The Correspondence of John Tyndall

VOLUME 13

THE CORRESPONDENCE OF JOHN TYNDALL

VOLUME 13

The Correspondence, June 1872–September 1873

Edited by
Michael D. Barton, Joseph D. Martin,
Gregory Radick, and Roy MacLeod

General Editors
Roland Jackson, Bernard Lightman, and Michael S. Reidy

Associate Editors
Michael D. Barton, Erin Grosjean, and Michael Laurentius

UNIVERSITY OF PITTSBURGH PRESS

Published by the University of Pittsburgh Press, Pittsburgh, Pa., 15260

Manufactured in the United States of America
Printed on acid-free paper
10 9 8 7 6 5 4 3 2 1

Cataloging-in-Publication data is available from the Library of Congress

ISBN 13: 978-0-8229-4742-4
ISBN 10: 0-8229-4742-0

CONTENTS

ACKNOWLEDGMENTS

Our first note of thanks must go to Alana Duggan, the research assistant for this volume, who completed the biographical register and timeline, and who was instrumental in cracking many of the numerous puzzles that this volume threw at us. We are grateful too to the general editors of *The Correspondence of John Tyndall,* Roland Jackson, Bernard Lightman, and Michael Reidy, for their guidance; to members of the wider Tyndall Correspondence Project team, notably James Elwick, Erin Grosjean, and Michael Laurentius, for their genial support; to Jennie De Protani, at the *Athenaeum*, and Jane Harrison and Charlotte New, at the Royal Institution of Great Britain, for information and advice in connection with their institutions; and to James Braund and Ella Coulter for expert help with translations of German and French letters. In common with all of the editors contributing to this series, we owe an immense debt to the army of graduate students and others who originally identified, scanned, transcribed, and ordered the letters now presented. Finally, we thank the many colleagues who have generously lent us their expertise in responding to queries on specific issues: Richard Bellon, Glenn Branch, Janet Browne, Richard Carter, Christine Chua, Ken Corbett, Gowan Dawson, Diarmid Finnegan, Anne Dewitt, Kathy Gorman (research associate in the Institutional History Division of the Smithsonian Institution Archives), Anita Guerrini, Piers Hale, Roland Jackson, Frank A. J. L. James, Adrian Kirwan, Elizabeth Neswald, Alison Pearn (Darwin Correspondence Project), Jon Roberts, Kathleen Sheppard, Amanda Y. Su (former president of the *Harvard Crimson*), James Ungureanu, and Matthew Wale.

LIST OF ABBREVIATIONS

AAAS	American Association for the Advancement of Science
ADB	*Allgemeine Deutsche Biographie*
Alum. Cantab.	J. A. Venn, *Alumni Cantabrigienses, 1752–1900*, 6 vols. (Cambridge: Cambridge University Press, 1940–54)
ANB	*American National Biography*
Ascent of John Tyndall	Roland Jackson, *The Ascent of John Tyndall: Victorian Scientist, Mountaineer, and Public Intellectual* (Oxford: University of Oxford Press, 2018)
BAAS	British Association for the Advancement of Science
BL	British Library
Brit. Assoc. Rep.	*Reports of the British Association for the Advancement of Science*
Brock and MacLeod, *Hirst Journals*	William H. Brock and Roy M. MacLeod, eds., *Natural Knowledge in Social Context: The Journals of Thomas Archer Hirst FRS* (London: Mansell microfiche, 1980)
CDSB	*Complete Dictionary of Scientific Biography*, 27 vols. (Detroit, MI: Charles Scribner's Sons, 2008)
CUL	Cambridge University Library
Darwin Correspondence, vol. 20	Frederick Burkhardt et al., eds., *The Correspondence of Charles Darwin*, vol. 20, *1872* (New York: Cambridge University Press, 2013)
Darwin Correspondence, vol. 21	Frederick Burkhardt et al., eds., *The Correspondence of Charles Darwin*, vol. 21, *1873* (New York: Cambridge University Press, 2013)
DBF	*Dictionnaire de biographie française*
DCP	Darwin Correspondence Project
Desmond, *Huxley*	Adrian Desmond, *Huxley: From Devil's Disciple to Evolution's High Priest* (Reading, MA: Perseus Books, 1994)
DeYoung, *Vision of Modern Science*	Ursula DeYoung, *A Vision of Modern Science: John Tyndall and the the Role of the Scientist in Victorian Culture* (London: Palgrave Macmillan, 2011)
Diamagnetism	John Tyndall, *Researches on Diamagnetism and Magne-crystallic Action Including the Question of Diamagnetic Polarity* (London: Longmans, Green, 1870)

Endersby, *Imperial Nature*	Jim Endersby, *Imperial Nature: Joseph Hooker and the Practices of Victorian Science* (Chicago: University of Chicago Press, 2008)
Finnegan, *Voice of Science*	Diarmid A. Finnegan, "Science, Speech and Character: John Tyndall's Lectures on Light," in *The Voice of Science: British Scientists on the Lecture Circuit in Gilded Age America* (Pittsburgh: University of Pittsburgh Press, 2021), 30–63
ForMemRS	Foreign Member of the Royal Society
Forms of Water	John Tyndall, *The Forms of Water in Clouds & Rivers, Ice & Glaciers* (London: H. S. King, 1872)
Fragments of Science	John Tyndall, *Fragments of Science for Unscientific People: A Series of Detached Essays, Lectures, and Reviews*, 2nd ed. (London: Longmans, Green, 1871)
FRS	Fellow of the Royal Society
Glaciers of the Alps	John Tyndall, *The Glaciers of the Alps* (London: John Murray, 1860)
HLS	*Historisches Lexikon der Schweiz*
Hours of Exercise in the Alps	John Tyndall, *Hours of Exercise in the Alps*, 1st ed. (London: Longmans, Green, 1871)
IC HP	Imperial College of Science, Technology and Medicine, College Archives, Huxley Papers
In den Alpen	John Tyndall, *In den Alpen* (Braunschweig: F. Vieweg und Sohn, 1872)
Life and Letters of Forbes	John Campbell Shairp, Peter Guthrie Tait, and A. Adams-Reilly, eds., *Life and Letters of James David Forbes* (London: Macmillan, 1873)
Life and Letters of Hooker	Leonard Huxley, ed., *Life and Letters of Sir Joseph Dalton Hooker*, vol. 2 (London: John Murray, 1918)
LT	Louisa Tyndall, John Tyndall's widow
Molecular Physics	John Tyndall, *Contributions to Molecular Physics in the Domain of Radiant Heat. A Series of Memoirs Published in the 'Philosophical Transactions' and 'Philosophical Magazine,' with Additions* (London: Longmans, Green, 1872)
Mountaineering in 1861	John Tyndall, *Mountaineering in 1861: A Vacation Tour* (London: Longman, Green, Longman, and Roberts, 1862)
Nature's Government	Richard Drayton, *Nature's Government: Science, Imperial Britain, and the "Improvement of the World"* (New Haven, CT: Yale University Press, 2000)
NDB	*Neue Deutsche Biographie*
Notes and Records	*Notes and Records: The Royal Society Journal of the History of Science*
ODNB	*Oxford Dictionary of National Biography*

OED	*Oxford English Dictionary*
Papers of Joseph Henry	Marc Rothenberg et al., eds., *The Papers of Joseph Henry*, vol. 11, *January 1866–May 1878: The Smithsonian Years* (Sagamore Beach, MA: Smithsonian Institution with Science History Publications, 2007)
Phil. Mag.	*Philosophical Magazine*
Phil. Trans.	*Philosophical Transactions of the Royal Society of London*
PMG	*Pall Mall Gazette*
Proceedings at the Farewell Banquet	*Proceedings at the Farewell Banquet to Professor Tyndall given at Delmonico's, New York, February 4, 1873* (New York: D. Appleton, 1873)
RGO MS.RGO 6	Royal Greenwich Observatory Archives, Papers of George Airy
RI	Royal Institution of Great Britain
RI MS JT	Royal Institution, London, John Tyndall Papers
Roy. Inst. Proc.	*Proceedings of the Royal Institution of Great Britain*
Roy. Soc. Proc.	*Proceedings of the Royal Society of London*
RS	Royal Society of London
Six Lectures on Light	John Tyndall, *Six Lectures on Light Delivered in America in 1872—1873* (London: Longmans, Green, 1873)
Sound	John Tyndall, *Sound: A Course of Eight Lectures Delivered at the Royal Institution of Great Britain* (London: Longmans, Green, 1867)
Traveller and Plant Collector	Ray Desmond, *Sir Joseph Dalton Hooker: Traveller and Plant Collector* (London: Royal Botanic Gardens, Kew, 1999)
Tyndall Correspondence, vol. 3	Ruth Barton, Jeremiah Rankin, and Michael S. Reidy, eds., *The Correspondence of John Tyndall*, vol. 3, *The Correspondence, January 1850–December 1852* (Pittsburgh: University of Pittsburgh Press, 2017)
Tyndall Correspondence, Volume 6	Michael D. Barton, Janet Browne, Ken Corbett, and Norman McMillan, eds., *The Correspondence of John Tyndall*, vol. 6: *The Correspondence, November 1856–February 1859* (Pittsburgh: University of Pittsburgh Press, 2019)
Tyndall Correspondence, Volume 9	Michael D. Barton, Iwan Rhys Morus, and James Ungureanu, eds., *Correspondence of John Tyndall*, vol. 9, *The Correspondence, of John Tyndall*, vol. 9, *The Correspondence, February 1865–December 1866* (Pittsburgh: University of Pittsburgh Press, 2022)
Tyndall Correspondence: Volume 10	Michael D. Barton, Ken Corbett, and Roland Jackson, eds., *The Correspondence of John Tyndall*, vol. 10, *The Correspondence, January 1867–December 1868* (Pittsburgh: University of Pittsburgh Press, 2021)

Tyndall Correspondence, Volume 11	Adrian Kirwan and Elizabeth Neswald, eds., *The Correspondence of John Tyndall,* vol. 11, *The Correspondence, January 1869–February 1871* (Pittsburgh: University of Pittsburgh Press, 2022)
Tyndall Correspondence, Volume 12	Anne DeWitt and Kathleen Sheppard, eds., *The Correspondence of John Tyndall,* vol. 12, *The Correspondence, March 1871–May 1872* (Pittsburgh: University of Pittsburgh Press, 2023)
The X Club	Ruth Barton, *The X Club: Power and Authority in Victorian Science* (Chicago: University of Chicago Press, 2018)

This photograph of John Tyndall is one of several taken by the photographer Frederick Gutekunst (1831-1917) in November 1872 while Tyndall was in Philadelphia, Pennsylvania, during his American lecture tour. Two letters from after the lecture tour refer to this photograph. In a letter to Tyndall of 26 April 1873, Joseph Henry remarked, "Mr. Childs of Philadelphia, has presented me with a portrait of yourself which is an excellent specimen of the art of photography and the artist, Mr. Gutekunst, has presented another to the Institution. The latter occupies a prominent place on the walls of my office; the former will be displayed in the parlor of my house" (see letter 4073). Lady Mary Egerton also wrote about the photograph, on 16 March [1873], "How good of you to give me that beautiful Photograph. It is really a splendid specimen as well as being an excellent likeness. At first I thought it too grave, almost sad—a sort of—'Oh! That the desert &c'—expression; but sitting for one's picture is apt to produce a certain solemnity of countenance, and certainly it is you down even to the eye glass! Very many thanks for it" (see letter 4014).
Credit: National Portrait Gallery, Photograph Collections, NPG x13250

INTRODUCTION TO VOLUME 13

The 476 letters in the thirteenth volume of *The Correspondence of John Tyndall* document the period from 1 June 1872 to 28 September 1873, much of which was consumed by Tyndall's lecture tour of the United States from late September 1872 to early February 1873. These fifteen months of Tyndall's life were marked by controversy. We meet him in the midst of the Ayrton affair, which saw Tyndall coming to the defense of his friend and fellow X Club member Joseph Dalton Hooker against the first commissioner of works, Acton Smee Ayrton, in an acrimonious dispute over the governance of the Royal Botanic Gardens, Kew. Tyndall's tour of the United States was a rousing success by many measures, but he was not long on American shores before his well-documented skepticism of the efficacy of prayer stoked the waspish ire of the faithful in the pulpits and in the religious press. Tyndall's return to England in mid-February 1873 saw him begin preparations for his 1874 Belfast Address, in which he accepted the presidency of the BAAS and articulated a defense of materialism that scandalized many of his contemporaries. As we leave him in September 1873, Tyndall is engaged in sharp-elbowed jostling with Scottish physicist Peter Guthrie Tait in the pages of *Nature* over James David Forbes, whose theory of glacial motion Tait had defended against Tyndall's attacks, in a scientific disagreement that evolved into a personal one. Amid the tumult of controversy, though, these letters reveal a man of science riding high on widespread esteem, wielding the influence it brought him with gusto, and moving with ease through the rarefied social and intellectual circles into which he had climbed.

The Ayrton Affair, the Alps, and American Ambitions

Tyndall, in his correspondence of June 1872, appears to relish throwing his hard-won weight around in Hooker's defense.[1] The affair consumed a great deal of his attention. Over half of the letters Tyndall wrote or received in June, July, and August mention the Ayrton affair, either directly or obliquely, offering a crude measure of the extent to which British men of science were collectively aghast at Ayrton's efforts to strip Kew Gardens of its research role. Tyndall coordinated the signatories, including Charles Darwin and Charles

Lyell, of a lengthy letter to Prime Minister William Gladstone decrying Hooker's treatment and imploring Gladstone to intercede on his behalf, and on behalf of the British scientific community at large.[2]

The letter leaned heavily on Hooker's expertise, outlining his qualifications, experience, and accolades, before stating: "With this sketch of the early training of Dr. Hooker for his present post before you, you will be able to compare with it the early training of Mr. Ayrton for the position which, by your favour, he occupies as Dr. Hooker's master." Tyndall and his compatriots invoked the authority of expertise over the forces of tradition and political patronage, in confidence that British men of science had amassed sufficient authority to sway affairs of state. Tyndall's growth into and evident enjoyment of his position and influence are thus a microcosm of the evolving status of British science as it entered the late nineteenth century.

The bustle of London life kept Tyndall's mind off his impending lecture tour of the United States until he departed for the Alps in late July. As he was visiting some of his best-loved peaks in the vicinity of Chamonix and Belalp (or Bel Alp, as it was then known) from 25 June to 10 August, his "little book on glaciers," *Forms of Water*, in part a chronicle of his previous visit to the region, was in the final stages of production. The book was a further contribution to his growing reputation as a popularizer, but also to the tension in his dual role as producer and expositor of scientific knowledge.

At the age of fifty-one, Tyndall had collected his share of intellectual grudges, and he took the opportunity the book presented to air again his disagreement with James David Forbes over the mechanism of glacial motion.[3] For this he earned a rebuke from an anonymous reviewer in *Nature*, who conceded that "whatever comes from Dr. Tyndall's pen is sure to be vivid and clear," but maintained that, in a popular exposition, "it was not necessary to rake up again the . . . controversy."[4] Tyndall, however, perceived no such cleavage in his role. Although he found it difficult, as he confessed to Henry Bence Jones, "to pick out of the mass of science a portion which can be at once popularly and thoroughly treated,"[5] his popular writings and lectures aimed to guide and instruct as much as to entertain and inform.

Tyndall began the task of framing the course of instruction he would deliver in America while in Switzerland, amid the glaciers that had sparked his disagreement with Forbes. His preparations for America, and for confronting Americans in the flesh, included reading Alexis de Tocqueville's *Democracy in America*, which he found "filled with practical wisdom."[6] He was struck in particular by Tocqueville's notes on American science, which suggested that the short-term incentives imposed by democratic capitalism had created an environment ill suited to theoretical reflection. "I want to do much more than make experiments before the American people," he wrote to Henry Bence; "I want to impress upon them the means by which science flourishes."[7]

Columbian Peregrinations

On 28 September 1872, Tyndall set sail from Liverpool aboard the SS *Russia*, bound for New York, where he landed on 9 October, greeted by his cousin Hector Tyndale, Edward L. Youmans, and William Appleton, among others. Over a grueling four months in the United States, Tyndall would lecture in Boston, Philadelphia, Baltimore, Washington, DC, Manhattan, Brooklyn, and New Haven, Connecticut. The topics ranged widely, and included demonstrations of many electromagnetic and optical phenomena, but they were united in their mission of driving home the importance of abstract research for the health of science, and for the prestige of a nation.[8]

The America Tyndall encountered was in many ways a fragile nation. The wounds of the Civil War, which had ended just eight years prior—including the assassination of President Abraham Lincoln—were fresh. These would have been plainly visible to Tyndall, whose cousin Hector Tyndale had been a Union major general. The president, Ulysses S. Grant, who himself had been the commanding general of the Union Army, was in the midst of enacting reconstruction policies designed to integrate Southern states back into the Union on terms that would ensure stability. The United States was also a growing country. The flag Tyndall would have seen flying over civic buildings in the cities he visited had thirty-seven stars. Eight more would be added before the century was out.

The United States was thus a nation substantially different from the one Tocqueville had visited in 1831. Tyndall nevertheless kept Tocqueville in mind when delivering his lectures, and his rhetoric is striking. In a letter to Bence Jones, he reported finishing his final lecture in Boston by quoting Tocqueville as saying: "I refuse to regard these peoples as two, the one is the vast outgrowth and development of the other."[9] This line, however, is not to be found in Tocqueville, and the spirit of the Frenchman's words often cut sharply against the sentiment. In the 1862 translation of *Democracy in America*, by Henry Reeve—the version most likely at Tyndall's disposal—Tocqueville wrote of England and the United States: "It is remarkable that two nations, so recently sprung from the same stock, should be so opposite to one another in their manner of feeling and conversing."[10]

It is improbable that Tyndall would have knowingly fabricated the quote; it is more likely a misattribution, or a misremembering. By the time of his lectures in New York, which were committed to print, he makes the vaguer claim that Tocqueville "refuses to separate America from its ancestral home."[11] In any event, it is a revealing fiction. Like Tocqueville, Tyndall was struck most frequently by the differences between the United States and the old world. He did appreciate the European quality of Boston, and felt at home in its high-society halls, but he also remarked on the oddity of railway cars in which different classes of passenger intermingled.[12] He was taken aback by the froideur of the Philadelphia Quaker audience, who were indisposed to applause (and he took singular pride in wearing down their reserve).[13] He marveled at the grandeur of the edifices of commerce: "Think of a marble palace devoted to retail hosiery," he wrote to Thomas Archer Hirst. "[T]here are such in New York."[14]

By referring to the United States as the "vast outgrowth" of England, then, Tyndall was not so much recording his impressions—or Tocqueville's—as he was articulating an aspiration, which he hoped his audience might claim as their own. The practical ingenuity for which America was famed, Tyndall insisted—following Tocqueville more accurately in this instance—drew from the reservoir of abstract scientific knowledge cultivated in Europe, and which the United States did not yet produce domestically in any great measure. Tocqueville's observation that Americans, animated by the short-term incentives of democracy and capitalism, excelled in the practical but not the abstract sciences, rang true to Tyndall, who took it as his mission to inspire appreciation for the value of science for its own sake.

In pursuing this mission, he found ready allies in the budding American scientific community. Foremost among them was Joseph Henry, the first secretary of the Smithsonian Institution, who was then in his late seventies. Henry—who along with the science writer, editor, and publisher Edward Youmans, played a decisive role in bringing Tyndall to America—and his institution, the Smithsonian, represented a link with Britain. The Smithsonian had been founded with funds bequeathed to the US government in the will of James Smithson (1765–1829), a British chemist who himself had never set foot in the United States, but who wanted his fortune to contribute "an establishment for the increase and diffusion of knowledge among men" across the Atlantic.

Henry, despite his own advanced age, represented a youthful but ambitious American scientific community. Its scientific institutions, such as the Smithsonian, like its universities, were comparatively young and still looking east across the Atlantic for recognition. For Henry, Louis Agassiz, Asa Gray, and other American scientists with whom Tyndall interacted, his visit was an opportunity to show that their labors merited the attention of European luminaries and to advance the fortunes of their nation's young scientific community.

Not content with his words alone, Tyndall arranged upon his departure to establish a scholarship fund with the monies he had accrued during his tour. The scholarship, administered through the Smithsonian, would allow a promising student of natural science to travel to Europe to pursue advanced study.[15] In many ways, the Tyndall scholarship was a precursor to the philanthropic fellowship programs that established crucial groundwork for the tremendous growth of American science in the first decades of the twentieth century.

The endowment also attests to the great affection Tyndall developed for America during his visit. He effused to his friends in England of its rich natural beauty and the warmth with which he was received. Whatever controversy he would stoke in the public sphere, he became invested, literally and figuratively, in the American scientific community.

Major Themes of the American Tour

THE INVISIBLE TECHNICIAN

The major misfortune of Tyndall's trip struck in November, when his young assistant, George Miller, contracted typhoid fever from his brother-in-law while visiting his sister in Ontario. Tyndall evidently cared for the young man. He retained prominent Philadelphia physicians to attend to him during his illness, at considerable personal expense, and frequently commented on his condition to his correspondents in England. But despite their efforts, Miller succumbed to his disease on 30 November.

Along with a second assistant, John Cottrell, Miller had been a key component of the Tyndall roadshow. The *New York World* extolled his skill preparing and executing Tyndall's experimental demonstrations, writing: "Mr. Miller had become by long experience, as one may say, a part of Professor Tyndall's scientific self."[16] Such statements could easily be dismissed as hyperbole from reporters eager to eulogize Miller. But such a dismissal would miss the importance of the work Miller performed.

Miller was the quintessential invisible technician.[17] His given name, in fact, never appears in Tyndall's letters or notebooks.[18] Even more thoroughly erased from the historical record is the Black man Tyndall hired to replace Miller, having Cottrell train him to assemble his battery for the lectures in Philadelphia and Baltimore, for whom we lack any name at all but whose work was "very useful," as Tyndall noted in a letter to Youmans.[19] But the incident is a reminder of the importance of such labor in the conduct of science. Miller was, among other things, a demonstrator, and his facility with the apparatus of Tyndall's demonstrations was integral to creating the impression of smooth mastery on which Tyndall's claims to authority were founded.

THE PRAYER TEST DEBATE

In July 1872, Tyndall wrote a preface to an article his friend the physician Henry Thompson published anonymously in the *Contemporary Review*, suggesting that the efficacy of prayer could be tested by directing prayers toward the sick at one hospital while withholding them from those at another, and comparing the rates of recovery at each institution.[20] The notion that prayer could be subjected to empirical test hung over Tyndall's American tour, despite his efforts to avoid it at the lectern.[21]

Whereas Tyndall garnered praise and adulation from the American scientific community, his reception in the American religious world was somewhat frostier. The American Methodists organized meetings to pray for Tyndall in the cities in which he was lecturing, and he evidently received (though did not save) several letters expressing concern about his soul.[22] The religious press compared him unfavorably with William Tyndale (1494–1536), who had translated the Bible into English.

For the most part, Tyndall bore the adverse reactions of the American devout graciously. At the end of his journey, though, the controversy undercut his ambitions. His hosts arranged a farewell banquet at Delmonico's restaurant in New York on 4 February 1873, several days before he sailed for England.[23] Tyndall hoped that the event would serve as a pretext for assembling the luminaries of American science, and for illustrating the potential it had to proceed along the lines he had advocated in his lectures. Henry was originally meant to preside but was unable to attend, and so the duty fell to Agassiz. When he too was called away on that date, the former attorney general William Evarts was placed in charge. The organizers saw fit to allow the long-running prayer test controversy to guide the program. Rather than, as Tyndall had hoped, a celebration of the potential of American science, the evening became a discussion of the relationship between science and religion.

The theme, though doubtless a frustration to Tyndall, reflected the appetites of his American audience, who had been given ample opportunity to trace the furore of the prayer-gauge controversy in the press. The speakers included the Reverend Henry Ward Beecher, who discussed the unity of religious and scientific truth. But also on the program were John W. Draper and Andrew Dixon White, who, before the dawn of the new century, would establish the conflict thesis, insisting on enduring warfare between science and religion.[24] Whatever Tyndall's own frustrations, then, the event crystallized a growing tension in American culture the repercussions of which would echo through the twentieth century.

THE PURE SCIENCE IDEAL

Tyndall took the occasion of his own banquet speech, on the eve of his departure, to implore any young American drawn to science to turn away from the siren song of vocationalism, and to pursue it for its own sake. The necessity to advance pure science above all else was the leitmotif running through the lectures Tyndall delivered, whatever the area of science on which they focused. It represented his own view of his American legacy, embodied in the form of his scholarship fund, which was designed to steep young Americans in a European ethos of pure science as much as it aimed to provide for the training of promising young investigators.

When the likes of William Whewell used the term through the mid-nineteenth century, *pure science* referred to areas of inquiry that could be advanced a priori, without reference to the fickle empirical realm. And so, Whewell referred to the science of motion as "pure" because it "does not depend upon observed facts, but upon the Idea of motion."[25] But for Tyndall, pure science was opposed not to empirical, inductive science but rather to science pursued for the sake of profit or glory—the purity, for Tyndall, resided in the heart of the investigator.

This term's shifting meaning reflects both the rising authority of science, a trend to which Tyndall contributed, and the professionalization of science, a trend he deplored—despite making his own living from his scientific labors. The new pure-science ideal was designed to advance the former while forestalling the latter. It articulated the importance of science as a form of culture while regarding those who pursued science as an avocation as both more prestigious than and morally superior to those who pursued it for profit.

In advancing this particular vision of pure science, Tyndall placed himself at the vanguard of a philosophical agenda that would unfold on both sides of the Atlantic. Fellow X Club member Thomas Henry Huxley would position himself as the spokesperson for pure science in Britain, and Henry Augustus Rowland, who was a twenty-three-year-old instructor at the University of Wooster in Ohio when Tyndall first lectured in the United States, would put a similar case to the United States.[26] No evidence suggests that Rowland witnessed any of Tyndall's lectures himself, but Rowland's own "Plea for Pure Science," published a decade after Tyndall returned to England, both advanced much the same argument and proved influential for the development of American science.

DARWINISM IN AMERICA

Charles Darwin's *On the Origin of Species* (1859) was published thirteen years before Tyndall's American tour, and *The Descent of Man, and Selection in Relation to Sex* (1871) just a year earlier. Darwinism remained a subject of fierce debate in both Britain and the United States. It would be reasonable to expect, therefore, that Tyndall, one of Darwin's foremost advocates in Great Britain, who would hold up Darwin as the foremost exemplar of scientific virtue in his 1874 Belfast Address, would express the same fulsome support for evolution by natural selection when given an American stage.

But it would be wrong. Here, the letters speak most loudly through their silence. Tyndall interacted regularly with some of Darwin's great American supporters, whom he would have perceived as allies, including Joseph Henry, who was instrumental for disseminating Darwinian ideas, and Asa Gray, whose qualified acceptance—despite his preference for a splash of divine providence in his evolutionary theory—also helped Darwinian thinking take root in American soil. Tyndall also crossed paths with vocal opponents, such as Louis Agassiz, who was keen to engage him in the subject.

But Tyndall, though an enthusiastic proselyte of Darwin's, for the most part remained circumspect when moving within the American scientific community. He appears at times, for instance in a letter to Thomas Archer Hirst discussing Asa Gray, to be counting the house for American Darwinians.[27] But, so far as the records at our disposal indicate, he avoided the topic as best he could. His journals describe both Agassiz and his host in Boston, John Amory Lowell, as eager to debate the topic, but when we see it broached, it is not Tyndall doing the broaching, and he emphasizes the amicability with which he carried on such discussions.

Whatever Tyndall's own views on natural selection at that time—and room remains for thinking that latent teleology, linked to pantheism, was part of the package for Tyndall—he in fact had good reason to skirt such a topic, given that his ambitions were larger than the success of a specific theory. He was likely cautious of the fervor with which some Americans might oppose Darwin and would not have wanted to jeopardize his quest to build momentum for pure science, as he saw it, by alienating potential allies in this mission, such as Agassiz. The issue tended to inflame the same passions that Tyndall had already stoked with the prayer-gauge issue, which would have given him good reason to set the topic aside. The absence of evolution from Tyndall's American agenda is thus both a testament to his political skill and an indication of the extent that trans-Atlantic engagements tolerated diversity of opinion on evolution by natural selection in service of the larger aims of science.

The Return to England

"Now, Prof. Tyndall is on the point of leaving us. When he gets back to Albemarle Street, he will remember Broadway. I am sure that you will all join me in wishing him a pleasant voyage over the Atlantic. But I wish him something better than that, I will add—a safe return to America. There is a great deal for him to do here yet."[28] Thus spoke the chemist, historian, and champion of Darwinism John W. Draper at Tyndall's farewell banquet. The hopes of Tyndall's American admirers notwithstanding, he never did return. Even so, his American visit featured often in his correspondence once he was back in Britain. He had loose ends to tie up in relation to the fund that he established with the profits from his lecture tour in order to help young American scientists advance their educations. Fan mail required attention, including a letter from a Louisiana physicist who, via Joseph Henry, sent Tyndall a photograph of the fernlike plumes produced after the freezing of water in a basin where paintbrushes were resting.[29] Tyndall liked the image so much that he used it as the frontispiece for *Six Lectures on Light Delivered in America in 1872-1873*, published in the summer of 1873. And new questions demanded pursuit, most immediately about the geological past and future of Niagara Falls. He discussed these with Charles Lyell in March,[30] lectured on them at the RI in April, and subsequently published his conclusions on both sides of the Atlantic. But larger questions also loomed about the nature of matter and spirit and science—topics that Tyndall would take up so notoriously in his address as president of the BAAS at Belfast in August 1874.

The most conspicuous challenges awaiting him at Albemarle Street, however, were more mundane. As he wrote to Lady Emily Peel on 19 March: "I have been hard pressed since I came here. First of all we have been committed to new buildings: the sacred walls of our old laboratories have been pulled down and new laboratories have been erected. This has thrown the place into a chaotic condition. Then I had to deal with about six cubic feet of papers and correspondence which had accumulated during my absence."[31] By mid-July, the backlog still dogged him, forcing him to delay his summer trip to the Continent—incidentally, the trip that would kindle the relationship with his future wife, Louisa Hamilton. The noisy construction work at the RI was still ongoing in late September. Moreover, he had to face this challenge without the help of his close friend Henry Bence Jones, who had been overseeing the renovations, and who died on 20 April 1873.

Yet despite these trying circumstances, and the occasional distress signal that he emitted ("I find as I grow older," he wrote to an unknown correspondent on 17 April, by way of turning down a request, "the multiplication of small demands upon my time tending to take all dignity & wholeness of purposes out of life, reducing it in fact to driblets"[32]), Tyndall kept up the usual punishing schedule of lecturing, writing, examining, reporting on innovations in the design of lighthouses and associated systems, and scientific-social engagements. Below we shall come back to the case for seeing this period as when Tyndall began thinking about the Belfast Address and what he wanted to do with it. We will first consider two other developments that bulk especially large in the surviving correspondence: first, Tyndall's deepening involvement in an increasingly bitter controversy with the Scottish physicist Peter Guthrie Tait, whom he loathed, about the late Scottish glaciologist James Forbes; second, Tyndall's friendship with fellow X Club member Thomas Henry Huxley, author of the most famous (and also funniest) line in this volume.

CONTROVERSY WITH TAIT OVER FORBES

Tyndall's American lecture tour coincided with the publication there of the *Forms of Water*. The book was the inaugural volume in the new International Scientific Series, masterminded by Youmans and eventually to be, in Bernard Lightman's words, "one of the most famous of all Victorian attempts to codify and popularize scientific knowledge in a systematic fashion to a wide reading public across national boundaries."[33] The evolutionary naturalism so overt at Belfast was subtly present in *Forms of Water* too, from the opening emphasis on how observations of natural phenomena cannot satisfy until embedded in further knowledge of causes and effects, to the rejection of the idea that water becomes less dense when passing from liquid to solid form in order to ensure that fish can survive through winter, as per the wise Creator's intentions (for if ice were denser than water, then ice at the top of a lake would sink, displacing water lower down, which, being less dense, would then rise to the top, where it would in turn freeze and sink, with the whole cycle repeating until the whole lake was frozen, and the life within it extinguished).[34]

What plunged Tyndall back into a public dispute with Tait, however, was the coverage, toward the end of the book, of different theories of glacier movement. Recounting the history of the theory that glacier ice is not statically hard but, like honey, dynamically viscous—the theory that Forbes made his name by advancing—Tyndall praised Forbes while nevertheless making it plain that another man, the French Catholic priest Louis Rendu, had stated the theory first, and furthermore that Forbes had known what Rendu had written. Tyndall wrote in a similar spirit about blue veins within glaciers, a phenomenon discovered and then valuably publicized by Forbes, but discovered independently in the same year by the Swiss naturalist Louis Agassiz, and several years before both of them by another Swiss naturalist, Arnold Henry Guyot.[35]

It was hardly fighting talk on Tyndall's part. But it was enough to antagonize Tait, who, with two collaborators, was finishing work on the literary equivalent of a monument to Forbes—a nearly six-hundred-page "life and letters" volume—when, in early 1873, *Forms of Water* was published in Britain. Tyndall somehow got wind of the Forbes project, guessing that Tait had complained to the editor of *Nature*, Norman Lockyer, about Tyndall's swipes at Tait's hero, and that Lockyer in turn had transmitted that complaint in the short, unsigned review of his book that appeared in the 27 March issue ("surely it was not necessary to rake up again the Forbes-Rendu controversy, nor to renew the claims of Agassiz and Guyot").[36] On 6 April, in a letter to Henry thanking him for passing on the plumey photograph, Tyndall reported that he had inadvertently provoked an old adversary:

> By the way that little book [*Forms of Water*], which I imagined to be without offence, is likely to bring me into trouble. You remember I dare say some years ago a passage of versus between Professor Thomson & Tait and myself regarding Mayer and the Mechanical Theory of Heat. I thought all ill feeling had ended for we met each other as friends in Dundee; and I was delighted to be assured that no further quarrels could arise between us—I was too sanguine. For while it was my earnest wish in the "Forms of Water" that the pupil should /emerge/ from its perusal with no other feeling than one of respect for all of those who had wrought at the glaciers, Prof. Tait I am told has taken deep offence at my references to Forbes, and intends to call me to account in a Life of Forbes on which he is now engaged. It will be an ungracious thing to quarrel over a grave, but I may be freed to that which will be exceedingly disagreeable to me.
>
> In a brief review the editor of Nature has given expression to Mr. Tait's feeling: certainly no charge had ever less justification than that I wished to "rake up" a buried controversy.[37]

Tyndall's prophecy was borne out: the new controversy with Tait and Forbes's other allies absorbed a huge amount of time and energy from spring to autumn 1873, along the way drawing in Huxley as well as Agassiz's son Alexander and prompting Tyndall to publish a lengthy defense of his position in the August *Contemporary Review*, under the title "Principal Forbes and His Biographers." Unsurprisingly, the episode brought Tyndall's customary combativeness in intellectual disputes to the fore.

Tait seemed to get under Tyndall's skin during this period. Though "a very able fellow," Tait was, Tyndall groused to the German physicist Rudolf Clausius in late June, while laboring away on the *Contemporary Review* response, "a quarrelsome, vulgar man"—and about as different from the reasonable and genial James Clerk Maxwell as two people could be. "It is very disgusting having to occupy time in this way," Tyndall continued, "and I shall take occasion to tell the public that I do not feel called upon to notice Tait for the future."[38] Yet when Tait responded in the letters column of *Nature* in September, Tyndall felt once again called upon to take notice, this time writing something whose impassioned vehemence, when he saw it in print, he regretted, and he wrote again to retract the harshest comments. "[God] forsook me for a moment when I allowed that burst of anger to overtake me as regards Tait," Tyndall wrote to Huxley on 24 September.[39] But then again, the irrepressible Tyndall continued, no less a figure than Goethe counseled that anyone who is never overtaken by big emotions, and so never guilty of big mistakes, may as well be dead already.

FRIENDSHIP WITH HUXLEY

In the *Oxford Dictionary of Scientific Quotations*, Tyndall merits thirteen entries. Only one of them is from his correspondence: a passage from a December 1875 letter to Huxley about Tyndall's experimental study of the fate of bacteria on substances covered with *Penicillium* (Roland Jackson has described Tyndall's near-miss in discovering the antibacterial properties of penicillin as "one of those 'what ifs' of history").[40] In our view, however, another, earlier letter to Huxley, included in this volume, shows the quotable Tyndall, as a scientific man of letters and a literary man of science, at his singular best.

The context was the brewing controversy with Tait over Forbes in spring 1873. After the publication of the anxiously anticipated *Life and Letters of James David Forbes*, Huxley was dismayed to find in its pages what he judged to be "a regular onslaught on Tyndall."[41] The volume printed a hitherto private letter from November 1859 in which Forbes wrote of missing out on the RS's Copley Medal that year because, he surmised, Tyndall, Huxley, and their London allies had thrown doubt on his claim to priority in the viscous theory as well as on "his straightforwardness in making that claim," as the volume's editors put it in their commentary on the letter.[42] But however painful the sting of the accusations, Forbes went on, he took comfort from the support of the friends who had rallied around him, and also from the apparently chastening impact on his enemies. "I believe that the effect of the struggle [on the part of his friends]—though unsuccessful in its immediate object—will be to render Tyndall and Huxley and their friends more cautious in their further proceedings. For instance, Tyndall's book, again withdrawn from Murray's 'immediate' list, will probably be infinitely more carefully worded relative to Rendu than he at first intended."[43]

Huxley set about preparing a riposte for *Nature*. "I am thinking of printing in 'Nature' the document I submitted to the Council of the R.S. and of which I sent you a copy some time back," Huxley wrote to Tyndall on 6 May. "Can you let me have that copy again[?]"[44] Huxley had in mind a statement that he had drawn up in autumn 1859 for the benefit of council members who were due to consider Forbes's candidacy for the RS's greatest honor but were not, in Huxley's estimation, sufficiently well informed about the state of glacial science, and Forbes's role in it, to come to a just view. The statement was Huxley's work alone; indeed, Tyndall had never laid eyes on it until, in mid-May 1872, Huxley had sent it to him, after coming across it while looking for something else. Tyndall replied a few days later, saying that he would send the document on whenever Huxley wished, but advising against printing it straight off. "Let the fight go on for a time," wrote Tyndall, "and then you will see your way more clearly."[45] A week later, 17 May, Tyndall returned the document,[46] and Huxley wrote his letter, which appeared in the 22 May issue under the title "Forbes and Tyndall."[47]

In the end, Huxley took Tyndall's advice and merely described the document, insisting that Tyndall had known nothing about it at the time, and so

Forbes's notion that a new cautiousness had taken hold of the battle-scarred Tyndall when it came to public statements about the rival claims of Forbes and Rendu was all the more ludicrous. "Forbes and Tyndall" is, as with everything Huxley ever wrote, sharp and stylish. But it was the cover note that Tyndall included with the document that deserves quotation. "I could hardly have believed," he began, "that a year has elapsed since you placed this document in my hands." And then a wise and witty one-liner: "But like a falling stone life moves with accelerated velocity as we make our final approach to the earth."[48]

To read the correspondence between Huxley and Tyndall is to be in the company of two fascinating and fascinated friends who loved each other and, even better, loved amusing each other. For evidence of the former, see especially the letters in this volume from April 1873 detailing the extraordinary efforts that Tyndall, Darwin, Spencer, and others in Huxley's circle made when they were concerned that, due to overwork, Huxley's final approach to the earth was imminent. He needed, they reckoned, a long, proper holiday, and urgently. To force him to take one, they donated funds for the purpose, and then agonized about how to offer the money to him so that he would accept it. In the end they succeeded, and by the end of summer 1873 Huxley seemed to have a new lease on life. The best-remembered example of their delight in making each other laugh comes from a letter that Huxley sent from Baden Baden on 30 July. After a hugely funny recounting of the scatological misadventures that he and his companion, Joseph Hooker, had suffered of late, Huxley turned to the news that the Bishop of Oxford, Samuel Wilberforce, with whom Huxley had famously skirmished over the theory of evolution, had died, killed when, out riding, he hit his head on a rock after being thrown from his horse. Huxley's immortal comment: "Poor Sammy! His end has been all too tragic for his life. For once reality and his brains came into contact and the result was fatal."[49]

BEGINNINGS OF THE BELFAST ADDRESS?

The Tyndall section of the *Oxford Dictionary of Scientific Quotations* concludes with a bonus quotation, not by Tyndall but about him. In August 1861, Huxley wrote to Spencer about a "favourite problem" of Tyndall's: "Given the molecular forces in a mutton chop, deduce Hamlet or Faust therefrom." "He is confident," Huxley added, "that the Physics of the Future will solve this easily." Huxley's testimonial is a vivid reminder that the materialism expressed in Tyndall's Belfast Address went a long way back. But when, exactly, did Tyndall begin planning to use that address to that remarkable end? In particular, might he have started as early as summer or even spring 1873, not long after his return from America, where his position on matters of matter and spirit had attracted so much controversy?

The traditional view is that, right up to the Bradford meeting of the BAAS in September 1873, Tyndall was undecided about whether to accept the invitation to serve as president at the 1874 meeting in Belfast, and so he cannot have begun thinking about his presidential address before then. Furthermore, over the next year he was so secretive about what he was planning to say that not even Huxley knew what the address would address until just before Tyndall delivered it. Nevertheless, there are signs, in letters in this volume and elsewhere, of an earlier start. The clearest is in a letter from Lady Mary Egerton dated 4 August 1873.[50] She did not include the year, but—the Tyndall Correspondence editors agree—it must be 1873, since she mentions her eagerness to read Tyndall's just-published *Contemporary Review* article on Forbes. Later in the letter she writes: "Are you really already beginning preparations for Belfast? It seems a long time beforehand! but I can well conceive you would have no rest in your next year's holiday if it was left till the last." At a minimum, then, it appears that, in early summer 1873, Tyndall had told Lady Mary, at least, that he had made a serious start on preparations for something big at the Belfast BAAS meeting.

Might he have done more, and shared with her the ambition of saying something more philosophically challenging than the usual survey of scientific achievements? That possibility becomes tantalizing when we notice another passage from the same letter, as well as passages from two other letters from Lady Mary. In the letter of 4 August 1873, she wrote: "I wonder whether the little book has even been opened? I do not want you to agree with me intellectually, only to observe how intellects on a par with your own, have thought and felt spiritually. And I know you can appreciate beautiful thoughts, even though you may not agree in the views they imply." In a letter dated only 19 April, she wrote to Tyndall after a dinner party at the Spottiswoodes', following up their conversation on that occasion by sending him a volume of Johann Gottlieb Fichte to reinforce what she had told him about Fichte on personal immortality. ("Of course I do not ask you to believe a thing because another man, however noble, has believed it; I know how your independent spirit recoils from the bare idea of 'imitation'! But that a man like Fichte, quite independent of all theological bias, should have found in the needs of his own moral and intellectual nature, a deep conviction of the truth of a life beyond this, surely shows the idea not to be one which the Philosopher is bound to [crush] out, when it arises, as arise it must, in every heart full of tenderness & human sympathy?") Toward the end of the letter, she encouraged him to take the volume with him on his upcoming trip, and said how much she was looking forward to his "address." Finally, in an undated letter from her that—the Tyndall Correspondence editors agree—was written in late August or early September 1873, she wrote, apropos of Tyndall's recent Swiss travels: "I was so glad to see that your old friend Fichte had been again one of your travelling companions."[51]

If the 19 April letter comes from 1873, then it is tempting to see in the above a sequence that begins with a dinner-party discussion in which Tyndall in an exploratory way had shared with Lady Mary his incipient thoughts on matter and spirit, science and souls, as he meant to address them in Belfast if he accepted the invitation,[52] and she responded, as thoughtful people will, by pressing on him a book that she reckoned he needed to read. On 4 August, she checked to see how he was getting on with the volume she had sent.[53] And in late August or early September, she conveyed how pleased she was by, it seems, his affirmative reply to her query. Looking beyond Tyndall's correspondence with Lady Mary, we can notice too that, in September 1873, Tyndall, back from Switzerland, was holed up in a hotel on the English coast, where he was reading, and wrestling with, another work of far-from-godless materialism: Spencer's *First Principles.*

At this point, however, another joke of Huxley's, at Spencer's expense, comes to mind: that for Spencer, a tragedy was a beautiful theory killed by an ugly fact. There are reasons to prefer 1874 over 1873 for the 19 April letter, which is why the letter is included not in this volume but in the next volume in the series. Be that as it may, and whatever the debt Tyndall owed to a late-night chat at a posh dinner party, the Fichtean element in the Belfast Address is inarguable, as Ruth Barton brought out in her classic "rereading." And anything in this volume that might help present-day readers be a bit more curious about it is to the good.[54]

Notes

1. For detailed background on the Hooker-Ayrton affair, see the Introduction to *Tyndall Correspondence*, vol. 12, as well as, in this volume, letter 3755, n. 1.
2. Letter 3755.
3. See the introduction to *Tyndall Correspondence*, vol. 6; and Nanna Katrine Lüders Kaalund, "A Frosty Disagreement: John Tyndall, James David Forbes, and the Early Formation of the X-Club," *Annals of Science* 74 (December 2017): 282–98.
4. "Tyndall's Forms of Water," *Nature* 7 (27 March 1873): 400–401.
5. Letter 3796.
6. Letter 3785.
7. Letter 3836.
8. Tyndall's tour in the United States brought him to Boston (six lectures from 15 to 25 October at the Lowell Institute), Philadelphia (six lectures from 11 to 22 November at the Horticultural Hall), Baltimore (three lectures from 26 to 30 November at the Peabody Institute), Washington, DC (six lectures from 3 to 12 December at Lincoln Hall), New York (six lectures from 17 to 31 December at the Cooper Institute), Brooklyn (six lectures from 4 to 18 January at the Academy of Music), and New Haven, Connecticut (two

lectures from 22 to 23 January at Yale College). He took with him two laboratory assistants from the RI, John Cottrell and George Miller, and only Cottrell would return, as Miller contracted typhoid fever in Philadelphia and died. Embarking on the SS *Russia* (of the Cunard Line) from Liverpool on 28 September 1872, Tyndall arrived in New York on 9 October after an eleven-day journey across the Atlantic. He returned to England on the *Cuba* on 19 February 1873 after eleven days at sea, arrivolving in London later the same day. His five months in America brought him in before numerous notable Americans and dazzled audiences, involved him in a debate on the efficacy of prayer, took him on a side trip to Niagara Falls (see letter 3738, n. 3), led to the opportunity to set up a fund for pure science research in America using his profits from the lecture tour (see n. 15), and ended with a lavish farewell banquet in New York before his departure from American soil (see n. 23). He published the lectures as *Six Lectures on Light*, with the speeches from the banquet published as *Proceedings at the Farewell Banquet*. The banquet was also reported in the *New York Times*, 5 February 1873, 5. The American trip has been discussed in: *Ascent of John Tyndall*, chapter 18; Michael D. Barton, "The 'Efficient Defender of a Fellow-Scientific Man': John Tyndall, Darwin, and Preaching Pure Science in Nineteenth-Century America" (MA thesis, Montana State University, 2010); I. Bernard Cohen, "Some Reflections on the State of Science in America during the Nineteenth Century," *Proceedings of the National Academy of Sciences* 45 (1959): 666–77; A. S. Eve and C. H. Creasey, *Life and Work of John Tyndall* (London: MacMillan, 1945), chapter 14; DeYoung, *Vision of Modern Science*, 46–47; Finnegan, *Voice of Science*, 30–63; Thomas F. Gieryn, "John Tyndall's Double Boundary-Work: Science, Religion, and Mechanics in Victorian England," in *Cultural Boundaries of Science: Credibility on the Line* (Chicago: University of Chicago Press, 1999), 54; Bernard Lightman, "The Victorians: Tyndall and Draper," in *The Warfare between Science and Religion: The Idea That Wouldn't Die*, ed. Jeff Hardin, Ronald L. Numbers, and Ronald A. Binzley (Baltimore: Johns Hopkins University Press, 2018), 71–73; Albert E. Moyer, *American Physics in Transition: A History of Conceptual Change in the Late Nineteenth Century* (Los Angeles: Tomash, 1983), 72–73; Edward J. Pfeifer, "United States," in *The Comparative Reception of Darwinism*, ed. Thomas F. Glick (Chicago: University of Chicago Press, 1988), 196–97; Katherine Sopka, "An Apostle of Science Visits America: John Tyndall's Journey of 1872–1873," *Physics Teacher* 10 (1972): 369–75; and Katherine Sopka, "John Tyndall: International Populariser of Science," in *John Tyndall: Essays on a Natural Philosopher*, ed. William H. Brock, Norman D. McMillan, and R. Charles Mollan (Dublin: Royal Dublin Society, 1981), 193–203.

9. Letter 3885.
10. Alexis de Tocqueville, *Democracy in America*, trans. Henry Reeve (London: Longman, Green, Longman and Roberts, 1862), 2:269.
11. *Six Lectures on Light*, 226.
12. Journal, 27 October 1872, RI MS JT/2/12.
13. See letters 3896, 3900, 3902, and 3903.
14. Letter 3873.
15. Before Tyndall even left England for the United States, he decided that the profits from his lecture tour would be donated to establish a fund for pure science research in America. In total, $13,033 was donated and invested in bonds, through the trustees Tyndall named—Joseph Henry, Hector Tyndale, and Edward L. Youmans—for "the advancement of theoretic science, and the promotion of original research, especially in the department of physics, in the United States," in the form of scholarships to send American students to German universities, not the University of Chicago, as Tyndall originally intended. This proved difficult, and following the death of Henry and Tyndale, in 1884 the $32,000 trust was equally divided among Columbia College, Harvard University, and the University of Pennsylvania. Funds from the trust continued to support students well into the twentieth

century. Tyndall also donated a small sum ($250) to the Yale Scientific Club. See *Ascent of John Tyndall*, 302–3, 312–15; Finnegan, *Voice of Science*, 59; Howard S. Miller, *Dollars for Research: Science and Its Patrons in Nineteenth-Century America* (Seattle: University of Washington Press, 1970), 119–24; Sopka, "Apostle of Science Visits America," 74; and "The Tyndall Trust for the Promotion of Science in the United States," *Annual Report of the Board of Regents of the Smithsonian Institution, showing the operations, expenditures, and condition of the Institution for the year 1872*, Serial Set Vol. No. 1573, Session Vol. No. 3, 42nd Congress, 3rd Session (Washington: Government Printing Office, 1873), 104–6 (this was also published in *Popular Science Monthly* 3 [May 1873], 100–101).

16. Quoted in the *Standard* (London), 30 December 1872, 5.
17. Steven Shapin, "The Invisible Technician," *American Scientist* 77, no. 6 (1989): 554–63.
18. We thank Alana Duggan for discovering the contemporary press reports that used his full name.
19. Letter 3919.
20. Letter 3779; "The 'Prayer for the Sick'—Hints towards a Serious Attempt to Estimate Its Value," *Contemporary Review* 20 (1872): 205–10.
21. See Finnegan, *Voice of Science*, 45.
22. See letter 3902.
23. On 4 February 1873, several days before his departure from New York on the SS *Cuba*, a lavish farewell banquet was given in Tyndall's honor at Delmonico's, to fête him before returning to England. In attendance were some two hundred leading men in science and public life, bidding farewell to one of Britain's leading men of science. Speeches were given by prominent academic and religious figures, while letters from those unable to attend were published in the banquet's proceedings. The choice of some of these speakers reflected the controversy Tyndall's comments about religion during his time in America had incited, and many of the remarks addressed the relationship of science and religion. A letter from Joseph Henry opened the evening; he noted that one object of the lecture tour was to "advocate the claims of abstract science to higher appreciation and more liberal support," and remarked that Tyndall was "a distinguished laborer in the line of original research." The chemist John Lawrence Smith brought up Tyndall's conflict with American religious communities over prayer and the politician William Evarts discussed the utility of pure science. Oliver Wendell Holmes shared in a letter that "We cannot over-estimate our debt to the men who give their lives to the sincere study of Nature. They are becoming more and more our leaders in this morning twilight of knowledge, this period of transition, when the eye is getting the better of the ear, the reason of the imagination, our trust in the established order of the universe of those terrors of the supernatural out of which have been forged the chains of thought and the weapons of tyranny." The chemist and historian John William Draper shared what he thought America still had to show Tyndall. Besides western mountain peaks and the Pacific Ocean, America could show Tyndall how it had turned Irish and German immigrants into English-speaking citizens. Further, "we will show him how we have purged the African, the wooly-headed black man, of the paganism of his forefathers, and are now trying our hand at Darwinizing him into a respectable voter. . . . We are mixing European and Asiatic, red Indians and black Africans together, and I suppose certain English naturalists will tell us that the upshot of thing will be a survival of the fittest." Clergyman Henry Ward Beecher expressed his view that there is "no antagonism between Nature and the Bible." Tyndall stated in his own speech, "Hitherto their efforts have been directed to the practical side of science, and this is why I sought in my lectures to show the dependence of practice upon principles. On the ground, then, of mere practical, material utility, pure science ought to be cultivated." He then announced his gift of the net profits of his lecture tour ($13,033) to a fund for "the advancement of theoretic science and the promotion of original research, especially

in the department of physics" (see n. 15). Reverend Henry Whitney Bellows concluded the evening's speeches: "Give our loving compliments to Huxley, and Maudsley, and Mill, and Spencer, and the rest, and tell them that we expect there will be good Christians left after they have got to the lowest bottoms of their explorations! May your voyage home be less stormy than we fear (we are sorry you won't let us pray for it, yet you can't prevent it), and your welcome there be as warm and grateful as this earnest and heart-felt farewell!" The speeches were published as *Proceedings at the Farewell Banquet*, as well as in the April 1873 issue of *Popular Science Monthly*—the "Tyndall number," as Tyndall called it in letter 4046, to its editor, Edward L. Youmans. The banquet was also reported, with partial transcripts, in the *New York Times*, 5 February 1873, 5. See also Barton, "'Efficient Defender of a Fellow-Scientific Man'"; and Finnegan, *Voice of Science*, 57–60.

24. Draper and White went on to write seminal books promoting the idea that the history of science and religion was a history of warfare, *History of the Conflict between Religion and Science* (1874) and *History of the Warfare of Science with Theology in Christendom* (1896), respectively. On Draper, White, and their "military metaphor," see James. R. Moore, *The Post-Darwinian Controversies: A Study of the Struggle to Come to Terms with Darwin in Great Britain and America, 1870–1900* (Cambridge: Cambridge University Press, 1979), chapter 1. More recent scholarship, however, reassesses the conflict thesis; see Bernard Lightman, ed., *Rethinking History, Science, and Religion: An Exploration of Conflict and the Complexity Principle* (Pittsburgh: University of Pittsburgh Press, 2019); Lightman, "Victorians," 65–83; and James C. Ungureanu, *Science, Religion, and the Protestant Tradition: Retracing the Origins of Conflict* (Pittsburgh: University of Pittsburgh Press, 2019).
25. William Whewell, *Aphorisms Concerning Ideas, Science, and the Language of Science* (London: Harrison, 1840), 8.
26. Graeme Gooday, "'Vague and Artificial': The Historically Elusive Distinction between Pure and Applied Science," *Isis* 103 (2012): 546–54.
27. Letter 3886.
28. John W. Draper, "English and American Science," *Popular Science Monthly* 2 (April 1873): 734–36, on 735.
29. Letter 4021.
30. Letters 4017, 4019, and 4023.
31. Letter 4018.
32. Letter 4058.
33. Bernard Lightman, "The Popularization of Evolution and Victorian Culture," in *Evolution and Victorian Culture*, ed. Bernard Lightman and Bennett Zon (Cambridge: Cambridge University Press, 2014), 286–311, on 294.
34. Tyndall, *Forms of Water*, 1–14, 122–25; Bernard Lightman, "The International Scientific Series and the Communication of Darwinism," *Journal of Cambridge Studies* 5, no. 4 (2010): 27–39, on 29 (quotation) and 33–34.
35. Tyndall, *Forms of Water*, 158–62, 183–87.
36. "Tyndall's Forms of Water" (see n. 4).
37. Letter 4033.
38. Letter 4120.
39. Letter 4171. On emotional volatility and approval seeking as features of Tyndall's personality, see *Tyndall Correspondence*, vol. 3, introduction.
40. William F. Bynum and Roy Porter, eds., *Oxford Dictionary of Scientific Quotations* (Oxford: Oxford University Press, 2005), 590–602; *Ascent of John Tyndall*, 344.
41. See Thomas H. Huxley to Charles Darwin, 26 April 1873, Darwin Correspondence, vol. 21: 190–92; this is letter 8882 in the online edition of the Darwin correspondence.
42. *Life and Letters of Forbes*, 384.
43. *Life and Letters of Forbes*, 387.
44. Letter 4081.

45. Letter 4086.
46. Letter 4095.
47. Thomas H. Huxley, "Forbes and Tyndall," *Nature* 8 (22 May 1873): 64.
48. Letter 4095.
49. Letter 4151.
50. Letter 4153.
51. Letter 4181.
52. Which was very much on his mind that month; see letter 4033.
53. Gowan Dawson has suggested that the volume in question was *Johann Gottlieb Fichte's Popular Works: The Nature of the Scholar; The Vocation of Man; The Doctrine of Religion*, trans. W. Smith (London: Trübner, 1873).
54. Ruth Barton, "John Tyndall, Pantheist: A Rereading of the Belfast Address," *Osiris* 3 (1987): 111–34.

EDITORIAL PRINCIPLES

Our aim is to include all letters to and from John Tyndall that are currently extant. Some of the letters are from published sources, such as letters written by Tyndall to newspapers or journals, but the majority are either the originals or typescripts produced by Louisa Tyndall from originals. One central editorial principle has been to transcribe, wherever possible, from original letters rather than Louisa's typescripts. This is because of Louisa's conscious and unconscious editorial interventions, a point discussed in more detail below. We have also aimed to reproduce, as accurately as possible, the text of the letters as they were written. Spelling mistakes that have appeared in the handwritten letters have therefore been preserved; mistakes appearing where only the typescript letter has survived are silently corrected, as in our judgment they are almost always typographical errors. But, in general, non-standard spellings (e.g., the use of dialect) has been retained even when in a typescript-only letter (though there is the possibility that the spelling is Louisa's, not her husband's). Where the transcription is based on a typescript transcription by Louisa (or one of her assistants), this will be indicated by "LT Transcript Only."

We have included illustrations that are part of the letter. In cases where it was possible to reprint the original handwritten image, we have done so. In other cases, volume editors have reconstructed drawings because it was not possible to reprint an image from the handwritten letter (due, for example, to deterioration of the original document).

The letters are presented in chronological order of writing. Where more than one letter has the same date, letters from Tyndall (in alphabetical order of addressee) come first, followed by letters to Tyndall (in alphabetical order of writer), with the exception that if the order of writing can be determined, that order takes precedence. Where letters cannot be dated precisely they are placed at the latest likely date of writing. (In some cases if it is possible, but unlikely, that they were written later, notes indicate such possibilities.) To indicate the beginning of a new year, the date of that year has been inserted in a large font before the first letter from that year. Readers should note that although volumes are arranged in chronological order, letters that are discovered from years already covered in published volumes will appear in the final volume of the correspondence.

For non-English-language letters, both the transcription of the original and the English translation are presented. The transcription of the non-English letter appears first, followed by the English translation in a smaller font. Editorial notes are inserted in the English translation. Some foreign letters are in an original hand, while others exist only as transcriptions by Louisa or others who often did not have a full command of the language. In the latter case these letters contain frequent spelling, grammatical, and interpretation errors that do not seem to have been committed by the original author. In these typescript-only cases, and only in these cases, spelling and grammatical errors have been corrected, as have poor interpretations where a better word choice was obvious from the context.

There is a standard format for each letter. The first line lists the author, if it is to Tyndall, or the recipient, if it is from Tyndall; the date; and the letter number. The second line lists any information supplied by the writer in their salutation, such as the place, date, or time of day when the letter was written. (In some cases, though, where the writer put the date at the foot of the letter we have left it there in order to reproduce the text accurately.) The opening salutation, the text of the letter, and closing then follow. Then come the source and editorial notes, the latter indicated in the text by arabic numerals.

Editorial notes are intended to provide the contextualization necessary to understand the contents of each letter. There are basically six types of editorial notes. Many of them are informative notes on persons mentioned in the letters. If a person referred to in a letter is not described in a note, they have been mentioned more than twice in the volume and they will have an entry in the biographical register at the back of the volume. For the rest—those mentioned once or twice—biographical information will appear in the editorial note appended to the letter in which he or she is first mentioned. When there is a second reference to that person a note will refer back to the letter in which he or she was first mentioned. Biographical notes on well-known figures such as Faraday or Huxley stress what that person was doing at the time when the letter was written rather than providing a full overview of the person's life.

A second type of editorial note identifies allusions and quotations. In the case of quotations, where possible, the first edition of the work is cited unless a different edition is mentioned in the letter. If a quote is not in English a translation is given. Glosses on obscure places, abstruse words, and words used in an unfamiliar sense constitute the third type of editorial note. The fourth type provides information about the context of the letter, drawing on such sources as contemporary publications, scholarly sources, Tyndall's personal journals and field notes, and other letters in the *Tyndall Correspondence*.

The fifth type of editorial note is cross-references to other letters in the correspondence, referring to the letter number. The last type is information on significant textual changes that Tyndall made to the original that is thought to be relevant to the meaning of the letter.

There are a series of conventions for indicating when editors have given additional information about the text of a letter:

When Louisa Tyndall has made an annotation or marginal insertion that the editors have deemed important enough to retain, we have included her annotation in an editorial footnote.

When editors other than she have made insertions, they are indicated by unitalicized square brackets; for example, [word].

When certain words are ambiguous and we have had to conjecture their meaning, they are indicated by italicized square brackets; for example, *[*words*]*; when we have considered such words illegible the number of illegible words is indicated in full italicization and square brackets; for example, *[3 words illeg]*.

If a word or series of words are not present or have been destroyed—by inkblots, holes, or mold, for example—they are indicated as missing, italicized and in angular brackets; for example, *<3 words missing>*.

In extremely rare cases where it is clear that Louisa herself has intentionally destroyed a few words, they are indicated as excised, italicized and in angular brackets; for example, *<3 words excised>*.

Finally, a brief word about Louisa. She was responsible for collecting together many of Tyndall's letters in order to write a biography of her late husband. This biography was not completed. It is important to recognize that Louisa was not always reliable as a transcriber. She often threw out the manuscript letter when she had made her transcription, but some original manuscripts remain, which indicate her editing processes. Like any transcriber, she was sometimes inaccurate, but corrections show that she proofread her transcriptions. However, she also edited Tyndall's writing to make his letters seem more consistent or grammatically correct. For example, she made punctuation more formal, corrected grammar, and, occasionally, corrected quotations. She consistently corrected Tyndall's usual (deliberate) lowercase for "god" to "God." Her typewriter was unable to insert the superscripts that Tyndall used in abbreviations, or the "&c" symbols that stood for an address at the foot of the letter. (Modern typefaces are unable to deal with all the variant scrawls that the "etc." symbol took and they are all formalized here to "&c.") Louisa also left out details that she considered too private or too salacious for public consumption. She equated postmark and date of writing and overlooked the time taken to write some letters. When she inferred dates from internal evidence, her reasoning was sometimes faulty. Similarly, her identifications of people are sometimes in error. Independent evidence has always been sought for her dates and identifications. Lastly, the patterns of punctuation differ between LT sources and manuscript sources. Tyndall himself used dashes more often than commas, semicolons, and stops, and he used dashes of many different lengths, which are all standardized here to em-dashes. It is often difficult to distinguish between very short dashes and commas or very short dashes and stops. Similarly, capitalization is often ambiguous; distinguishing *S*/*s*, for "science," for example, is a particular problem. Therefore, scholars for whom punctuation and capitalization are significant will need to consult the manuscripts. Users should also be aware that throughout this project the transcribers and editors have generally worked from scans and photographs rather than the original manuscripts.

NOTE ON MONEY

In the nineteenth century the currency used in England was the pound sterling, usually abbreviated to £ (or, occasionally, to *L*). A pound consisted of 20 shillings (abbreviated *s*) and each shilling contained 12 pennies (*d*). A guinea was 21 shillings; that is, £1 1s. A crown was 5 shillings (5s), and half a crown equaled two shillings and sixpence (2s 6d). The smallest coin was a farthing, one quarter of a penny (¼d).

Several forms of representation were in contemporary use. Sometimes the numbers for pounds, shillings, and pence were separated by dots, forward slashes, or spaces; thus, for example, 3 pounds, 4 shillings, and 5 pennies could be written as £3.4.5, £3/4/5, or £3 4 5. A zero could be represented by a dash; for example, 7 shillings was often written as 7/- or just 7/. A half crown was often written as 2/6.

US dollars feature in this volume. As today, the dollar, abbreviated $, was a decimal currency, with 100 cents (abbreviated ¢) to one dollar. US coinage in common circulation before the Coinage Act of 1873 included one- and two-cent pieces, a three- and a five-cent nickel, a ten-cent dime, and a twenty-five-cent quarter. In 1872–73, one British pound was worth approximately 5.5 dollars, as it returned to a fairly stable exchange rate of approximately five dollars to the pound after historic lows in the value of the dollar during the American Civil War.

French currency is also occasionally mentioned in this volume. The franc was the national currency in France from 1795 until 1999, when France adopted the Euro. In the mid-nineteenth century the exchange rate between British and French currencies was approximately 25 francs per £. A centime was a coin worth one-hundredth of a franc.

TIMELINE OF JOHN TYNDALL'S LIFE

Year	Event
c. 1822	John Tyndall born at Leighlinbridge, County Carlow, Ireland
c. 1836	Attends John Conwill's National School in Ballinabranna
1839	Begins employment as civil assistant with Ordnance Survey of Ireland in Carlow
1840	Becomes civil assistant in the Ordnance Survey Office at Youghal, Cork County
1841	First appearance in print (poem in *Carlow Sentinel* under pseudonym "W[alter] S[nooks]")
1842	Joins English Ordnance Survey in Preston
1843	Leads written criticism of Ordnance Survey in *Liverpool Mercury*; dismissed November
1844	Home in Ireland, mostly unemployed, until obtaining a position with the firm of Nevins and Lawton, Surveyors, of Manchester
1845–47	Works for Richard Carter, surveyor, of Halifax
1845	Meets Thomas Archer Hirst in Halifax
1847	Begins teaching mathematics and surveying at Queenwood College, Hampshire
1848–50	At University of Marburg, working for PhD with Robert Bunsen and Friedrich Stegmann
1850	Returns to England; meets Michael Faraday and William Francis and attends his first British Association meeting, before returning to Marburg
1851	Moves from Marburg to Berlin (April–June); then returns to Queenwood
1852	Elected fellow of the Royal Society
1852	Friendship developed with Thomas Henry Huxley
1853	First lecture at Royal Institution (RI)
1853	Meets Herbert Spencer
1853	Appointed professor of natural philosophy at RI
1853	Awarded Royal Medal for magnetic work; declines when controversy arises over award
1856	Meets Thomas Carlyle
1857	Climbs Mont Blanc for first time with Hirst
1859	Demonstrates existence of greenhouse gases and climatic implications
1859	Joins Government School of Mines as professor of natural philosophy
1860	*The Glaciers of the Alps* published
1860	Meets John Lubbock and his wife, Ellen Lubbock

1861	First ascent of the Weisshorn
1861	Delivers his first Christmas Lectures at RI
1862	*Mountaineering in 1861* published
1862	Begins to champion Mayer's priority in discovery of conservation of energy
1863	*Heat Considered as a Mode of Motion* published
1864	Awarded Royal Society's Rumford Medal
1864	First meeting of X Club
1865	*On Radiation* published
1865	Engages in public controversy over the efficacy of prayer
1866	Succeeds Michael Faraday as scientific adviser to Trinity House
1867	Faraday dies; Tyndall becomes superintendent of the RI
1867	*Sound: A Course of Eight Lectures* published
1868	Successfully climbs Matterhorn
1868	Discovers the cause of light scattering, to be known as "Tyndall Effect"
1868	*Faraday as a Discoverer* published
1868	"Scientific Materialism" lecture at the British Association
1868	Resigns from Royal School of Mines
1869	Joins Metaphysical Society
1870	"On the Scientific Use of the Imagination" lecture at British Association
1870	*Three Scientific Addresses* published
1870	*Researches on Diamagnetism and Magne-crystallic Action* published
1871	Meets Louis Pasteur for first time while in Paris
1871	*Fragments of Science* published
1871	*Hours of Exercise in the Alps* published
1871	*Light and Electricity* published
1872	Debates over how to measure the efficacy of prayer, aka the "Prayer-Gauge Debate"
1872	*Contributions to Molecular Physics in the Domain of Radiant Heat* published
1872	*The Forms of Water in Clouds and Rivers, Ice and Glaciers* published
1872–73	USA Lecture Tour
1873	*Six Lectures on Light* published
1874	As president of the BAAS, delivers the "Belfast Address"
1876	Marries Louisa Charlotte Hamilton
1877	Develops "Tyndallization" (discontinuous heating and cooling process to destroy heat-resistant spores)
1877	Tyndall and Louisa build summer cottage, Alp Lusgen, at Belalp, northern side of Valais, above Brig
1877	*Fermentation and Its Bearings on Phenomena of Disease* published
1879	*Fragments of Science*, which had gradually expanded, first published in two volumes
1881	*Essays on the Floating Matter of the Air* published
1885	Tyndall and Louisa build Hindhead retreat, Surrey Downs
1887	Resigns from RI
1890	Fight with Gladstone in the *Times* over Irish Home Rule
1892	Hirst dies
1892	*New Fragments* published
1893	Dies at Haslemere, Surrey, accidentally poisoned

TIMELINE OF EVENTS IN JOHN TYNDALL'S LIFE SPECIFIC TO VOLUME 13

Month and year	Biographical details	Notable social, political, and cultural events
June 1872	Delivers the last of a course of nine weekly lectures at the RI, "On Heat and Light," which began on 11 April (6th) Finishes drafting the memorial in support of Joseph Hooker in his dispute with Acton Ayrton over his role as director of the Royal Botanic Gardens, Kew; begins to seek signatories Memorial presented to Prime Minister William Gladstone by John Lubbock (20th) Begins journey to Switzerland (25th); in Paris, visits with Jules Jamin and Louis Pasteur (26th–27th); travels to Geneva (28th) Climbs the Glacier Montanvert (Mer de Glace) on the northern slopes of Mont Blanc (30th) Publishes *Contributions to Molecular Physics* Publishes "Action of Dark Radiations" in the new American magazine *Popular Science Monthly*	American newspaper publisher James Gordon Bennett dies (1st) William Odling delivers a Friday Evening Discourse at the RI, "On the History of Ozone" (7th) Women's Suffrage Convention held at Mercantile Hall, Philadelphia (18th) Ballot Act, England (18th) American painter James McNeill Whistler exhibits *Whistler's Mother* at the Royal Academy of Arts, London
July 1872	Based in Chamonix (1st–7th) Attempts to climb the Aiguille du Géant with Thomas Kennedy (5th) Based at the Bel Alp (8th–at least 5th August) House of Lords discusses Hooker-Ayrton dispute (29th)	Mexican president Benito Juárez dies (18th) Attempted assassination of the king and queen of Spain
August 1872	House of Commons debates Hooker-Ayrton dispute (8th)	BAAS holds its forty-second annual meeting at Brighton (14th–21st)

Month and year	Biographical details	Notable social, political, and cultural events
August 1872	Presumably returns to London from his trip to Switzerland (10th) Attends only one day of the BAAS meeting at Brighton; attends William Spottiswoode's lecture (17th) Appointed a metropolitan gas referee in London (29th)	Australian overland telegraph completed and first message sent (22nd) Spain's general election (24th)
September 1872	Prepares his American lectures while in Ascot and Folkestone before departing England Publishes *Forms of Water* Has dinner with Thomas Archer Hirst and Heinrich Debus (26th) Travels from London to Liverpool; has dinner with George Warren (27th) Embarks on the SS *Russia* from Liverpool (28th)	Charles XV, King of Sweden and Norway, dies; Oscar II crowned king (18th) Karl Marx speaks in Amsterdam (10th) Russian composer Anton Rubinstein gives first concert on American tour at Steinway Hall (23rd)
October 1872	Arrives in New York on the SS *Russia* (9th) Travels from New York to Boston (12th) Delivers six lectures at the Lowell Institute in Boston (15th–25th) Travels on overnight train to Niagara Falls (31st) Publishes "On Prayer" in the *Contemporary Review*	Bloomingdale's department store opens in New York (3rd) American politician William H. Seward dies (10th) Prince Albert of Prussia dies (14th)
November 1872	Travels to and explores Niagara Falls (1st–8th) Travels from Niagara Falls to Philadelphia (8th) Delivers six lectures at the Horticultural Hall in Philadelphia (11th–22nd) Travels from Philadelphia to Baltimore (25th) Delivers three lectures at the Peabody Institute in Baltimore (26th–30th)	Ulysses S. Grant defeats Horace Greeley in US presidential election (5th) American suffragist Susan B. Anthony votes for the first time (in defiance of the law) (5th) Great Boston Fire of 1872 destroys a significant portion of the city (9th–10th) British chemist Baldwin Francis Duppa dies (10th) Charles Darwin publishes *The Expression of the Emotions in Man and Animals* (26th) Scottish science writer Mary Somerville dies (29th)

Month and year	Biographical details	Notable social, political, and cultural events
November 1872	Publishes "Science and Religion" in *Popular Science Monthly*; the magazine publishes "Sketch of Professor Tyndall" in the same issue Tyndall's younger assistant, George Miller, dies of typhoid fever in Philadelphia (30th)	RS awards its Copley Medal to Friedrich Wöhler for "his numerous contributions to the Science of Chemistry; and more especially for his researches on the products of the decomposition of Cyanogen by Ammonia; on the Derivatives of Uric Acid; on the Benzoyl Series; On Aluminum (discovered by him), Yttrium, and Beryllium; on Boron, Silicon, & their compounds; on Titanium; and on Meteoric Stones" (30th) RS awards its Royal Medal to Henry John Carter for "his researches in Palaeontology and Zoology, on the Infusoria and Rhizopoda, and the root-cell of the *Chara*; but more particularly for his inquiries into the Natural History of the Spongiadae"; and Thomas Anderson for "his investigations on the Organic Bases of Dippell's Animal Oil; on Codeine; on the Crystallized Constituents of Opium; on Piperin and on Papaverin; and for his researches in Physiological and Agricultural Chemistry" (30th) RS awards its Rumford Medal to Anders Jonas Ångström for "his researches on Spectral Analysis" (30th)
December 1872	Travels from Baltimore to Washington, DC Delivers six lectures at the Smithsonian's Lincoln Hall in Washington, DC (3rd–12th) Travels from Washington, DC to New York (14th) Appleton reception (14th) Delivers six lectures at the Cooper Institute in New York (17th–31st) Visits Stevens Institute of Technology in Hoboken, NJ (24th)	Abandoned American ship *Mary Celeste* found in the Atlantic (4th) Springwall Pit disaster in Dawley, England (6th) French naturalist and proponent of spontaneous generation Félix-Archimède Pouchet dies (6th) HMS *Challenger* (1858) departs Portsmouth, England, beginning its four-year scientific expedition (21st) Scottish engineer William Rankine dies (24th)

Month and year	Biographical details	Notable social, political, and cultural events
January 1873	Travels from New York to Brooklyn (4th) Delivers six lectures at the Academy of Music in Brooklyn (4th–18th) Travels from Brooklyn to New Haven, CT Delivers two lectures at Yale College in New Haven (22nd–23rd) Travels from New Haven to New York (25th) Travels from New York to Philadelphia, visits with cousin Hector Tyndale (27th)	Napoleon III, last emperor of France, dies (9th) First Battle of the Stronghold in the "American Indian Wars" fought: Modoc defeat the US Army (17th) William Spottiswoode delivers a Friday Evening Discourse at the RI, "On the Old and New Laboratories at the Royal Institution" (17th) English politician and writer Edward Bulwer-Lytton (1st Baron Lytton) dies (18th) British geologist Adam Sedgwick dies (27th)
February 1873	Attends banquet in his honor at Delmonico's in New York (4th) Embarks on the SS *Cuba* from New York (8th) Arrives in Liverpool, England, on the SS *Cuba*; arrives in London later the same day, at Euston Station, where Thomas Archer Hirst and Heinrich Debus greet him (19th)	American oceanographer Matthew Fontaine Maury dies (1st) Earthquake on Greek Island of Samos (4th) Four hundredth anniversary of the birth of Nicolaus Copernicus celebrated in Thorn, Prussia (19th) Geological Society of London awards its Wollaston Medal to Philip de Malpas Grey Egerton for his forty years of service to geology, especially for "the light you have thrown on the structure and affinities of fossil fishes and reptiles" (21st) James Clerk Maxwell delivers a Friday Evening Discourse at the RI, "On Action at a Distance" (21st)
March 1873	Tyndall congratulated on his American lecture tour at a meeting of the RI (4th) Takes short trip to Bath with Thomas Henry Huxley (15th–17th) Dinner at the Athenaeum Club to celebrate Tyndall's return from America and Thomas Archer Hirst's appointment as Director of Studies at the new Royal Naval College, Greenwich (22nd)	Ulysses S. Grant is sworn in as president of the United States (his second term) (4th) American botanist John Torrey dies (10th) Charles Lyell delivers a Friday Evening Discourse at the RI, "On the Mythology and Temples of India" (21st) Emancipation day for Puerto Rico (22nd)

Month and year	Biographical details	Notable social, political, and cultural events
March 1873		William Kingdon Clifford delivers a Friday Evening Discourse at the RI, "The Meaning of Force and Energy" (28th) James Joseph Sylvester delivers a Friday Evening Discourse at the RI, "On the Meaning of Force and Energy" (28th) Cambridge wins the Oxford and Cambridge boat race (29th) Leo Tolstoy's *Anna Karenina* released in serial installments in the *Russian Messenger*
April 1873	Delivers a Friday Evening Discourse at the RI, "Some Observations of Niagara" (4th) Concludes gas is superior to paraffin or colza oil for lighthouse illumination in a report to the Irish Inspector of Lights (10th) Attends luncheon with Ralph Waldo Emerson, Thomas Archer Hirst, and Thomas Henry Huxley (23rd) Donates, along with seventeen other friends, to a fund to support Huxley in a much-needed health-restoring holiday	British ocean liner RMS *Atlantic* sinks off Nova Scotia, killing 547 (1st) Second Battle of Stronghold in the Modoc War, United States (15th–17th) German chemist Justus von Liebig dies (18th) English physician and chemist Henry Bence Jones dies (20th) English actor and manager William Charles Macready dies (27th) Scottish mathematician James Clerk Maxwell publishes *A Treatise on Electricity and Magnetism* Maskelyne & Cooke inaugurate their "Home of Mystery" magic show in the Egyptian Hall, Piccadilly, London
May 1873	"Tyndall Trust for Original Investigation" established Publishes his paper on Niagara Falls in *Macmillan's Magazine*	Opening of Vienna's World Fair (1st) British philosopher John Stuart Mill dies (8th) Drummond Mine explosion in Westville, Nova Scotia (13th) William Spottiswoode delivers a Friday Evening Discourse at the RI, "On Spectra of Polarised Light" (23th) Thoroughbred racehorse Doncaster wins the Derby Stakes at Epsom, Surrey (28th) Thomas Henry Huxley publishes "Forbes and Tyndall" in *Nature*

Month and year	Biographical details	Notable social, political, and cultural events
May 1873		US Marines attack Panama Heinrich Schliemann discovers "Priam's Treasure," a cache of gold
June 1873	Receives an honorary Doctor of Civil Law (DCL) degree from Oxford (18th) Publishes *Six Lectures on Light* Publishes "Some Observations on Niagara" in *Popular Science Monthly*	Modoc War ends with the capture of Chief Kintpuash (1st) Sultan of Zanzibar signs new treaty and closes slave market under foreign pressure (5th)
July 1873	Travels to Ireland for lighthouse work (8th) King Oscar II of Sweden makes Tyndall, Joseph Dalton Hooker, and Thomas Henry Huxley each a Knight of the Order of the Polar Star (18th) In Folkestone to finish some work (28th–30th)	Bishop Samuel Wilberforce dies (19th) Third Carlist War (Spain): Battle of Alpens (9th) French poet Paul Verlaine shoots and wounds Jean Nicolas Arthur Rimbaud (10th) Jesse James and James Younger's first train robbery in the American West (21st)
August 1873	Begins journey to Switzerland via Paris with Thomas Archer Hirst (1st) Arrives in Paris (1st); leaves Paris (5th) Arrives in Switzerland (6th); arrives in Brieg (7th); arrives at the Bel Alp (8th) Visits with the two daughters of the late Mary Somerville (10th) Various alpine excursions with the Hamiltons, Hirst joins for some (12th–19th) Hirst departs from Tyndall (20th), travels in Italy, returns to London (1st September) Publishes "Principal Forbes and His Biographers" in *Contemporary Review*	Peace treaty signed by imperial Russia and Khanate of Khiva, making khanate a protectorate of Russia (12th) First ascent of Mount Whitney, California (18th) Charles II, Duke of Brunswick, dies (18th)
September 1873	Returns to London (about the 11th, according to Hirst) Dispute in pages of *Nature* with Peter Guthrie Tait over whether James David Forbes plagiarized Louis Rendu's work on the movement of glaciers	Last German troops leave France upon the completion of payment of indemnity for the Franco-Prussian War (16th) BAAS holds its forty-third annual meeting at Bradford (17th–24th)

Month and year	Biographical details	Notable social, political, and cultural events
September 1873		New York stock market crash, "Panic of 1873" triggers the Long Depression (18th) Czech composer Antonín Dvorák writes String Quartet no. 5 First guests welcomed at Anton Dohrn's Stazione Zoologica in Naples, Italy

The Correspondence of John Tyndall

1872

From Carlo Cadorna[1] 1 June 1872 3707

20 Hill Str. 1 Juin 72

Monsieur le Professeur

Je dois m'excuser si à cause des circonstances particulières je prends la liberté de vous envoyer à la veille le billet d'invitation ci-inclus, et j'espère que vous voudrez bien l'accepter, si vous n'êtes pas engagé.

Tout dévoué | Ca Cadorna

20 Hill Str.[2] 1 June 72

Professor,

I must apologize that, due to unusual circumstances, I am taking the liberty of sending you the enclosed invitation[3] at the last minute. I hope that you will accept it, if you are not otherwise committed.

Yours truly | Ca Cadorna

RI MS JT/1/C/2

---

1. *Carlo Cadorna*: Carlo Cadorna (1809–91), an Italian politician and ambassador to Great Britain from 1869–75 (*Dizionario Biografico degli Italiani*, 100 vols (Istituto dell'Enciclopedia Italiana, 1973), vol. 16, pp. 97–104; A. Litta Visconti Arese, *The Birth of Modern Italy: Posthumous Papers of Jessie White Mario* (New York: Charles Scribner's Sons, 1909), p. 5).
2. *20 Hill Str.*: at Berkeley Square in London.
3. *enclosed invitation*: enclosure missing.

From Joseph Dalton Hooker [1 June 1872][1] 3708

Royal Gardens, Kew. | Aug. 29th 1872.

Dear Tyndall,

There is a consideration respecting Gladstone's conduct[2] which though of no practical moment, I may as well mention, which is this:—that when he decided to "alter my position in relation to the First Commissioner"[3] he was not actuated solely by the desire to remove the difficulty which Ayrton's[4] rule over Kew had given rise to; but, we may hope, by the desire to place this Institution (if not its Director) in a better official position in relation to the Government Offices with which it had really most practical relations.

The reference to my duties, in my letter to him of Aug. 21[5] (of which you have a copy) may have influenced him to this course.

For myself I can truly say, that I looked forward to this move, as the reward, and the only one I ever should care for, for my father's[6] and my own devotion to this Institution. Though too good to hope for, even when the prospect of its being carried out was brightest, the abandonment of "the plan" was the bitterest disappointment I ever experienced. It has coloured my official life ever since; and to be frank, it keenly intensifies the repugnance with which I regard any attempt to force a retraction out of me. I believe that in offering (if demanded under the circumstances) to apologise, or express regret, for the tone wording or manner (call it what is right) of my letter to Mr West;[7] I am doing right; the so doing will be a bitter but not dishonouring mortification, if not humiliation. Contrasted with the alternative position which was so long in prospect for me, it cannot but be a cruel blow, and one which might well make resignation or dismissal a welcome alternative, and it is one which Mr Gladstone should ward off.

It might be well to state at bottom of p. 12 in a footnote that Dr Hooker's letter adducing the five distinct acts,[8] was never answered or acknowledged by Mr Ayrton (I had not told him I was sending a copy to Mr Gladstone)

P.21 I would suggest the following addition at middle of page beginning "and your secretary proposed that, as an official answer would be sent to Dr Hooker's official application, the letter from which etc"[9]

Mr West made a great point, in his subsequent letter, of my having accepted a private message as an answer to an official appeal, quite forgetting that Mr Gladstone's private note from Balmoral[10] was intended to be a full and final answer to my two official applications[11]

Ever yours | J. D. Hooker.

RI MS JT/1/TYP/8/2708
LT Typescript Only

1. *[1 June 1872]*: the date provided by LT on the typescript is suspect, as August 1872 would be too late for Hooker to be writing to Tyndall with suggested edits for the Memorial (see n. 3). We date this to [1 June 1872] by relation to a letter dated 29 May 1872 (letter 3706, *Tyndall Correspondence*, vol. 12), in which Hooker wrote 'I will see to the corrections the first thing after my return from Edinburgh', and hoped to be back on Saturday (1 June). Tyndall's next letter to Hooker on 3 June (3710) has him taking the proof of the Memorial to the printer.
2. *Gladstone's conduct*: see letter 3755, n. 1.
3. *"alter my position in relation to the First Commissioner"*: the words 'alter his position in relation to the First Commissioner' are from the Memorial in support of Joseph Hooker in his dispute with Acton Ayrton over his role as Director of the Royal Botanic Gardens, Kew (see letter 3755, esp. n. 1).
4. *Ayrton's*: Acton Ayrton.
5. *my letter to him of Aug. 21*: Hooker may be mistaken with the date, and referring to his letter to Ayrton of 19 August 1871. See 'Copies of Correspondence between the Board of Works and Dr. Hooker relating to Changes proposed to be introduced into the Direction and Management of the Gardens at Kew', HC 25, Parliamentary Papers (1872), p. 28.
6. *my father's*: William Jackson Hooker.
7. *my letter to Mr West*: on this letter to Algernon Edward West (1832–1921), an English civil servant who was private secretary to William Gladstone from 1861–94 (*ODNB*), see 'Copies of Correspondence between the Board of Works and Dr. Hooker relating to Changes proposed to be introduced into the Direction and Management of the Gardens at Kew', HC 25, Parliamentary Papers (1872), p. 47.
8. *a footnote that Dr Hooker's letter adducing the five distinct acts*: this footnote was included in the final version of the Memorial (see letter 3755).
9. *following addition at . . . letter from which etc"*: the words 'as an official answer would be sent to Dr Hooker's official application' were included in the final version of the Memorial (see letter 3755).
10. *Mr Gladstone's private note from Balmoral*: for this letter from Balmoral Castle in Scotland, see letter 3755; or 'Copies of Correspondence between the Board of Works and Dr. Hooker relating to Changes proposed to be introduced into the Direction and Management of the Gardens at Kew', HC 25, Parliamentary Papers (1872), p. 46.
11. *my <u>two</u> official applications*: probably referring to his letters to Ayrton of 12 July 1871 and 19 August 1871 (see 'Copies of Correspondence between the Board of Works and Dr. Hooker relating to Changes proposed to be introduced into the Direction and Management of the Gardens at Kew', HC 25, Parliamentary Papers (1872), pp. 28, 106).

From Joseph Dalton Hooker [2 June 1872][1] 3709

Royal Gardens, Kew. | [Sunday][2]

Dear Tyndall,

I got back this forenoon pretty well tired with two night journeys and two days of examining witnesses and *[1 word illeg]* at Museum.[3] There never was such a muddle as we found things to be in.

I am at your service whenever you wish, but what is Lord Derby about? Has he had any answer from Lord Granville? If not we must get Lubbock to call for the papers without delay. The session is fast running away and whatever else is done, I fancy this must go before Parliament.[4]

I should like Lubbock (if Lord Derby cannot act) to call for the correspondence with Gladstone and the papers about the accident to the heating.

I shall now have a tough job with this Edinburgh Report.[5]

Ever yours | J. D. Hooker.

RI MS JT/1/TYP/8/2643
LT Typescript Only

1. *[2 June 1872]*: the date is given by relation to a letter from Hooker dated 29 May 1872 (letter 3706, *Tyndall Correspondence*, vol. 12). Hooker wrote to Tyndall: 'I will see to the corrections the first thing after my return from Edinburgh, whither I go to-night, to enquire into this squabble concerning the Museum. I hope to be back on Saturday'.
2. *[Sunday]*: this was handwritten on the typescript, indicating that it might not have been on the actual letter.
3. *examining witnesses and [1 word illeg] at Museum*: not identified. Possibly having to do with the Edinburgh Museum of Science and Art and Hooker's role with the Royal Commission on Scientific Instruction and the Advancement of Science (see letter 3747, n. 6).
4. *this must go before Parliament*: see letter 3755, esp. n. 1.
5. *Edinburgh Report*: see n. 3.

To Joseph Dalton Hooker 3 June [1872][1] 3710

Royal Institution of Great Britain | 3rd. June.

My dear Hooker,

I dare say you imagined my last telegram[2] to be a kind of response to your last note.[3] It had however been sent off before your note arrived.

I have taken all precautions with Spottiswoode; but I really do not see what risk you mean. As a public document[4] it would not lose a whit of wight through the fact of my putting the thing, on my own responsibility, and purely for the sake of convenience & despatch, into print. I suppose it will be signed by all our best men, and they are scattered over England. My notion was that copies should be sent to them so as to secure their signatures; for it would take a long time for the same account to filter through the whole of them.

Besides it will expedite matters to send copies to those who more immediately advise you in these matters—so that where they come together they shall be prepared with their comments & observations.

I intend to walk on with the proof myself to Spottiswoode's this morning—and I hope that tonight the complete memorial will be in my hands. Then I think it would be well to call a meeting tomorrow afternoon at the Athenaeum.[5] I trust & think that but little alteration will be necessary.

I blew off some steam apropos of Ayrton[6] (without naming him) at the Chemical dinner on Friday.[7] Lowe was there, and when I spoke of the crass ignorance of men in authority (always excepting Lowe) and their doing things before High Heaven that would make the Angels' weep, Lowe said "That's Ayrton I quite agree with Tyndall".

I had a talk with Huxley yesterday—I wish I could carry him off to Switzerland. But he has six week's work before /being/ still at Kensington.[8] I must make another attempt, for he is not himself.

Immediately after the memorial has passed muster I will have it fairly copied, and made ready for sending in.

Yours ever | John Tyndall

RBG, Kew Papers Relating to Kew 1867–1872, Ayrton Controversy, volume 1, f. 222

---

1. *[1872]*: the year is given based on reference to the Memorial in support of Joseph Hooker in his dispute with Acton Ayrton over his role as Director of the Royal Botanic Gardens, Kew (see n. 4).
2. *my last telegram*: telegram missing.
3. *your last note*: letter missing.
4. *public document*: see letter 3755, esp. n. 1.
5. *Athenaeum*: the Athenaeum Club, a private members' club located at 107 Pall Mall in London, particularly for those with intellectual interests, and accomplishments in art, science, or literature. The club was founded in 1824, at the instigation of John Wilson Croker. Tyndall became a member in 1860, Thomas Huxley in 1858, Joseph Hooker in 1851, and Thomas Hirst in 1866. For Tyndall's election to the club, see *Ascent of John Tyndall*, pp. 142–3. On the Club in general, see M. Wheeler, *The Athenaeum: More Than Just Another London Club* (New Haven, CT: Yale University Press, 2020).
6. *Ayrton*: Acton Ayrton.

7. *Chemical dinner on Friday*: a dinner on 31 May 1872 organized by the Chemical Society of London in honor of Italian chemist Stanislao Cannizzaro (1826–1910).
8. *at Kensington*: see letter 3711, n. 7.

To Thomas Henry Huxley 3 June [1872][1] 3711

Royal Institution. | 3rd. June.

My dear Huxley,

Emerson's daughter, Mrs Forbes,[2] and her husband Col. Forbes[3] are here until Thursday and they wish exceedingly to see you before they leave. They are staying at the Albemarle Hotel Albemarle Street.

I went over to Jermyn Street[4] to day but learned that you were engaged at South Kensington.[5] I had some notion of rigging up a luncheon, but I do not know how your times would agree. At all events by hook or crook[6] they must manage to see you before they go—

Just drop me a line to say how it could be arranged.

Speaking of South Kensington. The word "impossible" which you employed yesterday is obviously <u>relative</u>. For were you as I was 12 years ago the impossibility would dissolve and you would be forced to ground arms. The sole question is whether it is <u>wise</u>, not whether it is possible. You may think me somewhat wild about money matters, but I have had sufficient prudence to render myself perfectly secure. So that if I struck work tomorrow I should be able to eat the bread of independence. I declare to you I would willingly devote £500 to pay the expenses of those schoolmasters in London if you could displace those lectures[7] and come with me for a month to Switzerland. Look at the case of Spencer. He has never recovered his strength of brain; and though you are very different from him in point of toughness you can far less afford to run any risk. The thought of your being in any way crippled in future is simply intolerable to me. I say then unless you have absolute confidence in your powers of endurance, it is your bounden duty to your family to me to the world not to entangle yourself in this work just at present. To only one or two in the world could I write in this way without impertinence. But a brother will not take my language amiss. Mind I mean what I say. I am quantitatively accurate in my statement regarding the displacement of the lectures.

Yours ever | J. Tyndall.

RI MS JT/1/TYP/9/3013
IC HP 8.120
Typed Transcript Only

1. *[1872]*: the year is given by the addition of '1872' on the letter by LT, and by relation to this letter's reply, dated 4 June 1872 (letter 3712).
2. *Mrs Forbes*: Edith Emerson Forbes (1841–1929), the third child of Emerson and his second wife Lidian Emerson (née Jackson, 1802–92). Their other children were: Waldo Emerson (1836–42), Ellen Tucker Emerson (1839–1909), and Edward Waldo Emerson (1844–1930) (*ANB*).
3. *Col. Forbes*: William Hathaway Forbes (1840–97), an American businessman who served for the Union Army in Massachusetts during the American Civil War (1861–5). He married Edith Emerson in 1865 ('William Hathaway Forbes', *The Norfolk Virginian*, 24 October 1897, p. 10).
4. *Jermyn Street*: Huxley was Professor of Natural History at the Royal School of Mines, located on Jermyn Street in London, from 1854–85.
5. *South Kensington*: in the early 1870s, the Royal School of Mines moved to new buildings at South Kensington.
6. *hook or crook*: an English phrase meaning 'by any means necessary'.
7. *those lectures*: a short summer course of lectures Huxley delivered to science teachers at the Royal School of Mines at South Kensington (previously on Jermyn Street), which he first gave in 1871, focusing on his view of teaching biology through the laboratory study of physical specimens. See T. H. Huxley and H. N. Martin, *A Course of Practical Instruction in Elementary Biology* (London: Macmillan and Co., 1876), pp. 376–81. This laboratory guide went through numerous editions.

From Thomas Henry Huxley 4 June 1872 3712

June 4th. 1872.

My dear Tyndall,

I <u>must</u> be at work on examination papers[1] all day today, but to-morrow I am good to lunch with you (and abscond from the Royal Commission[2] which will get on very well without me) or to go and call with you on your friends, whatever may be most convenient.

Many thanks for all your kind and good advice about the lectures,[3] but I really think they will not be too much for me and it is of the utmost importance I should carry them on.

They are the commencement of a new system of teaching which, if I mistake not, will grow into a big thing and bear great fruit, and just at this present moment (nobody is necessary very long) I am the necessary man to carry it on. I could not get a supplicant if I would, and you are no more the man than I am to let a pet scheme fall through for the fear of a little risk of self. And really and truly I find that by taking care I pull along very well. Moreover it isn't my brains that get wrong, but only my confounded stomach.

I have read your memorial[4] which is very strong and striking, but a difficulty occurs to me about a good deal of it, and that is that it won't do to quote Hooker's official letters before they have been called for in Parliament, or otherwise made public. We should find ourselves in the wrong officially, I am afraid, by doing so. However we can discuss this when we meet. I will be at Athenaeum[5] at 4 o'clock.

Ever yours faithfully | T. H. Huxley.

IC HP 8.121
Typed Transcript Only

---

1. *examination papers*: probably from students at the Royal School of Mines, where Huxley was Professor of Natural History from 1854–85.
2. *Royal Commission*: Huxley served on the Royal Commission on Scientific Instruction and the Advancement of Science (see letter 3747, n. 6).
3. *your kind and good advice about the lectures*: see letter 3711, esp. n. 7.
4. *your memorial*: see letter 3755, esp. n. 1.
5. *Athenaeum*: the Athenaeum Club (see letter 3710, n. 5).

To Thomas Henry Huxley 4 June [1872][1] 3713

Royal Institution | 4th. June.

My dear Huxley

I expect great results from what you are now engaged upon,[2] and I expect to be able to lend you an active helping hand by and by.

Indeed I think you will regard my little glacier book[3] as a step in this direction.

All I want and desire now is that you should take care of yourself, and not by the unreasonable outlay of power diminish the power of the future.

....

Yours ever | John Tyndall

My notion was, the moment the terms of the Memorial[4] are fixed to urge upon Lord Derby the necessity of moving at once for the correspondence, and thus include the Memorial which as a summary will assist hazy or busy members in coming to a conclusion.

I am walking out to Kew but shall be at the Club[5] at 4. P.M.

Hooker by the way may not be able to come in sooner than by the train which reaches Waterloo at 4.3. Please find Lubbock and Bentham at the Club and ask them for a quarter of an hour's grace.

RI MS JT/1/TYP/8/2645
IC HP 8.122
Typed Transcript Only

1. *[1872]*: the year is given by a note from LT on the letter, as well as content—the Memorial in support of Joseph Hooker in his dispute with Acton Ayrton over his role as Director of the Royal Botanic Gardens, Kew (see letter 3755, esp. n. 1).
2. *now engaged upon*: see letters 3711 and 3712 regarding lectures Huxley was planning to give.
3. *my little glacier book*: *Forms of Water* was published just before he left England for the United States in September. This book was published as the first title in Edward L. Youmans's International Scientific Series (see letter 3871, n. 3). An American edition was published the same year by Appleton and Co. in New York.
4. *Memorial*: see letter 3755, esp. n. 1.
5. *the Club*: the Athenaeum Club (see letter 3710, n. 5).

From Heinrich Rudolf Vieweg[1] 4 June 1872 3714

Braunschweig 4tn Juni 1872 | Herrn Profeßor John Tyndall | London.

Hochgeehrtester Herr Professor.

In Ihrem gefälligen Schreiben vom 15 Febr. zeigten Sie uns an dass Sie M^r^. Longman beauftragt hätten, uns sowohl die zweite Auflage Ihres Werkes „über den Schall" sowie die neue Auflage der Fragments zu senden. Wir haben beides nicht erhalten und sind derselben jetzt sehr benöthigt, da die neue Ausgabe des *[ersteren]* jetzt sofort in Angriff genommen werden muß und die Uebersetzung der Fragments vollendet, nach der neuen Auflage abgeändert werden muß.

Sie würden uns sehr verbinden, wenn Sie die shleunigste Zusendung der Bücher veranlaßen wollten.

Mit der ausgezeichnetsten Hochachtung | ergebenst | Friedr Vieweg & Sohn.

Braunschweig 4th June 1872 | Herr Professor John Tyndall | London.

Most esteemed Herr Professor.

In your kind letter of 15 Feb.[2] you indicated to us that you had instructed M^r^. Longman[3] to send us both the second edition of your work "On Sound"[4] as well as the new edition of the Fragments.[5] We have not received either of them and now require them very much, as the new edition of the *[former]*[6] now has to be tackled straight away and the translation of the Fragments[7] has to be completed and revised in accordance with the new edition.

We should be very obliged to you if you would arrange for the books to be sent as quickly[8] as possible.

With the most excellent esteem | your humble servants | Friedrich Vieweg & Son.

RI MS JT/1/V/11

1. *Heinrich Rudolf Vieweg*: while the letter is signed 'Friedr Vieweg & Sohn', we believe the sender is Heinrich Rudolf Vieweg, who took over the family publishing firm, Friedrich Vieweg und Sohn, from his father Eduard Vieweg (1797–1869).
2. *kind letter of 15 Feb.*: letter missing.
3. *M^r. Longman*: either Thomas Longman (1804–79) or William Longman (1813–77), brothers who ran the publishing firm Longmans, Green, and Co., publisher of Tyndall's *Sound* (see n. 4).
4. *"On Sound"*: *Sound.*
5. *new edition of the Fragments*: the second edition of *Fragments of Science* was published in the same year as the first edition (1871).
6. *new edition of the [former]*: J. Tyndall, *Der Schall: Acht Vorlesungen Gehalten in der Royal Institution von Grossbritannien*, trans. G. Wiedemann and H. Helmholtz (Braunschweig: F. Vieweg und Sohn, 1874).
7. *translation of the Fragments*: J. Tyndall, *Fragmente aus den Naturwissenschaften*, trans. A. Helmholtz and H. Helmholtz (Braunschweig: F. Vieweg und Sohn, 1874). 'A. Helmholtz' is Anna Helmholtz, wife of Hermann Helmholtz who previously translated some of Tyndall's books into German without receiving publication credit.
8. *quickly*: our translator noted that the word in German for 'quickly'—*schleunigste*—was written by Vieweg without the 'c'.

From Joseph Dalton Hooker 5 June 1872 3715

Royal Gardens, Kew. | June 5th 1872.

My dear Tyndall,

I am indeed glad to hear that your Address[1] was so fully agreed to, with the alterations and modifications which you yourself suggested.

I have just cleared off a heavy piece of work, and am at your service for details.[2] We shall I hope meet to-morrow at the Club.[3] I see but two points on which I am susceptible, and on these I am as to the one childish, as to the other impregnable. The first is the "injured in health".[4] I think this is a trifle undignified, and an acknowledgment that Ayrton[5] has bored a hole into me. This wound I would as anxiously cover as if it had been in my back, instead of my front! and it rather dulls the effect of the next passage, which is crushing as an argument, and which refers to impaired scientific usefulness. The other is "natural capacity of the highest order".[6] Kindly temper this superlative to "high", which is all too much.

As to the rest, I have really nothing of importance to suggest even, nothing but trifles, altogether in detail, none that will weaken, and most that strengthen the position. Of course I regard the position we have taken as unassailable; and your fort bristles with ordnance, the shot and shell do rattle.

There is no time to be lost, I dine with Lubbock to-day and shall speak to him urgently.

I have reason to believe that the Treasury are again check-mated by Gladstone. They see well enough the impending disgrace to themselves as regulators of official duties in the Departments. When this comes out, the first thing the official mind will say is, "What were the Treasury officers about?"

But there is no time to be lost. The state of parties makes it most urgent, and I hope that L[or]d Derby will act, or give up without parley. Depend on it, he has got a put-off from Granville.

Telegraph, and I am with you an hour after.

Yours | J. D. Hooker

RI MS JT/1/TYP/8/2646
LT Typescript Only

1. *your Address*: by 'Address', Hooker likely was referring to the Memorial in support of him concerning his dispute with Acton Ayrton (see letter 3755, esp. n. 1).
2. *at your service for details*: regarding the Memorial.
3. *the Club*: the Athenaeum Club (see letter 3710, n. 5).
4. *"injured in health"*: these words do not appear in the Memorial sent to Gladstone.
5. *Ayrton*: Acton Ayrton.
6. *"natural capacity of the highest order"*: these words do not appear in the Memorial sent to Gladstone; the following words might have replaced them: 'in whom natural capacity for the post he occupies has been developed by a culture unexampled in variety and extent'.

From Mary Adair — 6 June [1872][1] — 3716

63. Portland Place.[2] | Friday 6th June[3]

Dear Dr Tyndall,

I must thank you for the great pleasure you have given me by permitting me to attend your Lectures.[4] They were so clear that a Child might learn from them, and so profound that they guided us on towards limitless ranges of Thought.

My Mother[5] would, I am sure, have desired me to add her Thanks to mine, but she has left London for the country. We are very glad that you will soon be amongst the Mountains[6] and rest and renew your strength—and hope that the American expedition[7] will prove in all ways delightful. Only don't allow our Transatlantic brethren to tempt you over en permanence with their claims Direct or Indirect—for the "consequential Losses" to British Science would be irreparable.

My sister[8] gave me all your kind messages and pray do not think of troubling yourself to send any answer to,

yours ever truly | H. M. Adair

RI MS JT/1/A/11

---

1. *[1872]*: the year is given by a note on the letter from LT. The year is also established by Adair's mention of Tyndall's 'American expedition' (see n. 7).
2. *63. Portland Place*: residence in London.
3. *Friday 6th June*: in 1872, 6 June fell on a Thursday, so either Adair mistakenly wrote 'Friday', or she got the date wrong and she wrote this letter on 7 June.
4. *your lectures*: Tyndall delivered a course of nine lectures, 'On Heat and Light', at the RI on Thursdays from 11 April to 6 June 1872 (*Roy. Inst. Proc.*, 6 (1870–72), p. 384).
5. *My Mother*: Harriet Eliza Adair.
6. *amongst the Mountains*: Tyndall left London for Switzerland on 25 June 1872, and presumably returned to London by 10 August 1872 (as he remarked in letters 3771, 3794, and 3806).
7. *American expedition*: for Tyndall's American lecture tour, see Introduction, n. 8.
8. *My sister*: Harriet Camilla Adair (d. 1909), eldest daughter of Alexander and Harriet Adair (J. B. Adair, *Adair History and Genealogy* (Los Angeles: Published by author, 1924) p. 24).

From Edward Stanley — 6 June 1872 — 3717

23 St James's Square[1] | June 6th 1872.

Dear Professor Tyndall

I had not overlooked Dr Hooker's affairs[2] and have obtained Lord Granville's consent to the papers being "laid on the table".[3] I shall give notice to-night, move for them to-morrow and they ought to be in our hands in three or four days. There cannot be the least objection to quoting the official correspondence, whether published or not. You need not fear that I shall neglect the matter. I fear we are a long way from the end of the session still!

Very faithfully yours | Derby.

RI MS JT/1/TYP/8/2647
LT Typescript Only

1. *23 St James's Square*: residence in London.
2. *Dr Hooker's affairs*: Stanley was referring to the Memorial in support of Joseph Hooker in his dispute with Acton Ayrton (see letter 3755, esp. n. 1). In letter 3713, Tyndall wrote to Huxley that his 'notion was, the moment the terms of the Memorial are fixed to urge upon Lord Derby the necessity of moving at once for the correspondence'.
3. *papers being "laid on the table"*: as referred to in other letters, calling for the correspondence and papers related to Hooker and Ayrton's dispute.

To George Biddell Airy 7 June 1872 3718

Private. | Royal Institution[1] | 7th June 1872

My dear Sir,

A memorial[2] signed by a few of our leading men of science is about to presented to Mr. Gladstone. It has reference to the very arbitrary treatment of Dr. Hooker, the Director of the Botanical Establishment at Kew by Mr. Ayrton,[3] the First Commissioner of Works. I should be exceedingly glad to place a condensed summary of the case before you if you could give me reason to hope that you would be likely to sign the memorial.[4]

Most faithfully yours | John Tyndall

G. B. Airy Esq &c &c

RGO MS.RGO 6/438

1. *Royal Institution*: while this is written on the letter, the stationary bears a stamp with: 'ATHENAEUM CLUB | PALL MALL'. For the Athenaeum Club, see letter 3710, n. 5.
2. *memorial*: see letter 3755, esp. n. 1.
3. *Mr. Ayrton*: Acton Ayrton.
4. *you would be likely to sign the memorial*: while Airy declined to sign the Memorial on account of his being a servant of the British government (see letter 3726), Tyndall did not stop trying (see letters 3730, 3759, 3763, and 3772).

To Mary Adair 8 June 1872 3719

Royal Institution of Great Britain | 8th. June 1872.

My dear Miss Adair.

You must allow me to tell you that your note[1] has gratified me much; but I will not ask you to allow me to tell you how pleased I was to see you here week after week[2]

Certainly if good will could induce me to change England for America I should have a strong inducement to go there. They are wonderfully kind to me—They have named mountains and associations after me,[3] and their private expressions are still kinder than their public acts.

The first strong inducement to my going amongst them was found in one of the letters addressed to me,[4] wherein it was said that the growth of good will between England and America would be greatly promoted if the non-political scientific men of both countries could be induced to interchange friendly visits

They have a wonderfully organised system of lecturing in America and I have agreed to lecture in five of their cities.[5] The people of Chicago have suffered so much that they have many wants to supply before they can turn their thoughts to science. So my notion is to hand over the proceeds of the lectures, which they tell me will be large, to the help of science in Chicago.[6]

This is one of the benefits of bachelorhood[7]—one is free to do such things!

The whole matter is in the hands of my scientific friends; Professor Henry the Director of the Smithsonian Institution, the leading physical philosopher of the United States, having consented to take the whole organisation into his hands.

I have been gathering my memoirs[8] together into a volume which I wish you to have, but not to read. It will reach you today.

And now goodbye—say all manner of kind things to your Mother[9] for me, & believe me

always yours | John Tyndall

RI MS JT/1/T/87
RI MS JT/1/TYP1/12–3

---

1. *your note*: letter 3716.
2. *see you here week after week*: Adair attended a lecture course delivered by Tyndall (see letter 3716, n. 4).
3. *named mountains and associations after me*: part of the California Geological Survey's explorations in the Sierra Nevada included an ascent and measurement of a peak (on 6 July 1864) that Clarence King named Mount Tyndall (it rises to just over 14,000 feet). This peak is part of the Mount Whitney range, named after Josiah Dwight Whitney (1819–96), an American geologist who was director of the California Geological Survey. King recounts the ascent and descent of Mount Tyndall in chapters 3 and 4 in his *Mountaineering in the Sierra Nevada* (Boston: James Osgood & Co., 1872), pp. 49–93. See also letter 2224, *Tyndall Correspondence*, vol. 9; and W. H. Brewer, 'Discovery of Mount Tyndall', *Popular Science Monthly*, 2 (April 1873), pp. 739–41. Admirers in Columbus, Ohio, set up a 'Tyndall Association" for the discussion of science and public lectures (see letter 3920; *Ascent of John Tyndall*, p. 310; *Constitution and By-Laws of the Tyndall Association of Columbus, Ohio, As Adopted June 6, 1874* (Columbus, OH: Paul & Thrall, 1875)).
4. *one of the letters addressed to me*: letter not identified, perhaps missing. Tyndall was invited to visit the United States at least twice before: by Edward L. Youmans in 1870 (see letter

3175, *Tyndall Correspondence*, vol. 11); and by John Amory Lowell in 1864, to lecture for the Lowell Institute in Boston (see letters 2160 and 2188, in M. D. Barton, P. J. Hale, N. Kapoor, and E. Neswald (eds), *The Correspondence of John Tyndall, Volume 8: The Correspondence, June 1862–January 1865* (Pittsburgh: University of Pittsburgh Press, 2021)).

5. *I have agreed to lecture in five of their cities*: his lecture tour in the end encompassed seven cities (see Introduction, n. 8).
6. *to the help of science in Chicago*: see Introduction, n. 15.
7. *benefits of bachelorhood*: Adair had previously turned down Tyndall's offer of marriage in November 1869 (see letter 3059, *Tyndall Correspondence*, vol. 11).
8. *my memoirs*: *Molecular Physics*.
9. *your Mother*: Harriet Eliza Adair.

To Charles Darwin 8 June [1872][1] 3720

Royal Institution of Great Britain | 8th June.

My dear Darwin.

The enclosed[2] is now being copied & you will be asked to sign it in a day or two.

We purpose confining the signatures to 10 or 12 good names.

I hope you will think that we have made out a good case. I have boiled down the correspondence so as to enable outsiders to form a quick judgment

Yours ever | John Tyndall | Char[le]s. Darwin Esq

CUL GBR/0012/MS DAR 106: C9[3]
RI MS JT/1/TYP/9/2817

1. *[1872]*: the year is given by reference to the Memorial (see n. 2).
2. *enclosed*: enclosure missing, but presumably the Memorial in support of Joseph Hooker in his dispute with Acton Ayrton (see letter 3755, esp. n. 1).
3. This letter has been published in *Darwin Correspondence*, vol. 20, pp. 249–50; this is letter 8375 in the online edition of the Darwin correspondence.

To Joseph Dalton Hooker 8 June 1872 3721

Royal Institution of Great Britain | 8th. June 1872

My dear Hooker

I have visited Sir H. Holland, Paget; Lyell; Burrows & Busk, and all of them are ready to deliver the blow.[1] Holland was very courageous, saying that this was to time to go in at them with all our might. I have written to Airy[2] and expect his reply[3] today.

This afternoon I shall place the Memorial in the hands of all the London signatories, and also post one to Darwin.[4]

I must ask you to go once more over it paragraph by paragraph & see that no flaw is in any of them.

May I ask did Ayrton[5] ever pay you a friendly visit at Kew? If you can effectually bar all shuffling on his part it is in the highest degree desirable that you should do so.

My young secretary[6] is already engaged in copying: but the Revises have not yet reached me from Spottiswoode.

Yours ever | John Tyndall

RBG, Kew Papers Relating to Kew 1867–1872, Ayrton Controversy, volume 1, f. 224

---

1. *deliver the blow*: the Memorial in support of Joseph Hooker in his dispute with Acton Ayrton (see letter 3755, esp. n. 1).
2. *written to Airy*: letter 3718.
3. *his reply*: letter 3726.
4. *post one to Darwin*: see letters 3720 and 3732.
5. *Ayrton*: Acton Ayrton.
6. *My young secretary*: not identified.

To Edward Stanley 8 June [1872][1] 3722

Royal Institution | 8th June.

Dear Lord Derby

The enclosed[2] will inform you of our next move.

We purpose restricting the signing of the memorial to a dozen names. If we cared we could get the Science of England to sign it. That may be necessary afterwards.

I trust you will find nothing in the memorial of which you disapprove: it is now being copied in manuscript and will be sent in early next week.

I cannot tell how glad I am to find that you really can throw your force into this matter without misgiving.[3]

Most truly yours | John Tyndall

RI MS JT/1/TYP/8/2648
LT Typescript Only

---

1. *[1872]*: the year is given by reference to the Memorial (see n. 2).

2. *enclosed*: enclosure missing, but presumably the Memorial in support of Joseph Hooker in his dispute with Acton Ayrton (see letter 3755, esp. n. 1).
3. *throw your force into this matter without misgiving*: see letter 3717.

From Mary Adair 9 June [1872][1] 3723

63 Portland Place.[2] | June 9th.

Dear Dr. Tyndall,

The motives which induce you to visit America[3] are indeed high and disinterested. May they be abundantly accomplished! but do not forget in the enthusiasm of International Fraternity that England cannot relinquish altogether the Chief of her physical Philosophers—

A thousand thanks for the splendid memoirs.[4] I must own that at first sight they appear some what appalling—as if one's spiritual vision would be "blinded by excess of light"[5]—however I hope in the quiet leisure of country life to disregard your friendly caution and read them throughout.

Yours ever truly. | H. M. Adair—

RI MS JT/1/A/12

---

1. *[1872]*: the year is given by reference to Tyndall's upcoming trip to the United States.
2. *63 Portland Place*: residence in London.
3. *motives which induce you to visit America*: see Introduction, n. 15. For Tyndall's American lecture tour in general, see Introduction, n. 8.
4. *splendid memoirs*: *Molecular Physics*; in letter 3719, Tyndall wrote to Adair: 'I have been gathering my memoirs together into a volume which I wish you to have, <u>but not to read</u>'.
5. *"blinded by excess of light"*: a common phrase in theological writings in the nineteenth century.

From Henry Bence Jones 9 June 1872 3724

Dear friend

I would not have a word altered of what you wrote.[1] Oh no. I have not written to thank you for it & for the book[2] partly because I thought to see you every moment & partly because I wanted to look at the new work in the book before I said anything. I ought to have thanked you however at once & left Gladstones book[3] which came at the same moment until I had at least thanked you

Your memorial of my work for the Institution[4] will always remain: & that is a pleasant thought whilst I live: & when I am gone the praises of the men who would best estimate what I did ought to be a pleasant thought to those I leave behind.

Many thanks dear friend very many

Ever Yours | H Bence Jones | June 9. 1872

RI MS JT/1/J/66
RI MS JT/1/TYP/2/704

---

1. *what you wrote*: Tyndall reviewed Jones's *The Life and Letters of Faraday*, 2 vols (London: Longmans, Green and Co., 1869) in *The Academy: A Monthly Record of Literature, Learning, Science, and Art*, 1 (14 May 1870), pp. 204–7; and (11 June 1870), pp. 232–4. He included this in *Fragments of Science*, pp. 349–72.
2. *the book*: *Molecular Physics*.
3. *Gladstone's book*: J. H. Gladstone, *Michael Faraday* (London: Macmillan and Co., 1872).
4. *Your memorial of my work for the Institution*: possibly Tyndall's dedication of *Molecular Physics* to Bence Jones, which reads: 'TO HENRY BENCE JONES, M.D., D.C.L., F.R.S. | HON. SEC. R. I. | IF unswerving devotion to the ROYAL INSTITUTION, firstly, and above all, as a school of original enquiry, and secondly as an organ for the diffusion of scientific knowledge, merit the grateful recognition of its Members and its Professors, then justice ought to require no stimulus from friendship, in associating these Researches with your name. They were one and all conducted on the spot, whence, during sixty years, issued in unbroken succession the labours of YOUNG, DAVY, and FARADAY. Would that they were more worthy of their immortal antecedents!' (p. v). The book (which bears the full title *Contributions to Molecular Physics in the Domain of Radiant Heat: A Series of Memoirs Published in the 'Philosophical Transactions' and 'Philosophical Magazine,' with Additions*) was a collection of ten memoirs and several additional papers on the subject of radiant heat. It may also be worth noting that the year before Bence Jones published *The Royal Institution: Its Founder and Its First Professors* (London: Longmans, Green, and Co., 1871).

From Charles Lyell 9 June 1872 3725

73 Harley Street W[1] | 9th June 1872

My dear Tyndall

I have read the printed paper,[2] which is not a bit too strong, and makes one's blood boil and may well help to turn out a ministry.

I would alter a word or two on p. 4, as some readers will think we are setting up a scientific expedition as more important than Nelson's victories.[3] The word exploring might be introduced. At p. 6 I have put a query to the passage about the career being likely to terminate in a manner discreditable to the country.[4] It would be better to assume that it is impossible that the country should allow his career so to terminate, and anything like a threat of resignation should be avoided.

I shall be happy to sign whenever you direct me.

Yours very faithfully | Charles Lyell.

RI MS JT/1/TYP/8/2649
LT Typescript Only

---

1. *73 Harley Street W*: residence in London (from 1854–75).
2. *printed paper*: the Memorial in support of Joseph Hooker in his dispute with Acton Ayrton (see letter 3755, esp. n. 1).
3. *scientific expedition as more important than Nelson's victories*: Lyell was referring to lines describing the Antarctic expedition that Hooker was a part of. The final wording was as follows: 'the most perilous, perhaps, that ever sailed from these shores, and the scientific results of which exceeded in importance those of any other naval exploring expedition of this century' (see letter 3755). Lyell was also referring to Admiral Horatio Nelson (1758–1805), a British naval commander who won numerous naval battles during the Napoleonic Wars (*ODNB*).
4. *passage about the... discreditable to the country*: Lyell was referring to the following line: 'But we think it highly desirable that you and England should know as much of his career as will enable you to decide whether its arbitrary interruption by your First Commissioner be creditable to the Government of this country' (see letter 3755).

From George Biddell Airy 10 June 1872 3726

(Address Greenwich)[1] | Loch Earn Head, Perthshire. | 1872 June 10.

My dear Sir

Your note of 7th.[2] reached me at Crieff[3] this morning. As I know nothing of Dr. Hooker's affair[4] I am unable to answer specifically, and can therefore only put before you some general ideas:—

1. With regard to any personal or official question between Mr. Ayrton[5] and Dr. Hooker, I think that we have nothing to do. Nor do I see that any body else can interfere below a Member of Parliament.

2. If science, abstractedly from persons, is concerned, we can interfere.

3. My position, as that of an officer under the Government, makes it very difficult for me to take any steps affecting another officer.[6]

I am, my dear Sir, | Yours very truly | G. B. Airy.

Professor Tyndall

RGO MS.RGO 6/438
RI MS JT/1/TYP/8/2664

1. *(Address Greenwich)*: this letter was written on letterhead for the Royal Observatory, Greenwich.
2. *Your note of 7th.*: letter 3718.
3. *Crieff*: a town in Scotland, a tourist destination in the nineteenth century.
4. *Dr. Hooker's affair*: referring to the Memorial in support of Joseph Hooker in his dispute with Acton Ayrton (see letter 3755. esp. n. 1).
5. *Mr. Ayrton*: Acton Ayrton.
6. *My position... affecting another officer*: on Airy's resistance to signing the Memorial, see letter 3718, n. 4.

From Joseph Dalton Hooker 10 June [1872][1] 3727

Royal Gardens, Kew. | 10th June.

Dear Tyndall,

I have again and again revised the Address,[2] and can find no flaw. Helps glanced through it, and all he noticed was that the word "instructed" at top of p. 19 should be replaced by "asked," or some other that does not imply that Gladstone had a right to order Lord Ripon to do this. This is a technical point, the fact being, that the Cabinet has no existence as a part of the Constitution,—has no Secretary, keeps no minutes, and all are equal at the table. It is a self-constituted and self-regulated assembly. I went straight to Lubbock, but he was at High Elms.[3]

I am now writing to him again, begging him to see Lord Derby and make out clearly what he will ask for. I would suggest that this should be the correspondence with Gladstone, and the subsequent correspondence with the Board[4] (or Treasury which received copies of it) on the subject of the accident in the Orchid house. I have again offered to Lubbock to come up if telegraphed for.

It is most important to know precisely what correspondence Lord Derby will call for. Could you see him? I again put the case of the visit to Helps. He sees no escape from the conclusion, that the visit alluded to by Gladstone was the one, and only possible one.

The case stands thus:—

In a letter to himself, I accuse Mr Ayrton[5] of paying a clandestine visit. I send a copy of this letter to Mr Gladstone. Mr Ayrton never answers my letter. Mr Gladstone answers for him, that he is assured by Mr Ayrton that he (Ayrton) paid the visit, (that in which "he failed to see me") for "the very purpose" etc.

Therefore Ayrton either lied to or cheated Gladstone.

RI MS JT/1/TYP/8/2651
LT Typescript Only

---

1. *[1872]*: the year is given by reference to the dispute between Joseph Hooker and Acton Ayrton (see letter 3755, n. 1).
2. *Address*: meaning the Memorial in support of Joseph Hooker (see letter 3755, esp. n. 1).
3. *High Elms*: the home of John Lubbock, which he inherited on the passing of his father John William Lubbock (1803–65). High Elms is near Charles Darwin's home Down House.
4. *Board*: Board of Works.
5. *Mr Ayrton*: Acton Ayrton.

To Joseph Dalton Hooker 10 June [1872][1] 3728

Royal Institution of Great Britain | 10th June

My dear Hooker.

The alterations and additions you suggest[2] have been attended to.

One or two others made by Sir H. Holland & Sir C. Lyell have also been introduced.

Now let me be perfectly definite. There is no use in my seeing Lord Derby: but there is every use & necessity that Lubbock and he should come to a definite understanding. Therefore, remember, I know the onus upon you of making perfectly clear to Lubbock what you wish him to do.

I think with you that our case is unassailable.

Lyell's letter[3] is so satisfactory & so vigorous that I send it to you—pray send it back.

I am going finally to Spottiswoodes to get the last revises.

Yours ever | John Tyndall

RBG, Kew Papers Relating to Kew 1867–1872, Ayrton Controversy, volume 1, f. 225

---

1. *[1872]*: the year is given based on reference to the Memorial (see n. 2).
2. *alterations and additions you suggest*: to the Memorial in support of Joseph Hooker in his dispute with Acton Ayrton (see letter 3755, esp. n. 1). Hooker also had Arthur Helps look over the Memorial (see letter 3727).
3. *Lyell's letter*: letter 3725.

From Edward Stanley 10 June 1872 3729

23 St James's Square[1] | June 10th 1872.

Dear Professor Tyndall

I moved on Thursday for the papers,[2] finding that no notice was required when there was no opposition.

I am glad you are printing the case separately.[3] The official correspondence will be printed for us, but not any private note or letter. I have not yet read your printed case, being only just come up out of the country, but shall do so to-day or to-morrow.

Very truly yours | Derby.

RI MS JT/1/TYP/8/2650
LT Typescript Only

---

1. *23 St James's Square*: residence in London.
2. *I moved on Thursday for the papers*: see letter 3717.
3. *printing the case separately*: the Memorial in support of Joseph Hooker in his dispute with Acton Ayrton (see letter 3755, esp. n. 1).

To George Biddell Airy 11 June 1872 3730

Royal Institution of Great Britain | 11th. June 1872

Dear Sir,

I beg to forward you the accompanying confidential document,[1] and to ask you whether you would have any objection to sign it when presented to you in manuscript?

The Presidents of all the scientific Societies that have been consulted have agreed without hesitation to sign the memorial.[2] Those interested in the matter are naturally anxious to add the name of the President of the Royal Society.[3]

I am dear Sir | very faithfully yours | John Tyndall

G. B. Airy Esq | Pres. R.S.
RGO MS.RGO 6/438

---

1. *confidential document*: enclosure missing, but presumably the Memorial in support of Joseph Hooker in his dispute with Acton Ayrton (see letter 3755, esp. n. 1).
2. *Presidents of all... sign the memorial*: of the eleven signatories of the Memorial, five were society presidents: George Bentham (Linnean Society, 1861–74); George Burrows (Royal

College of Physicians, 1871–6); George Busk (Royal College of Surgeons, 1871–2); Henry Holland (RI, 1865–73); and Henry Creswicke Rawlinson (Royal Geographical Society, 1871–3; 1874–6). The other six signatories were Charles Lyell, Charles Darwin, James Paget, William Spottiswoode, Thomas Huxley, and of course, John Tyndall.

3. *President of the Royal Society*: Airy was President of the RS from 1871–3, and had Tyndall and other X Club members to thank for it (see *The X Club*, pp. 230, 269, and 272). Tyndall had already written to him about signing the Memorial (letter 3718), but Airy declined (letter 3726). He tried again, hoping the signatures of other scientific society presidents would sway Airy (see letters 3759, 3763, and 3772).

To Joseph Dalton Hooker 11 June [1872][1] 3731

Royal Institution of Great Britain | 11th. June

My dear Hooker.

I send you Lord Derby's last note:[2] be good enough to send it back to me.

I saw Sir H. Rawlinson yesterday—He is quite ready to sign the Memorial[3]—A copy was forwarded to him last night.

The copying has been delayed a little through the alterations; but the document will be ready for signature this evening.

I forward by this post a Copy to Airy:[4] telling him that every other President applied to had agreed to sign without hesitation[5]—but that we were naturally anxious to include the President of the R.S.

I see he is on the point of being made a K.C.B.[6] I trust his honour may not [daunt] him.

Yours ever | John Tyndall

RBG, Kew Papers Relating to Kew 1867–1872, Ayrton Controversy, volume 1, f. 227

---

1. *[1872]*: the year is given based on mention of the Memorial (see n. 3).
2. *last note*: letter 3729.
3. *Memorial*: the Memorial in support of Joseph Hooker in his dispute with Acton Ayrton (see letter 3755, esp. n. 1).
4. *Copy to Airy*: George Biddell Airy, who was knighted (Knight Commander of the Order of the Bath) by Queen Victoria at Osborne on 30 July 1872, just seven weeks after this letter, declined to sign the memorial in support of Hooker, despite Tyndall asking him more than once (see letters 3718, 3726, 3730, 3734, 3759, 3763, and 3772). Note that Airy had previously declined a knighthood three times, in 1835, 1847, and 1863 (W. Airy (ed), *Autobiography of Sir George Biddell Airy* (Cambridge: Cambridge University Press, 1896), pp. 111–13, 187, 254–6, 296).
5. *every other President applied to had agreed to sign without hesitation*: see letter 3730, n. 2.
6. *K.C.B.*: see n. 4.

From Charles Darwin 11 June 1872 3732

Bassett, Southampton,[1] | June 11th 1872.

My dear Tyndall

I write one line to give you my present address. I am delighted to hear of the proposed step.[2] If the document[3] has been already sent to Down, it will be forwarded to me directly, but there will unfortunately be a day or two's loss of time.[4]

In haste | Yours very sincerely | Charles Darwin.

P.S—I saw Lady Derby[5] the other day and talked to her about Hooker's case, and she said she would repeat all I said to Lord Derby, in reference to the House of Lords. I mention this that you may remember to send a copy of the letter[6] to Lord D.

The memorial seems to me to be admirable.

RI MS JT/1/TYP/8/2654[7]
LT Typescript Only

1. *Bassett, Southampton*: according to the DCP, Darwin was staying with his son William Erasmus Darwin (1839–1914) in Southampton.
2. *proposed step*: in letter 3720, Tyndall wrote to Darwin in regard to the Memorial in support of Joseph Hooker in his dispute with Acton Ayrton (see letter 3755, esp. n. 1), that 'We purpose confining the signatures to 10 or 12 good names'.
3. *document*: see letter 3755.
4. *day or two's loss of time*: in a letter to Hooker dated 14 June [1872], Darwin wrote that 'I signed with real pleasure yesterday the memorial'. He went on to write that the 'memorial seems to me very clear & good; but I cannot help fearing too severe, not against justice, but for policy. I enjoyed the severity much; but on reflexion became fearful about it' (*Darwin Correspondence*, vol. 20, pp. 255–6; this is letter 8385 in the online edition of the Darwin correspondence). Hooker replied to Darwin on 15 June 1872, sharing that with regard to the Ayrton dispute, he was trying to keep a level head and longed for rest (*Darwin Correspondence*, vol. 20, pp. 256–7; this is letter 8386 in the online edition of the Darwin correspondence).
5. *Lady Derby*: Mary Catherine Stanley (1824–1900), wife of Edward Stanley and political hostess (*ODNB*). Some of her correspondence was published in *A Great Lady's Friendships: Letters to Mary, Marchioness of Salisbury, Countess of Derby, 1862–1890* (London: Macmillan and Co., 1933). The earl and countess of Derby were renting Holwood House, Beckenham, near Down (*Darwin Correspondence*, vol. 20, p. 244).
6. *letter*: it is unclear as to what letter this refers to.
7. This is letter 8383F in the online edition of the Darwin correspondence; it will be published by the DCP in a forthcoming supplement of *The Correspondence of Charles Darwin*.

From Joseph Dalton Hooker 11 June [1872][1] 3733

Royal Gardens, Kew | June 11th

Dear Tyndall

Again thanks galore.[2] I am much relieved by Bentham's approval,[3] for he is cautious to a fault, and knows my career in all its aspects better than I do myself; and I am morbidly sensitive about all the good thoughts and words you have lavished on me in the address.[4] I wrote again to Lubbock offering to go in and meet him wherever he pleased, but he did not answer by the messenger, but wrote saying that he would see Lord Derby as yesterday and arrange with him.

I hope that he will keep me up to the Parliamentary moves when the time comes for action. I will gladly take up my abode in London for the time being if necessary. All this is my affair proper. It will be a case of crooked answers in the Commons[5] and requires adroit handling.

Lord Derby is a man of action indeed. I am to see Lord John Manners[6] to-morrow, who was my chief when the Hot Houses were handed over to my charge absolutely,[7] and he will, I doubt not, lift up his voice.

I hope Airy is above being affected by K.C.B.[8] except perhaps a little annoyance at having to take what he cannot want. I suspect it is a Gladstonian sop to science.[9] I sometimes wonder whether he was annoyed at my declining knighthood.[10]

I find that though Parliament cannot remove Kew from the Board of Works except by a bill, they can by '<u>Resolution</u>' prepare the way for it, and such a resolution would checkmate Ayrton[11] most effectually. I shall try this on. The worst of it is that I am utterly cut off from <u>official</u> guidance by the P.M.,[12] the Cabinet, and the Treasury too, and it is this that makes it so difficult for me to pick my way to a solution.

Ever, my dear fellow, | Yours | J. D. Hooker.

RI MS JT/1/TYP/8/2653
LT Typescript Only

---

1. *[1872]*: the year is given by reference to the dispute between Hooker and Acton Ayrton (see letter 3755, esp. n. 1).
2. *Again thanks galore*: this letter is a reply to letter 3731.
3. *Bentham's approval*: George Bentham was one of eleven signatories of the memorial (see. n. 4).
4. *address*: the Memorial (see letter 3755, esp. n. 1).
5. *in the Commons*: the case was debated in the House of Commons on 8 August 1872 ('The

Royal Gardens, Kew—Dr. Hooker, and the First Commissioner of Works', HC Deb 08 August 1872 vol. 213 cc710–758, *HANSARD 1803–2005* (online, accessed 30 May 2021)).

6. *Lord John Manners*: John James Robert Manners (1818–1906), a British politician (*ODNB*).
7. *my chief when the Hot Houses were handed over to my charge absolutely*: Manners served as First Commissioner of Works in all three of Lord Derby's terms as Prime Minister, in 1852, in 1858–9, and in 1866–8, the last of which coincided with the first few years of Hooker's role as Director of the Royal Botanic Gardens, Kew.
8. *being affected by K.C.B.*: see letter 3731, n. 4.
9. *Gladstonian sop to science*: meaning that by doling out knighthoods to scientists, Gladstone was looking for support for his administration.
10. *my declining knighthood*: after declining on several occasions, Hooker accepted a knighthood in 1877 (Endersby, *Imperial Nature*, p. 308; *Life and Letters of Hooker*, vol. 2, pp. 145–51).
11. *Ayrton*: Acton Ayrton.
12. *P.M.*: Prime Minister William Gladstone.

To Joseph Dalton Hooker 11 June [1872][1] 3734

11th. June.

My dear Hooker,

Can you not take the memorial[2] down to Darwin tomorrow (Wednesday) It will be ready for you.

Yours ever | John Tyndall

Shall you be at home on Thursday afternoon? I was thinking of accompanying Lady Mary Egerton—a very choice spirit in a drive to Kew on that day

I have just had a note from Darwin[3] which I enclose

Ball | Bentham | Holland | Busk | Spottiswoode | Huxley | Paget | Lubbock[4]

RBG, Kew Papers Relating to Kew 1867–1872, Ayrton Controversy, volume 1, f. 226

1. *[1872]*: the year is given based on reference to the Memorial in support of Joseph Hooker in his dispute with Acton Ayrton (see n. 2).
2. *memorial*: the Memorial in support of Joseph Hooker in his dispute with Acton Ayrton (see letter 3755, esp. n. 1).
3. *note from Darwin*: letter 3732.
4. *Ball | Bentham | Holland | Busk | Spottiswoode | Huxley | Paget | Lubbock*: written on the back of the letter. Of these eight names, seven signed the Memorial (see n. 2): George Bentham, Henry Holland, George Busk, William Spottiswoode, Thomas Huxley, James Paget, and John Lubbock. John Ball was not a signatory; the other signatories were Charles Darwin, Charles Lyell, Henry Rawlinson, and, of course, Tyndall.

To Heinrich Rudolf Vieweg 11 June 1872 3735

Royal Institution of Great Britain | 11th June 1872

Gentlemen.[1]

I requested the Messrs. Longman[2] to send the 'Fragments'[3] & the 'Lectures on Sound".[4]

Please in the title omit the words for unscientific people & call it simply "Fragments of Science"

I have cut out the introductory part of the essay on Imagination but on looking at it again I would advise you to retain it in the translation

I have also desired the Messrs Longman to send you a new volume entitled "Contributions to Molecular Physics in the Domain of Radiant Heat".[5]

I shall probably ask you to publish a translation of this work,[6] and also of the first half of Diamagnetism[7]—excluding the letters and shorter papers. This will be the best practical answer to the unprovoked attack recently made on me by Herr Zöllner.[8]

Yours faithfully | John Tyndall

TU Braunschweig/Universitätsbibliothek/UABS V 1 T : 63

---

1. *Gentlemen*: while this letter is addressed to 'Gentlemen', we believe the recipient is Heinrich Rudolf Vieweg, who took over the family publishing firm, Friedrich Vieweg und Sohn, from his father Eduard Vieweg (1797–1869).
2. *Messrs. Longman*: Thomas Longman (1804–79) and William Longman (1813–77), brothers who ran the publishing firm Longmans, Green, and Co., publisher of many of Tyndall's books.
3. *'Fragments'*: the German translation of the second edition of *Fragments of Science* was published as: J. Tyndall, *Fragmente aus den Naturwissenschaften: Vorlesungen und Aufsätze*, 2 vols, trans. A. Helmholtz (Braunschweig: Friedrich Vieweg und Sohn, 1874).
4. *'Lectures on Sound"*: Vieweg first translated *Sound* in 1869, and followed with a second edition in 1874: J. Tyndall, *Der Schall: Acht Vorlesungen gehalten in der Royal Instiution von Grossbritannien*, 2nd edn, trans. H. Helmholtz and G. Wiedemann (Braunschweig: F. Vieweg und Sohn, 1874).
5. *"Contributions to Molecular Physics in the Domain of Radiant Heat"*: *Molecular Physics*.
6. *translation of this work*: Vieweg declined to publish a translation of this work (see letter 3790).
7. *first half of Diamagnetism*: Vieweg declined to publish a translation of Tyndall's *Diamagnetism* (see letter 3790).
8. *attack recently made on me by Herr Zöllner*: see letter 3740, n. 2 and n. 3.

To C. Cecil Trevor 12 June 1872 3736

Royal Institution 12th June 1872.

Sir,

I have had a lengthened interview with Mr. Wigham, during which he explained to me in great detail, and by means of models, the principles and performance of the so-called Triform Light.

This interview justifies me in giving renewed expression to the high opinion I entertain of Mr. Wigham's labours in relation to the application of gas in Lighthouses.

The present proposal is, in my opinion, well worthy of being worked fully out. A visit to Ireland could teach me no more than I have already learned from Mr. Wigham regarding the "Triform Light," in its present condition. The experiments which have yielded such satisfactory results were executed with a mere temporary arrangement, unfit to solve more than a portion of the problem under consideration, and even that imperfectly. It is, in my opinion, highly desirable that arrangements should be made for its complete solution.

One strong motive which actuated me in recommending gas at Rockabil[1]l was, that practical experiments could be made there on Revolving Lights, while Howth Baily[2] offered similar facilities with regard to Fixed Lights. The problem now in hand falls under the former of these two heads.

Mr. Wigham proposes to employ three lenses placed vertically over each other with a light in the focus of each, the number of lights employed to be determined by the state of the weather. That the use of gas renders this possible has been demonstrated by the experiments already made at Howth Baily with the temporary arrangement above referred to.

This arrangement consists, firstly, of a piece of the refracting belt of a fixed apparatus; secondly, of the lens, the purchase of which was authorised by the Board of Trade[3] some time ago; and thirdly of a broken lens, surmounting the two other pieces.

What I now beg to recommend is that the Board of Irish Lights[4] be empowered to purchase one additional lens with a view to the completion of these experiments. The two extra-lenses, together with the octagonal belt already existing at Rockabill, will enable Mr. Wigham, during the present summer, to carry out his arrangements in a complete manner; and on the completion of the work I shall be in a condition to lay before the Board of Trade[5] a Report[6] founded on data which do not yet exist, but which are necessary to the solution of the problem.

The more I consider the subject, the more thoroughly I am persuaded that an important future awaits the application of gas in Lighthouses; and the more I feel justified in recommending to the Board of Trade the encouragement of experiments of such obvious importance, and involving such definite and moderate outlay, as those now contemplated by Mr. Wigham. I am of opinion that the labours of this zealous and intelligent mechanician will eventually profoundly influence for good the lighting of our Coasts.

With regard to the gas gun[7] it is my intention to arrange with Mr. Wigham a series of experiments, comparative and otherwise, which will furnish me with the necessary groundwork of a Report.[8]

Under these circumstances I propose to postpone my contemplated visit to Ireland until the experimental arrangements have assumed a more perfect form.

I have the honour to be, Sir | Your obedient Servant. | John Tyndall

C. Cecil Trevor Esq | &c &c[9]

National Archives MT 10/165. Fog signals-gun flashes. H4380[10]
Transcript Only

---

1. *Rockabill*: a lighthouse on the 'The Rock', one of two islands making up Rockabill in the western Irish Sea, northeast of Skerries, County Dublin, Ireland. The other island is named 'The Bill'.
2. *Howth Baily*: a lighthouse on the southeastern part of Howth Head in County Dublin, Ireland.
3. *Board of Trade*: a British government department concerned with making policy and instituting regulations regarding economic activity in Britain and its empire. Under its direction was Trinity House, the lighthouse authority for England and Wales, established by Royal Charter in 1514. See A. Adams and R. Woodman, *Light Upon the Waters: The History of Trinity House 1514–2014* (London: The Corporation of Trinity House, 2013). On Tyndall as scientific adviser to Trinity House, see letter 3869, n. 4.
4. *Board of Irish Lights*: or Commissioners of Irish Lights, the lighthouse authority for Ireland. There was also Commissioners of Northern Lighthouses, the lighthouse authority for Scotland.
5. *Board of Trade*: in the copy of this letter to Robin Allen (see n. 10), Tyndall wrote 'Elder Brethren' instead of 'Board of Trade'. Trinity House is governed by a council of thirty-one Elder Brethren who are drawn from the four hundred Younger Brethren.
6. *Report*: not identified.
7. *gas gun*: an apparatus that uses a controlled gas explosion to warn ships of rocks in dense fog, as a means to replace the use of actual cannon fire. A description of the apparatus can be found in 'Sound-Signals', *Popular Science Monthly*, 19 (July 1881), pp. 428–9.
8. *Report*: not identified.
9. At the end of the letter someone, possibly Trevor, wrote the following words: 'The purchase of the additional lens was at once authorised—'.
10. A copy of this letter to C. Cecil Trevor of the Board of Trade was sent to Robin Allen of Trinity House, as indicated at the end of this letter.

From George Burrows[1] 12 June 1872 3737

18 Cavendish Square[2] | June 12th 1872.

My dear Sir,

I have read with much pain the document containing the account of the various affronts that have been put upon Dr Hooker.[3] I deeply sympathise with him, for it must be most galling to a high-minded man of science to be under the subjection of such a vulgar, unfeeling, "Jack in Office".[4] Clear and able and convincing as is the statement to any unprejudiced mind, I much fear that nothing very satisfactory will be obtained from the Prime Minister.[5] It is clear that he wishes to put aside the consideration of the conduct or misconduct of his colleague in office, and not be forced to take any decided steps against him. However, I shall be happy to cooperate with you and others, if it is thought that good will result from our proceeding as is proposed. May we not put Dr Hooker in such a position that he cannot any longer with self-respect retain his appointment under such a chief? Is he prepared for such a result?

At page 24, line 3, of the concluding paragraph I find the expression "entirely maimed", and would suggest the consideration of the expression "seriously crippled",[6] as more correctly expressing the effect of Mr Ayrton's proceedings. Also should we intimate at line 11 of same paragraph that Mr Gladstone is oppressed by his public duties? Might we not better say, "engrossed by" or absorbed in"?[7] Oppressed would seem to signify that the duties were greater than Mr Gladstone could bear. Excuse this verbal criticism.

Do you propose any conjoint consideration, of the course to be pursued before the signatures are affixed.

Yours very faithfully | G. E. Burrows

Professor Tyndall.

RI MS JT/1/TYP/8/2656

LT Typescript Only

---

1. *George Burrows*: see letter 3730, n. 2.
2. *18 Cavendish Square*: residence in London.
3. *document containing the account of the various affronts that have been put upon Dr Hooker*: the Memorial in support of Joseph Hooker in his dispute with Acton Ayrton (see letter 3755, esp. n. 1).
4. *vulgar, unfeeling, "Jack in Office"*: Acton Ayrton.
5. *Prime Minister*: William Gladstone.
6. *"entirely maimed"... "seriously crippled"*: the final version shows that the word 'entirely' was removed, leaving 'maimed'.

7. *oppressed by... "engrossed by" or absorbed in"*: the final version shows that 'oppressed' was replaced with 'engrossed'.

From Joseph Henry 12 June 1872 3738

June 12 1872

My */dear/* Sir

Your letter of the 28th. of May[1] has just been received and I hasten */to/* inform you in reply that I have */had/* interviews with your cousin Gen[era]l Tyndale,[2] Prof[esso]r. Lesley in Phil[adelphi]a., and, also, with Prof[esso]r. Youmans since I last wrote you. Prof[esso]r. Youmans visited me at the Light House Depot on Staten Island, at my request, and I have made an agreement with him and with the other gentlemen */1 word illeg/* will be duly arranged! Mr. Lesley and Gen. Tyndale will take charge of the business in Philadelphia and Prof[esso]r. Youmans in New York while I will attend to that in Washington and Baltimore

We regret to learn that you cannot make arrangements to visit California. We think the 3000 miles of sea and an equal number of miles travel by land would give you a sufficient change */of view/* to obviate the necessity of an Alpine excursion. Man, however, is the creature of habit and perhaps your physical organization has materially become as much adapted to occult influences found only in Switzerland that you cannot recuperate in any other locality. Be this as it may, we hope you will come on as early in October as possible that you may have some time for seeing the country, for example visiting Niagara,[3] before you */commence/* your lectures.

The Congress of the United States does not meet again until December and I do prefer therefore that lectures be given in Washington at some time in this */month/*.

Your first course might be given in New York; the second in Boston; the third in Philadelphia; the fourth in Baltimore and the fifth in this city, the number of lectures at each place can be */determined/* by your time and inclination. The preparation for each city may be the same although you may, in certain cases, vary the program if absolutely necessary, by omitting a lecture.[4]

To save labor while you are in the United States the experiments for each lecture should be carefully studied and apparatus fitted up so that the */load/* may be safely transported by rail from one place to another; and in making this preparation you need not */concern/* yourself */on/* account of the expence.

I was much pleased to */make the acquaintance/* of your relative, Gen. Tyndale who is a gentleman, I find, highly respected in Philadelphia. He had a very active part in favor of the North during the late civil War.

I have been selected as Chairman of the Light House Board[5] which office, without interference with my duties as the Director of the Smithsonian Institution, gives me facilities for excursions along our Coast. I shall visit New York and Philadelphia in the course of a week or so on Light House business and will then make more definite arrangements as to your lectures.

Very truly | Your friend | Joseph Henry | Sec[re]t[ar]y Smith[sonia]^{n} Inst[itutio]^{n}

Prof John Tyndall, | London

Smithsonian Institution Archives. RU 33, Vol. 29, Page 477

---

1. *Your letter of the 28th. of May*: see letter 3699, *Tyndall Correspondence*, vol. 12. In this letter Tyndall agreed to come to the United States for a lecture tour (see Introduction, n. 8).
2. *your cousin General Tyndale*: Hector Tyndale.
3. *visiting Niagara*: Tyndall visited Niagara Falls from 1–8 November 1872, after his lectures in Boston and before Philadelphia, and during his time wrote of his visit to Edward L. Youmans (3884), Henry Bence Jones (letter 3885), and Thomas Hirst (letter 3886). After his return to London (on 19 February 1873), he delivered a Friday Evening Discourse at the RI on 4 April, 'Some Observations on Niagara' (*Roy. Inst. Proc.*, 7 (1873–75), pp. 73–91). This lecture was also published as J. Tyndall, 'Niagara', *Macmillan's Magazine*, 28 (May 1873), pp. 49–62; J. Tyndall, 'Some Observations on Niagara', *Popular Science Monthly*, 3 (June 1873), pp. 210–26 (minus the final paragraph and map); and later in J. Tyndall, *Fragments of Science for Unscientific People: A Series of Detached Essays, Lectures, and Reviews*, 5th edn (London: Longmans, Green, and Co., 1876), pp. 218–45.
4. *Your first course... omitting a lecture*: see letter 3719, n. 5.
5. *Chairman of the Light House Board*: Henry was a member of the United States Lighthouse Board since 1851, and chairman from 1871 until his death in 1878. This role later brought him into a dispute with Tyndall in 1875 over Tyndall's criticisms of the Lighthouse Board, specifically concerning conceptual matters regarding sound and fog signals. On this, see *Ascent of John Tyndall*, pp. 338–9; and J. Tyndall, 'On the Motions of Sound', *Popular Science Monthly*, 7 (August 1875), pp. 410–26. Letters about this will be published in the fourteenth and fifteenth volumes of the Tyndall correspondence.

From James Paget 12 June 1872 3739

1 Harewood Place | Hanover Square[1] | 12th June 1872.

My dear Tyndall

Probably I need not write to you, but for form and surety I do, that I entirely agree with the letter to Mr Gladstone.[2]

Sincerely yours | James Paget.

RI MS JT/1/TYP/8/2655
LT Typescript Only

1. *1 Harewood Place | Hanover Square*: residence in London.
2. *letter to Mr Gladstone*: the Memorial in support of Joseph Hooker in his dispute with Acton Ayrton (see letter 3755, esp. n. 1). Paget was one of eleven signatories.

To Hermann Helmholtz 18 June [1872][1] 3740

18th June

My dear Helmholtz,

The accompanying letter was written on the 13th of April.[2] I put it aside not wishing to trouble you with a matter unworthy of your attention; but on second thought I send it on to you.

I am on the point of starting for the Alps where I shall forget Zöllner[3] and all similar nuisances.

Yours ever | John Tyndall

BBAW, NL. Helmholtz, Nr. 477, BI_19
RI MS JT/1/TYP/2/506

1. *[1872]*: the year is given by the addition of '[1872]' on both the handwritten letter and the typescript. Also, Helmholtz's reply (letter 3770) is dated 23 June 1872.
2. *accompanying letter was written on the 13th of April*: see letter 3987, *Tyndall Correspondence*, vol. 12. In this letter, Tyndall shared with Helmholtz his views on being attacked by Zöllner (see n. 3), writing: 'When I turned over the leaves of his book I could hardly credit my senses; the discharge of animosity against me was so virulent, and the cause of it all so inadequate. From Germany I had small expectation of such an attack as this. I did not read all that Zöllner said regarding me, but handed the volume over to Bence Jones, who informs me that it is "savage and venemous" to the last degree'.
3. *Zöllner*: in his book *Über die Natur der Cometen, Beiträge zur Geschichte und Theorie der Erkenntniss* (Leipzig: Engelman, 1872), Zöllner attacked not only Tyndall (on cometary theory and his popular writings), but numerous other scientific practitioners such as Hermann Helmholtz and Peter Guthrie Tait (*Ascent of John Tyndall*, p. 277; *CDSB*). Tyndall had delivered a lecture to the Cambridge Philosophical Society on 8 April 1869, 'On Cometary Theory' (published in *Phil. Mag.*, 37 (April 1869), pp. 241–5).

To Hector Tyndale 13 June 1872 3741

June 13th 1872.

My dear Hector

Whatever be the time decided upon for the commencement of the lectures I shall be in the United States at that time, always supposing that the campaign[1] does not begin before the middle of October.

And whatever the arrangements are as to the places of lecturing I shall do the utmost to conform to those arrangements

In my letters to Prof. Henry[2] I agreed to give 6 lectures at each of the five cities[3] he mentioned—of course I am equally willing to give a less number than six. Indeed if the numbers of places be extended the numbers of lectures will probably have to be diminished.

I have agreed to give three lectures a week—this would detain me at least a fortnight in each city. The time of transit from city to city must also be taken into account.

I think I told you that I intended to take two assistants; the one highly skilled and trained by myself[4]—the other an intelligent young fellow[5] who helps me in my lectures, and sometimes acts as secretary.

In every city I shall need a little further assistance; that of a carpenter will probably be in all cases necessary—there will be no difficulty I should say in obtaining the local assistance. It will simply be required for the setting up of apparatus and perhaps for the suspension of diagrams.

It must be borne in mind that I have to be in London by the end of the year. What my poor brain and body can do within the specified limit I am willing to do. I place myself entirely in the hands of my scientific friends, as I have no other object or desire than to conform to such arrangements as they consider will be most gratifying to the American people.

Since this letter was begun I have seen Froude, who is also going out on a lecturing campaign.[6] We shall probably arrange to quit England together towards the end of September.

To close up—you have me from the middle of October to the middle of December.[7] Do with me what seemeth to you good during this interval; fix my times and places of lecturing as you please, I shall willingly conform to your arrangements.

Yours affectionately, | John Tyndall.

Smithsonian Institution Archives. RU 32, Box 1, Folder: Tyndall, J.
Transcript Only[8]

1. *campaign*: Tyndall's American lecture tour (see letter Introduction, n. 8).
2. *my letters to Prof. Henry*: for previous letters between Tyndall and Joseph Henry regarding the American lecture tour, see letters 3504, 3510, 3542, 3669, and 3684, *Tyndall Correspondence*, vol. 12.
3. *five cities*: his lecture tour in the end encompassed seven cities (see Introduction, n. 8).
4. *highly skilled and trained by myself*: John Cottrell.
5. *intelligent young fellow*: George Miller.
6. *lecturing campaign*: Froude delivered a series of lectures in the United States beginning in the fall of 1872, attempting to sway Irish Americans' negative views of British policies toward Ireland. Opposition to his lectures caused Froude to end his trip earlier than planned, returning in February or early March 1873 (R. M. Bennett (ed), 'Letters of James Anthony Froude', *The Journal of Rutgers University Library*, 12 (1949), pp. 38–53, on p. 45; H. Paul, *The Life of Froude* (New York: Charles Scribner's Sons, 1906), pp. 203–27). Froude indeed sailed to America on the SS *Russia* with Tyndall.
7. *to the middle of December*: in the end Tyndall's American trip went into February 1873.
8. A note on this letter indicates that it is a 'copy', and thus not in Tyndall's hand.

From William Carpenter 15 June 1872 3742

[University of London, | Burlington Gardens, W.] | June 15th [18]72

My dear Tyndall

I am sorry that you have not given me more time to consider the propriety of attaching my signature to the Memorial[1] you have forwarded to me. As a Scientific Man I cannot but feel the greatest indignation at the treatment which Hooker has received at the hands of Mr Ayrton,[2] who has not hesitated to snub the Senate of the University of London in the same style.

But as a quasi Public Servant, I doubt how far it would be becoming in me to join in so personal an attack upon Mr Ayrton; and though I stand at present in such a relation to him that I could scarcely suffer by his increased animosity, I cannot but apprehend that my uniting in it might be held to compromise the University, of which I am the official representative in its communications with the Office of Works. At any rate I should feel that I ought not to do it without the sanction of the Vice-Chancellor;[3] and you have given me no time to consult him on the matter.

It might be thought that I am as free as Huxley to act in such a matter; but Huxley is not (I imagine) in anything like the direct representative relation with Mr Ayrton that I am. And I notice that no Officer of the British Museum has joined in the Memorial, which I understand that Prof. Airy has declined to sign,[4] on grounds similar to my own.

yours faithfully | Will[ia]m B Carpenter | Prof. Tyndall

RI MS JT/1/T/690
RI MS JT/1/HTYP/587

1. *Memorial*: the Memorial in support of Joseph Hooker in his dispute with Acton Ayrton (see letter 3755, esp. n. 1).
2. *Mr Ayrton*: Acton Ayrton
3. *Vice-Chancellor*: the Vice-Chancellor of the University of London at this time was John Lubbock, a good friend of Tyndall's and fellow member of the X Club, who would present the Memorial to Gladstone (see letter 3755, n. 1). On the X Club, see letter 3757, n. 2.
4. *Prof. Airy has declined to sign*: see letters 3718, 3726, 3730, 3759, 3763, and 3772.

To Thomas Archer Hirst [15 June 1872][1] 3743

Dear Tom.

This is my reward for wishing to pay a compliment to your Chef.[2] How did he get the notion about Airy?[3] It is perhaps hardly worth while to disabuse him—Airy has not declined—probably has not yet read the Memorial.[4] However we need neither, and the document will go in today.

Yours ever | J. T.

RI MS JT/1/T/690
RI MS JT/1/HTYP/587

1. *[15 June 1872]*: the date is given by relation to letter 3742, which was enclosed with this letter.
2. *your Chef*: probably intending to write 'Chief', Tyndall was referring to William Carpenter, who was Registrar of the University of London, which encompassed among other universities the one where Hirst was Chair of Mathematics, University College London.
3. *notion about Airy*: see letter 3742.
4. *Airy has not declined—probably has not yet read the Memorial*: the Memorial in support of Joseph Hooker in his dispute with Acton Ayrton (see letter 3755, n. 1). On Airy, see letters 3718, 3726, 3730, 3759, 3763, and 3772.

To Joseph Dalton Hooker 15 June 1872 3744

Royal Institution of Great Britain | Saturday | June 15 1872

My dear Hooker

I have written to Lords Salisbury & Stanhope & sent them copies[1]

I have also told the new K.C.B. ab[ou]t my mind[2]

Yours ever | John Tyndall
What an intelligent man your friend Brandis[3] is!

RBG, Kew Papers Relating to Kew 1867–1872, Ayrton Controversy, volume 1, f. 230

1. *copies*: of the Memorial in support of Joseph Hooker in his dispute with Acton Ayrton (see letter 3755, esp. n. 1).
2. *told the new K.C.B. ab[ou]t my mind*: letter to George Biddell Airy missing (see letter 3731, n. 4).
3. *Brandis*: Dietrich Brandis (1824–1907), a German-British botanist and forester who worked with the British Imperial Forestry Service in colonial India for nearly three decades (*ODNB*).

From Elie Wartmann[1] 15 June 1872 3745

[ACADÉMIE DE GENÈVA | Genève, le] 15 juin 1872
[Le Recteur,] | A Monsieur le Professeur Tyndall

Cher Monsieur et ami,

Je viens de recevoir le beau volume que vous m'avez adressé et j'ai hâte de vous en remercier. J'y ai trouvé avec intérêt la série de vos recherches sur le rayonnement et les résultats aussi variés qu'importants auxquels vous êtes parvenus. Dès la publication de votre premier travail sur l'absorption due à la vapeur d'eau, j'avais montré quelles conséquences cette découverte avait en météorologie. Ce que vous dites, à cet égard, page 137 et suivantes, est parfaitement fondé. Les négations de Magnus n'ont jamais diminué ma confiance dans vos résultats.

Je lis, dans le Catalogue de la Bibliothèque de la Société Royale que Henry Goode a publié dans les Transactions of the Cambridge Philos. Society, tome VIII, page 493, 1849 un petit mémoire sous le titre: On a peculiar defect of vision. Nous ne possédons malheureusement pas cette collection à Genève, et, s'il vous était possible de jeter un coup d'œil sur cet article, vous me feriez plaisir en me faisant savoir s'il y est question de « Colour-blindness ».

Croyez, mon cher Tyndall, à mes sentiments bien affectueux. | E. Wartmann

Academy of Geneva | Geneva, 15th June 1872
Chairman | To Professor Tyndall

Dear Sir and friend,

I have just received the beautiful volume that you sent me[2] and I could not wait to thank you. In it I found, with great interest, the series of your research on radiation and the results that are as various as they are important. Since the publication of your first work on absorption due to water vapour,[3] I have shown the consequences of this discovery in meteorology.[4] What you say on this matter, page 139 and following, is perfectly founded.[5] Magnus' negations[6] have never diminished my confidence in your results.

I read, in the Catalogue of the Library of the Royal Society,[7] that Henry Goode[8] published in the Transactions of the Cambridge Philos. Society, tome VIII, page 493, 1849 a small essay under the title: On a peculiar defect of vision.[9] Unfortunately, we do not have this collection in Geneva, and I would be so pleased if you could have a look at this article and tell me if it is about "Colour-Blindness".

Yours most affectionately, my dear Tyndall, | E. Wartmann

RI MS JT/1/W/8

---

1. *Elie Wartmann*: Elie Wartmann (1817–86), a Swiss physicist and Professor of Natural Philosophy at the Academy of Geneva (*HLS*).
2. *beautiful volume that you sent me*: *Molecular Physics*.
3. *your first work on absorption due to water vapour*: probably the third memoir in the collected volume: J. Tyndall, 'On the Relation of Radiant Heat to Aqueous Vapour. Third Memoir', *Phil. Trans.*, 153 (1863), pp. 1–12.
4. *I have shown the consequences of this discovery in meteorology*: if published, this has not been identified.
5. *What you say on this matter, page 139 and following, is perfectly founded*: in this section Tyndall described the large range of solar radiation and temperature while ascending a mountain, or at different times of the day in other locations as due to the lack of aqueous vapour (pp. 139–41).
6. *Magnus' negations*: for a detailed account of the differences between Tyndall's and Heinrich Magnus's views regarding absorption of heat due to water vapour, see I. M. McCabe, *Second Best as a Researcher, Second to None as a Populariser? The Atmospheric Science of John Tyndall* (PhD dissertation, University College London, 2012), pp. 146–65. This friendly dispute between them faded on Magnus's death in 1870. Even so, as late as the 1880s, Tyndall felt the need to demonstrate further evidence for the absorption of heat by water vapour, given its importance for meteorology, as described in *Ascent of John Tyndall*, pp. 393–4.
7. *Catalogue of the Library of the Royal Society*: possibly *Catalogue of Scientific Papers (1800–1863.). Compiled and Published by the Royal Society of London* (London: Eyre and Spottiswoode, 1868), vol. 2, p. 934.
8. *Henry Goode*: Henry Goode (1819–1918), the second son of a lawyer from the Isle of Wight, matriculated at Pembroke College at the University of Cambridge when he was 19. He graduated B.A. in 1842, Bachelor of Medicine (M.B.) in 1846, and gained the Medical Licentiate (M.L.) in 1849. Goode's chief subject of study was medicine (*Alum. Cantab.*; J. R. Levene, 'The True Inventors of the Keratoscope and Photo-Keratoscope', *British Journal for the History of Science*, 2 (1965), pp. 324–42, on pp. 327–9).

9. *On a peculiar defect of vision*: H. Goode, 'On a Peculiar Defect of Vision', *Transactions of the Cambridge Philosophical Society*, 8 (1849), pp. 493–6. This was not about colour blindness, on which Wartmann was an authority.

From James Colvile 18 June 1872 3746

8 Rutland Gate[1] | June 18th

My dear Professor,

I imagine that it is to your kindness that I owe a copy of the letter to Mr Gladstone[2] which reached me this evening. It is marked "confidential" and I presume I am, for the present at least, to treat it as such.

I have just read it with much interest. It is a very pungent document and I thought I could recognise your "fine Roman hand" in many a sentence, but I agree with every word of it; and feel that it is an additional ground for regretting that I am not a man of science, that I cannot sign it. I have tried to stir up sundry members of either house in this matter.[3]

Yours always sincerely | J. W. Colvile

RI MS JT/1/TYP/8/2657
LT Typescript Only

---

1. *8 Rutland Gate*: residence in Hyde Park, London.
2. *letter to Mr Gladstone*: the Memorial in support of Joseph Hooker in his dispute with Acton Ayrton (see letter 3755, esp. n. 1).
3. *I have tried... in this matter*: as a civil servant, Colvile probably had some sway with Members of Parliament.

From Joseph Dalton Hooker 18 June 1872 3747

Royal Gardens, Kew | June 18th. 1872.

Dear Tyndall

I greatly fear that Lubbock has been prevailed against. Campbell,[1] who saw him yesterday, and questioned him, has an idea that he has had an interview with Ayrton;[2] meanwhile what can Lord Derby be about?

Up to 6 p.m. yesterday it was not settled "in what form" Lubbock's motion for the paper was to be worded! This looks black: he was to see Lord Derby again before doing any thing.

Rollo Russell[3] will dine with us on Friday, and Dr Brandis,[4] who I am sure you will like to know

Ever yours aff[ectionate]ly | Jos. D. Hooker.

I shall be in town all tomorrow—all forenoon at S. Kensington,[5] afternoon at Science Commission[6] till three or four I suppose, when I shall go up to Athenaeum.[7]

RI MS JT/1/TYP/8/2658
IC HP 8.124
Typed Transcript Only

---

1. *Campbell*: probably George Douglas Campbell (1823–1900), Duke of Argyll, a Scottish politician and scientist. In the peerage of Scotland he was eighth Duke of Argyll and was first Duke of Argyll in the peerage of the United Kingdom. Campbell served over ten years as Lord Privy Seal and from December 1868 to February 1874 was Secretary of State for India. He was a proponent of educational reform and led the Argyll Commission (1865–8) into the Scottish school system. As an amateur scientist he wrote on ornithology, geology, and evolution. In his work on evolution he defended the special creation of humans apart from other species, publishing *The Reign of Law* in 1867 (*ODNB*).
2. *Ayrton*: Acton Ayrton.
3. *Rollo Russell*: Francis Albert Rollo Russell.
4. *Dr Brandis*: Dietrich Brandis (see letter 3744, n. 3).
5. *S. Kensington*: probably the South Kensington Museum. On the South Kensington Museum, see B. Robertson, 'The South Kensington Museum in context: an alternative history', *Museum and Society*, 2 (2004), pp. 1–14.
6. *Science Commission*: the Royal Commission on Scientific Instruction and the Advancement of Science (or the 'Devonshire Report') convened from 1870–75 to 'make Inquiry with regard to Scientific Instruction and the Advancement of Science and to Inquire what aid thereto is derived from Grants voted by Parliament or from Endowments belonging to the several Universities in Great Britain and Ireland and the Colleges thereof and whether such aid could be rendered in a manner more effectual for the purpose' (*Royal Commission on Scientific Instruction and the Advancement of Science, Vol. 1: First, Supplementary, and Second Reports, Minutes of Evidence and Appendices* (London: Eyre and Spottiswoode, 1872), p. 3). Seeking to define the scientific health of the nation, the commission produced eight reports from 1872–5 (A. J. Meadows, *Science and Controversy: A Biography of Sir Norman Lockyer* (Cambridge, MA: The MIT Press, 1972), pp. 75–111).
7. *Athenaeum*: the Athenaeum Club (see letter 3710, n. 5).

From Elizabeth Rutherford 18 June 1872 3748

Dalhousie Castle[1] | Lasswade—| Scotland.—| 18th June 1872

Many thanks My Dear Friend, for remembering to write to me,[2] & for giving me your Swiss address—

You have taken my joke too seriously it was merely a little badinage,[3] and instead of being annoyed, at what you said, I thought it a great compliment, only it amused me because it is not quite true. Never mind.—

I hope your Swiss tour will in every way prove satisfactory to you, and prepare you for your longer absence in America.[4] Froude ought to be an agreeable compagnon de Voyage[5] and therefore I hope he will be with you[6]—You must much require a change of scene and of thought after the laborious work you have lately undergone—in preparing your last Vol. for publication.[7] Your dedication[8] is very elegant—

Now, I must say Adieu—to be in time—and with every kindest wish for your welfare, Believe me, | Very sincerely & Affectionately, | Yours, | Elizabeth Rutherford—

Our weather has become very warm and very fine—in consequence, I feel better—

I am not personally acquainted with Professor Oakeley,[9] altho' I have often seen him. Poor man he is much to be pitied, for his hobby is the Organ—which requires limbs in good order for the Pedals. I don't think he is a real Musician. He is not the right man in the right place—By the way why were you not a Musician?—

RI MS JT/1/R/86

---

1. *Dalhousie Castle*: the estate of the Earl of Dalhousie, who at the time was the British politician Fox Maule-Ramsay (1801–74), 11th Earl of Dalhousie (*ODNB*).
2. *write to me*: letter missing.
3. *badinage*: humorous, witty, or trifling discourse; banter; frivolous or light-hearted raillery (*OED*).
4. *longer absence in America*: for Tyndall's American lecture tour, see Introduction, n. 8.
5. *compagnon de Voyage*: a traveling companion (*OED*).
6. *be with you*: see letter 3741, n. 6.
7. *last Vol. for publication*: *Molecular Physics*.
8. *Your dedication*: Tyndall dedicated the volume to Henry Bence Jones (see letter 3724, n. 4).
9. *with Professor Oakeley*: with this letter Rutherford shared a clipping from the *Times*, titled 'ACCIDENT IN SWITZERLAND', which read: 'Sir—Travellers should, I think, be warned in the present state of the roads here against taking a carriage from Zermatt to this place. A frightful accident occurred yesterday, when a small carriage, containing three persons, was precipitated into a rocky turnout. As one of those unfortunate persons—I might almost say fortunate, in a merciful preservation from death—I venture to dictate these few lines to you while lying helpless here with an unset broken leg and other injuries. | I remain, Sir, your obedient servant, | H. S. OAKELEY, Prof. Mus[ic]., Univ., Edinburgh. | Grand Hotel, St. Nicolas, Zermatt Valley, | Switzerland, June 12.' (*Times*, 17 June 1872, p. 7). Herbert Stanley Oakeley (1830–1903) was an English composer and Professor of Music at the University of Edinburgh (*ODNB*).

To James Colvile 19 June 1872 3749

Royal Institution of Great Britain | 19th. June

My dear Sir James

It is a source of strength as well as a source of pleasure to me that you should write to me thus.[1] I tried to tell a round unvarnished tale,[2] & right glad I am to find that in your estimation I have succeeded.

I myself see no reason why the memorial sh[oul]d not be generally known but I will consult my friends upon the subject.

Yours ever | John Tyndall

[Envelope] Sir James Colvile | 8 Rutland Gate | Hyde Park.[3]

[Postmark] LONDON W | JU19 | 72

BL Add MS 60634, ff. 133–4

1. *write to me thus*: letter 3746.
2. *to tell a round unvarnished tale*: the Memorial in support of Joseph Hooker in his dispute with Acton Ayrton (see letter 3755, esp. n. 1).
3. *8 Rutland Gate | Hyde Park*: residence in London.

From Edward Stanley 19 June 1872 3750

23 St. James's Square[1] | June 19th 1872.

Dear Professor Tyndall

I am only waiting to give notice until the official papers for which I asked are out.[2] There is not the least reason why the memorial[3] should not be circulated at once. The signatures will ensure its being attended to by the press.

Very truly yours | Derby.

RI MS JT/1/TYP/8/2660
LT Typescript Only

1. *23 St. James's Square*: residence in London.
2. *official papers for which I asked are out*: Derby requested in the House of Lords for the official correspondence between Hooker and government officials including Acton Ayrton, later debated on 29 July 1872 (HL Deb 29 July 1872 vol. 213 cc. 3–23, *HANSARD 1803–2005* (online, accessed 27 November 2022); see also letter 3755, n. 1).
3. *memorial*: the Memorial in support of Joseph Hooker in his dispute with Acton Ayrton (see letter 3755, esp. n. 1).

To Thomas Henry Huxley 19 June [1872][1] 3751

Royal Institution, 19th. June.

My dear Huxley

I wrote to Lord Derby[2] to day regarding the propriety of circulating the Memorial.[3] He replies.[4] "There is not the least reason why the Memorial should not be circulated at once. The signatures will ensure its being attended to by the press."

Some body told me that you had the Command of the Daily Telegraph. I will make a dash at the Times.[5]

Yours ever | John Tyndall

RI MS JT/1/TYP/9/2962
IC HP 8.123
Typed Transcript Only

1. *[1872]*: the year is given by a note from LT on the letter, as well as content (the Memorial; see n. 3).
2. *wrote to Lord Derby*: Tyndall's letter to Edward Stanley is missing.
3. *Memorial*: the Memorial in support of Joseph Hooker in his dispute with Acton Ayrton (see letter 3755, esp. n. 1).
4. *He replies*: letter 3750.
5. *Some body told me... dash at the Times*: it does not appear that the Memorial was printed in either the *Times* or *The Daily Telegraph*.

To Thomas Henry Huxley 19 June [1872][1] 3752

19th. June

My dear Huxley,

I hope Hooker's surmise[2] is not correct. If it were it would damage our friend very much in my estimation. I entrusted the sending forward of the Memorial[3] to him; he consented to send it on with alacrity, and to send with it a note explaining to Mr Gladstone that the sentiments expressed in the Memorial would, we had reason to believe, be those of the Scientific world as soon as it had been made aware of the facts.

Yours ever | John Tyndall.

I am on the point of starting for Switzerland:[4] So I fear you must come in for the further conduct of this affair.

You know, I take it for granted, that immediately after seeing you on

Saturday I had the cash transferred to the London & Westminster St. James's Square.

RI MS JT/1/TYP/9/2962
IC HP 8.125
Typed Transcript Only

---

1. *[1872]*: the year is given by reference to Joseph Hooker's dispute with Acton Ayrton (see n. 3).
2. *Hooker's surmise*: probably Joseph Hooker's surmise that 'Lubbock has been prevailed against' (see letter 3747). Tyndall would have forwarded Hooker's letter to Huxley or told him.
3. *Memorial*: the Memorial in support of Joseph Hooker in his dispute with Acton Ayrton (see letter 3755. esp. n. 1).
4. *I am on the point of starting for Switzerland*: see letter 3716, n. 6.

From Joseph Dalton Hooker 19 June 1872 3753

Royal Gardens, Kew. | June 19th 1872.

Dear Tyndall

Thanks for your note and telegram[1] and the copies. I intended to have seen you to-day but was so tired that when I got to the Athenaeum[2] (after working from 10 to 2 in town) and into a chair, I could not get up! (I have had neuralgia[3] over half the head for a week now)

I must be in town again to-morrow and could see you after 5.0 p.m. Shall we dine at the Athenaeum? If not let me know by a line addressed to me at the Linnean Society Burlington House, and I will seek you elsewhere.

Everyone approves of the address,[4] but I value Colvile's opinion[5] more than <u>all the rest put together</u>. I will see Helps tomorrow morning anent circulation and bring you the result.

Lubbock's motion for the papers[6] comes on to-morrow I am told. If it does not include the correspondence with Gladstone,[7] it will be futile.

Ever, dear Tyndall, | Yours | J. D. Hooker.

I do not understand whether Lubbock presents the address <u>before</u> calling for the correspondence or after. This I hope to know tomorrow as also <u>what Lord Derby has got</u>.

I have just seen Helps, he is quite averse to circulation previous to the correspondence being produced, but thinks that I should have no voice in the matter, but rely on the wisdom of those who have so gallantly and warmly and kindly taken up the affair, who should always consult Derby. What says Huxley?

RI MS JT/1/TYP/8/2659
LT Typescript Only

1. *your note and telegram*: letters missing.
2. *Athenaeum*: the Athenaeum Club (see letter 3710, n. 5).
3. *neuralgia*: pain, typically stabbing or burning, in the area served by a nerve; (also) an instance, type, or case of this (*OED*).
4. *the address*: the Memorial in support of Joseph Hooker in his dispute with Acton Ayrton (see letter 3755, esp. n. 1).
5. *Colvile's opinion*: see letter 3746.
6. *Lubbock's motion for the papers*: in the House of Commons on 20 June 1872 (HC Deb 19 July 1872 vol. 212 cc. 1417–30, *HANSARD 1803–2005* (online, accessed 27 November 2022)), where it was later debated on 8 August 1872 (see letter 3755, n. 1).
7. *correspondence with Gladstone*: Edward Stanley (Lord Derby) had been working to have the official correspondence included with the Memorial (see letters 3713, 3717, 3729, and 3750).

From John Lubbock 19 June 1872 3754

15 Lombard St[1] | 19 June 1872

My dear Tyndall

I have received the memorial[2] & will with pleasure present it to Mr Gladstone.[3]

Till this is done the document must of course be regarded as confidential

Y[ou]rs very faithfully | John Lubbock

Kew RBG Papers Relating to Kew 1867–1872, Ayrton Controversy, Volume 1, f. 240
RI MS JT/1/TYP/8/2662

1. *15 Lombard St*: address of the bank Robarts, Lubbock & Co. in London.
2. *memorial*: the Memorial in support of Joseph Hooker in his dispute with Acton Ayrton (see letter 3755, esp. n. 1).
3. *present it to Mr Gladstone*: see letter 3755, n. 1.

To William Gladstone 20 June 1872[1] 3755

<u>*CONFIDENTIAL*</u>.] | TO THE RIGHT HON. W. E. GLADSTONE, | *First Lord of the Treasury, &c. &c.*

We, the undersigned, deeply interested in the condition of English Science, and viewing with special concern the treatment which the eminent Director of the Botanical Establishment at Kew[2] has systematically received at the hands of Mr. Ayrton,[3] since his appointment to the office of First Commissioner of Works, do most respectfully beg your attention to the following statements and observations.

In the year 1840 the private Botanic Gardens of Kew,[4] which had previously been in the possession of the Royal Family, were handed over by the Queen to the Government.

A Commission then appointed to report on their condition recommended that they should be enlarged and maintained as a national scientific establishment, which should form a centre of reception for the useful products of the vegetable kingdom, a centre of reference and distribution for England, India, and the Colonies, and a means of augmenting the rational pleasure, increasing the knowledge, and refining the taste of the English public.

The late Sir William Hooker[5] was at that time Professor of Botany in the University of Glasgow. The founding of an establishment like that contemplated at Kew harmonised so completely with his scientific tastes and power of organisation, that at a sacrifice of more than half his income he offered to undertake the superintendence of Kew Gardens. His offer was accepted, and he was appointed Director of Kew, at a salary of £300 a year.

Sir William Hooker was at that time the possessor of an excellent private Herbarium and of a scientific Library, both of which were wanting at Kew. To provide houseroom for these an additional £200 was granted by the Government. No allowance, however, was made for the maintenance or increase of either the Herbarium or the Library. The expense of both fell upon the Director.

During his residence in Glasgow the excellence of his collections had attracted to the house of Sir William Hooker various active investigators, the number of which increased materially after his arrival at Kew. Fourteen rooms of the house he occupied were devoted to his Herbarium, which, for twelve years, was the resort of the scientific botanists of Europe. Unaided by the Government, save to the extent above mentioned, Sir William Hooker devoted his private means to the purchase of new books and specimens, and opened a correspondence with botanists of all lands. He thus made his house the most extensive botanical laboratory in this country, and the most important centre of reference regarding systematic, economic, and descriptive botany as illustrated by his Herbarium.

The gardens expanded equally under his vigorous and enlightened supervision; in ten years after his appointment they became the first in the world.

For twenty-five years he had been collecting textile fabrics, drugs, gums, dyes, and other products to illustrate the structure, uses, and physiognomy of plants. With these collections, made at his private cost, Sir William Hooker founded in Kew Gardens the first museum of the kind that had ever been established. Of such museums there are now three at Kew. They contain upwards of 50,000 named objects of scientific and economic interest, views of tropical vegetation, and maps illustrating the distribution of plants over the globe. These museums constitute concrete courses of instruction, unrivalled in concentration and completeness; and the public interest in them is proved by the number of persons who avail themselves of the stores of information thus provided.

The contributions of Sir William Hooker to these museums were his free gift to the country, for which he never received a farthing of remuneration.

In 1852 the Director's salary, which had previously been raised to £600 a year, was augmented to £800, together with a house which had become vacant at the time. The Herbarium was then lodged in a separate building, and immediately afterwards donations and legacies (some to the Director, some to the Government of the day) poured into it. The labour of naming the collections of expeditions, and of drawing up botanical reports, became at length so excessive that the public need of the Herbarium was still further recognised by the Government. The Director had previously borne the expense both of assistance and maintenance: of these he was now relieved, though he still continued to bear the cost of books for his Library and of new specimens of plants.

Without the personal devotion on the part of the Director, the development of Kew would have been a simple impossibility. For five-and-twenty years his purchases were made and his collections elaborated at his own expense and risk, though they were constantly employed in the work of the country. Before his death, knowing that his son[6] could not afford to be as regardless of pecuniary considerations as he had been himself, he gave directions to have his Herbarium valued by competent persons, and offered to the Government at the lowest valuation. On these terms the collections which had previously been devoted to the nation's use became the property of the nation itself.

This is a brief but sufficient statement of the relationship of Sir William Hooker to Kew Gardens. It shows him to have been their virtual creator.

The antecedents and achievements of the present Director of Kew may be thus sketched. In 1839 Dr. Joseph Hooker was appointed Assistant-Surgeon and Naturalist to the Antarctic Expedition,[7] the most perilous, perhaps, that ever sailed from these shores, and the scientific results of which exceeded in importance those of any other naval exploring expedition of this century. During the voyage Dr. Hooker received from the Government the pay of his rank as a medical officer. His outfit, his books, his instruments, were provided by his father. The expenses of travelling and collecting ashore during his four years' voyage of circumnavigation were defrayed from the same source, though this work was done with the express object of enriching a public establishment.

On his return, he waived his claim to promotion in the navy, and devoted four additional years to the classification and publication of the results of the voyage.[8] He also aided his father as an unpaid volunteer in the development of the scientific branches of the Kew establishment.

In 1847 Dr. Hooker was sent to India, to explore, in the interests of Kew, an unknown region of the Himalaya;[9] and he was directed to proceed subsequently to Borneo to report on its vegetable resources.[10] His outfit, both for India and Borneo, which embraced a large collection of expensive instruments, cost the Government nothing. To cover all expenses incidental to his three years' travelling and collecting, including the cost of assistants and specimens, a sum of £1,200 was received, while the real disbursements of Dr. Hooker during this time amounted to £2,200. The difference was contributed by Sir William Hooker and his son, in the interest of the establishment to which they had consecrated their best energies.

On his return from India, Dr. Hooker again devoted himself to the work of aiding his father in the scientific development of Kew. He was also employed by the Admiralty, during the nine years from 1851 to 1860, in publishing the botanical discoveries of various naval and other voyages, from Captain Cook's downwards, to parts of the world visited by Dr. Hooker himself.[11] For this service he received three years' pay as a medical officer in the Navy, together with a sum of £500, which was accompanied by 'the expression of their Lordships' approbation of the zeal, perseverance, and scientific ability displayed in bringing to a successful completion this great botanical work.' For three years he was occupied with the arrangement and distribution of his Indian collections, and with the publication of his journals.[12] To cover the expenses incidental to these labours, an allowance of £400 a year was granted by the Government.

Besides the voyages and travels above adverted to, Dr. Hooker has made journeys to various parts of Europe,[13] to Western Asia,[14] and to North Africa.[15] The expenses of these journeys, though they were made with the express object of adding to the interest and completeness of Kew, have been borne by himself, and the results given to the establishment of which he is Director.

We place these data before you, not with a view of founding on them either censure or complaint. The labours of Dr. Hooker, and the heavy drain upon his father's purse which his unexampled education as a botanist involved, constituted the discipline which made him the man he now is. But we think it highly desirable that you and England should know as much of his career as will enable you to decide whether its arbitrary interruption by your First Commissioner be creditable to the Government of this country.

In 1855, Sir William Hooker being then 70 years of age, Dr. Hooker was appointed his Assistant-Director, at a salary of £400 a year, without a house; and from this time his share in the duties of the Garden were added to his more purely scientific ones. In 1858 his salary was increased to £500 a year, with a house, and in 1865, on the death of his father, he succeeded to the Directorship without an assistant.

The liberality of his father and his own self-denying life in the public service have, we think, been sufficiently demonstrated. We will therefore ask permission to place before you only one additional specimen of his conduct. As regards the Floras of Asia, Africa, and America, the Herbarium at Kew had been long unrivalled. Europe, however, was but scantily represented. Three years ago, a collection embracing the very Flora needed for the completion of Kew was offered for sale in Paris. At his own private cost, Dr. Hooker purchased this collection for £400, and presented it to the Kew Herbarium.[16]

His income at Kew is £800 a year, and here is one-half of it voluntarily devoted to the establishment which it had been the continual object of his father and himself to raise to the highest possible perfection. Had these things been known to the Parliament and public of England, the First Commissioner of Works would, we imagine, have hardly ventured to inflict upon the Director of Kew the unnecessary toil, worry, indignity, and irredeemable loss of time against which this memorial is a remonstrance.

Under the auspices of his father and himself, Kew Gardens have expanded from 15 to 300 acres. They have long held the foremost rank in Europe. In no particular does England stand more conspicuously superior to all other countries than in the possession of Kew. The establishment is not only without a rival, but there is no approach to rivalry as regards the extent, importance or scientific results of its operations. Upwards of 130 volumes on all branches of Botany, including a most important series of Colonial Floras,[17] but excluding many weighty contributions to scientific societies and journals, have issued from Kew. To these are to be added guide books and official papers. The vast literature has been produced and published through the efforts of the Directors of Kew, for the most part at no expense whatever to the nation.

To these labours is to be added the correspondence of the Directors[18] with all parts of the world, a mere selection from which, now bound together at Kew, embraces some 40,000 letters addressed to the Directors, and for the most part answered with their own hands.

Of the popularity of the Gardens, which has been attained without prejudice to their scientific use and reputation, it need only be stated that from

9,000 visitors in 1841, the numbers have risen to an average of nearly 600,000 a year. What they have done towards the elevation and refinement of the tastes and conduct of the working classes may be inferred from the fact that last Whit Monday[19] 37,795 visitors entered and quitted the Gardens without a single case of drunkenness, riot, theft, or mischief of any kind being reported.

Since Dr. Hooker's accession the Gardens have been to a great extent remodeled, and the establishment wholly reorganised. A great saving in outlay has been thus effected, without any sacrifice of efficiency. During the ten years from 1863 to 1872 inclusive, the annual number of living plants sent from Kew to various parts of the world has been doubled, amounting on an average to eight or nine thousand annually. Of seeds ripened at Kew, or obtained by the Director from various parts of the world, the annual average distributed amounts to about seven thousand.

Of the practical value of these labours, the introduction of the Cinchona plant[20] into India, Ceylon,[21] and Jamaica, the commercial success of which is established, constitutes one of many illustrations. The introduction of Ipecacuanha[22] is another. This will be corroborated by Her Majesty's Secretaries of State for India and the Colonies. We would add, that there is scarcely a horticultural establishment at home or abroad which would not be willing to acknowledge its indebtedness to Kew.

In India upwards of thirty gardeners trained at Kew are now employed in forestry, cotton, tea, and cinchona plantations, Government gardens, &c., and a far greater number are usefully employed in other parts of the world.

By the joint effort of the Directors, a series of complete Floras of India and the Colonies[23] was set on foot at Kew, of which those of the West Indies, all the Australian Colonies, New Zealand, Tropical Africa, the Cape Colonies, and British India are completed or in progress. These are standard works of inestimable value in the countries whose plants they describe, as well as to scientific travelers and institutions in Europe.

We have hitherto confined ourselves to a statement of Dr. Hooker's services in relation to Kew, and have said nothing of his labours in geology, meteorology, and other sciences, nor of his researches while Botanist of the Geological Survey. During his single year of office he contributed to the Records of the Survey two memoirs, which are to be regarded as landmarks in the history of fossil botany.[24] In presenting the Royal Medal to Dr. Hooker in 1854, the President of the Royal Society spoke of these memoirs as 'one of the most important contributions ever made to fossil botany.'[25] We may add a reference to his adventurous explorations of the northern frontier of India, in regions never visited by a European before or since.

It is not likely that a man of these antecedents, accustomed to the respect which naturally follows merit of the most exalted kind, would in any way expose himself, and more especially in matters relating to the welfare of Kew, to the just censure of his official superiors. Until the advent of the present First Commissioner, he had never been the object of a censure, and was never interfered with in the practical discharge of his duties by the Board of Works. His proposals and suggestions were rightly scrutinised, and his estimates regulated by the opinions of the Board, but the current duties were left entirely to his conduct and supervision; the extension and improvement of the establishment being always the origination and work of the Director.

With this sketch of the early training of Dr. Hooker for his present post before you, you will be able to compare with it the early training of Mr. Ayrton for the position which, by your favour, he occupies as Dr. Hooker's master. You will be able to judge how far the First Commissioner is justified in treating the Director of Kew with personal contumely, and in rudely upsetting the arrangements which he had made with reference to the invaluable collections for which he is responsible, not to Mr. Ayrton alone, but to his conscience and his country.

Neither you, sir, nor the English public have forgotten the speech of the First Commissioner on presenting himself for re-election at the Tower Hamlets, when he went out of his way to insult 'architects, sculptors, and gardeners.'[26] That speech was a warning to every cultivated man who held office under the Board of Works, and it was, as you know, duly laid to heart by the Director of Kew. His desire to avoid all cause of offence was thus expressed in a letter addressed to yourself on the 31st of August 1871: 'Having regard to the tenour of the sentiments Mr. Ayrton is reported to have expressed in public accepting office, I felt it incumbent on me to be especially circumspect in my conduct and demeanour under his rule.'

Circumspection, under the circumstances, was of small avail, and one of Mr. Ayrton's first acts, after taking office, was to send a reprimand to Dr. Hooker. It was a new experience to the Director of Kew. During his thirty years of public service such a thing had never once occurred; indeed, the very reverse of it had always occurred, the respect due to intellectual eminence and moral worth having been always cheerfully accorded to Dr. Hooker by his official superiors. The first reprimand of his life was, moreover, not due to any fault of his, but arose entirely from the First Commissioner's own misconception.

The responsibility of the warming and ventilation of the plant-houses had, by special order, devolved upon the Director. After a searching enquiry, Dr. Hooker had been entrusted by a previous First Commissioner with the task of remodeling the heating apparatus throughout the establishment; and this led to the construction at Kew, in accordance with the Director's plans and estimates, of the most complete range of hot-houses for scientific purposes in existence. In 1871, however, he accidentally discovered that he had been superseded in this duty, without notice given or reason assigned. He wrote a respectful letter of enquiry to the First Commissioner,[27] and received the short—we are persuaded you will agree with us in adding, insolent—intimation that he had been superseded and would have 'to govern himself accordingly.'[28]

He would, in our opinion, have been equally unfaithful to the science of which he is a leader, and the public which he had so long served, if he had bowed in silence to this rebuke. He wrote a second letter of remonstrance to the First Commissioner,[29] in which he expressed himself as follows: 'The matter, therefore, stands thus: Several months ago I was, unknown to myself, deposed from the discharge of a function of great importance. I was left to hear this accidentally and, I have now to add, through one of my own subordinates.

'I do not for a moment question the First Commissioner's power to exercise arbitrary authority over the Director of Kew, but I do submit that there has been hitherto no plea whatever for such action as regards myself, and that the repetition of such acts, and the leaving me to be informed of them, on each occasion, by my subordinate, constitute a grievous injury to my official position, and tend to the subversion of all discipline in this department.'

At this point, sir, Dr. Hooker turned in the fullest confidence to you. He had undoubting trust in your will and power to protect both Kew and him from the arbitrary and, we would add, ignorant acts of the First Commissioner. He respectfully claimed the privilege of bringing the matter under the cognizance of the Right Honourable the First Lord of the Treasury.

You doubtless remember the letter addressed to you by Dr. Hooker on the 19th of August, 1871.[30] You could not fail to remark the reluctance with which he appealed to you, and his previous anxiety to take all possible measures to avert the necessity of such an appeal. 'I cannot express to you sir, sir,' he writes 'the anxiety that this step costs me, nor how earnestly I have endeavoured, by suppressing all personal feelings, to conduct my duties here under Mr. Ayrton to his and to my own satisfaction.

'After upwards of thirty-two years spent in the public service at home and abroad without a suspicion of mistrust on the part of my many previous superiors, I have had since Mr. Ayrton's accession to submit to various arbitrary measures, which, though compromising my position and authority, have been concealed from myself and become known to my subordinates, through whom alone I have first been made cognizant of them.'

From you, sir, the Director of Kew received no direct reply to this communication; but by the First Commissioner he was requested to 'furnish the dates and particulars of the conspicuous proofs of disregard to his office, and the particular occasions and facts, with dates, of his being left to be informed through his subordinates of acts of arbitrary authority of the First Commissioner, and the dates and particulars of those acts.'

To this challenge the Director of Kew replied by adducing five distinct acts of arbitrary interference, with their 'dates,' 'proofs,' and 'particulars.' * Among them was included what we have a right to call clandestine tampering with the subordinates at Kew. For example, the Curator of the Gardens[31] was tempted by Mr. Ayrton, personally, to leave Kew by the offer of a higher position, involving authority over works at Kew;[32] and he was requested by Mr. Ayrton to keep the fact from the knowledge of Dr. Hooker. To the loyalty of this man to a master whom he trusted and loved, the Director of Kew owes the discovery of proceedings which under any previous First Commissioner would have been impossible.

Your attention, sir, was drawn to this reply in a letter addressed to you by the Director of Kew on the 31st of August, 1871. It is in every respect so excellent, and so sure to be appreciated by all who know the real meaning of scientific work, and the baneful effect upon such work of this harassing conflict with your First Commissioner, that we do not hesitate to reproduce it here *in extenso*.

ROYAL GARDENS, KEW: | *August* 31, 1871.

Sir,—I beg most respectfully to submit copies of my further correspondence with the Right Honourable the First Commissioner of Her Majesty's Works, &c.

The acts detailed in the accompanying letter are, I believe, correctly described. I trust that I do not exaggerate in characterising them as grievously injurious to my official position, and tending to the subversion of discipline in this establishment; and I have evidently no protection from a repetition of them, except through the intervention of a higher authority.

Of these acts, those referred to under 1, 2, 3, formed the subject of a prolonged correspondence between the First Commissioner and myself; that under 4, I brought to the notice of Mr. Stansfeld,[33] and the result was the abandonment of the proposal; that under 5 will, I venture to hope, be revoked by your authority. I refrain from commenting on these acts of the First Commissioner in reference to their seriously interfering with the execution of my peculiar and multifarious duties here.

These include the labours of a Scientific Botanist, a Horticulturist, and the administration of Public Gardens, Museums, and Pleasure Grounds, frequented annually by upwards of 600,000 persons.

Besides the living collections, I have the direction of the largest and by far the most frequent Herbarium (by botanists and amateurs) in existence, and a very extensive Library.

I conduct, without a secretary, a responsible and onerous correspondence with Foreign and Colonial Gardens, as also with the Admiralty and Indian and Colonial Offices, on all subjects connected with Horticulture, Forestry, Botany, and the appointment of officers to duties in connection with these matters, and the introduction of useful plants everywhere.

I have, further, the editorship or control of various botanical works now being published, by order of Government, at Kew; and I have to devote every moment that

I can spare from my duties to maintaining, by researches and publications of my own, a position as a Scientific Botanist.

Until the accession of Mr. Ayrton, I have been enabled to fulfill these duties with satisfaction to myself, having been treated with uniform confidence, consideration, and courtesy by my superiors. I was invariably consulted on all prospective changes affecting my own and my subordinates' positions and duties. On the revision of my estimates at the Board, before their transmission to the Treasury, I was referred to; and amongst my other current duties was the control of the construction and repairs of the hot-houses and heating apparatus throughout this establishment.

Subsequently to Mr. Ayrton's accession, my position has been materially changed in all these respects. He had hardly entered on his duties when he hastily administered to me a wholly unmerited reprimand (the first I ever received), and his last act (known to me) has been to take from me the above-mentioned control, without pretext, warning or subsequent intimation.

I venture to hope that this may be restored to me, if the reasons I have adduced in the enclosed letter to the First Commissioner are satisfactory to you. These I would add, that in all similar establishments with which I am acquainted, in England or abroad, the opinion of the cultivator is entitled to the first consideration in all matters relating to plant-houses and heating apparatus; that to trust him with the care and treatment of invaluable collections, and make him amenable to the opinions of another in respect of the apparatus he requires, is as obviously wrong in principle as to refuse a surgeon his choice of instruments and hospital appliances. Nor would it be candid in me to withhold from you my conviction, that I have by this arbitrary act of the First Commissioner been lowered in the eyes of those who know no more of the circumstances than that I am deposed from the full control of buildings and apparatus which I was entrusted to erect and have still to use.

Let me assure you, sir, that I am unconscious of any feelings of personal animosity against Mr. Ayrton.

Having regard to the tenour of the sentiments he is reported to have expressed in public, on accepting office, in respect of professional duties such as mine, I felt it incumbent on me to be especially circumspect in my conduct and demeanour under his rule. And in evidence of this let me add, that when still smarting under his unprovoked reprimand, I, at his special request, devoted many nights to examining and reporting upon various books and pamphlets on the public parks of England, France, and America, for his guidance—a labour not very congenial and wholly beyond my province as Director of Kew, and which I further undertook in the hope that it might lead the First Commissioner to judge more generously of the acquirements and duties of some of the officers of the department he controls.

I am, Sir, | Your most obedient Servant, | JOS D. HOOKER, | *Director*.

To the Right Honourable | The First Lord of the Treasury.

To this letter Dr. Hooker was honoured by a reply from yourself, couched in kind and considerate terms. You had communicated with Mr. Ayrton, and had received his explanations, which you forwarded to the Director of Kew, 'in the hope that they would convey to his mind the assurance that there has been no intention on Mr. Ayrton's part to disregard his feelings or withhold the consideration due both to his person and his office.'

Stress of public duty is quite sufficient to account for the fact of your overlooking the serious omissions and inaccuracies of the First Commissioner's 'explanations.' These, however, were immediately pointed out to you by Dr. Hooker. His object in addressing you was not simply to complain of personal discourtesy on the part of the First Commissioner, but of five official acts subversive of discipline in the Kew establishment, and fraught with mischief to the public service. Had these acts merely affected him personally, he would have been perfectly willing to accept the assurance of Mr. Ayrton's consideration, though he failed to discover any trace of it either in his explanations or in the treatment which official papers sent from Kew continued to receive from the office of Works. He regarded it, however, as his duty, as an officer in the public service, to the Government and the scientific public, to spare no effort to procure a reversion of the policy introduced into the management of each and all of the departments of Kew (the Gardens, Museum, and Herbarium) by the present First Commissioner.

The specific acts enumerated by Dr. Hooker in answer to the request of the First Commissioner are thus summarised in a letter to yourself:—

'1. A transaction with my subordinate of a nature so new to my long experience of official life, and so repugnant to my principles, that I refrain from characterising it.

'2. Removing the Curator from his duties under me, without any communication with me.

'3. Empowering the Curator to act independently of me in regard to the times he should consider himself under my orders, and instructing me to make my arrangements in deference to his, and in concert with him.

'4. Submitting to the Treasury plans and estimates for extensive alterations to the Museums at Kew, without even informing me of his intentions; which works would have most seriously embarrassed me, as Director of the Museums, and would have involved a large expenditure for which I believe no estimate was submitted, and which would have been in every respect detrimental.

'5. Superseding me, without previous or subsequent communication, in duties for the execution of which I held the Board's authority, and which I am of opinion should unquestionably be performed by the Director, i.e. the control of the heating apparatus of the hot-houses &c.'

In Mr. Ayrton's letter of 'explanations,' which you considered so satisfactory, the first three of these charges are skillfully ignored, and the other explanations contain statements which are demonstrably at variance with fact.

In the very considerate note above referred to, which was addressed by you from Balmoral[34] to Dr. Hooker, you say: 'There must be some mistake about Mr. Ayrton's failing to see you at Kew, as he assured me that he paid the visit there for the very purpose of personal and friendly communication.'

We respectfully ask you to consider how this 'purpose' was carried out. Dr. Hooker was at home when the first Commissioner paid his visit to Kew. He omitted to enquire for the Director at his house, or at the Gardens, or of his subordinate, to whom Mr. Ayrton's visit was really paid. He held a conversation with this subordinate seriously compromising the Director's position and authority, which conversation he subsequently desired should not be communicated to the Director. The result of this conversation, moreover, was a communication to the Treasury affecting Kew, which was also kept from the knowledge of the Director. You will learn from these facts what the First Commander understands by 'personal and friendly communication.'

From its effect upon himself, Dr. Hooker could infer how disturbing the continual intrusion of this subject upon your attention must be. He was anxious to reduce this disturbance to a minimum, and therefore ventured to suggest that he should be put in communication with one of your private secretaries, to whom he might explain his position. To this request you, in the kindest manner, assented, and placed Mr. West[35] in communication with the Director of Kew.

Sorely against his inclination, but driven to it by the necessities of the case, Dr. Hooker, at an interview with Mr. West on the 30th of October, distinctly pointed out the grave errors and omissions contained in the 'Explanations' given by the First Commissioner to the First Lord of the Treasury.

The end of the year approached without any answer being made to these communications and representations, and towards the close of December Dr. Hooker wrote again to Mr. West, who thereupon replied that a plan was under the consideration of the Government which would materially alter his position with reference to the First Commissioner of Works. He was subsequently informed semi-officially that the scheme was maturing, and the hope was expressed that he would take no step likely to embarrass the Government. This was far from his wish or intention. But after waiting till the 21st of February, the Director was semi-officially informed that the Government plan for his relief, and for the protection of Kew, had been abandoned.

In the hope of a satisfactory settlement, the matter was subsequently placed by you in the hands of the Marquis of Ripon,[36] and on the 13th of March, 1872, before a Committee of the Cabinet, consisting of the Marquis, Lord Halifax,[37] and Mr. Cardwell, Dr. Hooker, by the desire of the Committee, handed in a Memorandum[38] containing a statement of the points wherein his relations to the Government required definition and correction.

The upshot of these friendly efforts was this: On the 15th of April, 1872, Lord Ripon was asked to convey the following verbal message from yourself to Dr. Hooker, which, the noble Marquis added, was to be regarded by the Director of Kew as a final answer to his appeal:—'Mr. Ayrton has been told that Dr. Hooker should in all respects be treated as the head of the local establishment at Kew: of course in subordination to the First Commissioner of Works.'

At this time the controversy had, unhappily, reached a pitch far too serious to be stilled by such a message. In a letter to your private secretary, written immediately subsequent to the interview on the 30th of October, Dr. Hooker put his case thus: 'I am at a loss what to say as to my future position under a Minister whom I accuse of evasion, misrepresentation, and misstatements in his communications to the First Minister of the Crown, whose conduct to myself I regard as ungracious and offensive, and whose acts I consider to be injurious to the public service, and tending to the subversion of discipline. Granting,' he continues 'that the functions of a Director are restored to me, how am I to act when ordered to undertake works that involve wasteful expenditure, or are otherwise detrimental? I should be thankful for Mr. Gladstone's instructions on this head.'

With great deference, we submit that the verbal intimation conveyed from you to Lord Ripon, and from Lord Ripon, through Mr. Helps, to Dr. Hooker, by no means met the issues here raised by the Director of Kew. He had suffered from the secret tampering of the First Commissioner with his subordinates; he had successfully resisted extravagant and foolish proposals made by the same Minister. His duties and responsibilities as regards the warming of the plant-houses had, to the imminent jeopardy of plants of the rarest value, been transferred, without notice or justification, to the *Director* of Works. Another class of duties had, in the same secret manner, been transferred to the *Secretary* of the Board of Works. Surely, sir, your message through Lord Ripon, to all intents and purposes, empowered the First Commissioner to continue his course of studied indignity? Wrong upon wrong has been committed, which your answer left unredressed. No wonder that, notwithstanding his esteem and regard for you personally, and his respect for all authority rightly exercised, the Director of Kew should be driven to address to you, on the 22nd of April, a letter containing the following remonstrance:—

The fact is, that the Directorship of Kew, which was formerly subordinated to the First Commissioner alone, has been by Mr. Ayrton *officially* subordinated to the Secretary of the Board and the Director of Works in London, and this surreptitiously, without fault found or notice of any kind given, the Director being left to discover his altered position as best he could; and the Director has further been subjected to a series of arbitrary and offensive measures on the First Commissioner's part, against which he could not defend himself. These measures being destructive of discipline, and injurious to this establishment, the Director felt it to be his duty to bring both their nature and consequences officially under your notice, and to seek from you that justice which (as he has been assured by the officers of the Treasury) could be obtained only through an official appeal to the Prime Minister.

After eight months' interval, during which further arbitrary measures have been resorted to by the First Commissioner (and four of which were passed under the assurance that a measure for effectual relief was under consideration), the position of the Director of Kew is not better, but worse than when it was first brought under your notice; for within that period his views with regard to the scientific and other appointments in the establishment have been absolutely set at naught.

These circumstances are well known to the Director's subordinates. They know that he has been virtually deprived of authority and responsibility, and that his official appeals have been unanswered, and his complaints ignored. The basis of all order and discipline in the establishment is thus sapped, and the position of the Director rendered so anomalous, that his desire and determination to uphold the interests of science at Kew, strengthened as they are by the moral and material support guaranteed to him, hardly suffice to render that position endurable.

Your own practical wisdom will enable you to judge whether such a state of things is to be remedied by the curt and vague announcement (and such you must allow me to call it) which you have been good enough to make to me through Lord Ripon.

I have the honour to be, etc. | (Signed) J. D. HOOKER.

Your verbal announcement through the Marquis of Ripon was subsequently defined by Mr. West as a private and friendly communication; and your secretary proposed that, as an official answer would be sent to Dr. Hooker's official application, the letter from which the foregoing extract is made should be considered as *non avenue*.[39] Dr. Hooker, however, had shown his letter to friends whose counsel he had sought in this matter, and he therefore pleaded that you ought to see that which had been seen by others. With regard to the character of the verbal communication, Dr. Hooker had been given distinctly to understand that it was *official* and *final*. He, however, cheerfully accepted the assurance of your secretary, and awaited the official reply.[40] It came, and we hereby respectfully submit it to your calm interpretation.

TREASURY CHAMBERS: | *April* 25, 1872.

SIR,—I am directed by the Lords Commissioners of Her Majesty's Treasury to acquaint you that their Lordships have been in communication with the First Commissioner of Works as to the matters contained in the letter which you have addressed to the First Lord of the Treasury.

Their Lordships find that there is no difference of opinion upon the question of your position, which may be briefly defined as that of head of the local establishment at Kew, of course in subordination to the First Commissioner, and they anticipate no difficulty in the future regulation of the relations of that important establishment to the office of the Board of Works, in which the duties and powers of management are vested by statute.

The present form of estimate for the Kew Gardens laid by their Lordships before the House of Commons cannot now be altered, but it will be acted upon, and will in future be framed in accordance with this letter.

I am, Sir, | Your obedient Servant, | (Signed) CHARLES W. STRONGE,[41] | *Principal Clerk.*

Dr. Hooker, | Director of Royal Botanical Gardens, Kew.

The concluding paragraph of this document, which is evidently the really important one, has been submitted to various persons accustomed to the language of official life, and we do not believe that a single one of them is sure of its meaning. Dr. Hooker, while willing to put the best construction upon it, thought it necessary to make a final enquiry,[42] which was preceded by these remarks:

I am most desirous of giving their Lordships no further trouble, and am, of course, prepared either cheerfully to submit to their decision, whenever it is clearly given, or to resign the office which I hold; but I am unable to find in your letter any judgment whatever upon the points contained in the accompanying Memorandum, which have all been submitted to the First Lord of the Treasury in my letters of August 19 and 31, or to the Committee of the Cabinet which I had the honour of attending on March 13 at Lord Ripon's residence.

I trust that their Lordships will observe, that in seeking their decision on these several questions, I am raising no superfluous difficulties, but that it is impossible for me to understand my position until it receives their Lordships' authoritative definition in respect of the above matters.

I am, Sir, | Your obedient Servant, | (Signed) JOS D. HOOKER, *Director.*

Charles W. Stronge, Esq.

MEMORANDUM.

1. Up to the date of the appointment of a Director of Works (under the Board of Works) in 1870, I was entrusted, by a special warrant of the Board, with the duty of preparing the estimates for the construction and repairs of the plant-houses,

museums, and warming apparatus in this establishment. This warrant has been cancelled, without fault found, enquiry made, or even intimation given, and the duty transferred to the Director of Works.

Will you be good enough to inform me if it is their Lordships' decision that the powers conferred upon me by that warrant be restored to me?

2. Previous to the accession of the present First Commissioner to office, I was consulted whenever changes were made in the estimates which it is my duty to submit to the Board, prior to their transmission to the Treasury.

Am I to understand that hereafter the estimates will not be altered by the Board without giving me an opportunity of stating my views?

3. I was entrusted with the custody and distribution to scientific bodies &c. of the copies of the first volume of the 'Flora of Tropical Africa,'[43] a work the publication of which I am officially instructed to superintend at Kew. On the publication of the second volume,[44] the undistributed copies of the first were withdrawn, without enquiry, from my custody, and sent, together with those of the second volume, to the Stationery Office for sale.

Would you be so good as to state, whether I am in future to be entrusted with the custody and distribution of scientific works of which I (the unpaid Editor) am entrusted with the publication by the Board?

4. Previous to the accession of the present First Commissioner to office, I was consulted in all cases of prospective changes in the position and duties of my subordinates, and in all cases of proposed works that might affect my duties and responsibilities.

I shall be glad to know whether I am in future to be consulted in regard to such matters.

5. The Department of Works having been brought under the rules of the Civil Service Commissioners, all candidates for employment at Kew are liable to be chosen by open competition, except in cases where the qualifications required 'are wholly or in part professional, or not ordinarily to be acquired in the Civil Service,' as set forth in Clause VII. of the Commissioners' Rules. The present First Commissioner of Works refuses to allow me to take advantage of Clause VII. in cases both of purely botanical and horticultural appointments.

Am I hereafter to be allowed to avail myself of this clause, when it is of importance to the public service that I should do so?

6. The Director of Works having been given power to interfere in matters for which I am still in part responsible, I am anxious to know—

Whether I am to consider myself subordinate to the Director of Works in such matters, and to submit to his control in respect of them.

JOS D. HOOKER. | Kew: *May* 1, 1872.

To this letter no answer has been received.

It but rarely falls in either with our duties or our desires to meddle in public questions; and not until we found Dr. Hooker maimed as regards his scientific usefulness—not until we saw the noble establishment of which he has hitherto been the living head in peril of losing services which it would be absolutely impossible to replace; not, indeed, until we had observed a hesitation upon your part which we believe could only arise from a lack of information—did the thought of interference in this controversy occur to us. Knowing how difficult it must be for one engrossed in the duties of your high position to learn the real merits of a conflict like that originated by the First Commissioner of Works, we venture to hope that you will not look with disfavour on an attempt to place a clear and succinct statement of the case before you.

That statement invites you, respectfully, to decide whether Kew Gardens are, or are not, to lose the supervision of a man of whose scientific labours any nation might be proud; in whom natural capacity for the post he occupies has been developed by a culture unexampled in variety and extent; a man honoured for his integrity, beloved for his courtesy and kindliness of heart; and who has spent in the public service, not only a stainless, but an illustrious life. The resignation of Dr. Hooker under the circumstances here set forth would, we declare, be a calamity to English science and a scandal to the English Government. With the power to avert this in your hands, we appeal to your justice to do so. The difficulty of removing the Directorship of Kew from the Department of Works cannot surely be insuperable; or, if it be, it must be possible to give such a position to the Director, and such definition to his duties, as shall in future shield him from the exercise of authority which has been so wantonly abused.
CHARLES LYELL.

CHARLES DARWIN.
GEORGE BENTHAM, Pres. Linn. Society.
HENRY HOLLAND, Pres. Royal Institution.
GEORGE BURROWS, Pres. Roy. Coll. of Physicians.
GEORGE BUSK, Pres. Roy. Coll. of Surgeons.
H. C. RAWLINSON, Pres. Roy. Geogr. Society.
JAMES PAGET.
WILLIAM SPOTTISWOODE.
T. H. HUXLEY.
JOHN TYNDALL.

*Dr. Hooker's letter containing these changes was never answered, or even acknowledged, by Mr. Ayrton.

RI MS JT/1/TYP/8/2648A

1. *20 June 1872*: while the date is not given on this letter to Gladstone, when John Lubbock was asked to forward the letter to Gladstone following the procuring of signatories during the first couple weeks of June, he did so on 20 June 1872. Lubbock's cover letter bearing this date is published in the Darwin correspondence, and reads: 'My dear Mr. Gladstone, | I have been requested to forward to you the accompanying memorial from some of our most eminent scientific men, on the subject of the changes recently introduced as regards

the botanical establishment at Kew. The signatures have been intentionally restricted to a few well-known names, but I have reason to know that the opinions expressed in the memorial would be shared not only by the science of England, but by scientific men throughout the world. | I am, &c. | John Lubbock' (*Darwin Correspondence*, vol. 20, pp. 262–78; this is letter 8403F in the online edition of the Darwin correspondence). The letter was published in 'Copies of Correspondence between the Board of Works and Dr. Hooker relating to Changes proposed to be introduced into the Direction and Management of the Gardens at Kew', HC 25, Parliamentary Papers (1872), pp. 41–9; and *Nature*, 6 (11 July 1872), pp. 211–16. After the Memorial was submitted to Parliament, there was difficulty in obtaining letters from the government, despite the support from several members of Parliament. The issue was discussed in the House of Lords on July 29 1872 (The Royal Gardens, Kew—Dr Hooker and The First Commissioner of Works', HL Deb 29 July 1872 vol. 213 cc3–23, *HANSARD 1803–2005* (online, accessed 30 May 2021); this was also published as 'Dr. Hooker and Mr. Ayrton', *Times*, 30 July 1872, p. 5), and debated in the House of Commons on 8 August 1872 ('The Royal Gardens, Kew—Dr. Hooker, and the First Commissioner of Works', HC Deb 08 August 1872 vol. 213 cc710–758, *HANSARD 1803–2005* (online, accessed 30 May 2021)). In the end, Gladstone essentially evaded the issue, Hooker did not retract any of his complaints and remained Director of the Royal Botanic Gardens, Kew, and Ayrton, despite Parliamentary and press support in favor of Hooker, made no apology and received no immediate consequences for his treatment of Hooker. However, the support of the scientific community around Hooker resulted in his election to the presidency of the RS in November 1873, while a few months before, in August, Gladstone placed Ayrton in a new position within his administration and thus no longer in control of public institutions such as the Royal Botanic Gardens, Kew.

2. *eminent Director of the Botanical Establishment at Kew*: Joseph Hooker.
3. *Mr. Ayrton*: Acton Ayrton. On Hooker's dispute with Ayrton, see *Ascent of John Tyndall*, pp. 295–9; L. H. Brockway, *Science and Colonial Expansion: The Role of the British Royal Botanic Gardens* (New Haven, CT: Yale University Press, 2002), pp. 96–9; Desmond, *Huxley*, pp. 421–3; *Nature's Government*, pp. 211–20; Endersby, *Imperial Nature*, pp. 282–93; J. Endersby, 'Odd Man Out: Was Joseph Hooker an Evolutionary Naturalist?', in G. Dawson and B. Lightman (eds), *Victorian Scientific Naturalism: Community, Identity, Continuity* (Chicago: University of Chicago Press, 2014), pp. 157–85, on pp. 161–3; *Life and Letters of Hooker*, vol. 2, pp. 159–77; and R. MacLeod, 'The Ayrton Incident: A Commentary on the Relations of Science and Government in England, 1870–73', in A. Thackray and E. Mendelsohn (eds), *Science and Values: Patterns of Recognition and Change* (New York: Humanities Press, 1974), pp. 45–78.
4. *Botanic Gardens of Kew*: on the history of the Royal Botanic Gardens, Kew, see L. H. Brockway, *Science and Colonial Expansion*; R. Desmond, *The History of the Royal Botanic Gardens Kew* (London: Royal Botanic Gardens, Kew, 2007); *Nature's Government*; and K. Teltscher, *Palace of Palms: Tropical Dreams and the Making of Kew* (London: Picador Books, 2020).
5. *Sir William Hooker*: William Jackson Hooker.
6. *his son*: Joseph Hooker.

7. *Antarctic Expedition*: Hooker was assistant to the ship's surgeon Robert McCormick on HMS *Erebus*, which under the command of James Clark Ross went on a voyage of exploration with HMS *Terror* from 1839–43.
8. *classification and publication of the results of the voyage*: J. D. Hooker, *Flora Antarctica: the Botany of the Antarctic Voyage of HM Discovery Ships* Erebus *and* Terror *in the years 1839–1843, under the command of Captain Sir James Clark Ross* (London: Lovell Reeve, 1844–7). Hooker would further publish volumes on the botany of New Zealand and Tasmania: J. D. Hooker, *Flora Novae-Zelandae: The Botany of the Antarctic Voyage of HM Discovery Ships* Erebus *and* Terror*, in the years 1839–1843, under the command of Captain Sir James Clark Ross*, 2 vols (London: Lovell Reeve, 1853–5); and J. D. Hooker, *Flora Tasmaniae: The Botany of the Antarctic Voyage of HM Discovery Ships* Erebus *and* Terror*, in the years 1839–1843, under the command of Captain Sir James Clark Ross*, 2 vols (London: Lovell Reeve, 1855–60).
9. *to explore... of the Himalaya*: on Hooker's travels in India and the Himalayas from November 1847 through March 1851, see *Traveller and Plant Collector*, chapters 6–10; J. D. Hooker, *Himalayan Journals; or, Notes of a Naturalist in Bengal, the Sikkim and Nepal Himalayas, the Khasia Mountains, &c.*, 2 vols (London: John Murray, 1854); and *Life and Letters of Hooker*, vol. 2, chapters 11–18.
10. *to Borneo to report on its vegetable resources*: while Hooker was originally to spend two years in India followed by one year in Borneo, the Borneo trip was cancelled and he spent a third year in India (*Traveller and Plant Collector*, pp. 94, 161).
11. *publishing the botanical discoveries... Dr. Hooker himself*: for his publications between 1851 and 1860, see *Life and Letters of Hooker*, vol. 2, pp. 491–6.
12. *arrangement and distribution of his Indian collections, and with the publication of his journals*: *Himalayan Journals* (see n. 9 above); and J. D. Hooker and T. Thomson, *Flora Indica: Being a Systematic Account of the Plants of British India, together with Observations on the Structure and Affinities of their Natural Orders and Genera* (London: W. Pamplin, 1855).
13. *journeys to various parts of Europe*: Hooker traveled to Europe in early 1845 for the purpose of meeting with continental botanists and visiting museums and herbaria in Paris, Brussels, and cities in the Netherlands (*Life and Letters of Hooker*, vol. 1, pp. 177–89). In 1867, he traveled to Paris to serve as a juror in the botanical section of the Exposition Universelle (*Life and Letters of Hooker*, vol. 2, p. 85). In May and June 1869, he traveled to St. Petersburg, Russia, visiting museums and herbaria in major cities along the way (*Life and Letters of Hooker*, vol. 2, pp. 85–9).
14. *to Western Asia*: in the autumn of 1860, Hooker traveled to Syria and Palestine with Daniel Hanbury (1825–75), a British botanist (*ODNB*), and published some botanical papers based on that trip. On his travels in the Middle East, see *Traveller and Plant Collector*, pp. 210–21; and L. Huxley (ed), *Life and Letters of Hooker*, vol. 1, pp. 528–33.
15. *to North Africa*: from April to June 1871, Hooker traveled to Morocco with John Ball, George Maw (1832–1912), a British botanist (*ODNB*), and Edward Crump (d. 1827/8), a young gardener from Kew (R. Desmond, *Dictionary of British & Irish Botanists and Horticulturalists*, 2nd edn (London: Taylor and Francis/Natural History Museum, 1994), p.

182). Hooker and Ball later published *Journal of a Tour in Marocco and the Great Atlas* (London: Macmillan and Co., 1878), in which Edward Crump is only referred to as 'Crump' (Desmond mistakenly identified him as William Crump (1843–1932) in *Traveller and Plant Collector*). On his travels in Morocco, see *Traveller and Plant Collector*, pp. 222–40; and L. Huxley (ed), *Life and Letters of Hooker*, vol. 2, pp. 90–4.

16. *purchased this collection for £400, and presented it to the Kew Herbarium*: Hooker purchased Jacques Gay's herbarium in 1868 (L. Huxley (ed), *Life and Letters of Thomas Henry Huxley*, 2 vols (London: Macmillan and Co., 1900), vol. 1, p. 48).
17. *series of Colonial Floras*: colonial floras were published for New Zealand, Tasmania, British India, the British West Indies, Hong Kong, the Cape of Good Hope, and Australia. On the colonial floras, see C. Bonneuil, 'The Manufacture of Species: Kew Gardens, the Empire and the Standardisation of Taxonomic Practices in late 19th Century Botany', in M.-N. Bourguet, C. Licoppe, and O. Sibum (eds), *Instruments, Travel and Science: Itineraries of Precision from the Seventeenth to the Twentieth Century* (New York: Routledge, 2002), pp. 189–215; *Nature's Government*, pp. 204–6; and Endersby, *Imperial Nature*, pp. 163–5.
18. *correspondence of the Directors*: since 2004, the archives of the Royal Botanic Gardens, Kew have been digitizing one of its largest collections, the Directors' Correspondence, making scans of the letters available through JSTOR Global Plants (*https://plants.jstor.org/collection/DIRCOR*). See K. Harrington, 'The Digitisation of Kew's Directors' Correspondence Collection', *NATSCA News*, Issue 23 (2012), pp. 43–9.
19. *Whit Monday*: a traditional holiday for workers, which became an official holiday in 1871. John Lubbock, Liberal MP for Maidstone, was responsible for the 1871 Bank Holidays Act, which created Whit Monday and three other holidays.
20. *Cinchona plant*: a genus of evergreen trees or shrubs with fragrant white or pink panicled flowers, growing in the tropical valleys of the Andes, and now extensively cultivated in India and Java for the sake of the bark (*OED*); the medicinal bark of species of Cinchona, Peruvian bark; also the drug prepared from it (*OED*). See *Traveller and Plant Collector*, p. 221; *Nature's Government*, pp. 206–11; and Endersby, *Imperial Nature*, p. 234.
21. *Ceylon*: now Sri Lanka.
22. *Ipecacuanha*: the root of *Cephaëlis ipecacuanha*, a South American small shrubby plant, which possesses emetic, diaphoretic, and purgative properties; also popularly applied to various forms in which the drug is employed (*OED*). In *Life and Letters of Hooker*, Leonard Huxley noted that the introduction of ipecacuanha to India from seeds raised at Kew 'was unsuccessful' (vol. 2, p. 1, n. 1). See also Endersby, *Imperial Nature*, pp. 298, 308.
23. *complete Floras of India and the Colonies*: see n. 17 above.
24. *two memoirs, which are to be regarded as landmarks in the history of fossil botany*: Hooker published three memoirs on fossil botany in *Memoirs of the Geological Survey of Great Britain*: 'On the Vegetation of the Carboniferous Period, as compared with that of the present day', 2 (1848), pp. 387–430; 'On some Peculiarities in the Structure of Stigmaria', 2 (1848), pp. 431–9; and 'Remarks on the Structure and Affinities of some *Lepidostrobi*', 2 (1848), pp. 440–56.

25. *Royal Medal to Dr. Hooker... made to fossil botany'*: Hooker received the RS's Royal Medal on 30 November 1854 for his botanical researches during and stemming from his travels on the Antarctic expedition and to the eastern part of the Himalayan range (*Roy. Soc. Proc.*, 7 (1856), pp. 261–2).
26. *speech of the First Commissioner... 'architects, sculptors, and gardeners'*: Ayrton's re-election speech was reported in *Times*, 9 November 1869, p. 4. His bid for re-election as Liberal MP for Tower Hamlets was successful.
27. *respectful letter of enquiry to the First Commissioner*: for Hooker's letter of 12 July 1871, see 'Copies of Correspondence between the Board of Works and Dr. Hooker relating to Changes proposed to be introduced into the Direction and Management of the Gardens at Kew', HC 25, Parliamentary Papers (1872), p. 106.
28. *'to govern himself accordingly'*: for the 'insolent' reply of 17 August 1871 from Robert John Callendar, Assistant Secretary at the Office of Works, see 'Copies of Correspondence between the Board of Works and Dr. Hooker relating to Changes proposed to be introduced into the Direction and Management of the Gardens at Kew', HC 25, Parliamentary Papers (1872), p. 107.
29. *second letter of remonstrance to the First Commissioner*: for Hooker's letter of 19 August 1871, see 'Copies of Correspondence between the Board of Works and Dr. Hooker relating to Changes proposed to be introduced into the Direction and Management of the Gardens at Kew', HC 25, Parliamentary Papers (1872), p. 28.
30. *letter addressed to you by Dr. Hooker on the 19th of August, 1871*: the DCP notes that this letter was not included in 'Papers Relating to Kew Gardens', HC 335, Parliamentary Papers (1872), pp. 41–9—nor is it in 'Copies of Correspondence between the Board of Works and Dr. Hooker relating to Changes proposed to be introduced into the Direction and Management of the Gardens at Kew', HC 25, Parliamentary Papers (1872)—and that the letter is in the Hooker collection at Kew (*Darwin Correspondence*, vol. 20, p. 278, n. 18).
31. *Curator of the Gardens*: John Smith.
32. *offer of a higher position, involving authority over works at Kew*: Ayrton had offered John Smith the post of superintendent of works in Hyde Park (*Life and Letters of Hooker*, vol. 2, p. 162; *Nature's Government*, p. 214).
33. *Mr. Stansfeld*: James Stansfeld (1820–98), a British Liberal politician who at the time of the Ayrton dispute was Financial Secretary to the Treasury (1869–71) (*ODNB*).
34. *Balmoral*: Balmoral Castle, a Royal estate in Aberdeenshire, Scotland, since 1852.
35. *Mr. West*: Algernon Edward West (see letter 3708, n. 7).
36. *Marquis of Ripon*: George Frederick Samuel Robinson.
37. *Lord Halifax*: Charles Wood (1800–85), a British Whig politician who served, among a number of positions, as MP for four different constituencies from 1826–66, Chancellor of the Exchequer (1846–52), and at the time of the Ayrton dispute, Lord Keeper of the Privy Seal (1870–74) (*ODNB*).

38. *Memorandum*: Hooker's Memorandum is reproduced at the end of this letter to Gladstone.
39. *non avenue*: not having happened; (*spec.* of a legal agreement) annulled, completely invalidated (*OED*).
40. *official reply*: see 'Copies of Correspondence between the Board of Works and Dr. Hooker relating to Changes proposed to be introduced into the Direction and Management of the Gardens at Kew', HC 25, Parliamentary Papers (1872), p. 21.
41. *CHARLES W. STRONGE*: Charles Walter Stronge, Principal Clerk to the Treasury.
42. *final enquiry*: see 'Copies of Correspondence between the Board of Works and Dr. Hooker relating to Changes proposed to be introduced into the Direction and Management of the Gardens at Kew', HC 25, Parliamentary Papers (1872), p. 23.
43. *'Flora of Tropical Africa'*: D. Oliver, *Flora of Africa. Vol. I. Ranunculaceæ to Connaraceæ* (London: L. Reeve and Co., 1868). The whole series of nine volumes was completed in 1919.
44. *second volume*: D. Oliver, *Flora of Africa. Vol. II. Leguminosæ to Ficoideæ* (London: L. Reeve and Co., 1871).

To Joseph Dalton Hooker 20 June 1872 3756

June 20/72

I have written to Delane of the Times, & to the Editor of the Daily News[1] & Pall Mall Gazette[2]—sending to each of them a copy of the memorial.[3]

You ought to learn the <u>truth</u> regarding Lubbock & not just content with surmises[4]—If he falters it must be taken out of his hands.

<u>J. T.</u>

RBG, Kew Papers Relating to Kew 1867–1872, Ayrton Controversy, volume 1, f. 237

---

1. *Editor of the Daily News*: Frank Harrison Hill (1830–1910) was editor of the *Daily News* from 1869–86 (*ODNB*). See 'Mr. Ayrton and Dr. Hooker', *The Daily News*, 10 July 1872, p. 5.
2. *Pall Mall Gazette*: Frederick Greenwood (1830–1909) was editor of the *PMG* from 1865–80 (*ODNB*). See 'Mr. Ayrton and Dr. Hooker', *PMG*, 11 July 1872, pp. 10–11.
3. *memorial*: the Memorial in support of Joseph Hooker in his dispute with Acton Ayrton (see letter 3755, esp. n. 1).
4. *surmises*: an idea formed in the mind (and, often, expressed) that something may be true, but without certainty and on very slight evidence, or with no evidence; a conjecture (*OED*).

To Thomas Henry Huxley 20 June [1872][1] 3757

Royal Institution. | 20th. June.

My dear Huxley,

I shall be at the X,[2] where we can make our final arrangements.[3]

Nothing could exceed the distinctness of my question to Lord D.[4] or the distinctness of his replies.[5]

Yesterday in answer to a direct question regarding the circulation of our case, he declared it to be perfectly unobjectionable to circulate it at once.

I do not like this five per cent business, as my average interest does not exceed four and a half. However have your way, I will circumvent you yet.

Yours ever | John Tyndall.

RI MS JT/1/TYP/9/2962
IC HP 8.126
Typed Transcript Only

1. *[1872]*: the year is given by a note from LT on the letter, as well as content (the Memorial; see n. 3).
2. *the X*: a meeting of the X Club, this time a country excursion to Windsor (Brock and MacLeod, *Hirst Journals*, p. 1943). While letter 3778 from Huxley implies Tyndall did not attend, letter 3783 from Tyndall to Thomas Hirst implies he did, so it is uncertain if he in fact did or not. In 1864 Tyndall joined George Busk, Edward Frankland, Thomas Hirst, Joseph Hooker, Thomas Huxley, John Lubbock, Herbert Spencer, and William Spottiswoode as members of the X Club, a monthly dining club in which they discussed the promotion of science in British society. On the X Club, see *The X Club*.
3. *our final arrangements*: for the circulation of the Memorial in support of Joseph Hooker in his dispute with Acton Ayrton (see letter 3755, esp. n. 1).
4. *my question to Lord D.*: the letter to Edward Stanley (Lord Derby), for which 3750 is a reply, is missing.
5. *distinctness of his replies*: see letter 3750.

To John Lubbock 20 June 1872 3758

Royal Institution of Great Britain | 20th. June 1872

My dear Lubbock

I have read your note[1] with the very greatest concern.

It did not occur to me that you would think of delivering the Memorial[2] committed to you for transmission to Mr. Gladstone.[3]

That memorial I am sorry to say has already been treated as a public document. Might I beg of you therefore to send it to me so that I may at once place is in the hands of M^r^. Gladstone?

A note received from Hooker[4] this morning contains the following passage:—"Lubbock's motion for papers comes on tomorrow I am told. If it does not include the correspondence with Gladstone it will be futile."

Yours sincerely | John Tyndall

RBG, Kew Papers Relating to Kew 1867–1872, Ayrton Controversy, volume 1, f. 240

---

1. *your note*: letter 3754.
2. *Memorial*: the Memorial in support of Joseph Hooker in his dispute with Acton Ayrton (see letter 3755, esp. n. 1).
3. *to M^r^. Gladstone*: see letter 3755, n. 1.
4. *note received from Hooker*: letter 3753.

From George Biddell Airy 20 June 1872 3759

Royal Observatory. | Greenwich. | 20 June /72.

My dear Sir,

I have lately returned from Scotland, and find your note of 11th.[1] accompanying the printed Memorial to Mr Gladstone concerning Dr. Hooker and Mr Ayrton.[2] I have read the memorandum with due care.

I cannot sign it.[3]

A misunderstanding has arisen between a Head of Department and an officer in an important position, which has after a time been referred to the highest recognised authority by whom decision has been given. I am far from saying whether the decision is just or unjust; but I am clear that uninterested persons ought not to interfere with the hope of aiding one party.

It happened to me once to think myself aggrieved by my immediate superiors (the Admiralty) and I did thereon appeal to the Treasury. But it never occurred to me that under any circumstances whatever could any outsiders put in a word between us. I did not even ask my own Board of Visitors to interfere.

I am sorry that it happens that our views disagree so much.

I am, my dear Sir, | Yours very truly | G B Airy

Professor Tyndall

RGO MS.RGO 6/438
RI MS JT/1/TYP/1/46

1. *your note of 11th.*: letter 3730.
2. *Memorial to Mr Gladstone concerning Dr. Hooker and Mr Ayrton*: the Memorial in support of Joseph Hooker in his dispute with Acton Ayrton (see letter 3755, esp. n. 1).
3. *I cannot sign it*: this was the second time Tyndall asked Airy. For the first, see letters 3718 and 3726. See also letters 3730, 3763, and 3772.

From Juliet Pollock 20 June [1872][1] 3760

59 Montagu Square W.[2] | June 20th

I don't know whether I shall see you again so I write a good bye, & good wishes for both your expeditions.[3] The American one I in a manner envy you. That is, I should like to go there.

Fred[erick][4] & I met by chance quite curiously among the Americans on Tuesday at the Century.[5]

Col. Higginson's[6] speech[7] was very good and he is a really interesting man. He & I have become friends in a short space of time but we are likely to remain so for long; we shall keep it up by letters, (shaking hands across the vast) till he sees me again in London or till I see him at Boston.

Poor little Dot[8] has had troubles about her sister's marriage[9] the peculiarities of the Lake Harris Government being in some ways very unfortunate. However it is of no use to war with the inevitable. We must accept & hope Laurence & Alice have been uniform in love both to me and to Dot and I don't know why their affection for us should ever experience an interruption. They are now both at Paris.

Yours ever | Juliet Pollock

RI MS JT/1/P/189
RI MS JT/1/TYP/6/2137

1. *[1872]*: the year is given by mention of Tyndall's upcoming trip to the United States.
2. *59 Montagu Square W.*: the Pollocks' residence in Marylebone, London.
3. *both your expeditions*: firstly, his summer excursion to the Swiss Alps (see letter 3716, n. 6); secondly, his lecture tour in the United States (see Introduction, n. 8).
4. *Fred[erick].*: William Frederick Pollock.
5. *Century*: the Century Club.
6. *Col. Higginson's*: Thomas Wentworth Higginson (1823–1911), an American clergyman, author, abolitionist, and soldier. He served from 1862–4 as colonel of the first African American regiment (1st South Carolina Volunteers) during the American Civil War (*ANB*). He traveled to London in 1872, where he met the Pollocks (T. W. Higginson, *The*

Writings of Thomas Wentworth Higginson, vol. 1: *Cheerful Yesterdays* (Cambridge: The Riverside Press, 1900), p. 272). For his reminisces of Juliet Pollock, see *Cheerful Yesterdays*, pp. 280, 292; and T. W. Higginson, *Part of a Man's Life* (Boston: Houghton, Mifflin, and Company, 1906), p. 244.

7. *speech*: a speech the Century Club. From the 22 June 1872 issue of the American newspaper *The Index*: 'Last Sunday evening he put in an appearance at the Century Club, and the young men there for once in their lives ceased to pair off for separate conversations, and all gathered around him, while he gave them his impressions of London. Before he left the room he was down for an infinitude of breakfasts and dinners, and he had received an invitation, which I hope he will accept, to meet the Anglo-American Committee and give them an address "on any subject whatsoever." Professor Tyndall has taken him to dine with the Royal Society, and he has been welcomed in the palatial rooms of the Athenaeum Club' (p. 198).
8. *Poor little Dot*: not identified, but an undated letter from Juliet Pollock refers to 'dear little Dot' as 'Mrs H Finch' (RI MS JT/1/P/216).
9. *her sister's marriage*: the marriage of Alice le Strange (d. 1886) to the traveler and Christian mystic Laurence Oliphant (1829–88), who was a follower of the spiritualistic prophet Thomas Lake Harris (1823–1906). By 'Lake Harris Government', Pollock was probably referring to peculiarities of marriage within the Christian utopian communities that Lake Harris created. They were married in London on 8 June 1872.

From Charles d'Almeida[1] 21 June 1872 3761

[JOURNAL DE PHYSIQUE | THEORIQUE ET APPLIQUÉE | 31 rue Bonaparte, Paris. | Paris,] le 21 Juin [*187*]2

Cher Monsieur,

J'ai reçu votre excellent recueil de mémoires sur la chaleur, et je me suis empressé de les faire connaître à nos lecteurs. Vous le verrez en lisant le numéro de juin qui vous parviendra dans quelques jours.

Avec votre permission, je ferai quelques extraits que j'insérerai bientôt. L'étendue trop restreinte du journal ne nous permet pas des développements suffisants; mais j'espère qu'ils exciteront le désir de lire les mémoires en entier.

Votre heureuse prédiction à notre égard s'est accomplie déjà. Si l'an prochain tout continue comme cette année-ci, le journal est très solidement fondé. Je vous remercie.

Veuillez agréer l'assurance de mon respect et de mon dévouement. | Ch. d'Almeida

[JOURNAL OF PHYSICS | THEORETICAL AND APPLIED | 31 Bonaparte Street, Paris. | Paris,] 21 June [*187*]2

Dear Sir,

I have received your excellent collection of memoirs on Heat,[2] and I hurried to make it known to our readers. You will see it by reading the issue of June[3] that will reach you in few days.

With your permission, I will make some abstracts that I will insert soon. The restricted length of the journal will not allow us sufficient developments; but I hope that they will arouse the desire to read the whole memoirs.

Your happy prediction about us has already been accomplished. If next year everything goes like this year, the journal is solidly founded.[4] I thank you.

Please accept the insurance of my respect and devotion. | Ch. d'Almeida

RI MS JT/1/D/7

---

1. *Charles d'Almeida*: Joseph-Charles d'Almeida (1822–80), a French physician and founder of the *Journal de physique théorique et appliquée* in 1872, as well as the Societé de Physique in 1873 (*Dictionnaire de biographie française*).
2. *excellent collection of memoirs on Heat*: *Molecular Physics*.
3. *see it by reading the issue of June*: this notice in the *Journal de physique théorique et appliquée* has not been identified.
4. *the journal is solidly founded*: the *Journal de physique théorique et appliquée* ran from 1872 to 1919.

From William George Valentin — 21 June 1872 — 3762

Schools of Science | South Kensington | 21st. June 1872
Prof. J. Tyndall

Dear Sir,

I called on you this morning to have some conversation with you about the enclosed papers.[1] As you are likely to leave Town on Monday next,[2] I hasten to write what I think about the subject matter of the communication from the Trinity House.[3]

I think the whole can be put in a nut shell. If all were perfect, i.e. if no oils were ever likely to be delivered of a dangerous nature, no precautions need be necessary. But as we have found the flaming points of the same kind of oil to vary considerably in—different samples, (I need only remind you of oil 5[a] or of Young's oils[4])—it stands to reason that some oil might be delivered that might be of a dangerous nature. I do not think the precautions adopted by the Trinity House are at all too onerous or too troublesome. Altho' I agree with D[r]. Stephenson M[c]Adam[5] that no fear of explosions need be entertained if the oils are all that is required: I yet think the precautions adopted are both judicious and to the purpose. For why should Paraffin be kept in large tanks necessitating this refilling of smaller vessels with the days supply of oil, instead of having such supply storred up in 5. gallon vessels <u>at once</u>.

Great care should be bestowed at first upon the testing of the oils for it appears from what has come under my notice that barrels of oil taken from the same assignment and coming from the same distillers, differ considerably when the flashing point is taken. Oil 5 varied the other day from 100° to 126° F. I believe there would be[6]

National Archives MT 10/143. H3088 2545–6
Transcript Only[7]

---

1. *enclosed papers*: enclosures missing.
2. *leave Town on Monday next*: to travel to Switzerland (see letter 3716, n. 6).
3. *Trinity House*: see letter 3736, n. 3. On Tyndall as scientific adviser to Trinity House, see letter 3869, n. 4.
4. *Young's oils*: James Young (1811–83), a Scottish chemist who developed a method of distilling paraffin from coal and oil shales and founded the Young's Paraffin Light and Mineral Oil Company, though he was not active within the company (*ODNB*).
5. *D[r]. Stephenson M[c]Adam*: Stevenson Macadam (1829–1901), a Scottish chemist, author, and lecturer at the Edinburgh School of Medicine and University of Edinburgh ('Dr. Stevenson Macadam', *Pharmaceutical Journal*, 12 (2 February 1901), p. 121).
6. Fragment only; letter incomplete.
7. Notes at the top of the letter, not in Tyndall's hand, read: '<u>Copy</u> | Enclosure 2 in N° <u>4</u>.'.

To George Biddell Airy 22 June 1872 3763

Royal Institution of Great Britain | 22nd. June 1872.

Dear Sir,

I am sorry to have given you the trouble of reading the Memorial to M[r]. Gladstone.[1] It was sent to you prior to the receipt of your first note from the North.[2] Otherwise the doubt as to your willingness to sign it would have caused me to hold it back.

Once sent, however, I did not think it worth while to countermand it.

I now beg to send you a corrected copy, not with a view to its perusal, but that you may see the signatures attached to it.[3] The memorial, thus signed, had been already placed in the hands of M[r]. Gladstone when your second note declining to sign it[4] arrived.

I would not trouble you with either this note or the memorial, were it not for a single expression in your last letter to me. You there say "I am clear that uninterested persons ought not to interfere with the hope of aiding either party."

If by "uninterested persons" you mean persons such as yourself, I have no objections to this phraseology; but if you mean to apply it to such men as have signed the memorial, then, I think, you have unwarrantably extended your preposition. The signers of the Memorial feel, one and all, the most profound interest in this subject; and they are not only justified in giving, but determined to give expression to that interest; in the usual constitutional way.

I also am as sorry as you can be to find our views so widely divergent.

I am dear Sir | very faithfully yours | John Tyndall

Sir George Biddel[5] Airy Bart. &c &c

RGO MS.RGO 6/438
RI MS JT/1/TYP/1/47

1. *Memorial to Mr. Gladstone*: the Memorial in support of Joseph Hooker in his dispute with Acton Ayrton was presented to William Gladstone (see letter 3755, esp. n. 1).
2. *your first note from the North*: for previous correspondence between Tyndall and Airy about the memorial, see letters 3718, 3726, 3730, and 3759. See also letter 3772.
3. *signatures attached to it*: see letter 3755.
4. *second note declining to sign it*: letter 3759.
5. *Biddel*: 'Biddell' misspelled by Tyndall.

To Robin Allen 22 June 1872 3764

22nd. June 1872

Sir,

Before giving the opinion requested in your communication № 2865,[1] I thought it well to communicate with Mr. Valentin.[2] His conclusions as set forth in the accompanying note agree with those which I had previously formed myself. He approves, as I do, of the precautions proposed to be taken by the Elder Brethren,[3] with regard to the storage of mineral oils.

I have the honour to be | Sir | Your obedient Serv[an]t. | (signed) John Tyndall | Robin Allen Esq | &c &c &c

National Archives MT 10/143. Papers re Fog Signals. H3088[4]
Transcript Only

1. *your communication № 2865*: letter missing.
2. *communicate with M^r. Valentin*: while Tyndall's letter to William Valentin is missing, Valentin's reply is letter 3762.
3. *Elder Brethren*: see letter 3736, n. 5.
4. This letter is a handwritten transcript, not in Tyndall's hand.

To Robert Arthur Talbot Gascoyne-Cecil — 22 June [1872][1] — 3765

Royal Institution | 22d June.

Dear Lord Salisbury,

Through the spontaneous kindness of Lord Derby the subject of this memorial[2] will be brought before the House of Lords. I trust when it is brought forward that a man of science may rely upon your willingness to lend him a helping hand.

Most faithfully yours | John Tyndall

RI MS JT/1/TYP/8/2666
LT Typescript Only

1. *[1872]*: the year is given by reference to the Memorial (see n. 3).
2. *this memorial*: the Memorial in support of Joseph Hooker in his dispute with Acton Ayrton (see letter 3755, esp. n. 1).

To Margaret Ginty — 22 June 1872 — 3766

Royal Institution | 22nd. June 1872.

My dear Mrs Ginty,

I only know the place you mention by its repute.[1] Your Dr.[2] may be right, but in sending people to such places they are certainly often not right. The best plan perhaps would be to consult Dr. Farre,[3] telling him frankly that if you could you would willingly avoid the expense.

Dr. Bence Jones is very sagacious and would give you good advice.

Ever yours faithfully | John Tyndall.

RI MS JT/1/TYP/11/3669
LT Typescript Only

1. *place you mention by its repute*: letter from Margaret Ginty missing, and thus the 'place' is not identified.
2. *Your Dr.*: not identified.
3. *Dr. Farre*: possibly Frederic John Farre (1804–86), an English physician at St. Bartholomew's Hospital, London and Royal London Ophthalmic Hospital. He was elected a fellow of the College of Physicians in 1838, for which he served in a number of positions for nearly a half century (*ODNB*).

To Philip Henry Stanhope 22 June [1872][1] 3767

Royal Institution | June 22d

Dear Lord Stanhope

Lord Derby has been good enough to introduce the subject of this memorial[2] to the notice of the House of Lords. We may I trust rely upon your sympathy and support on that occasion.

Most faithfully yours | John Tyndall

RI MS JT/1/TYP/8/2668
LT Typescript Only

1. *[1872]*: the year is given by reference to the Memorial (see n. 2).
2. *this memorial*: the Memorial in support of Joseph Hooker in his dispute with Acton Ayrton (see letter 3755, esp. n. 1).

From Herbert Spencer 22 June 1872 3768

37 Queen's Gardens | Bayswater, W.[1] | 22nd June 1872.

My dear Tyndall,

I forgot to thank you on Thursday at the Athenaeum[2] for the copy of your new volume.[3] Let me do so now. Though my reading powers will not serve me to do more than look through parts of most interest to me, I hope to do this presently; & if I can cull some illustrations serving the purposes of my sociological arguments, so much the better.

Very sincerely yours, | Herbert Spencer

RI MS JT/1/S/99
RI MS JT/1/TYP/3/1116

1. *37 Queen's Gardens | Bayswater, W.*: Spencer's London residence since September 1866.

2. *Athenaeum*: the Athenaeum Club (see letter 3710, n. 5).
3. *your new volume*: *Molecular Physics*.

From Robert Arthur Talbot Gascoyne-Cecil 23 June 1872 3769

20 Arlington Street[1] | June 23d 1872.

Dear Professor Tyndall

I have just received your letter.[2] I will carefully read and consider the memorial[3] you have sent me and shall be very glad if it should be in my power to render you any assistance.

Yours very truly | Salisbury.

RI MS JT/1/TYP/8/2667
LT Typescript Only

1. *20 Arlington Street*: residence in London.
2. *your letter*: letter 3765.
3. *memorial*: the Memorial in support of Joseph Hooker in his dispute with Acton Ayrton (see letter 3755, esp. n. 1).

From Hermann Helmholtz 23 June 1872 3770[1]

BERLIN, 23 Juni, 1872.

Verehrter Freund,

Zöllner's Buch hat auch uns in Deutschland in das äusserste Erstaunen gesetzt. Der erste Eindruck war allgemein dass er geisteskrank sei, um so mehr so, als man wusste, dass in seiner Familie mehrere Fälle von Geisteskrankheit vorgekommen sind, und namentlich bei einem Bruder seiner Mutter die Krankheit mit der Herausgabe eines ähnlichen Buches *[anfing]*. Indessen ist er vorläufig noch nicht so weit gekommen, dass man ihn hätte einsperren können in ein Irrenhaus. Im Gegentheil, die Leipziger haben ihn noch erst nach der Veröffentlichung dieses Buches zum Professor *[Ordinarius]* gemacht, und er soll viele Freunde in Leipzig haben, die ihn bewundern. Übrigens besteht gegenwärtig eine Art von Rivalität zwischen den kleineren Geistern der Universität Leipzig und Berlin, und man rechnet es ihm dort vielleicht als *[Verdienst]* <an> auf die Berliner losgeschlagen zu haben.

Ich muss sagen, dass mir trotzdem und trotz eines gewissen Gedanken-Zusammenhangs in dem tollen Buche, doch die Hypothese, er sei geisteskrank immer noch die wahrscheinlichste ist. Der giftige Neid gegen den äusseren Erfolg Anderer und der verbissene Grimm wäre auch ohne das zu erklären bei einem Menschen, der sich in seine Arbeiten vergraben hat, ein gewisses receptives Talent in sich fühlt, aber doch mit dem Fluche der Unproductivität geschlagen ist, und nicht vorwärts kommen kann. Aber seine eigene Cometentheorie ist zu unsinnig, als dass ein gesunder Mensch, der so viel Kenntnisse und Talente hat, wie Zöllner anfangs jedenfalls gehabt hat, sie hätte vorbringen können. Er lässt aus den Cometen und der Sonne positive E. in das Unendliche entwickeln, ohne auch nur ein einziges Mal zu fragen, wo die negative bleibt. Er lässt sie entwickeln durch Zerstäubung von Wasser wodurch, wie wir genau wissen, sich keine E. entwickelt. Es ist sogar ein Fehler gegen den Satz vom zureichenden Grunde, dass Wasser von Wasser sich trennend electrisch werden soll. Warum soll denn eines positiv, das andre negativ werden. Er glaubt ruhende E. einen durchsichtigen Körper überziehend im Gleichgewichtszustande, leuchte. Das ist ein Fehler gegen die Erhaltung der Kraft. Und endlich seine Rechnung über <die> ungeheure Geschwindigkeit mit der die vom Cometen ausgestossenen Theilchen fliehen, beruht auf einer Verwechselung der elektrostatischen Einheit mit dem Milligr.

Ich habe es deshalb verschoben ihm zu antworten. Man macht sich lächerlich, wenn man ihm geantwortet hätte, und er dann für wahnsinnig erklärt würde. Ich habe also vorläufig nur eine Antwort verfasst, gegen die von ihm benutzten Einwände von C. Neumann gegen mich wegen des Weberschen Gesetzes. Ich will aber, falls er bis dahin nicht eingesperrt ist, eine Vorrede für die zweite Abtheilung von W. Thomson's Natural Philosophy, deren Übersetzung unter meiner Durchsicht erscheint, *[beliefern]* für die Scientific Fragments, *[worin]* ich dem nicht wissenschaftlichen Publikum auseinandersetzen will, was es von Herrn Zöllner zu denken hat.

Übrigens ist der Kern von Zöllner's Widerspruch gegen uns doch eigentlich die Feindschaft des Metaphysikers gegen die Naturwissenschaft und leider zeigt es sich bei dieser Gelegenheit, dass die alten naturwissenschaftlichen (die metaphysischen) Schwärmereien noch in vielen deutschen Gemüthern verborgen liegen, und nur nicht gewagt haben laut zu werden. Solche Leute sind geneigt, an Zöllner zu glauben, und ich selbst habe schon erfahren müssen, wie viel Neid auch gegen meine, wahrlich sehr mässigen äusseren Erfolge im Leben in manchen Herzen geschlummert hat. Zöllner gilt den Philosophen als sachverständige Autorität für Naturwissenschaft, und sie benutzen dies gern, um ihren gepressten Herzen Luft zu machen.

Dann haben wir seit dem Kriege unter uns eine Anzahl von Chauvinisten, namentlich unter den Leuten, die nichts gethan und nichts gelitten haben, aber eines schönen Morgens als die Sieger von Sedan aufgewacht sind, und sich nun vor Nationalstolz nicht zu lassen wissen. Für diese hat Zöllner sein Buch sehr geschickt zurecht gemacht.

Entschuldigen Sie nicht Ihre populären Bücher, die auch in Deutschland den grössten Beifall gefunden haben. Hätten sie es nicht gethan, so würde Herr Z. wohl geschwiegen haben. Und geniessen Sie froh der Alpen.

In treuer Freundschaft | Ihr | H. Helmholtz.

BERLIN, 23 June 1872.

Dear friend,

Zöllner's book[2] has caused us in Germany extreme astonishment too. The first impression generally was that he was mentally ill, and all the more so, once one knew that several cases of mental illness have occurred in his family, and, in particular, that with a brother of his mother the illness */began/* with the publication of a similar book. However, he has not, for the time being, reached the point that one would have been able to lock him away in a lunatic asylum. On the contrary, the people in Leipzig still made him a */full/* professor only after publication of this book, and he is supposed to have many friends in Leipzig who admire him. Incidentally, there is presently a kind of rivalry between the smaller minds of the University of Leipzig and Berlin, and one perhaps thinks it */to his credit/* there for having lashed out at the people in Berlin.

I have to say, that despite this and despite a certain coherence of ideas in the mad book, the hypothesis that he is mentally ill is still the most plausible to me. Toxic envy towards the outward success of others and dogged wrath would also be otherwise explicable with a person who has buried himself in his works, perceives a certain receptive talent in himself, but is nevertheless struck with the curse of unproductiveness, and cannot move forward. However, his peculiar comet theory is too absurd for a healthy person with as much knowledge and talent as Zöllner at least initially had to have been able to propose it. He has positive e[nergy]. being generated from comets and the Sun infinitely, without asking only a single time where the negative remains. He has it being generated through the vaporisation of water by which, as we know precisely, no e[nergy]. is generated. It is even an aberration against the law of sufficient reason that water should become electrically charged when separating from water. Why on earth should one become positively charged and the other one negatively. He believes that resting e[nergy]. glows when it suffuses a transparent body in a state of equilibrium. This is an aberration against the conservation of force. And finally, his calculation of the tremendous speed with which particles ejected from the comet fly off, is based on a confusing of the electrostatic unit with the milligr.

I have therefore put off replying to him. One will make a fool of oneself if one had replied to him, and he was then declared insane. I have therefore, for the time being, written only one reply, against the objections used by him that he took from C. Neumann[3] against me in relation to <u>Weber's</u> Law.[4] However, in the event that he is not locked away by then, I shall */supply/* a preface for the second part of W. Thomson's Natural Philosophy,[5] the translation of which[6] is appearing under my supervision, for the Scientific Fragments, *[in which]* I shall explain to the non-scientific public what it has to think about Herr Zöllner.

Incidentally, the nucleus of Zöllner's opposition to us is really just the enmity of the metaphysician to the natural sciences, and unfortunately it turns out on this occasion that the old scientific (metaphysical) passions still lie hidden in many German minds, and have only not dared to become known. Such people are inclined to believe in Zöllner, and I myself have already had to learn how much envy, even towards my truly very middling outward successes in life, has slumbered in some hearts. For the philosophers, Zöllner is regarded as an expert authority for the natural sciences, and they gladly use this in order to get everything off their constrained hearts.

Then, since the war,[7] we have had a number of chauvinists among us, and in particular among people who have not done anything and have not suffered anything, but have woken up one fine morning as the victors of Sedan,[8] and now are delirious with national pride. For these people, Zöllner has very cleverly prepared his book.

Do not apologise for your popular books, which have met with the greatest approval in Germany too. Had they not done so, then Herr Z. would probably have remained silent. And do enjoy the Alps.[9]

In loyal friendship | Your | H. Helmholtz.

RI MS JT/1/H/48
LT Typescript Only

---

1. Our German translator notes that while LT's typescript is missing numerous umlauts, he supplied them here in the German transcription.
2. *Zöllner's book*: see letter 3740, n. 3.
3. *C. Neumann*: Carl Gottfried Neumann (1832–1925), a German mathematician (*NDB*).
4. *Weber's Law*: named for the German physician Ernst Heinrich Weber (1795–1878), a law stating that the smallest perceptible increase in a stimulus is a constant proportion of the initial stimulus (*OED*). It is also called the Weber-Fechner law.
5. *Natural Philosophy*: W. Thomson and P. G. Tait, *Treatise on Natural Philosophy* (Oxford: Clarendon Press, 1867).
6. *translation of which*: W. Thomson and P. G. Tait, *Handbuch der theoretischen Physik*, 2 vols (Braunschweig: F. Vieweg und Sohn, 1871–4).
7. *war*: the Franco-Prussian War (19 July 1870–28 January 1871), a six-month territorial war between France and Prussia; a Prussian victory resulted in the the unifcation of Germany and the end of the Second French Empire under Napoleon III.
8. *victors of Sedan*: the Battle of Sedan from 1–2 September 1870, in which the Prussians captured French emperor Napoleon III and 100,000 of his soldiers, was a defining point of the Franco-Prussian War.
9. *enjoy the Alps*: see letter 3716, n. 6.

To Joseph Henry 24 June 1872 3771

24th. June 1872

My dear Professor Henry.

I am here upon my last legs. Tomorrow morning I start for Switzerland.[1] I have already made arrangements for the taking of a cabin to New York.[2]

I shall be there on the 8th of October if all go well.[3]

For the subsequent two months I am under your orders—My pleasure will be to carry out all arrangements that you may make.

Renewed thanks for your kind letter[4] received this morning.

Should you write during my absence your letter will be sent on to my Alpine house.[5]

I shall be back here on the 10th of August or thereabouts; and employ the subsequent five or six weeks in preparations.

Yours ever | John Tyndall

I am willing to make the courses as short as you please.

John Tyndall to Joseph Henry, 24 June 1872, mssRH 3966, box 43, William Jones Rhees papers, The Huntington Library, San Marino, California

1. *Tomorrow morning I start for Switzerland*: see letter 3716, n. 6.
2. *the taking of a cabin to New York*: on the SS *Russia*. For Tyndall's lecture tour in America, see Introduction, n. 8.
3. *be there on the 8th of October if all go well*: Tyndall arrived in New York on 9 October 1872, after 11 days' travel across the Atlantic.
4. *your kind letter*: letter missing.
5. *my Alpine house*: the Bel Alp Hotel in the village of Bel Alp (now Belalp), above Brieg (now Brig), Switzerland. Tyndall and his future wife Louisa would build a summer cottage in Bel Alp in 1877, which they named Alp Lusgen. On Tyndall and Bel Alp, see T. Braham, 'John Tyndall (1820–1893) and Belalp', *Alpine Journal*, 98 (1993), pp. 193–8.

From George Biddell Airy 24 June 1872 3772

Royal Observatory, Greenwich, | SE | 1872 June 24

My dear Sir

In my expression about "uninterested persons" in my note of 20th,[1] I had not the least intention of blaming any one; I have no title to criticise the actions of others.—I ought to have said "an uninterested person", as describing specially my own position and the duty which it seems to impose on me.

I am, my dear Sir, | Yours very truly | G B Airy

Professor Tyndall[2]

RGO MS.RGO 6/438
RI MS JT/1/A/54
RI MS JT/1/TYP/1/48

---

1. *my note of 20th*: letter 3759. For previous correspondence between Tyndall and Airy about the Memorial in support of Joseph Hooker in his dispute with Acton Ayrton, see letters 3718, 3726, 3730, and 3763. On the Memorial in general, see letter 3755, esp. n. 1.
2. At the top of Airy's letter, Tyndall wrote: 'To be preserved. J.T. & given to me on my return'.

From James Joule 24 June 1872 3773

343 Lower Broughton Road | Manchester 24/6/72

Dear Tyndall

Many thanks for your kind present of the handsome volume of your researches[1] which came to hand some days ago. I hope to restudy your discoveries on radiation and am glad to have them placed in my library in one volume.

Yours ever truly | James P. Joule

RI MS JT/1/J/142
RI MS JT/1/TYP/2/667

---

1. *handsome volume of your researches*: *Molecular Physics*.

From Joseph Dalton Hooker [before 25 June 1872][1] 3774

Royal Gardens | Kew

My dear Tyndall,

The enclosed[2] is indeed considerate in the best sense.

I will dine with you at the Athenaeum[3] on Monday with pleasure and bring some papers.

Ever yours | J. D. Hooker.

RI MS JT/1/TYP/8/2607
LT Typescript Only

---

1. *[before 25 June 1872]*: the date is given based on Tyndall working with Hooker in June 1872 on the Memorial in support of Hooker in his dispute with Acton Ayrton (see letter 3755, esp. n. 1). Tyndall completed writing the Memorial before he left for Switzerland on 25 June 1872.

2. *enclosed*: not identified, but possibly having to do with Hooker's dispute with Acton Ayrton (see letter 3755, n. 1).
3. *Athenaeum*: the Athenaeum Club (see letter 3710, n. 5).

To Thomas Carlyle [25 June 1872][1] 3775

Tuesday Morning

My dear Friend

I turned my face towards Cheyne Row[2] last night when that terrible thunderstorm came down upon me. I wished to shake your hand before going to the Alps,[3] and to see with my own eyes that good condition which your friends report regarding you. I fear I shall be a long time now without seeing you for if you are not at home in September I shall have started for America[4] before you return. There I labour for two months—I hope a labour of good will—as no other motive carries me to America. I am now on the point of starting for the mountains.

Goodbye and believe me with old fidelity & affection | John Tyndall

RI MS JT/1/T/160
RI MS JT/1/TYP/1/192

1. *[25 June 1872]*: the date is given as Tyndall left for Switzerland on 25 June 1872, a Tuesday.
2. *Cheyne Row*: 5 Cheyne Row was Carlyle's home in Chelsea, central London since 1834.
3. *going to the Alps*: see letter 3716, n. 6.
4. *started for America*: for Tyndall's American lecture tour, see Introduction, n. 8.

To Thomas Archer Hirst 26–8 June 1872 3776

Paris 26th June 1872

My dear Tom,

Thus far all is right, and while waiting for breakfast I while away the time thus. The tidal boat enabled me to get a good start from London. The passage was pleasant, many sick however, but many not, and among the latter your friend.[1] I seem to have become indurated to the sea of late years, and this makes me look to my American trip[2] without anxiety. All my anxieties in America converge upon their hospitality: I must set my face like a flint against it, and have one influence, and that not the least personally what is destroyed. What a drawback it is! How I should enjoy the faces & conversation of my fellow man, that is the pick of them, were it not for this. It is too bad that a man with amicable instincts towards humanity should be thus crippled.

Well though the wind and tide were against us we reached Boulogne[3] in good time. What pretence it is to ask a man's name on leaving the steamer, the pretense being doubled by pretending to write it down as a flood of passengers are passing the writer. But it is a part of a system of centralization which is not without its good points. Still it cripples <u>men</u> though it may make nations in the presence of outside foes temporarily strong. I fear the pursuit of individual well-being with all its base concomitants will for ages yet be the motive force of human development, and it is only where people are left within certain wide limits to themselves that this force can have free play. Here I am waiting for my tea, expostulating with the waiter for keeping it away so long, in fact pursuing my own well being like the rest, not too regardful of the comfort of the waiter who would take pleasure in doing nothing. Thus we act & react upon each other.

But we came to Boulogne, and here the base pursuits of self manifested itself strikingly—I had some bouillon, then a slice of what they called roast beef: very small; and very leathery. I took a little of it, put it aside, and asked for a little fowl. The relation was preserved. With reference to English articles of the same name I might say as beef is to beef so is fowl to fowl. Both were atrociously bad. Then I had half a bottle of indifferent St. Julian—and the sum total of my bill, including waiter was 8 francs! A respectable French gentleman equally let in protested. He went and consulted the tariff, and found that the waiter had stuck on half a franc to each. He wrote under the 'Avis'[4] 'Le Buffet vole':[5] and then he questioned the waiter about the extortion of half a franc. He was ready with his reply: it was pour les legumes[6]—the 'legumes' being two small potatoes which nobody had demanded. Well I may possibly expose this, reach upon the Boulongese, and thus promote *[1 word illeg]*. It is a weary mode of development to a man of quick aspirations.

Well I got comfortably to Paris, sharing a carriage with an elderly lady & two young men, whom she fed and cared for as a hen for her chickens. She must have been an old traveller, for she had the most extraordinary appliances for holding meat and serving it out. She also made a cup of aromatic tea despite the motion of the train, which by throwing in synchronous vibrations now & then often brought her water over the edge of her "Etna" or boiler.[7] They were pleasant people—At least her affection for them, and theirs for her were very beautiful to see. I got easily through the Douane,[8] and put up at my old place, the hotel de St. James.

I had a cutlet for supper and half a bottle of Chateau Lafitte. It cost 8 francs a bottle—but it did <u>not</u> agree with me. My sleep was bad and disturbed by the most wonderfully consistent dream about my assistant Cottrell;[9] who had left me for some cause or other. There goes your theory. I walked out after supper and into the Palace Royal.[10] It was only a quarter past 10, still many of the shops were shut, some had their brilliants[11] covered with paper, and very few indeed were illuminated. One side was almost wholly dead. You would as easily recognize the Palais of old in the Palais of today. The soul seems squeezed out of the people. There may be purer energies at work elsewhere, let us hope there are.

I dine with Jamin today: as yet I have not sallied forth or seen anybody.

27th.

I wish you would send a copy of my addresses—I mean 'Chamounix'[12] and 'Bel Alp'[13]—to Bence Jones.

Would you also look in from time to time & see whether the mutinous builders have begun their work of demolition on the Institution?[14]

Tell Spottiswoode that I have already spared an hour or two with Hofmann,[15] and have selected a good number of crystals and <u>verres trempés</u>.[16] But as I am informed that he (Spottiswoode) selected pieces altogether 'exceptionally', I may come down on him for a loan on my return to London.

I called on several friends yesterday, but saw only a few of them. Old Becquerel[17] at 85 is as brisk and active as ever; full of his experiments; some of which are very pretty indeed.

I spent a joyous evening with Jamin. He his wife, his son (a pupil of the Polytechnic School) and his daughter, a very handsome maiden, were my entertainers.[18] He talked much of France and Germany. France he thinks might have borne the loss of Strasburg[19] without an eternal longing for war afterwards, but not the loss of Metz.[20] It was purely French.

Yours ever | John Tyndall

28th. I start at 8.40 P.M. for Geneva

Send me a copy of Thomson's[21] article in the Contemporary:[22] I think the Editor[23] will send a separate copy to the Institution.[24]

BL Add MS 63092, ff. 114–17

---

1. *your friend*: Tyndall referring to himself.
2. *my American trip*: see Introduction, n. 8.
3. *Boulogne*: a northern French city on the English Channel.
4. *'Avis'*: 'Notice' (French), probably meaning the restaurant bill.
5. *'Le Buffet vole'*: 'Le Buffet steals' (French). Le Buffet appears to be the name of the restaurant.
6. *pour les legumes*: for vegetables (French).
7. *"Etna" or boiler*: any of various kinds of spirit lamp or stove, esp. a portable one. In later use chiefly *attributive*, as Etna lamp, Etna stove (*OED*).
8. *Douane*: Customs (French).
9. *my assistant Cottrell*: John Cottrell.

10. *Palace Royal*: the Palais-Royal was a royal palace in Paris, built in the 1630s and now seated across from the Louvre.
11. *brilliants*: diamonds of the finest cut and brilliancy (*OED*).
12. *'Chamounix'*: Chamonix (or Chamouni), a town in the valley north of Mont Blanc, located in Upper Savoy, France.
13. *'Bel Alp'*: see letter 3771, n. 5.
14. *their work of demolition on the Institution*: while Tyndall was on his American lecture tour from October 1872 to February 1873, Henry Bence Jones oversaw the construction of a new two-story laboratory in the east wing of the RI building at 20 and 21 Albemarle Street, London (F. A. J. L. James and A. Peers, 'Constructing Space for Science at the Royal Institution of Great Britain', *Physics in Perspective*, 9 (2007), pp. 130–85, on pp. 162–3). Letters 3796 and 3801 indicate that the work was begun even earlier while he was on his summer excursion to Switzerland (see letter 3716, n. 6). Following this renovation, Spottiswoode delivered a Friday Evening Discourse at the RI on 17 January 1873, 'On the Old and New Laboratories at the Royal Institution' (*Roy. Inst. Proc.*, 7 (1873–75), pp. 1–10).
15. *Hofmann*: August Wilhelm Hofmann (1818–92), a German organic chemist who had studied with Justus Liebig at the University of Giessen. On the recommendation of Liebig and Prince Albert he was appointed director of the newly-established Royal College of Chemistry in Oxford Street in 1845. In 1853 the College was incorporated into the Metropolitan School of Science Applied to Mining and the Arts which the British government had founded in 1851 as the Government School of Mines and of Science Applied to the Arts. His work on the "ammonia type" kick-started the British dyestuffs industry. The large laboratory at Berlin University was built under his direction and was completed in 1869 (*CDSB*; *NDB*).
16. *verres trempés*: tempered glasses (French).
17. *Becquerel*: Antoine César Becquerel (1788–1878), a French scientist (*CDSB*). Two years later his son would marry the daughter of Jules Jamin (see n. 18).
18. *were my entertainers*: Jamin's wife Theresa Josephine Eudoxia Jamin (née Lebrun, 1832–80); his son Paul Joseph Jamin (1853–1903), a painter who married Augustine Marie Caroline Bastien (1852–1933) in 1882; and his daughter Lucie Zoé Marie Jamin (1857–78), who married the physicist Antoine Henri Becquerel (1852–1908) in 1874.
19. *loss of Strasburg*: the French city of Strasbourg became under German control in 1871 following the Franco-Prussian War.
20. *loss of Metz*: the French city of Metz became under German control in 1871 following the Franco-Prussian War.
21. *Thomson's*: Henry Thompson (1820–1904), a London physician and surgeon (*ODNB*).
22. *article in the Contemporary*: 'The "Prayer for the Sick"—Hints towards a Serious Attempt to Estimate its Value', *The Contemporary Review*, 20 (1872), pp. 205–10. This article was published anonymously, with a covering letter from Tyndall, and is letter 3779 in this volume.
23. *Editor*: the second editor of *The Contemporary Review* from 1870–77 was the English architect and editor James Thomas Knowles (1831–1908), who founded the Metaphysical

Society in 1869. After leaving the *The Contemporary Review* in 1877, he founded another periodical, *The Nineteenth Century* (*ODNB*).

24. *Institution*: the RI.

From Joseph Dalton Hooker [29 June 1872][1] 3777

Dear Tyndall

The above is all that has occurred. I immediately took the Treasury letter[2] to Spottiswoode who will lay it before Lord Derby and the rest and I am to meet them on Monday upon the subject. Spottiswoode suggested my writing to Ayrton[3] on the strength of it, but this would be a regular knock under... I could not do that, and let Parliamentary action proceed, and after all the Treasury letter gives me no real basis of operations.

I hope my dear fellow that you are enjoying your outing.[4]

Ever your aff[ectiona]te | J. D. Hooker.

RI MS JT/1/TYP/8/2672
LT Typescript Only

---

1. *[29 June 1872]*: the month and year are given by relation to other letters discussing Hooker, Acton Ayrton, and the Memorial to William Gladstone (see letter 3755, esp. n. 1); the day is given by the suggestion of LT in a note she wrote on the typescript.
2. *Treasury letter*: Hooker enclosed with this letter one he received from Ralph Robert Wheeler Lingen (1819–1905), an English civil servant who at this time was secretary of the treasury (*ODNB*). The letter from 26 June 1872 reads, 'In reply to your letters of the 1st, 14th, and 16th ult[imo]. and of the 3rd inst[ant]., wherein you bring under the notice of the Treasury various questions which have arisen between yourself and the First Commissioner of Works, I am directed by the Lords Commissioners of H. M. Treasury to inform you that My Lords have communicated their views to the First Comm[issione]r, and that they trust that no further difficulty may be found in reconciling the authority which must always belong to his office with such a degree of liberty and consideration as is due to the Director of the Royal Gardens. My Lords are not of opinion that this end would be furthered by prolonged discussions, They entertain a hope that the particular questions pending between yourself and the First Comm[issione]r may now be found to admit of satisfactory explanations and settlement within the Department of Public Works' (RI MS JT/1/TYP/8/2672).
3. *Ayrton*: Acton Ayrton.
4. *your outing*: to Switzerland (see letter 3716, n. 6).

From Thomas Henry Huxley 29 June 1872 3778

26 Abbey Place[1] | June 29th 1872

My dear Tyndall

I have received your two notes about Hooker's affair[2]—and had a long talk with Hooker himself yesterday—It is rather a tangle but I will do my best to keep the threads straight in your absence.

I have no command of the Telegraph that I know of—but I will see what I can do with that paper *[in]* the Daily News[3]—I hope Lord Derby may be right but I have a lingering suspicion that the letters quoted in the memorial ought not to have been published before they had been obtained in Parliament.

I thought you were coming to the X[4] and am disgusted you are off[5]—many thanks for the dib[6]—I send a receipt on the other page—put it among your Archives.

Ever yours | T H Huxley

26 Abbey Place NW | June 29th 1872

Received from John Tyndall for the sum of £1000 to bear interest at 5 per cent until repaid—Interest to date from June 1. 1872—

T H Huxley

RI MS JT/1/H/525

1. *26 Abbey Place*: the Huxleys lived at 26 Abbey Place, St. John's Wood from 1861–72.
2. *your two notes about Hooker's affair*: see letters 3751, 3752, and 3757.
3. *I have no command... the Daily News*: in letter 3751, Tyndall had asked Huxley to see about getting the Memorial in support of Joseph Hooker in his dispute with Acton Ayrton (see letter 3755, esp. n. 1) published in *The Daily Telegraph* while Tyndall tried for the *Times*. It does not appear that the Memorial was published in either of these newspapers.
4. *the X*: the X Club (see letter 3757, n. 2).
5. *you are off*: to Switzerland (see letter 3716, n. 6).
6. *dib*: donation of money.

From Henry Thompson[1] June 1872 3779[2]

DEAR PROFESSOR TYNDALL,

SINCE our conversation the other night, when you were good enough to listen to a suggestion I made relative to a means of determining the value of prayer to the Deity, it occurred to me to put the idea into writing, and to ask you to do me the further kindness of looking at it in this shape.

It seems to me impossible at the present day to find ourselves in contact with a source of power available for human ends, or affirmed to be so on high authority, without recognising a necessity—or even that it is a duty—to estimate its value. And especially if the power be one which is effective for the production of physical results, is it desirable to examine its nature, and to measure its extent, and the conditions under which it works.

The value of prayer to the Deity has been recognized in all ages and by all nations, not merely by the ignorant and superstitious, but by the more cultivated portions of the human race. And I think it may be said that among the great body of religious people of all denominations in this country, a belief in its efficacy is almost universally professed. As to the objects which it is believed are attainable by prayer, they are almost without limit as to kind. Taking as an authority that well-known compendium which none will dispute to be the national epitome of English religious idea on the subject, "The Book of Common Prayer,"* the legitimate objects of supplication to God may be classified as follows:—

Class A. Spiritual improvement; moral superiority; intellectual power.

" *B.* National supremacy. Preservation from pestilence, famine and battles. The fertility of the soil; weather suitable for the growth and preservation of vegetable products. The health, wealth, and long life of the chief national ruler. A special share of grace and wisdom for the Nobility, and for members of the Legislature and of the Executive.

" *C.* For all that are in danger; for the preservation of travellers, of sick persons, of young children, prisoners, orphans, and widows; protection against murder and sudden death.

" *D.* Comprehends *special* forms for occasional use, *e.g.*, for "moderate rain and showers," &c.; that "scarcity and dearth may be turned into cheapness and plenty;" that "this plague and grievous sickness may be withdrawn;" and the prayer for "sick persons," which is not precise in its requests on their behalf.

From all the foregoing it is impossible to resist the conclusion already more than hinted, that a very ample belief exists in the Christian Church, in the efficacy of prayer to God to avert dire physical evils, which without it might occur; such, for example, as disease and death. Were any one, however, hardy enough to question this, it would suffice to point out that the custom of offering prayers for the recovery of sick persons when in great danger is almost universal here. And it may be added that, in the larger and more ancient section of the Church, prayer still continues on behalf of the deceased, a custom, perhaps, not less pious and reasonable than the first-named.

Now, I propose to examine this subject from one point of view only, in the endeavor to discover a means of demonstrating, in some tangible form, the efficacy of prayer. I commence by remarking, however, that the objects of prayer in Class *A* clearly present inordinate difficulties, and are obviously unfitted for our purpose. Class *B* furnishes subjects which might be examined, which are less easy of treatment than some of those to be found in Classes *C* and *D*. But even here, elements of disturbance present themselves; thus, in reference to the influence of prayer on states of the weather in limited localities, that food may be cheapened, that travelers may be preserved from accident, &c., it is certain that considerable difficulty would arise in any systemic attempt to arrive at accurate conclusions. But this leads me to remark that there appears to be one source from a study of which the absolute calculable value of prayer (I speak with the utmost reverence) can almost certainly be ascertained. I mean its influence in affecting the course of a malady, or in averting the fatal termination. For, it must be admitted that such an important influence, manifestly either does, or does not exist. If it does, a careful investigation of diseased persons by good pathologists, working with this end seriously in view, must determine the fact. The fact determined, it is simply a matter of further careful clinical observation to estimate the extent or degree in which prayer is effective. And the next step would be to consider how far it is practicable to extend this benefit among the sick and dying. And I can conceive few inquiries which are more pregnant with good to humanity when this stage has been arrived at.

You will naturally next say, What practical shape does the method take by which you propose to attain your end? The method has its difficulties, but I see none that are insuperable. If I may reckon on the active co-operation of those who most believe in the value of such prayer, and I think I have a right to do so, the enquiry will be easy. For few more interesting subjects of enquiry can exist for the honest believer than the extent of a man's influence with Heaven, at the most momentous crisis in his personal history.

Before entering on the details demanded, it is first necessary to remark that prayer for the recovery of sick persons exists in two distinct forms, or, if I may use the term, in two orders or degrees of quality. For, first, there are the general prayers for the sick, made without distinction as to individuals, or to numbers, on most occasions of public worship. These prayers are offered by, perhaps, thirty thousand congregations every Sunday in our country, since it is no less the practice of the Dissenter than of the Churchman to remember devoutly the sick in the weekly supplication. But besides these, there are the special prayers for individual sick persons, which are by general consent deemed also necessary; and thus it is that when the patient holds a very high place in society, a special form of petition is sometimes ordained to be used throughout the national churches for his recovery. It is one of the advantages of rank and gentle birth in England, that special prayers are made for such every week at least, in most churches throughout the country.

The first kind, or general prayer, then, must be held to have a certain value not inconsiderable, since it is this kind which is relied on against the dangers of travel, of murder, and of sudden death, and respecting which no other or special petitions are provided. This general prayer for the recovery from sickness is constantly ascending, if I may use the term, in a broad stream to Heaven. Yet its objects, "all men," being so numerous, it is not held to suffice for all individual cases. Hence the second kind, or special prayer; and the object sought by those who are interested in the recovery of the sick, obviously is to concentrate the special prayers of many on the recovery of one, in the belief that by this means the malady may be more certainly checked than were the patient's fate to depend only on the influence of the "general prayer." With this end it is that the special prayers of a congregation are asked for A or B, or a special prayer-meeting is held to offer the one object of petition. I have been myself present at such meetings, and have witnessed the number, the minuteness, and the length of the petitions.

Now the latter kind, or special prayer, is that which readily lends itself to the earnest enquirer in this matter, and it is by its means, if carefully and conscientiously pursued, that we may certainly arrive, if at all, at a solution of the great question I have proposed.

The following appears to me to indicate the manner of conducting the enquiry. It should be pursued on a system somewhat analogous to that which is pursued by the Faculty when a question arises as to the value of any particular mode of treating disease. For example, a new remedy has been proposed, or is said on high authority to be efficacious, and as authority does not suffice in medicine further than to recommend a given course, and never to prescribe it, the remedy is carefully tested. Usually a hospital or a ward is assigned for the purpose. All the patients suffering from the disease to be treated are, during a certain period, divided into two classes, and all are subjected, as far as possible, to the same conditions, that single one of treatments alone excepted. The ages, sexes, and many other particulars of the patients are taken into account, and duly noted. The one class is treated by the old system, and the other by the new remedy. When a very large number—for in large numbers only is there truth—has been thus dealt with, the results are compared, and the value of the remedy can be definitely expressed; that is, its influence above or below that of the old treatment, as the case may be, will appear in the per-centage of recovery, or of other results.

Now, after much thought and examination of the various questions and objections which may possibly be urged, I do not hesitate to propose an analogous arrangement, in order to estimate and rightly appreciate the influence of special prayer to check disease, or to avert death.

We possess unquestionable data in reference to certain well-known maladies, particularly the fevers of eruptive type; such as small-pox, typhoid, scarlet fever, &c. Of some local acute disorders, such as pneumonia, we know what is termed the natural history pretty well, their duration, and probable termination at different ages, &c. The mortality which follows the great surgical operations at different ages is a matter known and determined; for example, after lithotomy and lithotrity,[3] amputations of the limbs, hernia, &c. The very large records of past cases which exist, and the very wide and careful researches which have been made, have had for their result the production of known numerical mortality-rates per cent, and applicable to future patients of different ages and conditions. Indeed, the whole system of life assurance is, all the world over, based solely on the accuracy of such data, and on the certainty with which they will reproduce themselves. Whatever these numerical results have been—whether the mortality rates deduced belong to healthy lives or to diseased lives—all have been necessarily made, subject to the conditions of human life as it now exists, and including, among a thousand other influences, that most important one of "general prayer" by the whole Christian Church for "all men," as it has been already described, and influencing as it does, whatever may be its extent, the sick, the suffering, those exposed to murder and sudden death, &c., throughout the whole world. Subject to this influence is that of every drug prescribed. Influenced by this is the result of every surgical operation.

Now, for the purpose of our inquiry, I do not propose to ask that one single child of man should be deprived of his participation in all that belongs to him of this vast influence. But I ask that one single ward or hospital, under the care of first-rate physicians and surgeons, containing certain numbers of patients afflicted with those diseases which have been best studied, and of which the mortality rates are best known, whether the diseases are those which are treated by medical or by surgical remedies, should be, during a period of not less, say, than three or five years, made the object of special prayer by the whole body of the faithful, and that, at the end of that time, the mortality rates should be compared with the past rates, and also with that of other leading hospitals, similarly well managed, during the same period. Granting that time is given, and numbers are sufficiently large, so as to ensure a minimum of error from accidental disturbing causes, the experiment will be exhaustive and complete.

I might have proposed to treat two sides of the same hospital, managed by the same men; one side to be the object of special prayer, the other to be exempted from all prayer. It would have been the most rigidly logical and philosophical method. But I shrink from depriving any of—I had almost said—his natural inheritance in the prayers of Christendom. Practically, too, it would have been impossible; the unprayed-for ward would have attracted the prayers of believers as surely as the lofty tower attracts electric fluid. The experiment would be frustrated. But the opposite character of my proposal will commend it to those who are naturally the most interested in its success;

those, namely, who conscientiously and devoutly believe in the efficiency against disease and death of special prayer. I open a field for the exercise of their devotion. I offer an occasion of demonstrating to the faithless an imperishable record of the real power of prayer.

Athenaeum Club, Pall Mall, | June, 1872.

*Although not used by Dissenters, they do not reject it on account of its contents, since its very phraseology is often employed by them, for the most part because all *forms* are deemed by them undesirable.

'The "Prayer for the Sick"—Hints towards a Serious Attempt to Estimate its Value', *The Contemporary Review*, 20 (1872), pp. 205–10

---

1. *Henry Thompson*: while this letter to Tyndall was published in the the *Contemporary Review* anonymously (Tyndall did not include the name of the author when he forwarded the letter for publication), its authorship was later attributed to Henry Thompson (1820–1904), a London physician and surgeon (*ODNB*). This letter, including Tyndall's introduction, was later published in J. Tyndall, F. Galton, et al., *The Prayer-Gauge Debate* (Boston: Congregational Publishing Society, 1876), pp. 9–19. See also Stephen G. Brush, 'The Prayer Test', *American Scientist*, 62, no. 5 (1974), pp. 561–3.
2. This letter was forwarded by Tyndall to the *Contemporary Review* (see letter 3780).
3. *lithotomy and lithotrity*: lithotomy is the operation, art, or process of cutting for stone in the bladder (*OED*); lithotrity is the operation of crushing a stone in the bladder by means of a lithotrite (*OED*).

To the Editor of the *Contemporary Review*[1] June 1872 3780

[THE "PRAYER FOR THE SICK:" | HINTS TOWARDS A SERIOUS ATTEMPT TO ESTIMATE ITS VALUE.]

THE following suggestive letter has been placed in my hands, with a view to publication. It is sure, I think, to interest the thoughtful readers of the *Contemporary Review*. It deals, indeed, with a subject which interests everybody, and regarding which all manner of men, from the Prime Minister downwards, have given the public the benefit of their views.

If such be attainable, it is surely desirable to have clearer notions than we now possess of the action of "Providence" in physical affairs. Two opposing parties here confront each other—the one affirming the habitual intrusion of supernatural power, in answer to the petitions of men; the other questioning, if not denying, any such intrusion. The writer of the letter wishes to bring these opposing affirmations to an experimental test.[2] He considers the subject to be accessible to experiment, and makes a proposal which, if faithfully carried out, would, he thinks, displace assertion by demonstration as regards the momentous point in question.

It was justly stated by the Archbishop of York[3] at a recent meeting of the supporters of the Palestine Exploration Fund,[4] that the progress of the human mind is from vagueness towards precision.[5] The letter before us seems an illustration of this tendency. Instead of leaving the subject to the random assertions of half-informed sceptics on the one hand, and hazy lecturers of the Victoria Institute[6] on the other, the writer seeks to confer quantitative precision on the action of the Supernatural in Nature. His proposal is so fair, and his mode of stating it so able and conciliatory, that I could not, when asked to do so, refuse to give it the support implied by these few lines of introduction.

JOHN TYNDALL. | ATHENAEUM CLUB,[7] June, 1872.

[...][8]

'The "Prayer for the Sick"—Hints towards a Serious Attempt to Estimate its Value', *The Contemporary Review*, 20 (1872), pp. 205–6

---

1. *Editor of the Contemporary Review*: James Thomas Knowles (see letter 3776, n. 23).
2. *an experimental test*: Tyndall had previously written about the efficacy of prayer in a chapter entitled 'Reflections' in *Mountaineering in 1861*, pp. 33–40. As Tyndall biographer Roland Jackson has noted, this was an odd topic to discuss in an otherwise mountaineering-focused book (*Ascent of John Tyndall*, p. 213). This chapter was also published with a different title, 'Thoughts on Prayer and Natural Law', in *Fragments of Science*, pp. 29–37 (on p. 38 is a note explaining what prompted him to write the chapter). In October 1865, Tyndall wrote two letters to the *PMG* in response to the call for prayer against a cholera epidemic (see letters 2314 and 2318, *Tyndall Correspondence*, vol. 9). On Tyndall and the debate over prayer, see: *Ascent of John Tyndall*, pp. 212–14, 219–20; N. K. C. Bossoh, 'Scientific Uniformity or "Natural" Divine Action: Shifting the Boundaries of Law in the Nineteenth Century', *Zygon*, 56 (2021), pp. 234–53; S. G. Brush, 'The Prayer Test', *American Scientist*, 62:5 (1974), pp. 561–3; R. B. Mullin, 'Science, Miracles, and the Prayer-Gauge Debate', in David C. Lindberg and Ronald Numbers (eds), *When Science & Christianity Meet* (Chicago: University of Chicago Press, 2003), pp. 203–24; and R. Ostrander, *The Life of Prayer in a World of Science: Protestants, Prayer, and American Culture*, 1870–1930 (New York: Oxford University Press, 2000), pp. 17–34. The 'prayer-gauge debate', as it came to be known, was lively from July 1872 to March 1873. Many of the contributions are collected in J. O. Means (ed), *The Prayer-Gauge Debate* (Boston: Congregational Publishing Society, 1876).
3. *Archbishop of York*: William Thomson (1819–90) was Archbishop of York from 1862–90 (*ODNB*).
4. *the Palestine Exploration Fund*: a British society founded in 1865 by Biblical archeologists

and clergymen to carry out surveys of the topography and ethnography of Palestine in order to 'capture the high ground of imperial religious control' by way of 'the scientific exploration of the Holy Land, its reduction to maps and photographs, its exploration and archaeological excavation, and the identification of its sacred sites' (J. J. Moscrop, *Measuring Jerusalem: The Palestine Exploration Fund and British Interests in the Holy Land* (London: Leicester University Press, 1999), p. 2).

5. *progress of the human mind is from vagueness towards precision*: a quotation has not been identified.
6. *Victoria Institute*: the Victoria Institute, or Philosophical Society of Great Britain, was founded in 1865 by a group of London evangelicals who while skeptical of Darwinian evolution, their chief concern was Biblical criticism. On the Institute, see R. L. Numbers, *The Creationists: From Scientific Creationism to Intelligent Design*, 2nd edn (Cambridge, MA: Harvard University Press, 2006), pp. 162–6, 175–9; S. Mathieson, *Evangelicals and the Philosophy of Science: The Victoria Institute, 1865–1939* (New York: Routledge, 2020); and S. Mathieson, 'The Victoria Institute, Biblical Criticism, and *The Fundamentals*', *Zygon*, 56 (March 2021), pp. 254–74.
7. *ATHENAEUM CLUB*: see letter 3710, n. 5.
8. Tyndall copied letter 3779 here.

To Thomas Archer Hirst 30 June–1 July 1872[1] 3781

Chamounix[2] 1st. July 1872.

My dear Tom,

To the left of my writing table there is a bundle of drawers and to the left of the bundle a series of small ones. In the second drawer from the top you will find a lot of photographs. Choose from them the one you consider best and send it to Mr. Dayton[3] (Appleton & Co) 16 Little Britain, with a line to say that Prof. Youmans wishes to have it "cut by the best artist in London".[4] These are his words.

There are two to which your choice will probably be confined; the one with my head leaning upon my hand,[5] the other the original from which the cut in the Illustrated News was taken.[6] This last Youmans considers an excellent likeness: if he thinks so, why let him have it. But if <u>you</u> think the other better let him have <u>it</u>. Both are by Elliott and Fry.[7]

Again, Lovell Reeve[8] (now dead, but the firm exists) of Henrietta Street Covent Garden published the little biographical notice that we concocted when at Waverly Place.[9] Get two of these: cut my part out of both numbers & send one to Youmans, (Appleton & Co New York) and the other to me to the Bel Alp.[10] Among a bundle of Newspapers to the right of the large bookcase facing the fire in my room (at the bottom) are two or three biographical notices; mine might be found by M^rs^. Tunstal[11] and if you think these worth anything they might also be sent to Youmans. It is just as well to strike out that line about the Irish peasantry, for it is not correct.

I am very loth to bother you with these things; and it disturbs me to have any thing to do with them. Had I apprehended anything of the kind it would certainly have been an element in weighing whether I ought to go to America.[12] But I am committed to it, and to check misstatements Youmans says the thing must be done.

I hope you read Helmholtz's letter.[13] It is very interesting.

I had a very instructive two hours with Jamin in his laboratory[14]—it is admirably organized—organized that is for practical earnest work. He is an excellent fellow. He has a son in the Polytechnic School, and a very charming daughter. Both were there.[15]

I went to see Pasteur,[16] and spent a most interesting hour with him. He is cultivating germs, and has numberless bottles containing special samples of these smaller forms of vegetable life. He is sifting them, the one from the other, and cultivating them separately. He aims at detaching those which produce fermentation from those which do not, and also at detecting and warding off those which ruin the pure work of fermentation and spoil our beer. Pasteur strikes me the most philosophical head in Paris; Indeed any where he would stand forth as a remarkable man.

I reached this place on Saturday having caught a glimpse of DelaRive[17] on Friday at Geneva. Thank the Arbitrators, we have been to all appearance rescued from the degrading position into which we were thrusting ourselves regarding this Alabama affair.[18] Gladstone is a striking example of a man who is all <u>outworks</u>. Able & subtle as regards the fear of his intellect, the rock of character on which intellect ought to repose is in him a rare quicksand: I think considering the times in which we live, he is the most dangerous statesman that England has had hitherto at the head of her affairs. I said this years ago, and the experience of these years had only confirmed my diagnosis.

Yesterday I went up to the Montanvert,[19] and thence to that singular cliff which you and I have often seen from the Montanvert. I had some severe climbing in ascending but in coming down I mistook the way, if there be a way, and to escape from my difficulties had the hardest bit of climbing to accomplish that ever I went through in my life. Six or seven times I found myself in places from which I am not ashamed to say it required both skill & strength to rescue me. As obstacle after obstacle met me, and as I felt the work beginning to tell upon me, my thoughts were very serious indeed. I did not get home till between 9 & 10 oC.

Nothing could exceed in magnificence the sunset of yesterday. I was at the Montanvert when a cloud banner dyed in Tyrian Crimson rolled away from the Aiguille de Dru.[20] The Jorasse,[21] the Periades,[22] the Géant,[23] the Aiguille de Charmoz[24] all glowed like fire. Today the heavens are absolutely cloudless, and the skies of Mont Blanc[25] purer than I have ever seen them before.

Do not neglect the little commissions referred to at the outset. Should you see Lady Mary Egerton give her my best regards. I will write to her from the Bel Alp. A purer sweeter soul I do not know than hers. The years have only excelled my opinion of her. She is one of those who do much to reestablish one's shaken confidence in humanity. Our friend M^rs^. Spottiswoode[26] is such another.

July 1st.—

Your last note[27] makes me pray that the Gods may inspire you in the will and power to join me at the Bel Alp.

Kiss Lilly[28] for me as the proper form of thanks for writing me so long a resumé.

Send one of the porters of the R.I. to Henrietta Street for the number of Lovell Reeves biographies which contains our sketch. Your letter makes me doubly upset putting you to this trouble.

Yours affectionately J. Tyndall

BL Add MS 63092, ff. 118–21

---

1. *30 June–1 July 1872*: although Tyndall wrote '1st. July' at the top of this letter, he also wrote 'July 1st.' in the body of the letter, indicating a separation of dates in which he composed this letter. We suggest that he began the letter the day before, 30 June, and finished it on 1 July.
2. *Chamounix*: see letter 3776, n. 12.
3. *M^r^. Dayton*: not identified.
4. *"cut by the best artist in London"*: a letter from Youmans including these words is missing. Referring to a photograph to be used to cut an engraving.
5. *one with my head leaning upon my hand*: the RI holds in its archive a photograph of Tyndall taken by Elliott & Fry in 1869 (RIIC 3619). Tyndall is seated in a chair, his right arm leaning on the right arm of the chair, his right fist resting upon his right cheek, with his left hand rested on the right arm, holding some papers.
6. *other the original from which the cut in the Illustrated News was taken*: the RI holds in its archive a photograph of Tyndall taken by Elliott & Fry in 1872 (RIIC 3630). This one is a more standard forward-facing portrait, and the engraving based on it appeared in *The Illustrated London News*, 6 January 1872, p. 9 (it accompanied a short biographical piece on p. 10).
7. *Elliott and Fry*: photographers Joseph John Elliott (1835–1903) and Clarence Edmund Fry (1840–97) established their firm, Elliott & Fry, in 1863 at 55 Baker Street (J. Hannavy (ed), *Encyclopedia of Nineteenth-Century Photography*, 2 vols (New York: Routledge,

2008), vol. 1, pp. 479–80). The firm was responsible for thousands of portraits between 1863 and 1962, when the firm was acquired by Bassano & Vandyk. The National Portrait Gallery owns all the surviving negatives.

8. *Lovell Reeve*: Lovell Augustus Reeve (1814–65), a British conchologist and publisher (*ODNB*).
9. *little biographical notice that we concocted when at Waverly Place*: Hirst contributed a brief life of Tyndall to L. Reeve (ed), *Portraits of Men of Eminence in Literature, Science, and Art, with Biographical Memoirs. The Photographs from Life, by Ernst Edwards, BA* (London: Lovell Reeve & Co., 1864), vol 2, pp. 25–32.
10. *the Bel Alp*: see letter 3771, n. 5.
11. *M^rs. Tunstal*: Jane Tunstall.
12. *go to America*: for Tyndall's American lecture tour, see Introduction, n. 8.
13. *Helmholtz's letter*: possibly Hermann Helmholtz's letter of 23 June 1872 (3770), which discussed Tyndall's issue with Johann Zöllner (see letter 3740, n. 2 and n. 3).
14. *Jamin in his laboratory*: on his way to Switzerland, Tyndall visited Jules Jamin in Paris.
15. *Both were there*: for Jamin's family, see letter 3776, n. 18.
16. *Pasteur*: Louis Pasteur (1822–95), a French microbiologist who is most well-known for his work on microorganisms. His work helped to advance the prevention of diseases, especially rabies and anthrax, through vaccination. He also impacted the manufacture of beer and wine through his research on fermentation (*CDSB*; *DBF*).
17. *DelaRive*: Auguste Arthur de la Rive.
18. *Alabama affair*: a reference to the Confederate ship *Alabama* during the American Civil War, which had been built in English territory and sent to sea despite unsuccessful intervention by the British government, much to the resentment of Abraham Lincoln's administration. A later tribunal deemed the British held responsibility for the maritime losses the United States suffered due to the *Alabama*, and awarded the United States damages of $15,500,000 ('Alabama Claims', *New American Supplement to the latest edition of the Encyclopaedia Britannica* (New York: The Werner Company, 1902), vol. 25, pp. 110–11; H. Jones, *Union in Peril: The Crisis over British Intervention in the Civil War* (Chapel Hill, NC: University of North Carolina Press, 1992), pp. 146–7).
19. *Montanvert*: a lookout point above the Mer de Glace, a large glacier on the north side of Mont Blanc.
20. *Aiguille de Dru*: a mountain in the Mont Blanc massif.
21. *Jorasse*: Grandes Jorasses, a mountain in the Mont Blanc massif.
22. *Periades*: a mountain ridge in the Mont Blanc massif.
23. *Géant*: Dent du Géant, a mountain in the Mont Blanc massif.
24. *Aiguille de Charmoz*: Aiguille des Grands Charmoz, a mountain in the Mont Blanc massif.
25. *Mont Blanc*: the highest mountain in the Alps, on the border between France and Italy, which Tyndall first summited on 13 August 1857 (see *Tyndall Correspondence*, vol. 6).
26. *M^rs. Spottiswoode*: Eliza Spottiswoode.
27. *Your last note*: letter missing.
28. *Lilly*: Emily Hirst.

From Henry Bence Jones 2 July 1872 3782

84 Brook St | W[1] | July 2 1872

My dear friend

I have little to say but still you may like to hear what we did yesterday. The chief thing was to adjourn for a fortnight when M[r] Berkley is to make another report to the managers M[r] Harwood had written to me "no doubt if Mess[rs] Smith & Co[2] cannot proceed with the work[3] in 3 or 4 weeks time we could find other builders & our contract provides for this emergency". M[r] Harwood was unable to be at the meeting & M[r] Berkley knew nothing about this possibility of another builder. However we shall know more in another fortnight about the strike. I suspect from what I hear the masters will give in.

I have made arrangements about the chemical lectures so that the Laboratory will not be absolutely requisite until Jan[uar]y or even Feb. Odling will prepare the Juveniles[4] somehow even if there is no Laboratory finished & Debus will give the first three chemical lectures on Thursdays beginning Jan[uar]y 16[5] & after him Armstrong[6] will give four lectures[7] which will carry us to the end of Feb[ruar]y then Harcourt will begin in March & go on till Easter.[8]

I have put your Friday evening down for the 4th of April[9] the last before Easter

& your lectures after Easter[10] will be April 24 May 8 15 22 19 & June 5 The 1st of May being the annual meeting you can give no lectures. I hope all this will suit you. There are 16 essays for the Actonian Prize[11] one Bastian's[12] two volume book!![13] Some however are very short.

I saw your M[rs] Tunstall[14] who I heard had a wounded leg, but I found it all healing kindly & she will soon be all right again

I hope this fine weather extends to Switzerland[15] if Mr Arnold[16] is with you make him promise a Friday evening in Jan[uar]y or Feb[ruar]y[17] I shall want some substitute for your attractive powers sadly

I go on well all my dropsy[18] is gone thanks to drugs I only fear to lay aside my crutches. I feel like a miserable dog escaped from drowning but who will be drowned again as soon as he is caught.

Good bye my dear friend may you get all health & heartiness in the mountains

& Believe me | yours most truly | H Bence Jones

RI MS JT/1/J/67

1. *84 Brook St | W*: residence in London.
2. *Mess[rs] Smith & Co*: the building company of George Smith—George Smith & Co Ltd,

Sun Foundry, Glasgow, which expanded to London and Dublin in 1863—was hired to build the new laboratory at the RI (F. A. J. L. James and A. Peers, 'Constructing Space for Science at the Royal Institution of Great Britain', *Physics in Perspective*, 9 (2007), pp. 162–3).

3. *the work*: renovation of the RI's laboratory (see letter 3776, n. 14).
4. *the Juveniles*: the RI's annual juvenile Christmas Lectures began in 1827 and still continue (see F. A. J. L. James (ed.), *Christmas at the Royal Institution: An Anthology of Lectures* (Hackensack, NJ: World Scientific, 2007)). Odling delivered the upcoming season's course, 'Six Lectures on Air and Gas', on 28 and 31 December 1872, and 2, 4, 7, and 9 January 1873 (*Roy. Inst. Proc.*, 6 (1870–72), p. 559).
5. *first three chemical lectures on Thursdays beginning Jan[uar]y 16*: Debus delivered 'Three Lectures on Oxidation' at the RI on Thursdays from 16 to 30 January (*Roy. Inst. Proc.*, 6 (1870–72), p. 559).
6. *Armstrong*: Henry Edward Armstrong (1848–1937), a British chemist and educational reformer (*ODNB*).
7. *four lectures*: Armstrong delivered 'Four Lectures on the Artificial Formation of Organic Substances' at the RI on Thursdays from 6 to 27 February 1873 (*Roy. Inst. Proc.*, 6 (1870–72), p. 559).
8. *Harcourt will begin in March & go on till Easter*: Augustus Vernon Harcourt delivered 'Five Lectures on the Chemistry of Coal and its Products' at the RI on Thursdays from 6 March to 3 April 1873 (*Roy. Inst. Proc.*, 6 (1870–72), p. 559).
9. *your Friday evening down for the 4th of April*: Tyndall delivered a Friday Evening Discourse at the RI on 4 April 1873, 'Some Observations on Niagara' (see letter 3738, n. 3).
10. *your lectures after Easter*: Tyndall delivered a course of six lectures at the RI, 'On Light', on Thursdays from 24 April to 5 June 1873 (*Roy. Inst. Proc.*, 6 (1870–72), p. 560; *The Athenaeum*, 19 April 1873, p. 489). These were essentially a redelivery of his lectures on light in America (see Introduction, n. 8).
11. *Actonian Prize*: a prize awarded every seven years by the RI beginning in 1844 to a 'person who in the judgement of the committee of managers for the time being of the Institution, should have been the author of the best essay illustrative of the wisdom and beneficence of the Almighty, in such department of science as the committee of managers should, in their discretion, have selected'. The announcement in 1871 for the 1872 prize, published in *Nature*, *The Athenaeum*, *Journal of the London Institution*, *The Academy*, *The Quarterly Journal of Science*, *The Lancet*, and other periodicals, called for submissions to address the subject of 'The Theory of the Evolution of Living Things'. The winners were announced at the 2 December 1872 general monthly meeting of the RI: 'The two ACTONIAN PRIZES of 105l. each for Essays "On the Theory of the Evolution of Living Things," were this day awarded by the Managers to the REV. GEORGE HENSLOW and to B. THOMPSON LOWNE, Esq' (*Roy. Inst. Proc.*, 6 (1870–72), p. 561). George Henslow (1835–1925), a British botanist, Lamarckian evolutionist, and a son of Charles Darwin's Cambridge mentor John Stevens Henslow (1796–1861) (*ODNB*), published his essay as *The Theory of Evolution of Living Things and the Application of the Principles of Evolution to*

Religion (London: Macmillan and Co., 1873). Benjamin Thompson Lowne (1839–1925), a British physician and naturalist (F. Burkhardt, et al. (eds), *The Correspondence of Charles Darwin* (Cambridge: Cambridge University Press, 2015), vol. 23, p. 651), published his essay as *The Philosophy of Evolution (An Actonian Prize Essay)* (London: John van Voorst, 1873).

12. *Bastian's*: Henry Charlton Bastian (1837–1915), a British physician and neurologist. He studied at University College, London, completing his MD in 1866. Following a post as assistant physician and lecturer in pathology at St Mary's Hospital, Paddington, Bastian returned to University College as professor of pathological anatomy in 1867. He was appointed physician there in 1878, and chair of medicine from 1887 to 1898. An early interest in flora and fauna saw him elected FRS in 1868. Bastian was a supporter of the theory of evolution and was a prominent promotor of spontaneous generation. This led to confrontational disputes with numerous scientists and clinicians, including Tyndall, Thomas Huxley, Joseph Lister, and Louis Pasteur (*ODNB*).
13. *two volume book*: H. C. Bastian, *The Beginnings of Life: Being Some Account of the Nature, Modes of Origin and Transformations of Lower Organisms*, 2 vols (London: Macmillan and Co., 1872).
14. *Mrs Tunstall*: Jane Tunstall.
15. *Switzerland*: see letter 3716, n. 6.
16. *Mr Arnold*: Matthew Arnold (1822–88), an English poet who was professor of poetry at Oxford from 1857–67 (*ODNB*).
17. *Friday evening in Jan[uar]^y^ or Feb[ruar]^y^*: Arnold did not deliver a Friday Evening Discourse at the RI in 1873.
18. *dropsy*: a morbid condition characterized by the accumulation of watery fluid in the serous cavities or the connective tissue of the body (*OED*).

To Thomas Archer Hirst — 3–4 July 1872 — 3783

Chamounix[1] 3rd July 1872

My dear Tom

After my gymnastics on Sunday[2] my arms were chiefly tired: today it is rather my legs and brain that feel stiffness. When I open my eyes in the morning the pure snows of Mont Blanc[3] and of the Dôme du Gouter[4] are right before me. I can see the mountain from the Grands Mulets[5] upwards without quitting my bed. Yesterday I had a scamper to the Grands Mulets. I desired the exercise, and I wished to see the ameliorations[6] introduced there of late years: Joseph Tairraz,[7] our hosts[8] at the Montanvert,[9] and Couttet my present host[10] here accompanied me—They wished to make an excursion with me, and as the wish fell in with mine it was fulfilled. We scrambled along the usual slopes with the route first pursued by us from the Montanvert overhead; found at the Pierre pointue[11] a good cottage with a good cook; good mutton & good potatoes à la Coco. We had taken a bottle of truly good wine with us from Chamouni. At the Pierre l'Eschette,[12] where we put on our gaiters fifteen years ago we repeated the process. The snow was plentiful & very soft, so the gaiters were needed. We made the usual detours among the séracs,[13] crossed the usual crevasses, ascended the usual slopes, and finally at 1 oClock found ourselves in the cabin of the Grands Mulets.

Beds were there, and it was very agreeable to stretch onesself[14] on the soft mattress. The heat outside was great. The sun seemed to lay his beams like something solid against the naked neck; but the temperature within the cabin might be called wintry—It was only a few degrees above the freezing point. The detachment of the air, from the rays which traversed it could not be more strikingly illustrated. I had taken some concentrated beef from Fortnum & Mason,[15] so we made some excellent soup and enjoyed it. The cabin is moreover a store of wine, so we uncorked a bottle of good champagne & quaffed it. The snows as I said in a former letter[16] are of exceptional purity this year; so the journey was in the highest degree pleasant and exhilarating.

We descended at a good pace. Halted for a moment at the Pierre pointue and then scrambled slowly down the rhododendron covered slopes. The Alpenrose[17] is splendid; its purple pure, & fresh, and deep. The day was wound up by a dinner here, when we allowed ourselves to regale upon the best the house could afford.

Today, as a draw-back I feel a singular absence of mental flexibility. My brain seems as stolid as that of an ox. The memory discharges accomplished yesterday for purely physical ends seem to have left the mind without stock in hand; and I must wait for the recovery of the lost energy.

It is exceedingly cool on the part of Routledge[18] to send me a book[19] without a word of intimation; to evoke my courteous thanks, and then to tell me that I must send them fifteen shillings—I do not want the book; but I suppose the simplest solution is to send them the money. What mud some people are made of!

I hope you read Helmholtz's letter:[20] it was very interesting.

As yet I have done next to nothing for America;[21] but I will throw myself wholly into the work by and by and get it clustered into order.

Will you give my kind remembrances to Lady Lubbock[22] when you see her? I thought her weak on the day of the X meeting;[23] and it is curious what an effect the appearance of illness has upon me. When my friends are in the enjoyment of health and pleasure, I do not feel to care about them. But the moment they appear to suffer my heart becomes human. In other words when there is no need of friendship I do not care to be a friend; but in the day of need I think, I should be one.

(July 4th)

We had a splendid sunset again last night. I notice this regarding the crimson of the clouds and mountains, that it does not appear when the sky is most clear. In dropping a solution of mastic[24] into water, the quantity necessary to give you the blue of reflexion is totally insufficient to give you the red of transmission. To produce the red you must make the liquid visibly turbid. And so with the air. On the occasion of late where the red was most pronounced the sky though free from cloud was visibly turbid so that after sunset it was only the light of the larger stars that could struggle through.

I have been feeding you with correspondence since I came away, but I know not how the food may agree with you. Today I wander up to the Montanvert: tomorrow I have a scramble among the Séracs,[25] and a look at the Aiguille du Geant.[26] A very strong man named Kennedy (of Leeds)[27] is here with two first rate guides, so I link myself on to him for the inspection of the Géant. I shall aim to get over the Col de Balm[28] on Saturday or Sunday, and then reach my perch on the Bel Alp.[29]

Love to Lilly[30] | Yours affectionately | J. Tyndall

Nothing I suppose has as yet been done regarding Hooker.[31]

BL Add MS 63092, ff. 122–5

---

1. *Chamounix*: see letter 3776, n. 12.
2. *my gymnastics on Sunday*: in his letter to Hirst of 1 July 1872 (a Monday), Tyndall described his descent of the Montanvert the day before, when he mistook his way, as 'the hardest bit of climbing to accomplish that ever I went through in my life' (see letter 3781).
3. *Mont Blanc*: see letter 3781, n. 25.
4. *Dôme du Gouter*: a mountain in the Mont Blanc massif.
5. *Grands Mulets*: a rock island on the north side of Mont Blanc.
6. *ameliorations*: improvements (*OED*).
7. *Joseph Tairraz*: Joseph Tairraz (1827–1902), a French pioneer of alpine photography (M. Dessemond, *Tairraz père et fils: photographes de montagne* (Paris: Contrejour, 1992)).
8. *hosts*: not identified.
9. *Montanvert*: see letter 3781, n. 19.
10. *Couttet my present host*: the alpine guide François Couttet, owner of le Grand Hotel Couttet in Chamonix.
11. *Pierre pointue*: a large rock outcrop on Mont Blanc.
12. *Pierre l'Eschelle*: a large rock outcrop on Mont Blanc.
13. *séracs*: a tower of ice on a glacier formed at a crevasse (*OED*).
14. *onesself*: an alternative spelling, according to *OED*.
15. *Fortnum & Mason*: a high quality grocery store in Piccadilly, London (founded in 1707, it remains in business as a department store).

16. *former letter*: letter missing. The only letter written from Switzerland previous to this one (3781) does not mention anything about the snow.
17. *Alpenrose*: an alpine species of rhododendron (*Rhododendron ferrugineum*), which produces pinkish-red flowers in the summer.
18. *Routledge*: possibly George Routledge (1812–88), a British publisher who founded the publishing house Routledge (*ODNB*).
19. *book*: not identified.
20. *Helmholtz's letter*: see letter 3781, n. 13.
21. *America*: for Tyndall's American lecture tour, see Introduction, n. 8.
22. *Lady Lubbock*: Ellen Frances Lubbock (née Hordern, 1834–79), daughter of an Anglican minister, who married John Lubbock in 1856. Together they had three sons and three daughters.
23. *X meeting*: the X Club (see letter 3757, n. 2).
24. *mastic*: an aromatic gum or resin which exudes from the bark of the lentisk or mastic tree, *Pistacia lentiscus*, used chiefly in making varnishes and, formerly, in medicine (also mastic gum) (*OED*).
25. *Séracs*: a region on Mont Blanc consisting of seracs (see n. 13).
26. *Aiguille du Geant*: or Dent du Géant, a mountain in the Mont Blanc massif. See letter 3785, and *Ascent of John Tyndall*, p. 298.
27. *Kennedy (of Leeds)*: Thomas Kennedy.
28. *Col de Balm*: Col de Balme, an alpine pass on the border of France and Switzerland, between the summits of Tête de Balme and Les Grandes Otanes.
29. *my perch on the Bel Alp*: see letter 3771, n. 5.
30. *Lilly*: Emily Hirst.
31. *regarding Hooker*: referring to the Memorial in support of Joseph Hooker in his dispute with Acton Ayrton (see letter 3755, esp. n. 1).

From Joseph Dalton Hooker [3][1] July 1872 3784

July 2d 1872.

Dear Tyndall

At Spottiswoode's and Huxley's suggestion I have written the enclosed to Ayrton, and the Treasury.[2] I was very loth to do it, but they saw no escape from the propriety of my thus addressing Ayrton under the circumstances. They further proposed that I should send Ayrton copies of my correspondence with the Treasury (to which the Treasury letter refers) together with a copy of the Treasury's letter itself; but this I overruled, on the grounds that (1) It was showing Ayrton my whole hand, whilst neither he nor the Treasury had shown me theirs (2) The first letter to be copied would be that in which I say that "I will cheerfully abide by their Lordships' decision, or resign", and as in writing to Ayrton at all, I am accepting him as in some sort the exponent of the Treasury's decisions (which are communicated to him) I think this would have been most unwise. (3) I should have had to send a copy of the Treasury's letter to me, asking me to give them information which they should have asked from the Board of Works, and which they asked of me with the private intimation that they did not want Ayrton to know what they were about. (4) There are various passages in my letters to the Treasury, upon which Ayrton would call for explanations, and get up a correspondence.

The course I have pursued further recommended itself to Huxley and Spottiswoode as enabling me to inform the Treasury, that whereas Ayrton has made no move, I have gone out of my way to advance towards him, in the direction the Treasury wish the matter settled; which as I am indirectly informed is in my favour on all points:—and I must never lose sight of the fact, that, whether by this ministry or another, the ultimate settlement must be wholly arranged by, or in concert with, the permanent officers of the Treasury.

Before taking the above course, I asked and obtained H. M. T.'s assurance that my action and its results would in no way influence the course of the Memorialists,[3] whose object is a permanent settlement; or that to be taken in Parliament, which I assume to be the exposure of Mr Ayrton's conduct.

The papers called for by L[or]d Derby and Lubbock are still being printed. I have urged action but I suppose that the ministers, or Gladstone, are throwing every impediment in the way of the production of the most important papers. I have not heard a whisper of Gladstone's proposed action in regard to the Memorial. He surely cannot ignore it! Huxley is perfectly disgusted with Lowe, and altogether Science under Government is in a pretty mess.

Ever your aff[ectiona]te | J. D. Hooker.

<I> have just received a telegraph to the effect that Mr Gladstone is coming to the Garden at 6 this evening.[4] The enclosed letter has not yet been sent so there is no sort of connexion.

RI MS JT/1/TYP/8/2673
LT Typescript Only

1. *[3]*: LT left a note on the letter stating that the date should have been '3d' since the enclosures were dated 3 July.
2. *written the enclosed to Ayrton, and the Treasury*: enclosed letters from Hooker to Acton Ayrton and the Treasury are missing.
3. *Memorialists*: referring to the signatories of the Memorial in support of Joseph Hooker in his dispute with Acton Ayrton (see letter 3755, esp. n. 1). The names of the signatories are listed at the end of the Memorial (as well as letter 3730, n. 2).
4. *Mr Gladstone is coming to the Garden at 6 this evening*: Hooker described this visit in a letter to Tyndall of [7 July 1872] (see letter 3786).

To Thomas Archer Hirst [7 July 1872][1] 3785

Chamounix[2] Sunday

My dear Tom.

You have long since ceased to send my letters to Chamounix, so that I have long since ceased to hear from you. I have been here just over a week, and I leave tomorrow morning for Martigny,[3] overtaking, I trust, the train at 2.23 P.M. and getting on, perhaps as far as Brieg[4] tomorrow night. I am well tanned—too much so in fact—I am burned and scorched and reduced to blisters and scurf. A few days ago a gentleman named Kennedy[5] (a machine maker in Leeds) whom I had previously met in the Alps came over here from the Italian side across Mont Blanc,[6] accompanied by two first rate guides.[7] He wished me to join him in an excursion, and I did so. We went up to the Montanvert[8]—I along the Chapeau side,[9] employing a day in making observations on the wasted glacier, & its moraines: he by the ordinary route. On Friday morning we were called at 4. Unhappily there were others there who had been called two hours earlier & who made the dawn hideous by their noise. Kennedy intended to cross the Col du Géant[10] to Courmayeur[11] and the project was to examine the Aiguille du Géant[12] on the way, and to ascertain whether it was accessible. I had been promised a good man at the Montanvert for I knew I should have to repass the Séracs[13] after the ice-cascade on my return: the 'good man'[14] was a mere youth who knew next to nothing of the Séracs. However our party in going up was sufficiently strong to surmounts all obstacles; and though we happened to choose the most difficult side of the glacier, the skill of our leading guide carried us through successfully. We reached the upper snows of the Col; and near the Col itself the guides abandoned their knapsacks; and we turned toward the Aiguille. We were soon upon rock where I generally find the rope intolerable, so I shook myself out of its coils, and followed for a time alone. Prudence however caused me again to attach myself: the youth from the Montanvert electing to sling below. We had some very rough work indeed to reach the highest accessible point of the Aiguille, and where these two hundred feet of smooth granite jutted above us like a perpendicular tower. It was not to be touched so we sat down, at times looking at the clouds and mountains & blue sky, at times closing our eyes and taking slumber. For London legs it was a heavy piece of work; but the tough old pins did not fail me. We descended—the work certainly needed thought and the quality of a first rate guide. We picked up the man who had remained upon the lower buttress of the mountain, and were soon beside our knapsacks. The snow which was fine in the morning was as soft as flour—no tenacity or toughness in the mass—I thought of my former escapade with young

Balmat[15] among the Séracs and had no pleasure in the thought of renewing it. The rope, moreover, taken singly was thin and old, so we doubled it and tied ourselves together. It then appeared that one of the guides (Carrel of Val Tournanche[16]) was about to be released by Kennedy, and that a party awaited him at Chamounix: So it was proposed that I should take Carrel back, there being no difficulty on the Courmayeur side of the Col. I did so; but Carrel, who is first rate upon rocks, is a mere child upon the glaciers. The Val Tournanche does not furnish the necessary discipline to its guides. Seeing a track on the further side of the snow field, I decided on reaching it and following it as far as possible. We did so, and thus got through the Séracs more rapidly than we otherwise should have done. Still it was weary, cautious, plodding work, the snow was so very soft, and the state of the bridges so uncertain. Then came the long pull along the glacier du Géant & the Mer de Glace[17] to the Montanvert where I had a bit of food & descended afterwards to Chamounix in the dark. It was 10 oClock when I reached the hotel.

It is I think certain that there is something in the quality of the solar beams at these great heights which produces chemical changes & consequent irritation in the skin. I had no veil: my eyes were protected but not my face, and as a consequence I find myself rather seriously burned. I hardly think I shall do any more heavy mountain work: I need all my nervous energy for my American work, and one cannot have this energy and squander it at the same time upon the mountains. This letter, for example, illustrates the stupidity of which a man who has been engaged in severe physical labour is capable.

I have been reading De Tocqueville[18] at intervals and since I read Wilhelm Meister[19] have not met a book so filled with practical wisdom.

I wish you w[oul]d write one line to M[e]ssrs. King and Co. publishers Cornhill[20] asking them to let me know whether they can spare room for a few words of preface regarding the present state of the Mer de Glace.[21]

Yours affecti[o]n[ate]ly | John Tyndall

Say to Dr. Bence Jones that I will write to him soon after my arrival at the Bel Alp.[22]

I see nothing in the papers about Kew ?[23]

BL Add MS 63092, ff. 133–4
RI MS JT/1/HTYP/594–5

---

1. *[7 July 1872]*: while LT noted on the typescript that this was July 1871, the content of the letter (climbing in the Alps with Thomas Kennedy, reading De Tocqueville before his American trip, and the preface about the Mer de Glace) points to 1872. The 'Sunday' following the climbs with Kennedy on 4 and 5 July 1872 (see letter 3783) was 7 July 1872.
2. *Chamounix*: see letter 3776, n. 12.

3. *Martigny*: a city in southwestern Switzerland, a junction of roads joining Italy, France, and Switzerland.
4. *Brieg*: Brig, a town in the Rhone Valley in the Swiss canton of Valais.
5. *Kennedy*: Thomas Kennedy.
6. *Mont Blanc*: see letter 3871, n. 25.
7. *two first rate guides*: for one of the guides, see n. 16 below.
8. *Montanvert*: see letter 3781, n. 19.
9. *Chapeau side*: referring to the refuge set up in 1855 for those crossing the Mer de Glace, Le Chapeau, on the opposite side of the glacier from the Montanvert.
10. *Col du Géant*: the high glacier pass between Chamonix and Courmayeur.
11. *Courmayeur*: an Italian town at the foot of the south side of Mont Blanc.
12. *Aiguille du Géant*: or Dent du Géant, a mountain in the Mont Blanc massif.
13. *Séracs*: a region on Mont Blanc consisting of seracs, towers of ice on a glacier formed at a crevasse (*OED*).
14. *'good man'*: not identified.
15. *former escapade with young Balmat*: referring to Edouard Balmat (b. 1846), the son of Chamonix porter and guide Alexandre Balmat (1815–75). Rather than hiring a guide in the summer of 1857, Tyndall explored the Mer de Glace and climbed the Col du Géant with Edouard Balmat, 'a little boy, who can climb well' (letter 1419, *Tyndall Correspondence*, vol. 6; *Glaciers of the Alps*, pp. 47, 67). Tyndall referred to Balmat as his 'little demon', Known as 'le petit Balmat', he was also among the porters who assisted Tyndall in his research on the Mer de Glace in December 1859 (*Glaciers of the Alps*, pp. 195–219).
16. *Carrel of Val Tournanche*: Jean-Antoine Carrel (1834–91) of Valtournenche (an Italian alpine town). He was an Italian soldier, mountaineer, and alpine guide who Tyndall had hired as a porter during his failed attempt on the Matterhorn in 1862. Carrel made the second ascent of the Matterhorn on 17 July 1865 just three days after Edward Whymper (1840–1911) made the first ascent (*Hours of Exercise in the Alps*, pp. 157–65; C. D. Cunningham and W. W. Abney, *The Pioneers of the Alps*, 2nd edn (London: Sampson Low, Marston, Searle, and Rivington Ltd., 1888), pp. 126–31).
17. *Mer de Glace*: a large glacier on the north side of Mont Blanc.
18. *De Tocqueville*: a translation of Alexis de Tocqueville's two-volume *De La Démocratie en Amérique* (1835–40); probably *Democracy in America*, trans. Henry Reeve (New York: George Dearborn & Co., 1838–40). Tyndall read this in preparation for his American lecture tour (see Introduction, n. 8), and referred to de Tocqueville several times in his lectures. Tyndall took note of section 1 of chapter 10 in the second volume, 'Why the Americans Are More Addicted To Practical Than To Theoretical Science', where de Tocqueville comments at length on the typically American proclivity for the practical and neglect of the cultivation of abstract knowledge. Tyndall's comments on de Tocqueville were reprinted in 'Professor Tyndall in America', *Nature*, 7 (23 January 1873), pp. 224–5. See also *Six Lectures on Light*, pp. 214, 216, 223–5. During the lecture tour, Tyndall wrote to Edward L. Youmans that 'I repeated one of De Tocqueville's hardest sayings with

reference to the poverty of their achievements in the higher walks of science' (see letter 3903).

19. *Wilhelm Meister*: *Wilhelm Meister's Apprenticeship*, the second novel of Johann Wolfgang von Goethe, first published in 1795–96. Tyndall was probably familiar with Thomas Carlyle's translation: J. W. von Goethe, *Wilhelm Meister's Apprenticeship: A Novel from the German of Goethe*, 3 vols, trans. T. Carlyle (London: G. & W. B. Whittaker, 1824).
20. *M[e]ss^rs^. King and Co. publishers Cornhill*: Henry S. King and Co. was a London-based publisher with offices at 65 Cornhill & 12 Paternoster Row.
21. *a few words of preface regarding the present state of the Mer de Glace*: in 1872, just before leaving England for the United States in September, Tyndall published *Forms of Water*, which included a preface about the wasting away of glaciers including the Mer de Glace, as well as the torrent caused by the breaking of the ice barrier at the Märgelin See (pp. vii–x; on the Märgelin See, see letter 3810). Tyndall's *Forms of Water* was the first title in Edward L. Youmans's International Scientific Series (see letter 3871, n. 3). An American edition was published the same year by Appleton and Co. in New York.
22. *Bel Alp*: see letter 3771, n. 5.
23. *Kew*: referring to the dispute between Joseph Hooker and Acton Ayrton, and the Memorial in support of Hooker (see letter 3755, esp. n. 1).

From Joseph Dalton Hooker [8 July 1872][1] 3786

Dear Tyndall,

Beyond a brief note from Lubbock saying that the "papers were not yet distributed to members" I have no news regarding the movement in the House.[2]

The first of these letters on the other side is the Board's[3] answer to mine of which I sent you a copy.[4] The second has just been received. Some progress has been made, I find, inasmuch as Ayrton[5] has himself had to discharge the Clerk! and this must have been by Treasury order. He has also given in about one of the Herbarium appointments which I am to fill up without reference to the C[ivil]. S[ervice]. Comm[issione]rs.[6]

What do you think of Gladstone's having paid me a visit with wife and family! and going all over the Gardens, Houses and Museums with me. Either he has not read the memorial[7] or he is a magnanimous fellow after all! I had a telegraph in the morning that he would arrive at 6. driven by Mr Tollemache[8] (M.P. for Cheshire) an old friend of mine to whom I had sent the memorial, in a Drag and four-in-hand![9] G[ladstone]. was surprised and delighted with what he saw. I do not think he had any previous notion of the place, its extent and variety of interests. No-one could be more pleasant, though he was evidently under some restraint at times. I saw Huxley last week He is, I think better on the whole and enjoys his post and his work at S. Kensington.[10] He is furious with Lowe for his neglect of science and the Institution.[11]

The Spectator has an article[12] on your share of the Contemporary writer's views on Prayer.[13] The article is a confused one and until I see the Contemporary I cannot follow it. It is mild enough—and inconsequential as far as I could follow it.

I assume that you have had 2 communications from me June 29th, July 2d.[14] Neither required any answer, and I allude to them only lest my direction should have proved illegible!

Ever yours aff[ectionate] | J. D. Hooker.

RI MS JT/1/TYP/8/2675–6
LT Typescript Only

---

1. *[8 July 1872]*: in letter 3788, dated 11 July 1872, Hooker wrote that his last letter to Tyndall was from 'Monday 8th'. This undated letter from Hooker evidently fits.
2. *movement in the House*: referring to the Memorial in support of Joseph Hooker in his dispute with Acton Ayrton (see letter 3755, esp. n. 1).
3. *Board's*: Board of Works.
4. *I sent you a copy*: this letter from George Russell (1830–1911), Secretary of Her Majesty's Office of Works (M. H. Port, 'A Contrast in Styles at the Office of Works. Layard and Ayrton: Aesthete and Economist', *The Historical Journal*, 27 (1984), p. 155, n. 33), dated 3 July 1872, reads: 'Sir, I am directed by the First Comm[issione]r of H.M. Works [Acton Ayrton] to acknowledge the receipt of your letter of this day's date and in reply I am to inform you that the Board, after fully considering the representations you have made will communicate to you the conclusion at which it may arrive. | I am, Sir, | Your obedient servant | George Russell' (RI MS JT/1/TYP/8/2675).
5. *Ayrton*: Acton Ayrton.
6. *C[ivil]. S[ervice]. Comm[issione]rs*: the Civil Service Commission was established in 1855 by William Gladstone to regulate recruitment to the Civil Service as a means to ensure impartiality.
7. *memorial*: see n. 2 above.
8. *Mr Tollemache*: John Jervis Tollemache (1805–90), 1st Baron Tollemache, a British Conservative Member of Parliament and a major landowner in Cheshire (*ODNB*). His son Wilbraham Frederic Tollemache (1832–1904), 2nd Baron Tollemache, would marry Mary Stuart Hamilton, the youngest sister of Tyndall's future wife Louisa, in 1878.
9. *Drag and four-in-hand*: a 'drag' is a private vehicle of the type of a stage coach, usually drawn by four horses, with seats inside and on the top (*OED*); 'four-in-hand' refers to a vehicle with four horses driven by one person (*OED*).
10. *his post and his work at S. Kensington*: see letter 3711, n. 7.
11. *neglect of science and the Institution*: Lowe, 'a friend of science' who supported Hooker in the Ayrton controversy (J. Parry, 'Gladstone, Liberalism and the Government of

1868–1874', in D. Bebbington and R. Swift (eds), *Gladstone Centenary Essays* (Liverpool: Liverpool University Press, 2000), pp. 94–112, on p. 107), was not a friend of Huxley's when it came to whether or not the government should spend money on well-rounded public education. Lowe, an educationalist always looking to cut expenditures, argued against the government footing the bill for classical education, instead promoting a practical system of science instruction with standardization as the key. Huxley, on the other hand, favored a culture of education that could have science and literature under the same roof. See P. White, 'Ministers of Culture: Arnold, Huxley and Liberal Anglican Reform of Learning', *History of Science*, 43 (2005), pp. 115–38, on pp. 120–1; P. White, *Thomas Huxley: Making the "Man of Science"* (Cambridge: Cambridge University Press, 2003), pp. 88–9; and *The X Club*, p. 327.

12. *The Spectator has an article*: 'The Proposed Prayer-Gauge', *Spectator*, 45 (6 July 1872), pp. 846–7.
13. *your share of the Contemporary writer's views on Prayer*: see letter 3779, esp. n. 1 and n. 2.
14. *2 communications from me June 29th, July 2d*: letters 3777 and 3784.

To Peter Guthrie Tait 10 July [1872][1] 3787

Bel Alp.[2] 10th. July

My dear Tait

I remember Mr. Dewar very well and have forwarded your note regarding him to Dr. Bence Jones who has the Friday evening arrangement in his hands.[3]

There will be no more lectures this Summer: the last Friday evening was on the 7th of June.[4]

Well as you are refractory[5] I hope sincerely that Clausius will not continue the discussion:[6] Such discussion does no good.

Yours faithfully | John Tyndall

University of Edinburgh GN 237 Coll-234 Gen. 2169, Letter 230

---

1. *[1872]*: the year is given by relation to letters 3793 (16 July) and 3801 (22 July), both from Henry Bence Jones.
2. *Bel Alp*: see letter 3771, n. 5.
3. *Friday evening arrangement in his hands*: as Secretary of the RI (1860–73), Bence Jones was in charge of scheduling lecturers for Friday Evening Discourses and lecture courses. Dewar delivered his first Friday Evening Discourse at the RI on 7 March 1873, 'On the Temperature of the Sun and the Work of Sunlight' (*Roy. Inst. Proc.*, 7 (1873–75), p. 57).
4. *last Friday evening was on the 7th of June*: William Odling delivered a Friday Evening Discourse at the RI, 'On the History of Ozone', on 7 June 1872 (*Roy. Inst. Proc.*, 6 (1870–72), p. 546–55).
5. *refractory*: obstinate, stubborn; unmanageable, rebellious (*OED*).

6. *discussion*: in 1872 Clausius fell into dispute with Tait over who was the originator of thermodynamics when James Clerk Maxwell published *Theory of Heat* (London: Longmans, Green, and Co., 1871). With barley a mention of him by Maxwell, Clausius claimed that the British were attempting to claim too much credit, for which Tait disagreed. Clausius sought the credit wholly for himself (see C. Smith, *The Science of Energy: A Cultural History of Energy Physics in Victorian Britain* (Chicago: University of Chicago Press, 1998), pp. 255–60). Their exchanges in the *Phil. Mag.*—the articles and responses Clausius referred to, and more—are: R. Clausius, 'A Contribution to the History of the Mechanical Theory of Heat', *Phil. Mag.*, 43 (February 1872), pp. 106–15; P. G. Tait, 'Reply to Professor Clausius', *Phil. Mag.*, 43 (May 1872), p. 338; R. Clausius, 'On the Objections raised by Mr. Tait against my Treatment of the Mechanical Theory of Heat', *Phil. Mag.*, 43 (June 1872), pp. 443–6; P. G. Tait, 'On the History of the Second Law of Thermodynamics, in reply to Professor Clausius', *Phil. Mag.*, 43 (Supplement for 1872), pp. 516–8; R. Clausius, 'A necessary Correction of one of Mr. Tait's Remarks', *Phil. Mag.*, 44 (August 1872), p. 117; and P. G. Tait, 'Reply to Professor Clausius', *Phil. Mag.*, 44 (September 1872), p. 240.

From Joseph Dalton Hooker 11 July 1872 3788

Kew | July 11th 1872.

Dear Tyndall

The enclosed from Lubbock[1] will disappoint you: this is the third time he has been put off.[2] I have told him plainly that he is being trifled with. I saw Spottiswoode and Huxley together to-day. In spite of all my exertions no-one, not even Lubbock has seen Lord Derby. I cannot but suppose that in Lubbock's political position he dares not. I left Sp[ottiswoode]. and H[uxley]. en route to Lord Derby.

I send the Daily News.[3] It is fuller and more explicit than the Times of Monday,[4] in which the murder came out. Who fired the mine I do not know, nor do H[uxley]. or S[pottiswoode]. I suspected Derby. It was the only thing to be done and I am right glad it is done. Lubbock's question of today would have brought it all before the house and Huxley, Spottiswoode and I parted full of hopes and expectations. Huxley has made out (through Lockyer) that Ayrton[5] and Galton[6] have laid their heads together and that A[yrton]. complains to Galton that Gladstone never showed him my letters of accusation! This is just what I suspected—that Gladstone sent me a garbled copy of Ayrton's "explanations", and that it was upon this garbled document that I accused A[yrton]. of evasions, misstatements, and misrepresentation in his communications to the First Minister of the Crown. No wonder that G[ladstone]. wanted the matter hushed up anyhow.

Huxley seems also to have made out that Ayrton and Owen are in the business—Owen persuading Ayrton that I am working to have the Brit[ish]. Mus[eum]. Botanical collections sent to Kew. This also confirms my suspicions that Ayrton's design was, to disestablish Kew as a scientific Institution, first worrying me out of it, then transferring the collections to the British Museum. I send you the first-fruits of the Treasury's communication to Ayrton.[7] My answer to the Treasury must convince them that they were deceived by Ayrton as to the terms of publication. Gladstone, I need hardly say has never answered the memorial. I am urging its being sent to every member of both Houses without delay. Huxley has sent it to "Nature"[8] with a short heading.

The Globe[9] comes down like a sledge hammer on poor Lord Ripon, for not helping me in the Cabinet. The poor soul did what he could I know. The Botanists and Horticulturists are about to memorialise Gladstone.[10]

The memorial is pronounced to be splendid. The Telegraph alone has not noticed it, <u>on dit</u>[11] that Glyn[12] has gagged it. Huxley promised to write to the Editor[13] at once, he has written to the Pall Mall[14] and Spectator.[15] Another friend[16] is stirring up the Edinburgh and Glasgow papers.[17]

It is quite understood by the Government that Lord Derby is to make an onslaught and they are preparing a reply. So there is no fear but that all will go right in the evening *[run]* so far as the Upper House is concerned, but it is now clear that there are but 3 settlements, Ayrton's resignation, mine, or the removal of Kew. I cannot now accept Ayrton's withdrawal of every point. He is utterly false and with Galton to back him—I should be at his mercy the moment Parliament is up.

Ever yours | J. D. Hooker

My last to you was Monday 8th.[18] I send Daily News[19] and Pall Mall.[20]

RI MS JT/1/TYP/8/2677–8
LT Typescript Only

---

1. *The enclosed from Lubbock*: enclosure missing.
2. *third time he has been put off*: in his attempts to put forward in the House of Lords of Parliament the Memorial in support of Joseph Hooker in his dispute with Acton Ayrton (see letter 3755, esp. n. 1).
3. *Daily News*: 'Mr. Ayrton and Dr. Hooker', *The Daily News*, 10 July 1872, p. 5.
4. *Times of Monday*: *Times*, 8 July 1872, p. 11.
5. *Ayrton*: Acton Ayrton.
6. *Galton*: Douglas Strutt Galton (1822–99), a British engineer, best known for his work in sanitary engineering. He was educated at the Royal Military Academy, Woolwich and was commissioned in the Royal Engineers in December 1840. Galton was elected FRS

in 1859. He was a member of the Royal Commission on Railways in 1866 and the Director of Public Works and Buildings from 1869–75. Starting in 1871 Galton served as a general secretary for the BAAS, and was elected its president in 1895. Galton advocated for improving education for women and for children with learning disabilities. He also received recognition for his work with the sick and wounded on both sides of the Franco-Prussian War (*ODNB*). Hooker and Galton had been in dispute since 1870, when Galton insisted that Hooker should seek multiple bids for the building of new plant houses at Kew, instead of Hooker simply using his own contractors (Endersby, *Imperial Nature*, p. 284).

7. *I send you the first-fruits of the Treasury's communication to Ayrton*: enclosure missing.
8. *"Nature"*: the Memorial was published in *Nature*, 6 (11 July 1872), pp. 211–16. Huxley's heading reads as follows: 'THE comments which have been made by several London journals upon the circumstances referred to in the subjoined memorial, seem, in the interest of all persons concerned, to render its publication in its entirety desirable. We cannot doubt that any delay which has occurred, or may occur, in the delivery of a reply to the memorialists, is due to the desire of the Prime Minister to deal with the important questions at issue in a manner which, while it is not inconsistent with the interests of the public service, is not incompatible with a due regard for the interests of Science, and for fair dealing towards a very eminent scientific man' (p. 211). In the same issue is the following editorial note: 'We reprint in another column a document which will be read with the greatest indignation by scientific men in every part of the world, and with shame by all Englishmen. We refer to the remonstrance presented to the Treasury by eleven of our most distinguished scientific men against the treatment which the eminent Superintendent of the Royal Gardens at Kew has received at the hands of his official superior, the Chief Commissioner of the Board of Works. Though we might take exception to some of the arguments brought forward by the memorialists, yet, there can be but one opinion that the systematic discourtesy and want of consideration with which Dr. Hooker has been treated is entirely incompatible with the efficient conduct of the department of the public service with which he has been entrusted, and deserves the most searching inquiry on the part of the Government' (p. 209).
9. *The Globe*: probably 'Mr. Ayrton's Last', *The Globe*, 11 July 1872, p. 2. This short article does not mention Lord Ripon by name, but as Lord President of the Council at this time, he would be implicated with the following words from the article, which were quoted from the *Standard*: 'the remonstrance, which is signed by Mr. Darwin, Sir Charles Lyell, Professor Huxley, Sir James Paget, Professor Tyndall, and others, was addressed to Mr. Gladstone many months ago, and the matter was brough before the cognizance of the Cabinet. Yet, practically, nothing has been done, and Dr. Hooker has been put off with half promises never fulfilled. Verbal messages of an unsatisfactory and shuffling character, tardily reduced to writing by an obscure underling, constitute the only redress that has been offered him'.

10. *Botanists and Horticulturists are about to memorialise Gladstone*: at the 17 July 1872 meeting of the Royal Horticultural Society, the following resolution was made: 'That the members of the Scientific Committee of the Royal Horticultural Society, having regard to the present question between Dr. Hooker and Mr. Ayrton, desire to express their sense of the importance of Dr. Hooker's efforts to maintain unimpaired the scientific character of Kew, as an establishment indispensable to Horticultural and Botanical Science; and are further unanimous in giving their support to the memorial which has already been forwarded to Mr. Gladstone' (*Journal of the Royal Horticultural Society of London*, 4 (1877), p. xxx). The Royal Botanic Society's support came in the form of the following resolution: 'That the Council of the Royal Botanic Society of London desire to express their deep sympathy with Dr. Hooker in his present position as Directors of the Royal Gardens at Kew, and they earnestly hope that nothing that has hitherto occurred will induce him to resign an office which he has filled with such advantage to the country and with so much honour to himself'. Further still, 'That the Council of the Royal Botanic Society of London, having the fullest confidence in Dr. Hooker as the Director of the Royal Gardens at Kew venture to express their hope that Mr. Gladstone will give his personal attention to the important object of preventing the public from being deprived of Dr. Hooker's valuable services. His scientific attainments and practical knowledge, as evinced in his directorate, have been of the highest order, and the Council of the Royal Botanic Society would, therefore, learn with great regret that any steps had been taken calculated to endanger the connexion of Dr. Hooker with the Royal Gardens at Kew, or his control in scientific matters connected with his office' (*Medical Times and Gazette*, 27 July 1872, p. 99). The names of botanists and naturalists who could not attend a meeting of the Committee of Botanists and Horticulturalists in support of Hooker are given in *Gardeners' Chronicle*, 27 July 1872, p. 1002.
11. *on dit*: an item of gossip; a report based on hearsay (*OED*).
12. *Glyn*: probably George Grenfell Glyn (1824–87), a British Liberal politician who was Member of Parliament for Shaftesbury from 1857–73 (then to the House of Lords), and Parliamentary Secretary to the Treasury under Gladstone from 1868–73 (then to Privy Council). Glyn served in Gladstone's second and third administrations as well (*ODNB*).
13. *Editor*: the editor of *The Daily Telegraph* at this time was Thornton Leigh Hunt (1810–73).
14. *Pall Mall*: 'Mr. Ayrton and Dr. Hooker', *PMG*, 11 July 1872, pp. 10–11.
15. *Spectator*: 'Mr. Ayrton and Dr. Hooker', *Spectator*, 13 July 1872, pp. 873–4.
16. *another friend*: not identified.
17. *Edinburgh and Glasgow papers*: see e.g., 'Dr Hooker and Mr Ayrton', *Glasgow Herald*, 22 July 1872, p. 5.
18. *My last to you was Monday 8th*: letter 3786.
19. *Daily News*: see n. 3.
20. *Pall Mall*: see n. 14.

From Edward Stanley 12 July 1872 3789

Holwood, Beckenham.[1] | July 12th 1872.

Dear Professor Tyndall

The long delay that has occurred in Dr Hooker's case[2] leads me to send you this line of explanation.

I thought that the discussion would take place at great disadvantage to us, if we had not before us the papers containing Ayrton's[3] side of the question: and I decided accordingly to wait for them. I have asked Lord Granville for them two or three times, he promised to give them but could not tell me when. I see now that they are promised for next week and as soon as we have them I am ready to go on.

Very truly yours | Derby.

RI MS JT/1/TYP/8/2681
LT Typescript Only

---

1. *Holwood, Beckenham*: Lord Derby and his wife were renting Holwood House, Beckenham, near Down at this time (see Charles Darwin to Joseph Dalton Hooker, 14 June [1872], *Darwin Correspondence*, vol. 20, pp. 255–6; this is letter 8385 in the online edition of the Darwin correspondence).
2. *Dr Hooker's case*: the Memorial in support of Joseph Hooker in his dispute with Acton Ayrton (see letter 3755, esp. n. 1).
3. *Ayrton's*: Acton Ayrton.

From Heinrich Rudolf Vieweg[1] 13 July 1872 3790

Braunschweig den 13 Juli 1872. |
Herrn Professor John Tyndall | London

Hochgeehrter Herr Professor.

Wir haben uns von mehreren unsrer wißenschaftlichen Freunde Rath erbeten, ob die Veröffentlichung einer deutschen Ausgabe Ihrer Werke über Diamagnetismus und die Contributions to molecular physics in the domain of radiant heat der wißenschaftlichen Welt erwünscht sein würde.

Alle Antworten lauten gleichmäßig dahin, daß diese Werke wesentlich Zusammenstellungen von schon früher erschienenen Aufsätzen in wißenschaftlichen englischen Zeitschriften seien, daß die wichtigeren in deutsche Journale in *[*ziemlicher*]* Anzahl übergegangen seien, daß ein anderer Theil der [englischen] Bücher mehr Intereße für dortige als für deutsche Leser habe, so namentlich die abgedruckten Briefe.

So werthvoll die Abhandlungen an sich seien, so müße doch hervorgehoben werden, daß dieselben den deutschen Gelehrten bereits durch die englischen Journale wie durch Uebersetzung aus denselben in deutsche Journale so zugänglich seien, dass ein Absatz der Reproduction und [Zusammenstellung] in Deutschland nicht zu erwarten sei, wo überhaupt der einzelne Gelehrte im Allgemeinen spärlich im Ankauf von Büchern sei da Bibliotheken in fast allen Städten ihm zu Gebote ständen. Da nun in den Bibliotheken von einiger Bedeutung die wißenschaftlichen englischen Zeitschriften nicht zu fehlen pflegten, so würden auch diese Bedenken tragen die Reproduktion der Abhandlungen über strahlende Wärme und Diamagnetismus zu kaufen.

Wir sind leider nicht in Stande dieser Deduction zu widersprechen. Aus diser Ansicht geht aber auch die Unmöglichkeit hervor die Ausgabe einer deutschen Uebersetzung dieser Werke zu versuchen.

Da wir glauben, daß es Ihnen angenehm gewesen wäre, wenn wir den Verlag übernommen hätten, so müßen wir Ihnen unser Bedauern darüber mittheilen, daß nach so sorgfältiger Ueberlegung wir dazu nicht im Stande zu sein, uns überzeugt haben.

Die Uebersetzung Ihres Buches „die Alpen" ist bis auf Titel und Inhaltsangabe gedruckt, erscheint also in 14 Tagen.

H. Helmholtz wird das Mspt des Buches, Ueberstzg der Fragments... noch einmal durchsehen und der Druck <wird> nach seinem Wunsch im October begonnen und rasch beendigt werden

Unter Versicherung unsrer größten Hochachtung | ergebenst | Friedr Vieweg & S.

Braunschweig 13 July 1872. |
Herr Professor John Tyndall | London

Esteemed Herr Professor.

We have requested advice from several of our scientific friends as to whether the publication of a German edition of your works on diamagnetism and the Contributions to molecular physics in the domain of radiant heat[2] would be desirable to the scientific world.

All the replies consistently say something to the effect that these works are essentially compilations of essays that have appeared previously in English scientific periodicals, that the more important ones have gone into a fair number of German journals, and that another portion of the [English] books, particularly the printed letters, may have more interest for readers there than for German ones.

As valuable as the papers may be by themselves, it would nevertheless have to be emphasised that these are already so accessible to German scholars through English journals as well as through translations from them into German journals that sales of a reproduction and compilation may not be expected in Germany, where at the best of times the individual scholar is generally frugal in purchasing books, as libraries would be at his disposal in almost all cities. As English scientific periodicals are not normally wanting now in libraries of some significance, then these too would have reservations about buying reproductions of papers on radiant heat and diamagnetism.

We are, unfortunately, not in a position to dispute this conclusion. From this[3] view, however, there also emerges the impossibility of us attempting the publication of a German translation of these works.

As we believe you would have preferred it if we had assumed responsibility for distributing the books, then we have to inform you of our regret that after such careful consideration, we have convinced ourselves that we are not in a position to do this.

The translation of your book "In den Alpen"[4] has, with the exception of the title page and contents, been printed, and will therefore appear in a fortnight.

H. Helmholtz is going to look through the MS[5] of the book and trans[lation].[6] of the Fragments[7]... one more time, and printing will[8] be commenced at his wish in October and be completed quickly.

Assuring you of our greatest esteem | your humble servants | Friedr[ich]. Vieweg & S[ohn].

RI MS JT/1/V/12

1. *Heinrich Rudolf Vieweg*: while the letter is signed 'Friedr. Vieweg & S[ohn]', we believe the sender is Heinrich Rudolf Vieweg, who took over the family publishing firm, Friedrich Vieweg und Sohn, from his father Eduard Vieweg (1797–1869).
2. *German edition of your works on diamagnetism and the Contributions to molecular physics in the domain of radiant heat*: in letter 3735, Tyndall wrote that he may ask Vieweg to publish German translations of *Diamagnetism* and *Molecular Physics*.
3. *this*: in the German, Vieweg misspelled the word 'dieser' (meaning 'this') as 'diser'.
4. *"In den Alpen"*: the German translation of Tyndall's *Hours of Exercise in the Alps* was J. Tyndall, *In den Alpen* (Braunschweig: Friedrich Vieweg und Sohn, 1872).
5. *MS*: in the German, Vieweg wrote 'Mspt', presumably an abbreviation for 'manuscript'.
6. *trans[lation]*: in the German, Vieweg wrote 'Ueberstzg', which is an abbreviation for 'Uebersetzung', the German word for 'translation'.
7. *translation of Fragments*: see letter 3735, n. 3.
8. *will*: in the German, Vieweg left out the word 'wird' which would be necessary in this clause.

From John William Draper 14 July 1872 3791

University | Washington Square | New York | July 14 1872

My Dear Dr Tyndall

I have sent today to the Philosophical Magazine a paper "on the Distribution of heat in the spectrum",[1] and as you have worked so much on that subject[2] I think it cannot fail to interest you. So I have requested Dr Francis[3] to send you proof sheets of it, that you may look at it at your leisure.

Don't you remember the day when I had the Pleasure of dining with you at the Athenaeum[4] and I told you what my views on the subject were.

Youmans tells me that you are to be in America this autumn[5]—I hope so. We will try to make your visit pleasurable. Let me know when your arrangements take on a definite state, and I will be on the look out for you.

Yours very truly | J W Draper

Library of Congress, John William Draper Family Papers

---

1. *"on the Distribution of heat in the spectrum"*: Draper's paper, originally published in the *American Journal of Science*, 4 (September 1872), pp. 161–75, was also published as J. W. Draper, 'Researches in Actino-Chemistry.—Memoir First. On the Distribution of Heat in the Spectrum', *Phil. Mag.*, 44 (August 1872), pp. 104–17. The second part appeared later the same year as: J. W. Draper, 'Researches in Actino-Chemistry.—Memoir Second. On the Distribution of Chemical Force in the Spectrum', *Phil. Mag.*, 44 (December 1872), pp. 422–43.
2. *worked so much on that subject*: Tyndall of course was no stranger to the subject of heat, conducting a variety of experiments and delivering lectures over the course of his career. He published two books on heat: J. Tyndall, *Heat Considered as a Mode of Motion: Being a Course of Twelve Lectures Delivered at the Royal Institution of Great Britain in the Season of 1862* (London: Longman, Green, Longman, Roberts, & Green, 1863), which went through several editions; and *Molecular Physics*, a collection of scientific articles on radiant heat. For more on Tyndall's work on heat, see *Ascent of John Tyndall*, chapters 9, 10, and 24.
3. *Dr Francis*: William Francis.
4. *dining with you at the Athenaeum*: Thomas Hirst recorded in his journal that Draper dined with him, Tyndall, and Herbert Spencer at the Athenaeum Club on 25 November 1870 (Brock and MacLeod, *Hirst Journals*, p. 1887). His visit with Tyndall also included lunch at the RI and viewing experiments in the laboratory. Tyndall, Draper wrote, 'had two assistants who performed the experiments and they were very skillful men. He simply stood by and directed. I could not help thinking how much more I should have done if I had had

such advantages!' (D. Fleming, *John William Draper and the Religion of Science* (Philadelphia: University of Pennsylvania Press, 1950), p. 123). On the Athenaeum Club, see letter 3710, n. 5.

5. *in America this autumn*: for Tyndall's American lecture tour, see Introduction, n. 8.

From Joseph Dalton Hooker — 15 July 1872 — 3792

15th July 1872.

Dear T.

The resolution I proposed[1] was that "This House regards with dissatisfaction the conduct of the First Comm[issione]r of Works as to the administration of the Royal Botanical Gardens of Kew." Helps approved, and said he could not better it, so I sent it to Spottiswoode and Huxley for approval, and forwarding to Gladstone and Lubbock if approved.

Hirst is sending you the papers—which are bursting with the subject. The "Observer" I hear is gagged, like the "Telegraph"[2]

What can Ayrton[3] mean by telling Lubbock in the House that my correspondence requires careful investigation before it can be produced! The "Globe" has taken up that answer, and asks if it means "garbling"[4]—which no doubt it does. Bruce[5] and family spent yesterday afternoon with us. He told me that Gladstone's plan was the removal of Kew: but we must not quote him for this! I told Bruce that there was now no option between this and Ayrton's and my resignation.

Not a whisper of the proposed action or attitude of the Gov[ernmen]t has reached me, but I cannot fancy their retaining Ayrton under the storm of indignation he has raised apropos of Kew, and yet quite independent of it.

How I wish it were all over.

Love to Ball[6] if he be with you.[7] | Ever your aff[ectionate] | J. D. Hooker.

RI MS JT/1/TYP/8/2682–3
LT Typescript Only

---

1. *The resolution I proposed*: regarding the Memorial in support of Hooker's dispute with Acton Ayrton (see letter 3755, esp. n. 1).
2. *like the "Telegraph"*: in letter 3788, Hooker wrote: 'The memorial is pronounced to be splendid. The Telegraph alone has not noticed it, on dit that Glyn has gagged it'.
3. *Ayrton*: Acton Ayrton.
4. *"Globe" has taken up that answer, and asks if it means "garbling"*: 'The Ayrton Case', *The Globe*, 13 July 1872, p. 4. The writer of this article did not use the word 'garbling'—meaning

'to make selections from (a statement, fact, piece of writing, etc.), typically with an unfair or malicious purpose; to purposefully distort or corrupt (a statement, a piece of writing, etc.) so as to misrepresent or mislead (*OED*)'—but instead used 'cooking'—meaning 'to alter (something) in a dishonest, underhand, or surreptitious manner; to tamper with, to doctor; to falsify' (*OED*).

5. *Bruce*: probably Henry Austin Bruce (1815–95), a British politician and Gladstone's Home Secretary. Bruce first entered as a Liberal Member of Parliament for Merthyr Tudful, in Wales, in 1852; during the 1860s he was particularly involved in issues relating to education. He became Baron Abedare in 1873 (*ODNB*).
6. *Ball*: John Ball.
7. *he be with you*: in letter 3796 from Henry Bence Jones, Tyndall wrote that 'John Ball (an Alpine man) was here when I arrived', in Chamouni (now Chamonix).

From Henry Bence Jones 16 July 1872 3793

84 Brook St | W[1] | July 16 1872

My dear friend

Many thanks for your letter from Chamouni[2] I am very glad to hear that you are well & I hope you will get into the best of health for your American campaign.[3]

We had a meeting of managers yesterday, but we did not get any further excepting that Mr Berkley told us the lock out was over[4] & was not likely to be renewed, so that any day sufficient men may return to work & as soon as they do our Institution will begin Meanwhile the season at the Clarendon[5] is passing & they will not trouble us. They might have put us to great trouble. They would not let the screen be put up. If we can only get the work finished by the time you return from America we shall do.

I wish we could give you all the arrangements that Jamin has in his Laboratory.[6] Is there nothing we could do; when you come back Berkley & Harwood will be here & you can have anything done that you like. It will not be too late at all

Pray think of this sometimes in your walks. could we not get drawings of Jamin's laboratory for Mr Harwood

I have written to Professor Tate[7] about Mr Dewar if he has a good subject for an experimental lecture it will be well to have him.[8] I think I told you in my last that I had got a course from Mr Freeman[9]

Foley[10] is making a very good bust of Sir Faraday.[11] It is far beyond Nobles bust[12] already.

I forgot to say that we adjourned the managers meeting indefinitely leaving every thing in the power of the building committee & when some one asked who formed the building committee I said Mr Berkley. So that in fact we have left the matter in his hands & unless he finds that we must get some non-struck-against-firm to take the contract I suppose we shall hear no more until November of the managers & as the lectures are all but settled for the year there will be nothing for them to do then.

I have no news to tell you L[ord]' Derby next Monday is going to bring Hookers case before the house of Lords.[13] The article for which you stood sponsor in the Contemporary[14] seems to have made some talk. Who was your friend I wish it had been some grand original discovery by some unknown man that you had brought before the public.

I hear of Dumas[15] making fresh original research sounds good. Some months since I urged Odling to take up fermentation as a work & to keep at it until he got something worth having I hear Miss *[1 word illeg]* will have some thousands a year some day

But my gossip must end. Goodbye
dear friend I wish you all health & happiness & Believe me
Ever yours most truly | H Bence Jones

RI MS JT/1/J/68

1. *84 Brook St | W*: residence in London.
2. *your letter from Chamouni*: letter missing; probably a reply to Bence Jones's letter of 2 July 1872, 3782. For Chamouni, see letter 3776, n. 12.
3. *your American campaign*: for Tyndall's American lecture tour, see Introduction, n. 8.
4. *lock out was over*: the renovation of the RI's laboratory (see letter 3776, n. 14) was awaiting the end of a labor strike. See letter 3782.
5. *Clarendon*: the Clarendon Hotel at 169 New Bond Street in London, directly behind the RI. See letter 3912, n. 12.
6. *Jamin has in his Laboratory*: on his way to Switzerland, Tyndall visited Jules Jamin in Paris (described in a letter to Thomas Hirst; see letter 3781).
7. *Professor Tate*: while LT's typescript reads 'Tate', we believe this is in error and Bence Jones was referring to Peter Guthrie Tait. Dewar had been a student at the University of Edinburgh, where Tait was a professor, and at this time they conducted research together.
8. *good subject for an experimental lecture it will be well to have him*: Dewar delivered his first Friday Evening Discourse at the RI on 7 March 1873, 'On the Temperature of the Sun and the Work of Sunlight' (*Roy. Inst. Proc.*, 7 (1873–75), p. 57).
9. *course from Mr Freeman*: Edward Augustus Freeman (1823–92), an English historian and architectural artist (*ODNB*), delivered a course of 'Six Lectures on the Comparative Political Institutions of Different Nations' on Saturdays from 18 January to 22 February 1873 (*Roy. Inst. Proc.*, 6 (1870–72), p. 560). He published these lectures in E. A. Freeman,

Comparative Politics: Six Lectures Read Before the Royal Institution in January and February, 1873, with The Unity of History; the Rede Lecture Read Before the University of Cambridge, May 29, 1872 (London: Macmillan and Co., 1873).

10. *Foley*: John Henry Foley (1818–74), a prominent British sculptor (*ODNB*). Foley would later design a full statue of Faraday, but when he died in 1874, it was completed in 1877 by his pupil Thomas Brock (1847–1922). The statue is displayed in London outside the Institution of Engineering and Technology, Savoy Place.
11. *Sir Faraday*: Michael Faraday was never given a title.
12. *Nobles bust*: Matthew Noble (1817–76), a British portrait sculptor, produced a bust of Michael Faraday in 1854, copies of which reside at the RI and the RS. The RI's copy was at first owned by the civil engineer James Walker (1781–1862), who gave it to the RI upon his death. The RS's copy was donated by Bence Jones. See *Royal Society Minutes of Council*, Printed, v.4 1870–77, meeting 20 March 1873, p.168' (*https://pictures.royalsociety.org/image-rs-13594*). A photograph of the bust graces the cover of the fourth volume of *The Correspondence of Michael Faraday*.
13. *Hookers case before the house of Lords*: the Memorial in support of Joseph Hooker in his dispute with Acton Ayrton (see letter 3755, esp. n. 1).
14. *article for which you stood sponsor in the Contemporary*: see letter 3779.
15. *Dumas*: Jean Baptiste André Dumas (1800–84), a French chemist and politician who worked with Haussmann on the modernization of Paris (*CDSB*).

To Joseph Henry 17 July 1872 3794

Bel Alp,[1] Brieg Switzerland | 17th. July 1872.

My dear Professor Henry.

You would render me still further your debtor if by the 10th. of August when I return to London you would kindly inform me when my lectures are to commence, and on what city they are to commence.[2]

The papers I see are full of the Presidential election[3] & I am quite willing, if you think well of it to postpone matters, until after the presidential election—I am equally willing if you prefer it to commence operations earlier—

I have taken my passage in the Russia which leaves England on the 28th of September.[4] But the organization of my work must depend on the period of its commencement—therefore I should feel much obliged to you if you would enlighten me on this head.

Pray write to me to London | Yours ever faithfully | John Tyndall

John Tyndall to Joseph Henry, 17 July 1872, mssRH 3967, box 43, William Jones Rhees papers, The Huntington Library, San Marino, California

1. *Bel Alp*: see letter 3771, n. 5.
2. *when my lectures are to commence, and on what city they are to commence*: for Tyndall's American lecture tour in general, see Introduction, n. 8. Tyndall's lecture tour commenced in Boston on 15 October 1872 at the Lowell Institute (*Ascent of John Tyndall*, p. 305).
3. *Presidential election*: the 22nd Presidential election of the United States occurred on 5 November 1872, resulting in the incumbent Ulysses S. Grant defeating Horace Greeley to gain a second term.
4. *my passage in the Russia which leaves England on the 28th of September*: Tyndall previously informed Henry that he had arranged his travel across the Atlantic (see letter 3771).

To Thomas Archer Hirst 18 July 1872 3795

18th July 1872

My dear Tom.

The remark about the cabin[1] renders me very anxious. I shall have a good deal of head work to do on board,[2] & if I am not well placed I cannot do it. In short the success of my visit to America depends upon this very point.

My young assistants[3] can if necessary rough it a little; but I am not up to that.

I am willing to pay any price for a good cabin. If necessary I will change the ship. But surely Mr. Burns can arrange for the 28th. I have given him a long notice

Would you kindly send me
Lord Derby's letter[4]
Vieweg's letter[5]
Henry Tyndalls letter?[6]

RI MS JT/1/T/691
RI MS JT/1/HTYP/597

1. *cabin*: Tyndall arranged passage on the SS *Russia* from Liverpool to New York, and left on 28 September 1872 (see Introduction, n. 8).
2. *head work to do on board*: presumably the preparation of his lectures on light for America.
3. *My young assistants*: John Cottrell and George Miller.
4. *Lord Derby's letter*: possibly letter 3789 from Edward Stanley, dated 12 July 1872.
5. *Vieweg's letter*: probably letter 3790 from Heinrich Vieweg, dated 13 July 1872.
6. *Henry Tyndalls letter*: letter missing. Tyndall had at least two relatives named Henry Tyndall. In his journal entry for 4 February 1866, he wrote: 'About two years ago I received a letter from a Mr Henry Tyndall who wished me to help him to a situation On enquiry I found that he was the son of a half brother of my father's I could not help him by asking for an improvement of his position as telegraphic operator, and told him frankly that I was

too proud to ask favours' (RI MS JT/2/13c/1303). Tyndall continued the journal entry, however, by describing his efforts to help Henry to improve his position, until he received a letter from a woman in Ireland stating that Henry had broken their engagement and taken her money. Henry admitted breaking the promise of marriage but denied taking her money. The woman continued to write and hired a lawyer, but Tyndall ignored her letters (RI MS JT/2/13c/1303–5). A Henry Tyndall is also mentioned in two letters in this volume, but it does not appear the two are the same person (see letters 3963 and 4004).

To Henry Bence Jones 19 July 1872 3796

19th. July. 1872.

My dear Friend

Previous to quitting Chamouni[1] I wrote you a longish letter[2] on such paper as I then possessed. Soon afterwards I crossed the Col de Balme[3] to Martigny[4] thence by rail to Sierre,[5] and thence by diligence[6] to Brieg[7] where I spent the night, and the following day, mounted to this my perch upon the hills.[8] Since I came here I have had very little exercise. In fact this American work[9] is slow to the last degree. It is so difficult to pick out of the mass of science a portion which can be at once popularly and thoroughly treated. If I fill my time with illustrations the whole appears superficial, and if I take up too much time in exposition the whole will appear dry. So that I am much exercised by these lectures and have only begun to make a little progress.

John Ball[10] (an Alpine man) was here when I arrived: and we have had a floating population, some known to me others unknown ever since. Bradley[11] and his wife[12] are here (now Master of University College Oxford: formerly Head Master of Marlborough). Various other clergymen are and have been here. In fact in relation to their total number the fraction of clergymen who visit the Alps is a high one.[13] Perhaps they come like myself to repair the hurts of the brain.

I dare say you are now pulling to pieces that sacred old edifice in Albemarle Street.[14] I saw by the Swiss Times,[15] a very scanty rag, that the masons and the masters had come to terms,[16] and this I suppose was all you wanted. I wrote to Smith[17] asking him to send me any paper that contained anything interesting. He knows I should have felt interest in the debate about Governor Eyre,[18] but he has not sent me that or anything else. I have had the Times, Daily News and Pall Mall, about Ayrton and Hooker:[19] but the Globe contains an article[20] I should like to see, and probably other papers also.

I learn also from the Swiss Times[21] that the good orthodox people are hitting me hard apropos of that letter in the Contemporary.[22] When the time arrives I shall have my say to the good orthodox people. At present my hands are full. On this head I have had the Spectator[23] and that only.

I hope you continue well. I think my best plan will be to halt at Folkstone[24] and have a talk with you. This will economise time, and as the crisis approaches I am getting into a condition of half anger and half fear regarding my American work. I did not think it would bother me so much; and I am enraged with myself that it should do so.

But I shall strike into it by and by, and then all the difficulty will disappear as a trifle.

I rise early every morning and have a sponge in a local stream. Then I work for about 5 hours, permitting myself usually a scamper afterwards. The weather however for the most part has been cloudy and clammy.

My health is good. Were it not for my confinement within doors it would be robust.

I hope Hooker's affair is thriving. I do not know whether I sent you a copy of that Memorial[25] before I came away. But as I send you everything it is possible that I sent you this.

19th Noon

Just had your letter,[26] many thanks. How you hunger for y[ou]r discoveries. They will not come a bit the quicker!

Yours ever and always | John Tyndall

RI MS JT/1/TYP/3/807
LT Typescript Only

---

1. *Chamouni*: see letter 3776, n. 12.
2. *I wrote you a longish letter*: letter missing.
3. *Col de Balme*: see letter 3783, n. 28.
4. *Martigny*: see letter 3785, n. 3.
5. *Sierre*: a city in southwestern Switzerland (Siders in German).
6. *diligence*: a public stage-coach (*OED*).
7. *Brieg*: see letter 3785, n. 4.
8. *my perch upon the hills*: the hotel at Bel Alp (see letter 3771, n. 5).
9. *American work*: his lectures on light for his American lecture tour (see Introduction, n. 8).
10. *John Ball*: John Ball.
11. *Bradley*: George Granville Bradley (1821–1903), an Anglican clergyman and schoolmaster, who was headmaster of Marlborough College (1861–70), Master of University College, Oxford (1870–81), and later Dean of Westminster (1881–1902) (*ODNB*).
12. *his wife*: Marian Jane Bradley (née Philpot, 1831–1910), married in 1849 ('Marriages', *Cambridge Independent Press*, 22 December 1849, p. 3).
13. *fraction of clergymen who visit the Alps is a high one*: Fergus Fleming has noted that given the spiritual or religious aspect of visiting the Alps, it 'was not surprising that of the 148

members who joined Britain's Alpine Club in 1857, thirty-four were clergymen' (F. Fleming, 'The Alps and the Imagination', in *Ambio: Special Report Number 13. The Royal Colloquium: Mountain Areas: A Global* Resource (2004), pp. 51–5).

14. *pulling to pieces that sacred old edifice in Albemarle Street*: the renovation of the RI's laboratory (see letter 3776, n. 14).
15. *Swiss Times*: not identified.
16. *masons and the masters had come to terms*: see letter 3782.
17. *wrote to Smith*: letter missing. Smith is presumably a porter at the RI (in letter 3797 to Heinrich Debus, Tyndall wrote: 'The Hall porter of the R.I. shall get a wigging on my return. He used to send me every thing likely to interest me. This year he has sent me nothing').
18. *debate about Governor Eyre*: a reference to the July 1872 Parliamentary debate over whether or not the British government should pay the legal expenses—they did—of Edward John Eyre (1815–1901), an English colonial governor, controversially known for his violent repression of the 1865 Morant Bay rebellion, while he served as governor of Jamaica (1862–5) (*ODNB*; R. W. Kostal, *A Jurisprudence of Power: Victorian Empire and the Rule of Law* (Oxford: Oxford University Press, 2005), p. 453; B. Semmel, *Jamaican Blood and Victorian Conscience: The Governor Eyre Controversy* (Boston, Houghton Mifflin, 1963), p. 178). In the years following the rebellion, London society was divided over whether Eyre had acted justly or criminally, resulting in the founding of the Jamaica Committee which sought to prosecute Eyre, and the Eyre Defence and Aid Committee which sought to defend Eyre's actions, of which Tyndall was a vocal member. In the end, Eyre would be unsuccessfully tried for private criminal charges brought by the Jamaica Committee. For Tyndall's support of Eyre, which included a speech and letters to the press, see *Tyndall Correspondence*, vol. 9; and *Tyndall Correspondence*, vol. 10. Joseph Hooker joined Tyndall in support of Eyre, while Charles Darwin and Thomas Huxley supported the effort to prosecute Eyre for his actions.
19. *Times, Daily News and Pall Mall, about Ayrton and Hooker*: 'Mr. Ayrton and Dr. Hooker', *Times*, 15 July 1872, p. 12; *Daily News* article not identified; the *PMG* article is probably either 'Dr. Hooker and Mr. Ayrton', 16 July 1872, p. 3, or 'Mr. Ayrton', 17 July 1872, p. 3.
20. *Globe contains an article*: possibly 'Mr. Ayrton's Last Exploit', *The Globe*, 16 July 1872, p. 6, but this only quotes from the *Daily Telegraph*.
21. *Swiss Times*: not identified.
22. *that letter in the Contemporary*: see letter 3779, esp. n. 1 and n. 2.
23. *Spectator*: see letter 3786, n. 12.
24. *Folkestone*: a town in Kent, England, on the English Channel, where Bence Jones had a home at 5 Albion Villas since 1851.
25. *Memorial*: the Memorial in support of Joseph Hooker in his dispute with Acton Ayrton (see letter 3755, esp. n. 1).
26. *your letter*: letter 3793.

To Heinrich Debus 19 July [1872][1] 3797

19th July.

My dear Heins

Send me the Spectator[2] like a good fellow. I never was so shabbily treated as regards papers in my life.

The Hall porter of the R.I.[3] shall get a wigging on my return. He used to send me every thing likely to interest me. This year he has sent me nothing.

I have had the Times, the Daily News & the Pall Mall about Hooker[4]—nothing more. and I know there is a great deal more.

I return the form signed:

I hear from the Swiss Times[5] that I have been attacked on all hands about the article in the Cotemporary;[6] Tom[7] has sent me one paper—the Spectator.[8]

I want a little light reading as a relief to my really heavy work,[9] and would thank you to send me some of the papers which treat of Hooker.

Yours ever | John Tyndall

I hope you will be in London when I return.

Here[10] the weather has been cold & clammy.

RI MS JT/1/T/316
RI MS JT/5/11/1371

---

1. *[1872]*: the year is given by content (some items of discussion are the same as in letter 3796).
2. *Spectator*: we are not certain if Tyndall was referring to the same issue of *The Spectator* referred to later in this letter (see n. 8 below).
3. *Hall porter of the R.I.*: 'Smith', according to letter 3796 to Henry Bence Jones.
4. *Times, the Daily News & the Pall Mall about Hooker*: see letter 3796, n. 19.
5. *Swiss Times*: not identified.
6. *article in the Cotemporary*: 'Contemporary' misspelled by Tyndall; see letter 3779, esp. n. 1 and n. 2.
7. *Tom*: Thomas Hirst.
8. *Spectator*: see letter 3786, n. 12.
9. *really heavy work*: his lectures on light for his American lecture tour (see Introduction, n. 8).
10. *Here*: at Bel Alp (see letter 3771, n. 5).

From Joseph Dalton Hooker 19 July 1872 3798

Kew | July 19th 1872.

Dear Tyndall

Thanks for yours of Monday.[1] I have no time except to tell you, (1) The Royal Horticultural Society have sent me a resolution[2] which I have forwarded to Spottiswoode to be sent to Lord Derby, "That the undivided control of the heating apparatus should rest with the Director, or whoever is responsible for the cultivation." (2) Lord Derby's movement[3] takes place on Monday, Lubbock's on the 30th. (3) Reginald Palgrave[4] writes me last night:—"I still have good hopes, on good authority, that the Gardens will be took away from under Mr Ayrton,[5] and that the Monday affair (Lord Derby's action) will not take place or be needed"

I doubt Derby letting them off so easy!

(4) The President (Christison)[6] and the Council of the R[oyal]. S[ociety]. Edinburgh have memorialised Gladstone in the sense of the address,[7] and sent copies to Lubbock and Lord Derby.

The papers continue to make a <u>prodigious</u> clatter.

I saw Lord Russell (also the Rosebud[8]) on Monday. He told me, with glee, that he had written to Lord Derby asking him to let him know when he would bring the matter forward, that he might attend. I have also seen Lord J. Manners[9] who entirely approved of the resolution I had proposed to Lubbock.

Also Lord Houghton and Sir P. Egerton[10] who promise to support me. I hear not a word of the course of action that the ministers will take. The Treasury asked me to send them a copy of the letter they had lost of Sept. 22d. The only letter I wrote of that date was to <u>Ayrton</u> (not the Treasury) urging the <u>selection</u> previous to or subsequent to competition of Herbarium assistants. Of course I sent it at once. The T[reasury]. sending to me for this is significant, and I cannot but think they are preparing for Ayrton's fall—not from the ministry, but from Kew.

I have not repeated the contents of Palgrave's letter to any-one: it is far too vague and favourable to be acted upon.

The papers are printed, I am told, a huge pile—including all the correspondence about the Clerk, and with the Treasury—<u>but not with the Prime Minister</u>! It however includes the address to him,[11] where the gist of that correspondence is quoted (thanks to your wisdom and foresight!)

I went with the Spottiswoodes to Combe Bank[12] on Wednesday, it was such a relief. It is a glorious place and you will enjoy it beyond measure and stint. You will revel in it.

They have made Helps a K.C.B,[13] I am glad to say, because he likes it, and because it disabuses my mind of an impression that he was looked on with disfavour for the uncompromising part he has taken in my favour. Poor Mrs H.[14] does not like it at all!

I am delighted to find that you write so heartily of yourself and work.

Ever your aff[ectiona]te | J. D. Hooker.

Ball[15] will be back to-night.

RI MS JT/1/TYP/8/2685–6
LT Typescript Only

---

1. *yours of Monday*: letter missing.
2. *The Royal Horticultural Society have sent me a resolution*: see letter 3788, n. 10.
3. *movement*: efforts to put forward in the House of Lords of Parliament the Memorial in support of Joseph Hooker in his dispute with Acton Ayrton (see letter 3755, esp. n. 1).
4. *Reginald Palgrave*: Reginald Francis Douce Palgrave (1829–1904), a British civil servant who was a clerk in the House of Commons from 1853–86, and Clerk of the House of Commons from 1886–1900 (*ODNB*). He was a cousin of Hooker's; Palgrave's mother Elizabeth Turner (1799–1852) was daughter to Dawson Turner (1775–1858), who was the father-in-law of Joseph's father William Jackson Hooker.
5. *Mr Ayrton*: Acton Ayrton.
6. *President (Christison)*: while the typescript showed 'Christiser', we believe this is in error and Hooker wrote 'Christison', for Robert Christison (1797–1882), a Scottish toxicologist and physician who served as President of the Royal College of Physicians of Edinburgh (1838–40 and 1846–8), President of the Royal Society of Edinburgh (1868–73); President of the British Medical Association (1875) (*ODNB*).
7. *memorialised Gladstone in the sense of the address*: not identified.
8. *Rosebud*: a pretty young woman (*OED*); possibly referring to John Russell's daughter Mary Agatha Russell (1853–1933), then aged about 18 or 19.
9. *Lord J. Manners*: John James Robert Manners (see letter 3733, n. 6).
10. *Sir P. Egerton*: Sir Philip de Malpas Grey Egerton (1806–81), a British paleontologist and politician (*ODNB*).
11. *the address to him*: the Memorial (see n. 3) was addressed to William Gladstone.
12. *Combe Bank*: or Coombe Bank, William Spottiswoode's home in Sevenoaks from 1864–83. His son Hugh (see letter 3855, n. 2) tenanted the estate until he sold it in 1906.
13. *made Helps a K.C.B*: Arthur Helps was made Knight Commander of the Order of the Bath on 18 July 1872 (*Times*, 20 July 1872, p. 7).
14. *Mrs H.*: Bessy Helps (née Fuller), married in 1836.
15. *Ball*: John Ball.

To Heinrich Debus [21 July][1] 1872 3799

Sunday 1872

My dear Heins

Just one word (as the post is starting) to say that the Hall porter of the Royal Institution has sent me in an envelope the articles[2] in most of the papers. I have written to him for those in the Globe and Spectator.[3]

I am very well, but much confined: when however I get the back of my work broken I will give myself a clear week's scampering.

Yours ever | J. Tyndall

RI MS JT/5/11

---

1. *[21 July]:* date is given from Tyndall's travels in Switzerland, and his earlier correspondence with Debus (see letter 3797).
2. *the articles:* probably a reference to the press response to the prayer-gauge debate. The letter Tyndall sent to the *Contemporary Review* (see letter 3779, n. 1) generated notices and commentaries in many British papers.
3. *Globe and Spectator:* 'Magazines for July', *The Globe*, 2 July 1872, p. 1; 'The Proposed Prayer Gauge', *The Spectator*, 45, no. 2297 (6 July 1872), pp. 846–7.

To the Editor of *Nature*[1] 22 July [1872][2] 3800

[Atmospheric Effect]

AN atmospheric effect, which is sometimes observed in England, displayed itself here in great beauty yesterday. The western sun had been cut off from us by an intervening ridge, while the upper atmosphere was still filled with his light. There was a good deal of opalescent haze in the atmosphere, which, had the sun shone upon it uniformly, would have presented a tolerably uniform hue. But besides the haze, small detached clouds floated in the air, and behind each of them was a sheaf of shadow, drawn through the haze. The density of these shadows varied with that of the clouds which produced them, nor was the density uniform for all parts of the transverse section of the same shadow. The parallel bars of graduated shade thus produced converged, through an effect of perspective, to a point *in the east*, exactly as if the sun were going to rise there. The display of the convergent glory was strikingly beautiful.

The same effect showed itself at Oran during one of the evenings spent there by the Eclipse Expedition.[3] I have seen it two or three times in England[4] (always, I think, near the coast), the last occasion being in company with Mr. Hirst at Margate.[5]

Faraday, if I remember aright, has described this effect.[6] There was also, I believe, a question asked regarding it some months ago in NATURE.[7] This brief account may interest the questioner.

Bel Alp,[8] Jul 22 | JOHN TYNDALL

Nature, 6 (1 August 1872), p. 260

1. *Editor of Nature*: Norman Lockyer.
2. *[1872]*: the year is given by the publication date of this letter in *Nature*.
3. *at Oran during one of the evenings spent there by the Eclipse Expedition*: Tyndall traveled to Oran, Algeria, to view the 22 December 1870 total solar eclipse, but poor weather kept the expedition from being able to view the eclipse (see 'Voyage to Algeria to Observe the Eclipse', in *Hours of Exercise in the Alps*, pp. 429–73); he published this again in *Fragments of Science for Unscientific People: A Series of Detached Essays, Lectures, and Reviews*, 5th edn (London: Longmans, Green, and Co., 1876), pp. 186–217.
4. *seen it two or three times in England*: Tyndall wrote in his journal on 22 July 1855: 'A drive on the omnibus to Chelsea and a walk home—noticed the shadows of the clouds on clouds. I don't know whether it has been noticed before' (RI MS JT/2/13c/779). Any other descriptions of this in his letters or journals have not been identified.
5. *at Margate*: Tyndall traveled to the seaside town of Margate in Kent, England, with Thomas Hirst in May 1869 (see letter 2996 in *Tyndall Correspondence*, vol. 11).
6. *described this effect*: M. Faraday, 'On a Peculiar Perspective Appearance of Aërial Light and Shadow', *Quarterly Journal of Science, Literature, and the Arts*, 22 (1827), pp. 81–3.
7. *question asked regarding it some months ago in NATURE*: possibly 'The Total Eclipse in Java', *Nature*, 6 (27 June 1872), pp. 160–1. This letter to the editor from the Dutch astronomer Jean Abraham Chrétien Oudemans (1827–1906) discusses shadow effects during a total solar eclipse.
8. *Bel Alp*: see letter 3771, n. 5.

From Henry Bence Jones — 22 July [1872][1] — 3801

84 Brook St W[2] | July 22

My dear friend

I have just got your letter[3] & wish the lectures for the Yankees[4] were easier. I shall be glad when all is over & you are back again

Tomorrow I expect the work of destruction[5] will begin & I expect your prayers have hitherto saved "the sacred old Edifice" as you call it.[6] I have just got a note from Mr Harwood to say he has given orders & I have had our own people taking down the cases and the strong chest to day & tomorrow I hope work will begin & be finished long before you come back from America. It shall not be my fault if it is not, but the expenditure will be great.

I think Max Muller[7] will give us a short course on Darwins views of language.[8] He gave a lecture on it I think, at Oxford.[9] I have heard from Prof Taits friend[10] to day. He makes no good proposal but I have written to him[11]

Hooker I saw on Saturday at Kew. Wild, quite wild with Ayrton.[12] To night L[or][d] Derby will make his speech.[13] I fear Lubbock will hardly get a chance. However some oil must be poured on the stormy waves before the adjournment. There is a rumor that Sir H. Thompson wrote that prayer test that you fathered.[14] I wish the sawbones or crushstones had burnt his own fingers. I heard a strange story that he & Paget will not speak or meet in consultation. How big guns fight!!

I expect to go to Folkestone[15] the end of next week the 3[d] of August & if you pass that way. You shall give me a line to Jamin as if I keep well I might go to Paris at the end of September. If I keep well!? I have great doubts but I have not gone wrong yet, but I have to take some preventative stuff. I will see about the papers for you at the Institution tomorrow. There has been nothing lately about the Prayer test except in the Press & the Hooker row will begin tomorrow only. You sent me the protest[16] & the Horticultural & Botanical Societies have also drawn up protests[17] I believe & more will follow[18]

If any lecturer comes in your way get one a Friday. I have written to Ruskin[19] to give the first.[20] He is in Italy. Max Muller at Baden Baden—Miss Emily Faithful!![21] is going with you to lecture in America. What say you to that oh Batchelor. If she should go in the same steamer[22] you may have a chance. I see nothing of Odling & hear nothing worth reporting to you. The Institution seems about deserted so it is a good time to begin. Would that all was finished. However we must like no stoppage may occur.

I am very glad to hear that you are "robust" it sounds oaky. I hope an ever green in non verdant sense

Good bye Dear friend | Ever yours most truly | H Bence Jones

RI MS JT/1/J/81

---

1. *[1872]*: the year is given by reference to Tyndall's upcoming lecture tour in the United States (see n. 4).
2. *84 Brook St W*: residence in London.
3. *your letter*: letter 3796.
4. *lectures for the Yankees*: his lectures on light for his American lecture tour (see Introduction, n. 8).
5. *work of destruction*: the renovation of the RI's laboratory (see letter 3776, n. 14).
6. *"the sacred old Edifice" as you call it*: see letter 3796.

7. *Max Muller*: Friedrich Max Müller (1823–1900), a German-born philologist and Sanskritist. In 1841 he entered the University of Leipzig to study classical languages and Sanskrit. Two years later, he began his doctorate there, and wrote his first book, a German translation of Hitopadeśa. Müller's greatest work of his lifetime was his *Rig Veda*, which he began while studying in Paris (1844) and finished after he moved to Oxford (1846), where the book was also published. A gifted pianist with a gregarious personality, he was a popular figure in Oxford society. He was appointed Taylorian professor (1851), and curator of the Bodleian Library (1856–63, 1881–94). In 1859 he married Georgina Adelaide (1834/5–1916) and they had four children, two of whom predeceased their parents. From 1869 onwards, Müller devoted most of his time to publishing and lecturing. He became a good friend of Tyndall's and his lectures at the RI on the science of language were especially popular. Müller's greatest legacy was his contributions to the fields of Vedic studies and comparative philology (*NDB*; *ODNB*).
8. *short course on Darwins views of language*: while Müller's course of three lectures at the RI, 'On Darwin's Philosophy of Language', was originally scheduled for Saturdays, 1, 8, and 15 March 1873 (*Roy. Inst. Proc.*, 6 (1870–72), p. 560), he actually delivered the lectures on 22 and 29 March, and 5 April 1873. These were published in *Fraser's Magazine*, 7 and 8 (1873), pp. 525–41, 659–78, 1–24; as well as F. M. Müller, *Lectures on Mr. Darwin's Philosophy of Language: Delivered at the Royal Institution in March and April 1873* (London: Longmans, Green, and Co., 1873).
9. *lecture on it I think, at Oxford*: not identified.
10. *friend*: James Dewar.
11. *no good proposal but I have written to him*: for a Friday Evening Discourse at the RI (see letters 3787 and 3793). Dewar delivered his first Friday Evening Discourse at the RI on 7 March 1873, 'On the Temperature of the Sun and the Work of Sunlight' (*Roy. Inst. Proc.*, 7 (1873–75), p. 57).
12. *Wild, quite wild with Ayrton*: referring to Hooker's dispute with Acton Ayrton (see letter 3755, esp. n. 1).
13. *make his speech*: see letter 3805, n. 16.
14. *Sir H. Thompson wrote that prayer test that you fathered*: see letter 3779, esp. n. 1 and n. 2.
15. *Folkestone*: see letter 3796, n. 24.
16. *protest*: the Memorial in support of Joseph Hooker in his dispute with Acton Ayrton (see letter 3755, esp. n. 1).
17. *Horticultural & Botanical Societies have also drawn up protests*: see letter 3788, n. 10.
18. *more will follow*: other societies that supported Hooker were the Meteorological Society, the Royal Zoological Society, and the Royal Society of Edinburgh (see letter 3798), as well as the West India Committee.
19. *Ruskin*: John Ruskin (1819–1900), a British art and social critic. While one of the leading art critics of his generation Ruskin also wrote and lectured on many subjects including geology, literature, botany, and economics (*ODNB*).
20. *the first*: the first Friday Evening Discourse of the upcoming season at the RI was delivered not by Ruskin, but by William Spottiswoode (see letter 3776, n. 14).

21. *Miss Emily Faithful*: Emily Faithfull (1835–95), a British women's rights activist and publisher (*ODNB*). The first of her three trips to America was in 1872, and she later published *Three Visits to America* (Edinburgh: David Douglas, 1884). She noted about her visit that a number of other British persons of note had made their way across the Atlantic: 'but that very season (1872–73) witnessed the *début* of Mr. Tyndall, Mr. Froude, Professor Huxley, Edmund Yates, George MacDonald, and several other Britons more or less distinguished' (p. 2).
22. *same steamer*: Tyndall traveled from Liverpool to New York on the SS *Russia*. Bence Jones later wrote with the impression that Faithfull was on the same ship, as was Anthony Froude (see letter 3880).

From Joseph Dalton Hooker — 22 July 1872 — 3802

Kew | 22d July 1872.

My dear Tyndall

I can hardly muster a dozen scratches of the pen for you, so heavy has been the strain of the last week. On Friday evening I received a letter from the Board[1] enclosing one from a fellow of known bad character[2] charging myself and the Curator[3] with maladministration, jobbery[4] and fraud; the charges though of the vulgarest and most ignorant description and known to the Board to be false, were forwarded without further comment than that I was to make any observations I pleased on them.

In the afternoon came another of a similar nature, this time from a poor half-witted soul,[5] who had been a foreman here, left under a delusion, and wrote to me only 6 weeks ago imploring my forgiveness, asking to be taken back, and expressing his confidence in me. This man has been in correspondence with Ayrton[6] <u>for weeks</u>. Then yesterday morning came a third letter, requesting me to send the names of every individual who had had plants from Kew since Jan[uar]y 1871! with no explanation of the demand.

The first two letters I am answering and am now about to take the answers to Lubbock. The third I shall insist upon an explanation of.

By the same post I receive 7 letters[7] from the Foreign Office, on the vine disease in France and Portugal,[8] referred to me to report on by Lord Granville <u>in his own hand</u>. So I have written him this morning, thanking him for his confidence; informing him of the three letters, with the last of which his despatches came, called his attention to the fact that this was the treatment I had been receiving for now 3 years—that I had appealed to the Government of which he is a distinguished member for relief, not redress, hitherto in vain—that hearing of this, a comparative stranger and member of the opposition,[9] had volunteered to a mutual friend[10] to bring my case before the Lords,[11] <u>this</u>

day, and so ended. I have just sent it by a messenger to his house with another to Mr Harwood[12] (the official secretary of the F[oreign]. O[ffice].) to call his attention to it, and a third to Ball,[13] to see Lord Enfield[14] (the Parliamentary Sec[retar]y, who forwarded Lord G[ranville]'s despatches) and get him to move Lord G[ranville]. I hear that Lord Granville is to answer Lord Derby tonight, and if only I can get that letter stuffed down his throat before-hand, I don't think he will be very fluent!

9.0 a.m. Post. A letter from Lubbock asking me to see him. The D[ean]. of St Albans[15] is I hear to reply to L[or]d Derby tonight; he is of course primed by Ayrton. I greatly fear that the whole affair will again be put off. I shall post this in town after I see Lubbock. Ball called yesterday.

2.0 p.m. I have seen Lubbock. His only news is that the papers are not yet presented. He fancies that L[or]d Derby's motion will be put off in consequence tonight. There is no notice of it in the Times!

Ever yours | J. D. Hooker

RI MS JT/1/TYP/8/2687
LT Typescript Only

1. *Board*: Board of Works.
2. *one from a fellow of known bad character*: probably referring to Richard Owen; see letter 3807, n. 3.
3. *Curator*: John Smith.
4. *jobbery*: the practice of turning a public office or position of trust to personal gain or political advantage (*OED*).
5. *poor half-witted soul*: not identified.
6. *Ayrton*: Acton Ayrton.
7. *7 letters*: this may be a mis-transcription in LT's typescript of '3 letters', as Hooker refers in the next sentence to 'the three letters', not 'seven'.
8. *vine disease in France and Portugal*: the invasive insect Grape phylloxera (*Daktulosphaira vitifoliae*), which feeds on the roots and leaves of grapevines, devastated most of the vineyards in Europe in the latter half of the nineteenth century. See G. D. Gale, *Dying on the Vine: How Phylloxera Transformed Wine* (Berkeley: University of California Press, 2011).
9. *comparative stranger and member of the opposition*: Edward Stanley (Lord Derby).
10. *mutual friend*: either Tyndall himself (in writing the Memorial in support of Hooker) or John Lubbock (in presenting the Memorial to Parliament).
11. *bring my case before the Lords*: efforts to put forward in the House of Lords of Parliament the Memorial in support of Joseph Hooker in his dispute with Acton Ayrton (see letter 3755, esp. n. 1).
12. *Mr Harwood*: not identified.
13. *Ball*: John Ball.
14. *Lord Enfield*: George Henry Charles Byng (1830–98), a British Liberal politician (*ODNB*).
15. *D[ean]. of St Albans*: Walter John Lawrance (1840–1914), Rector of St. Albans; he later became Dean in 1900 ('Death of the Dean of St. Albans', *Times*, 13 August 1914, p. 8).

From Joseph Dalton Hooker 23 July 1872 3803

Kew, July 23d 1872.

Dear Tyndall

I am tremendously hard at work, incensing Members of P[arliament]., sending letters detailing Ayrton's[1] last acts, and getting up estimates etc. for the row. Playfair and Ball[2] inform me that Ayrton will attack me on the excessive estimates and the contracts for the Heating,[3] and Playfair is coming tomorrow to examine the old accounts, for 3 years. I am delighted at this, as I can show such a tale of work and such a saving as will I hope charm the House into a vote according to their conscience![4]

Smith[5] had a long talk with Lord Derby yesterday, who is furious at these new accusations and delays, & for whom I am having copies made. Lord D[erby]. comes on for certain next Monday.

Meanwhile R. Palgrave[6] writes word:—"Papers will be issued on Wednesday containing some settlement"

Lubbock has seen Gladstone. I quote his letter.

22 July.

"My dear Hooker,

I have had some talk with Gladstone this evening. He says there is a Treasury minute just issued which he hopes may compose matters, but he added that the publication of the Times article[7] was a serious matter and that Ayrton felt much aggrieved about the "evasions" misrepresentation etc., and would probably assume the "offensive" When I see you again I must go carefully into these points.

Yours very sincerely | John Lubbock."

"4 Queensberry Place | S. Kensington | 22 July.

"My dear Hooker,

I propose to call on you on Wednesday about 11.0 clock if you are disengaged. Would you kindly have two or three estimates <s>ay from 1869 ready? I want to see inter alia[8] what Ayrton can say for himself on the ground of economy. Since the heating apparatus was taken in hand, has there been any saving

a. in coals

b. in labourers.

I am anxious to find out the weak points of your case if there are any, as possibly I may speak after Ayrton.

Yours truly | Lyon Playfair."

Hirst was here last night and I spoke to him about the newspapers. I really have not had time to read them, but Harriet[9] sends you a few slips by this post.

I have my confounded evidence on the Science Commission[10] to correct for tomorrow, a very difficult job, as between ourselves <I> only half understood what the Comm[issione]rs were driving at! Lockyer won't give me beyond tomorrow, which is not considerate.

Ball brought Monsell[11] out today who frankly said after going round, that he would not vote with the Ministry!

I am getting very worn with the work and alternate excitement and disappointment.

Ever aff[ectionate]ly yours | J. D. Hooker.

RI MS JT/1/TYP/8/2688–9
LT Typescript Only

---

1. *Ayrton's*: Acton Ayrton.
2. *Ball*: John Ball.
3. *for the Heating*: of Kew's conservatories (see letter 3755).
4. *charm the House into a vote according to their conscience!*: efforts to put forward in the House of Lords of Parliament the Memorial in support of Joseph Hooker in his dispute with Acton Ayrton (see letter 3755, esp. n. 1).
5. *Smith*: John Smith.
6. *R. Palgrave*: Reginald Palgrave (see letter 3798, n. 4).
7. *Times article*: probably *Times*, 8 July 1872, p. 11 (see letter 3788, n. 4).
8. *inter alia*: amongst other things (Latin; *OED*).
9. *Harriet*: probably Hooker's daughter Harriet Anne Hooker (1854–1945). In 1877 she married the British botanist William Turner Thiselton-Dyer (1843–1928), who after Hooker retired became the third director of the Royal Botanic Gardens, Kew (1885–1905). Note that Hooker's wife Frances Hooker had Harriet as her middle name.
10. *Science Commission*: see letter 3747, n. 6.
11. *Monsell*: William Monsell (1812–94), an Anglo-Irish Liberal politician (*ODNB*).

From Joseph Dalton Hooker — 29 July 1872 — 3804

Kew, | July 27th 1872.

Dear Tyndall

We are hard at it, deeper than ever in the mire. The papers are promised for today: meanwhile I have just received officially through the First Comm[issione]r,[1] the Treasury minute:[2] it defines my position with considerable skill and great liberality; rebukes Ayrton roundly, and states that my father[3] and I are "entitled to the gratitude of the country" for the Botanical Department: which it segregates from the Horticultural etc., and hence it takes from us any claim to credit for the organisation and administration of the whole affair.

It bears evidence of being based on information supplied by Ayrton, on the nature of the establishment and its departments, and relying on his statement separates the arboretum and its collections (commonly called the Pleasure Ground) from the rest of the Garden, and places it under Ayrton's uncontrolled authority, subject to communication with me or through me to Smith!![4] I have written a memorandum to Lord Derby (who certainly speaks to-morrow), showing that this is not to be suffered. The pleasure grounds of the old Palace were given by the Queen[5] to the public, five years after the Botanical Garden, on the suggestion of my father, at Prince Albert's[6] intercession to establish a <V>inetum and Arboretum worthy of the country. He laid it out and it contains, in classified series, miles of varied trees and shrubs from all temperate climates of the globe. Every successive First Comm[issione]r has fostered this idea and this use of the ground. It contains the great temperate House and nurseries for raising all hardy and half hardy plants, and I would as lief[7] give up direct control over the garden as over this. A more barefaced fraud was never practised by Minister on the Treasury and Parliament, and if Lord Derby manages it adroitly, I do not see how Ayrton can remain in office. I saw Lubbock and Cowper Temple[8] about it yesterday who took my view of it. Lord Derby was out when I called and yesterday I wrote begging him to give me a meeting by telegraph, but (4 p.m.) none has come. I shall send my memoranda to him to Holwood[9] by messenger tomorrow. I would go myself, but Huxley and Lubbock come here to talk over matters. Such a minute, flattering as it is, would compell my retirement. You shall have a copy of my memoranda to Lord Derby at once and of the Minute too.

I am tremendously busy. Poor Smith is broken down with anxiety and cant sign his name.

The Times article today written by <L>owe,[10] as he confessed, or rather boasted, to Lady D. Neville[11] (who told my wife)[12] is a gloss that must be shown up.

The Papers are out, a medley, without Gladstone's to me, or mine to Gladstone, as I expected. Ayrton does not appear to have contravened one of my statements, but has glozed them and dosed them I shall be at him again if Huxley and Co. do not prevent it.

Playfair spent several hours here going over the papers, estimates, and gardens, &c. He proposes to speak after Ayrton, and is staunch. Lord Derby writes:—"The more I look into the matter the more strongly I feel as to the injustice that has been done you"

I am overwrought with visitors and correspondence.

The Physical, Biological, and Chemical Lectureships of Jermyn St are at last transferred to S. Kensington[13] to Huxley's delight.

I must close. | Despatches will be sent you on Monday | Yours ever | J. D. Hooker.

RI MS JT/1/TYP/8/2690–1
LT Typescript Only

---

1. *First Comm[issione]r*: Acton Ayrton.
2. *Treasury minute*: an administrative regulation for any department under the Treasury (*OED*). A Treasury minute of 24 July 1872, heavily in favor of Hooker, concluded: 'My Lords see no reason why under these conditions there should be any serious difficulty in discharging the respective duties of the First Commissioner [Ayrton] and of the Director of the Gardens [Hooker] in a manner satisfactory to both, whoever may be the occupants of those offices, maintaining the proper authority of the First Commissioner, with due regard to the position and character of the Director of the Gardens' ('Copies of Correspondence between the Board of Works and Dr. Hooker relating to Changes proposed to be introduced into the Direction and Management of the Gardens at Kew', HC 25, Parliamentary Papers (1872), p. 176; and *Nature*, 6 (8 August 1872), pp. 280–1).
3. *my father*: William Jackson Hooker.
4. *Smith*: John Smith.
5. *Queen*: Queen Victoria.
6. *Prince Albert's*: Francis Albert Augustus Charles Emmanuel (1819–61), Prince Albert, husband of Queen Victoria (*ODNB*).
7. *lief*: willing, glad (*OED*).
8. *Cowper Temple*: William Francis Cowper-Temple (1811–88), a British Liberal politician (*ODNB*).
9. *Holwood*: Holwood House (see letter 3789, n. 1).
10. *Times article today written by <L>owe*: *Times*, 27 July 1872, p. 9.
11. *Lady D. Neville*: Dorothy Fanny Nevill (née Walpole, 1826–1913), a British hostess and horticulturist (*ODNB*).
12. *my wife*: Frances Hooker.
13. *Physical, Biological, and Chemical Lectureships of Jermyn St are at last transferred to S. Kensington*: see letter 3711, n. 5.

To Thomas Archer Hirst [28 July][1] 1872 3805

Bel Alp,[2] Sunday, 1872

My dear Tom.

I do not think you could do better than bring our friends[3] to this place.

Perhaps the most agreeable way would be to take the night train (at 8 P.M.) from Paris to Neuchatel,[4] and then come on to Sierre[5] by rail—Thence by diligence[6] to Brieg,[7] and then march to the Bel Alp.

But the weather is now broken and this may modify your desire for coolness. There is I am told an excellent train from Paris to Bâle,[8] that as you know would bring you into the Thun,[9] Interlaken,[10] Giessbach,[11] and Grindelwald[12] country.

But I confess I should much prefer this place to any of them.

If you would give me authority I would engage rooms from any day that you may fix upon. The landlord here will I am sure do every thing in his power to promote the comfort of any friend of mine.

I expect Lady Mary[13] here, with the two younger girls[14] and Hugh[15] tomorrow night.

Lord Derby, I suppose, comes on tonight. I hope he will succeed in knocking the breath out of Ayrton.[16]

I have more than once longed for one night's tussle with the bear in the House of Commons. With the weapon at our disposal he could be hewn in pieces before the Lord. I suppose you are aware of his last vile proceedings.

But Gladstone is more to blame than he for placing him in his present position.

I thank you for writing to M[r]. Burns. I am rather forced by the people on the other side[17] but I hope in my return to know exactly where they wish me to begin.

This will go via Paris in the hope that it will then reach you sooner.

Yours aff[ectionate][ly] | John Tyndall

BL Add MS 63092, ff. 130–1

---

1. *[28 July]*: the day and month are given by reference to Lord Derby (Edward Stanley) introducing the Memorial in support of Joseph Hooker (see letter 3755) into Parliament, which occurred on 29 July 1872, making this Sunday letter 28 July 1872.
2. *Bel Alp*: see letter 3771, n. 5.
3. *our friends*: not identified
4. *Neuchatel*: capital city of the Swiss canton of Neuchâtel.
5. *Sierre*: see letter 3796, n. 5.
6. *diligence*: a public stage-coach (*OED*).
7. *Brieg*: see letter 3785, n. 4.
8. *Bâle*: Basle, or Basel, a Swiss town close to the borders with France and Germany (Bale is the French spelling).
9. *Thun*: a town in the Swiss canton of Bern.
10. *Interlaken*: a Swiss town that serves as a gateway to the mountains and lakes of the Swiss Alps.
11. *Giessbach*: Giessbach Falls, located near the Brienzersee, Switzerland.
12. *Grindelwald*: a town in the Swiss canton of Bern.
13. *Lady Mary*: Mary Egerton.
14. *two younger girls*: probably the younger two of the four daughters of Mary Egerton and

Edward Christopher Egerton (1816–69), an English politician: Mary Alice Egerton (1848–1924), Charlotte Egerton (1850–1926), Emily Margaret Rosa Egerton (1851–94), and Georgina Renira Egerton (1853–1930).

15. *Hugh*: Hugh Egerton.
16. *succeed in knocking the breath out of Ayrton*: on 29 July 1872, Edward Stanley (Lord Derby) spoke before the House of Lords of Parliament about Hooker's dispute with Acton Ayrton over his role as Director of the Royal Botanic Gardens, Kew, and the Memorial in support of Hooker (see letter 3755, esp. n. 1). See The Royal Gardens, Kew—Dr Hooker and The First Commissioner of Works', HL Deb 29 July 1872 vol. 213 cc3–23, *HANSARD 1803–2005* (online, accessed 4 February 2022), columns 2–15.
17. *people on the other side*: Americans involved in bringing Tyndall over for a lecture tour (see Introduction, n. 8).

To Heinrich Debus 29 July 1872 3806

Bel Alp[1] 29th. July 1872.

My dear Heins.

I abstained from writing to you because I harboured the notion of sending on to you what I have been writing for America;[2] but it is too voluminous, so I must have your opinion of it when I return.

You know of course that I first visited Chamouni,[3] where I was well put up at the clean & comfortable Hotel Couttet[4]—(an old guide) I visited my old haunt on the Mer de Glace,[5] went up Mont Blanc[6] as far as the Grand Mulets,[7] and scorched my face upon the snow of the Col du Géant.[8] The action of the mountain sun upon the skin is very singular; it must be the quality rather than the quantity of the heat which inflames and indeed kills the skin. When thus burnt you have only to be patient till the outer skin dries, till the new one forms underneath—Then it becomes an amusement to peel of the old mask; but during the process of peeling the countenance is the reverse of beautiful.

From Chamouni I crossed the Col de Balme[9] to Martigny,[10] thence by train to Siders[11] where the railway ends. There I took the diligence[12] securing a seat on the Banquette[13] and carrying with me a bottle of wine and a crust, for I was very hungry. An elderly Catholic dignitary[14] with whom I had a good deal of pleasant talk, was my companion; we had no glass, so we finished the wine after the fashion of the loving cup[15] at City dinners, by taking each in turn a pull at the bottle.

I spent a night in Brieg[16] and on the following day mounted to the Bel Alp.

The weather for a time was very moist & cold. Raw fogs swathed us, and sometimes when I quitted the mountain stream in which I took my tub in the morning the cold wind seaping my body created a cloud behind me through the precipitation of the vapour produced by my own heat. Then the weather changed, and we had a fortnight of the most splendid days, the mountains coming out grandly day after day.

But it is again broken. I went yesterday to a place called the Nessel[17] where I had a grand view of the mountains. After some strawberries & milk furnished by a sunburnt Alpine maiden I lay down to read a paper & at intervals to look at the mountains. The atmosphere blackened gradually, and at length the mountains drew entirely away into the darkness. I could see a struggle between the north & south wind; and fearing rain I turned homewards. I was in sunshine for the most part; but at the opposite side of the valley of the Rhone[18] the thunder broke loose among the mountains. Local rains descended, and rainbows sprang into existence. At one point I thought I had received a shock of the electricity, but I am not sure.*

I have had various excursions on the Glaciers, and climbed (for the second time) with a friend[19] the peak of the Aletschhorn.[20] The ascent was rapid & pleasant, for the snow was frozen hard. As we returned it was so softened, & so deep, that we sometimes sank in it to the hips. We had one slip (on the part of the guide[21]) which might have been serious; but he was happily pulled up before any damage was done.

I am aiming to get home about the 10th of August: I hope to catch you before you start—at all events you will be in London before my departure for America.[22] Auf wider sehen[23] dear Heins

Yours ever | John Tyndall | over

*The lightning set a forest on fire which was very grand.

Tell Tom[24] I am extremely obliged to him for the good management of my affairs with Mr. Burns. I sh[oul]d relinquish the Russia[25] with regret because all my friends in America expect me to come by her.

RI MS JT/1/T/268
RI MS JT/1/TYP/7/2373–4

1. *Bel Alp*: see letter 3771, n. 5.
2. *writing for America*: his lectures on light for his American lecture tour (see Introduction, n. 8).
3. *Chamouni*: see letter 3776, n. 12.
4. *Hotel Couttet*: see letter 3783, n. 10.
5. *Mer de Glace*: see 3785, n. 17.
6. *Mont Blanc*: see letter 3781, n. 25.
7. *Grand Mulets*: see letter 3783, n. 5.
8. *Col du Géant*: see letter 3785, n. 10.
9. *Col de Balme*: an alpine pass on the border of France and Switzerland.

10. *Martigny*: see letter 3785, n. 3.
11. *Siders*: see letter 3796, n. 5.
12. *diligence*: a public stage-coach (*OED*).
13. *Banquette*: the long low bench behind the driver in a diligence or omnibus (*OED*).
14. *elderly Catholic dignitary*: not identified.
15. *loving cup*: a drink shared as a sign of friendship, unity, or goodwill (*OED*).
16. *Brieg*: see letter 3785, n. 4.
17. *Nessel*: a Swiss village with views across the Rhône valley, including the prominent peak Weisshorn (on 19 August 1861, Tyndall and his guides were the first to summit the Weisshorn; see *Mountaineering in 1861*, pp. 41–58).
18. *valley of the Rhone*: the valley formed by the Rhône river, which starts in Switzerland and passes through southeastern France.
19. *friend*: not identified.
20. *Aletschhorn*: Aletschorn, the second highest mountain of the Bernese Alps.
21. *guide*: not identified.
22. *my departure for America*: see Introduction, n. 8.
23. *Auf wider sehen*: Goodbye (German).
24. *Tom*: Thomas Hirst.
25. *Russia*: Tyndall traveled from Liverpool to New York on the SS *Russia* (see Introduction, n. 8).

From Joseph Dalton Hooker 31 July 1872 3807

Kew | July 31st 1872.

Dear Tyndall

It has been impossible for me to write to you for the last three days, much as I have had to say. The affair[1] is terribly complicated by Ayrton's[2] last dastardly action, of getting Owen to write a disparaging statement,[3] and censures on my father[4] and self, and actually appending this to the correspondence laid before Parliament, without affording me the opportunity of answering the charges it contains. It reminds me of certain animals that when they see that all is up, urinate in their enemy's face, or worse.

I ordered the Returns to be sent to you. Hirst assures me that he sends you all the newspapers. Huxley's in the Pall Mall[5] and in the Times of today (31st)[6] are killing. I hope he will not make the personal quarrel between myself and Owen too conspicuous. It would be beneath all our dignity, and divert attention from Ayrton and the main issues.

Derby's speech[7] was excellent, but not followed up. St Albans[8] wretched. By the way I have told you of all this.

The enclosed correspondence[9] explains itself. Gladstone's desiring the withdrawal of my letter accusing Ayrton of evasion etc., is a tremendous concession from a Prime Minister. It is an acknowledgment that the charges were not only true but that I could have proved them to the House! Gladstone has just lost his favourite brother-in-law which complicates matters.

Very many thanks for yours[10] received yesterday. My pile of unanswered letters from all round the world is gigantic. I shall let Ball[11] know of your exploit.

Ever yours aff[ectionate]ly | J. D. Hooker

RI MS JT/1/TYP/8/2692
LT Typescript Only

---

1. *affair*: Joseph Hooker was in dispute with Acton Ayrton over his role as Director of the Royal Botanic Gardens, Kew (see letter 3755, esp. n. 1).
2. *Ayrton's*: Acton Ayrton.
3. *disparaging statement*: as the Superintendent of the natural history collections at the British Museum, Owen had desired the botanical collections of Kew to be transferred under his control, for which Hooker vehemently disagreed (and even argued that the collection of botanist Joseph Banks should be moved from the British Museum to Kew). Knowing that they were not scientific friends, Ayrton solicited Richard Owen in writing a report about the Royal Botanic Gardens, Kew—titled 'Statement relative to the Botanical Departments respectively under the Trustees of the British Museum and the Commissioners of Works'—as a means to criticize Hooker and his management of the institution. In the report, which Hooker had not been made aware of before it was laid before Parliament and thus without a reply, Owen attacked Hooker and his father William Jackson Hooker for the competing manner in which specimens of dead plants had been obtained for the herbarium at Kew, for the mismanagement of specific species of trees in the grounds of Kew, for their systematic botany approach to organizing collections, and for the time Hooker invested in the herbarium and in publishing botanical works ('Copies of Correspondence between the Board of Works and Dr. Hooker relating to Changes proposed to be introduced into the Direction and Management of the Gardens at Kew', HC 25, Parliamentary Papers (1872), p. 169–75). Hooker's eventual reply of 6 August 1872, an almost point-by-point rebuttal of Owen's claims, was presented to Parliament and later published in *Nature* ('Dr. Hooker's Reply to Prof. Owen', *Nature* 6 (24 October 1872), pp. 516 –7).
4. *my father*: William Jackson Hooker.
5. *Pall Mall*: possibly, although off by a day, 'Mr. Ayrton and Dr. Hooker', *PMG*, 30 July 1872, p. 3.
6. *Times of today (31st)*: T. H. Huxley, 'Kew Gardens', *Times*, 31 July 1872, p. 7.
7. *speech*: see letter 3805, n. 16.

8. *St Albans*: Walter John Lawrance (see letter 3802, n. 15).
9. *enclosed correspondence*: enclosure missing.
10. *yours*: letter missing.
11. *Ball*: John Ball.

From Edward Stanley 31 July 1872 3808

Holwood, Beckenham,[1] | July 31st 1872.

Dear Professor Tyndall

I brought on Dr Hooker's case on Monday,[2] the papers referring to it having been delivered only on the Saturday morning preceding. These papers very considerably altered the state of matters, for though Mr Ayrton[3] is not removed, the Treasury minute[4] of July 24th does in effect, though in the guarded language familiar to official men, disavow him and directs in sufficiently plain terms the cessation of the annoyances to which Dr Hooker has been exposed. That this change of tone has been caused by the prospect of parliamentary interference cannot be doubted, and it has been so far successful that by separating Mr Ayrton's position from that of his colleagues, it gave the latter a much better defence than could otherwise have been made. In fact (no matter from what motive) they have met us half way. Under these circumstances, Sir J. Lubbock expressed himself in favour of withdrawing the motion. It seemed to me that this could not well be done without acknowledging, which I was not prepared to do, that the reparation made was sufficient. But I found it hopeless to make a good division in our house. The non-official peers are mostly gone, and ministers would certainly have secured a majority. I thought it better therefore, merely to raise a debate, in which expression could be given to the general feeling: the more so as even a successful division in the Lords could not have compelled Mr Ayrton's retirement; and an unsuccessful one would have injured the prospects of obtaining a similar vote in the Commons. There was scarcely time to consult Dr Hooker himself. In the House of C[ommons]. it was intended that the debate should come on last night, which it has not done, perhaps in consequence of Mr Gladstone's absence. Much will depend on what happens there. But in any event though entire success has not been obtained, certain results are secured:—(1) Dr Hooker's position is better defined (2) the Treasury have substantially sided with him, and disavowed Mr Ayrton (3) The expression of public opinion has been so strong and so unanimous, that we have a fairly good guarantee for the observance of the conditions imposed by the Treasury.

I regret much that Lord Russell who had promised to attend and speak, was not in the House. I suppose health was the cause. There was not a word said by either of the two Ministers who spoke, in defence of the First Commissioner, nor, I think, will there be in the House of C[ommons].

I hope you will have enjoyed your Swiss holiday.[5] I write to the <R>[oyal]. Inst[itution]. as you speak of returning on the 10th.

Believe me | Very sincerely yours | Derby.

RI MS JT/1/TYP/8/2698
LT Typescript Only

---

1. *Holwood, Beckenham*: see letter 3789, n. 1.
2. *I brought on Dr Hooker's case on Monday*: Stanley put forward in the House of Lords of Parliament (see letter 3805, n. 16) the Memorial in support of Joseph Hooker in his dispute with Acton Ayrton (see letter 3755, esp. n. 1).
3. *Mr. Ayrton*: Acton Ayrton.
4. *Treasury minute*: see letter 3804, n. 2.
5. *your Swiss holiday*: see letter 3716, n. 6.

From Joseph Dalton Hooker [1 August 1872][1] 3809

Kew.

Dear Tyndall

I have just returned, after a most fatiguing run in London after Huxley and Spottiswoode, in time to catch the post for you.

I found Lubbock most anxious that I should withdraw the charges of evasion etc., which I could not consent to. Spottiswoode agreed that it could not well be. Huxley suggested that I should propose to do so on condition that Ayrton[2] withdrew his memorandum and Owen's statement.[3] After long anxious discussion we agreed on this.

Dr H. to Sir J. Lubbock.

I am very glad to accept Mr Gladstone's assurance that Mr Ayrton has had no intention of giving me offence; and I am ready on my side to assure you that I am most willing to forget all personal grievances.

I have no wish to make any further reference to the charges of evasion etc. brought against Mr Ayrton, but I cannot declare myself satisfied so long as the new matter brought against me in Mr Ayrton's memorandum is unanswered, and before closing this controversy, I shall feel it my duty to my father[4] and myself to put upon official record a memorandum in reply to that of Mr Owen's, which I had never seen until it was laid before Parliament.

(This is Playfair's ending)

Huxley's is:—

"so long as no apology is offered for the publication in the Parliamentary papers of the document of Professor Owen of which I had no knowledge."

Playfair whom I lighted upon after Huxley and Spottiswoode had left objected to apologies as I do.

Huxley is off to-morrow and will be quite inaccessible in N. Devon,[5] 12 miles from railway. Spottiswoode is at Hastings.[6]

In great haste | Ever yours | J. D. Hooker.

RI MS JT/1/TYP/8/2699
LT Typescript Only

---

1. *[1 August 1872]*: LT wrote on the typescript 'about Aug 1. 1872', which indeed was a Thursday. At this time, Hooker was in a dispute with Acton Ayrton over the management of the Royal Botanic Gardens, Kew, and a Memorial in support of Hooker had recently been put forward before the House of Lords of Parliament (see letter 3755, esp. n. 1).
2. *Ayrton*: Acton Ayrton.
3. *Owen's statement*: see letter 3807, n. 3.
4. *my father*: William Jackson Hooker.
5. *N. Devon*: North Devon, a scenic coastal area of southwest England.
6. *Hastings*: a seaside town in southwest England.

To the Editor of the *Times*[1] — 3 August 1872 — 3810

[*THE ALETSCH GLACIER.*] |
TO THE EDITOR OF THE TIMES.

Sir,—At about the middle of its length the Aletsch glacier[2] once sent an arm eastward through a lateral valley. It seems to try to do so still, but the arm is perpetually broken off near the body of the glacier, ice-cliffs being thus exposed which rise sometimes to a height of sixty or seventy feet. A portion of the valley thus forsaken by the ice is filled by the well-known Märgelin See,[3] which receives the icebergs discharged from the glacier, and from which a considerable stream descends to the valley occupied by the Viesch glacier.[4] A portion of the drainage of the Aletsch escapes by turning aside into this lateral outlet; the main portion, however, continuing its course along the bed of the glacier, from which it escapes as an impetuous river named Massa.[5]

Yesterday a party of three of us[6] descended to the glacier and walked downwards along the ice to the savagely-picturesque gorge in which the glacier ends. Our way at one place lay along the hollow between the glacier and the mountain, which hollow three hours later was to be filled by a furious torrent.[7] We crossed a wooden bridge which spans the Massa, and ascended thence, mainly along the moraine of an extinct glacier, to the Bel Alp.[8] It was half-past 4 p.m. when we arrived. The domestics drew our attention to a sound like the roar of a cataract, which seemed to descend the Aletsch. For a time the sound was sub-glacial, but a yellow torrent at length appeared at the opposite side of the glacier, smoking and roaring as it tumbled down the declivities of ice. The front of the torrent soon appeared opposite to the Bel Alp, carrying every movable thing along with it. The ice-barrier which had dammed the Märgelin See had manifestly given way, and this was the result. Wishing to get near the torrent, I descended rapidly to the glacier, crossed it, and succeeded in getting quite close to the rushing water. Everywhere impetuous, it was divided into spaces of tolerably uniform slope, separated from each other by steep and broken declivities, down which the water plunged with tremendous fury. At the base of one of these falls it was met by a kind of reflecting surface, by which the rhythmic character of its motion was finely revealed. The water here was tossed upwards in a series of vast parallel fans, carrying with them ice-blocks and stones, and breaking above into a spray as fine as smoke. A bend of the glacier came in for the lateral portion of this spray, and over it the rounded blocks of ice and the stones were showered like projectiles. The sound of the torrent had not abated at bedtime, but this morning all is quiet, and no water is to be seen in the temporary channel. The Rhone[9] at Brieg[10] rose, I am told, considerably, but no loss of life has been reported.

Bel Alp Hotel,[11] Aug. 3. | I.T.[12,13]

Times, 8 August 1872, p. 5
RI MS JT/6/5/1/98–9

---

1. *Editor of the Times*: John Thadeus Delane.
2. *Aletsch glacier*: the longest glacier in the Alps, which flows past the Eggishorn.
3. *Märgelin See*: or Marjelin See or Märjelensee, a glacial lake that formed beside the Aletsch glacier, near the Eggishorn.
4. *Viesch glacier*: the Fiescher Glacier, on the south side of the Bernese Alps in the Swiss canton of Valais.
5. *Massa*: a river in the eastern Bernese Alps in the Swiss canton of Valais, fed by the melt-water from the Aletsch Glacier; its confluence is with the Rhône river.
6. *party of three of us*: Tyndall's companions on this excursion are not identified.
7. *a furious torrent*: Tyndall described the torrent caused by the breaking of the ice barrier at the Märgelin See in his *Forms of Water*, pp. vii–x.
8. *Bel Alp*: see letter 3771, n. 5.
9. *Rhone*: the Rhône river starts in Switzerland and passes through southeastern France.
10. *Brieg*: see letter 3785, n. 4.
11. *Bel Alp Hotel*: see letter 3771, n. 5.

12. *I.T.*: Tyndall's initials were mistakenly published as 'I.T.'.
13. In the handwritten transcript, the following was written under the letter: 'The three letters of Aug. 22, Aug. 23, & Sept 4, on the same subject, are only corroborative of the above account: the ice-cliffs bordering the See were heard and seen by a party to crack and fall, as the water in the lake gradually sank'.

To Joseph Dalton Hooker 5 August 1872 3811

5th. August 1872 | Bel Alp.[1]

My dear Hooker

I hope to with[2] you in a few days, and to hear form your own lips the progress of events.[3]

I am hardly competent to form any opinion as to the most recent phase of the question. I did not see why you should consider it a tremendous concession on the part of the Prime Minister[4] to ask you to withdraw the charge of evasion &c. It is certainly unusual—but it is at least as unusual for a Prime Minister to surround himself by circumstances which should prompt such a request. I am also unable to understand the meaning of Lubbock's advice. To save you from self stultification a great deal on the part of your antagonists would have to be done prior to any such withdrawal on your part. I can quite understand Lubbock's action, but were I in London I think with every wish for an amicable arrangement I should subscribe to the advice given to you by Spottiswoode & Huxley.

The debate in the Lords[5] surprised me—I thought they would make something of it—The speech of Lord Derby[6] was as you say excellent. But why he was not better backed I cannot understand. I read the speech[7] of Lord Stanley of Alderley[8] with blank amazement. Such a reckless perversion of truth I never encountered. I sent a line to the Times anent this speech, and I have since seen that Huxley has taken notice of it[9]

Is honour & fairness utterly extinct among the public men of England? By jove I have half a mind to settle down in America when I go there.[10]

How well the Times has behaved! That last article[11] was capital, and I have thanked the editor for it.[12]

I have tried to devise all manner of hypotheses to account for Owen's former behavior. I shrink from believing that a man who had worked so hard could be base: but I fear this last nerve of his[13] will force me to join the ranks of those who seem to have appreciated his character more truly than I did. A more rascally trick I have never heard of. On this subject I may have something to say by and by.

I am thinking of packing up & shall not be here long enough to receive anything from your after you receive this.

Yours aff[ectionate]ly | John Tyndall

RBG, Kew Papers Relating to Kew 1867–1872, Ayrton Controversy, volume 2, f. 415

---

1. *Bel Alp*: see letter 3771, n. 5.
2. *to with*: Tyndall left out the word 'be'.
3. *progress of events*: referring to the submission to Parliament of the Memorial in support of Joseph Hooker in his dispute with Acton Ayrton (see letter 3755, esp. n. 1).
4. *Prime Minister*: William Gladstone.
5. *debate in the Lords*: on 29 July 1872 (see letter 3755, n. 1).
6. *Speech of Lord Derby*: see letter 3805, n. 16.
7. *speech*: see 'The Royal Gardens, Kew—Dr Hooker and The First Commissioner of Works', HL Deb 29 July 1872 vol. 213 cc3–23, *HANSARD 1803–2005* (online, accessed 4 February 2022), columns 21–3.
8. *Lord Stanley of Alderley*: Henry Edward John Stanley (1827–1903), 3rd Baron Stanley of Alderley, a historian notable as the first Muslim member of the House of Lords, in which he served from 1869–1903 (*ODNB*).
9. *Huxley has since taken notice of it*: see T. H. Huxley, 'Kew Gardens', *Times*, 31 July 1872, p. 7.
10. *Is honour... when I go there*: these lines are crossed out in the original, but not enough to make them illegible. For Tyndall's American lecture tour, see Introduction, n. 8.
11. *last article*: Tyndall was probably referring to an article on p. 9 of the *Times* of 27 July 1872, which was supportive of Hooker and ended by stating, in reference to Acton Ayrton, 'it is greatly to be regretted that the administration of a Department should be gratuitously marred by such faults of temper and manner'.
12. *I have thanked the editor for it*: letter missing; the editor of the *Times* was John Thadeus Delane.
13. *this last nerve of his*: see letter 3807, n. 3.

To Heinrich Rudolf Vieweg — 5 August 1872 — 3812

The Bel Alp[1] 5th August 1872

Gentlemen[2]

Your letter[3] has been forwarded to me here in Switzerland.

You appear to have misinterpreted my last letter to you:[4] it had reference to the future, not to the present, and when I propose making arrangments with a publisher in Germany for a translation "of my memoirs" I shall endeavour to make such terms with him as shall secure him from loss.

Yours truly | John Tyndall

TU Braunschweig/Universitätsbibliothek/UABS V 1 T : 63

1. *The Bel Alp*: see letter 3771, n. 5.
2. *Gentlemen*: while this letter is addressed to 'Gentlemen', we believe the recipient is Heinrich Rudolf Vieweg, who took over the family publishing firm, Friedrich Vieweg und Sohn, from his father Eduard Vieweg (1797–1869).
3. *Your letter*: letter 3790.
4. *my last letter to you*: letter 3735. In this letter, Tyndall wrote that 'I shall probably ask you' to publish German translations of *Diamagnetism* and *Molecular Physics*, and in this current letter was stressing that he had not yet decided to ask him. Tyndall's short and pointed reply to Vieweg sharing that he would not publish these two volumes suggests he was offended to be told these works—indeed, compilations of previously published papers, as Vieweg stressed—were not important enough for translations.

From E. Diamilla Muller[1] 5 August 1872 3813

[Diamilla Muller | 32 Corso Magenta | Milano] | Milan 5 Août 1872

Monsieur,

J'ai l'honneur de vous expédier par la poste quelques études sur le magnétisme terrestre, et ma relation sur l'Eclipse totale du soleil de 1870. Veuillez les agréer comme un hommage respectueux de ma part.

J'ai lu et relu; j'ai étudié avec attention vos importantes leçons et je pense que c'est un dévoir impérieux de le repandre dans toutes les langues.

Or, je vous demande la permission de les traduire en italien, en commençant par votre charmant livre = Faraday as a Discoverer. Voulez vous m'accorder cette permission? Je vous en serai reconnaissant et plus que moi tous ceux qui étudient les sciences naturelles.

Je connais assez la langue anglaise pour accomplir ce travail avec conscience et devouement, mais je ne la connais pas assez pour oser vous écrire en anglais.

Si je dois juger votre caractère par votre style et par vos sentiments, je suis sûr d'avance que vous m'accorderez la permission demandée.

Parmi les voeux que je fais c'est de voir traité par vous la question des aurores polaires dans leur rapport avec le magnétisme terrestre.

Combien de fois j'ai été jaloux de votre science et de votre métode d'envisager les questions. Pour moi Sir John Tyndall est une émanation de Faraday;

et en lisant votre livre sur celui-ci on se trompe quelque fois et l'on *[croit]* de lire votre histoire.

Ne croyez pas que je sois un flatteur; je dis ce que je pense et ce que je sens.

Veuillez agréer, Monsieur, l'assurance de mon respect et de mon admiration.

Votre dévoué serviteur | E. Diamilla Muller

Diamilla Muller | 32 Corso Magenta | Milan | Milan 5 August 1872

Sir,

I have the honour of mailing you some studies on terrestrial Magnetism,[2] and my account of the total Eclipse of the sun in 1870.[3] Please accept them as a token of respect from me.

I have read and reread; I have carefully studied your important lessons, and I think that it is imperative to make them available in all languages.

Now, I ask your permission to translate them into Italian, starting with your charming book = <u>Faraday as a Discoverer</u>.[4] Would you grant me this permission?[5] I would be very grateful to you, and those who study natural science would be even more grateful.

I know English well enough to accomplish this work with dedication and devotion, but I do not know it well enough to dare to write to you in English.

If I had to judge your character by your style and your sentiments, I would be certain in advance that you will grant me the permission that I ask.

One of my desires is to see you address the question of polar auroras in their relation to terrestrial magnetism.

So many times I have been jealous of your science and your method of considering questions. For me, Sir John Tyndall is an emanation of Faraday; and while reading your book on him, you sometimes mistakenly think that you are reading your story.

Do not think that I am a flatterer; I say what I think and what I feel.

Please accept, Sir, the assurance of my respect and admiration.

Your devoted servant | E. Diamilla Muller

RI MS JT/1/H/150

---

1. *E. Diamilla Muller*: Demetrio Emilio Diamilla Muller (1826–1908), an Italian astronomer and numismatist (*Dizionario Biografico degli Italiani*, 100 vols (Istituto dell'Enciclopedia Italiana, 1991), vol. 39, pp. 642–5).
2. *studies on terrestrial Magnetism*: probably D. Diamilla-Muller, *Recherches sur le Magnétisme Terrestre* (Paris: Jean Polizzi et Comp., 1870).
3. *my account of the total eclipse of the sun in 1870*: possibly 'Rapporto dei signori Ing. Muller e Capitano Serra: Osservazioni Meteoriche e Magnetiche Fatte in Terrnova di Sicilia', in

Rapporti sulle osservazioni dell'ecclisse totale di sole del 22 dicembre 1870 eseguite in Sicilia dalla Commissione Italiana (Palermo: Stabilimento Tipografico, 1872), pp. 165–89; or D. Muller and L. Serra, *Eclisse totale del Sole del 22 dicembre 1870, Osservazioni meteoriche e magnetiche eseguite Terranova di Sicilia* (Milano: E. Treves, 1872).

4. *Faraday as a Discoverer*: J. Tyndall, *Faraday as a Discoverer* (London: Longmans, Green, and Co., 1868). Following Michael Faraday's death on 25 August 1867, Tyndall lectured at the RI on the life and work of Faraday in two Friday Evening Discourses on 17 and 24 January 1868, and he published these lectures as *Faraday as a Discoverer*. Preparing these lectures and transforming them into a book was a very heavy task for Tyndall (see *Tyndall Correspondence*, vol. 10).
5. *grant me permission?*: it does not appear that Diamilla Muller published a translation. This is the only letter from him in Tyndall's extant correspondence.

From William Rhees 5 August 1872 3814

Aug 5. 1872

Dear Sir

In the absence of Prof. Henry who has gone on a tour of Light House inspection,[1] I have the pleasure of acknowledging the receipt of your favor of the 17th July.[2]

We have made the following arrangements for your lectures.[3]

Oct 15. to Oct 26 – Boston. Lowell Institute[4]
" 28 " Nov 9 New York.
Nov 11 " Nov 23 Philad[elphi][a]
" 25 " Nov 30 Baltimore. Peabody Inst[itute].[5]
Dec 2 " Dec 14 Washington. Smithsonian

We have placed Washington last as we presume you would like to be here when Congress is in session.

The "Lowell Institute" is the principal establishment in Boston under whose auspices our first class lecturers appear.

Prof. Youmans will arrange in New York.

Your brother[6] and Prof Lesley have made arrangements in Philadelphia.

I do not think the Presidential election[7] will at all interfere with the proper appreciation of your lectures.

Trusting that you will have a safe and pleasant journey[8]

I am | Yours respectfully | W[ilia]m. J. Rhees | Secretary to Prof Henry.

Prof. John Tyndall. | Royal Institution. | London.

Smithsonian Institution Archives. RU 33, Vol. 30, Page 182

---

1. *tour of Light House inspection*: see letter 3738, n. 5.
2. *your favor of the 17th July*: letter 3794.

3. *your lectures*: for Tyndall's American lecture tour, see Introduction, n. 8.
4. *Lowell Institute*: Tyndall delivered six lectures from 15–25 October 1872 at the Lowell Institute, founded by a bequest from John Lowell (1799–1836) and opened in 1839 with the intent of presenting popular lectures and instruction in a range of subjects (*ANB*).
5. *Peabody Inst[itute]*: the Peabody Institute is an intellectual and arts center in Baltimore, founded by George Peabody (1792–1869) in 1857 but not opened until 1866 due to the American Civil War; since 1977, the Institute is part of Johns Hopkins University. Peabody, an American merchant banker and very wealthy philanthropist, moved from the United States to London in 1837 (*ANB*; *ODNB*).
6. *Your brother*: Hector Tyndale, who was Tyndall's cousin, not brother.
7. *Presidential election*: see letter 3794, n. 3.
8. *safe and pleasant journey*: Tyndall traveled from Liverpool to New York on the SS *Russia*, departing on 28 September 1872 (see Introduction, n. 8).

From Hector Tyndale 5 August 1872 3815

No: 1021 Clinton Street | Philadelphia August 5. 1872

My dear John

Professor Henry has placed the charge of your course of lectures in Philadelphia[1] in the hands of Prof[essor]: Lesley and myself. We have concluded to place the management of your business arrangements here with Mr: T. B. Pugh, a most respectable, useful and diligent man, who has made an annual course of lectures here highly successful. His "Star Course",[2] as it is called, embraces some of the first men in the country. Such as Senator Sumner,[3] Emerson, Hon[orable]: Geo[rge]: Boutwell[4]—Sec[re]'t[ar]'y of the Tres[uar]'y of the U.S.—Wendell Philips,[5] Curtis[6] and others. He hopes to have Mr Froude lecture in his course and certainly he could not have a better management than Mr Pugh's. I understand that Mr Froude will so lecture. I wrote to Prof Henry about it and received reply from Mr: W[illia]m J. Rhees chief clerk of the Smithsonian Institution who in Henry's absence has charge of the matters. He says "The whole matter so far as relates to Philadelphia is left with you (me) and Prof. Lesley and whatever arrangements you make with Mr: Pugh will be satisfactory to Prof. Henry". He further asks me to inform you of the programme as now arranged for your lectures. This I would suggest may be modified in some details, but the general tenor is right I should say—

" Programme

Oct 15 to 26 = 11 working days – Boston, at the Lowell Ins[titut]e[7]
" 28 to Nov 9 = 12 – " – " – New York, in charge Prof. Youmans
Nov 11 to 23 = 12 " – " Philadelphia " of {Gen. Tyndale & Prof Lesley}
" 25 to " 30 = 6 – " – " – Baltimore " Peabody Inst[itut]e[8]
Dec 2 to Dec 14 = 12 " – " – Washington " Smithsonian Inst[itutio]n "

Here in Philadelphia we have no public institution which undertakes to give courses of lectures, except such as are merely technical and of no width of popular reach. I saw Mr: Pugh—whom I have known for many years—and he agreed to undertake the management of your affairs here. He fully understands the matter and will, I have no doubt, satisfy you in every way. I agreed with him that he should give you five hundred (500) dollars for every night you lecture here. This is the highest amount ever paid in the United States to any body for lectures. Mr Pugh understands the conducting of the matter and will so act as to make your lectures here, perhaps, more brilliant and pleasant to you than in any other city. I believe so, at least I hope so. Inasmuch as the sum I have named is so much larger than paid to others it might be as well not to mention it as it might compromise Mr Pugh with other gentlemen. This is wholly my own suggestion and not Mr Pugh's. That gentleman requested me to ask you whether you would not please send him some general idea of your subjects and what you propose to do with them? He will take the finest place in the city—equal to any in America, and he wishes to please you, and of course himself, by giving you brilliant and large audiences. All the rooms or halls or theatres in which you may lecture in this country will be very much larger than your lecture room of the Royal Institution. Our Academy of Music,[9] for instance, will seat over 3000 persons, but not so many for distinctly hearing or rather seeing in lectures although for sound it is marvellously easy to speak in—but the great distance to the upper galleries render them, perhaps, ineligible for seeing nice experiments, such as you will make. This house is engaged by Pugh & others a year in advance. Now inasmuch as the Steamers are overcrowded in the Autumn, when thousands of Americans are returning home—I would suggest that you engage your passage[10] and that of your assistant at once on some good steamer direct for New York. The finest ships are said to be those of the White Star Line[11]—the "Oceanic"[12] and her mates. The Cunard Line[13] has a good reputation for safety and the ships are good, of these this line has two classes, one, mostly for freight, which is cheaper, and the other has rather more comfort. I would advise the best. But if not already done, I do advise taking passage at once. When you have taken passage let me know the vessel and the day of her sailing. You may have a good deal of trouble, possibly, with your luggage, apparatus &c on arrival and it is likely I can save you some of it by being in New York to arrange about it at the Custom House in advance. I certainly shall meet you in New York and shall be much hurt if you don't enable me to do it on shipboard by telling me beforehand the name of your ship & her day of departure from Liverpool.[14] You need not fear any trouble for it is but three (3) hours hence to New York. Besides you will really need some advice &c and having been a business man I can do you some service without trouble to myself and can influence by introductory letters from here a good deal.

By writing to the Agents of the Lines of Steamers in Liverpool you can have sent to you a list of the ships and their days of sailing and thus can engage your passage or some friend in Liverpool can do it. I think you had better start in September and thus have some days for arranging plans and for recruiting from the voyage before lecturing. And to this end you will come direct to my house—your rooms are now ready—and you can go easily to Washington to see Henry or elsewhere. Let me say, for fear you do not know, that from Boston to New York it is about 8 hours ride, from New York to Philadelphia 3 hours, thence to Baltimore 3½ hours, thence to Washington about 2 hours. So that you will lose but a small portion of your time in transit. You can leave New York by Steamer & rail for Boston and having a good night's sleep will arrive early the next morning or you can take rail all the way.

You will be here in all the excitement (popular) of a Presidential election[15] canvass. But I hope your lectures and the brilliant experiments, you so clearly make, will not be interfered with in point of attendance. I think our people are especially fond of experimental lectures.

Mr McLaughlin[16] has just been in and desires to send his respects, as does Mrs Tyndale.[17]

Affectionately yours | Hector Tyndale
Prof John Tyndall

RI MS JT/1/T/67
RI MS JT/1/TYP/5/1703–6

---

1. *your course of lectures in Philadelphia*: for Tyndall's American lecture tour, see Introduction, n. 8.
2. *"Star Course"*: Pugh established his Star Course in Philadelphia in 1869, organizing ten lectures annually at the Academy of Music, a concert hall and opera house (see P. Cherches, *Star Course: Nineteenth-Century Lecture Tours and the Consolidation of Modern Celebrity* (Boston: Sense Publishers, 2017), pp. 41–2).
3. *Senator Sumner*: Charles Sumner (1811–74), an American politician, lawyer, and reformer. Sumner served as Senator from Massachusetts from 1851–74 (*ANB*).
4. *Hon[orable]: Geo[rge]: Boutwell*: George Sewall Boutwell (1818–1905), an American governor and congressman who served as Secretary of the Treasury from 1869–73 during President Ulysses S. Grant's first term. Boutwell then served as Senator from Massachusetts from 1873–7 (*ANB*).
5. *Wendell Philips*: Wendell Phillips (1811–84), an American orator, abolitionist, and women's rights and labor advocate (*ANB*).
6. *Curtis*: George William Curtis (1824–92), an American writer, editor, and orator (*ANB*).
7. *Lowell Ins[titut]e*: see letter 3814, n. 4.
8. *Peabody Inst[itut]e*: see letter 3814, n. 5.
9. *Academy of Music*: a concert hall and opera house in Philadelphia, built in 1855–7, known as 'Grand Old Lady of Locust Street'.

10. *engage your passage*: Tyndall traveled from Liverpool to New York on the SS *Russia*, departing on 28 September 1872 (see Introduction, n. 8).
11. *White Star Line*: a British transatlantic shipping company founded in 1845, providing passenger and cargo services. Its most famous liner was HMS *Titanic* which sank in 1912, resulting in the death of more than 1,500 passengers and crew. The company ended in 1934 during the Great Depression.
12. *"Oceanic"*: RMS *Oceanic*, the White Star Line's first liner, was in operation for Atlantic crossings from 1871–5, then chartered to the Occidental and Oriental Steamship Company for the Pacific from 1875–95.
13. *Cunard Line*: a British transatlantic steamship company founded in 1840 which could not compete with its rival the White Star Line. The company remains in operation for pleasure cruises.
14. *day of departure from Liverpool*: see n. 10.
15. *Presidential election*: see letter 3794, n. 3.
16. *Mr McLaughlin*: John M. McLaughlin (b. 1826), who emigrated from Ireland to Philadelphia at a young age, served in the American Civil War in the Union Army in 1862–3 (J. H. Campbell, *History of the Friendly Sons of St. Patrick and of the Hibernian Society for the Relief of Immigrants from Ireland* (Philadelphia: The Hibernian Society, 1892), p. 485). McLaughlin was a friend of Tyndale's, and later published *A Memoir of Hector Tyndale* (Philadelphia: Collins, 1882).
17. *Mrs Tyndale*: Julia Tyndale.

From Hector Tyndale 6 August 1872 3816

Philadelphia Aug[ust]: 6. 1872

My dear John

Yours of the 17th July from Switzerland[1] came only yesterday—too late for reply so that you could have it by the 10th of this month, as you desired. I at once wrote to Mr Rhees, Chief Clerk of the Smithsonian Institution, and asked him to telegraph me his reply to day so that I could telegraph you to day whether any postponement of your lectures[2] could now be made. But up to this time I have received no reply, and probably will not until tomorrow by letter. The point of my letter to Mr: Rhees was to know whether the present arrangement was fixed and unchangeable and if in his opinion it was desirable to postpone your lectures on account of the Presidential election.[3] My own opinion is against a change but the whole matter having been put in Prof. Henry's hands he must decide the case. Lesley thinks the election canvass will make no difference at all, but I think it will make some though not any which should cause a change.

Yesterday before the receipt of yours I mailed you a long letter[4] (mine are all long somehow—arising, I suppose, from my desire to make things plain). In that I gave you the programme of your course of lectures in the different Cities; beginning in Boston on the 14th October. I also gave you some information as to this place having arranged with Mr: Pugh to conduct your course here. He will make it a success which no other man I know of can do. Lesley fully agrees with me in everything done. I trust you will like the arrangement & terms.

You have taken passage in an excellent ship. The Russia is a favorite. I hope you will have a pleasant voyage in her.[5] When Mr Rhees' reply comes I will probably telegraph you his decision—if there should be anything important to tell you. Your time of sailing, the times & places of lectures &c are well selected, but if you can remain here until January I would advise it. At any rate I advise you not to leave England with any definite date of return fixed in your mind—unless absolutely necessary—because when here you might, and probably will, find something to detain you and to make your visit more successful and pleasant. Your courses of lectures, will, as now arranged, occupy about two months and then there are other places to see and which you will desire to visit. Keep yourself free to act.

If I learn anything more to tell, I will write you details by next mail or soon thereafter.

I hope you enjoyed your Swiss trip[6] as much as usual.

As always affectionately yours | Hector Tyndale

I think Prof. Henry is now in California[7] and this may make some delay in answering.

RI MS JT/1/T/68
RI MS JT/1/TYP/5/1707–8

1. *Yours of the 17th July from Switzerland*: letter missing.
2. *your lectures*: for Tyndall's American lecture tour, see Introduction, n. 8.
3. *Presidential election*: see letter 3794, n. 3.
4. *long letter*: letter 3815.
5. *pleasant voyage in her*: Tyndall traveled from Liverpool to New York on the SS *Russia* (see Introduction, n. 8).
6. *Swiss trip*: see letter 3716, n. 6.
7. *now in California*: Tyndale was mistaken about California; Henry did travel to California the year before for investigations at lighthouse stations on the coast, as well as a visit to the California Academy of Sciences in San Francisco. In August 1872, however, Henry traveled to Maine on similar business for the Light-House Board (Kathy Dorman, Research Associate, Institutional History Division, Smithsonian Institution Archives, personal communication).

From George Biddell Airy — 8 August 1872 — 3817

[Royal Observatory, Greenwich, | London, S.E.] | 1872 August 8

My dear Sir

The n°. of Poggendorf[1] which has just reached me appears to me to be unusually interesting:—for its observations on the spectrum of radiant heat,[2]—but perhaps still more for M^r^. Königs[3] monometric flames.[4]

At some future time, I should like to know your opinion on the latter paper.[5]

I am, my dear sir, | Yours very truly | G B Airy

Professor Tyndall

RI MS JT/1/A/55
RI MS JT/1/TYP/1/49

1. *n°. of Poggendorf*: German physicist Johann Christian Poggendorff (1796–1877) edited the journal *Annalen der Physik und Chemie* from 1824–76. Here Airy is referring to the 13 July 1872 issue.
2. *observations on the spectrum of radiant heat*: S. Lamansky, 'Untersuchungen über das Wärmespectrum des Sonnen- und Kalklichtes', *Annalen der Physik und Chemie*, 222 (1872), pp. 200–31.
3. *M^r^. Königs*: Karl Rudolph Koenig (1832–1901), a German physicist and instrument maker whose main subject of study was acoustics (*CDSB*).
4. *monometric flames*: R. König, 'Die manometrischen Flammen', *Annalen der Physik und Chemie*, 222 (1872), pp. 161–99.
5. *your opinion on the latter paper*: a reply from Tyndall is missing.

From Elizabeth Rutherford — 12 August 1872 — 3818

Brodick Hotel—| Arran—| via Ardrossan[1] | 12th August 1872

My dear Friend,

I saw by your letter in the Times[2] that you were still at the Bel Alp[3] on the 3^d^ ins[tan]^t^ but as I imagine your laws to resemble those of the Medes & Persians,[4] & as you said you would be at home on the 10th[5] I believe you are there, and so I sit down in our little parlour & surrounded by Schoolboys, to welcome you back to your London den;[6] and what a multitude of letters, intelligence of all sorts, business & worry awaits you! My letter will add nothing to it and therefore, en avant![7]

When are you going to America?[8]—Mr Clarence King's Sierra Nevada[9] has afforded me much pleasure & amusement, and I think the having read some of Mark Twain's[10] clever sketches[11] previously, had accustomed me to the strong, pithy, rude language of the Americans—which has a certain zest & flavour of its own, very well suited to the description of the Sierra. The ascent of Mount Tyndall[12] is very interesting and I think you must have read with pleasure the last sentence of the chapter—"We reached the crest exactly at 12. oclock—I rang my hammer upon the topmost rock; we grasped hands, & I reverently named the grand Peak Mount Tyndall".[13] I do not think you will return from America—without having visited California—and if Mr King does not already know you I hope he will make your acquaintance in America.—

Your last work[14] came to me in a rather curious way and I was pleased to have it to place with the other works of yours I have—but it was of course all Greek and Hebrew to me—for I am utterly ignorant. You said if I reminded you, you would send me a copy of the Memorial sent to Mr Gladstone,[15] which Mr. Ayrton[16] designates "a scurrilous and calumnious libel"[17] Pray do—when you have leisure. The whole affair is infamous and I much fear Dr Hooker will resign, it is too, too bad—very disgusting.—Mr Ayrton's lengthy speech last Thursday[18] is truly ludicrous.—We shall remain here, I think, until the 22d. Arran is very beautiful—not quite so grand, as I thought it, in my early days—but there are reasons for that—Oh! I must tell you an anecdote of our clever boy[19] who is just 15—. I had promised him a Book, and if he could not find one to please him on my little Bookshelves, he was to name one for himself. The day before we left Dalhousie[20] he came to my room to make his selection; he looked at several—I said "There is Gibbon 12 vols,[21] a nice old copy of Milton 4 vols,[22] handsomely bound Warton's Virgil[23] ditto, & there is Sir W[ilia]m Hamilton 4. vols, Logic & Metaphysics".[24] "I have not yet dipped into Logic or Metaphysics" he replied—so I sat down to write, seeing he had not made up his mind—At length I looked up & saw him with his head buried in a vol. and oblivious of everything. I got up & looking over his shoulder found it was "Tyndall on Sound".[25] He started & said "This is the Book I should like to have. Please write my name in it". He then pulled out your lecture on Imagination[26] & exclaimed "What a very curious subject to write upon"!—Yes, I said, treated in a masterly manner.

Tom[27] is a very good Chemist and has a great turn for Science—I am afraid he is destined for the Bar—but of course his knowledge will be useful in that profession—

Now having occupied more than enough of your precious time, I will say Adieu—

Believe me, Always | Sincerely & Affectionately | Yours | Elizabeth Rutherford

RI MS JT/1/R/87

---

1. *Brodick Hotel—| Arran—| via*: hotel on the Scottish island of Arran, to the west of coastal town of Ardrossan in southwestern Scotland.
2. *your letter in the Times*: letter 3810.
3. *Bel Alp*: see letter 3771, n. 5.
4. *Medes & Persians*: the Medes were an ancient Indo-European people who established an empire in Media, southwest of the Caspian Sea and including most of Persia (now Iran), in the 7th cent. BC (*OED*); Persians were natives or inhabitants of Persia (now Iran) (*OED*).
5. *you would be at home on the 10th*: see letter 3716, n. 6.
6. *your London den*: Tyndall was living in Faraday's old rooms at the RI since 25 January 1868, following Faraday's death in August 1867.
7. *en avant*: onward (French).
8. *to America*: for Tyndall's American lecture tour, see Introduction, n. 8.
9. *Sierra Nevada*: C. King, *Mountaineering in the Sierra Nevada* (Boston: James Osgood & Co., 1872).
10. *Mark Twain's*: Mark Twain (1835–1910), an American humorist, author, and lecturer (*ANB*).
11. *clever sketches*: probably his 1867 collection of short stories, *The Celebrated Jumping Frog of Calaveras County and Other Sketches.*
12. *ascent of Mount Tyndall*: see letter 3719, n. 3.
13. *"We reached the crest... grand Peak Mount Tyndall"*: C. King, *Mountaineering in the Sierra Nevada*, p. 75.
14. *Your last work*: probably *Molecular Physics.*
15. *Memorial sent to Mr Gladstone*: the Memorial in support of Joseph Hooker addressed to Prime Minister William Gladstone (see letter 3755, esp. n. 1).
16. *Mr. Ayrton*: Acton Ayrton.
17. *"a scurrilous and calumnious libel"*: Ayrton stated before the House of Commons on 8 August 1872 that the Memorial addressed to Gladstone was 'a scurrilous and most calumnious libel upon him', and further that 'the charge made against him by Dr. Hooker was that he had been guilty of evasion and misrepresentation, and of all those errors which were used by a slave to escape from the anger of his master' (quoted in Endersby, *Imperial Nature*, p. 291).
18. *lengthy speech last Thursday*: on this, see n. 17, and Endersby, *Imperial Nature*, pp. 289–90.
19. *our clever boy*: not identified.
20. *Dalhousie*: Dalhousie Castle in Scotland.
21. *Gibbon 12 vols*: English historian Edward Gibbon's (1737–94) monumental six-volume *The History of the Decline and Fall of the Roman Empire* (1776–89) was published in twelve volumes in various editions in the early nineteenth century.
22. *Milton 4 vols*: English poet John Milton's (1608–74) *Paradise Lost.*

23. *Warton's Virgil*: the works of Roman poet Virgil (70–19 BC), as translated by the English academic Joseph Warton (1722–1800).
24. *Sir W[ilia]m Hamilton 4. vols, Logic & Metaphysics*: W. Hamilton, *Lectures on Metaphysics and Logic*, ed. H. L. Mansel and J. Veitch, 4 vols (Edinburgh and London: William Blackwood and Sons, 1860).
25. *"Tyndall on Sound"*: *Sound*.
26. *lecture on Imagination*: J. Tyndall, *Essays on the Use and Limit of the Imagination in Science* (London: Longmans, Green, and Co., 1870). This volume contains both Tyndall's 1868 BAAS lecture in Norwich ('Scientific Limit of the Imagination') and his 1870 BAAS lecture in Liverpool ('Scientific Use of the Imagination').
27. *Tom*: not identified; possibly the same son referred to in n. 19.

From Hector Tyndale 12 August 1872 3819

Philadelphia August 12th/72.

My dear John

On the 7th inst[ant]: I telegraphed to you at the Royal Institution[1] "First lecture Boston October fifteenth. Unchangeable". dated it only "Philadelphia" and signed "Hector"—to shorten it. I did this because I had not time to give you a written reply to yours of the 16th July,[2] from Switzerland,[3] by the 10th inst[ant]: as you desired. And my telegram gave you the information you wished viz: "the times & places" of your lectures[4] & it also disposed of your question as to the postponement of them—as the arrangements already made were "unchangeable".

I also wrote you, one on the 5th[5] another on the 6th inst[ant][6]—the last immediately on receipt of yours on the 17th July.[7]

A day or two since, speaking with a scientific gentleman of this city Doctor Morris,[8] he said he hoped you would give your courses of lectures in this Country (the whole United States he meant) "largely illustrated both by experiments and diagrams". He further said "The American people learn more through their eyes than ears". I have already given my opinion on this and obtruded my advice and that gentleman was right at least in the first expression of his hopes.

I presume however that the American lecture givers are like the same class everywhere and that our eyes are kings of our senses.

I did not state, in my telegram, the dates of other lectures for I deemed that unimportant. These I had given you in mine of the 5th.

I have also spoken of the large size of the halls you will speak in that you might regulate your diagrams &c by this. But the largest and best, I think, of the halls you will find the Academy of Music[9] here—if it can be had for your entire Philadelphia course Mr Pugh told me it could be had for only the last week of your being here and that he would try to get Baltimore or other places to exchange for one week.

Give my regards to Hirst and Debus—who I wish were coming with you.

When you write to Gorey[10] give my kindest regards to your sister Emma[11] and to Doctor John[12] and his family.

Affectionately yours | Hector Tyndale

I see Professor Lesley every day when we speak of you. He is one of the most generally learned men in the country and abreast of the Science of the day.

Prof[essor]: John Tyndall

RI MS JT/1/T/69
RI MS JT/1/TYP/5/1709–10

1. *7th inst[ant]: I telegraphed to you at the Royal Institution*: telegram missing.
2. *yours of the 16th July*: letter missing.
3. *from Switzerland*: see letter 3716, n. 6.
4. *your lectures*: for Tyndall's American lecture tour, see Introduction, n. 8.
5. *5th*: letter 3815.
6. *6th inst[ant]*: letter 3816.
7. *yours on the 17th July*: letter missing.
8. *Doctor Morris*: possibly Caspar Morris (1805–84), an American physician who was born and died in Philadelphia (*ANB*).
9. *Academy of Music*: see letter 3815, n. 9.
10. *Gorey*: in County Wexford, Ireland.
11. *Emma*: Emily Tyndall.
12. *Doctor John*: John Tyndall (of Gorey), Emily Tyndall's doctor.

From Heinrich Rudolf Vieweg[1] — 12 August 1872 — 3820

Herr Profeßor Tyndall | zu | Bel-Alp |
via Brig, Canton Wallis | Schweitz

Braunschweig 12tn August 1872.

Hochgeehrter Herr!

Wir erhielten Ihr gef. Schreiben datirt Bel-Alp 5tn Aug., aus welchem hervorgeht, daß wir Ihrem Brief vom 11 Juni nicht die von Ihnen beabsichtigte Deutung gegeben zu haben scheinen. Ein Mißverständniß ist ja bei einer Correspondenz in verschiedenen Sprachen [wohl] möglich; haben wir Sie nicht richtig verstanden, so bitten wir freundlichst, dieses entschuldigen zu wollen und überzeugt zu sein, daß wir außerordentlich großen Werth darauf legen, Ihre Verleger für Deutschland zu sein, und daß wir Ihnen mit dem größten Vergnügen dienlich sein werden, *[wo]* wir können. Wir sehen danach den in Ihrem letzten Briefe angedeuteten Vorschlägen in Betreff einer Übersetzung der ersten Hälfte Ihres Werkes „on Diamagnetism" entgegen.

Die Übersetzung Ihrer „Hours of Exercise" ist nun zur Versendung fertig. Sollen wir Ihnen vielleicht einige Exemplare nach der Schweitz senden, oder wünschen Sie dieselben wie früher nach London zu erhalten? Wir bitten eine gef. Nachricht darüber, ebenso, wohin wir das Honorar zu senden haben, welches zu Ihrer Disposition bereit liegt.

Mit größter Hochachtung | ganz ergebenst | Friedr. Vieweg & Sohn

Herr Professor Tyndall | at | Bel-Alp[2] | via Brig, Canton Wallis | Switzerland

Braunschweig 12th August 1872.

Esteemed sir!

We received your kind letter dated Bel-Alp 5th Aug.,[3] from which it emerges that we seem not to have given your letter of 11 June[4] the interpretation intended by you. A misunderstanding is of course *[no doubt]* possible with a correspondence in different languages; if we did not understand you correctly, then we ask most kindly that you would excuse this and be convinced that we place extraordinarily great importance on being your publishers for Germany, and that we shall be of service to you with the greatest pleasure *[where]* we can. We look forward accordingly to the proposals alluded to in your latest letter regarding a translation of the first half of your work "on Diamagnetism".[5]

The translation of your "Hours of Exercise"[6] is now ready for dispatch. Shall we send some copies to you in Switzerland perhaps, or do you wish to wish to receive them, as previously, in London? We ask for a kind confirmation of this, and, likewise, where we have to send the royalty which is ready at your disposition.

With the greatest esteem | your most humble servants | Friedr[ich]. Vieweg & Son

TU Braunschweig/Universitätsbibliothek/UABS V 1 T : 63
RI MS JT/1/V/13

---

1. *Heinrich Rudolf Vieweg*: while the letter is signed 'Friedr. Vieweg & Sohn', we believe the sender is Heinrich Rudolf Vieweg, who took over the family publishing firm, Friedrich Vieweg und Sohn, from his father Eduard Vieweg (1797–1869).
2. The original has the sender's note 'fr.', most probably an abbreviation for 'franco' (i.e.,

'postpaid'), on the same line as (but to the far left of) 'Bel-Alp'. For the Bel Alp, see letter 3771, n. 5.

3. *your kind letter dated Bel-Alp 5th Aug.*: letter 3812.
4. *your letter of 11 June*: letter 3735.
5. *first half of your work "on Diamagnetism"*: see letter 3735, n. 7.
6. *translation of your "Hours of Exercise"*: see letter 3790, n. 4.

To John William Draper 15 August 1872 3821

Royal Institution of Great Britain | 15th Aug 1872

My Dear Dr. Draper.

On my return from the Alps[1] I find your note[2] awaiting me here. A friend had forwarded your paper[3] to me in the Alps, where I read it, or rather I read it on my way home from the mountains.

The paper is marked by the skill and ability which attach themselves to all that you have done. Nevertheless I will not say that I intimately agree with it.

But I am so full of other matters now that I will not do more than make this brief reference to the paper.

It will give me great pleasure to see you in your country.[4]

I start in the Russia on the 28th of September[5]. Had I however known how difficult it is to put matters, which with my appliances here are easy enough, into due order I sh[ould][d] have shrunk from the responsibilities of lecturing in America.

But I am now in for it and must get through it as *[manageably]* as I can.

I thought I should have begun at New York, but a telegram[6] just informs me that I am to begin at Boston. This will bring me by an easier incline to my main trail.

I suppose there is no difficulty about acids in Boston. I dare say also I can have a good carpenter to assist me there. I hope finally that if it sh[oul][d] be needed I can get a good white screen for projection made at Boston.

Good bye till I see you | Always Yours | John Tyndall

Library of Congress, John William Draper Family Papers

---

1. *my return from the Alps*: see letter 3716, n. 6.
2. *your note*: letter missing.
3. *your paper*: see letter 3791, n. 1.
4. *in your Country*: referring to his American lecture tour (see Introduction, n. 8).
5. *Russia on the 28th of September*: see letter 3795, n. 1.
6. *telegram*: probably a telegram from Hector Tyndale that is missing (see letter 3819, n. 1).

To Thomas Archer Hirst [17 August 1872][1] 3822

Royal Institution of Great Britain | Saturday

My dear Tom.

The theatre of the School of Mines, not our own, is the scene of my operations at present.[2]

They are very heavy, and entirely shut out the thought of my joining you.[3]

I run down to Brighton to day[4]—but it is only for the night—to have some conversation with Spottiswoode Lubbock & Hooker, about the affairs of the last mentioned.[5]

A plain question was never so miserably mismanaged: & it will require the greatest skill to maintain dignity in what must be regarded as our defeat.

I wish heartily I could join you—it would be a great delight to me, but it is not to be hoped for.

Snowdon[6] from Capel Curig[7] is very fine. I hope you will have good weather for the ascent—today it is glorious here—May it be so there!

Yours ever aff[ectionate]ly | John Tyndall

RI MS JT/1/T/982
RI MS JT/1/HTYP/597/2

---

1. *[17 August 1872]*: the date is given by Tyndall's mention of attending only one day of the BAAS annual meeting in Brighton. In letter 3830, also to Hirst, he wrote that 'I was only one evening at Brighton: went down mainly on Hooker's business & to hear Spottiswoode lecture'. Spottiswoode's lecture occurred on 17 August 1872 (see letter 3830, n. 10).
2. *theatre of the School of Mines... my operations at present*: since the laboratory of the RI was under renovation (see letter 3776, n. 14), Tyndall was using the theatre at the Royal School of Mines to prepare his experiments for his American lecture tour (see Introduction, n. 8). Tyndall previously held the Chair of Natural Philosophy at the Royal School of Mines from 1859–68, in which he lectured to students during the week and in some years gave evening lectures for working men.
3. *joining you*: at this time Hirst was traveling in Wales.
4. *run down to Brighton to day*: see n. 1.
5. *affairs of the last mentioned*: Joseph Hooker's dispute with Acton Ayrton being taken up in Parliament (see letter 3755, esp. n. 1).
6. *Snowdon*: the highest peak in Wales.
7. *Capel Curig*: Hirst stayed at Capel Curig, a village in Wales, from 15–19 August 1872 (Brock and MacLeod, *Hirst Journals*, pp. 1947–8).

To Hector Tyndale 21 August 1872 3823

Athenaeum Club | Pall Mall S.W.[1] | 21st. August 1872

My dear Hector

Both your letters[2] have reached me & your telegram[3] preceded them—accept my best thanks for all.

I go, as stated to you, by the Russia,[4] and am exceedingly glad to have before me a fixed programme.[5]

Your account of Mr. Pugh[6] makes me feel that our relations will be amicable and agreeable.

One thing my dear Hector you must bear in mind. I shall illustrate my lectures sufficiently, I shall make as good experiments as under the circumstances I can make, but I have heard from American Scientific men accounts of lectures given in New York & elsewhere; brilliant to the last degree, but still such as I could by no means imitate.

In fact I must make the lectures sound science, aiming to address the intellect more than the eye, and only addressing the eye to make it the avenue to the intellect. Hence, you see, in comparison to the more fiery lectures to which I have referred, mine may seem cold.

I will go with you from New York to Philadelphia; but I must not halt there for more than a day. In fact the unpacking of my instruments and the arranging for the lectures will occupy some days.

Now let me be plain with you. Give me a garret[7] if you will, only let my bed room be quiet; and when you give me wine let it be really good claret—there is impudence for you.

My resulting weakness is a want of sleeping power—this power I must endeavour to keep intact. Nothing interferes with it so much as an error in point of quality of wine. Here my habit is to drink every day at dinner a pint of La Rose.

Remember you must prop your cousin up in every possible way; Keep him out of excitement, and see that his mind gets fair play from his miserable weed of a body.

Some funny Doctor has been writing about me calling me "a burley Englishman" who is always in robust health.[8] I am in the mountains, but never so in London.

The Messrs. Appleton[9] have been good enough to say that they will see my luggage through the hands of the officials in New York.

Give my kindest regards to your wife[10]

Give my warm remembrances to your friend MacLaughlin.[11]

Your affectionate Cousin | John

RI MS JT/1/T/1454
RI MS JT/1/TYP/5/1711–2

1. *Athenaeum Club | Pall Mall S.W.*: this letterhead is indicated on LT's typescript, but is not visible in the scan of the handwritten letter we had available for this transcription. For the Athenaeum Club, see letter 3710, n. 5.
2. *Both your letters*: previous letters from Tyndale are 3815, 1816, and 3819.
3. *your telegram*: telegram missing; see letter 3819, n. 1.
4. *the Russia*: Tyndall traveled from Liverpool to New York on the SS *Russia* (see Introduction, n. 8).
5. *fixed programme*: for Tyndall's American lecture tour, see Introduction, n. 8. In letter 3819, Tyndale sketched out the programme for the cities and dates of the lecture tour.
6. *Your account of M^r^. Pugh*: Tyndale informed Tyndall of the services of Thomas Burnett Pugh in letter 3815.
7. *garret*: a room on the uppermost floor of a house; an apartment formed either partially or wholly within the roof, an attic (*OED*).
8. *Some funny... in robust health*: letter missing.
9. *Messrs. Appleton*: William Henry Appleton (1814–99) and one or more of his four brothers with whom he ran the publishing firm Appleton and Co.
10. *your wife*: Julia Tyndale.
11. *MacLaughlin*: John M. McLaughlin (see letter 3815, n. 16).

From Edward Stanley 22 August 1872 3824

Private. | Knowsley, Prescot. | Aug. 22d 1872.

Dear Professor Tyndall

I join with you in regretting the very unsatisfactory issue of Dr Hooker's business in the House of C[ommons].[1] It is useless now to dwell on that, but I think it would have been better if no regular debate and division was to be taken, to have rested on the half apology (though inadequate) extracted by public opinion from the Treasury, which said as plainly as official language could be expected to say it, that Ayrton[2] had put himself in the wrong. Now the matter has passed into a new stage and Ayrton, as I gather from your letter,[3] assumes the tone of an injured person.

I should certainly say that no further concession can be, or ought to be, made by Dr Hooker. I am equally of opinion that he ought not, under any provocation, to resign, but to leave the discredit of removing him to his enemies. Probably they will not venture on that step: if they do there will be an outcry compared to which that of last session was very mild.

The papers were placed in my hands, and in those of others, so late that it was not easy to note their deficiencies; but it is obvious, on looking at them leisurely, that much has been suppressed which ought to have been included, and many things inserted (very skillfully) with the object of bringing the really important facts under a mass of irrelevant matter, and perhaps also of making the quarrel appear to turn on a multitude of wholly unimportant details.

I cannot conceive the existence of any reason why Dr Hooker should not use his own official letter to Mr Gladstone,[4] he having been challenged either to prove or to withdraw his charge.

Has any reference been made to the Chanc[ellor]. of the Exchequer[5] in this business? He is the real head of the Treasury department: he I believe drew, or at any rate revised, the minutes (there are two) supporting Dr Hooker against Ayrton,[6] and I know that he has a very strong opinion as to the conduct and official character of his colleague. In fact Ayrton left the Treasury because the two could not get on together. I am afraid at this distance I cannot be of much use, but pray consult me if there is any matter in which I can help your friend. I cannot understand the taking the view he has taken Prime Minister.[7] Hasty and violent he may be, but love of justice is his strong point.

Very sincerely yours | Derby.

RI MS JT/1/TYP/8/2702
LT Typescript Only

---

1. *Dr Hooker's business in the House of C[ommons]*: following discussion in the House of Lords about Joseph Hooker's dispute with Acton Ayrton over the management of the Royal Botanic Gardens, Kew, it was further discussed in the House of Commons, resulting in, as Stanley noted, a 'half apology' from the Treasury but no repercussions for Ayrton (see letter 3755, esp. n. 1).
2. *Ayrton*: Acton Ayrton.
3. *your letter*: letter missing.
4. *his own official letter to Mr Gladstone*: in a letter to the Board of Works dated 19 August 1871, Hooker wrote: 'It appears to me that acts so gravely concerning the subordination of the Director to the Board should be referred to a higher authority, and I beg respectfully to inform you that I claim the privilege of bringing them under the cognizance of the Right Honourable the First Lord of the Treasury' (that person also being the Prime Minister) ('Copies of Correspondence between the Board of Works and Dr. Hooker relating to Changes proposed to be introduced into the Direction and Management of the Gardens at Kew', HC 25, Parliamentary Papers (1872), p. 28). Hooker's letter to Gladstone, dated 31 August 1871, was published in 'Copies of Correspondence between the Board of Works and Dr. Hooker relating to Changes proposed to be introduced into the Direction and Management of the Gardens at Kew', HC 25, Parliamentary Papers (1872), p. 45–6; and *Nature*, 6 (11 July 1872), p. 213.
5. *Chanc[ellor]. of the Exchequer*: Robert Lowe.

6. *minutes (there are two) supporting Dr Hooker against Ayrton*: a Treasury minute is an administrative regulation for any department under the Treasury (*OED*). One of these two Treasury minutes is likely that of 24 July 1872, heavily in favor of Hooker, which concluded: 'My Lords see no reason why under these conditions there should be any serious difficulty in discharging the respective duties of the First Commissioner [Ayrton] and of the Director of the Gardens [Hooker] in a manner satisfactory to both, whoever may be the occupants of those offices, maintaining the proper authority of the First Commissioner, with due regard to the position and character of the Director of the Gardens' ('Copies of Correspondence between the Board of Works and Dr. Hooker relating to Changes proposed to be introduced into the Direction and Management of the Gardens at Kew', HC 25, Parliamentary Papers (1872), p. 176; and *Nature*, 6 (8 August 1872), pp. 280–1).
7. *Prime Minister taking the view he has taken*: after having evaded the issue for as long as he could, William Gladstone essentially stated at the House of Commons on 8 August 1872 that it was not within his capacity to intervene in the issues between Hooker and Ayrton, and that he expected Ayrton to exercise 'his rule with mildness', and for Hooker to do 'his duty in subordination' to Ayrton ('The Royal Gardens, Kew—Dr. Hooker, and the First Commissioner of Works', HC Deb 08 August 1872 vol. 213 cc710–758, *HANSARD 1803–2005* (online, accessed 30 May 2021)).

From Joseph Dalton Hooker 22 August 1872 3825

Royal Gardens, Kew. | August 22d 1872.

Dear Tyndall

I am much gratified by Lord Derby's letter.[1]

I am in the middle of a statement of my case for the prosecution[2] for Spottiswoode's guidance in which I hope I am bringing out the strong points in an unassailable form. I am in the hopes that this may be ready by to-morrow afternoon, and that you will be good enough to take it to Spottiswoode with you on Saturday night or Sunday morning. Please let me know the latest moment at which it should be delivered at Albemarle Street,[3] or elsewhere, to this end, and I will send it by messenger.

Helps thinks that it would be impossible to refuse Cardwell's offer[4] and augurs well from it.

Ever aff[ectiona]tely yours | J. D. Hooker.

Please telegraph if necessary so that my despatch to Spottiswoode may not miss you.

BL Add MS 63092, ff. 126–9
RI MS JT/1/TYP/8/2703

1. *letter*: letter 3824.
2. *a statement of my case for the prosecution*: it is not clear as to what document this refers to, but it is certainly related to the Memorial in support of Joseph Hooker in his dispute with Acton Ayrton (see letter 3755, esp. n. 1).
3. *Albemarle Street*: the RI.
4. *Cardwell's offer*: Edward Cardwell's offer is not identified.

From Joseph Dalton Hooker — 24 August 1872 — 3826

Royal Gardens, Kew, | Aug. 24th 1872.

Dear Tyndall

Thanks for telegraph.[1] I am overburdened with arrears.[2] Ayrton[3] sent his letter by special messenger with orders to wait for answer, or I should have brought it in to you.

Of course he will now write officially, offering me withdrawal of charges, or suspension from duties.

I am drawing up a careful document to send (if approved) to Gladstone.

The truth of this whole affair must be brought out somehow.

They have carried the recommendation in the section unanimously at Brighton[4] and in the Committee[5] too.

Lubbock has taken a written assurance from Galton[6] that "no change was contemplated at Kew" I do think that this is too bad, to accept D. G. as the scientific mouth-piece of either Ayrton or the Gov[ernmen]t, in the matter of an Institution of which I am Director, and with which he has nothing to do but superintend the execution of certain works.

Bentham comes here to-morrow to stay two days.

Ever yours | J. D. Hooker.

I assume that you know that Mr Glyn[7] was out of town.

RI MS JT/1/TYP/8/2704
LT Typescript Only

1. *telegraph*: in a letter to Tyndall of 22 August 1872 (letter 3825), Hooker wrote 'Please telegraph if necessary so that my despatch to Spottiswoode may not miss you'. The telegram from Tyndall is missing.
2. *arrears*: behind in the discharge of duties or liabilities; in indebtedness (*OED*).
3. *Ayrton*: Acton Ayrton.
4. *in the section at Brighton*: the following was published as the conclusion to a summary of John Lubbock's report to Section B (Biology) at the forty-second annual meeting of the BAAS held in Brighton from 14–21 August 1872: 'there was one matter which he could

not but touch upon, but which he yet could not properly treat at length. Great anxiety had been felt during the last few months lest changes should be made at Kew which would prove prejudicial to its scientific work, and lead to the retirement of Dr. Hooker. He felt sure that he only expressed the feeling of the scientific world when he said that such an event would be a misfortune to science, and when he stated his hope that the Government would do nothing to retard or impede the valuable scientific work now going on at Kew' (*Brit. Assoc. Rep. 1872*, p. 126).

5. *Committee*: probably the general committee of the BAAS. Perhaps also a reference to the Scientific Committee of the Royal Horticultural Society (see letter 3788, n. 10).
6. *Galton*: Douglas Strutt Galton (see letter 3788, n. 6).
7. *Mr Glyn*: George Grenfell Glyn (1824–87), 2nd Baron Wolverton, a British banker and politician. Glyn was Member of Parliament for Shaftesbury from 1853–73, then entered the House of Lords. He also served in Gladstone's first administration as Parliamentary Secretary to the Treasury, from 1868–73 (*ODNB*).

From Joseph Dalton Hooker 28 August 1872 3827

Kew, | Aug. 28th 1872.

My dear Tyndall

Your note[1] to Cardwell[2] is both right and good, I hope not too good for me. I do wish that he may respond.

Another American called on me today, an Emerson,[3] and I have just got a newspaper from Milwaukee Wisconsin[4] full of praise of Kew but <u>not</u> of Ayrton.[5]

Let me know if you are inclined for a walk next Sunday and shall we go to Down?[6]

Ever yours | J. D. Hooker—

RI MS JT/1/TYP/8/2707
LT Typescript Only

---

1. *Your note*: letter missing.
2. *Cardwell*: on Edward Cardwell, see letter 3825.
3. *an Emerson*: presumably a relative of Ralph Waldo Emerson.
4. *newspaper from Milwaukee Wisconsin*: the Library and Archives at the Royal Botanic Gardens, Kew holds a letter to Hooker from the American naturalist Increase Allen Lapham (1811–75), dated 12 Aug 1872, in which Lapham enclosed a clipping from the *Milwaukee Sentinel* newspaper that discussed Hooker positively, but the clipping is missing (Royal Botanic Gardens, Kew: Archives: Directors' Correspondence 197/467). The article was probably: 'A Director Under Discipline', *Milwaukee Sentinel*, 7 August 1872, p. 2.

5. *Ayrton*: Acton Ayrton.
6. *Down*: to Down House, Charles Darwin's home in the village of Downe in Bromley, Kent. Tyndall must not have agreed to go, since Darwin wrote to Hooker that he will be glad to have Hooker and his wife Frances at Down on 7 September 1872 (Charles Darwin to Joseph Dalton Hooker, 29 August [1872], in *Darwin Correspondence*, vol. 20, pp. 376–7; this is letter 8491 in the online edition of the Darwin correspondence).

From Frederick Pollock[1] 28 August [1872][2] 3828

Russischer Hof[3] | Heidelberg | Aug. 28.

My dear Tyndall,

The first thing that caught my eye in the bookseller's shops here was: In den Alpen.[4] Von John Tyndall. <*1 sentence excised*>. Whereat I was pleased. I have been three weeks in the Alps—I must have missed you by only a few days at the Bell Alp.[5] My companions were Pratt[6] and Leaf[7] of Trinity—good men and true both—My guide the excellent Peter Baumann.[8] Rubi[9] was engaged elsewhere. We had a noble day on the Finsteraarhorn[10]—and next day essayed to go up the N[orth]. side of the Aletschhorn.[11] It was determined to get up to the Dreieckhorn[12] from that side and then work along the ridge. But the climb of the Dreieckhorn proved much more difficult than it looked. Bad rocks, steep slopes, and much step-cutting in hard ice. Six hours from Faulberg[13] to Dreieckhorn—The connecting ridge to the Aletschhorn looked possible, but too long to be accomplished in the day. It might have been better to force a way straight up thro' the crevassed névé[14] and steep higher slopes closer under the main peak. But this also would certainly be long and only practicable in favourable conditions of snow. And so back to the Eggishorn.[15] Here my friends parted from me and I came to Grindelwald[16] by the Jungfrau Joch[17] with Baumann and Peter Schlegel[18]—a good rising guide. What glorious ice-cliffs and '<*1 word excised*>'. We cut down deep into a great Schrund[19] and came out through a natural archway of ice—and the loveliest moonlight walk from the Faulberg to the top of the pass. We were there at 4.25 pm and down safe on the rocks between the <*1–2 words excised*> and the Eiger[20] glaciers at 8. I should not have liked to be many hours later, yet I heard that Matthews[21] and Morshead[22] had a few days before ascended The Mönch[23] from the Wengern Alp[24] and came down the Jungfrau Joch quite late in the day: I don't understand how it can have been prudent. By the same token we found most of the steps ready made, which was a luxury.

This is the second time I have been disappointed of the Aletschhorn. You would hardly care to go up it a second or third time next year if I find you at the Bel Alp? May you have a good time and good listeners in the U.S.[25]

Yours ever | F. Pollock.

RI MS JT/TYP/6/2138
LT Typescript Only

---

1. *Frederick Pollock*: Frederick Pollock (1845–1937), third baronet, a jurist, and the eldest son of Juliet and William Frederick Pollock. He was schooled at Eton College (1858–63) before entering Trinity College, Cambridge, where he obtained his BA (1867) and MA (1870). He was admitted to Lincoln's Inn in 1868, was called to the bar in 1871, and married Georgina Harriet Deffell in 1873. Pollock was a distinguished jurist, author of many legal treatises, and became professor of jurisprudence at Oxford University (1883–1903). He was also an avid mountaineer (*ODNB*).
2. *[1872]*: the year is given by reference to Tyndall's upcoming American lecture tour (see n. 25), and that Pollock had later written about climbing in the Alps in August 1872 with Leaf and Pratt, and alpine guide Baumann, in his *For My Grandson: Remembrances of an Ancient Victorian* (London: John Murray, 1933), p. 213.
3. *Russischer Hof*: a hotel.
4. *In den Alpen*: the German translation of Tyndall's *Hours of Exercise in the Alps*—J. Tyndall, *In den Alpen* (Braunschweig: Friedrich Vieweg und Sohn, 1872)—includes a foreword by Gustav Wiedemann dated 19 July 1872, so the book must have just been published when Pollock saw it in a bookstore in August 1872.
5. *Bell Alp*: Bel Alp (see letter 3771, n. 5).
6. *Pratt*: John Henry Pratt (1809–71), a British clergyman and mathematician who was educated at Trinity College, Cambridge (*ODNB*).
7. *Leaf*: Walter Leaf (1852–1927), a British banker and classical scholar who was educated at Trinity College, Cambridge (*ODNB*).
8. *Peter Baumann*: Peter Baumann (1833–1921), an alpine guide based in Grindelwald ('Peter Baumann', *Alpine Journal*, 34 (1921), pp. 463–6).
9. *Rubi*: Peter Rubi (d. 1880), an alpine guide based in Grindelwald (F. Pollock, 'In Memoriam. Peter Rubi', *Alpine Journal*, 10 (1880), pp. 78–82).
10. *Finsteraarhorn*: the highest mountain of the Bernese Alps in the Swiss canton of Valais.
11. *Aletschhorn*: Aletschorn, the second highest mountain of the Bernese Alps in the Swiss canton of Valais.
12. *Dreieckhorn*: a mountain of the Bernese Alps in the Swiss canton of Valais, next to the Aletschorn.
13. *Faulberg*: the Faulberg Cave that was often used as a respite by climbers before attempting the Aletschorn or the Jungfrau, before any hut was built. Its location was probably near the Concordiaplatz, where later a hut was built (*Alpine Journal*, 31 (1917), p. 258).
14. *névé*: compacted or hardened snow, as found on the upper part of a glacier (*OED*).
15. *Eggishorn*: a mountain of the Bernese Alps in the Swiss canton of Valais.
16. *Grindelwald*: see letter 3805, n. 12.
17. *Jungfrau Joch*: Jungfraujoch, a saddle connecting the Jungfrau and the Mönch, both mountains of the Bernese Alps.
18. *Peter Schlegel*: an alpine guide (dates unknown).

19. *Schrund*: a crevasse (*OED*).
20. *Eiger*: a mountain of the Bernese Alps in the Swiss canton of Valais.
21. *Matthews*: Charles Edward Mathews (1834–1905), a British mountaineer and founding member of the Alpine Club (*ODNB*).
22. *Morshead*: Frederick Morshead (1836–1914), a British mountaineer and founding member of the Alpine Club (*ODNB*).
23. *The Mönch*: or Monk, a mountain of the Bernese Alps in the Swiss canton of Valais.
24. *Wengern Alp*: the slopes on the Lauterbrunnen side of the pass over Kleine Scheidegg towards Grindelwald.
25. *good listeners in the U.S.*: for Tyndall's American lecture tour, see Introduction, n. 8.

From François-Napoléon-Marie Moigno 30 August 1872 3829

[LES MONDES | Revue Hebdomadaire | DES SCIENCES | *et de leurs applications* | aux Arts et à l'Industrie | par | M. L'ABBÉ MOIGNO | RÉDACTION | 2, Rue d'Erfurth, 2.

Paris, *le*] 30 août 1872

Mon cher Mr Tyndall,

Mr Gauthier-Villars est prêt à remplir l'engagement qu'il a contracté avec vous, et, à mettre sous presse immédiatement la seconde édition française de La Chaleur. Envoyez moi, c'est-à-dire faites moi envoyer, la quatrième ou cinquième édition anglaise pour que je complette ma traduction. Tenez vous à corriger les épreuves?

Faut-il vous les envoyer en Amérique ou attendre votre retour?

Auriez vous le temps de les recevoir avant de partir.

A vous de coeur | l'abbé F. Moigno

[LES MONDES[1] | Weekly Review | OF THE SCIENCES | *and their applications* | to the Arts and the Industry | by | FATHER MOIGNO | EDITORIAL OFFICES | 2 Erfurth street, 2.

Paris,] 30 August 1872

My dear Mr Tyndall,

Mr Gauthier-Villars[2] is ready to fulfil the commitment that he made towards you, and to immediately start printing the second French edition of Heat.[3] Send me, that is to say, have sent to me, the fourth or fifth English edition[4] so that I will complete my translation. Do you want to correct the proofs?

Should I send them to you in America[5] or wait for your return?

Would you have the time to receive them before your departure?
Yours truly | Father F. Moigno

RI MS JT/1/M/121

1. *LES MONDES*: a scientific journal that Moigno founded in 1863.
2. *M<u>r</u> Gauthier-Villars*: Jean-Albert Gauthier-Villars.
3. *second French edition of <u>Heat</u>*: J. Tyndall, *La Chaleur: Mode de Mouvement*, 2nd edn, trans. F.-N.-M. Moigno (Paris: Gauthier-Villars), 1874.
4. *fourth or fifth English edition*: since the fifth edition of Tyndall's *Heat: A Mode of Motion* was not published by Longmans, Green, and Co. until 1875, for this translation the fourth edition of 1870 (also Longmans, Green, and Co.) was most likely used.
5. *in America*: for Tyndall's American lecture tour, see Introduction, n. 8.

To Thomas Archer Hirst 31 [August 1872][1] 3830

Royal Institution of Great Britain | Saturday 31st

My dear Tom.

Immediately on reading your note to me dated from Capel Curig,[2] and informing me that you would remain there a day a day[3] or two I wrote to that place.

I am very sorry that my note[4] did not reach you as it was somewhat out spoken on some of the points that interest us. Pray write to the Postmaster of Capel Curig.

I am well, though much bothered bothered[5] through perplexing changes of port, when a direct and courageous course is evidently the one thing needful. Perhaps I ought not to remark on this, for once upon a time in my own history I was vacillating, and to a certain extent cowardly. However we must do as well as we can.

I have been much occupied with my experiments:[6] but Hooker's affair[7] rather breakes me up. I must shut one or the other of them out from my mind for it is Chaos & ruin to attempt to drive both coaches at the same time.

I have resolved to abide by the Russia.[8] They have already improved my cabin & hope to improve it still more.

I was only one evening at Brighton:[9] went down mainly on Hooker's business & to hear Spottiswoode lecture.[10] <u>Your Chief</u>[11] is belauded by the Abbé Moigno![12]

I work daily at the School of Mines preparing for my lectures.[13]

London is very quiet which is a blessing. The Athenaeum[14] are at the Senior United Service.[15]

On the pressure of Farrer of the Board of Trade[16] I have agreed to be one of the gas referees[17] with Pole & Vernon Harcourt. He declares it will not occupy me more than two hours a month. I offered my advice on special occasions gratis—to this he demurred—so that I am forced to accept £200 a year.

Yours ever affe[ctionate]ly | John Tyndall

Give Lilly[18] love for love.

RI MS JT/1/T/981
RI MS JT/1/HTYP/598

---

1. *[August 1872]*: the month and year are given based on mention of Hirst being in Wales, 'Hooker's affair', Tyndall's upcoming American lecture tour, and the BAAS meeting at Brighton, and Tyndall's appointment as a Metropolitan Gas Referee.
2. *your note to me dated from Capel Curig*: letter missing. Hirst stayed at Capel Curig, a village in Wales, from 15–19 August 1872 (Brock and MacLeod, *Hirst Journals*, pp. 1947–8).
3. *a day*: Tyndall mistakenly wrote 'a day" twice.
4. *my note*: letter missing.
5. *bothered*: Tyndall mistakenly wrote 'bothered' twice.
6. *my experiments*: preparing the experiments for his upcoming lecture tour in America (see Introduction, n. 8).
7. *Hooker's affair*: Joseph Hooker's dispute with Acton Ayrton being taken up in Parliament (see letter 3755, esp. n. 1).
8. *the Russia*: Tyndall traveled from Liverpool to New York on the SS *Russia* (see Introduction, n. 8).
9. *Brighton*: the forty-second annual meeting of the BAAS was held in Brighton from 14–21 August 1872 (*Brit. Assoc. Rep. 1872*). Tyndall attended only on 17 August (a Saturday).
10. *to hear Spottiswoode lecture*: at the BAAS meeting, on 17 August, William Spottiswoode lectured on 'Sunshine, Sea, and Sky' for 'the Operative Classes of Brighton' (*Brit. Assoc. Rep. 1872*, p. lxviii).
11. *Your Chief*: William Carpenter.
12. *belauded by the Abbé Moigno*: soon after the BAAS meeting in Brighton, François-Napoléon-Marie Moigno gave a report in his journal *Les Mondes*, but this did not include anything laudatory about Carpenter's address ('Association Britannique pour L'Advancement des Sciences. Cloture de la Réunion de Brighton', *Les Mondes*, 29 (5 September 1872), pp. 124–32). Tyndall did not attend this address as he only attended one day of the meeting (see n. 9). Thus, it is not clear how Tyndall viewed Carpenter as 'belauded by the Abbé Moigno'.
13. *School of Mines preparing for my lectures*: see letter 3822, n. 2.
14. *Athenaeum*: the Athenaeum Club (see letter 3710, n. 5).
15. *at the Senior United Service*: the Senior United Service Club was a London gentleman's club founded in 1815 for members of the British Army and Royal Navy above a certain rank. It was located 116 Pall Mall in London, just down the street from the Athenaeum

Club at 107 Pall Mall. At this time members of the Athenaeum were welcome at the United Service Club. From the Athenaeum's General Committee minutes of 6 August 1872: 'Sir Lintorn Simmons [a member of the committee in attendance] conveyed an Invitation from the Committee of the United Service Club to the Members of the Athenaeum to resort to their House during the fortnight when the Coffee Room service is suspended. Resolved. That this Invitation be accepted the Secretary writing to express the acknowledgements of the Committee to that of the United Service Club for their courteous consideration and making all necessary arrangements. Sir Lintorn Simmons thanked for his obliging attention' (Jennie De Protani, Archivist, the Athenaeum Club, personal communication).

16. *Board of Trade*: see letter 3736, n. 3.
17. *gas referees*: the Metropolitan Gas Referees were appointed by the Board of Trade under the City of London Gas Act (1868). In 1872, Tyndall and Vernon Harcourt replaced J. H. Patterson and John Sampson Pierce, while William Pole had replaced Frederick John Evans (1815–85) in 1870. Tyndall served in this position until 1892, Pole until his death in 1900, and Harcourt until 1917 (K. Hutchison, 'The Royal Society and the Foundation of the British Gas Industry', *Notes and Records of the Royal Society of London*, 39 (1985), pp. 245–70, on p. 265; *Sulphur Compounds in Gas. Report of the Proceedings before a Committee of the House of Commons on the Crystal Palace District Gas and The Gaslight & Coke Company Bills. Session 1877* (London: William B. King, 1877), p. 87).
18. *Lilly*: Emily Hirst.

To Edward L. Youmans 31 August 1872 3831

August 31, 1872.

MY DEAR YOUMANS: I am in the midst of my preparations[1] here, and shall have them ready so as to enable me to start in the Russia on the 28th September.[2]

I shall need your friendly aid in getting my apparatus through the custom house....

With regard to the lecture rooms, in all of them I must be able to lower the lights promptly. Most of my experiments will be projected on a screen.

I purpose mixing experiment and philosophy in due proportions.

I deal with the illustrative phenomena of light: the laws of reflection and refraction, analysis and synthesis, the bearing and significance of theories. Spectrum analysis and its revelations regarding the constitution of the sun. The higher phenomena of optics, interference and polarization, reaction of crystals upon light. The building of crystals. The extension of radiation beyond the range of the eye. The identity of light and radiant heat. This is a rough sketch of the subjects which will probably occupy me. I shall not know for a certainty until my preparations are complete.

Do your audiences look down upon the lecturer?

I suppose I can borrow an air pump in New York if I need it.

I suppose if they do not possess ice in Boston I can have a clear block sent there from New York.

Acids, of course, are to be had everywhere.

Are they in the habit of using compressed hydrogen and oxygen in iron bottles in America, and, if so, could I borrow such bottles?

I am taking one screen with me, but I shall sometimes require two. Is such a thing to be borrowed?

Now, like a good fellow, answer these questions within twenty-four hours, and oblige

Yours, ever faithfully, | JOHN TYNDALL.

E. A. Youmans, 'Tyndall's American Visit', *Popular Science Monthly*, 44 (February 1894), pp. 502–14, on p. 509

---

1. *my preparations*: preparing the experiments for his upcoming lecture tour in America (see Introduction, n. 8).
2. *start in the Russia on the 28th September*: Tyndall traveled from Liverpool to New York on the SS *Russia* (see Introduction, n. 8).

From C. Cecil Trevor — 31 August 1872 — 3832

Board of Trade[1] (Railway Department), |
Whitehall Gardens, 31 August 1872.

Sir,

I AM directed by the Board of Trade to transmit to you herewith a copy of the Minute by which you are appointed to be one of the Gas Referees,[2] under the City of London Gas Act, 1868, and the Imperial and South Metropolitan Gas Companies Acts, 1869, for a period of one year from the 1st of September next, at a salary of 200l.

Offices have been provided for the use of the Referees at No 17, Buckingham street, Strand.

I am, &c. | (signed) *C. Cecil Trevor.*

John Tyndall, Esq., F.R.S. | Royal Institution.

'Correspondence between Board of Trade and Gas Referees, in reference to Mr. Patterson taking out Patent for Gas Purification', HC 260, Parliamentary Papers (1876), p. 17

---

1. *Board of Trade*: see letter 3736, n. 3.
2. *appointed to be one of the Gas Referees*: see letter 3830, n. 17.

From Emily Peel [August 1872][1] 3833

[Lady of the Lake][2]

The Times published a letter, dated from the Bel Alp & bearing the initials of J.T.[3] w[ich][h] I presume to have been from your pen, and which told me of y[ou][r] whereabouts. It gave me a hope that you might be tempted to pay y[ou][r] old friends here a visit, but we must now I fear, give up all hope of enjoying this pleasure. What have you done this season? Have you made any attempts to climb any hitherto impossible Peaks[4] or have you only quietly recruited your health on the lower lands?

I will no longer encroach on your precious time, but whilst wishing you success in your more serious winter avocations[5] ask you to believe me

yours very sincerely | Emily Peel

RI MS JT/1/P/74

---

1. *[August 1872]*: the date is given by reference to a letter in the *Times* (see n. 3 below).
2. *[Lady of the Lake]*: letterhead for the original steamer on Ullswater, a lake in England's Lake District.
3. *letter, dated from the Bel Alp & bearing the initials of J.T.*: J. T., 'The Aletsch Glacier', *Times*, 8 August 1872, p. 5.
4. *to climb any hitherto impossible Peaks*: in a very bold attempt, on 5 July 1872, Tyndall attempted, with Thomas Kennedy, to climb the Aiguille du Géant which remained unclimbed until 1882 (see letters 3783 and 3785; and *Ascent of John Tyndall*, p. 298).
5. *more serious winter avocations*: his upcoming American lecture tour (see Introduction, n. 8).

To Thomas Henry Huxley 5 September 1872 3834

Royal Institution | 5th. Sept. 1872.

My dear Huxley,

It seems like one of the Geological epochs since I heard directly from you or yours.[1] I spent a month at the Bel Alp[2] and as usual throve well there. The weather for half the month was very good, but for the remaining half it was atrocious. Still thanks to the presence of some agreeable people; and thanks also to the fact of having indoor work to do[3] I managed even during the bad weather to chase unhappiness away.

Hooker kept me well informed as regards his affairs[4]—and surely nothing could appear more promising. At Dijon[5] on returning I picked up a copy of the Daily Telegraph[6] which astounded me. A well planned campaign had turned out a fiasco. This came through putting qualities which would carry off the palm in a battle for the graces in a position which required to be filled by qualities needed in a battle for the strengths—Oh it did make me miserable.

But I soon shook that away, and was fit for another fight when I reached England. But then I found, or thought I found, that H[ooker]. himself had gone into utter collapse. In this however I was mistaken. He had acted in accordance with the advice of his friends, and after the first perplexity consequent on the debate he gathered himself up like a man for further action. But my first shock was so great, and my faith in all human things so shaken, that I fear I must have bothered poor H[ooker]. in my anxiety that every subsequent step should be marked by wisdom, strength and dignity.

No doubt he has told you how matters stand. That he has sent in a document[7] setting forth the grounds of the language which he used; at the same time apologizing for the tone, but holding firm to his complaint against the First Commissioner.[8] It was all that could be done, and it leaves the door open to future action on our part, if we wish to take it.

I have been organizing my staff and am about to begin my packing up. On the 28th. I start from Liverpool in the Russia.[9] I am booked for Boston on the 15th of October; and afterwards for New York, Philadelphia, Baltimore and Washington. These are already fixed,[10] but I can see that I shall be expected to do some outlying work besides.

Let me hear something about you. How are you? How is your wife?[11] How are the chicks?[12] I hope all flourishing. Give my love to them one and all.

There is a very nice and manysided American old gentleman named Cresson,[13] recommended to me by Lesley of Philadelphia. When the Athenaeum opens[14] he ought to be asked there for he is a man of distinction at home. He is Vice-president of the American Philosophical Society. I intend to put the matter in Hirst's hands, and should you see Mr Cresson at the Club I know you will speak kindly to him. He is a fine looking old fellow—tall, open-browed and white haired.

I think I have nothing more to talk about: one word about Owen. I never broke with that man, as you know, I hoped against hope that matters might in the end be healed up. But this last trick[15] makes me feel that those who broke with him knew him better than I did. It will greatly augment his isolation.

Yours ever and always, | John Tyndall.

RI MS JT/1/TYP/8/2712
IC HP 8.127
Typed Transcript Only

---

1. *yours*: Henrietta Huxley.
2. *Bel Alp*: see letter 3771, n. 5.
3. *indoor work to do*: preparing for his American lecture tour (see Introduction, n. 8).

4. *Hooker kept me well informed as regards his affairs*: Joseph Hooker's dispute with Acton Ayrton being taken up in Parliament (see letter 3755, esp. n. 1).
5. *Dijon*: a city in eastern France.
6. *copy of the Daily Telegraph*: not identified.
7. *document*: this may be the unidentified 'statement' Hooker referred to in letter 3825.
8. *First Commissioner*: Acton Ayrton.
9. *28th. I start from Liverpool in the Russia*: Tyndall traveled from Liverpool to New York on the SS *Russia* (see Introduction, n. 8).
10. *already fixed*: a general programme for Tyndall's American lecture tour was given to him by William Rhees (letter 3814) and Hector Tyndale (letter 3815).
11. *your wife*: see n. 1.
12. *chicks*: the Huxley children were: Jessie Oriana Huxley (1858–1927), Marian Huxley (1859–87), Leonard Huxley (1860–1933), Rachel Huxley (1862–1934), Henrietta (Nettie) Huxley (1863–1940), Henry Huxley (1865–1946), and Ethel Huxley (1866–1941). Their first child, Noel Huxley, died of scarlet fever in 1860 at the age of four.
13. *Cresson*: probably Charles M. Cresson (1828–93), an American chemist and engineer. He worked for the Philadelphia Gas Works (1849–64) and Philadelphia Board of Health, and was a member of the Franklin Institute (1849) and American Philosophical Society (1857). He further worked for several railroad companies, and was also a musician and amateur photographer (J. W. Jordan (ed), *Colonial Families of Philadelphia*, 2 vols (New York: Lewis Publishing Company, 1911), vol. 2, p. 958).
14. *When the Athenaeum opens*: see letter 3830, n. 15.
15. *this last trick*: see letter 3807, n. 3.

From Joseph Dalton Hooker — 5 September 1872 — 3835

Kew, | Sept. 5th 1872.

Dear Tyndall,

I have sent your letter[1] to Morthoe, Ilfracombe.[2] No answer yet. Helps and I had another long sitting over my letter on Monday before sending it. I was glad that I could report it "off" in answer to your telegram,[3] as I could not bear exacting more of your time, and my wife's[4] infinite patience in copying.

Helps dissuaded me from sending a copy to Gladstone—and this very decidedly. He argued that the reference to him as it were for approval, might get Gladstone's back up to no purpose, and so prolong the affair;—that if Ayrton[5] accepts it, good and well; if not then a copy should go to Gladstone as an appeal. I sent a copy to Spottiswoode for Glyn.[6]

We go to Darwin's on Saturday evening; will you not come down on Sunday and we will walk to Lubbock.[7]

Never, I beseech you, think that you bother me. I have been bereft of official support, or even advice, throughout this affair and but for you (before the public) and Helps (sub rosa)[8] I do not know what mess I might not have made long ago!

Ever yours aff[ectiona]tely | J. D. Hooker.

RI MS JT/1/TYP/8/2711
LT Typescript Only

1. *your letter*: letter missing.
2. *Morthoe, Ilfracombe*: or Mortehoe, on the north coast of Devon, England, where Thomas Huxley was staying, as shown in letter 3839 from Huxley to Tyndall.
3. *your telegram*: telegram missing.
4. *my wife's*: Frances Hooker.
5. *Ayrton*: Acton Ayrton.
6. *Glyn*: George Grenfell Glyn (see letter 3826, n. 7).
7. *We got to Darwin's... walk to Lubbock*: see letter 3827, n. 6. Darwin's home Down House was near the home of John Lubbock, High Elms. Lubbock inherited the estate on the passing of his father John Lubbock (1803–65).
8. *sub rosa*: privately, secretly, in strict confidence (Latin; *OED*).

To Henry Bence Jones 9 September 1872 3836

Royal Institution | September. 9th. 1872.

My dear Friend

I have been working hard through the tangle since my return to London;[1] and as far as the experiments go I now see my way.

But I want to do much more than make experiments before the American people.[2] I want to impress upon them the means by which science flourishes, and indeed to cast a certain amount of discredit upon the very work which they have called upon me to do. I want them and England to feel in their heart of hearts the importance of the original investigator and this requires work and thought of a totally different kind from that brought into play in the arrangement of experiment.

London is not a good place for work of this kind. So to day for the sake of the food air and water, and for the promotion of quiet meditation I go to Ascot Heath[3] and remain there for a few days. I want now to ask you whether you could engage for me a quiet bedroom and a second sitting room at Folkstone,[4] where I could carry forward the work begun at Ascot and get it complete.

I should much like to see you before I go; and if this can be arranged I shall be able to see you for several successive days. You */could/* give me a dinner whenever you feel hospitably inclined.

The Westcliff hotel[5] is by no means a bad place. But I had rooms once in Langhorne Gardens No 4,[6] I think, which were very comfortable, the people were decent people,—I believe from the North of England—and I should gladly go to them again for a few days if they could take me in.

The men are working away here.[7] They are now erecting the pillars to support the girders. I have spoken to Mr D[8] about that room, and he says he will bear it in mind. I have also had some correspondence with the Director of Works[9] regarding the time of completion; and he confidently expects that they will be cleared out by the end of the year. Then comes the period of fitting up; and here your genius will be of immense use.

The Presidential election[10] is a great draw back in America. I fear it will cause the postponement of my lectures in New York. Boston, New York, Philadelphia, Baltimore and Washington are fixed;[11] but the Interior Cities will I am told be sorely disappointed if I do not visit them. I must do the best I can.

Yours ever | John Tyndall

RI MS JT/1/TYP/3/804
LT Typescript Only

---

1. *my return to London*: see letter 3716, n. 6.
2. *make experiments before the American people*: for Tyndall's American lecture tour, see Introduction, n. 8.
3. *Ascot Heath*: a parish in East Berkshire, west of London.
4. *Folkestone*: see letter 3796, n. 24.
5. *Westcliff hotel*: a hotel in Folkestone.
6. *I had rooms once in Langhorne Gardens No 4*: in Folkestone. Tyndall wrote letters to Thomas Huxley and Heinrich Debus from here in June 1870 (see letters 3218 and 3220, *Tyndall Correspondence*, vol. 11).
7. *men are working away here*: the renovation of the RI's laboratory (see letter 3776, n. 14).
8. *Mr D*: not identified.
9. *correspondence with the Director of Works*: letters missing.
10. *Presidential election*: see letter 3794, n. 3.
11. *are fixed*: see letter 3834, n. 10.

To Joseph Henry 9 September 1872 3837

9th Sept. 1872.

My dear Prof. Henry.

Youmans, who is a man of real sagacity, writes me a startling account of the state of political excitement in New York.[1] He says it is the maddest that he has ever seen, and that if it continue, talking science[2] will be like crying in the wilderness.

I certainly should like to keep myself far aloof from any excitement of this kind, and I have telegraphed to Youmans[3] to say that I shall be prepared to lecture in New York in the latter half of December, or the first half of January.

Both he and my cousin[4] urge the desirability of extending the lectures to other places.[5] Youmans speaks of the disappointment that will be felt if this be not done. I can only say that I am willing to do anything in my power to avert dissatisfaction of every kind, and shall therefore remain in America as long as my duties here permit me to remain.

I think I might stretch my time till the middle of January.[6]

I have a very great esteem for Youmans; for he is a most able & useful man. He has been exceedingly obliging to many of us here in London and I am therefore very anxious to render him satisfied.

I shall be at my post in Boston on the 15th.

I sail on the 28th of this month in the Steamship Russia.[7]

If the postponement of the New York lectures be agreed upon then I shall be free for anything else from the 28th of Oct. to the 9th of November. But if nothing in the way of lecturing can be done in this interval I am quite willing to rest on my oars.

Sir Frederick Arrow, the Deputy Master of the Trinity House[8] is now in America. I suppose he will see you. You will find him a genial, vigorous & able man.

With best wishes | believe me | Yours ever faithfully | John Tyndall

John Tyndall to Joseph Henry, 9 September 1873, mssRH 3968, box 43, William Jones Rhees papers, The Huntington Library, San Marino, California

---

1. *writes me... in New York*: letter missing; Tyndall quotes from this in a letter to Hector Tyndale (3838).
2. *talking science*: for Tyndall's American lecture tour, see Introduction, n. 8.
3. *telegraphed to Youmans*: telegram missing.
4. *my cousin*: Hector Tyndale.
5. *extending the lectures to other places*: the initial programme for Tyndall's lectures included Boston, New York, Philadelphia, Baltimore, and Washington, DC (see letters 3814 and 3815), and added to this were Brooklyn and New Haven, CT.
6. *stretch my time to the middle of January*: Tyndall departed the United States from New York on 8 February 1873 (see letter 3928, n. 18).
7. *28th of this month in the Steamship Russia*: Tyndall traveled from Liverpool to New York on the SS *Russia* (see Introduction, n. 8).
8. *Trinity House*: see letter 3736, n. 3.

To Hector Tyndale 9 September 1872 3838

Royal Institution of Great Britain | 9th September 1872

My dear Hector.

Prof Youmans, a very sagacious man, and an everwilling helper in scientific matters, writes to me thus regarding the lecture arrangements.[1] "At all events they bear very hardly upon us. You are brought here on October 28th to Nov. 9th. being the week preceding, and the week including the election.[2] We are in the midst of the maddest and most furious presidential campaign[3] that I have ever seen, and if the excitement increases in the usual ratio, or even continues at its present intensity, talking science will be like crying in the wilderness." He also dwells upon other grave difficulties with regard to rooms and other matters.

I shall this day telegraph to Youmans[4] my willingness to postpone my lectures in New York until after the completion of those at Washington. No doubt he will communicate with you. This would leave the interval from the 28th to the 9th free—and I am willing to lecture somewhere else or to do nothing during this interval. Perhaps this gap might help Philadelphia, which I learn from Youmans suffers, though to a less extent, from a difficulty like that which besets New York. Rooms appear to be a great difficulty.

Youmans is so good a man—so self denying so helpful, and indeed so able, that I should not like to leave him any just cause of complaint. He has been urging me for many years to visit the States.[5]

I have thought a good deal over your advice as to lecturing in other places,[6] and have taken the builders[7] here into my counsels. They will have their work done at the end of the year. Still once in the States it would be a pity not to stretch a point if it would in any way gratify the people. I shall therefore do what in me lies to meet your views regarding lecturing in other places. Youmans also tells me that it will be a sore disappointment if I do not lecture in some of the interior cities. Rely upon my willingness to do as far as in me lies what you consider desirable. This I suppose may lie over until we meet: I hope our meeting will be at New York.

I have written to Mr: Lowell[8] under the assumption that I lecture at the Lowell Institute[9] on the 15th of October.

Give my kindest regards to Prof. Lesley. Both he & Prof. Henry appear to write with a little reserve which they did not show at first. I hope I have done nothing to account for this. Yours affectionately | John.

RI MS JT/1/T/1455
RI MS JT/1/TYP/5/1713–4

1. *writes to me thus regarding the lecture arrangements*: letter missing; Tyndall referred to this letter also in letter 3837 to Joseph Henry.
2. *election*: see letter 3794, n. 3.
3. *presidential campaign*: see letter 3794, n. 3.
4. *telegraph to Youmans*: telegram missing.
5. *He has been urging me for many years to visit the States*: see letter 3719, n. 4.
6. *lecturing in other places*: see letter 3837, n. 5.
7. *builders*: referring to the renovation of the RI's laboratory (see letter 3776, n. 14).
8. *Mr. Lowell*: John Amory Lowell.
9. *Lowell Institute*: see letter 3814, n. 4.

From Thomas Henry Huxley — 9 September 1872 — 3839

Morthoe, Ilfracombe,[1] | N. Devon | Sept. 9th 1872.

My dear Tyndall,

I was very glad to have news of you[2] and to hear that you are vigorous.

My outing hitherto has not been very successful, so far as the inward man is concerned, at least, for the weather has been good enough. But I have been worried to death with dyspepsia[3] and the hypochondrical[4] bedevilments that follow in its train until I am seriously thinking of returning to town to see if the fine air of St. John's Wood[5] (as the man says in Punch[6]) won't enable me to recover from the effects of the country.

I have been awfully disgusted with the course of Hookers affairs.[7] Hooker has kept me informed of all that was going on,[8] but I have had great hesitation about giving him any advice or indeed expressing any strong opinion.

So much depends in this matter not only upon what is to be done, but how it is to be done, that unless you can see an idea carried out it is dangerous to suggest it. The debate was a most unlucky one, but there was no help for it; and my judgment was clear that Hooker should have taken no steps whatever after it, but should have waited until he got some official statement from Gladstone of what was expected of him. Then there would have been something to go upon. At present he has acted upon his (or rather his friend's) interpretation of what Gladstone meant, and that most skilful of double-shufflers may escape from responsibility whatever be the result.

Put the other case. Suppose he had done nothing, and that Ayrton[9] had written to ask him what he meant to do. He could have replied that he had received no official intimation of what was expected of him from their common superior the treasury, and left Ayrton to appeal to the Treasury. The latter would have been cautious enough in their requirements and if they were satisfied a stopper would have been put on Ayrton.

However it is of no use crying over spilt milk. I wish I had stopped in town, as I was minded to do, until it was all over. I believe I might have mended matters.

What you say about Owen is consolatory to me for I have always been afraid that my personal quarrel with the man[10] might have made me paint him blacker than he is. But even my artistic powers could not put in the shadows more strongly than he has himself in that beautiful autotype memorandum furnished to Ayrton.[11] I don't know whether the folly or the baseness of it shocks me more. But he is not worth good ink and paper.

I wish I were going with you to Yankee Land,[12] not to do any lecturing, God forbid!, but to be a quiet spectator in a corner of the enthusiastic audiences. I am as lazy as a dog and the role of looker-on would just suit me. However I have a good piece of work to do in organizing my new work at South Kensington,[13] and I shall be content if my damnable bowels will only not interfere with my brains.

I have just asked my children[14] what message they have to send to you, and they send their love, very sorry they won't see you before you go, and hope you won't come back speaking through your nose!

I shall be in town this week or next and therefore <u>shall</u> see you.

Ever yours faithfully | T. H. Huxley.

I will look out for Mrs Cresson[15] and behave like an angel to her.

RI MS JT/1/TYP/8/2712
IC HP 8.128–9
Typed Transcript Only

---

1. *Morthoe, Ilfracombe*: see letter 3835, n. 2.
2. *very glad to have news of you*: see letter 3834.
3. *dyspepsia*: difficulty or derangement of digestion; indigestion: applied to various forms of disorder of the digestive organs, esp. the stomach, usually involving weakness, loss of appetite, and depression of spirits (*OED*).
4. *hypochondrical*: of, relating to, or affected by hypochondria (*OED*), which is a disorder characterized by low spirits and complaints of physical symptoms (real or imagined), esp. upper abdominal discomfort or dyspepsia (*OED*).
5. *St. John's Wood*: the Huxleys lived at 26 Abbey Place, St. John's Wood from 1861–72.
6. *Punch*: a British satirical magazine known for its humorous cartoons.
7. *Hookers affairs*: Joseph Hooker's dispute with Acton Ayrton being taken up in Parliament (see letter 3755, esp. n. 1).
8. *Hooker has kept me informed of all that was going on*: in letter 3835, Hooker noted to Tyndall that he forwarded a letter of Tyndall's (now missing) to Huxley at Morthoe, Ilfracombe.
9. *Ayrton*: Acton Ayrton.

10. *my personal quarrel with the man*: Huxley and Owen quarreled over a number of subjects, most notably man's relationship to apes, and whether or not there should be a national natural history museum. On Huxley and Owen, see N. Rupke, *Richard Owen: Biology Without Darwin* (Chicago: University of Chicago Press, 2009), chapter 6; and P. White, *Thomas Huxley: Making the "Man of Science"* (Cambridge: Cambridge University Press, 2003), pp. 43–55, 62–4.
11. *that beautiful autotype memorandum furnished to Ayrton*: see letter 3807, n. 3.
12. *to Yankee Land*: for Tyndall's American lecture tour, see Introduction, n. 8.
13. *my new work at South Kensington*: see letter 3711, n. 7.
14. *my children*: see letter 3834, n. 12.
15. *Mrs Cresson*: the spouse of Charles M. Cresson (see letter 3834, n. 13).

From Edward L. Youmans — 10 September 1872 — 3840

New York, Sept. 10th 1872.

My Dear Tyndall,

A thousand thanks for your yesterday's dispatch:[1] it has given us the relief for which we longed and makes it possible to do better justice to the very important enterprise of your lectures in this country.[2] I was not disposed to complain of Professor Henry's arrangement,[3] arbitrary and ill-advised as it was, but have gone to work heartily, and I must add desperately under it, yet my experience I can assure you has been anything but agreeable. There had been long expectation of your coming to us; much had been said about it and there was a great deal of pleasurable anticipation in connection with it. This I well knew, but I was not prepared for the amount of feeling which the announcement of the circumstances under which you were to lecture here elicited. Lord! how men swore! We have heard of the ingratitude of the Republics but it can't be equalled by their impiety. Men "clothed themselves with curses as with a garment".[4] "When is Professor Tyndall coming to lecture". "He will open a door of utterance to this community at election time".[5] The announcement in each case was like touching a trigger; the reaction was equally delicate and powerful. I had engaged a hall for two weeks continuously at an expense of a thousand dollars, but I reworked the arrangement yesterday upon the reception of your despatch, and we shall now be able to put things in better shape. There is great anxiety to hear you in some of the interior cities and a gentleman[6] came all the way from Chicago to see if you could not be induced to speak there. I shall go to Philadelphia in a day or two to counsel with your kinsman, General Tyndale[7] and without venturing to do much in the way of closing arrangements before you arrive I will inform myself as fully as possible of the general state of things, that you may take action upon them as soon as you get here. One thing I may mention as giving us some perplexity and that is your preference in regard to the number of lectures in a place. Henry arranged for six in each city consuming two weeks. It is felt that that is somewhat over-generous to the favored towns and under-generous to others. But we know not how far this is a preference of your own, not to be interfered with, I clearly remember your emphatic expressions in regard to exploring the work here, and knowing that it will be pleasanter not to change often, and more satisfactory to give the fullest possible course, I have assumed that there is to be no deviation from Henry's schedule. But if you can entertain the idea of giving two, three or four lectures in some places and thus enabling a larger number of our people to hear you[8]—the arrangements being within limits of convenience, it would be worthwhile for you to send me a despatch to that effect as soon as you receive this. There are many who would like to entertain you at their homes if you would allow it, but I have thought you would probably prefer the choice of one of our Hotels at first. Can I do anything for you in this direction? This will probably reach you in time to send me a line before you sail which we assume will take place on the 28th.[9]

If you come at that time you will be able to look over the ground and shape things very much to your mind.

Yours most truly | E. L. Youmans

RI MS JT/1/Y/1

---

1. *your yesterday's dispatch*: telegram missing. In letter 3838 to Hector Tyndale (dated 9 September 1872), Tyndall wrote that 'I shall this day telegraph to Youmans my willingness to postpone my lectures in New York until after the completion of those at Washington'.
2. *your lectures in this country*: for Tyndall's American lecture tour, see Introduction, n. 8.
3. *Professor Henry's arrangement*: for the programme of cities and dates for Tyndall's lectures as organized by Joseph Henry, see letters 3814 and 3815.
4. *"clothed themselves with curses as with a garment"*: Psalms 109:18.
5. *"He will open a door of utterance to this community at election time"*: the US Presidential election (see letter 3794, n. 3).
6. *gentleman*: not identified.
7. *General Tyndale*: Hector Tyndale.
8. *But if you can... people to hear you*: in the end, Tyndall delivered lectures in seven cities (see Introduction, n. 8).
9. *before you sail which we assume will take place on the 28th*: Tyndall traveled from Liverpool to New York on the SS *Russia* (see Introduction, n. 8).

From Joseph Dalton Hooker — 14 September 1872 — 3841

Kew | Sept. 14th 1872.

Dear Tyndall

The type of the Address[1] may be broken up I should say, it has done its duty splendidly. Let me beg of you not to let my friends be taxed for it, and other contingent expences. That should fall on me.

I am much puzzled what to do about Ayrton's last,[2] and Helps is as much so. Whether to answer it; or wait till next session and see whether anyone calls him to account, or questions Gladstone as to the termination of the affair; and then print the correspondence for the House of C[ommons]. Silence may look like assenting to his interpretation of the position: on the other hand what is a further correspondence likely to obtain from him, but a similar production to this last? "brute fulmen"[3] I have been getting through a lot of work, and have nearly concluded an arrangement for a Private Secretary which will relieve me, if carried out, of a mass of work.

I am very glad you are off for a holiday.[4] We must meet before <u>we</u> go to Amiens.[5]

Ayrton's "note of brutal triumph"[6] did not affect me so much as it has you. I am accustomed to it. And a reference to the letter that called it forth will, when both are published, render him ridiculous as well as hideously false and low.

Ever yours | J. D. Hooker.

RI MS JT/1/TYP/8/2714
LT Typescript Only

---

1. *Address*: the Memorial in support of Joseph Hooker in his dispute with Acton Ayrton (see letter 3755, esp. n. 1).
2. *Ayrton's last*: probably a letter from Acton Ayrton to Hooker dated 4 September [1872], which is quoted from in R. MacLeod, 'The Ayrton Incident: A Commentary on the Relations of Science and Government in England, 1870–73', in A. Thackray and E. Mendelsohn (eds), *Science and Values: Patterns of Recognition and Change* (New York: Humanities Press, 1974), pp. 45–78, on p. 68. Ayrton wrote that Hooker failed 'to appreciate the facts and to write about them in appropriate language'; that he was not 'guided by the better judgment of Mr. Gladstone'; and that his complaints regarding Ayrton were 'groundless'. 'It is desirable', Ayrton continued, 'that in future all communications respecting Kew Gardens should be carried on between the Board of Works in which the powers and duties

of administering them are vested by law and yourself as its subordinate officer in the manner which has been explained to you by the Secretary of the Board'. In the end, Hooker remained Director of the Royal Botanic Gardens, Kew until 1885, while in August 1873, Gladstone placed Ayrton in a new position within his administration and thus no longer in control of public institutions such as Kew (see letter 3755, n. 1).

3. *"brute fulmen"*: probably a mis-transcription for *brutus fulmen*, a mere noise; an ineffective act or empty threat (*OED*).
4. *off for a holiday*: as shown in letter 3836, first to Ascot Heath, and then to Folkestone, both to have time away from London to prepare for his American lecture tour (see Introduction, n. 8).
5. *we go to Amiens*: a city in northern France. By 'we', Hooker was probably referring to his family: his wife Frances Hooker, and their children, William Henslow Hooker (1853–1942), Harriet Anne Hooker (1854–1945), Charles Paget Hooker (1855–1933), Brian Harvey Hodgson Hooker (1860–1932), Reginald Hawthorn Hooker (1867–1944), and Grace Ellen Hooker (1868–1955). They had another daughter who died at age 6, Maria Elizabeth Hooker (1857–1863).
6. *Ayrton's "note of brutal triumph"*: see n. 2.

To Frederick Barnard 19 September 1872 3842

Royal Institution of Great Britain | 19th. Sept[embe]r.. 1872

My dear Sir,

Accept my heartiest thanks for your most kind offer of hospitality. I am however in the hands of my friends;[1] and my wild bachelor habits of life in England have accustomed me to hotels.

But I thank you much—and I would ask you to present my warm thanks to Mrs. Barnard.[2]

Yours most faithfully | John Tyndall

Columbia University Manuscripts; X973 C72 V01.2; Box 2 Folder 14; University Archives, Rare Book & Manuscript Library, Columbia University Libraries, Columbia University in the City of New York

---

1. *in the hands of my friends*: Joseph Peter Lesley, Hector Tyndale (Tyndall's cousin), and Edward L. Youmans assisted Tyndall with his arrangements during the American lecture tour; in New York, he stayed at the Brevoort Hotel.
2. *Mrs. Barnard*: Margaret Barnard (née McMurray, 1820–91), married in 1847 (*New York Times*, 6 October 1891, p. 3).

To John William Draper 19 September [1872][1] 3843

Folkestone[2] 19th Sept.

My Dear Dr. Draper,

My stay in New York after landing will be but short. I was strongly recommended to go Brevoort (I think that is the word) Hotel:[3] my sole desire is to have a quiet bedroom—and the hotel you recommended will probably be the best place to secure that.

But I am to be met—I hope by our friend Youmans—and a Philadelphia relative[4]—my movements will of course depend upon them.

The Russia[5] quits Liverpool on the 28th and I certainly go by that ship.

Many thanks to you, and regards the offer of apparatus. It is a comfort to know that such is there should I need it; but having to organize the lectures and prepare the experiments I thought it safest to make the acquaintance of my instruments. I have therefore purchased and tested some rather costly apparatus—in fact to the extent of more than £300. I dare say the half of it will not be used.

It will give me great pleasure to shake your hand upon the other side.

Yours ever faithfully | John Tyndall

I go first to Boston[6]—Can I calculate on Nitric & Sulphuric acid for my batteries? If not I will find two Carboys from New York.

Would you kindly ascertain this part for me? I carry with me 120 cells of Grove.[7]

Library of Congress, John William Draper Family Papers

---

1. *[1872]*: the year is given by reference to Tyndall's upcoming American lecture tour (see Introduction, n. 8).
2. *Folkestone*: while 'Folkestone' is written on the letter, the stationary bears the RI stamp. Tyndall made a visit to Folkestone before leaving for his American lecture tour.
3. *Brevoort (I think that is the word) Hotel*: on lower Fifth Avenue in New York, the Brevoort Hotel operated from 1854–1954.
4. *Philadelphia relative*: Hector Tyndale.
5. *The Russia:* Tyndall traveled from Liverpool to New York on the SS *Russia* (see Introduction, n. 8).
6. *to Boston*: Tyndall delivered six lectures in Boston from 15–25 October 1872 at the Lowell Institute.
7. *120 cells of Grove:* the Grove battery. See letter 3907, n. 7.

To Emily Peel 19 September [1872][1] 3844

Folkestone[2] | 19th September

My dear Lady Emily.

Your friendly letter[3] reached me a few days ago—accept my best thanks for it.

It never occurred to me that you were in the least degree likely to mistake a "Muff"[4] for a brave man. Still I am glad that your friend the Baron[5] has realised the hopes entertained of him by becoming a mountaineer of the first rate.[6]

I only went up the Alletschhorn this year,[7] and this to gratify a friend[8] rather than to please myself. We went up at a good speed, doing on the final arête[9] work in three quarters of an hour which it had taken a preceding party three hours to accomplish. The snow in going up was good, on returning it was abominable—It cost us nearly two hours to accomplish in coming down the same distance that required only three quarters of an hour in ascending.

My ice axe, moreover, in which I trusted, broke in my hand as I was leading the way down a steep ice-slope. The guide—or more correctly porter[10]—a good fellow but of little experience—once gave way & shot down an ice-slope. Happily I was firmly anchored among rocks and was able to pull him up.

I spent my holiday at the Bel Alp;[11] with two days at the AEggischhorn.[12] The emptying of the Märgelin See[13] was a most impressive sight. I also spent about eight days at Chamouni[14] this year. I passed through Geneva, saw my friend De la Rive, but learned that Lammermoor had not yet received its light & life.[15]

And now I am in the midst of my preparations for a journey to America.[16] We are pulling down our old laboratories & building new ones.[17] This enables me to accept an invitation which has been more or less pressed on me for the last six or seven years.[18] In fact the Americans have shown themselves exceedingly cordial friends of mine; and one of them writing to me at a time when the relations between America and England were very ticklish, urged upon me the consideration that as far as America was concerned nothing would be more calculated to promote kindly feelings than the occasional visits of men withdrawn from the arena of politics.[19] An invitation signed by 20 or 30 of their best men reached me last year;[20] and I accepted it.[21]

I shall certainly remain there for three months, having agreed to talk to them in Boston, New York, Philadelphia, Baltimore and Washington. The matter is in the hands of my friend Henry, the Director of the Smithsonian Institution in Washington.

And now good-bye. May peace and happiness take their residence in Lammermoor.

Yours ever faithfully, | John Tyndall.

I came down here to work and have done so till my head aches. I start on the 27th.[22]

RI MS JT/1/T/1118
RI MS JT/1/TYP/3/967

---

1. *[1872]*: the year is given by reference to Tyndall's travels in Switzerland and sharing details about his upcoming lecture tour in America.
2. *Folkestone*: as shown in letter 3836, Tyndall went first to Ascot Heath, and then to Folkestone (where Henry Bence Jones lived), both to have time away from London to prepare for his American lecture tour.
3. *Your friendly letter*: letter missing.
4. *"Muff"*: a foolish, stupid, feeble, or incompetent person; *spec.* one who is clumsy or awkward in some sport or manual skill (*OED*).
5. *your friend the Baron*: Albert Salomon Anselm Freiherr von Rothschild (1844–1911), an Austrian banker and member of the Rothschild banking family. He held the title of Baron. Rothschild had a number of interests outside of banking—such as chess, photography, and collecting art—and used his money to fund these. He owned the first car in Vienna, built a hospital, and had a private observatory built in one of the family's many properties in Vienna (*ADB*). Several letters in 1871 show Peel requesting Tyndall's assistance in having Rothschild elected a member of the Alpine Club in England (see letters 3525, 3531, and 3533, *Tyndall Correspondence*, vol. 12). He was elected as a member on 12 December 1871 ('Proceedings of the Alpine Club', *Alpine Journal*, 5 (1872), p. 336).
6. *first rate*: LT mistakenly transcribed this as 'first water' in the typescript.
7. *only went up the Alletschhorn this year*: see letter 3806, n. 20.
8. *friend*: not identified.
9. *arête*: a sharp ascending ridge or 'edge' of a mountain (*OED*).
10. *porter*: not identified.
11. *Bel Alp*: see letter 3771, n. 5.
12. *AEggischhorn*: Eggishorn (see letter 3828, n. 15).
13. *emptying of the Märgelin See*: see letter 3810.
14. *Chamouni*: see letter 3776, n. 12.
15. *Lammermoor had not yet received its light & life*: Emily and her husband Robert Peel (1822–95), son of former Prime Minister Robert Peel (1788–1850), had not yet stayed in their Geneva home, Villa Lammermoor, this year. The home was built on land (now Parc Barton) purchased in 1858.
16. *my preparations for a journey to America*: for Tyndall's American lecture tour, see Introduction, n. 8.
17. *pulling down our old laboratories & building new ones*: the renovation of the RI's laboratory (see letter 3776, n. 14).

18. *invitation which has been more or less pressed on me for the last six or seven years*: see letter 3719, n. 4.
19. *one of them... arena of politics*: Tyndall shared a similar sentiment in a letter to Mary Adair (letter 3719). The letter from the American is not identified, perhaps missing (see letter 3719, n. 4).
20. *invitation signed by 20 or 30 of their best men reached me last year*: see letter 3510, *Tyndall Correspondence*, vol. 12.
21. *I accepted it*: see letter 3542, *Tyndall Correspondence*, vol. 12.
22. *I start on the 27th*: towards Liverpool, to sail to the United States on 28 September.

To Edward L. Youmans 19 September 1872 3845

FOLKESTONE,[1] *September 19, 1872.*

MY DEAR YOUMANS: ... I hope they have clear ice in Boston,[2] also nitric and sulphuric acid; if not, I must stock myself from New York. I have written a line to Dr. Draper[3] on this point, but I should be truly thankful to you if you would make this point out for me, and if the acid is not to be had at Boston send there a carboy of nitric and one of sulphuric acid.

I am quite shocked at the mass of apparatus I have collected round me. Still I thought it best to take light apparatus—batteries, electric lamps, and costly optical apparatus—with me, having just given the experiments with them here.

Yours ever, | JOHN TYNDALL.

E. A. Youmans, 'Tyndall's American Visit', *Popular Science Monthly*, 44 (February 1894), pp. 502–14, on p. 514

---

1. *FOLKESTONE*: Tyndall made a visit to Folkestone before leaving for his American lecture tour (see Introduction, n. 8).
2. *in Boston*: Tyndall delivered six lectures in Boston from 15–25 October 1872 at the Lowell Institute.
3. *written a line to Dr. Draper*: letter 3843.

To Joseph Peter Lesley 20 September 1872 3846

Folkestone[1] 20th September 1872.

My dear Professor Lesley.

I came down here for the sake of quiet and good air to throw my work[2] into its finest organic form. But I have brought myself into a headache and done but little good. Still I now see my way with tolerable clearness through the tangle. Its general course is this.

I begun in a prefatory way and dwell upon the introduction of the experimental methods into science. I [s]peak of the ardour of investigators and of their rewards. I seek to share that most of them brought for the sake of knowledge and with no practical end in view, though their discoveries have led to the most astonishing practical applications. After dwelling on the importance of original enquiry I take up the real subject of the lectures. The instruments are explained and the principles on which they depend. I point out the proximate cause and action of the electric light which is to be used in our illustrations. The laws of reflexion are demonstrated, and one of two striking practical applications adduced. Then I go on to refraction. These elementary subjects are really touched upon in order to enable me in a subsequent lecture to reveal the workings of Newton's mind when he theorized upon the subject of light. Refraction is followed by an inquiry into the constitution of light—its analysis and synthesis. This occupies the first lecture.

In the second lecture the demonstrated constitution of light is applied to the doctrine of colours. I go very thoroughly and placidly into this matter making perfectly evident the causes on which ordinary colours depend, winding up by the experimental proof that yellow and blue lights, when mixed together, produce white and not green. Having exhausted the ordinary spectrum I describe the difference between the emissions from solids and their vapours. Metallic vapours are produced and shown with their characteristic colours. This light is then analyzed, and it is shown to be distinctive of the substance from which it comes. Spectrum analysis is dwelt upon and copiously illustrated. This brings me to the end of the second lecture.

I then deal with solar light, dwelling upon the distinctions between the bright lines of the metallic vapours and the dark lines of Fraunhofer.[3] The reciprocity gradiation and absorption is demonstrated, and it is shown experimentally that an incandescent vapour absorbs the light which it emits. This leads up to the theory of the physical constitution of the sun. Then I go on to show the extension of the spectrum beyond its visible range, performing with quartz prisms and lenses Stokes's experiments on Fluorescence,[4] and the rendering visible of invisible rays. Then the other side of the spectrum is handled—its extension as heat beyond the limits of light as demonstrated. Numerous experiments on the total heating power of the rays from the electric light are made; fusion and combustion being thus effected. Here I hope to perform the famous Florentine experiment of the ignition of a diamond in oxygen[5] using however a purely terrestrial source of radiant heat. I also hope to produce combustion by rays concentrated by a lens of ice. The heat rays are then filtered from the light rays, and it is shown that all the foregoing effects are produced by rays totally beyond the range of vision; fusion, combustion and explosion being produced at foci perfectly dark, and as far as the air is concerned, perfectly cold. It is also proved that these dark rays perform the work of evaporation in the tropical ocean, and the work of fusion upon the Alpine ice & snows. The rays moreover are shown to be competent to raise platinum to a white heat, so that by its interocution[6] you may extract from the dark rays all the colours of the spectrum. This brings me to the end of the 3rd lecture.

In the fourth lecture I show the irresistible tendency of the human mind to seek for governing principles which rule facts & earnest theory rendering them so to say organic. I dwell upon the exercise of the theorizing faculty, taking Newton as my example. I try to show how materially his optical theory grew out of his previous knowledge. The doctrine of colours is now extended by the introduction of the colours of thin plates, striated surfaces, &c. and I unravel the subtle addition which Newton made to his theory in order to fit is to these new facts. The theory of Emissions is then contrasted with the theory of undulation. The latter is rendered familiar to the mind by the preliminary considerations regarding water waves, and by experiments regarding sound. I dwell upon the labours of Thomas Young and the effect of Brougham's's[7] attacks in the Edinburgh Review.[8] This will be my most difficult lecture—but I have wrought hard to make it clear, and it is essential to the comprehension of the subsequent ones.

In the fifth lecture I enter upon the phenomena of crystallization & seek to give an intelligible explanation of crystalline architecture. The process of crystallization is experimentally illustrated. This is done with a view to the action of crystals upon light. In the first experiments, I deal with crystals solely with reference to the polarization of light. This is explained, and illustrated by numerous experiments. Double refraction and the state of the two halves of the divided beam are dwelt upon. Then come the chromatic phenomena of polarized light. Basing myself upon the principles explained in the fourth lecture I hope to make these effects comprehensible by all intelligent persons. The effects of mechanical strains and pressures in producing a quasi-crystalline structure are exhibited. Then the similar phenomena of unannealed glass. I hope to show these effects in a very splendid fashion. They will more than fill the fifth lecture.

The sixth lecture is devoted to the further illustration of the action of crystals upon light: uniaxial and biaxal[9] crystals. Circular polarization, and the chromatic effects produced by rock-crystal. The conferring of doubly refracting powers by sonorous vibrations. The magnetization of light. Although the syllabus is short it covers a good deal.

I have here sketched the course of six lectures, and the materials touched upon are ample to fill the six to overflowing allowing an hour & a half for each lecture. A seventh very striking lecture might be given on the identity of light and heat. Every experiment made in the optical lectures being shown capable of repetition with pure lightless radiant heat: the thermoelectric pile[10] & galvanometer being substituted for the eye. I have made an arrangement for the projection of the galvanometer[11] dial upon a scrim[12] which render it visible to any number of people.

As I worked at the subject the desire grew upon me to do it more and more thoroughly, and to spare no expense as regards apparatus. I have accordingly purchased between three & four hundred pounds worth of instruments; and have gone over all the experiments so as to render every thing sure. Thus I hope to do my duty in a manner worthy of the subject and the occasion.

Since I last wrote[13] to Hector[14] I have been in communication with our President & Secretary,[15] and the result is that I do not think it will be necessary to quit the United States so early as I at first intended. Indeed considering he costliness of the apparatus, and the labour incurred in testing them and packing them, and carrying them over it would by a pity to leave America without duly utilizing them. But until I see you it will not be necessary to make any further arrangements. I can only express my willingness so to act that nobody shall be disappointed. Indeed the people of the United States cannot desire to see me half so strongly as I desire to make these lectures all that they could wish for. Had I an additional month for preparation I should come nearer to my own ideal; but I trust my friends will be satisfied if I only do my best.

Believe me | Yours ever faithfully | John Tyndall

Mss.B.L56. American Philosophical Society

---

1. *Folkestone*: as shown in letter 3836, Tyndall went first to Ascot Heath, and then to Folkestone (where Henry Bence Jones lived), both to have time away from London to prepare for his American lecture tour.
2. *my work*: preparing for his American lecture tour (see Introduction, n. 8).
3. *dark lines of Fraunhofer*: the dark lines in the solar spectrum, caused by substances in the solar atmosphere absorbing radiation in specific parts of the continuous solar spectrum. They were noticed by the English chemist William Hyde Wollaston (1766–1828), then investigated by the German optician Joseph Fraunhofer (1787–1826) and named after him (*ODNB*; *NDB*; *OED*).
4. *Stokes's experiments on Fluorescence*: Stokes first introduced the term 'fluorescence' in G. G. Stokes, 'On the Change in Refrangibility of Light', *Phil. Trans.*, 142 (1852), pp. 463–562, on p. 479 (see endnote).
5. *famous Florentine experiment of the ignition of a diamond in oxygen*: see H. Davy, 'Some Experiments on the Combustion of the Diamond and Other Carbonaceous Substances', *Phil. Trans.*, 104 (1814), pp. 557–70.

6. *interocution*: Tyndall misspelled 'interlocution'.
7. *Brougham's*: Henry Peter Brougham (1778–1868), a British politician and one of the founders of the *Edinburgh Review* (*ODNB*).
8. *attacks in the Edinburgh Review*: [H. Brougham], 'The Bakerian Lecture on the Theory of Light and Colours. By Thomas Young, M.D.F.R.S. Professor of Natural Philosophy of the Royal Institution. From Philosophical Transactions for 1802. Part I', *Edinburgh Review*, 1 (1802–3), pp. 450–6; [H. Brougham], 'An Account of some Cases of the Production of Colours not hitherto described. By Thomas Young, M.D.&c. From Philosophical Transactions for 1802. Part II', *Edinburgh Review*, 1 (1802–3), pp. 457–60; and [H. Brougham], 'The Bakerian Lecture. Experiments and Calculations relative to Physical Optics. By Thomas Young, M.D.F.R.S. From Phil. Trans. 1804', *Edinburgh Review*, 5 (1804), pp. 97–103.
9. *biaxal*: Tyndall misspelled 'biaxial'.
10. *thermoelectric pile*: or thermopile, a thermo-electric battery, used in connection with a galvanometer, for measuring minute quantities of radiant heat (*OED*).
11. *galvanometer*: an apparatus for detecting the existence and determining the direction and intensity of a galvanic current (*OED*).
12. *scrim*: a screen or backcloth made from an open-weave fabric which appears opaque until lit from behind, typically used to create different spaces on a stage set (*OED*).
13. *last wrote*: letter 3838.
14. *Hector*: Hector Tyndale.
15. *in communication with our President & Secretary*: at this time, the President of the RI was Henry Holland (from 1865–73), and the Secretary was Henry Bence Jones (from 1860–73). It is not clear if Tyndall meant communication in person or by letter, although there is a letter to Bence Jones written on the same day as his last to Tyndale (letter 3836).

To Louisa Baring 22 September [1872][1] 3847

22nd. Sept.

On my return from Switzerland[2] I heard of your great loss.[3] Need I say how it grieved me?

Yours always | John Tyndall

I am on the point of starting for America[4] and shall not return for many months. Goodbye

National Library of Scotland Ashburton Papers Acc.11388 N0.101

1. *[1872]*: the year is given by reference to Tyndall's trip to the United States.
2. *my return from Switzerland*: Tyndall left London for Switzerland on 25 June 1872, and presumably returned to London by 10 August 1872 (as he remarked in letters 3771, 3794, and 3806).

3. *your great loss*: a fire at Baring's home Melchet Court in July 1872 (V. Surtees, *The Ludovisi Goddess: The Life of Louisa Lady Ashburton* (Michael Russell, 1984), pp. 147, 157–8, 164).
4. *starting for America*: for Tyndall's American lecture tour, see Introduction, n. 8.

To Frederick Barnard 22 September 1872 3848

22nd. Sept. 1872.

My dear Sir,

I ought to have thanked you earlier for the very beautiful volume on the "Metric system"[1] which you have been kind enough to send me.

Let me heartily wish you good speed in your enlightened efforts.

Even a cursory glance over your pages suffices to make evident the logical clearness and precision with which you have stated your case.

Most faithfully Yours | John Tyndall

Dr. Frederick Barnard | Pres[ident]. Col[umbia]. College | & & &

If fear much that[2]

Columbia University Manuscripts; X973 C72 V01.2; Box 2 Folder 13; University Archives, Rare Book & Manuscript Library, Columbia University Libraries, Columbia University in the City of New York

---

1. *volume on the "Metric system"*: F. A. P. Barnard, *The Metric System of Weights and Measures: An Address Delivered Before the Convocation of the University of the State of New York, at Albany, August 1, 1871* (New York: Columbia College, 1872).
2. Fragment only; letter incomplete.

To Margaret Ginty [22 September 1872][1] 3849

Royal Institution | Saturday

Dear Mrs Ginty,

I cannot understand your not wishing to come here. I will not ask you to come up stairs[2]—I will go down to you and we can have our conversation in the museum or some other room. Were I not so terribly pressed,[3] or were there any sufficient reason for it I would go to you. It is by no means as a punishment, but simply because if I do not set my face sternly against it my life will be dribbled away in London.

I shall expect you at 2 on Thursday.[4]

Yours faithfully | John Tyndall.

RI MS JT/1/TYP/11/3668
LT Typescript Only

1. *[22 September 1872]*: the date is given by relation to letter 3858.
2. *up stairs*: Tyndall's rooms at the RI.
3. *terribly pressed*: in preparing to leave for his American lecture tour on the upcoming Saturday, 27 September (see Introduction, n. 8).
4. *I shall expect you at 2 on Thursday*: in letter 3858 to Ginty, written on a Thursday and dated to [26 September] 1872, Tyndall wrote: 'At 1 o'clock I have an engagement at Brompton. Pray look in at me at 5 P.M. punctually'.

To Mr. Sedgwick[1] 22 September 1872 3850

22nd Sept. 1872

My dear Mr Sedgwick

I return your photographs[2] with the thousand apologies for keeping them so long—In fact I have been so immersed in my preparations for America[3] that they escaped my memory.

I have desired my publisher Mr H. S. King to send you a copy of my last little book about ice & water.[4] I hope you will not deem it unworthy of your acceptance.

faithfully yours | John Tyndall

RI MS JT/1/T/1275

1. *Mr. Sedgwick*: possibly John Bell Sedgwick (d. 1895), who appears in the list of Visitors of the RI. Sedgwick was also a member of the Council of the Photographic Society of London.
2. *your photographs*: subject of photographs not identified.
3. *my preparations for America*: for Tyndall's American lecture tour, see Introduction, n. 8.
4. *my last little book about ice & water*: *Forms of Water* was published in September just before Tyndall left England for the United States.

From Rudolf Clausius 22 September 1872 3851

BONN, 22/9/72.

Lieber Tyndall,

Ich habe Deinen freundlichen Brief aus der Schweiz nicht gleich beantwortet, weil ich fürchtete, mein Brief würde Dich in der Schweiz nicht treffen.

Jetzt darf ich aber nicht länger zögern, weil Du sonst vielleicht schon nach America abgereist bist.

Zunächst muss ich Dir mittheilen dass meine Frau wieder von einem Töchterchen entbunden ist. Es ist dieses das sechste Kind, und Du kannst Dir daher denken, dass es in unserem Hause lebhaft genug ist. Dieses Ereigniss hat mich verhindert, in diesem Sommer irgend eine grössere Reise zu machen. Ich bin froh, dass die Sache glücklich vorüber ist. Das Kind ist frisch und gesund, und meine Frau hat sich auch schon wieder ziemlich erholt.

Mein Streit mit Tait hat eine sonderbare Wendung genommen. Ich hatte in meinem ersten Artikel gesagt, dass er eine unrichtige Behauptung gemacht und diese durch ein ganz falsches Citat bekräftigt habe. Auf diesen bestimmten thatsächlichen Vorwurf erwidert er kein Wort, und greift statt dessen die Richtigkeit meiner Untersuchungen an, die er früher nie bestritten hatte. Darauf schreibe ich einen zweiten Artikel, worin ich seine Einwendungen rein wissenschaftlich widerlege. In der Antwort auf diesen Artikel übergeht er die Widerlegung der ersten Einwendungen mit Stillschweigen, macht aber wieder neue noch allgemeinere Einwendungen. Dabei spricht er von meinen Untersuchungen in einem höchst beleidigenden wegwerfenden Tone, und doch ergiebt <es> sich aus seinen eigenen gelegentlichen Angaben, dass er meine Arbeiten gar nicht, oder nur unvollständig kennt, denn er sagt selbst, dass er sich dasjenige was in seinem Buche über meine Arbeiten steht, von einem Freunde, der meine Arbeiten kennt, hat machen lassen. Endlich legt er mir mit voller Bestimmtheit Behauptungen in den Mund, die ich gar nicht gemacht habe, so sagt er z.B. mehrmals, ich hätte ihm vorgeworfen, dass er meinen Namen mit Absichtlichkeit unterdrückt hätte, und dabei versieht er das Wort Absichtlichkeit mit Anführungszeichen. Ich habe aber dieses Wort von ihm nie gebraucht, sondern nur von Maxwell. Unter diesen Umständen bin ich zu der Überzeugung gekommen, dass <es> sich mit einem solchen Menschen nicht streiten lässt. Es fehlt ihm an wissenschaftlichem Ernst und Aufrichtigkeit.

Mit grossem Interesse habe ich vernommen, dass Du nach America gehen und dort Vorträge halten wirst. Es ist ein kühnes Unternehmen, aber ich zweifele nicht, dass Du es mit Erfolg durchführen wirst. Ich wünsche Dir von Herzen Glück dazu. Diese Reise wird zu Deinen vielen interessanten Erlebnissen wieder neue von anderer Art hinzufügen und Deine Anschauungen noch mehr bereichern, so dass Du hoffentlich auch Freude und Genuss davon hast, während Du anderen Genuss und Belehrung bereitest. Meine Frau und ich haben in neuerer Zeit viel in Deinem jetzt deutsch erschienenen Buche, "In den Alpen" gelesen, und haben uns von Neuem an den schönen Schilderungen erfreut. Meine Frau ist aber oft über Deine Verwegenheit entsetzt gewesen.

Mit herzlichen Grüssen von meiner Frau und mir,

Dein | Clausius.

BONN, 22/9/72.

Dear Tyndall,

I did not reply to your kind letter from Switzerland[1] right away, because I feared my letter would not reach you in Switzerland. Now, however, I cannot hesitate any longer because you might otherwise have already departed for America.[2]

First of all, I have to tell you that my wife[3] has given birth again to a little girl.[4] This is our sixth child, and you can therefore imagine that it is lively enough in our household. This event has prevented me from going on any major trip this summer. I am glad that the matter was over happily. The child is bright and healthy, and my wife has also more or less recovered again already.

My quarrel with Tait[5] has taken a strange turn. I had said in my first article that he had made an incorrect claim and supported this by means of a quite false quotation. He does not say a word in response to this precise, factual accusation, and instead attacks the correctness of my investigations, which he had never disputed earlier. In response to all this, I write a second article in which I refute his objections in a purely scientific manner. In his reply to this article, he passes over my refutation of the initial objections in silence, though he does raise some new, even more general objections once again. In doing so, he speaks of my investigations in a highly insulting dismissive tone, and yet <it> turns out from his own, occasional statements that he does not know my works at all, or only incompletely, for he says himself that he has had what is said about my works in his book[6] done by a friend who knows my works. Finally, and with complete certainty, he puts claims into my mouth that I have not made at all, e.g., he says several times I had accused him of having deliberately suppressed my name, and in doing so he provides the word deliberately with quotation marks. However, I have never used this word referring to him, only to Maxwell.[7] Under these circumstances, I have become convinced that one cannot argue with such a person. He lacks scientific seriousness and honesty.

I have heard with great interest that you will be going to America and giving lectures there. It is a bold undertaking, but I do not doubt that you will carry it out successfully. I wish you good luck for this with all my heart. This trip will again add to your many interesting experiences new ones of a different kind and enrich your views even more, with the result that you will hopefully also gain enjoyment and pleasure from it, while giving enjoyment and guidance to others. Lately, my wife and I have been reading a great deal in your book "In the Alps",[8] which has now appeared in German, and have taken pleasure once again in the beautiful descriptions. My wife, however, has often been horrified by your daring.

With kind regards from my wife and me,

Your | Clausius.

RI MS JT/1/TYP/7/2327–8
LT Typescript Only

---

1. *to your kind letter from Switzerland*: letter missing.
2. *departed for America*: Tyndall traveled from Liverpool to New York on the SS *Russia* (of the Cunard Line), departing on 28 September 1872 (for his American lecture tour in general, see Introduction, n. 8).
3. *my wife*: Adelheid Clausius.
4. *little girl*: Else Clausius (1872–1954). Their five other children were: Mathilde (1861–1907), Helene (1863–1919), Rudolf John (1864–1915?), Hedwig (1866–1946), and Alfred (1868–1950).
5. *My quarrel with Tait*: see letter 3787, n. 6.
6. *his book*: P. G. Tait, *Sketch of Thermodynamics* (Edinburgh: Edmonston and Douglas, 1868). Tait discusses Clausius on pp. 24–45, 111–14.
7. *Maxwell*: James Clerk Maxwell.
8. *"In the Alps"*: *In den Alpen* was the German translation of Tyndall's *Hours of Exercise in the Alps* (London: Longmans, Green, and Co., 1871).

From Mary Egerton 22–3 September [1872][1] 3852

Mountfield Court, | Robertsbridge, |
Hawkhurst.[2] | Sept[embe]r. 22nd.

My dear Mr Tyndall,

I had so completely made up my mind that I had seen & heard the last of you, that your note[3] was a pleasant surprise. I do not apply the adjective to all its contents; "the end of January" or perhaps of February! that is a long long time to look forward to! I hoped you were coming back about Christmas![4]

Will you not be back for the opening lecture at the Institution?[5] And then you will be far too busy to write! May I ask Mr Hirst to show me a letter now & then that has "no secrets" in it, as you let me do at Chamouni?[6] You would not mind my seeing some of your remarks on people & things, sh[oul]d you? You know I would never repeat them, & I sh[oul]d be so much interested to know your impressions of it all. You can tell him whether I may or not. I cannot think how I let you go in London without telling me your programme.[7] I want so much to know how the lectures are to be arranged, whether different subjects for each place, &c.

I am so glad you have written that paper,[8] & I am so glad to have seen it before you went away, that I may tell you how much I like it: You have, as you say, made your own position clear, and it is one towards which all of us can go a long way with you. Of course we must still be at issue on the question of experiment,[9] which we hold to be impossible, just because it is not any "form of physical energy, or equivalent of such energy",[10] that we invoke, but a free & omniscient Will, the originator & director of energy; which Will must be guided in granting or denying, by a thousand circumstances beyond our knowledge. But even c[oul]d the experiment be possible, it w[oul]d never be convincing, because (seeing we ask no miracle) it w[oul]d always be open to the other side to say the event was mere coincidence, and would have happened all the same without the prayer.

The word "Christianity" standing alone, does not quite satisfy me, where you say, (what is so true in the sense you mean it), that it "varies with the nature on which it falls".[11] (like those autotype plates that bring out the image in blue or red or black!) Christianity is ever one & the same—but in some souls its image is blurred & defaced, in others the reflection, if feeble, is yet in some measure true. The "faith" that can "add to folly & ferocity"[12] is not the Christian faith, but a presumptuous superstition! As St James says on another subject, "Let not such a man think he shall get anything of the Lord"![13] Truly there have been but too many examples of it, from the Philips and Alvas[14] down to the present day!

But enough of this subject. May[15] bids me tell you how sorry she is not to have seen you before you go, and to give you her best wishes for the success of your expedition in which the others, especially y[ou]r alpine friends *[1 word illeg.]*. She & I have been reading Kant's[16] "Schöne und Erhabene"[17] together. It is wonderfully good & so like the grand noble spirit one fancies him to have been. Several passages show me why it was a favorite of yours.

I trust the voyage will blow away all the effects of hard work! You must have prepared a quantity of lectures! I think it was the New York Herald you said would be likely to give the fullest report of them?

I hope it was not wrong to keep the paper another post? thank you so much for sending it. I could not help lingering a little over this letter, because when it is closed and gone, it will be really "Good-bye"! You will have a cold voyage if the weather keeps like this! Mrs Rutherford's[18] jerseys will come in useful!

Well! it must go now! Fancy to yourself a warm "Händedruck"[19] and many good wishes, going with it.

Good-bye! | Believe me my dear friend, | Y[ou]r ever attached | M. F. Egerton | Monday[20]

I suppose you sail from Liverpool Saturday[21]—

RI MS JT/1/E/50

1. *[1872]*: the year is given by reference to Tyndall's American lecture tour.
2. *Mountfield Court, | Robertsbridge, | Hawkhurst*: the Egerton family estate.
3. *your note*: letter missing.
4. *coming back about Christmas!*: for Tyndall's American lecture tour, which went into February 1873, see Introduction, n. 8.
5. *opening lecture at the Institution*: William Spottiswoode delivered a Friday Evening Discourse at the RI on 17 January 1873, 'On the Old and New Laboratories at the Royal Institution' (*Roy. Inst. Proc.*, 7 (1873–75), pp. 1–10).
6. *Chamouni*: see letter 3776, n. 12.
7. *your programme*: see letter 3837, n. 5.
8. *that paper*: J. Tyndall, 'On Prayer', *Contemporary Review*, 20 (1872), pp. 763–77 (the pages following Tyndall's article (pp. 777–82) are a response from James McCosh (1811–94), a Scottish philosopher who was President of Princeton University from 1868–88 (*ODNB*).
9. *experiment*: an experimental test to determine the efficacy of prayer (see letter 3780, esp. n. 2).
10. *"form of physical energy, or equivalent of such energy"*: 'On Prayer', p. 764.
11. *"varies with the nature on which it falls"*: 'On Prayer', p. 766.
12. *"add to folly & ferocity"*: 'On Prayer', p. 766.
13. *"Let not such a man think he shall get anything of the Lord"*: James 1:7.
14. *Philips and Alvas*: a reference to Fernando Álvarez de Toledo y Pimentel (1507–82), 3rd Duke of Alba; and Philip II (1527–98), King of Spain. Both were strict defenders of Catholicism against Protestantism. The following is from an article in *Harper's* which Egerton may have been familiar with: 'The true monsters of history are the Philips and Alvas, not the Dantons and Robespierres' ('The Rise of the Dutch Republic', *Harper's New Monthly Magazine*, 13 (June 1856), pp. 83–96, on p. 85).
15. *May*: Mary Alice Egerton.
16. *Kant's*: Immanuel Kant (1724–1804), a highly influential German philosopher and Enlightenment thinker (*NDB*).
17. *"Schöne und Erhabene"*: I. Kant, *Beobachtungen über das Gefühl des Schönen und Erhabenen* (Königsberg: Bey Johann Jacob Kanter, 1764). In English, the title is *Observations on the Feeling of the Beautiful and Sublime.*
18. *Mrs Rutherford's*: not identified; possibly Elizabeth Rutherford.
19. *"Händedruck"*: Handshake (German).
20. *Monday*: Egerton began this letter on 22 September, a Sunday, and the 'Monday' here shows that she finished the letter on 23 September.
21. *sail from Liverpool Saturday*: Tyndall traveled from Liverpool to New York on the SS *Russia*, departing on 28 September 1872 (see Introduction, n. 8).

From Thomas Archer Hirst 24 September 1872 3853

University of London | Sep[tembe]r 24th 1872

My dear John.

Debus has yielded like a man to the invincible force of my argument It is with me that you ought, naturally, to take your last dinner before you cross the Atlantic.[1]

I am sure you will neither have the wish nor the cruelty to dispute my claim, so come here at 7 o'clock on Thursday and you shall have the most wholesome dinner, in every sense, that you ever enjoyed in your life.[2]

Ever, dear John, | Yours | Tom

RI MS JT/1/H/252

1. *cross the Atlantic*: Tyndall traveled from Liverpool to New York on the SS *Russia*, departing on 28 September 1872 (see Introduction, n. 8).
2. *most wholesome dinner... enjoyed in your life*: Hirst recorded in his journal entry for 26 September: 'Tyndall took his last dinner with me before crossing the Atlantic. Debus was the only one I invited to meet him, for we felt it to be in accordance with the fitness of things that we three old friends should pass the evening alone. John however left us early for he had still to say good bye to Thomas Carlyle' (Brock and MacLeod, *Hirst Journals*, p. 1952). He then recorded for the next day (27 September) that 'Debus and I took lunch with Tyndall and then accompanied him to Euston Square to see him off on his probably eventful journey. As I expected it cost me a struggle to say good bye with proper composure. He took his ticket for Edge Hill near Liverpool where he is to pass the night with a new friend whom he lately met in Switzerland. To-morrow at 3P.M. he is to sail from Liverpool in the Russia. The Equinoctial gales will I fear render the voyage a rough one. My only hope now is that he may come safely back to us' (p. 1952).

To Thomas Archer Hirst 25 September 1872 3854

Sep[tembe]r 25th 1872

My dear Tom.

Let it be considered to be in accordance with the eternal fitness of things.

ever affec[tionate]ly yours, | John[1]

RI MS JT/1/T/692
RI MS JT/1/HTYP/599

1. Hirst wrote on the letter: 'NB. The "it" here refers to his dining with me (and Debus) on the eve of his departure to America | TAH'. See letter 3853.

To William Spottiswoode 25 September 1872 3855

25[th]. Sept 1872.

My dear Spottiswoode

I waited on Tuesday till disappointed hope made the heart sick, thinking that when mathematics and physics set their heads together some means might be found of running down to you for a quarter of a day. I must now give up the thought attractive as it was, for I am sorely pressed.[1]

Again best love to the boys,[2] and best wishes to yourself & your most excellent wife[3]—A wife whose goodness wins a word of admiration from every body.

Good bye | always yours | John Tyndall

RI MS JT/1/T/1302
RI MS JT/1/TYP/3/1246

1. *I am sorely pressed*: Tyndall was in the final days of preparing for his American lecture tour (see Introduction, n. 8).
2. *boys*: William Hugh ('Hughie') Spottiswoode (see letter 3946, n. 1) and Cyril Andrew Spottiswoode (1867–1915), sons of William and Eliza Spottiswoode.
3. *your most excellent wife*: Eliza Spottiswoode.

From Henry Bence Jones [before 26 September 1872][1] 3856

Dear friend

Many thanks for all you have done to help me for the Institution.[2] I am much obliged for all

I hope your journey[3] will be prosperous what a kind man you are. You deserve kindness from every one

Y[ou]rs HBJ

I have had a good night & can eat a little & am going to be better again.

RI MS JT/1/J/98

1. *[before 26 September 1872]*: the date is given as this letter appears to come from just before Tyndall left London to journey to the United States for a lecture tour. Bence Jones wrote a farewell letter on 26 September 1872: 'I must write a last time to wish you every success'

(see letter 3857). Also, Bence Jones was ill before and during Tyndall's lecture tour, and died in April 1873.

2. *help me for the Institution*: before he left London for the United States, Tyndall helped Bence Jones with finding lecturers for the RI's upcoming season of Friday Evening Discourses and lecture courses, as he often did (see letters 3782, 3793, and 3801).
3. *your journey*: for Tyndall's lecture tour in America, see Introduction, n. 8.

From Henry Bence Jones 26 September 1872 3857

Folkestone[1] | Sep 26. 1872

Dear Friend

I must write a last time to wish you every success.[2] Pray all external & internal causes of disturbance be absent & may you come back well & happy in mind & body; & may I be near to see you then

Ever warmly yours | Goodbye | H Bence Jones

RI MS JT/1/J/69
RI MS JT/1/TYP/2/705

1. *Folkestone*: see letter 3796, n. 24.
2. *wish you every success*: for Tyndall's American lecture tour, see Introduction, n. 8.

To Margaret Ginty [26 September][1] 1872 3858

Thursday | 1872.

My dear Mrs Ginty,

At 1 o'clock I have an engagement at Brompton.[2] Pray look in at me at 5 P.M. punctually.

I am off to America to-morrow[3] and in the midst of my packing.

Yours ever | John Tyndall.

RI MS JT/1/TYP/11/3670
LT Typescript Only

1. *[26 September]*: the day and month are given since Tyndall left London on Friday, 27 September, to head toward Liverpool for his American trip, and then departed Liverpool aboard the SS *Russia* on 28 September.
2. *Brompton*: not identified, but possibly to see John Lubbock, who had a residence at 6 Foulis Terrace, Brompton.

3. *off to America to-morrow*: for Tyndall's American lecture tour, see Introduction, n. 8.

To Thomas Henry Huxley 27 September [1872] 3859

Royal Institution. | 27th. Sept.

My dear Hal,[1]

I would have run up to you if I could have shaken myself free from my entanglements here. I am off to day,[2] having accepted the offer of a dinner and bed from a Liverpool gentleman[3] whom I met fortuitously at the Aeggischhorn[4] last month.

It would be waste of words on my part to tell you to take care of your health, for your eyes must be wide open to its value. It will be a blessedness to me to hear that it is regaining its former vigour.

Goodbye. My love to your wife[5] and children.[6]

Your affectionate friend | John Tyndall.

IC HP 8.130
Typed Transcript Only

---

1. *Hal*: a nickname for Huxley.
2. *I am off today*: from London to Liverpool, from where on 28 September Tyndall departed England for the United States aboard the SS *Russia*. For his American lecture tour, see Introduction, n. 8.
3. *Liverpool gentleman*: George Warren (dates unknown), a British steamship owner and shipping merchant.
4. *Aeggischhorn*: Eggishorn (see letter 3828, n. 15).
5. *wife*: Henrietta Huxley.
6. *children*: see letter 3834, n. 12.

To Henry Bence Jones 27 September [1872] 3860

Royal Institution. 27th. Sep.

My dear Friend

Your note[1] was exceedingly pleasant to me; but it is not the cause of my writing this one. For I had already determined that my last words in England should be written to you. Need I say how ardently I wish that you may be strong and happy on my return.[2] Nothing will interest me more or give me greater pleasure than to hear from you from time to time. My head quarters, that is for letters, will be Philadelphia anything addressed to the care of

General Tyndale[3] will be sure to reach me.

And now Good bye There is no man of whom I shall think with more warmth of heart than yourself.

Yours ever | John Tyndall

I had a little dinner[4] with Hirst and Debus last night: the one I have known over 24, and the other over 27 years. It is so pleasant to stand well with one's friends after such an interval. And here you and I have been working together for well nigh twenty years without either of us becoming tired of the other. Well that is pleasant

Again good bye | Kindest remembrances to Lady Millicent[5] and all the chicks[6]

RI MS JT/1/TYP/3/803
LT Typescript Only

1. *Your note*: letter 3857.
2. *Need I say... on my return*: Tyndall knew that Bence Jones's health was declining. Fortunately, Tyndall was able to visit with him upon his return to England before Bence Jones died on 20 April 1873.
3. *General Tyndale*: Hector Tyndale.
4. *little dinner*: see letter 3853, n. 2.
5. *Lady Millicent*: Millicent Bence Jones.
6. *all the chicks*: the children of Henry Bence Jones and his wife Millicent were: Millicent Mary Bence Jones (1843–1932), Henry Robert Bence Jones (1844–1912), Olivia Mary Bence Jones (1845–1911), Edith Mary ('Edy') Bence Jones (1853–1919), and Archibald ('Archie') Bence Jones (1856–1937). Olivia's twin Ralph Noel Bence Jones died in 1866. Another daughter, Anabella Mary Bence Jones (1846–51), died at age 4.

To Thomas Archer Hirst 29 September 1872 3861

Sunday 11 A.M. | Sep[tembe]r 29th 1872

My dear Tom

I reached Edge hill[1] in a terrible rain-storm. Was met by Mr. Warren's[2] carriage, and driven with all my luggage to his country house. There I had a pleasant dinner with 8 pleasant and intelligent men. I had a capital bedroom: next morning at breakfast my host spoke of the ship's[3] claret in a discouraging tone; and he proposed & indeed insisted on putting up half a dozen bottles for me. All the more useful as my own wine sent to me from London has failed to reach me. He then drove me round the suburbs for an hour or so; then into the town:[4] halting at a tea merchants and buying half a pound of excellent tea which he told me I should enjoy more than that of the ship.

He drove me to MacIvers office, and there I saw M^r^. MacIver[5] himself. Every thing had been done to secure my cases from violence. Then I left my friend & wandered about for a time. But he met me at the landing stage: went on board the tender & accompanied me to the deck of the Russia. Greater and more courteous kindness it would be difficult to experience: so you are to remember this gratefully towards M^r^. Warren of Strawberry Field Woolton.[6]

I saw a portmanteau blown off a cab yesterday. The board for public notices had the words Fresh Gale printed on it. The paper placards bore the announcement "Heavy gale Critical position of Channel Steamers, foundering of a Welsh Schooner. Loss of Liverpool ship with 30 lives". The Mersey[7] was as angry as it could be, and things looked altogether uncomfortable. At 5 P.M. we had dinner; I confining myself to a slice of beef. At 10 I had a mouthful of grilled chicken and some brandy, and soon afterwards turned into bed. I did not sleep very soundly, but I rested very well, and towards morning the agitation subsided, and on rising we found a clear heaven & a keen clean wind. It is very pleasant. My two boys[8] appear to be in a state of high enjoyment, so that thus far things look well. Froude is here:[9] we have just spent an hour vibrating to and fro along the deck and talking over the affairs of Ireland.

It may be useful for you to know the address of D[r]. Tyndall[10] who takes care of my sister it is

D[r]. Tyndall | The Lodge | Gorey | Ireland.

We are now approaching Cork,[11] where this will be posted.

I do not think that I have anything more to say. Peace be with you my dear Tom.

Yours affect[ionatel]y | John

This bears evidence of the vibration of the screw.[12]

RI MS JT/1/T/693
RI MS JT/1/HTYP/599

---

1. *Edge hill*: a suburb of Lancashire near Liverpool, later incorporated into Liverpool.
2. *Mr. Warren's*: George Warren (see letter 3859, n. 3).
3. *ship's*: Tyndall traveled from Liverpool to New York on the SS *Russia* (see Introduction, n. 8).
4. *the town*: Liverpool.
5. *M[r]. MacIver*: probably David McIver (1840–1907), a British steam ship owner and Conservative politician. McIver was among the founders of the Cunard Line of steamships.
6. *Strawberry Field Woolton*: in Lancashire near Liverpool.
7. *Mersey*: the River Mersey passes Liverpool and Birkenhead as it meets the Irish Sea.
8. *My two boys*: presumably Tyndall's two assistants from the RI who joined him in the United States, John Cottrell and George Miller.

9. *Froude is here*: see letter 3741, n. 6.
10. *Dr. Tyndall*: John Tyndall (of Gorey).
11. *Cork*: a coastal city in southern Ireland.
12. *evidence of the vibration of the screw*: referring to his handwriting, affected by the vibrations of the propeller of the *Russia*.

From Joseph Dalton Hooker [September 1872][1] 3862

Kew.

My dear Tyndall

I should have liked to have shaken your hand and bid you God speed on your noble errand,[2] but I did not expect it, knowing all that you had to do.

I cannot say that I don't hope that an opportunity may arrive next session for publishing the concluding correspondence.[3] I enclose the Russellian verdict,[4] which please return. The "<R>osebud"[5] brought it with a poor account of Lady R.[6]

Her brother's[7] essay on storms[8] surprised me: it was a remarkably clear essay, and showed a very good grasp of the subject as treated hitherto. I have urged his holding on to physics and following Meteorology under the new lights of science. At Lady R's request I sent the essay to Sabine,[9] and am very curious to get his answer.

Now my dear Tyndall take care of yourself. You are very dear to us all; play no tricks with health. We shall think much of you and long to have you back.

With united affection | Ever yours | J. D. Hooker.

Huxley's looks greatly concerned me.

RI MS JT/1/TYP/8/2715
LT Typescript Only

1. *[September 1872]*: the date is given by reference to Tyndall's American lecture tour.
2. *your noble errand*: Tyndall's American lecture tour, for which he left London on 27 September, and departed from Liverpool to New York on the SS *Russia* on 28 September 1872 (see Introduction, n. 8).
3. *opportunity may arrive next session for publishing the concluding correspondence*: correspondence concerning the dispute between Hooker and Acton Ayrton that was presented before Parliament (see letter 3755, esp. n. 1).
4. *Russellian verdict*: John Russell, Prime Minister from 1846–52 and 1865–6, probably would have had a strong opinion about the Hooker-Ayrton dispute.
5. *"<R>osebud"*: probably Mary Agatha Russell, then aged about 18 or 19 years old and seemingly the definition of 'Rosebud' a pretty young woman (*OED*). Hooker also referred to her as such in letter 3798.

6. *Lady R.*: Frances Russell.
7. *Her brother's*: Rollo Russell.
8. *essay on storms*: probably either R. Russell, 'On the nature and character of European Winter Storms, with the best means of giving warning to their approach', *Proceedings of the Royal Society of Edinburgh*, 6 (1869), pp. 140–5; or R. Russell, 'On the regular current theory of storms, illustrated by two storms which passed over the United States, from the 13th to 19th March, 1869', *Proceedings of the Royal Society of Edinburgh*, 6 (1869), pp. 525–9.
9. *Sabine*: Edward Sabine (1788–1883), an Anglo-Irish astronomer, geophysicist, and military officer who led the British effort to establish magnetic observatories all over the globe. He accompanied J. C. Ross's Arctic expedition (1818) and W. E. Parry's expedition (1819). He received the RS's Copley Medal in 1821 for these magnetic observations in the Arctic. Sabine was closely connected with the administration of the RS, serving as its treasurer (1850–61) and president (1861–71) (*CDSB*; *ODNB*).

From [Robert Mayer][1] 1 October 1872 3863

1 Oct 72.

Hochverehrtester Herr!

Der Apotheker J. Fronheiser aus Philadelphia beabsichtigt seine chemisch-physikalische<n> Studien in nächster Zeit in London fortzusetzen und wäre Ihnen sehr dankbar, wenn Sie ihm hiebei mit Ihrem gediegenen Rathe an die Hand gehen wollten, wozu ich mir erlaube, Ihm *[deshalb]* zu empfehlen. Ihr neuestes *[ausgezeichnetes]* Werk über die Alpen, *[welches]* auch hier mit großem Interesse gelesen wird zeigt uns, daß Sie nicht aufgehört haben ein kühne<r> Tourist zu *[seyn]* und wir begleiten Sie *[umher]* ängstlich auf Ihren gefährlichen Wanderungen. Von meiner „Mechanik der Wärme“ habe ich gegenwärtig das Manuscript zur 2^{tn} Auflage zu anfertigen; im Wesentlichen will ich nichts ändern, als daß ich das mech<anische> W<ärme> Aeq<uivalent> statt von *[1 word illeg]* = 367 auf = 425 setze, was allerdings viele Redactionen erfordert

Unter besten Empfehlungen zeichne ich hochachtungsvoll | Ihr dankbar ergebenster

1 Oct 72.

Most esteemed Sir!

The apothecary J. Fronheiser[2] from Philadelphia intends to continue his studies in chemistry and physics in London in the near future and would be very grateful to you if you would lend him a helping hand *[in this connection]* with your sterling advice, for which I take the liberty *[of]* entrusting myself to you *[for this reason]*. Your newest *[excellent]* work about the Alps,[3] *[which]* is also being read here with great interest, shows us that you have not ceased to *[be]* a bold tourist and we accompany you nervously *[all around]* on your *[dangerous]* hikes. At the present time I have to prepare the manuscript of my "Mechanik der Wärme" for its 2nd edition;[4] basically, I do not intend to change anything, other than setting the mechanical equivalent of heat instead of at *[1 word illeg]* = 367 at = 425, which, though, requires a lot of editing

With the best compliments I sign respectfully | Your grateful and very humble servant[5]

Stadtarchiv Heilbronn, D032–155 no. 12

1. *[Robert Mayer]*: the sender of this letter is identified by the mention of the work 'Mechanik der Wärme'.
2. *J. Fronheiser*: James Jacob Fronheiser (1849–94), born of German parents in Johnstown, Pennsylvania, graduated from the Philadelphia College of Pharmacy in 1868. He then went to Germany to continue his education, first at the University of Berlin, and then earning his doctorate at Munich in 1873 (*Alumni Report* (Philadelphia: Alumni Association of the Philadelphia College of Pharmacy, 1893–4), pp. 188–9).
3. *Your newest [excellent] work about the Alps*: the German translation of Tyndall's *Hours of Exercise in the Alps* was J. Tyndall, *In den Alpen* (Braunschweig: Friedrich Vieweg und Sohn, 1872).
4. *"Mechanik der Wärme" for its 2nd edition*: J. R. Mayer, *Die Mechanik der Wärme in gesammelten Schriften,* 2nd edn (Stuttgart: J. G. Cotta'schen Buchhandlung, 1874).
5. No signature given at the end of the letter, which may indicate this was a draft.

From the Albany Institute[1] 2 October 1872 3864

[Albany Institute

This is to Certify that at a meeting of the ALBANY INSTITUTE held] Oct. 1, 1872 Prof. John Tyndall [was duly elected a Honorary member of the same.

Attest,] John V. L. Pruyn[2] [President.] | Daniel J. Pratt,[3] Rec[ording]. [Secretary. | Albany,] Oct. 2, [*18*]72.

RI MS JT/1/A/66

1. *Albany Institute*: a center of learning in Albany, NY, that had various organizations as predecessors going back to 1791, with its charter granted in 1829. It is now the Albany Institute of History & Art.

2. *John V. L. Pruyn*: John Van Schaick Lansing Pruyn (1811–77), an American lawyer and politician who was President of the Albany Institute from 1857–77 (*ANB*).
3. *Daniel J. Pratt*: Daniel Johnson Pratt (1827–84), an American educator and author. Pratt served as Recording Secretary of the Albany Institute from 1869–83, and was the author of *Manual of the Albany Institute: Prepared Under the Order of the Institute, March, 1870* (Albany: J. Munsell, 1870) (L. J. Bevan and W. H. Dannat Pell (eds), *Catalogue of Delta Upsilon* (New York: Delta Upsilon, 1917), p. 89; *Seventy-Third Annual Catalogue of the Officers and Students of Hamilton College, for the Academic Year, 1884–85* (Clifton, NY: Hamilton College, 1884), p. 64).

From Gustav Wiedemann 7 October 1872 3865

Leipzig, 7 October 1872

Mein lieber Tyndall!

Ihren so freundlichen und herzlichen Brief in Betreff meiner Vorrede zu Ihrer Alpenarbeit erhielt ich in Zermatt, mitten in dem Felde Ihrer Alpenarbeit und ich kann Ihnen mit aufrichtiger Freude mittheilen, dass der Genuss der herrlichen Gegend für meine Frau und mich durch Ihre Schilderungen und Betrachtungen ganz wesentlich erhöht worden ist. So haben Sie gleich einen experimentellen Beweis für die Trefflichkeit Ihres Werkes.—Sehr erfreut hat es mich, dass Sie auch aus meinem Vorwort ersehen haben, wie ich Ihnen in wahrer Freundschaft ergeben bin. Haben Sie nicht schon früher meine aufrichtige Zuneigung besessen, so hatten Sie sie durch die Art, wie Sie sich in Ihrem vorletzten Briefe über Zöllner ausreden, in vollstem Maasen erlangt, und meine Frau theilt vollständig meine Gesinnung. Dass ich Zöllner in der Vorrede nicht direct erwähnt habe, geschah, weil ich es für unpassend hielt, durch *[Nennung]* seines Namens auch besonders auf sein Scandalbuch aufmerksam zu machen; jeder Eingeweihte wird hoffentlich meine Absicht verstehen. Uebrigens bin ich, ebenso wie die übrigen, Zöllner früher wirklich aufrichtig ergebenen Freunde, denen er in seiner ganzen Entwickelung viel verdankt, ganz mit Zöllner *[auseinander]*. In seiner *[Eitelkeit]* hat er die Warnungen in Betreff seines *[Besuchs]* und namentlich in Betreff seiner letzten sehr leichtfertigen und bedenklichen Arbeiten nicht vertragen können und als Beleidigungen angesehen. Wir beklagen dennoch viel mehr, als wir darüber erzürnt sind. Wie Schade, dass ein so grosses Talent solche Wege gehen konnte! Das sind die traurigen Folgen der Beschäftigung mit der neueren pessimistischen Philosophie eines Schopenhauer, Hartmann, freilich begünstigt durch manche andere Umstände.

Wie ich höre, haben Sie eine weite amerikanische Reise gemacht. Bitte, erzählen Sie uns gelegentlich von Ihren Erlebnissen.

Meine Frau sendet Ihnen mit mir ihre herzlichsten Grüsse.

In der Hoffnung, recht bald viel Gutes von Ihnen zu hören, in aufrichtiger freundschaftlicher Ergebenheit

Ihr | G. Wiedemann

Leipzig, 7 October 1872

My dear Tyndall!

I received your so kind and warm letter[1] regarding my foreword to your work on the Alps[2] while in Zermatt, right in the middle of the field of your work on the Alps, and I can tell you with sincere pleasure that the enjoyment of the magnificent region has been quite substantially heightened for my wife[3] and me by your descriptions and reflections. You thus have experimental proof right away of the excellence of your work.—I was very pleased that you have also seen from my foreword how I am devoted to you in true friendship. If you did not have my sincere affection already earlier, then you had attained it in the fullest measure through the way in which you unburden yourself about Zöllner in your letter preceding the last,[4] and my wife completely shares my opinion. The fact that I did not directly mention Zöllner in the preface came about because I thought it inappropriate to draw special attention to his scandalous book[5] by *[naming]* his name; anyone who is privy will hopefully understand my intention. Incidentally, like all the others who earlier were really sincerely devoted friends with Zöllner, and to whom he owes much over his entire development, I have completely *[broken off]* with him. In his *[vanity]* he has not been able to bear the warnings regarding his *[visit]* and in particular regarding his recent very thoughtless and dubious works, and viewed them as insults. We nevertheless lament much more than we are angry about this. What a shame that so great a talent could take such paths! Those are the sad consequences of a preoccupation with the modern pessimistic philosophy of a Schopenhauer,[6] Hartmann[7]—aided, however, by a great many other circumstances.

You have, as I hear, gone on a long trip to America.[8] Please, tell us some time or other about your experiences.

My wife, along with me, sends you her warmest regards.

In the hope of hearing much good news from you quite soon, and in sincere friendly devotion

Your | G. Wiedemann

RI MS JT/1/W/53

---

1. *your so kind and warm letter*: letter missing.
2. *my preface to your work on the Alps*: Wiedemann wrote the preface for the German translation of Tyndall's *Hours of Exercise in the Alps*, published as J. Tyndall, *In den Alpen* (Braunschweig: Friedrich Vieweg und Sohn, 1872).

3. *my wife*: Clara Wiedemann.
4. *your letter preceding the last*: letter missing.
5. *his scandalous book*: for Zöllner's book and the dispute which it caused, see letter 3740, n. 3.
6. *Schopenhauer*: Arthur Schopenhauer (1788–1860), a German philosopher (*ADB*; *NDB*).
7. *Hartmann*: Karl Robert Eduard Hartmann (1842–1906), a German philosopher (*NDB*).
8. *long trip to America*: for Tyndall's American lecture tour, see Introduction, n. 8.

To Joseph Henry 9 October 1872 3866

9th Oct. 1872.

My dear Professor Henry

I take it as particularly kind of you to do so much as you have done to forward the arrangements for the lectures.[1] I have seen & conversed with Mr. Rhees, and he is now in Boston, resolved to make sure that I shall have a fair start.[2] I cannot more truly express my feeling than by saying that I feel myself thoroughly at home with my excellent American friends

I knew you would like Arrow.[3] He is genial, able, vigorous man, and he & his Brethren[4] are doing excellent work along the coast of England.

I hope you will not ask me to swerve too far from my original intention. Should the occasion arise I shall feel myself more free to tell both America & England their duty to Science, if I know that I am speaking without pay. The expenses already incurred are very high indeed, so that in any case I shall have to make a heavy draft upon you. I have laid hold of probably far more apparatus than I shall use: it is costly, and this added to other expenses will constitute a very serious subtraction[5] from the proceeds of the lectures.

With best wishes | Believe me | ever faithfully yours | John Tyndall

The night but one before I came away I went to see Mrs. Faraday[6] when she drew from a drawer the Daguerreotype of a very powerful looking man of about 40 or under. I recognized it to be your own.[7]

John Tyndall to Joseph Henry, 9 October 1872, mssRH 3969, box 43, William Jones Rhees papers, The Huntington Library, San Marino, California

---

1. *lectures*: for Tyndall's American lecture tour, see Introduction, n. 8.
2. *fair start*: Tyndall's very first lecture in the United States was in Boston on 15 October 1872.
3. *Arrow*: in letter 3837 to Henry, Tyndall wrote that 'Sir Frederick Arrow, the Deputy Master of the Trinity House is now in America. I suppose he will see you. You will find him a genial, vigorous & able man'.

4. *Brethren*: Elder Brethren of Trinity House (see letter 3736, n. 3 and n. 5). On Tyndall as scientific adviser to Trinity House, see letter 3869, n. 4.
5. *very serious subtraction*: Tyndall's total expenses ($10,066.66) and the net proceeds of his lecture tour ($13,033.34) were published in 'The Tyndall Trust for the Promotion of Science in the United States', *Annual report of the Board of Regents of the Smithsonian Institution, showing the operations, expenditures, and condition of the Institution for the year 1872.* Serial Set Vol. No. 1573, Session Vol. No. 3, 42nd Congress, 3rd Session (Washington: Government Printing Office, 1873), pp. 104–6 (this was also published in *Popular Science Monthly*, 3 May 1873, pp. 100–01). Tyndall donated the proceeds to establish a fund for pure science research in the United States (see Introduction, n. 15).
6. *M^rs^. Faraday*: Sarah Faraday.
7. *your own*: the Smithsonian Institution holds several daguerreotype portraits of Henry from *c.* 1830s or 1840s, and Princeton University Archives holds at least one as well.

To Frederick Barnard 11 October 1872 3867

11th. Oct. 1872.

My dear D^r^. Barnard

Since I came to New York every moment of my time has been occupied in inspecting the places where I shall have to lecture,[1] & in other matters. Were I to visit anybody in New York during my halt it would be you.

The whole of today I shall be out of town, and tomorrow early I start for Boston.[2] If I can secure a fragment of time I will endeavour to see you before I leave, but if this should be impossible it will give me true pleasure to meet you when I come here to lecture.[3]

Yours most faithfully | John Tyndall

Columbia University Manuscripts; X973 C72 V01.2; Box 2 Folder 91; University Archives, Rare Book & Manuscript Library, Columbia University Libraries, Rare Book & Manuscript Library, Columbia University in the City of New York

---

1. *to lecture*: Tyndall arrived in New York on 9 October. For his American lecture tour, see Introduction, n. 8.
2. *Boston*: Boston was the first stop on Tyndall's American lecture tour, six lectures from 15–25 October 1872 at the Lowell Institute.
3. *when I come here to lecture*: Tyndall returned to New York following his lectures in Washington, DC. His New York lectures ran from 17–31 December 1872 at the Cooper Institute (three were before Christmas Day, and three after).

From Theodore Negley[1] 12 October 1872 3868

Hall of Amer[ican]. Whig Society[2] |
Princeton N. J. | Oct. 12th 1872

Prof. Tyndall

Dear Sir

This is to inform you of your election to an honorary membership in the Amer[ican]. Whig Society—a literary organization founded in the year 1769 & composed of a portion of the students and Faculty of the College of New Jersey. An institution similar to our own called the Cliosophic Society[3] exists in the College between whom & us there is a treaty, to the following article of which we desire to call your attention

Art III

"If the same individual be elected an honorary member of both Societies, he shall be a member of that Society whose election he may have first accepted"

We number in our catalogue many of the most illustrious citizens of our country, & desire to add your name to the list.

Hoping for a favorable reply we remain

Yours Very Respectfully | Theo. S. Negley | By order of | The Amer[ican]. Whig Society.

RI MS JT/1/N/6

---

1. *Theodore Negley*: Theodore Shields Negley (1846–1911), of Pittsburgh, a student at the College of New Jersey (later Princeton University), who graduated in 1873, and then from Princeton Seminary in 1876. He went on to become a pastor for Presbyterian churches in Pennsylvania (G. T. Fleming (ed), *History of Pittsburgh and Environs: From Prehistoric Days to the Beginning of the American Revolution*, 6 vols (New York: American Historical Society, 1922), vol. 6, p. 61).
2. *Amer[ican]. Whig Society*: the Plain Dealing Club (later American Whig Society) and the Well Meaning Club (later Cliosophic Society) were founded about 1765 to promote literary and debating activities, the oldest such clubs in the United States. The two societies merged in 1928. See J. J. Looney, 'Useful without Attracting Attention: The Cliosophic and American Whig Societies of the College of New Jersey, 1765–1896', *The Princeton University Library Chronicle*, 64 (2003), pp. 389–424.
3. *Cliosophic Society*: see n. 2.

From Frederick Arrow 14 October [1872][1] 3869

[TRINITAS IN UNITATE[2] | TRINITY HOUSE[3] LONDON] |
5th. Avenue Hotel | New York | 14 October

My dear Tyndall.

I was so vexed to have missed you. I have been pushing on in the hope that we might meet & blame myself for not telegraphing you as I might then have got in to New York late on Friday whereas I took the early train from Philadelphia Saturday and so missed you by Two hours.

One of my objects in seeing you was to ask you on behalf of Light House Work, when you meet Professor Henry is to talk to him about his views on the Transmission of Sound as resulting from the experiments we made.[4] There were some very curious features, such as the lapse of sound, and thus its recontinuance at a greater distance than when it was lost, and I should be glad also if you visit Canada if you could see the oil region near new London[5] about half a days journey from Niagara[6] where the refineries are. This place would interest you & there is a good deal about Petroleum to be learnt. I have arranged for some to be *[tower]* distilled as a sample. They tell me it improves the illuminating power while it rises the Flashing point though I doubt it. We also at Home are unaware of the difference between Canadian & Pennsylvanian oil both being called American the latter being really much more volatile I can't say. I have learnt much from my visit here, but still there is always something to be picked up and I hope I have not neglected to do that if for nothing else. I am glad to have come here to make the acquaintance of Prof[esso]r. Henry who I feel quite an affection for, & his amicable family.[7] He is one the more you know him the more you like & respect his amiable disposition, his unobtrusive kindness, & his wide and varied knowledge. He has truly been more than kind to us. I wish all his Country men were like him *[1 or more pages missing]* or could through his spectacles which all the really able American gentlemen I have met *[1–2 words illeg]*. The Country is a noble one teeming with Gods best gifts. I wish I thought as much of its People, but the melange[8] of races seems though practical & enterprising to have lost many of mans higher attributes. Wishing you a pleasant *[sojourn]* and a safe return & if you have a leisure ½ hour to hear from you.

I am | always yours. | Fred Arrow

RI MS JT/1/A/116

1. *[1872]*: the year is given as Tyndall and Frederick Arrow were both in the United States in 1872.

2. *TRINITAS IN UNITATE*: the coat of arms for Trinity House bears this Latin phrase meaning 'Three in One', a reference to the Holy Trinity from which Trinity House took its name in 1514.
3. *TRINITY HOUSE*: see letter 3736, n. 3.
4. *experiments we made*: while Tyndall had assisted Michael Faraday since May 1866 in lighthouse matters (see letter 2423, *Tyndall Correspondence*, vol. 9), in July 1867 he officially took over for Faraday as Scientific Adviser to the Board of Trade (a British government department concerned with making policy and instituting regulations regarding economic activity in Britain and its empire) and Trinity House (the lighthouse authority for England and Wales, established by Royal Charter in 1514) (see letters 2613, 2625, and 2829, *Tyndall Correspondence*, vol. 10). In this role, Tyndall investigated and gave recommendations to Trinity House on matters of lighthouse illumination, and later, sound signaling. On Tyndall and his role with the Board of Trade and Trinity House, see R. M. MacLeod, 'Science and Government in Victorian England: Lighthouse Illumination and the Board of Trade, 1866–1886', *Isis*, 60 (1969), pp. 4–38; S. Courtney, '"A very diadem of light": Exhibitions in Victorian London, the Parliamentary Light and the Shaping of the Trinity House Lighthouses', *British Journal for the History of Science*, 50 (2017), pp. 249– 65; and *Ascent of John Tyndall*, pp. 282–5, 316–17, 345–6, 389, 406–13, 420–3, 433–4, and 437.
5. *new London*: it is not certain to which New London that Arrow refers, as there are several in the northeastern United States, in Connecticut, Ohio, and Pennsylvania.
6. *Niagara*: Tyndall did visit Niagara himself (see letter 3738, n. 3).
7. *his amicable family*: Joseph Henry married his cousin Harriet (née Alexander, 1808–82) in 1830. At this time, their three daughters were all adults, and may have been present during Arrow's visit: Mary Henry (1834–1903), Helen Henry (1836–1912), and Caroline Henry (1839–1920). A fourth child, their eldest and only son, died a decade before from jaundice: William A. Henry (1832–62).
8. *melange*: a mixture, combination, or blend (*OED*).

To Alfred Mayer 18 October [1872][1] 3870

Royal Institution of Great Britain | 18th. Oct.

My dear Sir,

A day under present circumstances is a very serious matter; but I certainly hope to be able to see your experiments.[2]

I predicted long ago that you would do some good work if not overloaded by antagonistic duties[3]

Yours faithfully | John Tyndall

Tyndall, John (1820–1893); Hyatt and Mayer Collection, C0076, Manuscripts Division, Department of Special Collections, Princeton University Library

1. *[1872]*: the year is given as this letter is during Tyndall's American lecture tour (see Introduction, n. 8).
2. *to see your experiments*: Tyndall was unable to visit Mayer's laboratory while in New York at the beginning of his time in the United States, and here writing from Boston, thought that maybe he would be able to when he returned to New York in December for lectures at the Cooper Institute. Again, a visit did not occur. Mayer later wrote, in letter 3985, 'I cannot but regret that I have seen so little of you and that the hope I had of your spending a day with me in my laboratory could not be realized'.
3. *I predicted long ago... antagonistic duties*: in a letter to Mayer of 11 March 1871, Tyndall wrote: 'Your papers have reached me and I can see in them, even from the superficial glance which up to the present moment I have been able to bestow upon the, firmness of intellectual fibre which is sure to make itself felt, and which it is of exceeding importance to cultivate, in America' (see letter 3377, *Tyndall Correspondence*, vol. 12).

To Edward L. Youmans [18][1] October 1872 3871

BOSTON, October 8, 1872.

MY DEAR YOUMANS: Thanks many for both your letters,[2] which interested and amused me greatly. This opening of the International Series[3] is very gratifying. I think in founding it you have entered upon a most important enterprise.

I have received a letter[4] from Mr. Winthrop,[5] of Boston, which interests me greatly. He is descended from John Winthrop,[6] who came over in 1630, and who married Margaret Tyndale,[7] whose great-grandson[8] was the most distinguished professor of natural science in New England of his time. Is not this curious?

The lectures[9] are going on, but they are difficult. The people crowd to hear them, and Mr. Lowell[10] declares that never previously were they held so fast in the lecture room.[11] I give them an hour and a half, though warned that they would not stand more than an hour. Still the lectures by no means please myself, one reason being that I am at sea as regards the intellectual level of my audience.*...

Go on and prosper, my dear Youmans, in the work you have undertaken. I hardly know any man in Europe or America who enjoys your opportunity of doing good, and the best of it is that it is an opportunity created by yourself.

Give my affectionate regards to the Methodists.[12] I will not return railing for railing, but, contrariwise, blessing[13]

Yours ever, | JOHN TYNDALL.

* An error which has played the mischief with more than one lecturer in America, notably the lamented Freeman, who thought it necessary to tell his

audiences in Boston and St. Louis simple outlines of English history known to every schoolboy. As I remember Tyndall's lectures, however, he did not go far astray in this direction. Any British lecturer will always be safe in addressing any American audience he is likely to meet in exactly the same way that he addresses his audiences at the Royal Institution in London or at the Philosophical Institution in Edinburgh.

J. Fiske (ed), *Edward Livingston Youmans: Interpreter of Science for the People* (New York: D. Appleton and Company, 1894), pp. 317–18

1. *[18]*: while the published letter bears the date of 8 October, this must be in error, as Tyndall refers to the first of his Boston lectures, which commenced on 15 October (see n. 9). 18 October is a likely date ('18' mis-transcribed' as '8'), and this would allow for the first two of the six lectures in Boston to have already occurred.
2. *both your letters*: letters missing.
3. *International Series*: the International Scientific Series was a popular series of scientific books founded by Youmans in 1871, with Tyndall, Thomas Huxley, and Herbert Spencer involved as a 'British Committee'. Tyndall's *Forms of Water* was the first title in the series (see L. Howsam, 'An Experiment with Science for the Nineteenth-Century Book Trade: The International Scientific Series', *British Journal for the History of Science*, 33 (2000), pp. 187–207; B. Lightman, 'The International Scientific Series and the Communication of Darwinism', *Journal Cambridge Studies*, 5 (2010) pp. 27–38; and R. M. MacLeod, 'Evolutionism, Internationalism and Commercial Enterprise in Science: The International Scientific Series 1871–1910', in A. J. Meadows (ed), *Development of Science Publishing in Europe* (Amsterdam: Elsevier, 1980), pp. 63–93).
4. *letter*: letter missing.
5. *Mr. Winthrop*: Robert Winthrop.
6. *John Winthrop*: John Winthrop.
7. *Margaret Tyndale*: Margaret Tyndal Winthrop.
8. *great-grandson*: John Winthrop (1714–79).
9. *lectures*: Boston was the first stop on Tyndall's American lecture tour (see Introduction, n. 8), six lectures from 15–25 October 1872 at the Lowell Institute.
10. *Mr. Lowell*: John Amory Lowell.
11. *in the lecture room*: of the Lowell Institute (see letter 3814, n. 4).
12. *Methodists*: American Methodists attacked Tyndall for his views on prayer, and organized prayer meetings in the major Eastern cities during his lecture tour to pray for him. On Tyndall and prayer, see letter 3780, n. 2.
13. *railing for railing, but, contrariwise, blessing*: 1 Peter 3:9.

From Robert Winthrop 19 October 1872 3872

Brookline, | near Boston, | 19 Oct[obe]r. 1872.

Dear Professor Tyndall,

Agreeably to my promise, after your Lecture,[1] on Thursday Evening I send you my "Life & Letters of John Winthrop",[2] who came over here as Governor of Massachusetts in 1630. After losing two wives[3] in very early manhood, he married Margaret Tyndale,[4] (daughter of Sir John Tyndale,[5]) who followed him to America, & lived in Boston until 1647. She was one of the loveliest women of our early Colonial period. I was myself descended from the Governor's first wife; but you may be interested to know that Margaret Tyndale's great grandson[6] was the most distinguished Professor of Natural Science in New England of his time,—the friend of Franklin,[7] a member of your Royal Society,[8] & whose Observations of the Transit of Mercury in 1740,[9] & of Venus in 1761,[10] were the subjects of great notice both in England & America.

I have put a mark at page 123 of the 1st. volume, & at page 363 of the 2d. volume, which will point you to all I have written as to the Tyndalls in our Country. You will hardly care to read even so much as this; but perhaps you will kindly take the work home with you as a Souvenir of your visit to Boston, & as furnishing evidence that your family name was not unknown or unhonored here in the olden time.

I send you, also, a copy of the Life of Count Rumford,[11] recently published by our American Academy of Arts & Sciences,[12] & which the Author, my friend the Rev. Dr. George E. Ellis, has requested me to present to you in his behalf. Rumford,[13] as you know, was born near Boston, & we are all proud in remembering the leading part he took in founding the Royal Institution,[14] to which Faraday & yourself have added so much renown.

Let me only say further, that it will afford me the greatest pleasure if you should be able to come out to luncheon on Thursday next, as you partly promised. But if any other day, after Monday, would suit you better, I shall gladly conform to your convenience. I will make arrangements for bringing you out & showing you something of our environs <u>en route.</u>

Believe me, Dear Sir, | Very faithfully, yours, | Rob[er]t. C. Winthrop.

Prof. Tyndall

P.S. Should you desire to name a different day, a Note addressed to "Brookline Mass[chuse]tts" would reach me by mail.

RI MS JT/1/W/84

1. *your Lecture*: the second of Tyndall's six lectures from 15–25 October 1872 at the Lowell Institute, on 17 October. For Tyndall's American lecture tour, see Introduction, n. 8.
2. *my "Life & Letters of John Winthrop"*: R. C. Winthrop, *Life and Letters of John Winthrop*, 2 vols (Boston: Ticknor and Fields, 1863–7).
3. *two wives*: first, Mary Winthrop (née Forth, 1582–1616), married in 1605 (*ODNB*); second, Thomasine Winthrop (née Clopton, 1583–1616), married in 1615 (*ODNB*). After his third wife Margaret Tyndal Winthrop's death in 1647, Winthrop married a fourth wife, Martha Rainsborough Coytemore Winthrop (*c.* 1617–60), married in 1647 (*ODNB*).
4. *Margaret Tyndale*: Margaret Tyndal (or Tyndale/Tyndall) Winthrop.
5. *Sir John Tyndale*: John Tyndal (or Tyndale/Tyndall) (*c.* 1552–1616), Master in Ordinary of the King's High Court of Chancery. He married Anne Egerton (1571–1620) in 1586 (K. G. Everingham (ed), *Royal Ancestry: A Study in Colonial and Medieval Families*, 5 vols (Salt Lake City, UT: 2013), vol. 5, p. 223).
6. *great grandson*: John Winthrop (1714–79).
7. *Franklin*: Benjamin Franklin.
8. *member of your Royal Society*: John Winthrop was elected a Foreign Member of the RS on 20 February 1766.
9. *Observations of the Transit of Mercury in 1740*: see J. Winthrop, 'A Letter from Mr. John Winthrop, Hollisian Professor of Mathematics and Astronomy at Cambridge in New-England, to C. Mortimer, M.D. Sec. R.S. concerning the Transit of Mercury Over the Sun, April 21. 1740. and of an Eclipse of the Moon, Dec. 21. 1740', *Phil. Trans.*, 42 (1742–3), pp. 572–8.
10. *Venus in 1761*: see J. Winthrop, 'Observation of the Transit of Venus, June 6, 1761, at St. John's, Newfound-Land: By John Winthrop, Professor of Mathematicks and Philosophy at Cambridge, New England', *Phil. Trans.*, 54 (1764), pp. 279–83.
11. *Life of Count Rumford*: G. E. Ellis, *Memoir of Sir Benjamin Thompson, Count Rumford, with Notices of His Daughter*, 2 vols (Boston: American Academy of Arts and Sciences, 1871).
12. *American Academy of Arts & Sciences*: a learned society in Cambridge, Massachusetts founded in 1780 by some of the Founding Fathers of the United States. It continues to operate.
13. *Rumford*: Benjamin Thompson.
14. *founding the Royal Institution*: on the founding of the RI in 1799, see F. A. J. L. James, 'Founders of the Royal Institution of Great Britain', *ODNB*; D. Knight, 'Establishing the Royal Institution: Rumford, Banks, Davy', in F. A. J. L. James (ed), *'The Common Purposes of Life': Science and Society at the Royal Institution of Great Britain* (New York: Routledge, 2002), pp. 97–118; and H. B. Jones, *The Royal Institution: Its Founder and Its First Professors* (London: Longmans, Green, and Co., 1871).

To Thomas Archer Hirst [20 October 1872][1] 3873

My dear Tom.

It was sunny weather when I last wrote to you,[2] but that soon changed and for days we had to contend against a gale which blew in our teeth. As long as the wind continued in this direction the ship[3] 'pitched' heavily: it veered at times to the north, taking us abeam, and then we 'rolled' heavily. But througout the voyage I was not for a moment sick, once or twice a little uncomfortable—that was all. The masses of beefsteak fronting Cottrell[4] on the breakfast table proved <u>him</u> all right. Miller[5] was disturbed for a time; but he also quickly shook away calamity and was as brisk as his companion.

The gleams of rainbows in the fine spray were beautiful; the foam with gleams of delicious green breaking through its whiteness was beautiful; the phosphorescence was at times beautiful; and the sea itself when the gale was at its highest was exceedingly grand. The ocean ridges were high, and where they crossed each other they formed <u>heaps</u> tufted with foam which was caught by the wind and carried like smoke through the air. I once rushed through the spray which swept the forward deck, and got just behind the bowsprit,[6] which seemed to divide the waves; seen from here the turmoil was finest, and Shakespeares[7] image of the wind

"taking the ruffian billows by the top"[8]

rose to my mind as a veritable description of the turbulence. I was warned away from the bow by one of the officers who assured me that he would not stand there for any consideration. For, he urged, the bowsprit itself dips under water, and the full force of the wave sweeps the deck sometimes without a moment's warning—Being wise I quitted my ocean observatory.

Planted on the maindeck of the Russia is an upper deck which in fine weather is pleasant to walk upon. It is carried quite from stem to stern of the vessel. Between this higher deck and the gunwale[9] was a space bounded on one side by the deck and on the other by the ship's side. Here ladies used to sit in shelter, choosing the side protected from the waves and spray. One day I was on deck watching the waves rising above the bow: curling over & sweeping the forward deck. The ladies occupied their usual chairs upon the sheltered side. I heard the swash of a wave, the crash of chairs, and immediately afterwards a scream. I looked over the deck railing. The wave had leaped the gunwale and carried the whole bevy along with it. I was of course instantly knee deep in water trying to assure and rescue the frightened fair. They differentiated themselves in the most beautiful manner into helpful & helpless; some picking themselves up practically, others screaming & doing nothing else. The men who joined in the rescue were similarly differentiated into helpful and confused.

Our speed during the voyage varied from about 6 knots to 15, the difference being due to the wind. On Wednesday the 9th we were called earlier than usual. There was a cloudless sky overhead, and the American coast was in sight. Under these favourable conditions we steamed between the forts which guard the entrance of New York harbour. I was not prepared for the view then opened out before me. It was splendid. To the left were the slopes of Staten Island,[10] fringed with coloured foliage. The ample bay was studded with small craft flying about with their white winglike sails. All round the bay the coalescence of land and water was picturesque. In fact the Bay of New York may fairly take its place beside that of Naples. At the so-called quarantine the Doctor came on board, and as we had a clean bill of health we were permitted to proceed without delay. The docks are rather shabby, looking like the neglected & decayed structures of an old country, rather than the vigorous production of a new one. A crowd of people awaited the arrival of the boat, and the first known face which appeared among the crowd was that of Hector.[11] Youmans was also there and Mr. Appleton and many others—all desirous of rendering me every possible aid. The Secretary of the Treasury[12] had given directions regarding my baggage, which was passed without halt or examination. The greeting was just such as I liked, quiet, cordial and practical. Thus ended my first voyage across the Atlantic.

I had suggested to Hector the possibility of a trip to Niagara[13] prior to my Boston lectures,[14] and he, good fellow, wishing to protect me from all strain & pressure in New York, had arranged to carry me off immediately. But this arrangement did not fall in with my notions. In fact the strain of the journey to Canada would have been far greater than the strain of the city.

I put up at an excellent, but exceedingly expensive hotel, called Brevoort House.[15] The hotels are often called Houses—I am now staying at the Revere House.[16]

Froude[17] is the great luminary in New York. His publishers, Messrs. Scribner & Co.[18] have arranged a magnificent banquet to welcome him.[19] One of the firm is I am told, a great newspaper man,[20] and he has arranged that nearly all the editors in New York shall meet the historian. I believe they have also secured Emerson, who I hear is on his way to England. It is amusing to hear the remarks of Youmans on the whole affair; but at all events it will make the harvest of Froude in the States a golden one, and it will be a glory to his publishers. Appleton I am told had contemplated something similar, but had postponed the matter on hearing of the action taken by Scribner. I bless the gods that he did so, for quiet at the commencement is very necessary.

On the afternoon of the day of my arrival I strolled through Broadway with Hector. It is a fine street, but less 'broad' than I had imagined. The retail trade has here received its deification—being carried on in temples worthy of the gods. Think of a marble palace devoted to retail hosiery—there are such in New York.

On the Thursday I visited the places where I am to lecture, the Cooper Institute[21] at New York, and the Academy of Music[22] in the adjacent city of Brooklyn. Many people called, and many friendly letters of welcome reached me. We had a stroll in the evening through the 5th Avenue, the paradise of New York capitalists—a really fine street and through other avenues. My bed room remained quiet until 4 A.M. when the clatter of pots & pans prevented further sleep.

On the Friday I accompanied Youmans and Mr. Appleton to West Point.[23] We wished to take a boat up the Hudson.[24] But the steamers had been discontinued, and we had to take the rail. The river was however almost always in view. The opposite side was bounded by a long reach of basaltic cliff which bears the name of 'the palisades'—by no means an inappropriate term. They disappeared and rounded Rhinelike[25] eminences clothed with woods took their place. We halted opposite West Point and took a ferry-boat across the Hudson. Mounted to a high and commanding plateau planed down to almost perfect horizontality, and on which the buildings of the Military Academy stand. I do not think I ever saw anything more charming than the scene from this plateau. Bold hills with valley inlets clothed with woods flanked the Hudson right and left. The river itself stretched like an indented lake between them. The colouring I should have pronounced gorgeous; though I am assured that the year is a bad one for colour owing to the great heat of the summer. Single trees however stood forth as miracles of colour. Their images upon my retina were sufficient to brighten up my whole mind, so startling was their beauty.

The General Superintendent was away, but we got hold of an excellent old professor—Kenrick[26] by name—who felt a pleasure in taking us over the whole place. It gratified me to find my books on the shelves of the libraries and their contents in the brains of the Professor. He was a hearty fellow who loves his work, and was evidently a man to sympathise with the young. I went though many of the rooms where the young fellows were at work under the supervision of young officers, who acted the part of assistant professors. We wound up at 3 P.M. with a glass of rye whiskey & water, and a biscuit. I afterwards dined with Appleton at his country house, and then returned with Youmans to New York.

Early on Saturday I was astir—paid my bill and started with Hector for Boston by the 9 P.M. train. An unfriendly land this New England must have been to the first settlers. It was hard & unfruitful. Still it has not been so hard as to break man's energies in a hopeless struggle with natture; but hard enough to call out all the energy they might possess. The foliage at times was of the character I have described—indescribably brilliant; and as those wondrous trees passed me each of them started its psychological equivalent of pleasure. The monkeys from which I came[27] must have been dandled amid such scenes.

I had some pleasant people in the train with me, who were also my companions in the Russia. M^r^. Peabody[28] a Boston Banker & his wife & children.[29] They greatly regretted the disorganization of their establishment, as they wished to take me with them. I put up at the Revere House.

Professor Henry had sent his principal assistant[30] from Washington to meet me in New York, and he on the day of my arrival started with Cottrell & Miller to Boston. He is now here. On Saturday night I saw Cottrell's disposition for the lectures. As I expected he was far forward. The excellent fellow leaves me very little to do in the way of preparation.

Yesterday after writing some letters and after a very instructive visit from D^r^. Beard[31] the second in command of the Smithsonian Institution, I walked out with Hector. When a boy I remember committing to memory a poem about Bunker Hill.[32] Yesterday I stood upon Bunker Hill, or rather Breed Hill, itself. There is a plain pillar on the top, built of granite and over two hundred feet high.

"Here the embattled Yeomen stood
& fired the shot heard round the world".

It is useless to cry over spilt milk but one wishes that wisdom, instead of George the third[33] had filled the throne of England in those days. Boston is in an entirely English town as far as its aspect goes. It is massive and wealthy, and full of active life.

D^r^. Cotting,[34] the curator of the Lowell Institute,[35] a practical sagacious man, evidently fears that I may prove prosy. He exhorts the limitation of the lectures to an hour; he exhorts the omission of all introductory matter, as he has known audiences to be killed by the first lecture. I laugh, and say that I have gone too far to retreat, and that now I must follow my own line.

During the last two nights I slept from 11 to 4 or thereabouts, but not more than four or five hours. The streets were within hail. But this morning I change my rooms, and doubt not that my rest henceforward will be sufficient for the work on hand.

Since writing the above nearly a week has passed and I have delivered three lectures.[36] This day week (Sunday) Hector and myself walked out for several hours—saw the famous Bunker Hill, one of the earliest fields of conflict between England and America. We walked towards Cambridge, but did not reach the place. On Monday we went to the cemetery of Mount Auburn,[37] one of the most beautiful places—in fact the most beautiful place of the kind that I have yet seen. From a high tower I had a view of the surrounding country, and a more lovely country I have never seen associated with a city. Boston itself impresses me greatly. Its size, solidity, and the massive vigour of its people are all such as you might expect in a first rate English town. Several persons called upon me, & Prof. Asa Gray wishes much that I should live with him. I decided however to cling to the hotel.

On Monday 14th night I imagined I had secured quiet sleeping quarters; but at 2 oC[lock] A.M., they began shovelling coals beneath my window further sleep was out of the question. I rose with a broken brain, and went again to Mount Auburn to mend it. Here I recovered a little energy & clearness and when the time came[38] I was fairly patched up—The reception given to me was warm, and after the lecture cordial greetings poured in upon me It was considered far more successful than I considered it myself. But the newspaper slips sent you by Hector will have told you all about the matter.

Emerson called on the Monday 14th on his way to New York. I missed him. On his return he & his daughter[39] called again: I missed them, but Hector was here. While I was taking luncheon on Thursday he came again & sat with me for some time. He would have us go to Concord[40] on the Friday. We went—He met us at the station & drove us through the village—the picture of perfect peace—to his ruined house which is now being restored.[41] It is the simplest little nest imaginable. Thence he drove us to the old Manse[42] where he took refuge when he was burnt out, and from the windows of which his own grandmother witnessed the first firing between the farmers & the troops.[43] We dined in a simple way—Saw his wife[44] & daughters[45] and were throughout not only happy but joyful. He took us to the Concord Monument;[46] where to use his own words the "embattled yeomen stood & fired the shot heard round the world". The grave of the two Englishmen shot there is still preserved. Two of the farmers were also shot & of course carried away by their friends. At 4 oC[lock] a pony carriage was ready to drive us to Lexington[47]—also a famous historic place. Emerson accompanied us, and here we parted from him & took our tickets to Boston. Nothing could be pleasanter than our visit.

He is a tall thin man; bald quite at the top & back of the head. Sometimes an expression of profound abstraction fills his eyes; but after dinner on the day of our visit, practical vigour took the place of abstraction. He urged us to go to a club[48] on the following day where he said we should meet numbers of the best people in Boston.

He gave us a note to Wendell Holmes & Hector presented it in the early part of the day. Thereupon Holmes, a most agreeable & intelligent fellow, called upon us; while he was here in came Emerson to explain a mistake that he had made regarding the day of meeting of the Club. Holmes had made the explanation previously. While they were here in came Agassiz—a few hours subsequent to his reaching Boston from his travels.[49] All were kind & hearty in the highest degree. Afterwards came a Spencerian[50] named Fiske,[51] a hard headed thinker who is librarian at Harvard College. Then came Longfellow. In fact a perfect galaxy of genius was squeezed into the two midday hours yesterday. I lectured last night.[52] The people crowd to hear me, but I do not at all satisfy myself—my main difficulty is a want of acquaintance with the intellectual level of my audience.

This is the only journal I shall write—so keep it—I may expand it when I have more time.

Youmans informs me that my little book the "Forms of Water"[53] is going "like thunder". One concern purchased 2000 copies of it at once. I am also told that prayer meetings are held with a view to my salvation.[54]

M^r^. Winthrop[55]—a member of one of the old historic families here has sent me some books accompanied by a letter which has interested me immensely. He writes to me thus:—

[...][56]

Here I must pause. Love to all my brothers of the X[57] and with steadfast affection.

believe me yours | John.

* Tell Lady Mary[58] that the wife of Sir John Tyndall[59] was Anne Egerton![60]

Posted 20th October, written in part earlier.

RI MS JT/1/T/999
RI MS JT/1/HTYP/600–04

---

1. *[20 October 1872]*: the date is given as Tyndall was on his American lecture tour (see Introduction, n. 8), and that he wrote at the top of the letter that he posted this letter on 20 October, while it had been written over a period of a week to ten days.
2. *when I last wrote to you*: letter 3861.
3. *the ship*: Tyndall departed for the United States aboard the SS *Russia* on 28 September 1872.
4. *Cottrell*: John Cottrell.
5. *Miller*: George Miller.
6. *bowsprit*: a large spar or boom running out from the stem of a vessel, to which (and the jib-boom and flying jib-boom, which extend beyond it) the foremast stays are fastened (*OED*).
7. *Shakespeares*: William Shakespeare (1564–1616), the English poet and playwright (*ODNB*).
8. *"taking the ruffian billows by the top"*: Shakespeare's *Henry IV* II.iii.22.
9. *gunwale*: the upper edge of a ship's side (*OED*).

10. *Staten Island*: at this time Staten Island was a town of its own, and not part of New York City until 1898.
11. *Hector*: Hector Tyndale.
12. *Secretary of the Treasury*: George Boutwell (see letter 3815, n. 4).
13. *trip to Niagara*: Tyndall visited Niagara Falls after his Boston lectures (see letter 3738, n. 3).
14. *Boston lectures*: Tyndall delivered six lectures in Boston from 15–25 October 1872 at the Lowell Institute.
15. *Brevoort House*: see letter 3843, n. 3.
16. *Revere House*: a Boston hotel in Bowdoin Square that operated from 1847–1912.
17. *Froude*: Anthony Froude traveled to the United States on the same steamship as Tyndall, the *Russia*.
18. *Messrs. Scribner & Co.*: Scribner and Company was an American publisher in New York, founded in 1846 by Charles Scribner I (1821–71) and Isaac D. Baker (1819–50). In 1871, upon Scribner's death, his son John Blair Scribner (1850–79) took over as President, and when his two brothers later joined him in the company, it was renamed Charles Scribner's Sons. The firm specialized in religious and historical texts, eventually adding significant British reprints and European translations. *Scribner's Monthly* magazine, established in 1870, became a leading reputable journal.
19. *magnificent banquet to welcome him*: the banquet was held on 15 October at Delmonico's in New York, the same venue where Tyndall's farewell banquet was held on 4 February 1873 ('The Froude Dinner', *New York Times*, 16 October 1872, p. 4). For Tyndall's farewell banquet, see Introduction, n. 23.
20. *great newspaper man*: probably referring to John Blair Scribner (see n. 18).
21. *Cooper Institute*: a private college in New York established in 1859, officially called the Cooper Union for the Advancement of Science and Art. The lectures were held in the Great Hall in its Foundation Building, at Third Avenue and Astor Place.
22. *Academy of Music*: see letter 3815, n. 9.
23. *West Point*: the United States Military Academy in West Point, New York, founded in 1802, educates cadets for commissioning into the United States Army.
24. *Hudson*: the Hudson River, flowing north to south through New York state and through New York City, emptying into the New York Bay.
25. *Rhinelike*: a reference to the Rhine River in Germany.
26. *Kenrick*: Henry Lane Kendrick (1811–91), at first a professor of chemistry, mineralogy, and geology at the United States Military Academy at West Point, relieved of duty in 1847 to participate in the war with Mexico followed by service in the western frontier. He returned to West Point in 1857 and remained a professor there until 1880 (*Col. Henry L. Kendrick, U.S.A.* (New York: E. P. Dutton & Company, 1892).
27. *the monkeys from which I came*: possibly a nod to Darwin's *The Descent of Man*, just published by John Murray in 1871.
28. *M^r^. Peabody*: possibly Francis H. Peabody or Oliver W. Peabody, both of the Boston firm Kidder, Peabody & Co., founded in 1865.

29. *his wife & children*: not identified.
30. *his principal assistant*: William Rhees.
31. *D^r. Beard*: not 'Beard', but 'Baird', for Spencer Fullerton Baird (1823–87), an American naturalist, museum curator, and Assistant Secretary of the Smithsonian Institution from 1850–78 (*ODNB*).
32. *Bunker Hill*: the battle of Bunker Hill on 17 June 1775, a British victory in the American Revolutionary War (1775–83).
33. *George the third*: George III (George William Frederick, 1738–1820), King of Great Britain and Ireland from 1760–1801, then of the United Kingdom of Great Britain and Ireland until 1820 (*ODNB*).
34. *D^r. Cotting*: Benjamin E. Cotting (1812–97), an American physician who was curator of the Lowell Institute from 1843–97. He was also president of the Massachusetts Medical Society (1874–6) and a fellow of the American Academy of Arts and Sciences ('Benjamin E. Cotting '34', *The Harvard Crimson*, 24 May 1897, p. 3).
35. *Lowell Institute*: see letter 3814, n. 4.
36. *delivered three lectures*: the first three of his six lectures in Boston, on 15, 17, and 19 October.
37. *cemetery of Mount Auburn*: Mount Auburn Cemetery, dedicated in 1831 and now a National Historic Landmark, is west of Boston on the line between Cambridge and Watertown.
38. *when the time came*: to deliver the first of his six lectures in Boston, on 15 October.
39. *his daughter*: Ellen Tucker Emerson.
40. *Concord*: a town in Massachusetts to the northwest of Boston, known as a significant site in the American Revolutionary War (1775–83), where "the shot heard round the world" on 19 April 1775 began the Battle of Lexington and Concord.
41. *his ruined house which is now being restored*: Coolidge Castle, the Emerson family home in Concord, Massachusetts, from 1835–82. The home is now a historic site operated by the Ralph Waldo Emerson Memorial Association.
42. *old Manse*: The Old Manse is a historic manse (or clergy house) in Concord, Massachusetts, built in 1770 for Emerson's grandfather William Emerson (1743–76). Ralph Waldo Emerson lived there in the mid-1830s, where he wrote the first draft of his famous essay 'Nature'.
43. *first firing between the farmers & the troops*: see n. 46.
44. *wife*: Lidian Emerson (née Jackson, 1802–92), married in 1835 (*ANB*).
45. *daughters*: Ellen Tucker Emerson and Edith Forbes (née Emerson, 1841–1929) (*ANB*).
46. *Concord Monument*: an obelisk commemorating the site where the American Revolutionary War began (see n. 46), dedicated in 1837 with a poem by Emerson titled 'Concord Hymn', from which the often repeated words 'the shot heard round the world' come from.
47. *Lexington*: a town to the northwest of Boston, between Boston and Concord, known as a significant site in the American Revolutionary War (1775–83), where on 19 April 1775 began the Battle of Lexington and Concord.
48. *club*: probably the Saturday Club, established in 1855, of which Holmes, Emerson, Longfellow, and Agassiz were members.

49. *his travels*: Louis Agassiz and his wife Elizabeth Cabot Cary Agassiz (1822–1907) were members of a scientific expedition aboard the USCSS *Hassler*, which from December 1871 to August 1872 dredged the seas around South America for marine specimens, as well as inland excursions to observe Andean glaciers (see C. Irmscher, *Louis Agassiz: Creator of American Science* (New York: Houghton Mifflin Harcourt, 2013), pp. 317–36). Many of the 7,000 specimens collected are in the collections of the Museum of Comparative Zoology at Harvard University and the Smithsonian Institution.
50. *Spencerian*: a follower of Herbert Spencer (*OED*).
51. *Fiske*: John Fiske (1842–1901), an American historian and popularizer of evolutionary science, and editor of *Edward Livingston Youmans: Interpreter of Science for the People* (New York: D. Appleton and Company, 1894) (*ANB*).
52. *I lectured last night*: the third of his six lectures in Boston, on 19 October.
53. *"Forms of Water"*: see letter 3785, n. 21.
54. *prayer meetings are held with a view to my salvation*: see letter 3871, n. 12.
55. *M^r^. Winthrop*: Robert Winthrop.
56. Tyndall copied letter 3872 here.
57. *all my brothers of the X*: for the X Club, see letter 3757, n. 2.
58. *Lady Mary*: Mary Egerton.
59. *Sir John Tyndall*: see letter 3872, n. 5.
60. *Anne Egerton*: see letter 3872, n. 5.

To Joseph Peter Lesley 22 October 1872 3874

Royal Institution of Great Britain[1] | 22nd. Oct. 1872

My dear Prof. Lesley

Many thanks to you for your note.[2] There is nothing in the least objectionable in the use you have made of my letter.[3] I may not quite adhere to the programme, but it is substantially correct.

The people are very kind to me here. They come to the lectures[4] in great numbers, and instead of a falling off a continually greater number is excluded every morning. Still I by no means do what I should like to do. In the first place, & particularly in the early lectures, I did not know the level of my audience, and besides that little mishaps arise through the strangeness of the situation. However I cannot complain. I am told the audience is more patient, more willing to sit still and give me their attention for an hour and a half than is usually the case.

In fact at the beginning the curator[5] of the Lowell Institute[6] warned me not to exceed the hour, but his fears are now dissipated.

I am thinking of rushing with Hector[7] to see Niagara[8] as soon as the lectures are over.

With best regards to M^rs^. Lesley.[9] | Believe me | Yours ever faithfully | John Tyndall

Mss.B.L56. American Philosophical Society

---

1. *Royal Institution of Great Britain*: while Tyndall wrote this letter on RI letterhead, he was in Boston at this time.
2. *your note*: letter missing.
3. *my letter*: letter 3846, in which Tyndall described the contents of each of the six lectures on light that he planned to deliver. Presumably Lesley had printed a programme for the Philadelphia lectures using Tyndall's own descriptions of each lecture.
4. *lectures*: Tyndall delivered six lectures in Boston from 15–25 October 1872 at the Lowell Institute.
5. *curator*: Benjamin E. Cotting (see letter 3873, n. 34).
6. *Lowell Institute*: see letter 3814, n. 4.
7. *Hector*: Hector Tyndale.
8. *Niagara*: see letter 3738, n. 3.
9. *M^rs^. Lesley*: Susan Inches Lesley (née Lyman, 1823–1904).

To Henry Bence Jones 23 October 1872 3875

Boston, 23rd. Oct. 1872.

My dear Friend

A squalling child and the rush of water in surrounding baths have urged me out of bed earlier than usual this morning and I take the top of the morning to write to you. This day I deliver the fifth lecture of my course.[1] The people are eager to the last degree, the stress for tickets is great; the room holds only 1200, and it is every night crammed. I am pressed on all sides to lecture in one of the great halls where the audience might amount to 3 or 4000. I keep myself almost wholly aloof from society—dine every day at the hotel, drink enormously dear Le Rose,[2] and am now after many fruitless efforts beginning to have a reasonable amount of sleep. I was very broken during the first lectures, still the people crowded to hear them, and the papers spoke kindly of them. Froude has been terribly feted in New York,[3] but I thank the Gods that I am out of the turmoil. They say my turn is to come[4] but I shall be a little hardened by the time it comes.

Lots of applications have poured in upon me regarding lectures elsewhere. It is stated that in two years I could make £20,000 by lecturing in America. Only think of a poor devil's self denial to resist that temptation. Agassiz has just returned.[5] He, Emerson, Longfellow and Wendell Holmes were all here within the same hour on Saturday.[6] They are friendly to the last degree. Had I been set down in Boston without knowing myself to be in America I should have thought it a thoroughly English town. The houses are English, the streets are English, and the tide of life rolling through the streets a strong English tide. The women here are very fine, and not of that fair and fading quality which I had thought characteristic of American ladies. Dr Ellis has sent me his life of Rumford[7] through Mr Winthrop:[8] and Mr Winthrop has sent me the life of his celebrated ancestor[9] who was Governor of Massachusetts when the revolution broke out. Oddly enough this governor was married to Margaret Tyndale,[10] and Margaret Tyndale's great grandson[11] was the most celebrated professor of natural science of his time in America! He was a fellow of our Royal Society, and of much consideration in England.

I spent one charming day with Emerson in his nest at Concord:[12] he is now off to Europe.[13]

And now I must start out and put my thoughts in order. I ran over the experiments last night, but I must regulate the thoughts which bind them during my ramble. I have thus prepared each lecture in the open air.

I find the harder I make the lectures the better they are liked. On Monday I gave them that terribly stiff one about Newton's rings and interference. They were delighted: I thought they would be terrified.

Take this scrap with my best regards.

Ever Your Friend | John Tyndall

RI MS JT/1/TYP/3/811
LT Typescript Only

---

1. *fifth lecture of my course*: of his six lectures delivered in Boston from 15–25 October 1872 at the Lowell Institute, the fifth lecture was on 23 October.
2. *La Rose*: Tyndall's favorite wine, which as he wrote to Thomas Hirst, was expensive (see letter 3886, n. 13).
3. *terribly feted in New York*: see letter 3873, n. 19.
4. *my turn is to come*: for the banquet held in Tyndall's honor before he departed the United States, see Introduction, n. 23.
5. *just returned*: see letter 3873, n. 49.
6. *all here within the same hour on Saturday*: in letter 3873, Tyndall wrote to Thomas Hirst of this gathering: 'In fact a perfect galaxy of genius was squeezed into the two midday hours yesterday'.
7. *life of Rumford*: see letter 3872, n. 11.
8. *Mr. Winthrop*: Robert Winthrop.
9. *life of his celebrated ancestor*: for the book about John Winthrop, see letter 3872, n. 2.
10. *Margaret Tyndale*: Margaret Tyndal (or Tyndale/Tyndall) Winthrop.

11. *great grandson*: John Winthrop (1714–79).
12. *one charming day with Emerson in his nest at Concord*: in letter 3873, Tyndall described his time with Emerson to Thomas Hirst.
13. *off to Europe*: in October 1872, Emerson traveled to Europe with his daughter Ellen Tucker Emerson and returned to the United States in May 1873.

From Joseph Henry[1] 23 October 1872 3876

[SMITHSONIAN INSTITUTION |
WASHINGTON,] Oct 23rd [187]2

My Dear Professor

Mr. Rhees has informed me of the great success of your first lecture[2] and of the warm reception you met with in Boston. I am not surprised at this since no one is more favourably known in this reading country than yourself. Indeed we are constantly receiving letters of inquiry as to the possibility of inducing you to visit almost all parts of the United States.

Had your object been money you might have realized a large sum; and in this case we would have given you in charge to a Lecture Bureau: But as we understood your object was not money we made an arrangement for your first appearance at the Lowell Institute, the most respectable lecturing establishment in this country. The remuneration however which you will receive, since the lectures are entirely free to the public, is not as great as that which you might have obtained had the plan been adopted which is proposed for New-York and Philadelphia. In making the arrangement with the Lowell Institute I was governed by what I considered your feelings, and I trust it will prove satisfactory. Although you will not be able to realize as much of what is considered among us as the great essential of life, I am sure you will do much good by your advocacy of the importance of abstract science in its relation to the advance of civilization, and the importance of making provision for original research.[3] The truth is that the teachers of science in this country have so much to do and are so poorly paid that they have neither the time nor the means for investigation. So low has been the estimate of the value of pure science and so great the demand for what is called practical skill that few among us have devoted themselves to the advancement of science.

It is only after twenty five years of constant struggle and entire devotion to the Smithsonian Institution that I am now beginning to bring the government, and the public generally into a proper understanding of the terms of the bequest: namely the "increase and the diffusion of knowledge among men" and preventing the income from being expended on local objects having no bearing on the advancement of science.

The distinction between the three important relations of scientific knowledge have not neither in this country as in England been properly reorganized: namely (1) The discovery of new scientific principles—(2[d]) The teaching of these principles and (3[d]) The applications of these to useful purposes in the arts of life. The 2nd of these relations, have been provided for in our colleges and universities; and the 3rd in our patent laws; but the first have been almost entirely neglected.

There are however many wealthy men in this country who may be induced to found establishments for original investigation and in advocating the importance of pure science you have a theme worthy of all your eloquence; the advocacy of which may tend to the realization of the conception of Bacon,[4] of a Solomon's House as put forth in his New Atlantis.[5]

I hope to meet you either in Phil[adelphia]: or in Baltimore[6] and to have you as my guest while you are in Washington:[7] I found Sir Frederic Arrow[8] a man of clear head and warm heart and shall ever retain for him a warm regard.

Truly your friend &——| Joseph Henry

Professor John Tyndall

RI MS JT/1/H/77

---

1. This letter has previously been published as document 198 in *Papers of Joseph Henry*, pp. 425–6.
2. *first lecture*: the first of six lectures Tyndall delivered in Boston from 15–25 October 1872 at the Lowell Institute.
3. *making provision for original research*: see Introduction, n. 15.
4. *Bacon*: Francis Bacon (1561–1626), an English natural philosopher who espoused that knowledge should be gained through empirical, inductive means. Often regarded as the father of the scientific method (*ODNB*).
5. *Solomon's House as put forth in his New Atlantis*: in his posthumously published novel *New Atlantis* (1627), Bacon envisioned 'Salomon's House'—or 'Solomon's House'—as an ideal institution of scientific research.
6. *Phil[adelphia]: or in Baltimore*: the second and third cities in Tyndall's American lecture tour (see Introduction, n. 8).
7. *in Washington*: Washington, DC, was the fourth city in Tyndall's American lecture tour (see Introduction, n. 8).
8. *Sir Frederic Arrow*: in letters 3837 and 3866, Tyndall informed Henry that he would likely meet Frederick Arrow who was also on a trip to the United States.

To Edward L. Youmans 24 October 1872 3877

BOSTON, *October 24, 1872.*

MY DEAR YOUMANS: The hall of the Lowell[1] still continues crowded, but I shoot above their heads sometimes.

In fact, this is my difficulty. I do not know the scientific level of my audience.

Still the people are most kind and attentive, and the newspapers, I believe, are very civil.

I go to Niagara next week,[2] so that the fortnight will be one of relaxation in part and in part of preparation.

I have pledged myself to lecture in New Haven in January.[3] They would be sorely disappointed if I did not do this.... I remain here till Wednesday, when I propose starting for Niagara.

E. L. Youmans, 'Tyndall's American Visit', *Popular Science Monthly*, 44 (February 1894), pp. 502–14, on pp. 509–10

1. *Lowell*: the Lowell Institute in Boston (see letter 3814, n. 4), where Tyndall delivered six lectures from 15–25 October 1872.
2. *I go to Niagara next week*: for Tyndall's visit to Niagara Falls, see letter 3738, n. 3.
3. *lecture in New Haven in January*: New Haven, CT, was the last city during his American lecture tour (see Introduction, n. 8), where he delivered two lectures for Yale College on 22 and 23 January 1873.

To Josiah Whitney[1] [25 October 1872][2] 3878

Friday

My dear Sir

I shall certainly do myself the pleasure of paying you a visit before I leave Boston & I hope soon to be able to arrange the time.

faithfully yours | John Tyndall

William Dwight Whitney Family Papers (MS 555). Manuscripts and Archives, Yale University Library

1. *Josiah Whitney*: Josiah Dwight Whitney (1819–96), an American geologist and professor of geology at Harvard University who was director of the California Geological Survey (*ANB*).

2. *[25 October 1872]*: the date is given since Tyndall left Boston for Niagara Falls on 31 October 1872 (a Thursday), making the Friday before 25 October, the date of his sixth and final lecture in Boston.

To Joseph Henry 26 October 1872 3879

26th Oct. 1872.

My dear Prof. Henry.

I finished my course in Boston[1] last night to the apparent satisfaction of my hearers. Nothing could have been kinder or more sympathetic than the behavior of the audience throughout, and last night especially many notable persons were present.

I was glad to receive your note[2] previously to the lecture, as it caused me to add emphasis to what I had to say regarding the claims of the original investigation. What I said was exceedingly well received. In neither America nor England have we any thing to expect from the governments, and indeed in their present state of ignorance their interference would do more harm than good. The only course open to us is so to[3] enlighten the public mind that those in a position to do so may be induced to help the work of original enquiry.

This morning the Cincinnati letter[4] reached me. I yesterday finally decided on the programme of New York and Brooklyn and the lectures in these places carry me on to the 20th of January.[5] I shall then be turning my thoughts homeward, so that for the present I fear the West will be beyond my range. I will however leave the question open.

With regard to other matters I am perfectly satisfied with every arrangement you have made. The kindly reception of the Boston people & the hope that my coming amongst them may do some good in promoting friendly feelings between England & America, are far more gratifying & important to me than the honorariums for the lectures. From some correspondence that has come to my knowledge I infer that Dr. Morison[6] of Baltimore[7] is much exercised regarding money matters—This I regret. I suppose it would not now be possible to omit Baltimore from your programme?

As Yale is a centre of education, and as the people there would be sorely disappointed if I quitted the country without seeing them I have agreed to lecture there in January.[8] I understood both Dr. Baird[9] & Mr Rhees to say that the stop is advisable.

It will give me great delight to meet you in Philadelphia[10]—I falter about inflicting myself on any of my friends. I am sometimes weary and stupid, and under these circumstances prefer in England the savage freedom of a hotel. But we will talk about the matter when we meet.

Meanwhile best regards & best thanks | always yours | John Tyndall

John Tyndall to Joseph Henry, 26 October 1872, mssRH 3970, box 43, William Jones Rhees papers, The Huntington Library, San Marino, California

---

1. *my course in Boston*: Tyndall delivered six lectures in Boston from 15–25 October at the Lowell Institute.
2. *your note*: letter 3875.
3. *so to*: it appears that Tyndall meant to write 'to so'.
4. *Cincinnati letter*: Tyndall had been invited to visit various inland cities in the United States.
5. *lectures in these places carry me on to the 20th of January*: Tyndall delivered six lectures in New York from 17–31 December at the Cooper Institute, and six lectures in Brooklyn from 4–18 January 1873 at the Academy of Music.
6. *Dr. Morison*: Nathaniel Holmes Morison (1815–90), First Provost of the Peabody Institute (1867–90) (*Memorial of Nathaniel Holmes Morison* (Baltimore: Privately Printed, 1892)).
7. *Baltimore*: Tyndall delivered three lectures in Baltimore from 26–9 November at the Peabody Institute.
8. *lecture there in January*: Tyndall delivered two lectures in New Haven, CT on 22 and 23 January at Yale College.
9. *Dr. Baird*: Spencer Fullerton Baird (see letter 3873, n. 31).
10. *Philadelphia*: Tyndall delivered six lectures in Philadelphia from 11–22 November at the Horticultural Hall.

From Henry Bence Jones [26–9][1] October 1872 3880

84. Brook St[2] | Oct 26 1872

Dear friend

Altho'. I have no news to send you yet I shall begin a letter to you to shew you that I am alive & to tell you how much I want you to send me a newspaper when Cotterell[3] has time to do so. I have got one posted the day after you arrived. You Froude & Miss Emily Faithful[4] if you were as I suppose all in the Russia[5] must have made an odd alliance. I hear Pepper[6] also is ghosting in America.[7] By this time your success must be determined & as the Presidential contest[8] has gone down to nothing you will have no interference from political attraction & the minds of most worth will be all free to attend to you. Sir H Holland came back all safely two or three days ago he was detained in Madeira[9] by the breaking down of some steamer. He immediately after saying how are you went off upon an invitation to Ceylon[10] & Brazil to the Governor[11] & Emperor[12] of course, & perhaps before you get back he will be off. He was or is 84 yesterday or today!!! my. my.

I have been all wrong & am so still. I have done a little in practice & down I go. Whither my days will put me up again or not seems doubtful

I have energy enough to lose my temper with Mr Dempsey[13] Harwood & Berkley & I ought to meet them this afternoon at the Institution[14] but the day is vile. Berkley leaves all to Harwood; Harwood all to Dempsey; & Dempsey leaves all alone!!! There is an awful mistake in the original plans as to lighting the lower laboratory. The stair case half swallows up the yard & so cuts off much reflected light from the theatre walls & until the state of the lower laboratory is quite clear your electric light house is postponed. Partly also it was postponed because there seemed no chance of getting the work in hand finished before the lectures begin[15] & the office, urinals & heating of the theatre must be finished if possible by the end of the year

Hitherto there has been little or no inconvenience to the members. Very few persons seem about the house & all I have spoken to have made no complaints

I went to the last meeting of visitors & found them recommending the managers to increase the salaries of the professors!! I told them you had not gone to America to make money quite the reverse.[16] The rumor is strong here that you are to have £300 each lecture. Lee[17] wanted a committee into the expenditures. He did not get it

Tuesday

So far I had got some days since, but I have caught a bad cold in the country doctoring & have been too unwell to do anything except to look at the fire. I have had a better night last night so I hope to finish my paper.[18] I have seen no one except Strzelecki[19] who comes to talk with me. He is going to fly off to Cannes.[20] I hope to be well enough to get to the managers meeting next Monday. I have got all the lectures in order[21] & I think the list of courses[22] as good as we ever had. Freeman[23] Morley,[24] Max Muller[25] & Clifford[26] for talkers ought to satisfy our audiences. The Friday evenings are not so forward. Saturday I had a meeting with Berkley Harwood *[Shustle]* & Dempsey & urged to the uttermost the finishing of the works. I had to be careful not to fall out with Mr Berkley who is ill and hardly looks after the place & as Harwood is scarcely ever there Mr Dempsey is really every thing & though he may be a good draftsmen his stuttering makes him appear at least slow & inefficient. However I hope to get them all out of the house in December & to have no trouble with them during the lecture time.

I have settled for one of the electric dials to be in your passage facing the green baize door. One in each laboratory the one in the theatre cannot be made to strike but an electric bell will be touched by a porter in the anteroom at the hour & he may do so twice at the quarter hour & then at the half hour after & so on.

I want much to hear how your first course of lectures[27] has done. We had the leading article in the Times on Froude's after dinner speech[28] certainly it did not read like an effective speech. The Yankees must see thro' flattery pretty sharply, & even if they could they w[oul][d] not put out a finger to cure Ireland or to stop the Irish from grumbling. But you will have too much to do to read all my scrawl.

I have no news to send you: Huxley I hear is looking ill Mrs Frankland[29] has to go to Cannes. Their eldest boy[30] has broken down also & goes with her. Oh health health what a blessing to have it. This is my abiding thought. Certainly no new or happy thought. Goodbye dear friend. How I hope this tour may not hurt your health & that you may come back safe & well is the hope of yours ever

HBJ

RI MS JT/1/J/70

---

1. *[26–9]*: the date range is given as Tyndall started this letter on 26 October, a Saturday, and finished it on a Tuesday, which was 29 October.
2. While Bence Jones wrote '84 Brook St' (his residence in London), this letter was written on RI letterhead.
3. *Cotterell*: John Cottrell.
4. *Miss Emily Faithful*: Emily Faithfull (see letter 3801, n. 21).
5. *Russia*: Tyndall traveled from Liverpool to New York on the SS *Russia*, departing on 28 September 1872 (see Introduction, n. 8).
6. *Pepper*: John Henry Pepper (1821–1900), a British popularizer of science, illusionist, and manager of the Royal Polytechnic Institution (*ODNB*).
7. *ghosting in America*: Pepper was most well-known for his on-stage ghost illusion which used a large sheet of glass to project onto a stage the 'ghostly' images of costumed actors. From 1872–8, Pepper traveled the United States and Canada on a lecture tour showing off the illusion. See H. Groth, 'Reading Victorian Illusions: Dickens's "Haunted Man" and Dr. Pepper's Ghost"', *Victorian Studies*, 50 (Autumn 2007), pp. 43–65; J. H. Pepper, *The True History of the Ghost: and All About Metempsychosis* (London: Cassell & Company, 1890); and B. Weeden, 'Pepper's ghost', in *The Education of the Eye: History of the Royal Polytechnic Institution 1838–1881* (Cambridge: Granta Editions, 2008), pp. 71–86.
8. *Presidential contest*: see letter 3794, n. 3.
9. *Madeira*: a British-controlled archipelago in the North Atlantic Ocean, to the southwest of Portugal and northwest of northern Africa. It is now an autonomous region of Portugal. Holland wrote to Charles Darwin in November 1872: 'I have myself had a somewhat busy autumn of travel; going round the North Cape, under the light of the midnight sun, & 200 miles beyond, to the borders of Russian Lapland—reaching 71"30 of Latitude. Since that time I have been to Lord Rosse's great Reflectors in Ireland, and still-more lately, have passed three weeks in the delicious Island of Madeira, warming myself there after the Arctic Seas: This is tolerable work for a gentleman in his 85th year' (*Darwin Correspondence*, vol. 20, pp. 484–5; this is letter 8604 in the online edition of the Darwin correspondence).

10. *Ceylon*: now Sri Lanka. Holland was very well-traveled, and published numerous accounts of his travels. He mentioned many of his travel destinations in *Recollections of Past Life* (London: Longman's Green, and Co., 1872), published the year before he died.
11. *Governor*: the Governor of British Ceylon at this time was William Henry Gregory (1816–92), an Anglo-Irish writer and politician, from 1872–7.
12. *Emperor*: Pedro d'Alcantara, Emperor Dom Pedro II (1825–91), also known as the Emperor of Brazil or Dom Pedro II 'the Magnanimous', was the last monarch of Brazil. He began his reign under regents at the age of five, and in 1840 at the age of fourteen assumed the throne. He was very popular in Brazil for most of his forty-nine year reign, and ruled until he was ousted by a military coup in 1889. He and his family were exiled to Paris two days after the coup, where he died two years later.
13. *Mr Dempsey*: not identified.
14. *meet them this afternoon at the Institution*: to discuss the current renovation of the RI's laboratory (see letter 3776, n. 14).
15. *lectures begin*: the RI's Friday Evening Discourses for the season, which began on 17 January 1873 with William Spottiswoode discussing the RI's laboratories (see letter 3776, n. 14). For the list of other Friday Evening Discourse lecturers that season, see *Roy. Inst. Proc.*, 7 (1873–75), pp. iii–iv.
16. *quite the reverse*: Tyndall donated the net profits of his lecture tour to a fund for pure science research in the United States (see Introduction, n. 15).
17. *Lee*: probably Thomas Lee.
18. *my paper*: not identified.
19. *Strzelecki*: Paul Edmund de Strzelecki (1797–1873), a Polish explorer, geologist, and philanthropist who became a British subject in 1854 (*ODNB*).
20. *Cannes*: a resort town on the French Riviera, on the Mediterranean Sea.
21. *lectures in order*: for the upcoming season's Friday Evening Discourses at the RI, from January to June 1873, see *Roy. Inst. Proc.*, 7 (1873–75), pp. iii–iv.
22. *list of courses*: for the lecture courses given at the RI before and after Easter 1873, see *Roy. Inst. Proc.*, 6 (1870–72), pp. 559–60.
23. *Freeman*: Edward Augustus Freeman (for his lecture course, see letter 3793, n. 9).
24. *Morley*: John Morley (1838–1923), a British politician and writer (*ODNB*), delivered a course of 'Three Lectures On the Limits of the Historic Method' at the RI on Tuesdays from 22 April to 6 May 1873 (*Roy. Inst. Proc.*, 6 (1870–72), p. 560).
25. *Max Muller*: Friedrich Max Müller (see letter 3801, n. 7; for his lecture course, see letter 3801, n. 8).
26. *Clifford*: William Kingdon Clifford's lecture course at the RI, 'Three Lectures on the Philosophy of the Pure Sciences', was delivered on Saturdays, 1, 8, and 15 March 1873 (*Roy. Inst. Proc.*, 6 (1870–72), p. 560).
27. *your first course of lectures*: Tyndall delivered six lectures in Boston from 15–25 October 1872 at the Lowell Institute.
28. *leading article in the Times on Froude's after dinner speech*: see letter 3873, n. 19.
29. *Mrs Frankland*: Sophie Frankland (née Fick, 1821–74), of Cassel, near Marburg in the state

of Hesse-Cassel. Edward Frankland met her in 1847, on his first visit to Marburg. They became engaged in 1849, and married in 1851. Two of her brothers were professors in the University of Marburg, and her father, Friedrich W. Fick (1783–1861), was chief engineer of the state (C. A. Russell, *Edward Frankland: Chemistry, Controversy and Conspiracy in Victorian England* (Cambridge: Cambridge University Press, 1996), pp. 83–91).

30. *Their eldest boy*: Edward and Sophie Frankland's children were: Margaret Nannie West (née Frankland, 1853–1941), Frederick William Frankland (1854–1916), Sophia Jeannette Frankland (1858–1946), and Percy Faraday Frankland (1858–1946). Another son, Lionel Edward, was born in 1863 but died the following year (Russell, *Edward Frankland: Chemistry, Controversy and Conspiracy in Victorian England*, pp. 86, 231, 344).

From Joseph Peter Lesley — 27 October 1872 — 3881

1008 Clinton Street | Phil[adelphi][a] |
Oct[obe][r] 27th 1872.

My dear Prof. Tyndall

If pleasure be heightened by surprise then I might be more pleased at your Boston successes,[1] for I certainly forknew them. Who can hope, however, that your potassium will burn as dazzlingly in our common air as in that jar of oxygen? What right have I to say this, who am a born Philadelphian, who knew it when it was only a great village, who have lived only of 53 years & yet long enough to see its size enlarged sixfold, and its quaker spirit vanish & be replaced by a generous public sentiment for science & fine art? Ah if you could remember, as I can, the lovely quaker village! If I live to be 80 I shall see Philadelphia as great & splendid a place as Paris was when I lived in it just, & only just, 30 years ago. Of that I am as sure as that I sit at this desk & write.

It is a sort of spiritual fortune we Americans have a chance to make, who, born in the beginnings of cities, fill long lives with memories all pertaining to growth on a large scale; the shooting out of streets in all directions like ice crystals; the erection of one public edifice after another undiminished by contrast with medieval magnificences; the introduction of new kinds of public feeling; the establishment of one manufacture more after another; the birth of the fine arts in succession, and the planting & watering of the sciences.

Philadelphia is now in the state of a healthy young fellow who has nearly got his growth and after a routine training at College is ready for his professional education in the higher courses. The close economics have given way to luxury and taste, and the enormous wealth accumulated here is ready to be poured out at the command of the first mastermind who, like Agassiz at Boston, will know how to conquer this society by his genial eloquence, unimpeachable knowledge, and attractive manners.

My impression is that no one can do this but—or so well as a stranger; and so far as I am concerned in the matter, I greatly desire to see the Chair of Physics,[2] just vacated by the death of Prof. John F Fraser,[3] filled from England rather than by a American. The Board of Trustees are resolved to obtain the very best incumbent the world can afford to their inquiries. It is the most important of all the Chairs in the two faculties of the University & every facility will be afforded to its occupant for making it both useful & splendid. We are just at the first spurt of the race; the competitors being the Inst[itute]. of Tech[nology]. at Boston, the Harvard, Yale, New York, Hoboken, Easton, Bethlehem & Troy Schools of Science and our University. We have the inside track, being in the second largest city, not commercial in the ocean shipping sense yet like New York; with more wealth at command in posse; at the Centre of American Manufactures; at the Centre of American Coal & Iron mining; at the old established Centre of American Literature, fine arts & especially science. Consequently, when our new organisation of the Department of Science opened this fall, instead of 35 or 40 students as we expected, we matriculated 104, a larger number than in any of the schools I have mentioned except perhaps Troy, which has been in existence 30 or 40 years.

We have 200 students in the building.

The Medical Department has about 450, and their special build[in]g will be finished in a year from now. A vast hospital will be attached to it & finished about the same time.

Certainly in a year or two wings or additions must be built for our Department of Sciences.

Fraser's classes are kept going by a temporary supply. But his successor must be chosen at or before New Year's day.

Now I know of but one man who has proved publicly his ability to do all that this Chair demands & and he is Pickering[4] of the Inst[itute]. of Tech[nology]. in Boston.

Next to him comes Meyer[5] of Bethlehem, now at Hoboken in the Stevens Institute of Tech[nolog]y under the Young Morton,[6] who was Fraser's assistant in our University. Meyer is an investigator & probably a good experimenter & lecturer.

If Prof Henry were 20 years younger I think his ties to Phil[adelphi]a would induce him to honor the University by accepting this Chair. From him we could have expected the most brilliant success.

This dearth of thoroughbred men of physical science proper in America makes the task of filling Fraser's place a most difficult one & the danger of its being allowed to lapse by default into improper hands imminent. We are all extremely anxious about the result.

In fact, the greatest of dangers already threatens us. Prof. Robert E Rogers[7] of the Medical Department is engineering a storming party and has good hopes of securing the place for himself. Once in it he retains it for life, of course.

You will see a good deal of this gentleman during your visit to Phil[adelphi][a] as he has attached himself to M[r]. Pugh and will be at your right hand everywhere & all the time.

My relations with M[r] Rogers are very friendly, but in common with every member (I believe) of both faculties (non medical) in the University I dread nothing so much as his installation as Prof. of Physics. It will be a stunning blow to the progress of the best sort of sentiment for real science in this community & cut down our hopes of seeing the organisation of a noble physical laboratory—the main standby of our Scientific Department.

If you desire my reasons for this opinion I can best state them viva voce,[8] when I have the pleasure of giving you a welcome here.

One of the Executive Committee of the Board of Trustees asked me earnestly if it were possible for them to offer you sufficient inducements to accept this, the principal Chair in our University; Having heard that at one time both Prof Huxley & yourself proposed to move to Toronto.[9] All I could say was, God grant it,—times are changed,—perhaps—who knows? How gladly I would give up my chair to Huxley or John Phillips,[10] and become their assistant, or step aside to enjoy the sight of my old Alma Mater glorified.

Then again, M[r] Crotch[11] of Cambridge, Eng. has talked to me about one of the Cambridge Fellows whose fellowship will determine in 3 years because he will not take orders, and who is attached to the University College London I have missed his name—Cummings[12] is it? or something like that—But Crotch spoke of him with enthusiasm, as a man of the greatest learning, power, originality, tact, & already famous in English Physics. You must know who he means, & can no doubt refute or confirm his statement.

But I have already trans<u>gressed</u> all bounds of your patience. Grant me pardon on the ground of the paramount importance of the subject to us here, and permit me to add to yourself also as an Apostle of Science[13] & especially Physics, on a mission tour through Macedonia.[14]

Wishing you a Niagara[15] of pleasure I remain ever Yours faithfully | Lesley.

P.S. M[r] Pugh called on me the other day to show me your letter about the programme.[16] It made me half regret the use I had made of your Folkstone letter;[17] but I think (& thought then) that such a thing being necessary, it had better be done at once, so that time might elapse for those who read it to forget all but the sentiments it might excite. No one will keep that notice as a programme; & therefore you will be as free to order the lectures as if it had not appeared. I might have spared the article, for the fame of your Boston lectures has gone abroad & they are common talk.

M[r] Pugh's programme is the ten or twelve lines you gave him.

He sent me a long skreed of his own to proof read; but I made no corrections, because it was entirely all Pugh, & that I suppose suits the multitudes, if it does not suit us. And after all such things are of no consequence.

RI MS JT/1/L/17

---

1. *your Boston successes*: Tyndall delivered six lectures in Boston from 15–25 October at the Lowell Institute.
2. *Chair of Physics*: at the University of Pennsylvania.
3. *Prof. John F Fraser*: John Fries Frazer (1812–72), first assistant geologist to the Geological Survey of Pennsylvania and Professor of Natural Philosophy and Chemistry at the University of Pennsylvania from 1844–72 (J. L. Le Conte, 'Memoir of John Fries Frazer. 1812–1872', in *National Academy of Sciences: Biographical Memoirs*, Vol. 1 (Washington City, 1877), pp. 245–56).
4. *Pickering*: Edward Charles Pickering (1846–1919), an American astronomer and physicist who was Thayer Professor of Physics at the Massachusetts Institute of Technology from 1868–77 (*ANB*).
5. *Meyer*: Alfred Mayer.
6. *Young Morton*: Henry Morton (1836–1902), an American scientist who was the first president of the Stevens Institute of Technology in Hoboken, NJ, from 1870–1902 (*ANB*; C. Sellers and A. R. Leeds, *Biographical Notice of Pres't Henry Morton, Ph.D., of the Stevens Institute of Technology* (New York: Engineering Press, 1892)).
7. *Prof. Robert E Rogers*: Robert Empie Rogers (1813–84), an American chemist and from 1852 professor of chemistry at the University of Pennsylvania, and from 1856 dean of its medical faculty (*ANB*; W. S. W. Ruschenberger, 'A Sketch of the Life of Robert E. Rogers, M.D., LL.D., with Biographical Notices of His Father and Brothers', *Proceedings of the American Philosophical Society*, 23 (1886), pp. 104–46).
8. *viva voce*: conveyed or expressed in speech instead of writing; given or stated by word of mouth; spoken; oral (*OED*).
9. *proposed to move to Toronto*: in 1851, Tyndall and Huxley both applied for positions at the University of Toronto in Canada, but were rejected (see *Tyndall Correspondence*, vol. 3).
10. *John Phillips*: John Phillips (1800–74), a British geologist (*ODNB*; J. Morrell, *John Phillips and the Business of Victorian Science* (Aldershot: Ashgate, 2005).
11. *M[r] Crotch*: probably George Robert Crotch (1842–74), a British entomologist who following his education and some time working at the University of Cambridge traveled to the United States in 1872 to collect insects. After arriving in Philadelphia and visiting some insect collections there, he traveled to California and the Pacific Northwest before taking a position at the Museum of Comparative Zoology at Harvard (J. Smart and B. Wager, 'George Robert Crotch, 1842–1874: a bibliography with a biographical note', *Journal of the Society for the Bibliography of Natural History*, 8 (1977), pp. 244–8).

12. *Cummings*: as Tyndall indicated in his reply (letter 3883), Lesley was probably referring to William Kingdon Clifford.
13. *Apostle of Science*: a moniker that Tyndall gained as a result of his American lecture tour.
14. *Macedonia*: possibly a reference to Macedon, New York (named after the birthplace of Alexander the Great), which may have been on the way via railroad from Boston to Niagara Falls.
15. *Niagara*: referring to Tyndall's upcoming visit to Niagara Falls (see letter 3738, n. 3).
16. *your letter about the programme*: letter to Thomas Burnett Pugh missing.
17. *your Folkstone letter*: letter 3846, in which Tyndall described the contents of each of the six lectures on light that he planned to deliver.

From Frederick Pollock and Juliet Pollock 27 October 1872 3882

59 Montagu Square | W.[1] | Sunday Oct. 27 1872

My dear Tyndall

We have only just ascertained that you have some comparatively fixed address not far from the Hub of Creation[2]—otherwise I should have written sooner. It was a loss to me, but perhaps a good thing for you, that you had already gone to Switzerland at a time which, as I am about to tell you, was critical for me. Had you been at the R.I. then, you might have had to listen as best you might to my variations on the theme Ich bin verliebt.[3] The past three months have wrought a happy change in the vowel, and you are now to hear of verliebt[4] and verlobt[5] all in one. Your first question will naturally be—Who is she? Her name is Georgina Deffell[6]—she is the only living child of her mother,[7] who is the widow of an Indian merchant.[8] They live very near us, but we made their acquaintance through the Ritchies[9] only three years ago—and I first learnt to know them really well last summer, when we met at Bonn and Heidelberg.[10] Ever since we worshipped Beethoven[11] together I have been gradually falling in love with her. (The continuous expression is strictly accurate, as I am since her decision still more in love with her every day). And so it came to pass that in the latter days of July, being certain in my own mind and having weighed all the difficulties, I told her so. She took time to consider: I saw her once more before I left England and wrote to her constantly during our travels. During this time it was of course not right to say anything even to real friends, as the result was quite in suspense: but it cost me something to suppress it in my letter to you from Heidelberg.[12] Finally, we met in Paris at the beginning of this month, and she gave me her answer, which has made me so happy that it has taken me quite these three weeks to believe it. Do you know her at all, I wonder? I fear not well, at any rate, and I am not strong in descriptions of character. It is a noble and sincere soul, true and clear as the light; with all a woman's tenderness, and yet a strength and wisdom which I am only beginning to know; wholly loveable and harmonious in all her ways. Besides being musical in the lofty Greek sense of the word, which I wish we had kept, she is a musician and plays divinely. I am almost impatient for you to come back to England and see her.

The difficulty I had to consider beforehand is this—that we must wait a long time, that is till I have made enough of this world's goods by the exercise of my wits to enable us to marry with reasonable prudence. We have therefore not made any formal announcement, but taken occasion to tell our real friends gradually—However we have trust in the future, and she has a faith and hope that are quite splendid and would keep mine alive against much greater obstacles than any we are likely to meet with. Indeed her only fault is a disposition to believe in me too much and in herself not enough. She has that rare high-mindness which is wholly unconscious of being anything out of the common way.

—I gave you the news of the Alps from Heidelberg. Since that the Saturday has done me the honour to print some thoughts of mine on Mountaineering,[13] with which I shall be curious in good time to know if you agree. At present I am finishing an article on Spinoza[14] which Morley[15] had undertaken to consider for the Fortnightly.

—For the news of the world, Roundell Palmer[16] is my Lord Selborne and Lord High Chancellor of Great Britain, and we shall see how his new broom sweeps in the great lumber room where the laws of England lie littered about because nobody has time or red tape to spare for putting them in order. Meanwhile the Bar will scramble for the practice he has left, and I suppose some few small crumbs may in the general commotion fall down to us juniors.

—I hope your lectures[17] leave you some space of time to enjoy yourself and see the best Americans. I think the really good sort of them delightful, and I should envy you the society of Boston if I were not so happy in London.

Yours ever | F. Pollock | turn over

Hear the Mother Speak.—

Georgie Deftell is all that the boy says & we all love her. She is quite pretty enough for any woman to be, but perhaps grace & picturesqueness are more characteristic in her than actual beauty. Her gentleness & modesty are quite remarkable & she seems capable of taking in any idea. She is clever but still more sympathetic than clever. Her manner recalls the Ritchies; with whom she lived in early days almost as a sister. She is very fond of me & of Sir F.[18] & her mother is also much our friend. There is then no drawback but that of having to wait.

—No one can say how long it may be because ability alone will not assure success: there must be circumstance wh[ich]. we call chance besides. However on the whole we all feel sure that it is a happy thing. The only other news is that the Henry Taylors[19] have taken a house in town & that Ant[hony]. Trollope[20] hopes to knock at our door on the 7.th Dec[embe]r.

Yours ever. | Juliet Pollock | I am afraid that Boxall[21] is breaking up.

RI MS JT/1/P/136

---

1. *59 Montagu Square | W.*: the Pollocks' residence in Marylebone, London.
2. *Hub of Creation*: a nickname for Boston.
3. *Ich bin verliebt*: I'm in love (German).
4. *verliebt*: in love (German).
5. *verlobt*: engaged (German).
6. *Georgina Deffell*: Georgina Harriet Deffell (1847–1935), whom Frederick Pollock married on 13 August 1873 (M. D. Howe (ed), *Holmes-Pollock Letters: The Correspondence of Mr. Justice Holmes and Sir Frederick Pollock, 1874–1932*, 2nd edn, 2 vols (Cambridge, MA: Belknap Press of Harvard University Press, 1961), vol. 1, p. xv).
7. *her mother*: Letitia Deffell (née Hill, 1818–97).
8. *Indian merchant*: John Deffell (1808–61), a merchant in Calcutta.
9. *Ritchies*: probably the family of William Ritchie (1817–62), a British lawyer and Advocate-General of Bengal from 1855–62. His widow was Augusta Charlotte Ritchie (née Trimmer, 1817–88), and they had four sons and four daughters.
10. *Bonn and Heidelberg*: cities in Germany.
11. *Beethoven*: Ludwig van Beethoven (1770–1827), the German composer and pianist (*NDB*).
12. *my letter to you from Heidelberg*: letter 3828.
13. *the Saturday has done me the honour to print some thoughts of mine on Mountaineering*: [F. Pollock], 'Mountaineering', *The Saturday Review*, 34 (7 September 1872), pp. 305–6.
14. *article on Spinoza*: F. Pollock, 'Spinoza: The Scientific Character of His Philosophy', *Fortnightly Review*, 19 (1873), pp. 567–85.
15. *Morley*: John Morley (1838–1923), who succeeded George Henry Lewes (1817–78) as editor of the *Fortnightly Review* in January 1867 (until 1882) (*ODNB*).
16. *Roundell Palmer*: Roundell Palmer (1812–95), 1st Baron Selborne (and Earl in 1882), an English lawyer and politician. Palmer first served as Lord High Chancellor of Great Britain from 1872–4 (*ODNB*).
17. *your lectures*: Tyndall had just completed six lectures in Boston from 15–25 October 1872 at the Lowell Institute. His next stop was Philadelphia, the first lecture commencing on 11 November.
18. *Sir F.*: William Frederick Pollock.
19. *Henry Taylors*: Henry Taylor (1800–86), a British poet, playwright, and diplomat who spent much of his career working in the Colonial Office (*ODNB*); his wife Theodosia Alice Taylor (née Spring-Rice, 1818–91), daughter of Thomas Spring-Rice (1790–1866),

first Baron Monteagle of Brandon in Kerry (they married in 1839); and their five children: Aubrey Ashworth Taylor (1845–70), who went to Balliol College, Oxford and died young; Eleanor Ashworth Towle (née Taylor, 1847–1911), a novelist; Ida Ashworth Taylor (1850–1929), a novelist and biographer (*ODNB*); Harry Ashworth Taylor (1855–1907), a captain in the British Army; and Una Ashworth Taylor (1857–1922), a novelist and author of *Guests and Memories: Annals of a Seaside Villa*, her memoir about her family's summer home and the guests they entertained there (H. Taylor, *Autobiography of Henry Taylor, 1844 to 1875* (London: Harrison and Sons, 1877); U. Taylor, *Guests and Memories: Annals of a Seaside Villa*).

20. *Ant[hony]. Trollope*: Anthony Trollope (1815–82), a British novelist (*ODNB*).
21. *Boxall*: probably William Boxall (1800–79), a British portrait painter. Initially aspiring to be a history painter he eventually focused more on portraits as a more stable source of income. A well-known painter already by the 1830s, he was elected Associate of the Royal Academy in 1851 and Royal Academician in 1863. He served as director of the National Gallery between 1866 and 1874 (*ODNB*).

To Joseph Peter Lesley 30 October 1872 3883

30th Oct. 1872.

My dear Prof Lesley.

I have just received your letter[1] which has my entire sympathy. With regard to myself I do not think it would be possible to tempt me from my nest in London.[2] As you surmise times have very greatly changed since Huxley and myself were candidates for chairs in Toronto.[3]

Your friend Crotch[4] must mean Clifford who is certainly one of the most remarkable young fellows I have ever met. If he had trust in himself—the stamina of persistence—that deeper force which cannot be expanded in flashes which produce more temporary astonishment and delight—he will become something extraordinary. He as a mathematician promises to be one of the first rate—if he only works—He has also, for one so young, made himself acquainted to a wonderful extent with modern philosophy. He now holds the chair of Applied Mathematics in University College,[5] sympathetic with you. On the use of original enquiry the intellectual of America especially requires expansion. He has the means in abundance & if they be only wisely applied *[he]* may wipe out the reproach of De Toqueville that democracies fail to find and foster the highest genius.[6]

Ever yours faithfully | John Tyndall

Hector desires his kind regards.

Mss.B.L56. American Philosophical Society

1. *your letter*: letter 3881, offering Tyndall the Chair of Physics at the University of Pennsylvania.
2. *my nest in London*: the RI.
3. *candidates for chairs in Toronto*: see letter 3881, n. 9.
4. *Crotch*: George Robert Crotch (see letter 3881, n. 11).
5. *chair of Applied Mathematics in University College*: Clifford was appointed professor of mathematics and mechanics at University College, London, in 1871, a post he held until his death in 1879.
6. *the reproach of De Toqueville... the highest genius*: on Alexis de Tocqueville, see letter 3785, n. 18.

To Edward L. Youmans 4 November 1872 3884

INTERNATIONAL HOTEL, NIAGARA FALLS,
Monday morning, November 4, 1872.

MY DEAR YOUMANS: I came here[1] on Friday afternoon and have been active ever since. The first impression made upon me by the Falls was tame, because my point of view was not a good one; but they have grown in strength and majesty as I have seen more of them. I had a somewhat exciting day on Saturday, penetrating into unexpected regions under the Horseshoe Fall.[2] I had a fine, strong fellow with me as guide; he had been put upon his mettle, and he led me into extraordinary places—into places, indeed, where no prudent man ought to be found....

I remain here doing some work until Thursday, when I start for Philadelphia.[3] If I find from my assistants[4] that matters are all right in Philadelphia, I may be induced to stay till Friday. There is nothing, I suppose, to be arranged regarding New York? If there were, I could go that way and have a word with you.

I am stronger than when I came, and my work will gradually become easier to me—at least I hope so. I quitted Boston[5] on Thursday, not completing all I wished to do, nor seeing all I wished to see. Still, my sojourn there was a most pleasant one. The only drawback was that many people—thousands I was told—wished to hear the lectures who were unable to hear them.

With kind regards to Mrs. Youmans[6] and your sister,[7] also to Mr. Appleton,
Believe me ever faithfully yours, | JOHN TYNDALL.

E. A. Youmans, 'Tyndall's American Visit', *Popular Science Monthly*, 44 (February 1894), pp. 502–14, on p. 510

1. *here*: Niagara Falls (see letter 3738, n. 3).

2. *Horseshoe Fall*: the largest of the three waterfalls forming Niagara Falls, which Tyndall described in his Friday Evening Discourse at the RI (see letter 3738, n. 3).
3. *Philadelphia*: Tyndall delivered six lectures in Philadelphia from 11–22 November at the Horticultural Hall.
4. *my assistants*: John Cottrell and George Miller.
5. *Boston*: Tyndall delivered six lectures in Boston from 15–25 October 1872 at the Lowell Institute.
6. *Mrs. Youmans*: Catherine Newton Lee Youmans (*c*. 1819–94).
7. *your sister*: Eliza Ann Youmans (1826–1914), who published several books on botany.

To Henry Bence Jones [26 October][1]–5 November 1872 3885

Begun in Boston ended at Niagara Falls[2] | 5th. Nov. 1872.

My dear Friend

The first ordeal has been passed, and with fair success. On Friday I gave my last lecture in Boston,[3] and nothing could have been more cordial than the finishing up. I spoke my mind with regard to original investigation, setting every thing else—lecturing included—beneath it; and what I said was exceedingly well received. This I am assured is the most pressing necessity of the United States at the present time. Agassiz, Henry and others assure me that no greater service could be done to science than the enlightenment of the public mind on this point. They do not understand it and they need to be taught. I closed the last lecture an allusion to the relations of America and England which was well received both by the audience and the press. These, are as nearly as I can recollect them, were my words

"And now ladies and gentlemen my course among you is nearly run. I began these lectures with anxiety, and I end them with regret. I have sometimes found them harder than I expected, and were it not for the devotion of my assistants, I should not have been able to get on. I shall ever gratefully remember my reception by my first American audience. Should I be fortunate enough to meet a similar reception elsewhere I shall assuredly return to the old country brimful of content. And I hope you will excuse me when I say that I have heard the expression 'the old country' oft repeated since I came to Boston. Of no other country than England can you speak thus. You cannot abolish your antecedents, and out of the loins of England you have come. I see its ancestry stamped everywhere upon this community. I see it in your mansions, in your streets in the activity of your commercial life, in the vigour of your population. De Toqueville says 'I refuse to regard these peoples as two, the one is the vast outgrowth and development of the other',[4] and I also refuse to regard myself as a foreigner here. Let me express the trust that the augmented intercourse between the two peoples, and especially of the intellectual men of both, will dissipate more and more the ignorance which is at the bottom of all our troubles until in the end each nation shall be willing to recognise and respect the individuality of the other while maintaining its own."

What I said was terser than this, but this was the substance of it, and I was glad to find it received with ringing applause. You know I was always a man of peace, but it is a new exaltation of my vocation to assume the character of a peace maker between nations.

In half an hour from the present time I accompany Dr Ellis to visit the birthplace of Rumford![5]

Dr Ellis came and we went together by rail to Woburn Centre, a sweet village, bathed when we arrived in sunshine, and adorned with bright foliage. A walk of two miles brought us to North Woburn, and there sure enough was Rumford's birthplace. It was midday and the people were just sitting down to dinner. The flavour of fried onions filled the room in which the hero was born. The sketch in Dr Ellis's book[6] is very nearly correct. It stands in a pretty rural place. In fact what struck me most in these new England villages is their extreme stillness and quietude. Dr. Ellis is a man charged with all manner of information which he willingly dispenses in a continued flow of conversation. I was a willing listener, so that we agreed well together. On our return we called in at the house of a Mr Baldwin[7] the descendant of one of Rumford's friends: close at hand is a kind of villa which had been built for Rumford when it was thought that he would return to New England; it was occupied by his daughter whom every body appears to have called "the Countess."[8]

Finding Dr Ellis so well informed and so obliging I gladly agreed to put myself under his tutelage for a run through Boston on the following day. We visited the birthplace of Franklin,[9] the burial place of his parents;[10] the Athenaeum,[11] the State House,[12] the Institute of Technology,[13] and during our march my companion gave me all manner of historic information. Boston is a truly fine and beautiful city; and the parties I met at Mr Lowell's[14] could not be distinguished by their manners from a party of thoroughbred English gentlemen.

I saw Agassiz and many other professors repeatedly at the parties to which I have just referred. At Mr Winthrop's[15] luncheon on Monday. At dinner at Prof. Cooke's[16] on Tuesday. Agassiz previously took me through his museum,[17] and I was much pleased with the thoroughness of the system of instruction adopted in it. He is evidently exceedingly anxious that his position as regards Darwinism should not be misconstrued I fear the subject may make his later days unhappy.[18] We had a great deal of earnest conversation regarding the general question of evolution. It evidently disturbs him much.

On Thursday I packed up and started at 3 P.M. for Niagara Falls. At Springfield, three hours from Boston, we changed to a sleeping carriage. When the time for retiring came it was interesting to see how rapidly the carriage was converted into a dormitory. I slept some and rested much; so that I rose fresh next morning.

I have had good weather and a good time at this place. I have seen the Falls thoroughly—nobody ever saw them more thoroughly. In doing this some danger was incurred, and places were visited which no traveller had previously seen; but I could not bear to quit the place without learning all it could teach me. Its beauty and grandeur grew upon me.

This day (Tuesday, 5th) I was amid the blinding spray in a boat, endeavouring to get right into the seething water underneath the Horseshoe Fall.[19] But after two attempts one of which threatened the boat with serious injury we had to be content with less than I wished for.

I remain here till Thursday: probably till Friday and then proceed to Philadelphia.[20] Cottrell[21] and Miller[22] came here with me, but their tastes lying in the direction of variety rather than thoroughness they swept off to Chicago and will probably reach Philadelphia to night. My course there begins on Monday evening next. I am recasting my lectures so as to avoid treading wholly in the old groove.

The President's election[23] has thrown me back seriously. I do not finish at Brooklyn[24] till the 20th of January, and I have promised the students and Professors of Yale College[25] to lecture in Newhaven afterwards. Deputations have come to me from St Louis and other places, applications from Cincinnati, Cleveland, Columbus and many other places.[26] I am also offered a temptation to remain here![27] They would make what is described as "a splendid position" in one of the chief cities for me. But they are not likely to draw me from the old nest at home.[28] I may however as you prophesied stay longer than I intended. I would not stay an hour longer, were I not persuaded that my staying will be productive of international good. Give me your notions on this point.

Good bye | Yours affectionately | John Tyndall

I shall be at Philadelphia for the coming fortnight. Afterwards for a week at Baltimore.[29] I think it would be best to write to the Smithsonian Institute Washington.[30]

RI MS JT/1/TYP/3/812–4
LT Typescript Only

---

1. *[26 October]*: The typescript notes that Tyndall finished this letter on 5 November but had begun it in Boston. He would have worked on it since 26 October at the earliest, the day following his last lecture in Boston on 25 October.

2. *Niagara Falls*: see letter 3738, n. 3.
3. *last lecture in Boston*: Tyndall delivered six lectures in Boston from 15–25 October 1872 at the Lowell Institute.
4. *"I refuse to regard these peoples as two, the one is the vast outgrowth and development of the other"*: possibly apocryphal, and conveniently at odds with much of what de Tocqueville writes in *Democracy in America*, a great deal of which is dedicated to observing the differences and animosities between Americans and the English. On de Tocqueville, see letter 3785, n. 18.
5. *birthplace of Rumford*: Benjamin Thompson (later Count Rumford) was born in Woburn, Massachusetts, a town approximately ten miles to the northwest of Boston.
6. *sketch in Dr. Ellis's book*: G. E. Ellis, *Memoir of Benjamin Thompson, Count Rumford, with notices of his daughter* (Boston: American Academy of Arts and Sciences, 1871). Ellis's sketch of Thompson's birthplace appears on p. 5.
7. *Baldwin*: George Rumford Baldwin (1798–1888), an American civil engineer, was the son of Loammi Baldwin (1744–1807), also an engineer, a childhood friend of Benjamin Thompson who gave his youngest son "Rumford" as a middle name in his honor (*ANB*). Tyndall likely visited the Baldwin Mansion in North Woburn, built in 1661.
8. *his daughter*: Sarah Thompson, Countess Rumford (1774–1852) inherited her father's title upon his death (*ANB*; *ODNB*).
9. *birthplace of Franklin*: Benjamin Franklin was born in a house on Milk Street in Boston, opposite the Old South Meeting House, where the organizers of the Boston Tea Party met.
10. *burial place of his parents*: Franklin's parents, Josiah Franklin Sr. (1657–1745) and Abiah Folger Franklin (1667–1752), are buried in the Granary Burying Ground, adjacent to Boston Common, which is also the resting place of a number of prominent figures from Colonial and Revolutionary America, including Samuel Adams, John Hancock, and Paul Revere (*ANB*).
11. *The Athenæum*: the Boston Athenaeum was established in 1807, and by the time of Tyndall's visit had moved into a purpose-build site at 10½ Beacon Street, directly adjacent to the Granary Burying Ground.
12. *the State House*: completed in January 1798; the dome was covered in copper by Paul Revere's firm in 1802 and at the time of Tyndall's visit, it would have been painted yellow. It sits just across Beacon Street from the Boston Athenaeum.
13. *the Institute of Technology*: the Massachusetts Institute of Technology (MIT) was chartered in 1861 and opened at its original location in Back Bay, Boston, at Boylston and Berkeley Streets, in 1866. It moved to its present location in Cambridge, Massachusetts, in 1916. The tour Tyndall described would therefore have been a short walk of about 1.5 to 2 miles, depending on the route.
14. *Lowell's:* the Lowells were a prominent Boston family, tracing their roots in America to the 1630s. This Mr. Lowell is likely John Amory Lowell (1798–1881), who at the time was the trustee of the Lowell Institute, at which Tyndall was lecturing (*ANB*). The high manners Tyndall described were used to deride the snobbery of Boston society by the early

twentieth century, as in the common piece of doggerel: 'Here's to good, old Boston | The town of the bean and the cod | Where the Lowells speak only to Cabots | And the Cabots speak only to God'.

15. *Winthrop*: Robert Winthrop.
16. *Cooke's*: Josiah Parsons Cooke (1827–94), an American chemist at Harvard University with an interest in atomic weights and spectroscopy (*ANB*).
17. *his museum*: Louis Agassiz founded the Museum of Comparative Zoology at Harvard University in 1859 (see M. P. Winsor, *Reading the Shape of Nature: Comparative Zoology at the Agassiz Museum* (Chicago: University of Chicago Press, 1991)).
18. *his position as regards Darwinism*: Agassiz was a prominent opponent of Darwinism in America, holding instead that organisms' adaptations were evidence of divine creation. Tyndall spoke of Agassiz in his 'Belfast Address' at the BAAS meeting in August 1874: 'One of the many times I had the pleasure of meeting him in the United States was at Mr. Winthrop's beautiful residence at Brookline, near Boston. Rising from luncheon, we all halted as if by a common impulse in front of a window, and continued there a discussion which had been started at table. The maple was in its autumn glory; and the exquisite beauty of the scene outside seemed, in my case, to interpenetrate without disturbance the intellectual action. Earnestly, almost sadly, Agassiz turned, and said to the gentlemen standing round, "I confess that I was not prepared to see this theory received as it has been by the best intellects of our time. Its success is greater than I could have thought possible"' (J. Tyndall, *Address delivered before the British Association assembled at Belfast* (London: Longman, Green, and Co., 1874), pp. 44–5). At the time Tyndall remarked in his journal that Agassiz 'was as vigorous as ever and as eager for the fray apropos of Darwin', and he 'tackled me about the doctrine of Evolution' (RI MS JT/2/12/18). On Agassiz and Darwin, see C. Irmscher, *Louis Agassiz: Creator of American Science* (New York: Houghton Mifflin Harcourt, 2013), pp. 4–5, 23–4, 91, 126–8, 128–9, 147–52, 282, and 265–6.
19. *Horseshoe Fall*: the largest of the three waterfalls forming Niagara Falls.
20. *Philadelphia*: Tyndall delivered six lectures in Philadelphia from 11–22 November 1872 at the Horticultural Hall.
21. *Cottrell*: John Cottrell.
22. *Miller*: George Miller.
23. *The President's election*: see letter 3794, n. 3.
24. *Brooklyn*: Tyndall delivered six lectures in Brooklyn from 4–20 January 1873 at the Academy of Music.
25. *Yale College*: Tyndall delivered two lectures in New Haven, CT, on 22 and 23 January at Yale College.
26. *Deputations have come... many other places*: a reference to further invitations to lecture Tyndall received after arriving in the United States.
27. *offered a temptation to remain here!*: an invitation to apply for the Chair of Physics at the University of Pennsylvania (see letters 3881 and 3883).
28. *old nest at home*: the RI in London.

29. *Baltimore*: Tyndall delivered three lectures in Baltimore from 26–9 November 1872 at the Peabody Institute.
30. *Smithsonian Institute Washington*: Tyndall delivered six lectures in Washington, DC, from 3–12 December 1872 at Lincoln Hall.

To Thomas Archer Hirst — 5 November 1872 — 3886

Niagara Falls[1] 5th. November 1872.

My dear Tom.

The ordeal at Boston[2] was successfully passed. As usual I fell below my own ideal, but the reception of all that I had to say was enthusiastic in the extreme. I wound up with an allusion to international relations[3] which was also exceedingly well received by the audience and by the press.

I saw some nice people in Boston. A dinner party at Mr. Lowell's;[4] a luncheon at Mr. Winthrops,[5] a dinner & reception at Prof Cookes.[6] Sundry visits to Cambridge: sundry rambles round Boston,[7] including a visit to the birthplace of Rumfor[8] at Woburn, made me feel pretty well acquainted with the place. I liked place and people much.—In fact it is a truly beautiful city, and its surroundings are surprisingly lovely.

I packed up on Thursday last and came here with Hector.[9] We quitted Boston at 3 P.M. changed at Springfield, 3 hours from Boston, into a sleeping "car".[10] It was amusing to see how rapidly when the time of rest came the carriage was converted into a dormitory. I slept some and rested much, so that I rose perfectly refreshed in the morning.

The first aspect of the Falls was rather tame, for the point of view accidentally chosen was not a good one. But their beauty and their majesty have grown upon me. I have seen them thoroughly—nobody more thoroughly, and in doing so have incurred some danger, for I went into places never visited by a traveller before. I have viewed them from top & bottom—from the leaps over which the water arches, to the blinding spray-storm at their base. This day I made an attempt to get right under the Horseshoe Fall[11]—I had a strong boat and strong & skilful men; we were tossed & twirled and bumped seriously against the rocks, being finally obliged to return before I had accomplished all I wished.

I find the people very civil—they all seem to know me.

We have reached the Falls before the bright foliage has entirely disappeared. Goat Island[12] is a lovely spot, and now that my muscular labours are ended I intend to make it my study for a couple of days. I am recasting the lectures, so as to avoid treading in the precise groove followed at Boston.

Hector is very good—after a few stormy onsets permanent calm appears to have settled down upon us. He attends to all business matters which is a great comfort to me.

Living is very expensive here. Good la Rose 16 to 18 shillings a bottle![13] Board, private room not included 4½ dollars (18 shillings) a day. My expenses at Boston for 19 days amounted to nearly 400 dollars—(80 pounds!)[14] In my efforts to find quiet they had landed me in the best rooms in the hotel, and made me pay for them. If cost adds to the relish of a wine, surely what I take here ought to be very enjoyable.

I am decidedly stronger than when I came; and look to the work before me without fear—the main difficulty is the transference of the apparatus from place to place, so as to get it unpacked in time for the first lecture. This difficulty has not hitherto shown itself, but it will, I fear, in the future.

Applications & diputations have reached me from various quarters, from St. Louis—from Cincinnati, from Providence, Cleveland, Columbus and many other places.[15] A temptation to remain here has been also thrown in my way[16]—I believe if I remained what is described as a "splendid position" would be made in one of the principal cities for me. But I am not to be seduced from the old nest,[17] and the old friends in that way. However, it shows the kindness of the people.

Cottrell[18] & Miller[19] came here, but remained only a single afternoon, utterly insufficient to see anything of the Falls. Preferring variety to thoroughness they have swept on to Chicago, and I hope they will reach Philadelphia tonight. My first lecture there is to be delivered on Monday the 11th.[20]

Had I time I would give you a description of one wild day spent with a splendid fellow[21]—himself however a little rash and wild amid the spray and torrents of the Horseshoe Fall. But I must defer doing this. Give my affectionate remembrances to the X:[22] Say to Hooker that I had a long talk with Asa Gray about him and that I like the move Gray has made.[23] It will strengthen our hand, if that be necessary, in the future.

Kindest regards to Heins.[24] | Ever aff[ectionatel]y yours John

Write a line to Gorey[25] saying that I am well

RI MS JT/1/T/694

---

1. *Niagara Falls*: see letter 3738, n. 3.
2. *the ordeal at Boston*: Tyndall delivered six lectures in Boston from 15–25 October 1872 at the Lowell Institute.
3. *an allusion to international relations*: Tyndall completed his Boston lectures with a possibly apocryphal quote from Alexis de Tocqueville to the effect that the English and American people were of a piece (see letter 3885, n. 4). The lecture was written up in the *Boston*

Globe, '"Light and Heat." Professor Tyndall's Last Lecture', 26 October 1872, p. 4. On de Tocqueville, see letter 3785, n. 18.

4. *M^r. Lowell's*: John Amory Lowell.
5. *M^r. Winthrops*: Robert Winthrop.
6. *Prof Cookes*: Josiah Parsons Cooke.
7. *sundry rambles round Boston*: see letter 3885 for descriptions of the locations visited.
8. *Rumfor*: misspelled by Tyndall; Benjamin Thompson (Count Rumford).
9. *came here with Hector*: to Niagara Falls with Hector Tyndale.
10. *sleeping "car"*: Tyndall would have regarded the use of 'car', as opposed to 'carriage', as a novel Americanism.
11. *Horseshoe Fall*: the largest of the three waterfalls forming Niagara Falls.
12. *Goat Island*: an islet in the Niagara River, just south of the Falls on the American side of the border.
13. *la Rose 16 to 18 shillings a bottle*: La Rose was Tyndall's favorite wine. The price of 16–18 shillings translates to about £71–80 (or $88–99) in 2023 currency.
14. *400 dollars*: approximately $9,730 in 2022 currency.
15. *Applications & diputations... many other places*: a reference to further invitations to lecture Tyndall received after arriving in the United States.
16. *A temptation to remain here has been also thrown in my way*: an invitation to apply for the Chair of Physics at the University of Pennsylvania (see letters 3881 and 3883).
17. *old nest*: the RI in London.
18. *Cottrell*: John Cottrell.
19. *Miller*: George Miller.
20. *My first lecture... Monday the 11th*: Tyndall delivered six lectures in Philadelphia from 11–22 November 1872 at the Horticultural Hall.
21. *splendid fellow*: Thomas Conroy (*c.* 1848–1917) ('Some Observations on Niagara' (*Roy. Inst. Proc.*, 7 (1873–75), pp. 73–91, on p. 78). Conroy was recognized for several heroic rescues at the Falls.
22. *the X*: the X Club (see letter 3757, n. 2).
23. *the move Gray has made*: it is unclear what specific move Tyndall had in mind, but it most probably refers to Gray's support for Darwinism. See J. G. Lennox, 'The Darwin/Gray Correspondence 1857–1869: An Intelligent Discussion about Chance and Design', *Perspectives on Science*, 18 (2010), pp. 456–79; and D. M. Porter, 'On the Road to the *Origin* with Darwin, Hooker, and Gray', *Journal of the History of Biology*, 26 (1993), pp. 1–38.
24. *Heins:* Heinrich Debus.
25. *Gorey*: Gorey, County Wexford, Ireland, where much of Tyndall's family still resided.

From Mary Adair 6 November 1872 3887

Heatherton Park[1] | Nov[em]ber 6th 1872.

My dear Dr. Tyndall

I have had a pleasant surprise in receiving another work of yours, through your Publishers.[2] I am sure that it is you that I must thank for giving me this charming volume which appears to be full of most pleasant reading of Rain, Snow, Ice, and the ever glorious glaciers. I took my time over "Molecular Physics", and reading leisurely found almost the whole was within my comprehension and very interesting.[3]

We are so glad to see that your visit has begun so auspiciously, by such brilliant success at Boston, and trust that your whole stay in America will be as agreeable; also that you will not overwork but as Socrates used to say "take care of your health".[4] We are all fairly well, but saddened by the grief of our friends the Douglas Hamiltons. Two of the Miss Hamiltons have died of typhoid fever, and others are ill with it.[5] I think you are acquainted with them.

Sir Charles[6] seems quite well again and gives good accounts of his cousins.

With our[7] kind regards and a great many thanks, believe me always yours truly

H. M. Adair

RI MS JT/1/A/5

---

1. *Heatherton Park*: the Adair family estate in Somerset, England. Mary's brother Alexander William Adair (1829–89) inherited the estate when their father Alexander Adair (1791–1863) died.
2. *another work of yours*: *Forms of Water*.
3. *Molecular Physics*: *Molecular Physics*.
4. *Socrates used to say "take care of your health"*: possibly Adair has in mind Plato's *Charmides*, which praises temperance, in which Socrates says, 'if the head and body are to be well, you must begin by curing the soul; that is the first thing' ('Charmides', in *The Dialogues of Plato*, trans. B. Jowett, 4 vols (New York: Charles Scribner and Company, 1871), vol. 1, pp. 1–33, on p. 11).
5. *the Douglas Hamiltons ... ill with it:* Alfred Douglas-Hamilton (1818–95); two of his daughters with Adelaide Black Douglas-Hamilton (1818–70) died in early November 1872: Gertrude Eleanor Douglas-Hamilton, on 2 November, aged 24; and Marie Adelaide Blanch Douglas-Hamilton, on 3 November, aged 26.
6. *Sir Charles:* probably Charles John James Hamilton (1810–92), a manager at the RI, and a relation of Mary Adair. In 1833, he married Katherine Emily Wynne, the second daughter

of William Wynne of Dublin (Edward Walford, *The County Families of the United Kingdom*, 5th edn (London: Robert Hardwicke, 1869), p. 450).

7. *our*: her mother Harriet Eliza Adair.

From Joseph Henry[1] 6 [November][2] 1872 3888

Oct 6 72

My Dear Professor

I have Washington on the 13th instant to meet you at a dinner to which my much esteemed friend Mr. Joseph Patterson[3] intends to invite you on the 14th[4]

I hope you will gratify my friend by accepting his invitation. He is a gentleman of wealth and influence with whose acquaintance I am sure you will be well pleased

I am glad to learn, by the news papers, that you have taken time to visit the Falls of Niagara,[5] and have enjoyed, as I am sure you have done, the emotions of the sublime which they never fail to produce in minds properly constituted to appreciate them

I spent a, few years ago, a day at Niagara in making physical observations and obtained some results which I thought of interest. I presume the sight of the falls suggested, in your mind, investigations of a similar character.

I am gratified to learn, through the papers, that the Bostonians and yourself have parted with mutual good feeling[6]—not that I had any doubt that you would be successful, but thought it possible that some incident might occur which would render your course less pleasant than it has been.

I hope nothing has occurred to induce you to give up the idea of lecturing in Baltimore. I think it an important point at which to make an impression for the good of science. It is a city of great wealth; and much refinement; but with little knowledge of physical principles. The late Mr Peabody[7] of London devoted one million of dollars to found the Institution which bears his name.[8] I was invited to give the opening course[9] and accepted the invitation in order that I might do some good in the way of suggestions as to the best plan of appropriating the money. I warmly advocated the adoption of the plan of the Royal Institution of London, which while it diffuses a knowledge of s[cience] among the higher classes of the community, it has enlightened the world with more new truths than were emanated from any other single establishment. The trustees as the time were not prepared to fully appreciate the value of the suggestion; although I think I did some good in checking the tendency to sink a larger part of the endowment in "[unattesting]" marble.[10]

I think however your lectures will produce the most marked effect in Washington where you will be heard by men of the most active minds in the country. There is a growing disposition in congress to make liberal appropriation for scientific objects and a few words from you on this point will I am sure produce a marked effect.

as ever truly yours | Joseph Henry

Dr. John Tyndall

Smithsonian Institution Archives. RU 7001, Box 4A, Folder 1, Pages 649–51

---

1. This letter has previously been published as document 199 in *Papers of Joseph Henry*, pp. 427–8.
2. *[November]*: while Henry clearly wrote 'Oct' for the month, this is in error and November is the correct month. As Henry noted, Tyndall had already lectured in Boston (six lectures from 15–25 October 1872 at the Lowell Institute).
3. *Mr. Joseph Patterson*: Joseph Patterson (1808–87), a Philadelphia merchant who from 1842 was President of Western National Bank in Philadelphia (*The Bankers' Magazine and Statistical Register*, 42 (October 1887), p. 315).
4. *invite you on the 14th*: Tyndall dined with Henry at Patterson's home outside of Philadelphia on 14 November 1872 (*Papers of Joseph Henry*, p. 428, n. 1).
5. *visit the Falls of Niagara*: see letter 3738, n. 3.
6. *Bostonians and yourself have parted with mutual good feeling*: Tyndall delivered six lectures in Boston from 15–25 October 1872 at the Lowell Institute.
7. *Mr Peabody*: George Peabody (see letter 3814, n. 5).
8. *Institution which bears his name*: see letter 3814, n. 5.
9. *opening course*: Henry delivered the very first lecture for the Peabody Institute on 20 November 1866, speaking on the philosophy of science (A. P. Molella, 'At the edge of science: Joseph Henry, "Visionary Theorizers", and the Smithsonian Institution', *Annals of Science*, 41 (1984), pp. 445–61, on p. 453, n. 21).
10. *"[unattesting]" marble*: either meaning a larger focus on art than on science, or the physical building of the Peabody Institute itself.

From Henry Bence Jones — 8 November 1872 — 3889

84 Brook St W.[1] | Nov. 8 1872 | Friday

Dear Friend

If for no other reason then to thank you for your letter[2] I must write this week. As far as your health goes I am tolerably satisfied but not quite. I shall like to hear soon of you again. As far as lectures are concerned all seems as well as can be.[3] Tell Cotterell[4] to send me more frequent papers and to keep an account against me I want as many as he pleases to send. They can not come too often tell him. All odd particulars in which American life strikes you as different from ours are very interesting. What you have to pay for things La Rose for example.[5]

What an hour that must have been with Agassiz Emerson Longfellow & Holmes.[6] They were too much for the time vastly too much.

We had our managers meeting on Monday & all went smoothly.[7] Your letter[8] had reached me that morning so I was able to give them & the members also who asked news of you the latest accounts. I had got a bad cold some days previously & feared I should not be at the meeting however I did it & without trouble. We shall have no other meeting till December—The works do not require it. I have hardly <u>seen</u> what is doing as I have been out very little but all I am told is going on well & I hope to get all but the Carpenters out by Xmas or sooner.[9] I am pressing the fitting of Odlings Laboratory so that the lecturers may have every convenience for the lectures.[10] The new stair case will be a big thing & certainly in the case of the fire upstairs or down should be vastly safer than the old one—Your floor of the laboratory is nearly done & your roof. The drains have all had to be new & that has delayed the lower works. There seems to be an awful lot more to do before the work will be finished however I live in hope—I hear nothing new of any kind. I go so little about that news must come to me. Gassiot[11] spoke as if the Hooker affair[12] was over & Hooker would stay on & nothing more take place about it. Owen this week again gives his opinion in an article in Nature.[13] I have had to direct Mr Wills to remove for a few days your salt specimens in the model room. There was fear that the pipes from the boiler in the cupboard might leak & spoil your things. He will take every care & all shall be put back again in a few days. We seem doing everything on a very grand scale. The heating is a monstrous affair & so indeed all the works are. They will last forever.[14] Such flags such gardens such stairs & such works I should have thought far beyond our needs. However I hope we shall all like it when done though that is much to expect. Spottiswoode will give the first Friday evening "on the old & new laboratories"[15] I should do it myself but he will do it much better & I shall probably not be able to go to hear him. Busk is going to keep order[16] at the Friday lectures[17] until you return to your old seat. There once was an Assistant Honorary Secretary I have not seen Huxley nor heard of him lately; Darwin's new book is out & appears very interesting.[18] He is again at work at some botanical paper & is going to put all his botanical papers together[19]

DuBois[20] has printed his speech at the German Association at Leipzig.[21] It is on matter & force body & soul & his ending is "Ignorabimus!" It is difficult German to me & I cannot make out some parts. I am asked whenever I go to the Institution by members how you are & when you will be home as everyone seems interested in your tour. I told the managers of the £20,000 that you might earn[22] by leaving us for 2 years & they opened their eyes. I had not long before told them that they had given Faraday £100 a year & in a committee on the expenses they reported that they would not cut down his salary, much as they wanted money, because he did so much for the Institution. Miss Barnard[23] was here not long since & reports Mrs Faraday[24] well. I have not been up there yet. I get home as soon as I can when I must go out. Sir H Holland is as lively as ever running about all day long. We all united in kind regards and in warmest wishes for your welfare. Perhaps you will bring back an American wife. They settle those matters quickly I believe & some of them are very charming when married to Englishmen & make very good wives. I reckon you are just finishing your course at Washington[25] & about next week to go to New York.[26] Where after that Baltimore?[27]

Goodbye dear friend may all go well with you in all ways

Ever Yours | H. Bence Jones

RI MS JT/1/J/71

---

1. *84 Brook St W.*: residence in London.
2. *your letter*: letter 3875.
3. *lectures are concerned all seems as well as can be*: probably meaning that from the American newspapers that Bence Jones was receiving in the mail from Tyndall's assistant Cottrell, the reception of Tyndall's American lecture tour thus far was positive. For Tyndall's American lecture tour in general, see Introduction, n. 8.
4. *Cotterell*: John Cottrell.
5. *La Rose for example*: see letter 3886, n. 13.
6. *Agassiz Emerson Longfellow & Holmes*: see letter 3875.
7. *mangers meeting*: at the RI, of which Bence Jones was secretary from 1860–73.
8. *your letter*: letter 3875.
9. *Carpenters out by Xmas or sooner*: the construction of a new laboratory at the RI (see letter 3776, n. 14).
10. *Odlings Laboratory*: the Chemical laboratory; William Odling was the Fullerian Professor of Chemistry at the RI (1868–72), but his time at the RI ended later in 1872 as he moved on to become the Waynflete Professor of Chemistry at Oxford University (1872–1912). Odling delivered the season's annual Christmas lectures, which went into early January 1873 (see letter 3782, n. 4).
11. *Gassiot*: John Peter Gassiot (1797–1877), an English electrician and wealthy wine merchant. He was one of the founding members of the London Electrical Society (1837) and served as the its treasurer until 1841, and he pursued electrical research in the laboratory he outfitted at his home in Clapham Common, near London (*ODNB*).
12. *Hooker affair*: on Joseph Hooker's dispute with Acton Ayrton, see letter 3755, n. 1.
13. *an article in Nature*: Richard Owen, 'The National Herbarium', *Nature*, 7 (7 November 1872), pp. 5–7.
14. *They will last forever*: the laboratory constructed in 1872 was demolished to make way for

a more modern facility in 1929 (F. A. J. L. James and A. Peers, 'Constructing Space for Science at the Royal Institution of Great Britain', *Physics in Perspective*, 9 (2007), pp. 130–85, on pp. 170–2).

15. *first Friday evening "on the old & new laboratories"*: see letter 3776, n. 14.
16. *Busk is going to keep order*: George Busk, treasurer of the RI, filled Bence Jones's duties for the upcoming season of Friday Evening Discourses.
17. *Friday lectures*: the series of Friday Evening Discourses beginning in January 1873 (see 3880, n. 21).
18. *Darwin's new book is out & appears very interesting*: C. Darwin, *The Expression of the Emotions in Man and Animals* (London: John Murray, 1872), published on 26 November.
19. *put all his botanical papers together*: Darwin's next four books would all be on botanical themes: *Insectivorous Plants* (1875); *The Effects of Cross and Self Fertilisation in the Vegetable Kingdom* (1876); *The Different Forms of Flowers on Plants of the Same Species* (1877); and *The Power of Movement in Plants* (1880).
20. *DuBois*: Emil du Bois-Reymond.
21. *printed his speech at the German Association at Leipzig*: du Bois-Reymond's 14 August 1872 lecture 'Über die Grenzen des Naturerkennens' (or 'The Limits of Science'), in which he finished with a single word, Latin for 'we will not know', to articulate his view that science would not be able to explain consciousness (E. d.-B. Reymond, *Über die Grenzen des Naturerkennens. Ein Vortrah in der Zweiten Öffentlichen Sitzung. Der 45. Versammlung Deutscher Natureforscher und Ärtze zu Leipzig AM 14. August 1872* (Leipzig: Verlag von Veit & Comp., 1876)).
22. *the £20,000 that you might earn…*: in letter 3875, Tyndall wrote that 'It is stated that in two years I could make £20,000 by lecturing in America'.
23. *Miss Barnard*: Jane Barnard.
24. *Mrs Faraday*: Sarah Faraday.
25. *Washington*: Tyndall delivered six lectures in Washington, DC, from 3–12 December 1872 at Lincoln Hall.
26. *New York*: Tyndall delivered six lectures in New York from 17–31 December 1872 at the Cooper Institute.
27. *Baltimore*: Tyndall delivered three lectures in Baltimore from 26–9 November 1872 at the Peabody Institute.

From George K. Winter[1] 11 November 1872 3890

2 Wilton Villas | Shepherds Bush[2] | 11 Nov[ember]: 1872

My dear sir

I trust you will pardon the liberty I am taking in thus addressing you.

You will probably have seen in the Phil[osophical]: Mag[azine]: for March, a paper, in which are given the results of my observations during the eclipse of last year.[3] The chief result was "the increase of the polarization of the corona with distance from the limb"[4]

There are three causes by which I have sought to explain this observed fact. One of these is given in the paper referred to, another in my letter to "Nature" of Sept. 5th[5] and a third,[6] as yet unpublished involving a question that I think may easily be answered by experiment—

I have unfortunately no apparatus with me in this country, but the subject of the polarization of light by small particles has already been so largely indebted to your beautiful researches that I feel sure you are able, and I trust you will be willing to help me

My explanation is briefly this

Let D and E be two particles in the corona, at different distances from the Sun, but in the same radial line DEG. Let these particles scatter the sun's light, which light, for the moment, we shall consider as coming from two points C & C′ on the limb at equal distances from G. Let DA and EB be the rays from the particles seen by an observer on the earth.

It is evident that the angles C'DA and CDA will be nearer to 90° the angle of the max: polarization than the angles C'EB, and CEB, respectively. The ray DA should therefore be more strongly polarized than the ray EB.[7] But I am met with the objection that in dealing with light scattered from particles, if the light come from different distances we must take the resultant of all the different directions of incidence. Let us suppose the sources C & C′ to be equal, then the common resultant direction of the light incident on D & E will be DEG and as DA and EB are to all intents parallel, the angles GEB and GDA are equal and the two rays DA and EB should be equally polarized—

I am not quite prepared to accept this without the evidence of experiment—

There would, I believe, be little difficulty in devising an experiment by which the question might be answered and I think I may say that it is not an uninteresting subject

I am going to Bristol for a few days but shall be back here about the end of the week.

Yours very truly | G. K. Winter
Prof[essor]: Tyndall FR.S | &.c &.c | Royal Institution

RI MS JT/1/W/83

---

1. *George K. Winter*: George Kift Winter (1842–98), an English civil engineer, was first Assistant Telegraph Engineer of the Madras Railway in India (in 1864), and then became head of the department until his death. He was interested in technical edition, and to this end was an examiner for the University of Madras. He published in the *Phil. Mag.*, and presented papers at BAAS meetings. He was a Fellow of the Royal Astronomical Society, and participated in eclipse expeditions in August 1868 and December 1871 (*Minutes of Proceedings of the Institution of Civil Engineers*, 133 (1898), pp. 412–13).
2. *2 Wilton Villas | Shepherds Bush*: in Middlesex, London.
3. *a paper... of last year*: G. K. Winter, 'Observations on the corona seen during the eclipse of December 11th and 12th, 1871', *Phil. Mag.*, 43 (1872), pp. 191–4.
4. *the increase ... from the limb*: the corona is the outermost segment of the sun's atmosphere, and the limb is the visual edge of the sun. Winter refers to an empirical phenomenon, whereby coronal light farther from the limb is radially polarized to a greater extent than coronal light from closer to the limb.
5. *my letter to "Nature" of Sept. 5th*: G. K. Winter, 'Radial Polarisation of the Corona', *Nature*, 6 (5 September 1872), pp. 371–2.
6. *third*: this paper does not appear to have been published.
7. *The ray DA ... than the ray EB*: Winter's proposed explanation, using ray optics, is that the degree of polarization depends on scattering angles, although here it does not appear to offer much explanatory power since his choice of the points C and C′ is arbitrary.

To William Jones Rhees 12 November 1872 3891

12th Nov.

My dear Mr. Rhees.

My cousin[1] & I have been considering your note.[2] I am anxious to do every thing that I can to please all my friends. New York has been already considerably a sufferer[3] & I cannot do less for it than I have promised to do—as it is they have to be content with three lectures before Christmas & three after.[4]

But in two or three days I shall be able to say whether it will not be possible to lecture on Monday the 9th Tuesday the 10th. & Thursday the 12th. You have no idea of the trouble of packing—it would be absolutely impossible to get *[at completely]* & the apparatus sent off on the Saturday.

Faithfully yours | John Tyndall

Best regards to Prof. Henry, whom I hope to meet on Thursday.

John Tyndall to William Jones Rhees, 12 November 1872, mssRH 3971, box 43, William Jones Rhees papers, The Huntington Library, San Marino, California

1. *My cousin:* Hector Tyndale.
2. *your note:* possibly letter 3814, in which Rhees layed out the proposed dates for each city in Tyndall's American lecture tour.
3. *New York has been already considerably a sufferer*: Tyndall was to have lectured in New York following Boston, but concerns of it coinciding with the US Presidential election (see letter 3794, n. 3) prompted delaying New York until later; thus, after Boston but before Philadelphia, Tyndall made an excursion to Niagara Falls (see letter 3738, n. 3).
4. *three lectures before Christmas & three after*: Tyndall delivered six lectures in New York from 17–31 December 1872 at the Cooper Institute.

To Frank Marx Etting[1] 13 November 1872 3892

13[th] Nov. '72

Dear Sir

The spokes of light & shade I know to be infrequent at sunset—there is then more haze in the air.[2]

You must be right as regards the Chandelier[3]—it shall be attended to.

faithfully y[ou]rs | John Tyndall

Thank you much for your cordial appreciation

[Envelope] Frank Etting Esq[ui][r][e] | 1315 Spruce Street. | City.

[Postmark: 'PHILADELPHIA']

Historical Society of Philadelphia, Etting Papers

1. *Etting*: Frank Marx "Francis" Etting (1833–90). After a decade in the army, Etting was director of the Philadelphia's public schools, chairman of the committee to restore Independence Hall, and director of the Historical Department of the Centennial (1876) (Historical Society of Pennsylvania).
2. *spokes... air*: presumably a reference to a question Etting asked about optics. In the 1860s Tyndall had undertaken research on the scattering of light in air and various liquids, leading to the discovery of what would later be called the 'Tyndall effect'—the scattering of light by a colloidal suspension.
3. *the Chandelier*: Etting was chairman of the committee to restore Independence Hall in preparation for the centennial of American independence in 1876. This is perhaps a reference to the glass chandelier in the assembly room (J. R. Young (ed), *Memorial History of the City of Philadelphia, from its First Settlement to the Year 1895* (New York: New York History Company, 1898), p. 107).

To Thomas Archer Hirst 13 November [1872][1] 3893

Philadelphia[2] U.S. | 13th. Nov. | 1872

My dear Tom.

Would you like a good fellow transfer the money now in your hands to the London Joint Stock Bank Pall Mall. It will then fructify,[3] and give me a right to draw on them.

Ask Mr. Hughes[4] to do the same with my Royal Institution cheque.

I gave them 2 hours all but 10 minutes on Monday night—not a sign of inattention or weariness from beginning to end.[5]

Hector[6] & his wife[7] are taking great care of me. I am stronger than I was in Boston.[8]

Always aff[ectionatel]y. | John

RI MS JT/1/T/965
RI MS JT/1/HTYP/607

---

1. *[1872]*: the year is given by reference to his lectures in Philadelphia and Boston. For Tyndall's American lecture tour in general, see Introduction, n. 8. '1872' is written on the letter, likely in another hand.
2. *Philadelphia*: Tyndall delivered six lectures in Philadelphia from 11–22 November 1872 at the Horticultural Hall.
3. *fructify*: to bear fruit, become fruitful (*OED*).
4. *Mr. Hughes:* not identified.
5. *Monday night*: Monday, 11 November 1872, was the evening of Tyndall's first lecture in Philadelphia.
6. *Hector*: Hector Tyndale
7. *His wife*: Julia Tyndale.
8. *Boston*: Tyndall delivered six lectures in Boston from 15–25 October 1872 at the Lowell Institute.

From William L. Faber[1] 15 November 1872 3894

Salt Lake City Utah Nov 15/72 | Prof. John Tyndall

My dear Sir

The writer had fondly hoped to use the privilege of an old fellow student[2] in being one of the first to welcome you on your arrival in New York;[3] but at that precise time an unforeseen call took him off to Utah. And finding that a month's stay will not suffice for his business here, but that a winter among the Mormons[4] is to be his lot, he now greets Prof. Tyndall being denied the pleasure of shaking hands with Wat Ripton.[5]

By this time you have recognized the writer—but I beg of you to think of him as somewhat different from the wild thoughtless Marburg student; and I am proud to say that much as a rather severe struggle with a rougher world than ever I was fitted for has had its due share in effecting in me a desirable change, a great deal is also to be ascribed to the influence you exercised over me during our intercourse. I have always desired an opportunity of personally expressing my gratitude toward you, who acted toward me always with the part of an elder brother than of a mere friend. The privilege of an hour's chat over old, old times, is now denied me, unless indeed you contemplate a trip across the continent. In such case I should be most happy to act as your cicerone[6] here and can promise you two items of interest: Alpine snow and scenery[7] and mines of extraordinary richness and extent,[8] out of one of which I am endeavoring to hammer a competency.[9]

And now I will venture one little request based on the supposition that you still take some interest in me. It is simply this: will you kindly devote one leisure hour to a call on my wife?[10] I am most happily married, and dare even to say that I wish you could return to England with the impression of my wife as a fair specimen of American women, although I am well aware that she is not such, and that her rare virtues and sterling qualities are not common. Mrs Faber is boarding with the children during my absense at No 227 Warren Street Brooklyn, where a call from you as one of my old friends would be a rare treat quite an era in her monotonous life.

She will give you, if you desire it a sketch of my chequered career during the past twenty years and in return she may claim some account of me at Marburg—which please give fully and freely, for I have no secrets from her. But enough for the present—I would write a much longer letter if I felt quite sure of its being acceptable; still I can not deny myself the pleasure of hoping that the friendship of nearly ¼ century ago is yet entertained by Wat Ripton as sincerely as it certainly is by

Yours truly | Wm L. Faber | Address: Bingham Cañon | Salt Lake County | Utah. | P. O. box 13

RI MS JT/1/F/1

1. *William L. Faber*: William L. Faber (d. 1873), an American who studied at the University of Marburg in Germany. Before he became a chemist, civil engineer, and metallurgist, Faber was a music teacher and artist (Arkansas State Archives).
2. *an old fellow student*: Tyndall and Faber studied together at the University of Marburg.
3. *your arrival in New York*: Tyndall arrived in New York on 9 October 1872 for his American lecture tour (see Introduction, n. 8).

4. *among the Mormons*: many members of the Church of Jesus Christ of Latter-day Saints migrated to Salt Lake Valley in the 1840s and continued to wield considerable influence in the Utah Territory. Utah gained statehood in 1896.
5. *Wat Ripton*: an alias Tyndall used in his Marburg days.
6. *cicerone*: a guide who shows and explains the antiquities or curiosities of a place to strangers (*OED*).
7. *Alpine snow and scenery*: probably referring to the Wasatch Range, part of the Rocky Mountains, which runs north-south along the Salt Lake Valley's eastern edge.
8. *mines of extraordinary richness and extent*: probably referring to the large copper deposits discovered in the Oquirrh Mountains to the southwest of Salt Lake City in the late 1840s. The site is now known as the Bingham Canyon Mine, both the largest man-made excavation and deepest open-pit mine in the world. See T. LeCain, *Mass Destruction: The Men and Giant Mines That Wired America and Scarred the Planet* (New Brunswick, NJ: Rutgers University Press, 2009).
9. *competency*: the condition of having a sufficient income (*OED*).
10. *my wife*: Georgia Lydia Knox (1847–1920). Tyndall delivered six lectures in Brooklyn from 4–20 January 1873 at the Academy of Music.

From Joshua Pusey[1] 15 November 1872 3895

22 North 7th st. | Philad[elphi]a. Nov[embe]r. 15th. 1872.

Prof. John Tyndall

Dear Sir:

I have attended your lectures in this city so far,[2] with great interest and pleasure.

In connection with this subject "Light" I beg leave to bring to your attention a curious experiment in Optics which I first made several years ago, and have since frequently repeated. It may not be new to you: but it is also possible that it has escaped your attention: If so I hope it will interest you. I will endeavor to explain the experiment as clearly as possible with the aid of the accompanying diagrams. I will relate it precisely as I made it.

I took a piece of wood, W., turned in the form shown, and painted its upper face, or disc I will term it, equally with the three primary colors.[3] In a socket in W, I inserted a stem S (unpainted) the whole forming a sort of a top. Fig. 1 is a plan view looking down upon this top. Now if S is perpendicular to the plan<e> of W or in the position of the red-ink line a.b. nothing peculiar is observable upon rotating the top except the intermingling of the colors of the disc to form a dirty white:[4] but if the stem, S, be inclined to the plane of the disc, as in Fig. 2., the colors on the latter (apparently) leave it, and two or more, or a mixture of them, mount the stem and appear thereupon, as appears in Fig. 3.; that is, provided the line drawn from the eye to the top of the stem, S, and protracted, falls within the periphery of the disc, as the red-ink line c.d. Fig. 3. In the present case I have supposed the stem, S, to be inclined towards the central red-ink line x.y. Fig 1. of the red portion of the disc, when, as the top rotates—the speed of rotation seems not to affect the result—, blue appears on the right side of S. and yellow on the left. Various beautiful combinations of colors may be caused to appear on the stem according as the latter is inclined towards different points of the disc. The dividing line between the colors is of course not so well defined as in Fig. 3. but is somewhat blurred. I have been unable to discover any satisfactory explanation of this curious effect, either from others, or from my own reasoning.

I hope you will find some matter of interest in what I have written herein.

Wishing you a pleasant visit in our country and a safe return to your native land, and that you will be spared for many years of continued usefulness. I am

Yours very respectfully | Joshua Pusey

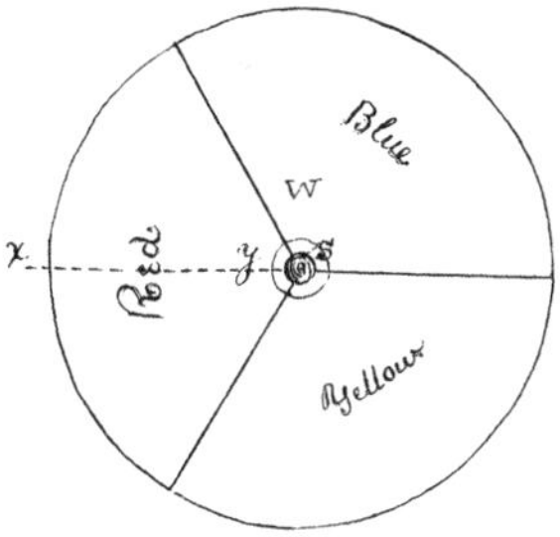

Figure 1

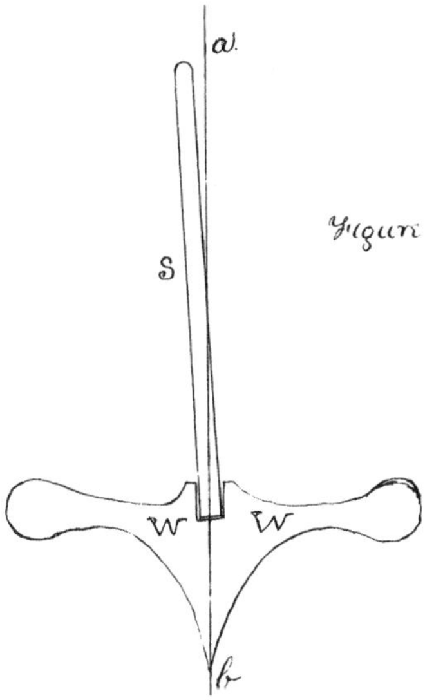

Figure 2

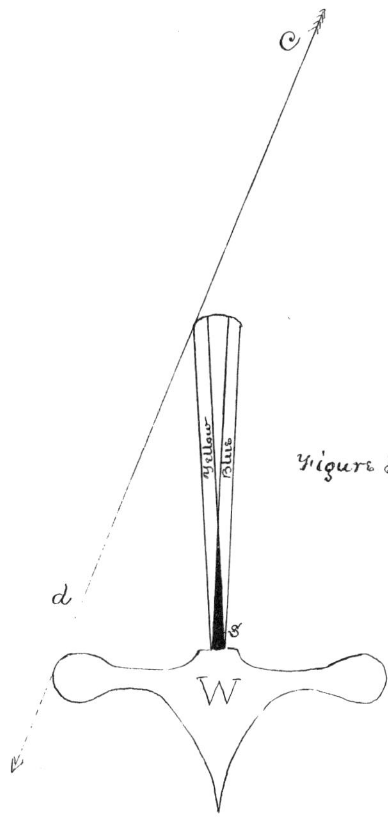

Figure 3

RI MS JT/1/P/260

---

1. *Joshua Pusey*: Joshua Pusey (1842–1906), a Philadelphia-based patent attorney and inventor. In 1892 he was granted a patent for the paper matchbook, which he later sold to the Diamond Match Company in Ohio (J. W. Jordan (ed), *A History of Delaware County, Pennsylvania and its People* (New York: Lewis Historical Publishing Company, 1914), vol. 3, p. 778; R. Welch and P. A. Lamphier (eds), *Technical Innovation in American History: An Encyclopedia of Science and Technology*, 3 vols (Santa Barbara, CA: ABC-CLIO, 2019), vol. 1, p. 96).
2. *your lectures in this city so far*: Tyndall delivered six lectures in Philadelphia from 11–22 November 1872 at the Horticultural Hall. At the time of this letter, Pusey would have attended either the first two, if he wrote this letter before the third lecture on 15 November, or three if he wrote this letter after the third lecture.
3. *three primary colors*: red, yellow, and blue.
4. *Now if S… dirty white*: Pusey describes a phenomenon known since antiquity, now referred to as 'Newton's disc', in which colored blocks or lines painted on a spinning surface blend to produce a uniform color.

To Henry Bence Jones 16 November 1872 3896

Philadelphia[1] | 16th. Nov. 1872.

My dear Friend

I am sorely grieved through your letter[2] which reached me yesterday to learn that you are ill.

And I am grieved that you should be worried by the laboratories.[3] This I suppose is the fashion of architects, and we must put up with it.

I have already written to you twice[4] and sent you one or two scraps of newspapers. Last night I delivered my third lecture here. The audience was at first reserved,—the quaker element[5] being strong here; but they came down with hot applause last night.

Henry of Washington[6] is here, and it is said that the audience is composed of the 'cream' of Philadelphia.

They are trying to keep me here and speak of large inducements—say 10000 dollars, or £2000 a year with lots of time for work and very few lectures. Were there many Mr Lees[7] at the Institution,—Indeed were the ruling party there not very different from Mr. Lee my course at present would be very plain and decisive.

After the Boston lectures[8] Miller [9] felt low. He then took a trip to Niagara, thence went to visit his sister. His brother-in-law had just recovered from a fit of typhoid fever, and Miller is now down with it. I have had the advice of two of the best physicians regarding him. They both say that he is doing perfectly well.[10]

Of course this is a draw back to me. I had some doubts as to the wisdom of bringing him with me, but he would have been unhappy had he been left behind, and so I brought him.

I transfer myself to Baltimore[11] a week hence and begin there on the Tuesday.

Take care of yourself dear friend and rejoice me by a letter announcing that you are better.

always Yours | John Tyndall

I have beautiful rooms here in a quiet street, and am fed like a fighting cock. I am staying with my cousin.[12]

RI MS JT/1/TYP/3/815
LT Typescript Only

---

1. *Philadelphia*: Tyndall delivered six lectures in Philadelphia from 11–22 November 1872 at the Horticultural Hall.
2. *your letter*: possibly letter 3880.
3. *the laboratories*: those under construction at the RI, under Bence Jones's supervision (see letter 3776, n. 14).
4. *written to you twice*: letters 3875 and 3885.
5. *the quaker element*: the colony of Pennsylvania was founded by William Penn, a member of the Society of Friends, and the Quakers remained prominent in Philadelphia, a city he helped to plan. The sect's emphasis on silent reflection made many reluctant to engage in noisy and demonstrative performances of approval.
6. *Henry of Washington*: Joseph Henry.
7. *Mr Lees*: probably Thomas Lee.
8. *Boston lectures*: Tyndall delivered six lectures in Boston from 15–25 October 1872 at the Lowell Institute.
9. *Miller*: George Miller.
10. *doing perfectly well*: two weeks after this letter, on 30 November at 7:15 a.m., Tyndall's younger assistant George Miller died from typhoid fever while in Philadelphia. Hector Tyndale received a telegram from Thomas Pugh, dated 30 November, that read: 'Mr Miller died this morning, advised his relatives in Hamilton' (RI MS JT/1/P/259). See letters 3902 and 3907.
11. *Baltimore*: Tyndall delivered three lectures in Baltimore from 26–9 November 1872 at the Peabody Institute.
12. *my cousin*: Hector Tyndale.

From Henry Bence Jones — 18 November [1872][1] — 3897

84 Brook St | W[2] | Nov 18

My dear friend

The end of your lecture last was in the paper you sent me, so I copied it out & sent it to "nature" they published it last week.[3] The globe had a notice of your ending which I send you with this.[4] I have not seen any thing else. There was a short notice of one lecture in the Times.[5]

Emerson is here[6] & gave Sir H. Holland a flourishing account of you but he thought you did not realise how much the Yankees know of science.

But now for the important question when? you will come back. No doubt you will have temptation upon temptation of all sorts.[7] You do not let me know what is in your own mind probably because you have not made it up. So I shall only write as seems to me for the good of the Institution the last Friday evening before Easter is April the 4th & you are announced to give it.[8] But there is no reason why this sh[oul]d not be changed. Your first lecture is the 24 of April & the 2nd not till May 8th.[9] We can stop this first lecture & so begin May 8th if you wish it so that you could stay in America until the middle of April easily. Every body would like to have you back in March if your plans admit of it but every body also will wish that you sh[oul]d do that which seems to you best & of which we here can not judge at all.

I hope you will have done all the international good you can do by the end of Feb[ruar]y. How much harm you may have done yourself by that time I shall judge if I am alive when you return. I tremble lower & lower & am almost confined to the house.[10] Busk is going to act on Friday evenings as ruler of the house.[11] I shall never get there.

I have not been at the Institution for a fortnight I shall do my uttermost to get to the managers meeting in Dec

Airy is going to announce his resignation next year.[12] The Duke of Devonshire[13] was spoken of. Grove[14] yesterday told me he thought Hooker would be the best man. Cambridge Miller[15] also is then going to resign the foreign secretaryship so the stream goes on.

I am sorry Dr Ellis did not send me a message by you; in some unconscious way I must have offended him

Pray let me know as soon as you have made up your own mind as to your return I like to look forward to seeing you again & as I don't look forward to our meeting in the other world my only chance is in this[16]

I believe all goes on well at the Institution I sh[oul]d hear if it were not so

Your account of Niagara interested us all but why should you always be trying to do things on the verge of danger? I suppose it is a part of your Irish inheritance.

I am glad you were not in Boston when the fire was there[17] you would have blown yourself up to save the houses.

Good bye | Dear friend | I am always yours | H Bence Jones | Write soon

RI MS JT/1/J/90

---

1. *[1872]*: the year is given by references to Tyndall's American lecture tour (see Introduction, n. 8). He delivered six lectures in Philadelphia from 11–22 November 1872 at the Horticultural Hall.
2. *84 Brook St | W*: residence in London.
3. *the paper you sent me ... last week*: the clipping from the *Boston Daily Advertiser*, 26 October 1872, was reprinted in *Nature*, 7 (14 November 1872), p. 34.
4. *The globe had... with this*: 'Tuesday Evening', *The Globe*, 12 November 1872, p. 1.
5. *short notice... in the Times*: 'Lectures on Heat at Boston', *Times*, 15 November 1872, p. 4.
6. *Emerson is here:* Ralph Waldo Emerson spent a day with Tyndall after his lectures in Boston, and traveled to England in October of 1872. See letter 3875.
7. *temptation upon temptation of all sorts*: a reference to further invitations to lecture Tyndall received after arriving in the United States (see letters 3885 and 3886).
8. *last Friday evening... give it*: on 4 April 1873, Tyndall delivered a Friday Evening Discourse at the RI on Niagara Falls (see letter 3738, n. 3).
9. *Your first lecture is the 24 of April & the 2nd not till May 8th*: see letter 3782, n. 10.
10. *if I am alive... confined to the house*: Bence Jones was in poor health and would die of heart failure on 20 April 1873.
11. *Busk... ruler of the house*: George Busk, treasurer of the RI, filled Bence Jones's duties for the upcoming season of Friday Evening Discourses.
12. *Airy ... resignation next year*: George Biddell Airy served as president of the RS from 1871–3. He was succeeded by Joseph Hooker, who served until 1878.
13. *The Duke of Devonshire*: William Cavendish (1808–91), the 7th Duke of Devonshire. He was a patron of the sciences who donated to the construction of the Cavendish Laboratory at the University of Cambridge and endowed the Cavendish Professorship. He studied mathematics at Trinity College, Cambridge, earning second wrangler in the mathematical tripos (*ODNB*).
14. *Grove*: William Robert Grove (1811–96), a Welsh physicist, member of the RI, and FRS (since 1840). In the 1840s, he made early contributions to thermodynamics that anticipated the principle of the conservation of energy (*CDSB*; *ODNB*).
15. *Cambridge Miller*: William Hallowes Miller (1801–80), a Welsh mineralogist and Fellow of St John's College, Cambridge. From 1832–70 he was professor of minerology at Cambridge, occupying the chair vacated by William Whewell. In the early 1870s, he served as Foreign Secretary of the RS.

16. *I don't look forward... chance is in this*: a reference to Tyndall being non-Christian. On Tyndall and religion, and his being labeled a materialist, pantheist, or agnostic, see G. Cantor, 'John Tyndall's religion: a fragment', *Notes and Records of the Royal Society of London*, 69 (2015), pp. 419–36. Bence Jones was not a believer himself, as he noted in letter 3912: 'Miss Barnard & Mrs Thistlethwayte try to convert me to the truth before I die but I fear they will give me up soon'.
17. *in Boston when the fire was there*: the Great Boston Fire of 1872 broke out on the evening of 9 November 1872, shortly after Tyndall departed on October 31. The fire resulted in the death of thirteen people and over $70 million of property damage.

To [Robert E. Rogers][1] 19 November 1872 3898

It will give me pleasure to join the party of Prof. Rogers on Thursday evening.[2]
John Tyndall | 19th. Nov. 1872

History For Sale 33672

1. *[Robert E. Rogers]*: probably Robert E. Rogers of Philadelphia (see letter 3881, n. 7). Tyndall was on his American lecture tour and delivered six lectures in Philadelphia from 11–22 November 1872 at the Horticultural Hall.
2. *on Thursday evening*: in his journal during the American lecture tour, Tyndall wrote, 'On the 21st walked some and joined a party at Professor Rogers's after dinner' (RI MS JT/2/12/44). Indeed, 21 November 1872 was a Thursday.

From Heinrich Rudolf Vieweg[1] 20 November 1872 3899

Braunschweig d. 20n Nov. 1872.

Herrn Professor John Tyndall. | London

Hochgeehrter Herr!

Es zeigt sich, dass wir an eine dritte Ausgabe der deutschen Bearbeitung Ihres Buches über die Wärme denken müßen, *[indem]* wir voraussehen, das ehe das Jahr 1873 zu Ende geht, alle Exemplare der zweiten Auflage, welche mit Berücksichtigung der dritten und vierten englischen Auflage bearbeitet werden, verkauft sein werden.

Wir erlauben uns schon jetzt bei Ihnen anzufragen, ob nach der vierten Auflage in Englischer Sprache, noch eine spätere mit wesentlichen Aenderungen erschienen ist, oder ob eine solche in den nächsten Zeit<en>

[zu erwarten ist]; vielleicht beabsichtigen Sie auch ohne dies der deutschen Auflage Zusätze oder Abänderung<en> zuzufügen. In solchem Falle hätten wir die Bearbeiter der deutschen Ausgabe auch um deren Mitwirkung zu ersuchen, während wenn nur ein der zweiten deutschen Ausgabe gleich lautender Abdruck zu machen sein würde gar keine Eile vorläge, da der Druck nach Bedürfniß rasch beschafft werden kann.

Ihren gefälligen Mittheilungen entgegensehend hochachtungsvoll *[&]* ergebenst

Friedr Vieweg & Sohn

Braunschweig, 20 Nov. 1872.

Professor John Tyndall. | London

Dear Sir!

It turns out that we have to think about a third edition of the German version[2] of your book on heat,[3] *[in that]* we anticipate that, before 1873 comes to an end, all copies of the second impression, which are being revised taking the third and fourth English impression into consideration, will be sold.

We permit ourselves to enquire with you right now whether, after the fourth English language impression, a further later one with substantial changes has appeared, or whether one *[is to be expected]* in the near future; perhaps, without this, you also intend adding further remarks or revisions to the German impression. In such a case, we would also have to request the editors of the German edition[4] for their cooperation, whereas if only a reprint identical to the second German edition would have to be done, there would be no hurry at all, as the printing can be arranged quickly according to need.

Looking forward respectfully *[&]* humbly to your kind advice,

Friedr[ich] Vieweg & Son

Vieweg-Archive der Universitatsbibliothek Braunschweig 663

---

1. *Heinrich Rudolf Vieweg*: while the letter is signed 'Friedr. Vieweg & Sohn', we believe the sender is Heinrich Rudolf Vieweg, who took over the family publishing firm, Friedrich Vieweg und Sohn, from his father Eduard Vieweg (1797–1869).
2. *third edition of the German version*: J. Tyndall, *Die Wärme betrachtet als eine Art der Bewegung*, trans. H. Helmholtz and G. Wiedemann (Braunschweig: Friedrich Vieweg und Sohn, 1875).
3. *your book on heat*: the most recent English edition of Tyndall's book on heat was: J. Tyndall, *Heat: A Mode of Motion*, 4th edn (London: Longmans, Green, and Co., 1870). Previous editions were published in 1863, 1865, and 1868.
4. *collaborators of the German edition:* Hermann Helmholtz and Gustav Wiedemann.

To Henry Bence Jones 21 November 1872 3900

21st. Nov. 1872.

My dear Friend

By this time you have had two or three letters from me;[1] and I am not going to write one now. But I cannot resist telling you of the pleasure which this last letter[2] (just received) gave me. Your first[3] made me terribly gloomy and anxious but this[4] cheers me up wonderfully.

And your reference to the managers[5] was very pleasant to me. It is surely a gratification after nearly 20 years of working together that this perfect and friendly understanding should exist between us. It is that which has welded, and which always will weld me to the Royal Institution.

I have given five out of my six lectures at Philadelphia.[6] The city is profoundly influenced by its quaker population[7] and a man expecting applause might be disappointed. But my audience have been unflagging in their attention, and they have steadily augmented in numbers. Could we have got a better hall the number would have been 3000. As it is about 1800 come to hear me. They warmed up moreover and I am assured such applause as that which they now show has been rarely, if ever, manifested in Philadelphia.

The people are as kind as possible to me.

My cousin[8] has a beautiful little house here and has lodged me royally. Had he a cooler, and more bending temper he might by this time be one of the foremost men in the United States. But his temper is as hot as fire; and his contempt for the politics of the day undisguised.[9]

I leave for Baltimore[10] on Monday. Then comes Washington[11] and then New York[12] and Brooklyn. [13]

Write to me when you have time I hope you will always be able to send me good news.

Ever Yours my dear friend | John Tyndall

My lectures here have occupied from 1.50 to *[1.30]* minutes

There is no mistake about the £20000.[14] It might be made almost in a single year!

RI MS JT/1/TYP/3/816
LT Typescript Only

1. *Two or three letters from me*: letters 3875, 3885, and 3896.
2. *this last letter*: letter 3889.
3. *Your first*: letter 3880.
4. *this*: letter 3889.

5. *your reference to the managers*: this presumably refers to Bence Jones's comment that news of Tyndall's success in America deterred the RI's managers from cutting salary costs.
6. *five out of my six lectures at Philadelphia*: Tyndall delivered six lectures in Philadelphia from 11–22 November 1872 at the Horticultural Hall.
7. *quaker population*: see letter 3896, n. 5.
8. *my cousin*: Hector Tyndale.
9. *contempt for the politics of the day undisguised*: Tyndale had been a Union general in the American Civil War (1861–5) and had spoken in support of Ulysses S. Grant's nomination at the Republican nominating convention in 1868. The specific source of his contempt here remains unclear, though it is possibly a reference to the 1872 presidential election, in which Grant defeated Horace Greeley to gain a second term.
10. *Baltimore*: Tyndall delivered three lectures in Baltimore from 26–9 November 1872 at the Peabody Institute.
11. *Washington*: Tyndall delivered six lectures in Washington, DC, from 3–12 December 1872 at Lincoln Hall.
12. *New York*: Tyndall delivered six lectures in New York from 17–31 December 1872 at the Cooper Institute.
13. *Brooklyn*: Tyndall delivered six lectures in Brooklyn from 4–20 January 1873 at the Academy of Music.
14. *the £20000:* in letter 3875, Tyndall wrote that 'It is stated that in two years I could make £20,000 by lecturing in America'.

To Henry C. Carey[1] 23 November 1872 3901

Philadelphia[2] 23rd Nov 1872

My dear M[r]. Carey.

I went in search of my pinch of snuff last night but the staff in which I trusted was not there to comfort me.

But you have been faithful in the matter of the book[3] which reached me last night just as I was I was[4] stepping out to the lecture. Let me thank you warmly for it. Whether I agree with it or not I am sure to find in it vigour of thought & utterance.

Ever faithfully yours | John Tyndall

H. C. Carey Esq[uire] | &c &c &c

Historical Society of Philadelphia, Dreer Collection

1. *Henry Charles Carey*: Henry Charles Carey (1793–1879), an Irish-American economist, and lifelong resident of Philadelphia, who was a prominent advocate of the American School of economics (*ANB*).

2. *Philadelphia*: Tyndall delivered six lectures in Philadelphia from 11–22 November 1872 at the Horticultural Hall.
3. *book*: probably H. C. Carey, *The Unity of Law; as Exhibited in the Relations of Physical, Social, Mental and Moral Science* (Philadelphia: Henry Carey Baird, 1872).
4. *I was*: these words repeated by Tyndall.

To Thomas Archer Hirst — 23 November 1872 — 3902

Philadelphia.[1] 23rd Nov. 1872.

My dear Tom.

Philadelphia is struck off the list. I gave them my last lecture last night. The quaker element[2] is strong in this city and at first the audience to a man accustomed to applause might have felt clammy and cold. But their attention was unflagging, and even from the first I am informed their demonstrativeness was far above the average. But they warmed up by degrees until in the end they behaved like other Christians—and a heartier burst[3] I never heard than that which greeted my farewell last night.

The best of it is that I am stronger after the dispatch of my first two cities[4] than when I began my work.

The best people in Philadelphia I am told attended the lectures; and hospitality has met me at every turn during my stay. But I have been very careful. And although obliged to go out oftener than I intended, I firmly resisted all temptations that were likely to seriously impair my strength. Last night I spoke to the audience in a manner which I am sure no Englishman had previously employed, and they listened to me without a murmur. I repeated to them de Toqueville's account of their poverty as regards genius in the higher sciences:[5] I repeated his explanation of this, that had they been alone in the world they would have found that practical science could not long be cultivated with success without abstract science, but that the[y] drew their intellectual treasury from England. I took to pieces the claims of their practical men, and opened out to them the region of antecedent discoveries to which the practical men were no contributors, but from which they drew their supplies. In fact I was as plain with them as I could be—Still I managed to say it all without exciting the slightest irritation; not only so but I was frequently interrupted by loud approval.

Leidey,[6] a most solid fellow was here this morning to tell me how the lectures gratified him. All the Professors I believe attended the lectures.

And there have been people praying for me,[7] and writing to me about my soul and its future prospects,[8] at such a rate. Some of the letters so full of indescribable sweetness, that it is a pain to give such people pain by differing from them. Half a dozen newspapers have been drawing contrasts between William Tyndale and myself; much to my disadvantage of course, but on the whole not ill-naturedly.[9] The press of Philadelphia, as far as science is concerned, appears dull, commonplace, and ill-informed. Perhaps I judge hastily as I have cast an eye over only one or two papers, and they may not be fair specimens.

I am now packing up, and the instruments will be forwarded today to Baltimore;[10] there I can give only three lectures: after Baltimore comes Washington[11] and then New York[12] & Brooklyn.[13]

Miller,[14] my second assistant, has as I dare say you know, given way, and is now *[down]* in typhoid fever. Two of the best physicians in Philadelphia see him twice daily. They both warned me not to see him myself, and for two or three days I contented myself with speaking to his attendant: yesterday however I rigged up a respirator of cotton wool and glycerine,[15] which would keep out all the plagues of Egypt, and paid Miller a long visit. Nobody knows how such a thing may turn out[16]—But as yet no alarming symptom has made its appearance. It was a great drawback to me to lose his assistance, but Cottrell[17] has already taught a coloured man[18] to put the battery together, and I take him with me to Baltimore.

Philadelphia has been quite a home to me. The kindness of the people is surprising—I seem to be better known here than I sh[oul][d] be in an English City. The houses here are wonderfully clean and luminous. As far as the Kerbstone all is beautiful—Beyond it all is—I had almost said <u>vile</u>. The street pavement is atrocious. Thus the private citizen seems admirable, while the public contractor leaves dirt and roughness as a foil to the perfect neatness, beauty, and in some cases magnificence, of the dwellings.

Love to Lilly;[19] and warmest regards to Heins.[20]

Ever aff[ectionatel]y yours | John Tyndall

RI MS JT/1/T/696
RI MS JT/1/HTYP/608–9

1. *Philadelphia*: Tyndall delivered six lectures in Philadelphia from 11–22 November 1872 at the Horticultural Hall.
2. *the quaker element*: see letter 3896, n. 5.
3. *burst*: a vehement outbreak (of emotion or its expression) (*OED*).
4. *first two cities*: Boston, where Tyndall delivered six lectures from 15–25 October 1872 at the Lowell Institute, and Philadelphia (see n. 1).

5. *de Toqueville's account of their poverty as regards genius in the higher sciences*: on de Tocqueville, see letter 3785, n. 18.
6. *Leidey:* probably Joseph Mellick Leidy (1823–91), professor of anatomy at the University of Pennsylvania. Leidy was a medical doctor, one of the founders of American vertebrate paleontology, and an early supporter of Darwinian evolution in the United States (*ANB*; *CDSB*).
7. *people praying for me*: see letter 3871, n. 12.
8. *writing to me about my soul and its future prospects*: such letters are missing.
9. *drawing contrasts... not ill-naturedly*: William Tyndale (1494–1536) was noted for his translation of the Bible into English. Tyndall created minor controversy during his lecture tour with the suggestion that the efficacy of prayer could be tested experimentally. See letter 3780, n. 2.
10. *Baltimore*: Tyndall delivered three lectures in Baltimore from 26–9 November 1872 at the Peabody Institute.
11. *Washington*: Tyndall delivered six lectures in Washington, DC, from 3–12 December 1872 at Lincoln Hall.
12. *New York*: Tyndall delivered six lectures in New York from 17–31 December 1872 at the Cooper Institute.
13. *Brooklyn*: Tyndall delivered six lectures in Brooklyn from 4–20 January 1873 at the Academy of Music.
14. *Miller*: George Miller.
15. *I rigged up a respirator of cotton wool and glycerine*: Tyndall became interested in methods for filtering air as recently as 1870, when he used cotton wool to filter particulate matter out of the air in which he was conducting optical experiments. On the basis of that line of inquiry, he developed a respirator for use in firefighting, mines, and other environments with contaminated air, employing the cotton wool and glycerine combination described here. See J. Tyndall, 'On the action of rays of high refrangibility upon gaseous matter', *Phil. Trans.*, 160 (1870), pp. 333–65; J. Tyndall, 'Foul Air in Mines and How to Live in It', *Nature*, 5 (7 March 1872), pp. 365–6; and J. Tyndall, 'On some recent Experiments with a Fireman's Respirator', *Roy. Soc. Proc.*, 22 (1874), pp. 359–61. See also *Ascent of John Tyndall*, pp. 274–5.
16. *how such a thing may turn out*: see letter 3896, n. 10.
17. *Cottrell:* John Cottrell.
18. *a coloured man:* not identified.
19. *Lilly:* Lilly Hirst.
20. *Heins:* Heinrich Debus.

To Edward L. Youmans 23 November 1872 3903

PHILADELPHIA, *November 23, 1872.*

MY DEAR YOUMANS: The second ordeal has been passed,[1] and I believe successfully. The audience at first might have damped[2] a person who reckoned on applause, for the Quaker element is strong in Philadelphia, and Quakers eschew the clapping of hands.[3] But the attention was unflagging throughout. I drew heavily upon their patience, occupying them sometimes for nearly two hours. I did not see one yawn in the assembly, nor one mark of weariness from beginning to end.

They warmed up, moreover, and behaved very much as other Christians in the end. I hardly think any Englishman ever spoke so freely to an American audience as I did to mine last night. I repeated one of De Tocqueville's hardest sayings with reference to the poverty of their achievements in the higher walks of science.[4] I took to pieces the claims of the so-called practical man, not attenuating his merit in the slightest, but opening to their view a region of antecedent discovery to which practical men were not contributors, but from which they drew their supplies. I managed to say all this and a good deal more without exciting a murmur; nay, I was frequently interrupted by expressions of approval, and when I ended the burst of applause was as hearty as I have ever heard.

So this matter is past, and I am now preparing for Baltimore.[5] I have received innumerable requests and invitations to lecture,[6] and could I hope to be able to respond to them or any of them, I should send them to you and ask you to select from the many those that you think most suitable. But I see no hope of being able to prolong my visit beyond the end of January. I dare say I shall be pretty well used up by that time.

As regards science, the newspapers that I have glanced at here are very dull and poorly reported. Perhaps I have not seen the best of them. ...

Always yours, | JOHN TYNDALL.

E. A. Youmans, 'Tyndall's American Visit', *Popular Science Monthly*, 44 (February 1894), pp. 502–14, on pp. 510–11

---

1. *the second ordeal has passed*: Tyndall delivered six lectures in Philadelphia from 11–22 November 1872 at the Horticultural Hall.
2. *damped:* to depress, deject, discourage, check (*OED*).
3. *the Quaker element... clapping of hands*: see letter 3896, n. 5.
4. *De Tocqueville's hardest sayings... higher walks of science*: see letter 3785, n. 18.

5. *Baltimore*: Tyndall delivered three lectures in Baltimore from 26–9 November 1872 at the Peabody Institute.
6. *I have received innumerable requests and invitations to lecture*: in previous letters, Tyndall shared that he received invitations to lecture in Cincinnati, Cleveland, Columbus, Providence, and St. Louis, among other cities (see letters 3885 and 3886).

From Robert Winthrop 15 November 1872 3904

Brookline (Boston) | 25 Nov. 1872.

Dear Professor Tyndall,

There is nobody in New York[1] who will be more sure to put you in the way of the things, & the places, & the persons you would like to see, & ought to see, than my friend Mr. Schuyler.[2]

He is himself descended from one of the oldest Dutch Founders of New Netherlands,[3] while his grandfather[4] was one of our most distinguished Officers in that Revolution (shall we not call it Evolution?), which made us a Nation.

His wife is a grand-daughter, meantime, of Alexander Hamilton,[5] the friend & Financier, par eminence, of Washington,[6] & the ablest Statesman, our hemisphere has produced.

Let me commend Mr. Schuyler to your friendly consideration.

Yours faithfully, | Rob[er]^{t}. C. Winthrop.

Professor | John Tyndall, &c. &c.

RI MS JT/1/W/85

1. *in New York*: Tyndall delivered six lectures in New York from 17–31 December 1872 at the Cooper Institute.
2. *Mr. Schuyler*: a member of the prominent Dutch family, which was influential in New York politics (F. A. Christoph, *Schuyler Genealogy: A Compendium of Sources Pertaining to the Schuyler Families in America Prior to 1800* (Friends of Schuyler Mansion, 1987)); probably George Lee Schuyler (1811–90), who was at the time married to Mary Morris Hamilton, Alexander Hamilton's granddaughter (*not* great granddaughter).
3. *one of the oldest Dutch Founders of New Netherlands*: Philip Pieterse Schuyler (1628–83).
4. *his grandfather*: Philip John Schuyler (1733–1804).
5. *Alexander Hamilton*: Alexander Hamilton (1755/7–1804), a prominent American statesman and first secretary of the treasury (*ANB*).
6. *Washington*: George Washington (1732–99), a Revolutionary War general and the first President of the United States (*ANB*).

From McFadden Alexander Newell[1] 26 November 1872 3905

[State of Maryland, | Department of Public Education. | OFFICE OF THE STATE BOARD OF EDUCATION, Baltimore,] Nov[embe]r 26th [*18*]72

John Tyndall Esq—

Dear Sir,

The Association of School Commissioners of Maryland will pay an official visit to the State Normal School[2] tomorrow (Wednesday) about noon; and I take the liberty of asking you to join them.

The deep interest which I know you take in educational matters, and the great need which we have of a revival, are the only apology I can offer for hinting that if you could find time to say a word to us on the importance of the scientific element in education, it w[oul]^{d} be a great favor and might be the means of doing much good. But if you do not desire to "talk", and will give us merely your presence the favor will be highly appreciated.

Yours, very respectfully, | M. A. Newell | President.

RI MS JT/1/N/8

1. *McFadden Alexander Newell*: McFadden Alexander Newell (1824–94), an Irish immigrant with a degree from Trinity College, Dublin, served as the first principal of the Maryland State Normal School from 1866–90 (Towson University Special Collections and University Archives).
2. *State Normal School:* Maryland State Normal School was established in Baltimore in 1866 as a teacher-training institute. In the early twentieth century it moved to Towson, Maryland, and underwent several name changes. It is now Towson University.

To Joseph Henry [27 November 1872][1] 3906

Mount Vernon Hotel Baltimore[2] Wednesday

My dear Professor Henry

I cannot tell you how heavily I agree with your estimate of M^{rs}. Russell Gurney[3] who is an old and an excellent friend of mine. I do not know a sweeter character.

And now with regard to lodging please hear me out.[4] I am terribly pressed by the lectures. They render my habits irregular. I am also not equal to the excitements of society. At home I steadfastly set my face against dining out either on the lecture evening itself or on the evening before. In the present sensitive state of my brain a glass of the wrong wine would upset me for a day. Now all these things taken together render me desirous to take up my quarters at a hotel, if I can only get a quiet one. I have not got into the American habit of dining early and taking tea in the evening, nor can I get into it. I dine at about half past six, and at that hour nourish my brain with a pint of La Rose.[5]

In Washington[6] of course I must relax my English rule; but if the lectures are to be successful I must be very careful. What I propose then is this, that you would be good enough to secure a perfectly quiet bedroom and sitting room for me in some hotel, regardless of expense. I should then arrange to join your parties whenever you wished me to do so.

Do not consider me churlish or unsocial. My reason for writing this is that I want to cook myself up so that my brain shall be strong and my temper free to do with effect the work that you would wish me to do.

I spoke out my mind emphatically in Philadelphia,[7] and what I said was surprisingly well received—No Englishman I think has spoke to an American audience so plainly as I did to mine. But they known that I spoke as a friend and tolerated me accordingly. I should like to speak equally strongly at Washington.

Yours ever | John Tyndall

Were my cousin[8] in the house he would thank you gratefully for your offer of hospitality.

John Tyndall to Joseph Henry, 27 November 1872, mssRH 3972, box 43, William Jones Rhees papers, The Huntington Library, San Marino, California

---

1. *[27 November 1872]*: the date is given by reference to Tyndall being in Baltimore during his American lecture tour.
2. *Baltimore*: Tyndall delivered three lectures in Baltimore from 26–9 November 1872 at the Peabody Institute.
3. *M^rs^. Russell Gurney*: Emelia Russell Gurney (née Batten, 1823–96), wife of Russell Gurney (see letter 3917, n. 5). She was a social reformer who helped to establish the Kensington Society, an influential group in the suffragist movement in the 1860s (*ODNB*).
4. *with regard to lodging please hear me out*: Tyndall declined Henry's invitation to stay with him while in Washington (see letter 3907 n. 2).
5. *La Rose*: Tyndall's favorite wine.
6. *Washington*: Tyndall delivered six lectures in Washington, DC, from 3–12 December 1872 at Lincoln Hall.
7. *Philadelphia*: Tyndall delivered six lectures in Philadelphia from 11–22 November 1872 at the Horticultural Hall.
8. *my cousin*: Hector Tyndale.

To Joseph Henry 30 November 1872 3907

Baltimore[1] 30th. Nov. 1872.

My dear Prof. Henry

I thank you much for your kind compliance with my desires,[2] though it would grieve me sincerely if you attributed my desires to any lack of sociability on my part.

I have received sad intelligence this morning from Philadelphia. My young assistant to whom I was much attached died there this morning at 7:15.[3]

Would you kindly see that there is an intelligent carpenter at hand when my assistant Mr. Cottrell[4] arrives at the Lecture Hall[5] on Monday morning?[6]

Would Mr. Rhees be good enough to take a comfortable room for him somewhere near the hall. I should like him to be comfortable, and above all things to have good food.

Ever faithfully yours | John Tyndall

Finally, would you kindly have a carboy of fresh nitric acid at the Hall by the time my assistant arrives. We employ Grove's battery.[7] In our first lecture here the battery sank rapidly[8] because of the defect of the nitric acid. I believe it had been used before.

Nothing could be kinder than my reception, nor more cordial than my leave-taking at Baltimore. I gave them a bit of my mind regarding original investigation.

John Tyndall to Joseph Henry, 30 November 1872, mssRH 3973, box 43, William Jones Rhees papers, The Huntington Library, San Marino, California

---

1. *Baltimore*: Tyndall delivered three lectures in Baltimore from 26–9 November 1872 at the Peabody Institute.
2. *compliance with my desires*: Henry invited Tyndall to stay with him in Washington; Tyndall indicated he was likely to decline, in favor of a hotel. See letters 3876 and 3906.
3. *My young assistant... morning at 7:15*: see letter 3896, n. 10.
4. *Cottrell*: John Cottrell.
5. *lecture hall*: of the Peabody Institute in Baltimore.
6. *an intelligent carpenter... Monday morning*: to assist in preparing Tyndall's demonstrations, and not to fashion a coffin, as the context might suggest. See letter 3741.
7. *Grove's battery*: a fuel cell invented by William Robert Grove (see letter 3897, n. 14) in 1838. It employed platinum and zinc electrodes with the former in nitric acid and the latter in zinc sulfate.
8. *the battery sank rapidly*: the battery quickly ceased to produce a current.

To William Rhees 30 November 1872 3908

[The Western Union Telegraph Company. | The rules of this Company require that all messages received for transmission, shall be written on the message blanks of the Company, under and subject to the conditions printed thereon, which conditions ha<v>e been agreed to by the sender of the following message. | Anson Stager, Gen[era]'l Sup[erintenden]'t, Chicago, Ill. 26} William Orton, Pres[iden]'t, O. H. Palmer, Sec[retar]'y} New York.][1]

Dated Balti[more] Nov 30 1872 | Received at N.E. cor[ner]. 14th St. and Penn[sylvani]a. Ave. Washington, D.C. 12 | To W[illia]m. J. Rhees | Smithsonian Inst[itution].

Please take a Comfortable room for Cottrell[2] in a good Hotel & telegraph the address.

John Tyndall
15 paid EJ

Smithsonian Institution Archives. RU 32, Box 1, Folder: Tyndall, J.

1. *[The Western... New York]:* boilerplate text on the telegram.
2. *Cottrell:* John Cottrell.

To William J. Rhees 1 December 1872 3909

1st Dec. 1872

Dear Mr. Rhees.

It is exceedingly important that the Carpenter,[1] and a handy man as well, should be there when my assistant[2] arrives. The labour of unpacking and arranging is immense.

Pray procure for us sometime tomorrow a small quantity of the cyanide of potassium.[3]

I have mentioned the acid to Prof. Henry.[4] Nitric & sulphuric will be both required for the battery.

Thankfully yours | John Tyndall

The death of my young assistant[5] is a grievous calamity. I suppose if advisable I can have Cottrell at Welckers?[6]

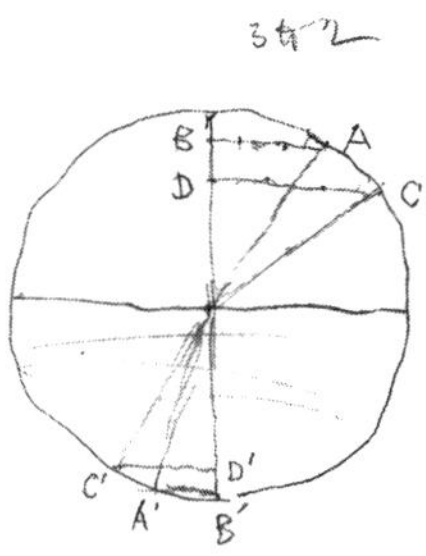

I want to refer historically to the law of optics & should feel exceedingly obliged if you would kindly get somebody to make the enclosed sample diagram for me. It must be large enough to be well seen—say a circle of three feet in diameter & strong lines.

Should the Hall contain a large white screen it will be useful

John Tyndall to William Jones Rhees, 1 December 1872, mssRH 3974, box 43, William Jones Rhees papers, The Huntington Library, San Marino, California

---

1. *Carpenter*: see letter 3907, n. 6.
2. *my assistant*: John Cottrell.
3. *cyanide of potassium*: in modern terminology, potassium cyanide, or KCN. It is unclear how Tyndall intended to use it, but it was often used as a fixing agent in the photographic development process and might have played a role in his optical demonstrations.
4. *mentioned... to Prof. Henry*: see letter 3907.
5. *death of my young assistant*: see letter 3896, n. 10.
6. *Welckers*: the Welckers Hotel in Washington, DC.

To Edward L. Youmans 1 December 1872 3910

BALTIMORE, *December 1, 1872.*

Nothing could be more genial and sympathetic than my reception at Baltimore.[1] They declare the lectures entirely successful. Both at Philadelphia and here I have spoken very strongly about their duty as regards scientific investigation.[2]

E. A. Youmans, 'Tyndall's American Visit', *Popular Science Monthly*, 44 (February 1894), pp. 502–14, on p. 511

1. *reception at Baltimore:* Tyndall delivered three lectures in Baltimore from 26–9 November 1872 at the Peabody Institute.
2. *their duty as regards scientific investigation:* a reference to Tyndall's mission to impress the importance of pure scientific investigation upon his American audiences.

From Thomas Archer Hirst 1 December 1872 3911

University of London | Sunday Dec[embe]r 1st 1872

My dear John.

I was rejoiced to learn from your letter, written at Niagara,[1] that you found your health improved after the exertions at Boston,[2] and that you felt quite equal to your new work at Philadelphia.[3] Yesterday your short letter, written on the 13th,[4] reached me and conveyed to me the first news of your having reached Philadelphia—for somehow or other no Philadelphia papers have yet arrived. English papers have made everyone acquainted with your great success at Boston, and most extravagant rumours are afloat about magnificent offers having been made to you with a view of enticing you away from the Mother Country.[5] I have no fear of their doing so. In the Pall Mall not long ago was the intelligence that Special Prayers were about to be offered up for you in New York.[6] I was asked more than once if you were ill or if any accident had happened to you at Niagara, or if they were merely praying for your conversion. Of course I answered that they were about to make an experiment just as impious as the celebrated "prayer guage" experiment[7] against which they were howling not long ago.

I will attend tomorrow to your request about the cheques.[8] I enclose all the letters which have come since I last wrote.[9] I have cut off your autograph from one of your letters to me and sent it to the lady at St Leonards.[10] I enclose also an important letter I received two days ago from Huxley. I spent a wakeful night thinking about it and longing to have you to go to for advice. I have replied to Huxley and Spottiswoode that although I shrink rather from the absorption of time the worry and the expenditure of force which such a position would involve I feel that I should not be justified in at once declining to entertain the offer and that I shall accordingly be glad to have an interview with Mr Goschen[11] with a view of learning more thoroughly what his intentions are and of ascertaining whether the Direction of the Studies at Greenwich would or would not be compatible with the prosecution of my own studies and researches. If not I should feel it my duty to decline the offer; if it should prove, however, that the Admiralty are willing to give me a sufficiently complete staff to work with and sufficient liberty as far as my taking an active part in lecturing &c is concerned I should be prepared to accept the office.[12]

Both Spottiswoode and Huxley urged me strongly yesterday, when we met at the Royal Society, to accept the position, which they consider to be unique in England and of the highest importance so far as the Higher Scientific training of naval officers is concerned.

How I wish you were at home again! I do not know how soon they may expect an answer from me, but if I can defer giving my final answer until I have had time to hear from you I will. I greatly fear however that for once I must take a very important step without advice or approval from you. The Royal Society Anniversary took place yesterday.[13] Huxley was unanimously elected to fill Sharpey's place, and the President gracefully alluded to the change *[in]* the permanent staff.[14] He also publicly announced his intention of resigning his own office[15] before this time twelvemonth. Already there is great talk of the Duke of Devonshire[16] as his successor. At the next X[17] however we are going to take into consideration whether the time has not come for bringing forward Hooker[18] for President. There would be an advantage if you would express your opinion on the matter in your next letter. Love to Hector[19] and kindest regards to Mrs Tyndale.[20]

Ever yours affectionately | T. Archer Hirst

I wrote to your sister[21] and had a letter in reply asking me to procure one of your respirators[22] for Dr Tyndall[23] who is not well. I have bought one at Bells and sent it to her.

RI MS JT/1/H/253

1. *your letter, written at Niagara:* letter 3886. On Tyndall's visit to Niagara Falls, see letter 3738, n. 3.
2. *exertions at Boston*: Tyndall delivered six lectures in Boston from 15–25 October 1872 at the Lowell Institute.
3. *new work at Philadelphia*: Tyndall delivered six lectures in Philadelphia from 11–22 November 1872 at the Horticultural Hall.
4. *your short letter, written on the 13th*: letter 3893.
5. *extravagant offers... Mother Country*: one such offer was an invitation to apply for the Chair of Physics at the University of Pennsylvania (see letters 3881 and 3883).
6. *the Pall Mall... New York*: 'Reuter's Telegrams', *PMG*, 22 November 1872, p. 6. The extent of the entry is: 'The New York papers report that special prayers are being offered for Professor Tyndall, who is now in America'. Tyndall created minor controversy during his lecture tour with the suggestion that the efficacy of prayer could be tested experimentally. See letter 3780, n. 2.
7. *"prayer guage" experiment*: a reference to a suggestion Tyndall made in one of his Boston lectures that the efficacy of prayer could be tested by scientific means, which caused considerable agitation in the American press. On Tyndall and prayer, see letter 3780, n. 2. The misspelling of 'gauge' is Hirst's.

8. *your request about the cheques*: see letter 3893.
9. *I enclose all... I last wrote*: letters not identified.
10. *the lady at St Leonards*: not identified.
11. *Goschen*: George Goschen (1831–1907), the First Lord of the Admiralty, 1871–74, who in this capacity was involved in establishing the Royal Naval College, Greenwich, in 1873 to provide advanced training for British Naval Officers (*ODNB*).
12. *accept the office*: Hirst was hired as the first Director of Studies of the Royal Naval College, Greenwich, serving in that position until February 1883 (R. Wilson and J. H. Gardner, 'Thomas Archer Hirst at Greenwich, 1873–1883', in R. Flood, T. Mann, and M. Croarken (eds), *Mathematics at the Meridian: The History of Mathematics at Greenwich* (New York: CRC Press, 2020), pp. 119–38).
13. *The Royal Society Anniversary*: the RS held General and Anniversary meetings each year, the Anniversary meetings occurring in November, as the Society was originally chartered on 28 November 1660. At these Anniversary meetings, which take place on 30 November each year, the President of the RS summarized the year in science and its annual awards were presented.
14. *Huxley... permanent staff*: Huxley became Secretary of the RS, succeeding the physiologist William Sharpey (1802–80). Hirst, at the time, was a Vice-President of the RS, under President George Biddell Airy.
15. *resigning his own office*: see letter 3897, n. 12.
16. *the Duke of Devonshire*: William Cavendish (see letter 3897, n. 13).
17. *the next X*: the next meeting of the X Club (see letter 3757, n. 2).
18. *Hooker*: Joseph Hooker, who indeed succeeded Airy as President of the RS.
19. *Hector*: Hector Tyndale.
20. *Mrs Tyndale*: Julia Tyndale.
21. *your sister*: Emily Tyndall.
22. *respirators*: see letter 3902, n. 15
23. *Dr Tyndall*: John Tyndall (of Gorey).

From Henry Bence Jones — 3 December 1872 — 3912

Tuesday | Dec 3 1872

My dear Friend

Your telegram to Spottiswoode reached him at the Royal Institution dinner on Saturday & Sir H. Holland came to me with the sad news[1] on Sunday I found nothing from Spottiswoode had reached the Institution on Sunday evening & so I begged Miss Savage[2] on Monday morning herself to go to Mrs Miller[3] & break the sad news. Miss Savage found them reading your letter of the 16th[4] saying he was ill & she made known to them the sad end. I have not heard of them since.

Your letter of the 16th to me[5] also came yesterday morning so I gave the managers & members news of you up to that day. At the members meeting when I said you w[oul]d not be home so soon as you had at first thought Mr Lee!![6] wanted that leave of absence sh[oul]d be granted you till 187<u>4</u>. I am sure he meant it in kindness & not that he might direct the servants &c as I think he tries to do in your absence but "one fears the Greeks" &c.[7] I had told the managers what I wrote to you to Washington[8] & they all hope you will do what you like best.

We all feel however that we want you home & that the place is nowhere without you. Busk is going to keep guard on Friday evenings when the lectures begin as my chance of being out is small.[9] Above all I miss you & how I am to do anything to forward your laboratories[10] I do not know. I crawled over them yesterday & stood mazed.[11] They look so splendid in size, such places. I felt overwhelmed. They have got on famously in the three weeks since I was here & promise to finish in another fortnight or three weeks at most, so far as bricks & mortar & such things are concerned. Really they will be noble places oh for you to live forever & to work there <u>always</u> & to make new & grand additions to human knowledge. At present I am overpowered with the size & look of the whole & the thought of what it must cost to keep such a machine in working order & then to provide for the washmen. If I had no children I should at once endow the place with some £30000 the Clarendon has been sold to Peter Robinson a great Haberdasher for 70000.[12] So I suppose we shall have an immense Redmayner[13] built there.

I have got some medicines to act at last & I have again got tolerably free from the worst of the dropsy[14] but I am still very swelled & very weak & down & cannot stop the *[drugs]* in the least without the trouble returning. Still I am decidedly better & I went thro' the managers meeting on yesterday quite as well as I would expect. There is nothing new to tell you but we all long for you to be safely back and glad shall we be to hear that you have taken your place. As to the American temptations[15] I cannot think of you ending your days there. You do not want money & you have no family to start in the world & you can do exactly what you please in London for the rest of your days. 2000 a year in America can hardly be more than 1000 here & tho' prices will fall there & rise here I tell you are too old now. So come back soon & never go away again.

We ought to open the new laboratories with a ball. In the lower room dancing & in the upper room refreshments & whist &c. My what! places they are & what! they must cost before they are done I am sure you will get so much glorification in America that you will find the old Country uncommonly stupid when you return. As I do not go out of course I hear very little indeed & all the Atheneum[16] talk is lost to me. Miss Barnard[17] & Mrs Thistlethwayte[18] try to convert me to the truth before I die but I fear they will give me up soon. Sir H. Holland dines out every day he can, but as far as I can hear never learns anything.

Goodbye dear friend | Ever yours | alive or dead | H B J

RI MS JT/1/J/72

1. *sad news*: see letter 3896, n. 10.
2. *Miss Savage*: Sarah Savage.
3. *Mrs Miller*: George Miller's mother.
4. *your letter of the 16th*: letter to the family of George Miller is missing.
5. *Your letter of the 16th to me*: letter 3896.
6. *Mr Lee*: probably Thomas Lee.
7. *"one fears the Greeks"*: a reference to the Latin phrase from Virgil's *Aeneid*, 'Timeo Danaos et dona ferentes' (I fear the Greeks, even when they bear gifts).
8. *I wrote to you to Washington*: letter missing.
9. *keep guard... being out is small*: see letter 3897, n. 11.
10. *forward your laboratories*: a reference to the laboratory facilities then under construction at the RI (see letter 3776, n. 14).
11. *mazed*: stupefied, dazed; insane, crazed; bewildered, confused (*OED*).
12. *the Clarendon... 70000*: *The Standard* for 23 November 1872, records the sale of the Clarendon Hotel and the German Gallery at 168–9 New Bond Street, immediately behind the RI, for £70,000. The report appears above a notice of the sale of a life insurance policy to 'Messers. Robinson, Son, and Fisher', evidently leading Bence Jones to believe that the building had been sold to Peter Robinson, a London merchant and clothier. 'Estate Exchange Report, Nov. 22', *The Standard*, 23 November 1872, p. 2. See letter 3793, n. 5.
13. *Redmayner*: not identified.
14. *dropsy*: see letter 3782, n. 18.
15. *American temptations*: one offer was an invitation to apply for the Chair of Physics at the University of Pennsylvania (see letters 3881 and 3883).
16. *Atheneum*: the Athenaeum Club (see letter 3710, n. 5).
17. *Miss Barnard*: Jane Barnard.
18. *Mrs Thistlethwayte*: not identified.

From Elizabeth Rutherford 3 December 1872 3913

13 Shandwick Place | Edinburgh | 3^d. December 1872.

My Dear Friend,

Your letter of the 16th ult[i]^m[o][1] which reached me yesterday, gave me great pleasure.—it arrived as all your letters have done, (there is a fatality in it) at a time when I stood in need of consolation.—and to tell you the truth,

I had wearied & longed for that letter. Well, I am glad you are having some enjoyment of Nature,[2] as well as prospering triumphantly in the object for which you went to America[3]—No one, who has the slightest knowledge of your character could for a moment suppose that your purpose was to make money. At the same time, I hope that your very handsome preparations for your Lectures will not cause you to be a loser.—The account of the 1st. lecture at Boston,[4] copied from an American paper into the Times,[5] amused me much—The description of yourself, your manner etc—was very good, but when it adds that you were "clad in full evening costume" I thought it very comical, being accustomed to read of those clad in Purple & fine Linen or in Velvet & Ermine.—and then the audience rushing out "to hunt up new adjectives with which to praise you" was delightful. Your English accent is commented on—they could hardly expect you to have an American one. I have to thank you, which I do heartily for sending my Boy[6] a beautiful copy of your last work.[7] He was as proud as Punch,[8] but very desirous of having your autograph in it, which I undertook for, in the course of time—

Had I been more happy, or less unhappy than I have been since before I received your kind letter of farewell in Sept[embe]r.[9]—I should have addressed a letter in grandiose style to "À mon Ami | Le Grand Physiciste d'Angleterre | en Amèrique".[10] but I had not spirit.

It would be worse than vain to oppress you, with a detail of my anxieties & sorrows—only now beginning to subside,—suffice it to say that the least of them was our leaving Dalhousie Castle[11] for if ever & for aye—I had 6 years ago furnished a very charming apartment there & I had hoped I might die with my household Gods around me—but my Son-in-law,[12] seeing that from different circumstances his family could only be resident there for 2 or 3 months during the year, and that the place was very expensive, resolved to give it up; I thought he was quite right & therefore did not allow him to see how much I regretted it—but it was a pang—however as dear little Johnnie says "A thing of Beauty is a joy for ever"[13] and I can remember it with pleasure. One of my dreams was, that in 1874, if I live so long, that when with the Association you would be in Scotland, you would have visited me there.[14] How I do detest the Quakers! Cold conceited, selfish sect![15]—I have just finished Hepworth Dixon's[16] new Life of William Penn,[17] an interesting and rather melancholy book. By the way, what a mistake that man has made in going to law with the Pall Mall.[18] Silence is often a great virtue. I read his Spiritual Wives, when it first came out, and gave him credit for a good intention in publishing it but I suppose I was wrong—Any-how, the Trial was worse than the Book—so disgusting one could not read it. I am now in the midst of Forster's[19] life of Dickens IId Vol[20]—I[t] interests me because I knew Dickens well—Forster also—But oh! how few people know how to write a Life or Biography. You do

& I do—and sometimes I wish that you had been as much my senior as I am yours, that I might have written "Tyndall, an *[*Idea*]*". You will say Thank ye for nothing. It was Sadi who said "Ce n'est qu'en laissant s'écouler un long espace de temps que l'on arrive à connaitre à fond la personne qu'on étudie".[21]—but I feel sure that I know you very well and Huxley or one of your other friends would have done the Scientific part—

I believe if you had remained much longer at Niagara,[22] you would have jumped into the fall—that sort of fascination is very dangerous—I wish you had time to go to the Sierra Nevada & to the top of Mount Tyndall & to see the Yosemite Valley[23]—& I hope you will meet your worshipper Clarence King.[24]—If, or when we meet again you will tell me all I want to know, but I hope you will write to me again & tell me when you are obliged to be at home in London—

I won't speak of the fire at Boston nor of the miserable French affairs, nor of the misdeeds of our own weak Government.[25] All these matters are in the Newspapers—and now, I will say Adieu & God Bless you—I am better, but I have been very unwell—and I am a little tired.

Believe me, | Yours Sincerely & Affectionately | Elizabeth Rutherford

Our weather is & has been deplorable, never very cold—but constant rain, the ground is saturated with water so that it is impossible to put the seed in the ground, or to do any farm work.

RI MS JT/1/R/88

---

1. *Your letter of the 16th ult[i]m[o]*: of the 16th of the previous month, 16 November 1872; letter missing.
2. *enjoyment of Nature*: probably referring to Tyndall's visit to Niagara Falls (see letter 3738, n. 3).
3. *object for which you went to America*: for Tyndall's American lecture tour, see Introduction, n. 8.
4. *Boston*: Tyndall delivered six lectures in Boston from 15–25 October 1872 at the Lowell Institute.
5. *the Times*: 'Lectures on Heat at Boston', *Times*, 15 November 1872, p. 4.
6. *my Boy*: Thomas Rutherford.
7. *your last work*: possibly *Forms of Water*.
8. *proud as Punch*: or 'pleased as Punch', an idiom meaning to be extremely pleased; this comes from the Punch and Judy puppet shows, Punch being short for Punchinello, a character fiendishly pleased with himself for his evil actions. The show was popular in Europe in the eighteenth and nineteenth centuries, and in England, the character's image was used by the satirical magazine *Punch*.

9. *your kind letter of farewell in Sept[embe]r*: letter missing.
10. "*À mon... en Amèrique*": To my friend | the great physicist of England | in America (French).
11. *Dalhousie Castle*: see letter 3748, n. 1.
12. *Son-in-law*: not identified.
13. *dear little Johnnie says "A thing of Beauty is a joy for ever"*: 'Johnnie' here is the poet John Keats (1795–1821), who published his poem *Endymion* in 1818, which begins with the line quoted here.
14. *with the Association... visited me there*: a reference to the possibility of Tyndall traveling though Scotland en route to the August 1874 meeting of the BAAS annual meeting in Belfast, where as President he delivered his famous Belfast address.
15. *the Quakers! Cold conceited, selfish sect!*: Tyndall's prior letter to Rutherford is missing, but in correspondence with others around this period he routinely referenced the characteristically quiet audiences in Philadelphia, which had a large Quaker population. See letter 3896, n. 5.
16. *Hepworth Dixon's*: William Hepworth Dixon (1821–79), an English journalist and historian (*ODNB*).
17. *Life of William Penn*: W. H. Dixon, *Life of William Penn: Founder of Pennsylvania* (London: Hurst and Blackett, 1872).
18. *a mistake... with the Pall Mall*: Dixon sued the *PMG* for libel after the paper accused him of indecency over his 1868 book *Spiritual Wives*. On 29 November 1872, a few days before this letter was written, Dixon won the judgement, but was awarded damages of only one farthing (the smallest unit of British currency at the time, equivalent to a quarter of one penny) ('Law Report', *Times*, 30 November 1872, p. 11).
19. *Forster's*: John Forster (1812–76), an English biographer and literary critic (*ODNB*).
20. *Life of Dickens II[d] Vol*: the second volume (1873) of J. Forster, *The Life of Charles Dickens*, 3 vols (London: Chapman and Hall, 1872–4).
21. *Sadi who said "Ce n'est... étudie"*: the quote translates: 'Only by letting a long span of time pass can you know the person you are studying inside-out'. It was attributed to the thirteenth-century Persian poet Saadi (1210–91/2), using this French phrasing, by the French literary critic Charles Augustin Sainte-Beuve (1804–89), in his *Nouveaux lundis*, vol. 13 (Paris: Michel Lévy Frères, 1872), p. 121.
22. *Niagara*: see letter 3738, n. 3.
23. *Sierra Nevada & to the top of Mount Tyndall & to see the Yosemite Valley*: in California.
24. *meet your worshipper Clarence King*: the American geologist Clarence King named a peak in the Sierra Nevada of California after Tyndall (see letter 3719, n. 3).
25. *fire at Boston... weak Government*: the references to other current events are ambiguous, but the fire is a reference to the Great Boston Fire of 1872 (see letter 3897, n. 17).

From Susan Lesley[1] [3 December 1872][2] 3914

Tuesday A.M.

Dear Professor Tyndall,

I send you this little Photograph, because I value it, I had nothing else I thought you would value as much. There were but six taken, but my niece has one of those, and at a future time, it can be reproduced.

I have no words to tell you of our grief & sorrow for your sorrow.[3] Mr Lesley[4] sends you his warmest remembrances & says "Tell him I am so sorry". And we look back with self reproach that the fear of being officious prevented us from trying to help you.—Dear Mrs. *[*dale*]*[5] was planning the most-devoted care, as soon as the Dr would allow her, and she has sorrowed as for a relative. But I will not grieve you by more words—Only be sure of our entire sympathy.

Your sincere friend | Susan I. Lesley

RI MS JT/1/L/18

1. *Susan Lesley*: see letter 3874, n. 9.
2. *[3 December 1872]*: the date is estimated from the timing of George Miller's death on 30 November 1872, a Saturday, making the following Tuesday 3 December. The news would have spread quickly in Philadelphia.
3. *your sorrow*: the death of his assistant George Miller (see letter 3896, n. 10).
4. *Mr Lesley*: Joseph Lesley.
5. *Mrs. [dale]*: not identified. If the name is 'Dale', Lesley did not capitalize it; it could possibly be the end of the name Tyndale, for Julia Tyndale, wife of Hector Tyndale who helped oversee Tyndall's visit in America. In a later letter to Hector, Tyndall wrote that he received 'an exceedingly interesting letter from Miller's father in which he speaks with boundless gratitude of Gen. Tyndale & his "loving lady"' (letter 3978).

From Asa Gray 4 December 1872 3915

[BOTANIC GARDEN, | CAMBRIDGE, MASS.] | Dec. 4, 1872

Dear Prof. Tyndall

My neighbour and friend Mr. Hubbard,[1] who has received kind attention from near friends of yours in England, desires to make your acquaintance, and I take pleasure in paving the way by this introduction, which he will deliver to you. I had to-day letters from Dr. Hooker, in which you were affectionately mentioned.

Believe me to be, Very sincerely yours | Asa Gray

RI MS JT/1/G/14

1. *Mr. Hubbard*: possibly Samuel Hubbard Scudder (1837–1911), an American entomologist and colleague of Gray's at Harvard, who was a supporter of Darwinian ideas (*ANB*; *CDSB*). Scudder lived at 156 Brattle St. in Cambridge, about ¾ mile walk from Gray's residence at 88 Garden St.

From Washington Chew van Bibber[1] 4 December 1872 3916

Baltimore, Wed. night, Dec 4th 1872, Prof. John Tyndall

My dear Sir,

What I am about to say you will please excuse, if it is in your judgment presumptuous because it is actuated only by the kindest feelings.

I regret the remarks made in the Sermons,[2] because I truly wish you to appear in the right light, &, in a pleasing light before the American public.

Now I really think the great cause of difference is the illustration you accidentally chose. You will find a different feeling here in this country, from what exists in Europe in the way of Democracy.

The bare idea of leaving patients with smallpox in a hospital without attention is a startling one here.[3] I know it would not be so elsewhere. From the comments I have heard more cant[4] concerning the subject, I conclude the matter of prayer is altogether a secondary idea. This would not have given rise to the ill-natured comments it has, had it not have been for the illustration. Therefore, I think your real feelings are misrepresented before the public. So consult with Prof Henry, & ask him if some explanation from you is not due to issue?

Let me give you some figures to show the power of the press here. This question originated in the Phil[adelphi]a Public Ledger, did it not? That sheet[5] claims a circulation of 84,000 daily. More than the N.Y. Herald, Times, Tribune, /Continent/. The Balt[imore] Sun claims 42,000 as its circulation. The two papers were once partners & are now great "chums". If you are misrepresented by these sheets I would really like you to be properly represented. You might therefore I think with great prosperity say in a note to the "Sun", the Ledger or both that "in the comments which had been made from the Pulpits & elsewhere concerning your remarks upon prayer. Your feelings had been misrepresented" or something to that effect.

The matter however need give you no further concern than this, that you, like everyone, would no doubt wish to be properly understood before a large audience.

With much respect | yours truly &c | W. C. van Bibber MD | 47 Franklin St.

RI MS JT/1/V/1

1. *Washington Chew van Bibber*: Washington Chew van Bibber (1824–92), a Baltimore physician who published widely in American medical journals in the late 1800s. He was Secretary of the Medical and Chirurgical Faculty of the State of Maryland from 1857–9, among many other appointments and memberships (E. F. Cordell (ed), *The Medical Annals of Maryland, 1799–1899* (Baltimore: Medical and Chirurgical Faculty of the State of Maryland, 1903), p. 603).
2. *remarks made in the Sermons*: probably a reference to the attacks from religious leaders Tyndall faced in the United States following his remarks about testing the efficacy of prayer. See letter 3780, n. 2.
3. *The bare idea... startling one here*: Tyndall proposed a 'prayer-gauge' experiment in which one group of ill patients would be provided with proven treatments, while a second group would only receive prayers for their improved health. See letter 3780, n. 2.
4. *cant*: the peculiar language or jargon of a class (*OED*).
5. *sheet*: contemporary slang for a newspaper.

To Henry Bence Jones 5 December 1872 3917

Washington. 5th. Dec. 1872

My dear Friend

I have time only for a few words.

I have been sorely smitten by the death of Miller.[1] The body has been embalmed and sent to England.

He was mourned like a relative in Philadelphia.

The lectures at Baltimore[2] were I think entirely successful.

I plunged into my work here[3] yesterday. The hall was crowded. I send you a newspaper scrap or two. I also send you the allusion referred to by the New York Times, to the establishment of friendly relations between England and America,[4] it was exceedingly well received.

Russell Gurney[5] and his excellent wife[6] are here, but I have not yet had time to see them. I dine with them tomorrow.

Health fair.

Weather thus far wonderful—clear, sunny, bracing. But we had a terrible downfall of temperature a few days ago—I liked it

I hope you are better. I shall be anxious until I hear from you.

Ever Your friend | John Tyndall

RI MS JT/1/TYP/3/814
LT Typescript Only

1. *death of Miller*: see letter 3896, n. 10.
2. *Baltimore*: Tyndall delivered three lectures in Baltimore from 26–9 November 1872 at the Peabody Institute.
3. *my work here*: Tyndall delivered six lectures in Washington, DC, from 3–12 December 1872 at Lincoln Hall.
4. *allusion referred to by the New York Times*: 'Washington Notes', *New York Times*, 4 December 1872, p. 1.
5. *Russel Gurney*: Russell Gurney (1804–78), a Member of Parliament representing Southampton. He was elected FRS in 1875 (*ODNB*).
6. *his excellent wife:* Emelia Russell Gurney (see letter 3906, n. 3).

From William Spottiswoode — 6 December 1872 — 3918

50 Grosvenor Place[1] | 6 December. 1872

My dear Tyndall

Your two telegrams[2] came safely to hand. The first arrived, sent on by Mrs Spottiswoode,[3] just at the close of the Royal Society Anniversary Dinner.[4] Unfortunately I have been shut up all this week with influenza, except that on Monday, I went to the R[oyal]. Inst[itu]t[io]n monthly meeting & to one or two other places of imperative business. Miss Barnard[5] was good enough to carry the sad intelligence to Mr & Mrs Miller;[6] & today, the first day that I have been about, I have been myself to see them. Your second telegram arrived last night, so I was able to tell them of your last act of thoughtful kindness. They were much touched by all that you have done for their boy, not only in this last stage of his life but all through the years during which he has been with you. And they begged me to express at the first possible moment their gratefulness. They begged me also to send word to Cotterill[7] how much obliged they were for his letters.

I doubt not that if there be anything more wherein I can be of use in these (as in all other) matters you will let me know.

Hirst is here teaching Hughie[8] the first steps of Mathematics; & I must go immediately & ask your address, as it is just post time.

We have all rejoiced at your success; & sympathize with you on this shadow thrown over its brightness. But beside what we gather from newspapers, & what you are able to tell in letters, we should like to know a great deal more of your doings & experiences. We should like to know in short what a school boy calls "all about it".

We shall have plenty to tell you about the Royal Society &c when you come back; but at present I need only mention (what you no doubt already know) that Huxley is fairly installed as secretary to the Society.[9]

Pray remember me kindly to Cotterill; & with benediction to the prisms, believe me

Ever your sincere friend | William Spottiswoode

M[rs] Spottiswoode & the boys[10] all send the warmest greeting; each after their kind.

RI MS JT/1/S/202

1. *50 Grosvenor Place*: residence in London.
2. *Your two telegrams*: telegrams missing.
3. *M[rs] Spottiswoode*: Eliza Spottiswoode.
4. *Royal Society Anniversary Dinner*: see letter 3911, n. 13.
5. *Miss Barnard*: Jane Barnard.
6. *M[r] & M[rs] Miller*: parents of Tyndall's recently deceased assistant George Miller. His father was W. Miller of Park Road, Dalstone, London.
7. *Cotterill*: John Cottrell.
8. *Hughie*: see letter 3855, n. 2.
9. *Huxley... the Society*: Thomas Huxley was elected Secretary of the RS.
10. *boys*: see letter 3855, n. 2.

To Edward L. Youmans [8][1] December 1872 3919

WASHINGTON, WELCKERIE[2] HOTEL, 15TH ST., *December, 1872.*

MY DEAR YOUMANS: ... The lectures here[3] are going off well. Lincoln Hall[4] is crowded, and I am assured that no such audiences ever assembled in Washington before. I was brief the first night, but gave them two hours the second night, and an hour and three quarters last night. By the way, when I came to the hall I found to my horror that I had put the wrong notes in my pocket, and so I had to speak for the hour and three quarters without once looking at a note.

No sign of weariness or inattention was to be seen in the audience from first to last.

You will not forget the taking of quiet rooms for me. Expense is quite a secondary matter, so if the Brevoort[5] be the best, please let me have rooms there. Quiet is the great thing—more precious than gold; yea, than much fine gold.

It is difficult to report these experimental lectures. Ordinary reporters can not possibly do it. Now, if you think the New York papers desire to report the lectures I might throw my notes into such a form as would help them, and let them have a copy of the notes of each lecture.[6] What do you say to this?

I hope you are all right again. I am well aided here, and have brought a colored man from Philadelphia, who is very useful.[7]

Always faithfully yours, | JOHN TYNDALL.

E. A. Youmans, 'Tyndall's American Visit', *Popular Science Monthly*, 44 (February 1894), pp. 502–14, on p. 511

---

1. *[8]*: the day is given as Tyndall refers to the third lecture in Washington, DC, as being 'last night'. The third lecture was on 7 December 1872.
2. *Welckerie*: this is a mis-transcription from the original letter; it is the Welckers Hotel.
3. *lectures here*: Tyndall delivered six lectures in Washington, DC, from 3–12 December 1872 at Lincoln Hall.
4. *Lincoln Hall*: Lincoln Hall was an auditorium located at 9th and D Streets, NW, in Washington, DC, widely regarded as a cultural nexus during its short period of operation. It was constructed in 1867 and burned down in 1886.
5. *The Brevoort*: see letter 3843, n. 3.
6. *if you think... notes of each lecture*: the *New York Times* reported on each of the six lectures in their issues of 18 December 1872, p. 8; 20 December 1872, p. 4; 22 December 1872, p. 1; 27 December 1872, p. 5; 29 December 1872, p. 5; and 1 January 1873, p. 5.
7. *a colored man... very useful*: Tyndall required an additional assistant after the death of his aide, George Miller. The man he hired in Philadelphia is not identified, but he is mentioned also in letter 3902.

From Thomas Corwin Mendenhall[1] — 8 December 1872 — 3920

Columbus Ohio. | Dec. 8 1872 | Prof. John Tyndall.

My Dear Sir,

A short time previous to your embarking for the country I addressed a letter to you at London which I suspect must have passed you on the way over. In it I returned my sincere thanks to you for your kindness in enclosing Card Photographs and your expression of an intention to procure a larger one for the rooms of our Association.[2] I also expressed a hope that at some time during your visit to this country we might enjoy the pleasure of meeting you in this City. It is in regard to the latter, more particularly, that I address you at the present time. I understand of course, how you are beset with invitations to lecture in all parts of the country, most of which you must necessarily refuse. I do not know indeed that I can induce you to think that we have any claim upon you, above that of many other similar Associations in the West. Two or three years ago four or five young men with a love for science began a series of meetings in a private room, gladly naming themselves the "Tyndall Association". As I said to you on a previous occasion the Association has grown in importance and usefullness, beyond our own sanguine expectations. We have rooms which we hope to make pleasant and worthy of one of the finest buildings in the City, in which they are situated. We give, annually, a course of Scientific lectures, this year—a course from Profs. Young of Dartmouth College,[3] Youmans, and Morse of Salem Mass;[4]—and we hold meetings every fortnight for the reading of scientific papers and discussions. We are poor but we notice with feelings of keen satisfaction the growth of interest among the people, in the objects of our association which will some day place us in better circumstances. I need not say that I think a visit from you and a couple of lectures, at some time during the winter, would arouse a healthy and much needed enthusiasm not only in our City and throughout our state but its influence would extend to adjoining states. In fact we, in the busy, hard working, I might say uncultivated West need help more than do our brethren in the East. Pardon me for occupying so much of your time in writing at so great length and if you should conclude it possible for you to come to Columbus I shall thank you to write me as soon as convenient concerning time, terms, or any thing else you may chose to mention.

With great respect | I am yours | T. C. Mendenhall | Pres[ident]. Tyndall Association.

RI MS JT/1/M/98

---

1. *Thomas Corwin Mendenhall*: Thomas Corwin Mendenhall (1841–1924), an American teacher, amateur physicist and soon to be appointed Professor of Physics and Mechanics at the Ohio Agricultural and Mechanical College (in 1873; now, The Ohio State University). He later held influential administrative positions, including superintendent of the US Coast and Geodesic Survey and President of Worcester Polytechnic Institute (*ANB*).
2. *our Association*: the Tyndall Association of Natural Science, established in 1870 to promote the study of science in Columbus, Ohio. See *Constitution and By-Laws of the Tyndall Association of Columbus, Ohio, As Adopted June 6, 1874* (Columbus, OH: Paul & Thrall, 1875).
3. *Young of Dartmouth College*: Charles Augustus Young (1834–1908), an American astronomer and solar spectroscopist and Professor of Natural Philosophy and Astronomy at Dartmouth (*ANB*).
4. *Morse of Salem Mass*: Edward S. Morse (1838–1925), a zoologist based at the Peabody Academy of Science in Salem, Massachusetts, of which he would later become the director (*ANB*).

From Charles A. Joy[1] 9 December 1872 3921

Columbia College, | New York, | December 9, 1872.

My dear Professor Tyndall,

I was deeply pained to hear from Dr Youmans the sad news of the death of one of your assistants.[2] I had hoped that everything connected with your visit to the United States would be of such an agreeable character that you would be tempted to repeat the visit at a time when you could extend your explorations to the far West & you would have leisure to study us better, but this melancholy event must cast a shadow across your path & I fear may deter you from soon to returning to us again. The loss of a trained assistant must add to the fatigue of preparation for your lectures[3] and I write to say that my assistant and my messenger will be at your service to "fetch & go" & procure such things as may be needed for your experiments. They know the city & can more easily find anything that is wanted than your own remaining assistant, who is a stranger to our streets & our ways. Do not hesitate to call upon me for any service of this kind.

I shall be glad to hear an account of your Niagara experience[4] when you come to town.

Very truly yours | Charles A. Joy

Prof. John Tyndall, | &c &c

RI MS JT/1/J/144

---

1. *Charles A. Joy*: Charles Arad Joy (1823–91), an American chemist. After graduating from Harvard Law School and working for the US Geological Survey, he studied chemistry in Berlin and Paris. He then held the chair of chemistry at Union College in Schenectady, New York, followed by a professorship in chemistry at Columbia College (1857–77; now Columbia University) (M. Benjamin, "Sketch of Charles A. Joy', *Popular Science Monthly*, 43 (July 1893), pp. 405–9).
2. *death of one of your assistants*: see letter 3896, n. 10.
3. *preparation for your lectures*: Tyndall delivered six lectures in New York from 17–31 December 1872 at the Cooper Institute.
4. *your Niagara experience*: see letter 3738, n. 3.

From Henry Bence Jones — 10 December 1872 — 3922

Tuesday | Dec 10. 1872

My dear friend

Your kind letter[1] and my leisure lead me to write to you again so soon although I have nothing much to say.

I hope all & each thing the managers do will be what you wish.[2] I am sure there is only one feeling to do what you would like. I think I told you we are going to have an unusual meeting on January 13 so as to decide on anything about building or anything else that might by chance press.

If you were there or if I was well my children[3] would certainly persuade as to have a ball on twelth night in the new laboratories.[4] Dancing below & refreshments above to promote a spirit of original investigation!! I went yesterday to see where the works have got to. The exit from the theatre down the new stair case is splendid & there ought never again to be a panic of fire. Certainly three & even four persons abreast might go down the stair case. It is far too magnificent. However there it is. I wish I could furnish your laboratories but alas I fear it must stay for you. The speaking tube to your rooms is done & like all the rest well done. The lavatories are to be nearly finished if not quite this week. The office is to be put in its former state this week & the antiroom & theatre will be all themselves. The carpenters work in the laboratories is the chief thing & that must get done as it can. Until you come very little progress as I have said can be made with your laboratory. Your letter of the 21 Dec[5] shews that you were in no anxiety then about Miller.[6] I have heard nothing of his people this week. The newspaper with your first lecture at Baltimore[7] was very interesting as shewing what interest your lectures cause £20000 is indeed a large sum.[8] It made Lecky's[9] mouth water I think last night when he thought of the small sums his books bring in, altho as he was telling me they are translated into Russian & Hungarian &c. He says Carlisle[10] is delighted with Froudes book because it glorifies Cromwell.[11] He Lecky is writing a strong Review against it.[12] In the Athenaeum[13] last week it was stated that Mr H Thompson & not Prof Tyndall had written the two articles in the Contemporary on Prayer.[14] It reads as if he had put it in himself. /

I have just had a letter from Dr Rigg[15] that old member we have; asking me whether you will want another assistant & suggesting his son an Oxford undergraduate?[16] Of course he thinks his son a genius & he seems to have done well.

Sir H Holland was for 24 hours full of Carpenters[17] dredging ship.[18] He spent all the day on board with some of the council of the RS &c.

I have no other news for you but as this sh[oul]d reach you soon after or at Christmas I must wish you every good wish & first & foremost that you were at home again as well & as strong & as happy as one who has done a good work should be. I am slowly getting better again but very weak & the weather too horrible. Good bye dear friend very many thanks for your kind words. I wish I had more life in me. But I am all water & no blood.

goodbye | goodbye | yours | H B J

RI MS JT/1/J/73

---

1. *Your kind letter*: letter 3900.
2. *the managers*: the governing body of the RI.
3. *my children*: see letter 3860, n. 6.
4. *new laboratories*: see letter 3776, n. 14.
5. *Your letter of the 21 Dec*: letter 3900, from 21 November, not December as Bence Jones mistakenly wrote.
6. *Miller*: see letter 3896, n. 10.
7. *Baltimore*: Tyndall delivered three lectures in Baltimore from 26–9 November 1872 at the Peabody Institute.
8. *£20000... large sum*: in letter 3875, Tyndall wrote that 'It is stated that in two years I could make £20,000 by lecturing in America'.
9. *Lecky's mouth... &c.*: William Edward Hartpole Lecky (1838–1903), an Irish historian who by 1872 had published *History of the Rise and Influence of the Spirit of Rationalism in Europe* (1865), *History of European Morals from Augustus to Charlemagne* (1869), and *The Leaders of Public Opinion in Ireland: Swift, Flood, Grattan, O'Connell* (1871). Later he published the monumental *History of England in the Eighteenth Century*, 8 vols (1878–90) (*ODNB*).
10. *Carlisle*: Thomas Carlyle.
11. *book... Cromwell*: Froude's *The English in Ireland in the Eighteenth Century* (London: Longman, Green, & Co., 1872), took a sympathetic view of English general Oliver Cromwell's (1599–1658) brutal military campaign in Ireland in 1649–50.
12. *strong review against it*: W. E. H. Lecky, 'Mr. Froude's English in Ireland', *MacMillan's Magazine*, 27 (1873), pp. 246–64.
13. *Athenaeum*: *The Athenaeum*, 7 December 1872, p. 733.
14. *two articles in the Contemporary on Prayer*: see letters 3779 and 3780; and letter 3852, n. 8.
15. *Dr Rigg*: probably Arthur Rigg (1812–80), an English clergyman and schoolmaster who was an RI member (F. E. Foden, 'The Rev. Arthur Rigg: Pioneer of workshop practice', *The Vocational Aspect of Secondary and Further Education*, 11:23 (1959), pp. 105–18).
16. *his son an Oxford undergraduate*: Edward Rigg (1850–1933), who studied at Queen's College, Oxford, and in 1873 became an Assistant Assayer of the Royal Mint until 1889, followed by chief clerk (1889–98), superintendent of the Operative Department

(1898–1918), and for a period was acting deputy master (*The Numismatist*, 46 (1933), p. 334).

17. *Carpenters*: William Carpenter.
18. *dredging ship*: Carpenter was one of the scientists organizing the soon-to-sail expedition of HMS *Challenger*, a ship which the RS obtained and had fitted for scientific purposes. It departed Portsmouth, England on 21 December 1872, and returned in 1876, having circumnavigated the globe, the crew studying the physical conditions of the sea, performing chemical studies of seawater, and collecting marine natural history specimens. See D. Macdougall, *Endless Novelties of Extraordinary Interest: The Voyage of H.M.S. Challenger and the Birth of Modern Oceanography* (New Haven, CT: Yale University Press, 2019).

From Charles Boyd Robertson[1] 11 December 1872 3923

[British Legation, | Washington.] | Dec 11. 1872.

Dear Mr. Tyndall,

Allow me to thank you very cordially for your "Contemporary" article on Prayer,[2] which reached me in due course of Post, & which has interested me greatly. It is perhaps almost presumptuous for me to notice the moderation and fairness of its tone, nevertheless I cannot refrain from doing so. On one or two cardinal points we differ,—of course—notably as regards the probability of physical results being produced, (without "miracle") by the agency of Prayer—yet not so widely as I had feared; but in many respects I venture to flatter myself that there is some identity of view between us.

Pray do not think me very egotistical for sending you, unsolicited, the inclosed letters[3] on the general subject, which I forwarded to the Editor of the "Spectator"[4] some time since, but which were courteously returned by him as having arrived "after the close of the controversy".

Should you be able to find time for their perusal you will gratify me much. Trifling & insufficient as they are, you will, I know, read them with the indulgence due to an inexperienced writer & thinker like myself. You can return them when next convenient to you.

I congratulate you cordially on the success of your Lectures here,[5] & thank you very sincerely for several pleasant evenings. Last night the experiments were especially interesting;—indeed, on the whole, "ignoramus" as I am, I have gathered more light from your teaching here than when I listened to you in London at the "R.I."[6]

Geo. Macdonald's Lecture tonight was "a treat" in a different style.[7] I am sorry that you should have been compelled to forego the pleasure of being present.

I look back with much gratification to our luncheon of last Sunday and trust that you will not have quite forgotten me should I turn up in London at some future day & pay you an unexpected visit.

Yours very sincerely | Charles Boyd Robertson.

RI MS JT/1/R/30

---

1. *Charles Boyd Robertson*: Charles Boyd Robertson (1843–1905), son of the minister Frederick W. Robertson (1816–53), was educated at Harrow and entered the Foreign Office in 1865 (*ODNB*). The United States Congressional Directory for 1872 lists him as Third Secretary of Great Britain's legation in Washington.
2. *your "Contemporary" article on Prayer*: see letter 3852, n. 8.
3. *inclosed letters*: enclosures missing.
4. *Editor of the "Spectator"*: at this time the *Spectator* had a joint editorship, Meredith White Townsend (1831–1911) and Richard Holt Hutton (1826–97).
5. *your Lectures here*: Tyndall delivered six lectures in Washington, DC, from 3–12 December 1872 at Lincoln Hall. For his American lecture tour in general, see Introduction, n. 8.
6. *The "R.I."*: the Royal Institution in London.
7. *Geo. Macdonald... different style*: the Scottish author and minister George MacDonald (1824–1905) delivered a lecture on Robert Burns in Washington, DC, on 11 December 1872 (J. Ricke, A. Chu, K. Smalley, K. Dwyer, and C. Hoelscher, 'George MacDonald: A Timeline of Lectures and Performance, 1855–1891', *North Wind: A Journal of George MacDonald Studies*, 37 (2018), pp. 107–79, on p. 135).

From John E. Davies[1] 12 December 1872 3924

[University of Wisconsin | President's Rooms, | *Madison,*] December 12th, [*187*]2.

Dear Sir

I am requested by a vote of the Faculty of the University to acertain if, in case you come as far west as Chicago, you could not be induced to come a little farther and favor us with your presence here.

We are all, Faculty and Students, very anxious to see and hear you; and, if you do come to Chicago and cannot come here, some of us are determined to go to Chicago where we may have that pleasure.

This place is about 100 miles from Chicago, but with the best of Railroad Facilities &c being the Capital of the state. I think the citizens also will take a hearty interest in the matter. There are excellent Lecture Halls here, easily obtained. Gov. Washburn[2]—present Governor of the State—has directed me to say that, if you come, he will be most happy to entertain you at his house while you remain in Madison.

If not asking too much, may I request that you will inform me whether you intend to visit Chicago ere you return from the United States; whether you would entertain a proposition to come on here in case you do visit C[hicago].,—and what will be the probable cost for a */series/* of three or four lectures; say upon Heat.

I have sent this letter through the Appletons,[3] hoping thereby to be sure to reach you. I have also ventured to send you a small Mathematics-Physical article of mine "On Potential Functions".[4] My object in writing and reading it was to arouse more interest in higher Science in this western country. I hope you may receive and accept it.

Trusting I may speedily hear from you, I am

With great respect | Your Obd[edien]'t Serv[an]'t | John E. Davies. | Sec[retar]y of the Faculty

D^r. John Tyndall | &c &c

RI MS JT/1/D/11

1. *John E. Davies*: John Eugene Davies (1839–1900), an Instructor of Agronomy and Physics and Chair of the Physics Department at the University of Washington, which had been established in 1868 (*Lawrence University. Alumni Record. 1857–1905* (Appleton, WI: General Education Board, 1906), pp. 137–8; C. W. Butterfield, *History of the University of Wisconsin, from Its First Organization to 1879; with Biographical Sketches of Its Chancellors, Presidents, and Professors* (Madison, WI: University Press Company, 1879), pp. 116–22).
2. *Gov. Washburn*: Cadwallader C. Washburn (1818–82), Governor of Wisconsin form 1872–4. Previously he was a Union general during the American Civil War and a mill owner thereafter, founding the company that would grow into General Mills (*Dictionary of Wisconsin Biography* (Madison: The State Historical Society of Wisconsin, 1960), p. 366).
3. *the Appletons*: the Appleton and Co. publishing firm (see letter 3823, n. 9).
4. *"On Potential Functions"*: J. E. Davis, 'On Potentials and their Application to Physical Science', *Transactions of the Wisconsin Academy of Sciences, Arts and Letters*, 1 (1870–72), pp. 155–64.

From Genealogical Society — 12 December 1872 — 3925

[LADIES and GENTLEMEN of energy, who desire the most lucrative and useful of all employments, are invited to correspond with us at once, giving in every letter their Post Office address, COUNTY and STATE; and applicants for territory will give the Township in which they desire to begin the canvass.

Genealogical Society, | PUBLISHERS OF | STANDARD WORKS AND STEEL ENGRAVINGS, | *Steel Engraving and Albert Types executed in the Best Manner, at Lowest Prices.* | SPECIAL ATTENTION GIVEN TO GENEALOGICAL WORKS, ENGRAVING ARMORIAL BEARINGS, FAMILY COATS OF ARMS, ETC. | New York Office, 945 & 947 Broadway, (near Fifth Avenue Hotel,) | *New York,*][1] December 12th [*18*]72

Prof Tyndall

Dear Sir

For our biographical work as explained in accompanying prospectus[2] and in which we design having a few good men from different states we would be pleased to to[3] have a sketch of your life from fifteen hundred to two thousand words well conceived and embodying the leading facts of your career.

Such sketch may be prepared by any literary friend of yours or will be written here by an eminent author from data provided by you.

The book although national in its character is designed as a family work.

If you will send us immediately a negative the size of an ordinary engraving and sufficient data we can give you a portrait that is almost as valuable as an oil painting. The biography can be made of great interest financially politically or as a family record.

The negative must not be albuminized.[4] Please send the data immediately as the work is ready for the binders.

Respectfully Genealogical Society | per C Northrop Sec[retary][5]

RI MS JT/1/G/1

1. *New York*: the text up to this point is typed letterhead.
2. *accompanying prospectus*: enclosure missing.
3. *to*: 'to' was repeated in the letter.
4. *albuminized*: coated or impregnated with albumen (*OED*).
5. *C Northrop Sec[retary]*: not identified.

From George F. Barker[1] 13 December 1872 3926

30 College St | New Haven, Dec. 13. 1872

My dear Professor Tyndall:—

Will you allow me to offer you my sincere congratulations upon the great success of your lectures[2] thus far, and my hearty thanks for the great assistance which you have rendered the cause of science in this country. Especially do I want to thank you for your earnest appeals in behalf of pure science and of original investigation. I am quite sure your words will not be lost, but will bear fruit of which you yourself will yet be proud.

Our citizens are looking forward to your lecture with great interest. Will it be convenient for you to let us know the dates in January when you can come to New Haven?[3] later than the 20th if convenient. The hall being a public one must be secured some time in advance. Moreover the College term closes next week to reassemble Jan. 3d. It appears desirable to announce your lectures before the term closes. Pardon our importunity, if it be such. "The young men of Yale" and the old men of Yale also, want to hear science from the lips of one of its very best expounders. Hence our desire to have the details arranged speedily.

Trusting you have not suffered in health from the severe tax upon one's strength required by such continuous lecturing, I remain

Very sincerely yours | George F. Barker

RI MS JT/1/B/14

1. *George F. Barker*: George Frederick Barker (1835–1910), Professor of Physiological Chemistry and Toxicology at Yale University. Later he served as President of the AAAS ('George Frederick Barker', *Report of the National Academy of Sciences for the Year 1910* (Washington: Government Printing Office, 1911), p. 24).
2. *your lectures thus far*: at this point in his American lecture tour, Tyndall had delivered lectures in Boston, Philadelphia, Baltimore, and Washington, DC. New York and Brooklyn would follow, with New Haven, CT, the last city.
3. *dates in January when you can come to New Haven*: Tyndall delivered two lectures in New Haven, CT, on 22 and 23 January at Yale College.

From Charles W. Eliot[1] 13 December 1872 3927

[HARVARD COLLEGE, | CAMBRIDGE, MASS.] | 13 Dec. [187]2.

My dear Sir,

I hope that you will return to Boston[2] for a few days at least, before you go home. If you would like to get a notion of this nascent University,[3] it would give me much pleasure to show it to you. Your easiest way to do this would be to come and spend a few days at my house in Cambridge. It would give me much pleasure, if you could accept this invitation *[in]* any moment & for any time which might suit your convenience.

With experimental lectures to give in strange places under difficult conditions you can have had little time to observe our social and political habits or our mode of life and ways of thinking. It always seems to me that these are the only things in this country which a travelled European needs to study—the only things indeed which can much interest him.

Hoping to have the pleasure of seeing you here; I am

Very truly yours | Charles W. Eliot

RI MS JT/1/E/101

---

1. *Charles W. Eliot*: Charles William Eliot (1834–1926), an American educator and President of Harvard University from 1869–1909 (*ANB*).
2. *return to Boston*: Tyndall had delivered six lectures in Boston from 15–25 October 1872 at the Lowell Institute.
3. *this nascent University*: Harvard University, although founded in 1636, only began to assume the form of a modern research university under Eliot's leadership.

To Henry Bence Jones [between 14 and 16][1] December 1872 3928

Brevoort House,[2] | Fifth Avenue, Cor[ner], of Eighth Street, | New York 17th. Dec. 1872.

My dear Friend

It is a good while since I wrote to you and calamities have since then occurred which for the time quite overshadowed me. Spottiswoode has no doubt informed you of the circumstances attending Miller's death.[3] It was a great blow to both myself and Cottrell.[4] Every thing that practical prudence could suggest was done, and when all proved unavailing every thing that could soften and beautify the event was done. Had he been at home he could not have been more tenderly cared for while living, nor treated with greater consideration when dead.

Since this event my lectures in Washington have begun continued and ended.[5] I had an audience which I was informed was without a parallel in Washington. President,[6] Cabinet, Ministers, Senators, Congress men, and all manner of minor lights; and they held on to me through many difficulties with unswerving attention. I kept my health in order by walking every day, though there are not many agreeable walks about Washington. Once only I dined out and that was with Russell Gurney and his excellent wife.[7] I met there our ambassador, Sir Edward Thornton,[8] Sumner[9] and others. The kindness of the people is not to be described. We had a special meeting of the Philosophical Society,[10] and a feast after the Meeting I have steadily avoided taking a mouthful of food or a drop of wine at these evening entertainments, and in this way have kept myself well.

Yesterday in quitting Philadelphia,[11] where I had spent the night I stumbled against Froude. He appeared to be well. He informed me that he had cut his bonds and given up the prosecution of his lectures.[12] In a few days he will be entirely free. I fear his time here has not been happy.[13] We had a walk together to day and he looks somewhat careworn. This however may be due to a bad night's rest, for he looked bright, I thought, yesterday.

I begin my lectures here on Tuesday,[14] and when New York is finished I repeat the lectures in the neighbouring city of Brooklyn.[15] This will detain me here until the 20th of January. Then I go according to promise to give two or three lectures to the students of Yale,[16] and then I steadfastly turn my face homewards. From numerous universities, and public Institutions I have received the most pressing and cordial invitations;[17] but I shall decline them all, for the yearning towards home is beginning to be very strong.[18]

The thought has occurred to me whether I ought not to do with myself as I do with Cottrell, and not accept any pay from the Royal Institution the Trinity House and the Board of Trade[19] for the time I have been away. Do you not think this reasonable?

My cousin[20] has sent you lots of papers from time to time I have been very busy and have left these matters to him. Not *[any]* English newspaper has been sent to me since I came away although Smith[21] well knows that a newspaper from time to time would be agreeable to me.

Mr Evarts has just left me; we have been talking about Sir Henry.[22]

Last night an entertainment was given by Mr Appleton, at which were present a great number of strong men of various kinds. They were all cordial in the extreme. I fear before I quit New York that I shall obtain a reputation for unsociability, for dinners and evening parties I am determined to eschew.

And now I think I have said my say, save that I am, as ever

Your friend | John Tyndall

The article from the Globe which you sent me was simply scandalous.[23]

RI MS JT/1/TYP/3/817–8
LT Typescript Only

---

1. *[between 14 and 16]*: we believe the date of 17 December on LT's typescript is in error. He finished his lectures in Washington, DC, on 12 December, was in Philadelphia for at least one night as I noted in this letter, and began his lectures in New York on 17 December, thus this letter would fall between 14 and 16 December. He may have sent the letter on 17 December, and LT got the date from the postmark.
2. *Brevoort House*: see letter 3843, n. 3.
3. *Miller's death*: see letter 3896, n. 10.
4. *Cottrell*: John Cottrell.

5. *my lectures in Washington have begun continued and ended*: Tyndall delivered six lectures in Washington, DC, from 3–12 December 1872 at Lincoln Hall.
6. *President*: at this time the President of the United States was Ulysses S. Grant, who the previous month defeated Horace Greeley in the election of 1872 to gain a second term.
7. *Russell Gurney and his excellent wife*: see letter 3917, n. 5, and letter 3906, n. 3.
8. *Sir Edward Thornton*: Edward Thornton (1817–1906), a British diplomatist. He was Envoy Extraordinary and Minister Plenipotentiary to the United States from 1867–81 (*ODNB*).
9. *Sumner*: probably Massachusetts Senator Charles Sumner (see letter 3815, n. 3).
10. *Philosophical Society*: the Philosophical Society of Washington, established in 1871 by Joseph Henry (F. Frenkiel, 'Origin and Early Days of the Philosophical Society of Washington', *Bulletin of the Philosophical Society of Washington*, 16 (1962), pp. 9–24).
11. *quitting Philadelphia*: presumably Tyndall visited Philadelphia between Washington, DC, and New York (see n. 1).
12. *his lectures*: Froude was also in the United States for a lecture tour; he and Tyndall traveled on the same ship across the Atlantic (SS *Russia*).
13. *He informed... not been happy*: Froude's lectures on his historical work on the English in Ireland (see letter 3922, n. 11) received a chilly reception from Irish-American audiences (H. Paul, *The Life of Froude* (London: Sir Isaac Pitman & Sons, 1905)).
14. *I begin my lectures here on Tuesday*: Tyndall delivered six lectures in New York from 17–31 December 1872 at the Cooper Institute.
15. *I repeat the lectures in the neighbouring city of Brooklyn*: Tyndall delivered six lectures in Brooklyn from 4–20 January 1873 at the Academy of Music.
16. *two or three lectures to the students of Yale*: Tyndall delivered two lectures in New Haven, CT, on 22 and 23 January at Yale College.
17. *the most pressing and cordial invitations*: in previous letters, Tyndall shared that he received invitations to lecture in Cincinnati, Cleveland, Columbus, Providence, and St. Louis, among other cities (see letters 3885 and 3886).
18. *yearning towards home is beginning to be very strong*: Tyndall departed New York on the SS *Cuba* on 8 February 1873, returning to England on 19 February and arriving in London later the same day, where he was greeted at Euston Station by Thomas Hirst and Heinrich Debus. The steamer was set to depart on 5 February, as Tyndall wrote in some letters (3951, 3959, 3967, 3975, and 3979), but was delayed, and thus previous sources have Tyndall departing New York the day after the farewell banquet in his honor on 4 February (e.g., in A. S. Eve and C. H. Creasey, *Life and Work of John Tyndall* (London: MacMillan & Co. Ltd., 1945), p. 173; and K. Sopka, 'An Apostle of Science Visits America: John Tyndall's Journey of 1872–1873', *The Physics Teacher*, 10 (1972), pp. 369–75, on p. 374). The *Cincinnati Daily Gazette* reported on 8 February 1873 that Tyndall was to depart on the *Cuba* that very day (p. 3), and the *Sunday Mercury* of 9 February 1873 reported that the *Cuba* sailed on 8 February and listed Tyndall as one of the 'distinguished passengers' on board (p. 5). As of 27 January, Tyndall still thought he was departing on 5 February (see

letter 3979), but in his speech at the banquet, he did not say 'tomorrow,' but that 'in a day or two I quit these shores' (*Proceedings of the Farewell Banquet*, p. 51).

19. *Trinity House and Board of Trade*: see letter 3736, n. 3. On Tyndall as scientific adviser to Trinity House, see letter 3869, n. 4.
20. *My cousin*: Hector Tyndale.
21. *Smith*: Smith is presumably a porter at the RI (in letter 3797 to Heinrich Debus, Tyndall wrote: 'The Hall porter of the R.I. shall get a wigging on my return. He used to send me every thing likely to interest me. This year he has sent me nothing').
22. *Sir Henry*: probably Henry Holland.
23. *The article from the Globe:* see letter 3897, n. 4.

From Hector Tyndale 17 December 1872 3929

Philad[elphi]a Dec[em]ber 17th 1872

My dear John

By this mail I have sent you an envelope, apparently enclosing several letters from Tom Hirst[1]—as, also, a copy of the Press of the 5th inst[ant][2,3]—the only one I can get. When you receive this your first New-York lecture[4] will be over and I trust this, as all the others, has gone off pleasantly enough. Do not forget that you have the Washington[5] check for $2000. in your pocket, nor that you should hand it to Appletons,[6] getting their receipt for it as before.

Please speak to Mr Cottrell[7] as to where I shall send Mr Miller's[8] trunk and bundle. The keys of the trunk, with others, I gave you when here. I suppose Cottrell will take charge of all.

My cold, though not well, is something better and a few days will bring it all right. Every body else is well and all send love to you. All hope for many visits from you, while you stay in New-York and Brooklyn,[9] as you can run over so easily. On the day before or on the morning of your coming please telegraph us of it.

I shall run over to see you, perhaps go back with you should you come and stay with us during the gap in your lectures from next Saturday the 21st to Thursday the 26th inst[ant].—so spending Christmas here and Christmas with us, as will be befitting. The photographs are being attended to.

Affectionately yours | Hector

RI MS JT/1/T/71

1. *several letters from Tom Hirst*: this may have included letter 3911; the letter from Hirst previous to 3911 is letter 3853 from September, so there are probably some missing letters from Hirst.
2. *inst[ant]*: of the current calendar month (*OED*).

3. *Press of the 5th inst[ant]*: *The Philadelphia Press* (or *The Press*) was a newspaper published from 1857–1920. The article has not been identified.
4. *your first New-York lecture*: Tyndall delivered six lectures in New York from 17–31 December 1872 at the Cooper Institute.
5. *Washington*: Tyndall delivered six lectures in Washington, DC, from 3–12 December 1872 at Lincoln Hall.
6. *Appletons*: see letter 3823, n. 9.
7. *Cottrell*: John Cottrell.
8. *Mr Miller's*: see letter 3896, n. 10.
9. *Brooklyn*: Tyndall delivered six lectures in Brooklyn from 4–20 January 1873 at the Academy of Music.

To Hector Tyndale 18 December [1872][1] 3930

18th Dec.

Thou art wise Oh Hector! While I was lecturing last night[2] a robber got into my rooms, rifled my portmanteaus and secured that purse of money which you recommended me to keep elsewhere.

It amounted to either 30 or 40 pounds—I do not remember which. He took nothing else.

The audience here is splendid and by far the most enthusiastic I have met

Love to all. Thanks for the letters.[3] | affectionately | John

RI MS JT/1/T/1472
RI MS JT/1/TYP/5/1715

1. *[1872]*: the year is given as this letter is a reply to letter 3929 from Hector Tyndale.
2. *lecturing last night*: Tyndall delivered six lectures in New York from 17–31 December 1872 at the Cooper Institute.
3. *letters*: letters from Thomas Hirst (see letter 3929, n. 1).

From Manton Marble[1] [20 December 1872][2] 3931

532 Fifth Ave. | 2nd door above | 44th St.

There, My dear Prof Tyndall, are the directions for your driver—or for yourself if you are so infatuated as to climb these Alpine snow peaks[3] on foot,—and at half past six to day the dinner-bell will ring.

I have asked to meet you Cha[rle]s. O'Conor[4] facile princeps[5] at the bar in these United States; Judge W[illia]m. D. Shipman,[6] of the U.S. Court, than whom no better man wears ermine;[7] Dr W[illia]m. H. Van Buren[8] the first surgeon among us; W[illia]m Butler Duncan,[9] the banker, whom you met Monday Evening; Prof. Youmans of whom we endorse all your praise, and M. B. Anderson,[10] my one teacher, the President of the University of Rochester, and if Youmans will not tell you that he is the foremost educator in this republic—then I forswear Y[oumans]'s friendship for the rest of my natural life, & stick to the proposition. Bless you, if he succeeded in educating me in any wise, doesn't that prove it?

Dr Anderson is a religious man (Baptist), but a Catholic thinker. Mr O'Conor & Dr Van Buren are R[oman]. Catholics. Judge Shipman resists classification. Youmans is a Pagan and I am a Buddhist, and Duncan does not worship Mammon.[11]

And this reminds me that I asked Henry Ward Beecher[12] who couldn't come because he has a prayer meeting Friday nights!—He writes—"you could hardly have done a kinder thing than to ask me to dine with Prof. Tyndall whose works are my familiar companions. It grieves me that I cannot come. x x I hope in some way to meet Prof. T[yndall]. before he returns[13]—"

Wherefore I now beg of you to give me another day—Tuesday or Wednesday or Thursday will be best for him—when I may gather a half dozen more of the elect—Evarts for one.

And I remain, dear Mr Tyndall, most faithfully yours | Manton Marble

RI MS JT/1/M/48

---

1. *Manton Marble*: a journalist who at this time served as editor of the *New York World*. He was a member, and later president, of the Manhattan Club.
2. *[20 December 1872]*: the date is given since Tyndall was in New York in December 1872 during his American lecture tour (see Introduction, n. 8). He arrived in New York on 14 December 1872, and left on 4 January 1873. This letter was written on a Friday, and during this time frame, Fridays fell on 20 December, 27 December, and 3 January. 20 December seems likely as Marble refers to the heavy snowstorm in New York, which Tyndall referred to in a letter to Hirst dated 22 December, sharing that 'We have had heavy snow, and the streets for a day or two were execrable' (3934), and in a letter to Miss Moore dated the same day, sharing that the roofs of houses were 'laden with snow which fell a few days ago, and which when it thawed rendered the streets of New York execrable' (3935). Also, in his letter of 23 December, a Monday, Marble referred to dinner in a room at the Manhattan Club, which very well may be this Friday evening gathering (3937).
3. *these Alpine snow peaks*: referring to the heavy snow that had recently fallen on New York.
4. *Cha[rle]s. O'Conor*: Charles O'Conor (1804–84), an American lawyer and politician who was a candidate in the 1872 presidential election (*ANB*).
5. *facile princeps*: a person or thing that is easily first; a person or thing considered to be the best or most notable; the acknowledged leader in a particular subject, field of expertise, etc. (Latin; *OED*).

6. *Judge W[illia]m. D. Shipman*: William Davis Shipman (1818–98), Federal Judge for the United States District Court for the District of Connecticut (J. W. White and D. Glenn, 'The Politics of Judging: The Civil War Letters of William D. Shipman', *Connecticut History Review*, 59 (2020), pp. 80–100).
7. *ermine*: the use of the fur of ermine (an animal of the weasel tribe) in the official robes of judges and the state robes of peers (*OED*).
8. *W[illia]m. H. Van Buren*: William Holme Van Buren (1819–83), an American surgeon (C. Herbermann (ed), 'Van Buren, William Home', in *The Catholic Encyclopedia*, 15 vols (New York: Robert Appleton Company/The Encyclopedia Press, 1907–12), vol. 15 (1912), p. 267).
9. *W[lilia]m Butler Duncan*: William Butler Duncan I (1830–1912), a Scottish-American banker and railroad executive ('Wm. Butler Duncan Dies in 82d Year', *New York Times*, 21 June 1912, p. 13).
10. *M. B. Anderson*: Martin Brewer Anderson (1815–90), an American clergyman and first president of the University of Rochester in Rochester, New York ('Dr. Martin Brewer Anderson', *New York Times*, 27 February 1890, p. 4).
11. *Mammon*: wealth, profit, possessions, etc., regarded as a false god or an evil influence (*OED*).
12. *Henry Ward Beecher*: Henry Ward Beecher (1813–87), an American preacher and social reformer (*ANB*).
13. *returns*: to London.

From Henry Bence Jones — 21 December [1872][1] — 3932

84 Brook St[2] | Friday Dec 21

Dear friend

Here is the end of the week that was to see our Royal Institution finished[3] & so I am going to send you a report I have just come from thence & oh how I want you. I cannot stir a step with your rooms through fear of stirring wrong & how to get them in order I do not know.

However the first thing was to get all right for the juveniles[4] & all will be ready tomorrow (the lectures begin tomorrow week). The passage to your rooms is thro Miss Savages[5] stopped by the theatre & the new door to the right hand of the lecturer is all open & the gas is all right. Then the office for Mr Hughes[6] will all be finished tomorrow & the books back & the anteroom to the theatre is done & also the urinals which are splendid. These will be usable on Monday. With water unlimited. So that every thing will look as it was formerly except that the door to your stair case is blocked up in the anteroom.

The new stair case is splendid but that is not finished as regards ballastrades. I wonder how it sh[oul]d be /left/. Mr Wills has not been able to prepare any thing in the lower laboratory & there is much to be done Your three laboratory rooms are being painted & I shall do as little as I can do for fear of going wrong. The whole thing has been done in a style I never anticipated.

Wills with Odling as soon as the juvenile lectures[7] are over will get that laboratory usable & when you come back Odling will be gone & there will be no one but you to use it until your own are to your liking.

Yesterday Mr Gassiott[8] made us a present of a marble bust of Mrs Somerville[9] by MacDonald of Rome.[10] I think its place should be in the upper Library. I think this makes up all my budget of news regarding the Institution. On the whole all is as well forward as we can expect & the members have had no trouble or disarrangement to signify & we shall start all right without delay of any kind.

I received your last letter[11] & the passage introduction[12] on Monday & I sent it with a note to Delane but it has not come out in the Times yet.[13] It may tomorrow. no. Some of the remarks regarding your lectures at Washington[14] I have sent to Nature & they will appear next week no doubt.[15] [What] odd folk they are in their newspapers. Your cousins[16] speech[17] was wonderfully lively & figurative. Certainly you have had a rich harvest of praise from every one & I fear you will think the old country very torpid when you come back. Success is such a [stimulant] no Ether comes near it & we shall all appear stupid & immobile when you get home

Poor Millers body[18] arrived on Monday! but as there was no medical certificate of death the customs suspecting! smuggling would not let it land. It seems I know Millers brother[19] & he came here & I wrote a certificate which has been taken & today or tomorrow the friends will see the poor fellow They seem all most grateful to you for all you have done & said & written in this matter. Their gratitude is unbounded

I must now thank you for myself also for your last kind letter.[20] It was very pleasant I wish I was not so badly broken down & could look forward to a longer time of our relationship. But tho' I am in no danger I am in a vile state kept alive by my medicines & even with them gradually growing worse & worse & I long for rest & wonder why I take the trouble to try & swim on when there is no hope of reaching the shore. However it looks as if I shall get on at least till you come home & and then—who knows. I would not work a day longer for my president[21] & indeed it is mighty little I can do. I have not been out for a week & then only to the Institution & back. Nothing more—

I am very glad that you keep well & that Cotterell[22] is well too. Come back as soon as you can & have a good rest before Easter[23]

Good bye Dear friend | I have no other news to send you | Every good Christmas & new years wish to you

RI MS JT/1/J/93

1. *[1872]*: the year is given by reference to the renovations at the RI while Tyndall was on his American lecture tour (see Introduction, n. 8, and letter 3776, n. 14).
2. *84 Brook St*: residence in London.
3. *Royal Institution finished*: see letter 3776, n. 14.
4. *juveniles*: see letter 3782, n. 4.
5. *Miss Savages*: Sarah Savage.
6. *Mr Hughes*: Henry C. Hughes, Clerk of Accounts and Collector at the RI (*Roy. Inst. Proc.*, 6 (1870–72), p. ii).
7. *juvenile lectures*: see letter 3782, n. 4.
8. *Mr Gassiott*: John Peter Gassiot (see letter 3889, n. 11).
9. *marble bust of Mrs Somerville*: Mary Somerville (1780–1872), a Scottish science writer who studied mathematics and astronomy. Primarily educated at home, her works included *On the Connexion of the Physical Sciences* (1834), *Physical Geography* (1848), and *On Molecular and Microscopic Science* (1869). In 1835, Somerville and astronomer Caroline Herschel (1750–1848) became the first women elected members of the Royal Astronomical Society (*ODNB*). The referred-to bust by Macdonald was sculpted in 1844 (*List of the Members, Officers, and Professors; with the Report of the Visitors, Statement of Accounts, and Lists of Lectures and Donations in 1873* (London: Royal Institution of Great Britain, 1874), p. xxxi). According to Somerville's daughter Martha, this likeness was 'by far the best' (M. Somerville (ed), *Personal Recollections. From Early Life to Old Age, of Mary Somerville. With Selections from her Correspondence* (Boston: Roberts Brothers, 1874), p. 61).
10. *MacDonald of Rome*: Lawrence Macdonald (1799–1878), a Scottish sculptor living and working in Rome (*ODNB*).
11. *your last letter*: letter not identified, perhaps missing.
12. *passage introduction*: not identified.
13. *in the Times yet*: it does not appear that anything from a letter to Henry Bence Jones was published in the *Times* in December 1872.
14. *your lectures at Washington*: Tyndall delivered six lectures in Washington, DC, from 3–12 December 1872 at Lincoln Hall.
15. *appear next week no doubt*: *Nature*, 7 (26 December 1872), p. 150.
16. *cousins*: Hector Tyndale.
17. *speech*: not identified.
18. *Poor Millers body*: see letter 3896, n. 10.
19. *Millers brother*: not identified.
20. *your last kind letter*: possibly letter 3917 (dated 5 December 1872), in which Tyndall wrote: 'I hope you are better. I shall be anxious until I hear from you'.
21. *my president*: Henry Holland.
22. *Cotterell*: John Cottrell.
23. *before Easter*: after Easter (13 April 1873), Tyndall delivered a course of lectures at the RI, 'On Light' (see letter 3782, n. 10).

To Henry Bence Jones 22 December 1872 3933

Brevoort House,[1] | Fifth Avenue |
New York Sunday, 22nd, Dec, 1872.

My dear Friend

Your last letter received two days ago greatly delighted me.[2] Those that preceded it made me gloomy;[3] out here the words "decidedly better" ring joyfully in my ears. I was glad to hear of your doings: amazed to hear of Mr Lee's proposition.[4] I believe the old man does not mean unkindness or discourtesy towards me; he has got a crotchet[5] and pursues it. That is all. I have tried to keep you supplied with papers since I came here. They wanted to get the Academy of music for the lectures,[6] but could not as I required it for a fortnight without interruption, so they fell back upon the Cooper Institute.[7] It is a large low room, ceiling groined and supported by numerous pillars; with a cellar-like aspect. The big people of New York had ceased to go to the Cooper Institute, but they now flow into it in full tide. I hear exclamations of surprise on all sides that such an audience could be gathered together. Not the wealthy alone, but the learned in literature science and art. They are the warmest people I have hitherto met in America. The applause is well-timed and cordial. I mean by well-timed that they choose bits of clear exposition for their heartiest approval. The same struck me on the delivery of my first lecture in Albemarle Street,[8] and at that time it greatly encouraged me. Deputations from various western cities[9] have come to me, but I now set my face firmly homewards.[10]

I am exceedingly interested in the account you give me of the laboratories.[11] It makes me wish myself twenty years younger. Could I but go back 20 years and carry with me the serenity of mind that experience has brought me, and which <u>you</u> know I now possess in such an eminent degree, I might do great things. As it is I am old and apathetic, and feel rather inclined to give up the beautiful place you describe than that those who built it should be disappointed in the meagerness of the results obtained in it.

I had no notion that I had so many friends in America. They crowd around me on all possible occasions—warm, cordial, hospitable. I fear they must consider me somewhat morose as the hospitality I rarely accept. Appleton told me last night that King[12] had informed him that 2000 copies of "Forms of Water" had been printed.[13] "Why" he said "we have printed 7000 copies here", and we shall have to print a far greater number. I keep on the whole well. I walk a good deal in the open air, and manage to lecture without notes and with some vigour. At all events the attention of the audience never flags even when I keep them nearly two hours. Three of my lectures here[14] have been reeled off. Three more are to be given after Christmas, and then the six are to be repeated in Brooklyn.[15] Two or three lectures subsequently at New Haven[16] finish my work and then hurrah for England! I am told however that the north Atlantic is atrocious in February, so that I may not after all get back to you. I wish I was now near you. I should claim my dinner from you on Christmas day.

My notion is to repeat the American course when I return to England.[17] I have bestowed some trouble upon it and its redelivery will relieve me of the toil of preparing another course.

I have had some experience of burglars since I came here[18] A wretch of this class watched my movements on the evening of my first lecture, and after I slipped out of my room he slipped in. On my return from the lecture, I found the hoziery of one of my portmanteaus strewn about the floor. I thought at first that a dog, or some romping child had gone into the room. Then I looked into my bedroom, and there saw a portmanteau which I had carefully locked wide open. Cheque book letters and other matters were strewn about, and a purse containing some 30 or 40 pounds had disappeared. I feel a kind of tenderness and respect toward the rascal: he flew high and disdained taking any thing but the money. A scrap of a gold watch chain which lay on my table he also captured, but he left me my socks and shirts the loss of which would have been serious to me.

And now good bye. Continue to grow better and better until I reach you
ever and always Yours | John Tyndall
Busk is a dear good fellow.[19]

RI MS JT/1/TYP/3/819
LT Typescript Only

---

1. *Brevoort House*: see letter 3843, n. 3.
2. *Your last letter*: letter 3912.
3. *Those that preceded it*: letters 3880, 3889, and 3897, in which Bence Jones complains of his ill health.
4. *Mr Lee's proposition*: that Tyndall be granted a leave of absence from the RI through 1874 (see letter 3912). For Mr Lee, see letter 3896, n. 7.
5. *crotchet*: a peculiar notion on some point (usually considered unimportant) held by an individual in opposition to common opinion (*OED*).
6. *Academy of music for the lectures*: see letter 3815, n. 9.
7. *Cooper Institute*: see letter 3873, n. 21.
8. *my first lecture in Albemarle Street*: the RI is located at 21 Albemarle Street, Mayfair, London. Tyndall delivered his first Friday Evening Lecture there, 'On the influence of Material Aggregation upon the manifestations of Force', on 11 February 1853 (*Roy. Inst. Proc.*, 1 (1851–54), pp. 254–9). Several months later, Tyndall was elected Professor of Natural Philosophy at the RI, a defining moment of his career (*Roy. Inst. Proc.*, 1 (1851–54), p. 339).

9. *Deputations from various western cities*: in previous letters, Tyndall shared that he received invitations to lecture in Cincinnati, Cleveland, Columbus, Providence, and St. Louis, among other cities (see letters 3885 and 3886).
10. *set my face firmly homewards*: see letter 3928, n. 18.
11. *account… of the laboratories*: see letter 3912. See also letter 3776, n. 14.
12. *King*: Henry S. King (1817–78), British publisher who established Henry S. King & Co. in 1871. Of all of Tyndall's books, only two were published by a firm other than Longmans, Green, and Co.: *Forms of Water* by King, and his first book, *Glaciers of the Alps*, by John Murray.
13. *2000 copies of "Forms of Water" had been printed*: a comparison between the print runs of Tyndall's *Forms of Waters*, published in the United States by Appleton and in England by Henry S. King.
14. *my lectures here*: Tyndall delivered six lectures in New York from 17–31 December 1872 at the Cooper Institute.
15. *Brooklyn*: Tyndall delivered six lectures in Brooklyn from 4–20 January 1873 at the Academy of Music.
16. *at New Haven*: Tyndall delivered two lectures in New Haven, CT, on 22 and 23 January at Yale College.
17. *repeat the American course when I return to England*: see letter 3782, n. 10.
18. *some experience of burglars since I came here*: see letter 3930 to Hector Tyndale.
19. *Busk is a dear good fellow*: George Busk took over Bence Jones's duties for the upcoming season of Friday Evening Discourses at the RI.

To Thomas Archer Hirst 22 December 1872 3934

Brevoort House,[1] | Fifth Avenue. |
Sunday, 22nd. Dec. 1872.

My dear Tom.

On receiving your last letter[2] I telegraphed to Huxley and from the tenor of the telegram[3] you might infer my notions regarding Greenwich.[4] But fearing that your insight might be at fault I telegraphed[5] subsequently to yourself. I find the strain upon me here[6] very great. It is so difficult to report experimental lectures; and ordinary reporters make such a hash of them, that I resolved in New York to write out an account of each of them, have it set up in type, so that the editors of the principal newspapers may have some guidance.[7] The manner in which the lectures are received is cordial in the extreme: in fact the audience is the most enthusiastic that I have met in the United States. And I am told it is altogether without a parallel as to quality: the whole mental force and refinement of the city—literature science & art being represented in it. I have kept myself strong thus far and hope to do so to the end. You remember my once giving a screed about Governor Eyre[8] in Willis's Rooms,[9] and you saying to me at the time, "I wish you had spoken your speech instead of reading it". That casual saying has influenced me in giving these lectures; I hardly ever look at my notes. I dine at 6½; take half a bottle of good claret[10]—sometimes costing as much as 25/-! usually about 20/-. In fact the price of wine here is atrocious. Every thing indeed is dear; but having lots of money expense is a matter of no moment to me. To keep my brain in order is the prime object of my life, and this by wisely partaking & by wisely abstaining I have managed to do thus far.

This Greenwich matter seems to me a very important one; but I hope they will not commit the error which I am preaching against here, and so overload you with official duties as to leave you no time for higher ones. It appears to me that their wisest plan would be to take you at your own terms. Your official duties ought to be those of organization and supervision, and you ought to lecture only when you deem it fit to do so. That fine establishment the International College will, I fear, go to ruin because the principal, though a first rate teacher in a class, has no faculty in the way of supervision.[11] It would pay well if he had that faculty to compel him to withdraw from tuition altogether and attend to the weightier matters. Of course at Greenwich you would have all our sympathies, and if any one *[of]* us could lend you a helping hand it would be a joy to do so. The retiring pension is also an element of moment.

My life since I came here has been the most unexciting possible. There was a grand reception at Appleton's on the evening of my arrival. That is about all thus far. I am so pressed by the preparation of the lectures that I have steadily declined all offers of hospitality. This week I give myself a little liberty.

You will probably have learned from the newspapers that I have had some experience of New York burglars.[12] Immediately after I went out to my first lecture a robber slipped into my rooms, and on my return I found the hoziery of a portmanteau in my sitting room strewn about the floor. I thought a dog or a romping child must have got into my room. But on looking into my bedroom I saw a second portmanteau that I had carefully locked wide open; the lock had been broken, such papers as the portmanteau contained strewn about, checkbooks, letters and other things formed a litter and a purse containing 30 or 40 pounds made away with. I feel a kind of respect for the scoundrel. He flew high, and disdained taking any thing of lower value than the precious metal. A scrap of a watch-chain which lay on the table he also captured.

The weather here is intensely cold, but bright and exhilarating. We have had heavy snow, and the streets for a day or two were execrable. There is a park—the Central Park about 2½ miles from here & thither I walk frequently.

Scaters[13] swarm like Ants over the ice of the ponds. I have had a letter from Faber[14] whom I think you knew in Marburg. It is dated from Utah and is very well written. He asked me to call upon his wife in Brooklyn.[15] The letter professes, and indeed seems to be that of a reformed man. Last night I had a letter[16] from January Searle:[17] dated from Jersey City, which is close to New York. Deputations from various western cities have come to me pressing me to visit them.[18] But I have set my face steadily homeward, and early February will see me on the Atlantic.[19] Best love to Lilly[20]—

always yours affectionately | John Tyndall

RI MS JT/1/T/697
RI MS JT/1/HTYP/610–11

1. *Brevoort House*: see letter 3843, n. 3.
2. *your last letter*: letter 3911.
3. *the telegram*: telegram missing.
4. *notions regarding Greenwich*: referring to Hirst's decision about whether to accept the position of Director of Studies at the new Royal Naval College at Greenwich (see letter 3911, n. 12).
5. *I telegraphed*: telegram missing.
6. *here*: Tyndall delivered six lectures in New York from 17–31 December 1872 at the Cooper Institute.
7. *I resolved... have some guidance*: see letter 3919, esp. n. 6.
8. *Governor Eyre*: Edward John Eyre (1815–1901), an English colonial governor, controversially known for his violent repression of the 1865 Morant Bay rebellion, while he served as governor of Jamaica (see letter 3796, n. 18).
9. *Willis's Rooms*: Tyndall spoke at a meeting of the Eyre Defence Committee, held at Willis's Rooms (also known as Almack's Assembly Rooms, a social club on King Street, St James's, London), on 18 February 1867. A transcript of the meeting, including Tyndall's speech, was published as Appendix C in H. Hume, *The Life of Edward John Eyre, Late Governor of Jamaica* (London: Richard Bentley, 1867), pp. 277–88, Tyndall's speech on pp. 278–85.
10. *good claret*: Tyndall's favorite wine was La Rose.
11. *International College... way of supervision*: London International College was a secondary school, catering to international students, that operated between 1867–89. Its headmaster was the classicist Leonhard Schmitz (1807–90).
12. *some experience of New York burglars*: see letter 3930 to Hector Tyndale.
13. *Scaters*: ice skaters.
14. *a letter from Faber*: letter 3894.
15. *Brooklyn*: Tyndall delivered six lectures in Brooklyn from 4–20 January 1873 at the Academy of Music.
16. *letter*: letter missing.
17. *January Searle*: George Searle Phillips (1816–89), a British writer and journalist who

moved to the United States and worked for various newspapers. He suffered a mental collapse in 1873 and was committed to an asylum in New Jersey. 'January Searle' was a pseudonym (*ODNB*).

18. *Deputations... to visit them*: in previous letters, Tyndall shared that he received invitations to lecture in Cincinnati, Cleveland, Columbus, Providence, and St. Louis, among other cities (see letters 3885 and 3886).
19. *early February will see me on the Atlantic*: see letter 3928, n. 18.
20. *Lilly*: Emily Hirst.

To Miss Moore[1] 22 December 1872 3935

Brevoort House,[2] | Fifth Avenue. | New York | 22nd Dec. 1872.

My dear Miss Moore.

The wind is whistling outside. I am near the top of this "house", whither I ascended for the sake of quiet, and overlook the roofs of many of the surrounding houses. They are laden with snow which fell a few days ago, and which when it thawed rendered the streets of New York execrable. But the frost has set in again and the weather is intensely cold. Overhead however is a blue sky, and the sun shines down upon the white roofs. I am at peace today & turn that peace to account in writing to you. You are probably pretty well acquainted by this time with my movements through America.[3] The presidential election[4] caused a postponement of two of the courses of lectures and this will detain me here longer than I at first intended to remain. I opened the campaign at Boston[5] where I was exceedingly well received. Then I allowed myself a little trip to Niagara[6] of which you have probably heard. Then I went on to Philadelphia,[7] and on my arrival there found my assistant M[r]. Miller[8] ill in bed—I will not dwell upon this subject which quite overshadowed me for a time. Suffice it to say that all that practical prudence care and tenderness could do was done for him. The quaker element was strong in the Philadelphia audience,[9] but they warmed up, and were positively enthusiastic at last. Then I went to Baltimore[10] where I met much kindness and cordiality—Thence to Washington[11] where I had all manner of big people among my pupils, the president,[12] Cabinett, judges senators and minor lights. I talked plainly to them and they received what I had to say kindly. I met at Washington the most cordial treatment possible. Then I came to New York,[13] where I found myself instantly surrounded by crowds of friends. At all the places that I have lectured the audiences have been, I am told, altogether unusual. At Philadelphia & Washington they were unparalleled. The surprise of those who know the city is still greater in New York. The audience here: certainly the warmest that I have met. On the evening of my first lecture I had some experience of New York burglars[14]—perhaps my robber may have been a

London thief—they, I am told, are the most accomplished here. However be that as it may, I found on my return from the lecture my hoziery strewn about the floor, the lock of my portmanteau broken and a purse containing 30 or 40 pounds taken away. The scoundrel left me my socks and shirts for which I respect him.

I wish you & your sister,[15] to whom remember me kindly, a happy new year. Tell M[rs]. Faraday[16] that I will write to her one of these days. I turn my face homeward early in February.[17]

Yours ever faithfully | John Tyndall

RI MS JT/1/T/1076

---

1. *Miss Moore*: either Harriet Jane Moore (1801–84) or Julia Moore (1803–1904), both daughters of Harriet Moore (1779–1866) and James Carrick Moore (1762–1860), a surgeon and promoter of vaccination. Julia was a friend of Tyndall's as was the rest of her family. Harriet Jane was a member of the RI since 1852 and was a watercolor artist who painted scenes of Faraday's apartment, study, and laboratory at the RI during the 1850s (F. A. J. L. James, 'Harriet Jane Moore, Michael Faraday, and Moore's Mid-Nineteenth-Century Watercolours of the Interior of the Royal Institution', in J. Hamilton (ed), *Fields of Influence: Conjunctions of Artists and Scientists, 1815–1860* (University of Birmingham Press, 2001), pp. 111–28). Another sister, Louisa, died in 1853. A son, John Carrick Moore (1805–98), was a field geologist and an active member of the Geological Society, becoming a fellow in 1838 and serving as secretary (1846–52); he was also a member of the RI from 1836, becoming a visitor in 1853 and a manager in 1857. The Moore family often entertained a number of well-known scientists at their home in Wimbledon, such as Charles Babbage and Tyndall (*Ascent of John Tyndall*, pp. 109, 144–5).
2. *Brevoort House*: see letter 3843, n. 3.
3. *movements through America*: for Tyndall's American lecture tour, see Introduction, n. 8.
4. *presidential election*: see letter 3794, n. 3.
5. *Boston*: Tyndall delivered six lectures in Boston from 15–25 October 1872 at the Lowell Institute.
6. *trip to Niagara*: see letter 3738, n. 5.
7. *Philadelphia*: Tyndall delivered six lectures in Philadelphia from 11–22 November 1872 at the Horticultural Hall.
8. *M[r]. Miller*: see letter 3896, n. 10.
9. *quaker element... Philadelphia audience*: see letter 3896, n. 5.
10. *Baltimore*: Tyndall delivered three lectures in Baltimore from 26–9 November 1872 at the Peabody Institute.
11. *Washington*: Tyndall delivered six lectures in Washington, DC, from 3–12 December 1872 at Lincoln Hall.

12. *president*: Ulysses S. Grant.
13. *New York*: Tyndall delivered six lectures in New York from 17–31 December 1872 at the Cooper Institute.
14. *experience of New York burglars*: see letter 3930 to Hector Tyndale.
15. *your sister*: Harriet Jane Moore or Julia Moore (see n. 1).
16. *M^rs^. Faraday*: Sarah Faraday.
17. *I turn my face homeward early in February*: see letter 3928, n. 18.

To Elizabeth Dawson Steuart 22 December 1872 3936

Brevoort House,[1] | Fifth Avenue | New York. | 22nd Dec. 1872.

My dear Mrs Steuart,

Before this reaches you the New Year will have come in—I wish you a happy one.

My campaign in America[2] has been thus far wonderfully pleasant and successful.

I return early in February.[3]

Yours ever | John Tyndall

RI MS JT/1/TYP/10/3407
LT Typescript Only

---

1. *Brevoort House*: see letter 3843, n. 3.
2. *My campaign in America*: for Tyndall's American lecture tour, see Introduction, n. 8.
3. *I return early in February*: see letter 3928, n. 18.

From Manton Marble[1] 23 December 1872 3937

532, Vth Ave. | Dec 23 '72

My dear Prof. Tyndall,

The Manhattan Club[2] is not far from your lodgings,[3] and you will find in the reading room, the Times & the best London weeklies, to say nothing of the American newspapers & weekly & monthly journals.

The room up stairs where we waited for dinner is a quiet corner all day at least, to read or write in.

The cards enclosed give you all the rights of membership until past the time of the next meeting of the Managing Committee, when your acceptance of them will be asked during the whole period of your stay in the country. And late may you return to England.

Some time to day you will receive Vol II of Middlemarch[4] which I have finished since I saw you & I cleave to my opinion that Dorothea[5] is the loveliest creature of fiction—the woman to hold a man to his best ideals & inspire him with better.

And along with the book is a box of the cigars I brought from Cuba.

Please don't bother to mention their arrival. My own servant will carry the parcel & I shall know that it reaches you.

Faithfully Yours | Manton Marble

Prof. John Tyndall

RI MS JT/1/M/47

---

1. *Manton Marble*: see letter 3931, n. 1.
2. *Manhattan Club*: a social club founded in New York in 1865.
3. *your lodgings*: Tyndall was staying at Brevoort House (see letter 3843, n. 3).
4. *Vol II of Middlemarch:* George Eliot's *Middlemarch, A Study of Provincial Life* was published in installments between 1871 and 1872.
5. *Dorothea*: Dorothea Brook, the protagonist of *Middlemarch*.

From Frederick Merrick[1] 23 December 1872 3938

Delaware, Ohio. Dec. 23d 1872.

Prof. Tyndall

Dear Sir

Will you excuse the impertinence of a stranger in addressing you through this note. Your name and writings have been familiar to me for some time. I am your debtor to no small amount. While you have afforded me much pleasure may I tell you in what you have made me sad? I have been made sad, my dear Sir, unspeakably sad, when I have read in your writings what tends to undermine the faith of men in the Bible as a revelation from God; and of course in the religion of the Bible—in Christ and his salvation.[2] I do not blame you for uttering your convictions. We must be honest. Truth is what we all need. None of us probably have it unalloyed. It becomes those to be tolerant who confess themselves fallible; and who may not err.

I have myself been a loving student of nature—more years, probably, than yourself but in an humble way. I love science still, tho' occupied less with its study than formerly. I accept as the thought of God, all its teachings. But I have learned that we may miss-read the thought. I cannot now teach as science what I once taught, in common with all, with confidence. Science may yet need "purifying", as well as religion.

I am also a student of the Bible; and, as with science, the more I study it the more I love it, and the more am I convinced it is from God. I cannot believe Christ was a vile imposter. I feel I grow better as I study his character and strive to become like him. I have had more than ten thousand students more or less under my care; and I find the more they think of Christ the truer and nobler they become. Nothing so holds them back from what would corrupt and degrade them; and nothing so awakens aspirations for all that tends to purify and exalt them, as faith in Christ. As an educator I would rather part with any other influence than this.

Laying down my pen after writing the above I picked up a volume of George MacDonald's[3] works and opened on these words "Then I thought of the wind which bloweth where it listeth, which is everywhere, and I quite forgot to open my Plato and thanked God for the Life of life whose story and whose words are in that best of books, and who explains everything to us, and makes us love Socrates and David, and all good men ten times more; and who follows no law but the law of love, and no fashion but the will of God; for where did ever one read words less like moralizing and more like simple earnestness of truth than all those of Jesus?"[4] And yet now utter untruthful he must have been if there is no supernatural intervention with the course of nature. He was constantly professing such intervention, and often under circumstances of peculiar solemnity where to trifle with men's most sacred beliefs would indicate a character despicable for its audaciousness and utter recklessness of truth. Can this be the character of one whose life and teachings are doing more than all else to purify and ennoble humanity?

Do not understand me as charging you with calling Christ an imposter: I am only saying how it seems to me I must regard him if I adopt your teaching on the subject of supernatural interference.

I am glad to find you saying: "It is no departure from scientific method to place behind natural phenomena a universal Father who in answer to the prayers of his children alters the currents of those phenomena"[5]—or of course for other sufficient reasons. I thank you in the interests of science and religion for these words. They are without doubt true. Would that you could see and take one step more in the same direction—accept the fact. But you must have proof, and proof by experiment. But, my respected Sir, are isolated facts to be proved by experiment? I beg you to reexamine this point. Have you not fallen into an error here?

I have only reached the point on which I did wish to venture a few suggestions for your consideration. But I have perhaps gone already too far. O, this touching the faith of men in reference to their relation to God is a serious thing. To mislead here is fearful indeed. Father of Light; "what in us is dark illume."[6] I have about done with this life: all my hope for the future is based on the Bible. As yet it is a rock unmoved.

Believe me, dear Sir, I have so much faith in my prayer that I daily pray for you, that you may be led into all truth.

Again begging you to excuse this intrusion, allow me to subscribe myself.

Most respectfully yours, | F. Merrick

RI MS JT/1/M/99

---

1. *Frederick Merrick:* Frederick Merrick (1810–94), second president of Ohio Wesleyan University from 1860–73. One of his legacies was to establish a lecture series at the university on the subject of experimental and practical religion (D. C. Thomas, 'Memoirs. Rev. Frederick Merrick', *Minutes of the Seventy-Ninth Session of the Ohio Annual Conference of the Methodist Episcopal Church* (Columbus, OH: Gazette Printing House, 1890), pp. 66–8).
2. *your writings... his salvation*: a reference to the prayer gauge controversy (see letter 3780, n. 2).
3. *George MacDonald*: see letter 3923, n. 7.
4. *"Then I thought... of Jesus"*: from G. MacDonald, *Annals of a Quiet Neighborhood* (New York: Harper and Brothers, 1867), p. 42.
5. *"It is no... those phenomena"*: from J. Tyndall, 'Science and Religion', *Popular Science Monthly*, 2 (November 1872), pp. 79–82, on p. 81.
6. *"what in us is dark illume"*: quote is adapted from the first book of John Milton's *Paradise Lost*, 'what in me is dark, | Illumine; what is now, raise and support | That to the highth of this great argument | I may assert Eternal Providence, | And justify the ways of God to men'.

To Heinrich Debus 26 December 1872 3939

Brevoort House[1] New York 26th. Dec. 1872. |
Corner of 5th. Avenue & 8th Street

My dear Heins

Christmas has come and gone and here we are in the midst of a New York winter. The cold is intense. I am writing in my bedroom which however possesses a comfortable fire, but outside the snow is whirling through the air, and the wind sounds mournfully through the chinks of the windows. Now and then the rustle of the snow, which is dry and granular, is heard against the window panes. Perhaps you asked yesterday how is John spending his

Christmas? Thus. Wrote during the early part of the day. In the afternoon had a 6 mile walk: returned and went to the Cooper Institute[2] where I had a battery of 40 calls waiting for me, and made sure of my experiments on the magnetization of lights which are to come off this evening. So you see I have not been spending the holiday time in dissipation. I might however do so if I would, for the people here are unspeakably kind and hospitable to me.

I reached New York on Saturday week, and on that evening there was a great reception at Mr. Appletons. Here I saw many of the celebrities of New York including General McClellan,[3] Bryant[4] the poet and the bishop.[5] I had met the kindly old man years ago at the seat of Lord Stanhope.[6] I have done my best since to keep out of dissipation and have in a great measure succeeded. There was a banquet on Monday at the Century Club.[7] On Tuesday there was a gathering at Hoboken where I met various church dignitaries. Among them the Bishop of America[8]—a most charming patriarchal old man. You cannot imagine the stir this prayer question[9] has excited in America. Still the clergy though some of them attack me, do not on the whole hold aloof from me. The bishop of New York[10] had been invited to Monday's banquet; but though full of kindness towards me personally he thought it better not to offend the weaker brethren. This is the only practical illustration I have had as to the manner in which the people look upon me here.

I have sent you papers from time to time which reveal to you how I am getting on. The audience is enthusiastic and I am assured without a parallel in the history of New York. Dr. Bellows[11] an eminent tractarian[12] minister said this in his speech at the "banquet", and everybody says the same. It has been thus throughout, so that I cannot complain of my reception. I really think it will do good in various ways. In my final lecture in which the course will be summed up I intend to tell some home truths regarding science. And the best of it is, they are prepared to hear such truths, and I believe to act upon them.

Tonight I deliver my fourth lecture, then two remain. Afterwards I give the whole six in the adjacent city of Brooklyn;[13] then two lectures to the students at Yale,[14] and then hurrah for England![15] Kind as the people are to me and pleasant as they have made my visit I long to be once more among my own old friends. If you write to me I think it would be best to address it to Appleton & Co Broadway New York.

goodbye dear Heins: a happy New Year! | Yours ever | John Tyndall

RI MS JT/1/T/269

---

1. *Brevoort House*: see letter 3843, n. 3.

2. *Cooper Institute*: Cooper Union, which hosted Tyndall's New York lectures (six lectures from 17–31 December 1872). See letter 3873, n. 21.
3. *General McClellan*: George B. McClellan (1826–85), general-in-chief of the Union forces during several months of the American Civil War (1861–5). In 1864 he ran for president on the Democratic Party ticket, losing to the incumbent, Abraham Lincoln. By the early 1870s, he was the chief engineer of the Department of Docks in New York City and president of the Atlantic and Great Western Railroad (*ANB*).
4. *Bryant*: William Cullen Bryant (1794–1878), an American poet and newspaper editor, noted for his poem 'Thanatopsis', written in the early 1800s. In the early 1870s, he was engaged in translating Homer's *Odyssey* (*ANB*).
5. *the bishop*: likely a reference to Horatio Potter (1802–87), professor of mathematics and natural philosophy at Washington College in Hartford, Connecticut, in the late 1820s and early 1830s. From 1861–7 he served as the Episcopal Bishop of New York, and traveled to England in the 1840s (G. E. DeMille, 'The Episcopate of Horatio Potter (1802–1887), Sixth Bishop of New York, 1854–1887', *Historical Magazine of the Protestant Episcopal Church*, 24 (1955), pp. 66–92).
6. *seat of Lord Stanhope*: Chevening House in Kent, home of Philip Henry Stanhope.
7. *Century Club*: the Century Association, a New York social club founded in 1847, of which William Cullen Bryant was a member.
8. *Bishop of America*: since the Catholic Church does not designate a bishop responsible for the whole of the United States, this might refer to Benjamin B. Smith (1794–1884), the presiding bishop of the Episcopal Church in the United States of America from 1868–84 (W. R. Insko, 'Benjamin Bosworth Smith, 1794–1884', *Historical Magazine of the Protestant Episcopal Church*, 22 (1953), pp. 146–228).
9. *this prayer question*: a reference to the prayer gauge controversy (see letter 3779, n. 2).
10. *The bishop of New York*: see n. 5.
11. *D^r^. Bellows*: likely Henry Whitney Bellows (1814–82), an American Unitarian minister, editor of a Unitarian newspaper, and co-founder of Cooper Union, which hosted Tyndall's lectures (*ANB*).
12. *tractarian*: distributing tracts (*OED*). Bellows does not appear to have had links to the Tractarian movement surrounding the Oxford Tracts of 1833–41, which sought closer links between Anglicanism and Roman Catholicism.
13. *Brooklyn*: Tyndall delivered six lectures in Brooklyn from 4–20 January 1873 at the Academy of Music.
14. *Yale*: Tyndall delivered two lectures in New Haven, CT, on 22 and 23 January at Yale College.
15. *hurrah for England!*: see letter 3928, n. 18.

From John Eaton[1] 26 December 1872 3940

[Department of the Interior, | BUREAU OF EDUCATION, | *Washington, D.C.,*] Dec. 26[, *1872.*]

Dear Sir.

In closing your lectures here[2] you gave some very excellent suggestions with regard to the aid that our country ought to extend to learning. We are closing our Annual Report on education for the country,[3] & I should esteem it a favor if you would take pains to send such an extract from those remarks as you would be willing I should insert in my representation of the educational progress of the year.[4]

I am confident it would be useful, as not only reaching the audience to whom it was addressed, but thoughtful men scattered over the entire nation.

Very truly yours. | John Eaton | Commissioner

Prof. John Tyndall | Care D. Appleton & Co. | New York City.

RI MS JT/1/E/1

---

1. *John Eaton*: John Eaton (1829–1906), appointed Commissioner of Education in 1870 under President Ulysses S. Grant, a post he held through 1886. Previously, Eaton served as a Union colonel and general during the American Civil War (1861–5) (E. O. Mason, 'John Eaton: A Biographical Sketch', in J. Eaton, *Grant, Lincoln and the Freedmen: Reminiscences of the Civil War* (New York: Longmans, Green, and Co., 1907), pp. ix–xxxiv).
2. *your lectures here*: Tyndall delivered six lectures in Washington, DC, from 3–12 December 1872 at Lincoln Hall.
3. *Annual report... country*: the report was subsequently published as *Report of the Commissioner of Education for the Year 1872* (Washington, DC: Government Printing Office, 1873).
4. *an extract... of the year*: while Tyndall's reply is missing, he did furnish Eaton with the conclusion of his lectures on light, which were published in part in the report cited in n. 3, on pp. xliv–xlvi. For Tyndall's conclusion, see *Six Lectures on Light*, pp. 207–26.

From Henry W. Matthews[1] 26 December 1872 3941

[CHIEF ENGINEER'S OFFICE. | Fire Department. | *Rochester, N.Y.,*] December 26th [*187*]2

Prof John Tyndall

Dear Sir. I have recently read in your "Fragments of Science"[2] and more lately in public prints[3] of a respirator for the use of Firemen, invented by you.[4]—

My experience of over fifteen years connection with the Fire Dep[artmen]t of this city has often illustrated the desirability of such an article and I have taken the liberty to address you in reference to it. First I wish to inquire whether the respirator can be obtained in this Country and if not I should feel under great obligation if you would give me more specific information than is contained in the description I have seen and which would enable me to construct one.

The points upon which I desire information are: the thickness or amount of cotton used, the manner of applying the Glycerine and Charcoal, whether the Charcoal should be animal or vegetable and how finely pulverized also the quantity—Can the cotton be filled with the Charcoal and moistened with Glycerine and in that condition be kept ready for use—In what is the preparation enclosed?

Would such a respirator serve as a protection against Carbonic Acid gas[5] to which a person using the Babcock[6] and other similar fire Extinguishers is exposed?

Having often noticed the liberality with which you have imparted valuable scientific information to the public, I have made bold to address these inquiries to you and any reply you may see fit to make will be thankfully received and heartily appreciated.

Very Respectfully | Your Obedient Servant | Henry W. Matthews | Ass[istan]t Engineer—| Rochester Fire Department

RI MS JT/1/M/56

---

1. *Henry W. Matthews*: Henry W. Matthews (dates unknown), Assistant Engineer at the Rochester Fire Department in Rochester, New York.
2. *"Fragments of Science"*: J. Tyndall, 'Dust and Disease', in *Fragments of Science*, pp. 289–343, on pp. 334–5. This is the published version of a Friday Evening Discourse that Tyndall delivered at the RI on 21 January 1870 (*Roy. Inst. Proc.*, 6 (1870–72), pp. 1–14).
3. *public prints*: not identified.
4. *respirator... invented by you*: see letter 3902, n. 15.
5. *Carbonic Acid gas*: carbon dioxide.
6. *Babcock*: a brand name for a fire extinguisher based on sodium bicarbonate dissolved in water, which would be exposed to an acid to produce carbon dioxide gas which could be used to smother a fire.

From Charles Walkem[1] 27 December 1872 3942

Woodston Villa | Kingston, Ontario | Canada | 27th. Dec[embe]r 1872

My dear Venerated Tyndall

Your letter[2] of March 1844 is before me dated Leighlin Bridge[3]—beginning with "my ever venerated "Charley" &c" which I shall ever esteem as a precious souvenir from my attached friend & shall I say "Companion" I need not recite the occasion or circumstance of your writing me that letter on my leaving for duty connected with the disputed boundary between Lower Canada & the State of Maine under the Ashburton Treaty[4] suffice it to say that you & Latimer[5] had decided to leave the Ordnance Survey (Preston) for the United States, and you had promised if you were within five hundred miles of us you would assuredly come & see me and Mrs Walkem. I hoped time & place has not damped your ardor in this respect & that before you return to England you will spend a little time with us & our family. I need not say that no one in this world will be more welcome. I shall say no more at present than wish you many happy returns of the Season and with Mrs Walkems sincere esteem & affection.

Believe me always | your devoted & sincere | servant & friend | Charles Walkem.

formerly RS&M[6] on Ordnance Survey

RI MS JT/1/W/3

---

1. *Charles Walkem*: Charles Walkem (b. *c.* 1805), a friend of Tyndall's from their days in the Ordnance Survey of Great Britain. He was a Corporal in the 14th Company of the Royal Sappers and Miners. He enlisted to the army in September 1822. He was a carpenter by trade. He worked in the A District of the Ordnance Survey of Ireland, and as a Private was involved primarily in surveying duties. He became Lance Corporal in August 1838, and he later became the Superintendent at the District headquarters at Waterford. In this post he was Captain Tucker's right-hand man. He left the Irish Survey, along with most of the workers in A District, at some point between August and October 1841; the records for these months are missing so there is no exact date of departure. He was transferred to the Survey in Preston, but following the disputes between the Preston civil assistants and the Survey leadership, Walkem was transferred to Quebec, Canada, in 1844. His departure, according to Tyndall's friend John Todd, was 'universally regretted' (see letter 0294, in M. Baldwin and J. Browne (eds), *The Correspondence of John Tyndall, Volume 2: The Correspondence, September 1843–December 1849* (Pittsburgh: University of Pittsburgh Press,

2016)). He appears on the 1871 Canadian census living in Quebec along with a wife, Mary Anne Walkem, and four children.

2. *your letter*: letter missing.
3. *Leighlin Bridge*: Leighlinbridge, in County Carlow, Ireland, was Tyndall's birthplace and childhood home.
4. *disputed boundary... Ashburton Treaty*: the northeastern border of the United States was a matter of dispute through the early 1800s. The Ashburton Treaty established the consensus borders separating the US State of Maine from New Brunswick to the east and Quebec to the north.
5. *Latimer*: George Latimer (dates unknown), a friend of Tyndall's from their days in the Ordnance Survey of Great Britain. He was a civil assistant on the Ordnance Survey. He worked in the 4th Division in the C District, under the command of Lieutenant Robert Fenwick. Latimer worked on the Irish Survey from 1837–42, then moved to the English Survey. Latimer had a brother named William who also worked on the Survey in the same district. In 1844 he was considering emigrating to America with Tyndall, but changed his mind at the urging of his family.
6. *RS&M*: Royal Sappers & Miners.

To the Editor of the *New York Times*[1] 29 December 1872 3943

[Prof. Tyndall and the Presbyterians.] | *To the Editor of the New York Times:*

Having far greater reason to congratulate myself on the friendliness than to complain of the enmity of Christian men, I should be exceedingly unwilling to utter a word which might be needlessly unpalatable to them. But I had been distinctly and circumstantially informed that, in one of the prominent pulpits of this great City,[2] not only were my notions of religion[3] handled with severity, to which I have no manner of objection, but that my visit to this country had been spoken of as a money speculation.[4] This I thought a case of striking below the belt, and it drew from me the mild remark on which you have commented in a recent number of the TIMES.[5]

The assurance, however, has been brought to me this day, that my informant indulged in an inference instead of stating a fact, and that no such language was used toward me. I hope, therefore, the reverent gentleman to whom I referred, without naming him, will permit me to offer him a friendly hand, and to express my regret that any word of mine, founded on misinformation, should have given him a moment's pain or uneasiness.

Your obedient servant, JOHN TYNDALL. | BREVOORT HOUSE,[6] Sunday, Dec. 29, 1872.

New York Times, 30 December 1872, p. 4

1. *Editor of the New York Times*: at this time the editor of the *New York Times* was Louis John Jennings (1836–93), a British journalist who moved to the United States in 1865 as a special correspondent for the London *Times*. He was editor of the *New York Times* from 1870–6 (*ODNB*).
2. *one of the prominent pulpits of this great City*: referring to Presbyterian minister John Hall of New York.
3. *my notions of religion*: a reference to the prayer gauge controversy. See letter 3780, n. 2.
4. *my visit to this country had been spoken of as a money speculation*: for Tyndall's American lecture tour, see Introduction, n. 8. For the donation of his lecture tour profits, see Introduction, n. 15.
5. *the mild remark ... of the TIMES*: the *New York Times* report of Tyndall's fourth lecture in New York reported on his response to American Presbyterians who reacted against his statements regarding religion: 'He did not propose to notice their attacks in any way, being resolved to act according to the precepts of an old book which he greatly revered, "not to render railing for railing, but contrariwise, blessing"'. ('Tyndall's Fourth Lecture', *New York Times*, 17 December 1872, p. 5). This article elicited a letter to the editor, signed 'A Presbyterian', published a day before Tyndall sent this letter to the *New York Times*, which began: 'I notice in the TIMES of this morning that Prof. Tyndall means to be very magnanimous in relation to the Presbyterians, who have been "railing at him," as he alleges'. ('Prof. Tyndall and the Presbyterians', *New York Times*, 28 December 1872, p. 2).
6. *BREVOORT HOUSE*: see letter 3843, n. 3.

From Henry Bence Jones — 31 December 1872 — 3944

Tuesday | Dec[embe]r 31 1872

Dear friend

What good news it is to hear that you are longing for home. Somehow a rumor had spread that you were coming in a fortnight but I could find no ground for it. It will be famous if you are back by the middle of Feb[ruar]y.[1] Every body wants you. Thanks to Miss Savage[2] every thing goes on superficially well but the masters eye is wanted & the mere fact of the head authority being always near makes a vast difference in the way things are done. However we shall go on quite right until the middle of Feb[ruar]y. As to your & Cotterells[3] income from the Institution you can say just what pleases you to the managers but they will not be affected by it in the least & will I am certain insist that you do not give up for him or yourself one farthing. Possibly tho' not probably you may find that an increase to your income has been voted by the time you return but you will I think be home too quickly for this to be done. But at the least the members & managers will not hear of your not taking what we can so well afford to give. I have not been able to settle anything about your laboratories to my satisfaction.[4] There will be plenty of room

I expect for a magneto electric light machine[5] & vertical air engine[6] in the new boiler room & I hope to have got all the information from Paris before you return as to size, price &c. By the middle of Feb[ruary] the lower laboratory will be quite usable & it is yours as much as the other. Odling will be at Oxford,[7] the workmen will be all gone, but any number will return to do all you wish in your rooms as quickly as possible & by Easter you will I hope have all things in order & to your liking.

I am very sorry I forgot to ask about the papers going to you. I have been at the Institution but seldom & I never spoke to Smith[8] about them. He goes foolish sometimes & seems up in a balloon or anywhere else than at the Institution.

I send the preface of your 1st Washington lecture[9] with a note to Delane asking him to put it in the Times but it never appeared. One or two amusing extracts from the Washington papers about y[ou]r lectures I sent to Nature & they were published therein.[10] Everywhere your success has been as great as it could be & must be very satisfactory to you. The leading article in the new york Herald for the 16th of Dec[embe]r on original research[11] which you have just this moment caused to reach me is all that you could wish & holds out a permanent advantage to the Yankees from your visit. But oh the difficulty is to find the man for discovery & I look at our new rooms with a fear almost greater than my hope of new investigations to be made there when we are feeding the grass or doing some other useful work elsewhere.

I have been going on moderately badly & my dropsy[12] will not yield to my drugs. However it does not overwhelm me yet. Archibald Smith[13] is just dead he was my chief Cambridge ally but when I became a doctor I ceased to see him & so our alliance ended. He leaves six sons & two daughters!! Think of that for a burden & be thankful that you have not a flock to consider & to educate.

I am so glad you are coming home. I shall be happy now in counting the days to the 10th of Feb[ruar]y which I conjecture may be your time. I am right glad also that you are well & Cotterell[14] also.

I have no news to send you. I see no one. The new members are coming in as they have done the last few years & our works are so finished no one coming to the Juveniles[15] perceives any difference. All has been done in great style. Too great: The greatest anxiety I have is that you should like the change when you come back.

& now dear friend many many happy years to you & every good wish to you comes from us all & welcome home from me especially

Believe me yours always | H B J

RI MS JT/1/J/75

1. *back by the middle of Feb[ruar]y*: see letter 3928, n. 18.
2. *Miss Savage*: Sarah Savage.
3. *Cotterells*: John Cottrell.
4. *settle anything about your laboratories to my satisfaction*: on the renovation of the RI's laboratories, see letter 3776, n. 14.
5. *magneto electric light machine*: an array of magnets, set into rotation in order to generate a direct current and produce arc lighting, invented by the Belgian electrical engineer Zénobe Théophile Gramme (1826–1901). Letter 4099 indicates that such a machine was obtained for the RI.
6. *vertical air engine*: a kind of steam engine.
7. *Odling... Oxford*: William Odling (see letter 3889, n. 10).
8. *Smith*: Smith (first name unknown) was a porter at the RI, and while Tyndall was in the United States, he was tasked with mailing to Tyndall British newspapers with items of interest, though, as Tyndall noted previously in letters 3796, 3797, and 3928, Smith was dropping the ball on this.
9. *1st Washington lecture*: Tyndall delivered six lectures in Washington, DC, from 3–12 December 1872 at Lincoln Hall.
10. *amusing extracts... published therein*: 'Notes', *Nature*, 7 (26 December 1872), pp. 150–2, on p. 150.
11. *The leading article... original research:* 'Professor Tyndall's Plea for Original Science—His Lectures on 'Light'", *New York Herald*, 16 December 1872, p. 7.
12. *dropsy*: see letter 3782, n. 18.
13. *Archibald Smith*: Archibald Smith (1813–72), a Scottish mathematician. He was Senior Wrangler in the Cambridge Mathematical Tripos in 1836, and maintained an interest in science and mathematics while pursuing a career in law. He died on 26 December 1872 (*ODNB*).
14. *Cotterell*: John Cottrell.
15. *the Juveniles*: see letter 3782, n. 4.

From Hector Tyndale 31 December 1872 3945

Philad[elaphi]a: December 31st 1872

Dear John

I wish you a happy New-Year—as do all my people. Since my return home on Christmas day I have not been out of the house, except for a time yesterday for which time I felt no better. The cold and lung-pains seem to be *[tough]* and settled. Still I am better on the whole. I found that the use of wine or spirits deepened the cold & sharpened the pains and so since my return have been abstemious—and sleepless. Enclosed[1] I send your receipted bills from Miller's nurse,[2] from La Pierre House,[3] from Doctors Murchison & McClellan.[4] Also two notes from Murchison, one to you.[5] Other bills receipted & unreceipted I left with you in New-York. These bills include everything I know of against you on account of Miller's death or other account. Had my health been better these matters should have been attended to before.

By this mail I send to your care a note for Mr Cottrell.[6] Some days ago I wrote him, to the care of the Cooper Institute,[7] asking his address in New-York—that I might send him Mr Miller's luggage now in my house. Having had no reply I presume my letter miscarried. Please hand him this one I refer to of to day.

Do not forget to add to the list of the drafts[8] drawn by you on Appletons & Co,[9] which list I left with you in New-York, that drawn by you on the 25th (or 24th Dec) for Six Hundred & Five (605) dollars, to settle all the bills for account of Miller's death.

I presume you will count as part of your expenses, while in this country, the loss of your money &c from your trunk in New-York, as it was an insuperable risk, incurred by travelling. If health permits, I shall run over to New-York for a day or two next week. But it is possible, in the mean time, that you may find time to come over to see us before that.[10]

Your foreign letters I suppose are sent direct to New-York as none have come for you for about a fortnight. Again wishing you a happy New-Year

Yours affectionately | Hector Tyndale

The letter from Dr: Murchison to you refers to the decease of Mr Miller and will be of interest to you and to the relations.

RI MS JT/1/T/72

---

1. *Enclosed*: all enclosures referred to missing.
2. *Miller's nurse*: caretaker for Tyndall's deceased assistant, George Miller (see letter 3896, n. 10). Not otherwise identified.
3. *La Pierre House*: a hotel adjacent to the Academy of Natural Sciences in Philadelphia.
4. *Doctors Murchison & McClellan*: presumably the two Philadelphia physicians who cared for Miller.
5. *one to you*: letter missing.
6. *Cottrell*: John Cottrell.
7. *Cooper Institute*: Cooper Union, which hosted Tyndall's New York lectures (six lectures from 17–31 December 1872). See letter 3873, n. 21.
8. *draft*: the 'drawing' or withdrawing of money from a stock by means of an order written in due form (*OED*).
9. *Appletons & Co*: Tyndall's American publishing company.
10. *you may find time to come over to see us before that*: Tyndall visited with Tyndale in Philadelphia in late January 1873, before finalizing his time in the United States in New York.

From Hugh Spottiswoode[1] [1872][2] 3946

Combe Bank, | Sevenoaks.[3]

dear J.O.T[4]

thank you for kindly sending me your book.[5] I will read it and try to under-stand all you say in it

I hope you like life in America[6] and the Yankees too and I am sure they must think your Lectures very jolly We are going to London next Saturday Combe is so jolly Good bye with love

from your affect[io]n[a]t[e] | Hugh

RI MS JT/1/S/196

1. *Hugh Spottiswoode*: William Hugh ('Hughie') Spottiwoode (1864–1916), the eldest son of Eliza and William Spottiswoode and an editor. He founded Printers' Pie and Winter's Pie, publications that supported the Printers' Pension Corporation. After his father's death he became a partner in the firm Eyre & Spottiswoode. In 1899 he donated his father's collection of physical apparatus to the RI where he also worked as a manager ('Literary Gossip', *Athenaeum*, 28 August 1915, p. 146).
2. *[1872]*: the year is given by reference to Tyndall's lecture tour in the United States (see Introduction, n. 8). It is possible this letter is from 1873, which would place it within this volume as well.
3. *Combe Bank, | Sevenoaks*: see letter 3798, n. 12.
4. *J.O.T*: Jolly Old Tyndall, a moniker given to Tyndall by Hugh and his brother Cyril Andrew Spottiswoode (1867–1915), the second son of Eliza and William Spottiswoode who later served as director of the printing firm Eyre & Spottiswoode.
5. *your book*: not identified; possibly *Forms of Water*, which was published just before Tyndall left England for the United States in September.
6. *life in America*: for Tyndall's lecture tour in the United States, see Introduction, n. 8.

1873

From Thomas Henry Huxley — 1 January 1873 — 3947

4 Marlborough Place | Abbey Road N.W.[1] |
January 1st. 1872.[2]

My dear Tyndall,

I cannot let this day go by without wishing you a happy New Year, and lamenting your absence from our customary dinner.[3] But Hirst and Spencer and Michael Foster[4] are coming and they shall drink your health in champagne while I do the like in cold water making up by the strength of my good wishes for the weakness of the beverage.

You see I write from the new house.[5] Getting into it was an awful job, made worse than needful by the infamous weather we have had for weeks and months, and by the stupid delays of the workmen whom we had fairly to shove out at last as we came in. We are settling down by degrees and shall be very comfortable by and bye though I do not suppose that we shall be able to use the drawingroom for two or three months to come. I am very glad to have made the change but there is a drawback to everything in "this here vale" as Mrs Gamp says,[6] and my present thorn in the flesh is a neighbour,[7] who says I have injured him by certain operations in my garden and is trying to get something out of me by Chancery proceedings. Fancy finding myself a defendant in Chancery![8]

It is particularly hard on me as I have been especially careful to have nothing done without Burton's[9] sanction and assurance that I was quite safe in law and I would have given up anything than have got into bother of this kind. But 'sich is life.'[10]

You seem to have been making a Royal Progress[11] in Yankee-land. We have been uncommonly tickled with some of the reports of your lectures which reached us, especially with that which spoke of you having 'a strong English accent.'[12]

The loss of your assistant[13] seems to have been the only deduction to be made from your success. I am afraid you must have felt it very much in all ways.

"My Lord" received your telegram only after the business of "securing Hirst" was done.[14] That is one of the bright spots in a bad year for me. Goschen[15] consulted Spottiswoode and me independently about the headship of the new Naval college and was naturally considerably surprised by the fact that we coincided in recommending Hirst—I doubted whether Hirst would leave the University[16] so I suggested Henry Smith[17] as an alternative man. Goschen asked Spottiswoode and me to meet Admiral Key[18] who is to be the naval head of the new Institution, and Dr. Woolley.[19] We had several hours' talk over the business and the upshot was that Goschen asked me to communicate with Hirst and see if he would be disposed to accept the offer. So I did and found to my great satisfaction that Hirst took to the notion very kindly. I am sure he is the very best man for the post to be met with in the three kingdoms,[20] having that rare combination of qualities by which he gets on with all manner of men, and singularly attracts young fellows. He will not only do his duty, but be beloved for doing it, which is what few people can compass.

I have little news to give you. The tail of the Ayrton Hooker storm[21] is drifting over the scientific sky in the shape of fresh attacks by Owen on Hooker. Hooker answered the last angelically and I hope they are at an end.[22]

The wife[23] has just come in and sends her love (but is careful to add "second-best") The chicks grow visibly and audibly and Jess looks quite a woman.[24] All are well except myself, and I am getting better from a fresh break-down of dyspepsia.[25] I find that if I am to exist at all it must be on strictly ascetic principles, so there is hope of my dying in the odour of sanctity[26] yet. If you recollect, Lancelot did not know that he should "die a holy man"[27] till rather late in life. I have forgotten to tell you about the Rectorship of Aberdeen.[28] I refused to stand at first, on the score of health, and only consented on condition that I should not be called upon to do any public work until after the long vacation. It was a very hard fight and although I had an absolute majority of over 50 the mode of election is such that one vote, in one of the four nations would have turned the scale by giving my opponent[29] the majority in that nation. We should then have been ties, and as the chancellor,[30] who has under such circumstances a casting vote, would have (I believe) given it against me, I should have been beaten.

As it is, the fact of anyone, who stinketh in the nostrils of orthodoxy, beating a Scotch peer at his own gates in the most orthodox of Scotch cities, is a curious sign of the times. The reason why they made such a tremendous fight for me, is I believe that I may carry on the reforms commenced by Grant Duff,[31] my predecessor. Unlike other Lord Rectors, he of Aberdeen is a power and can practically govern the action of the University during his tenure of office.

I saw Pollock[32] yesterday, and he says that they want you back again.[33] Curiously enough the same desire is epidemically prevalent among your friends and not least here.

Ever yours | T. H. Huxley.

IC HP 8.131
Typed Transcript Only

---

1. *4 Marlborough Place | Abbey Road N.W.*: the Huxleys moved to this London address in mid-1872, having purchased it in June. Previously, they were at 26 Abbey Place, St. John's Wood from 1861–72.
2. *1872*: Huxley wrote '1872' instead of '1873'.
3. *our customary dinner*: Huxley traditionally hosted a New Year's Day dinner, with X Club members and other friends. Tyndall could not visit as he was in the United States for a lecture tour (see Introduction, n. 8).
4. *Michael Foster*: Michael Foster (1836–1907), a British physiologist, was a student and close associate of Huxley, whom he succeeded as the Fullerian Professor of Physiology and Comparative Anatomy at RI, from 1869–72 (*ODNB*).
5. *the new house*: see n. 1.
6. *"this here vale"... says*: Sarah Gamp is a character in Charles Dickens's *Martin Chuzzlewit* (1844), who sprinkles her speech with the pronouns 'this here' and 'that there'.
7. *neighbour*: not identified.
8. *Chancery*: the Court of Chancery was a court in England and Wales that functioned in parallel with the common law courts, with which it was merged in 1873.
9. *Burton*: Edward F. Burton was Huxley's solicitor.
10. *'sich is life'*: Huxley again invokes Dickensian dialect. This line is spoken by Mr. Dennis in *Barnaby Rudge* (1841).
11. *Royal Progress*: a term used to describe a monarch's tour of his or her kingdom.
12. *'a strong English accent'*: the American press routinely commented on Tyndall's accent, evidently unaware of his Irish origins.
13. *your assistant*: see letter 3896, n. 10.
14. *"My Lord"... "securing Hirst"*: a reference to efforts to hire Thomas Hirst as Director of Studies of the new Royal Naval College, Greenwich. Tyndall reported telegramming Huxley regarding Hirst's appointment, so by "My Lord" Huxley perhaps refers to himself. See letters 3911 and 3934.
15. *Goschen*: George Goschen (see letter 3991, n. 11).
16. *University*: University College, London.
17. *Henry Smith*: probably Henry John Stephen Smith (1826–83), a British mathematician who at the time held the Savilian Chair of Geometry at Oxford University (*ODNB*).

18. *Admiral Key*: Astley Cooper Key (1821–88), the first president of the Royal Naval College, Greenwich (*ODNB*).
19. *Dr. Woolley*: Joseph Woolley (1817–89), an Anglican clergyman and fellow of the Royal Astronomical Society who was the Inspector General and Director of Studies at the Royal School of Naval Architecture and Marine Engineering, the predecessor institution to the Royal Naval College, Greenwich (*ODNB*).
20. *three kingdoms*: England, Scotland, and Ireland.
21. *Ayrton Hooker storm*: see letter 3755, n. 1.
22. *fresh attacks... at an end*: Joseph Hooker replied in print to Richard Owen's latest attacks in: 'Dr. Hooker's Reply to Prof. Owen', *Nature*, 6 (24 October 1872), pp. 516–17. See also letter 3807, n. 3.
23. *the wife*: Henrietta Huxley.
24. *the chicks... a woman*: Huxley and his wife had seven surviving children (see letter 3834, n. 12), the eldest of whom, Jessie Oriana Huxley (1858–1927), was on the cusp of turning 15.
25. *dyspepsia*: see letter 3839, n. 3.
26. *odour of sanctity*: a state of saintliness or moral rectitude (*OED*).
27. *"die a holy man"*: quote from 'Lancelot and Elaine' in A. Tennyson, *Idylls of the King* (London: Edward Moxon & Co., 1859), p. 222.
28. *Rectorship of Aberdeen*: Huxley became Lord Rector of the University of Aberdeen in 1873 (through 1880), having been elected by the students there on 14 December 1872. See letter 4041, n. 1.
29. *my opponent*: Huxley's opponent, Charles Gordon (1847–1937), 11th Marquess of Huntly, was later Rector of Aberdeen from 1890–99 (*Alum. Cantab.*). Adrian Desmond quoted from this letter in his biography of Huxley: 'Huxley came in over a clansman, the Marquis of Huntley; and a Sassenach "who stinketh in the nostrils of orthodoxy, beating a Scotch peer at his own gate... is a curious sign of the times"' (Desmond, *Huxley*, p. 439).
30. *the chancellor*: Charles Gordon-Lennox (1818–1903), 6th Duke of Richmond, was chancellor of the University of Aberdeen from 1861–1903 (*ODNB*).
31. *Grant Duff*: Mountstuart Elphinstone Grant Duff (1829–1906), a Scottish politician and civil servant, was Rector of the University of Aberdeen from 1866–72 (*ODNB*).
32. *Pollock*: William Frederick Pollock.
33. *want you back again*: to return from the United States; for Tyndall's American lecture tour, see Introduction, n. 8.

From Joseph Henry 2 January 1873 3948

January 2^d 73

My dear Sir:

Permit me to introduce to your acquaintance the bearer of this note, Mr. George Mayer, of New York. He is the Engineer in the Delamater Iron

Works[1] in charge of the manufacturing of the Daboll Fog Signals[2] used by the Light House Board. He is anxious to make your acquaintance and will, I trust, be able to give you some information on a subject in which you are much interested.[3]

I remain, very truly | yours, etc | Joseph Henry

Prof. John Tyndall, | *<1 word missing>*

Smithsonian Institution Archives. RU 33, Vol. 31, Page 711

---

1. *Delamater Iron Works:* a Manhattan foundry established by Cornelius Henry DeLamater (1821–89).
2. *Daboll Fog Signals*: a brand of compressed-air foghorn developed and patented in the United States by Celadon Leeds Daboll (1818–66).
3. *subject in which you are much interested*: see letter 3869, n. 4.

To Joseph Henry 2 January 1873 3949

Brevoort House[1] | corner of 5th Av. & 8th St | 2nd January 1873

My dear Professor Henry

The time is come for you & me to set our heads together regarding the upshot of these lectures. I find the enthusiasm, and willingness to act upon these very things in New York. In fact there are in this country numbers of wealthy men who are willing to promote the highest science if they only see their way. The great difficulty, as you well know, is not the building of institutions but the finding of <u>men</u>.

And here I want to counsel with you. My expenses, including cost of instruments, have been very weighty: Living in hotels in the United States is a very serious affair. The death of my assistant[2] also, involving as it did the charges of two physicians[3] for a long period; also the expenses of some of my assistant's friends from Canada to Philadelphia & back, finally the embalment and transference of the body to England, was attended with great expense. Still when all has been added up I think it not unlikely that a surplus of 10000 dollars may remain. I wish to consult you regarding the application of this surplus.[4]

Thinking much over the matter causes me to incline to the following plan:—which I think will impress the community here with a sense of the value that I attach to the choice of <u>men</u>. This sum divided into four parts would amply support four students for three years each at Heidelberg or Berlin. What I propose then is that the choice of <u>three</u> of these pupils; should be vested in three gentlemen in the United States, or if preferred a small committee of gentlemen. That after due inquiry the most promising pupils of their acquaintance should be selected and sent to Germany to be trained with reference to the prosecution of original research subsequently at home.

I would ask you to name one of those pupils; leaving you an absolutely unfettered choice; having however by preference to Physics as the subject of study. I should probably ask Prof. Youmans, who has been indefatigable here, to name another. A third gentleman I shall find out subsequently, and I think I should like to retain the power of having one pupil at the other side of the water myself. There is no use in going further into details at present. But the hour of my departure approaches, and I must have every thing arranged definitely before I go.

With best wishes for the new year, and kindest regards to M^{rs}. Henry and the young ladies[5] believe me

ever faithfully yours | John Tyndall

My cousin[6] who is I think the tenderest, most chivalrous, courageous, and at the same time, from a worldly point of view, the most impracticable man that I have ever met, accompanied me, cared for me & worked for me, paying his own hotel expenses & housing me when he had the opportunity of doing so; so that he has helped to make the outlay[7] less than it otherwise would have been.

John Tyndall to Joseph Henry, 2 January 1873, mssRH 3976, box 43, William Jones Rhees papers, The Huntington Library, San Marino, California

1. *Brevoort House*: see letter 3843, n. 3.
2. *death of my assistant*: see letter 3896, n. 10.
3. *two physicians*: see letter 3945, n. 4.
4. *application of this surplus*: see Introduction, n. 15.
5. *M^{rs}. Henry and the young ladies*: see letter 3869, n. 7.
6. *my cousin*: Hector Tyndale.
7. *outlay*: expenditure (*OED*).

To Henry Bence Jones 3 January 1873 3950

New York, 3rd. Jan. 1873.

My Dear Friend

I wish I were at home to "open the ball"[1] for the youthful ones. Last night your last note[2] reached me, and I was right glad to receive it. To day the mails are made up, and to morrow this letter will commence its journey to England.

I think I may now say that the back bone of my work in America is broken. I have passed the ordeal of New York,[3] and they say, with unqualified success. Yesterday I had my apparatus transferred to Brooklyn, where I begin on Saturday evening.[4] I hope to finish there upon the 18th: and the following week I hope to complete my labours in America.[5]

My friends here wish to give me some "entertainment" before leaving and this will detain me for a few days. On the 5th of February however the Cunarder[6] 'Cuba' sails, and as I am told she is a tough sound vessel, able to cope with the expected violence of the North Atlantic, I intend to take my passage in her.[7] I send you with this my last screed in New York, and a short editorial article apropos of the course. Both are from the Tribune.[8]

I am told by scientific men, and by religious men, that my audience was a wonderful one; and the curious thing is that we assembled in an Institute[9] which had been for some years forsaken by the magnates of New York. Still most of the magnates of wealth, science, literature and the professions put in an appearance. Night after night through the most inclement weather they trooped to the lecture hall. On the evening of the 31st for example, the streets were fearfully blocked with snow, and the air was full of the descending flakes. Still when I came near the door of the Institute I could hardly accomplish an entrance through the crowd of carriages. The best possible feeling is abroad, and wealthy people are willing to forward science in many ways if they only see their way clearly. I believe if the Royal Institution could be transplanted here, before the lapse of a fortnight you would have it endowed with half a million sterling!

It is the custom here—an ancient Dutch custom—to pay flying visits to the ladies on New Year's day. I spent some time thus three days ago. It is a curious and not unpleasant habit. It renews social ties, for people that do not see each other through the year see each other on new year's day. On my return with a friend[10] who was good enough to place his carriage and pair of horses at my disposal, he said to me "It is quite in your power to carry home with you a competence[11] for life." There is not a doubt of this: but what can I do when a friend at the other side of the Atlantic is calling me home; a friend who gives up all hope of seeing me in the other world, and therefore wishes to see me again in this one?[12] Well if I should end my days in a poorhouse I shall know in part to whom my decay is to be attributed.

As it is a balance will doubtless remain to me. My expenses, including those attendant on poor Miller's death,[13] have been terribly high. Still two thousand pounds will remain as a surplus. I wrote yesterday to Henry of Washington[14] regarding the application of this money.[15] Doubtless we shall come to some arrangement about it soon.

From beginning to end nothing has occurred to mar my visit, if I except a slight passage of arms with an illtempered Irish Presbyterian,[16] who does not know when he is well treated. But this is nothing. In fact as far as success was possible, success has been actual. I long however for the quiet of home. You will have time before I leave to answer that question regarding my income from Royal Institution: but address henceforth, not to Philadelphia; but to the care of Appleton & Co[17] Broadway New York.

Cottrell[18] has behaved exceedingly well and has won golden opinions from the people here.

I think I have no more to say: let the carpentering[19] stand over,[20] it can grow up slowly as it is needed. The next great point is that your magnificent body should have a living soul. Froude[21] sails tomorrow. I fear he returns disheartened. There are two parties in America; one of which care nothing about the Irish question and will never be got to take England's view of it; the other, the Irish themselves, and they as a friend of Froude's said to me yesterday, behave towards it as a bull towards a red rag.

always Yours | John Tyndall

RI MS JT/1/TYP/3/821
LT Typescript Only

1. *"open the ball"*: see letter 3922.
2. *your last note*: letter 3922.
3. *ordeal of New York*: Tyndall delivered six lectures in New York from 17–31 December 1872 at the Cooper Institute.
4. *where I begin on Saturday evening*: Tyndall delivered six lectures in Brooklyn from 4–20 January 1873 at the Academy of Music.
5. *complete my labours in America*: Tyndall's last stop in his American lecture tour was in New Haven, CT, for two lectures on 22 and 23 January at Yale College.
6. *Cunarder*: a ship belonging to Cunard Line (see letter 3815, n. 13).
7. *take my passage in her*: see letter 3928, n. 18.
8. *I send you... the Tribune*: 'Spectrum Analysis: Prof. Tyndall's Closing Lecture', *New York Tribune*, 1 January 1873, pp. 1–2; 'What Tyndall Has Done for Us', *New York Tribune*, 1 January 1873, p. 4.
9. *an Institute*: Cooper Union, which hosted Tyndall's New York lectures (six lectures from 17–31 December 1872). See letter 3873, n. 21.
10. *friend*: Abram S. Hewitt (see letter 3951, n. 11).
11. *competence*: a sufficiency of means for living comfortably (*OED*).
12. *friend who gives up... again in this one*: in letter 3897, Bence Jones wrote that 'I like to look forward to seeing you again & as I don't look forward to our meeting in the other world my only chance is in this' (see n. 16 in letter 3897).
13. *Poor Miller's death*: see letter 3896, n. 10.

14. *Henry of Washington*: Joseph Henry.
15. *application of this money*: see Introduction, n. 15.
16. *slight passage... Presbyterian*: see letter 3943.
17. *Appleton & Co*: Tyndall's American publisher.
18. *Cottrell*: John Cottrell.
19. *carpentering*: a reference to the ongoing renovations of the RI's laboratories (see letter 3776, n. 14).
20. *stand over*: to be left or reserved for treatment, consideration or settlement at a later date (*OED*).
21. *Froude*: see letter 3741, n. 6; and 3922, n. 11.

To Thomas Archer Hirst 3 January 1873 3951

New York 3rd. Jan[uar]y. 1873.

My dear Tom,

It is now nearly three weeks since I reached this city, and last Saturday evening, the last evening of the year I finished my work here.[1] I have tried to live a quiet life in New York, and have in great part succeeded, though I fear to the discontent of some of my friends. On the first days of my sojourn I used to get habitually away from the city to the so called Central Park, which is about two and a half miles from the hotel.[2] There I took pleasure in the keen brisk air, and in the bright, though ineffectual sunshine. When I first entered the park the rock masses which rose above its general level had a strangely familiar aspect to me; they appeared evidently glaciated; but where were the mountains from which the glacier could have come? Well they may be the mountains of the moon but the evidences of glaciations are conclusive. I found on seeking them the grooves and striae[3] as fresh as if they had been formed yesterday. All this northern Continent must have been encased in ice.[4]

Then came that terrible snowstorm;[5] preceded at intervals by "snaps" of terribly cold weather. Feeling a little shady one day I sought and found and had a Turkish Bath; but either in the bath or afterwards I must have been incautious as I was assailed by a violent cold. I had engaged to meet a party at dinner, but to the keen disgust of my host I had to give the dinner up. Only think of my splendid teeth being a bother to me. They were so however. A hard tumour-like swelling rose up in my left jaw, and the residue of it is still there. I was racked by headache, lay in bed all day with the thought of a lecture in the evening pressing upon me. I screwed myself up for the work and went through it. The scene of the lectures has been the Cooper Institute, as it was not possible to procure the Grand Academy of Music[6] for so long a period. The Cooper Institute has been forsaken for some years by the magnates of

New York; but thither they streamed, in spite of the most inclement weather, and a state of the streets almost indescribable. On the evening of the last lecture the snow blockade was fearful, and snow was freshly falling; still I found it difficult to accomplish an entrance to the Institute through the crowd of carriages. A poor fellow—one of the single ticket holders edged his way up to me and said—"Well Professor, you ought to be a proud man." "How so?" I said. "Well" said he "to draw such an audience on such a night!"

But Thank God! the work is thus far complete. I long for the quiet of home; But I see my way now to the end of the campaign. I shall be able to finish in Brooklyn on the 18th. Then I run down to New Haven,[7] and give two lectures mainly to the students of Yale College. Afterwards it is proposed to have a kind of farewell banquet,[8] and the day after this banquet viz. Wednesday the 5th of February, I hope to step on board the "Cuba."[9] The north Atlantic has been very ferocious of late; let us hope that its ill temper will have wasted itself by February. But I am told that I must calculate on heavy weather.

The weather at present, or rather the state of the streets arising from the weather is simply abominable. I walked out for an hour today and such slush! I am not as strong as I have been, and this imprisonment prevents me from gathering strength by exercise. The sleighs were ringing everywhere during the snow; I had a ride one day round the Central Park with Mr. Appleton. The ponds in the park were crowded with skaters;[10] regular anthills of human beings swarming every-where. Up to my arrival in New York I kept a journal, but not afterward. In fact I have kept so quiet that little of interest to a journal has transpired. On New Year's day the gentlemen run about to visit the ladies. My good friend Hewitt[11] came with carriage & pair at 2. P.M. and kept me at this work of visiting till 6 P.M. The usual plan is to rush in shake hands & rush out again. A hundred calls can be thus accomplished in a day. I saw some remarkable people & some very lovely girls; New York is particularly rich in feminine beauty. The blossoms that one meets in the streets & avenues are very lovely indeed.

I quit New York for Brooklyn[12] tomorrow—Pray write hence-forth to the care of Appleton & Co.[13] Broadway New York.

The lectures I am told have realized over 10 000 dollars. The expenses of it all etc. will be 2000; so that 8000 will remain. Only think of that my boy! £1500 for a course of six lectures. A gentleman said to me last night that I might, if I liked, carry from this Country a competence[14] for life. But I do not intend to do so—How glad I shall be when it is all ended! Ever affectionately Yours John Tyndall.

RI MS JT/1/T/698
RI MS JT/1/HTYP/612–3

1. *I finished my work here*: Tyndall delivered six lectures in New York from 17–31 December 1872 at the Cooper Institute.
2. *hotel*: Brevoort Hotel (see letter 3843, n. 3.).
3. *striae*: a small groove, channel, or ridge; a narrow streak, stripe, or band of distinctive colour, structure, or texture (*OED*).
4. *All this northern Continent must have been encased in ice*: Tyndall was very familiar with the ice age theory espoused by Louis Agassiz beginning in the 1840s and largely accepted by the 1860s (see C. Irmscher, *Louis Agassiz: Creator of American Science* (New York: Houghton Mifflin Harcourt, 2013), pp. 64–74, 77–8).
5. *terrible snowstorm*: a storm deposited eighteen inches of snow in New York on 26 December 1872.
6. *Grand Academy of Music*: Tyndall delivered six lectures in Brooklyn from 4–20 January 1873 at the Academy of Music.
7. *New Haven*: Tyndall's last stop in his American lecture tour was in New Haven, CT, for two lectures on 22 and 23 January at Yale College.
8. *farewell banquet*: see Introduction, n. 23.
9. *the "Cuba"*: see letter 3928, n. 18.
10. *skaters*: ice skaters.
11. *Hewitt*: Abram S. Hewitt (1822–1903), son-in-law of Peter Cooper, the founder of Cooper Institute where Tyndall delivered his New York lectures. Hewitt later served in Congress representing New York, and then as mayor of New York City (*ANB*).
12. *Brooklyn*: see n. 6.
13. *Appleton & Co*: Tyndall's American publisher.
14. *competence*: a sufficiency of means for living comfortably (*OED*).

To D. Edwin Conery[1] 3 January [1873][2] 3952

[PARK HOTEL. | *ON THE EUROPEAN PLAN.* |
New York,] | Jan[uar]'y 3$^{\underline{d}}$, [*18*]72

Prof. Tyndall: Among those who through lectures delivered in different parts of our common country by yourself, have learned that: Addison,[3] or any other man, may walk the earth a second time, beg leave without being considered impertinent, I trust: to tender my kindest acknowledgement for near matters, presented a la Blackstone,[4] six law, upon a dry subject, in a captivating manner.

Englishmen are noted for long wind and he who has sit out a night London or Dublin, knows that the listeners art is possessed but by few, of their speakers. You afford both wind and sunshine and hence all is pleasant sailing!

If I were you, (for one purpose), it occurs to me that I should edit a <u>cheap</u> pamphlet edition of the lectures, for a dual purpose, preservation of matter for the public, and due credit as well—Lardner[5] did!

Respectfully & kindly Y[ou]rs, | <u>D. Edwin Conery</u>

RI MS JT/1/C/41
RI MS JT/1/HTYP/616

1. *D. Edwin Conery*: D. Edwin Conery (dates unknown) was a New York–based lawyer.
2. *[1873]*: while Conery wrote in '2' for '1872', this is clearly in error as Tyndall was in the United States for his American lecture tour in January 1873 (see Introduction, n. 8).
3. *Addison*: possibly a reference to Joseph Addison (1672–1719), an English writer noted for both the clarity and informality of his prose style (*ODNB*).
4. *Blackstone*: William Blackstone (1723–80), an English legal writer and judge. He was the author of *Commentaries on the Laws of England* (Oxford: Clarendon Press, 1765–9) (*ODNB*).
5. *Lardner*: probably a reference to Dionysius Lardner (1793–1859), a prolific author of popular scientific works (*ODNB*).

From Mary Anne Coxe[1] 3 January 1873 3953

Kinellan | Edinburgh | January 3rd 1873.

1000 thanks my dear kind and never forgotten Friend, for your welcome and pleasant proof of remembrance.[2]

How delighted we <u>all</u> are to see how thoroughly you have been appreciated across the wide Atlantic,[3] and to read of the absolute ovation that has been paid to you. I assure you the mere reading of it "sent my old blood, bounding free through pulse and heart and vein"[4] how I wished I had been present!

We are all well and in our quiet vegetating way—very happy.

Mrs Rose Innes[5]—whom you designated "Frozen bones" is now in Naples—her sister Mrs Stuart[6] in Dinan[7]—both were with us in the summer.

The Rose Innes's[8] were so far on their way to Egypt, but cholera and quarantine prevented their proceeding.

Lord Arch. Campbell[9] the D[uke]. of Argyll's[10] 2nd son, is I am delighted to say, removed from Liverpool where he has been for 2 or 3 years in a Mercantile House[11] on no Salary—to London as junior partner into Coutts's—House[12] on a commencing Salary of £2000 per an[num].

Something good I doubt not will ere long, come the way of Lord Walter[13] who was in America.

I must not take up your time with my twaddle, but I could not help just saying—May God bless you, through this and many succeeding years

I am thine | M. A. C. | I had your letter[14] yesterday

RI MS JT/1/TYP/1/298
LT Typescript Only

1. *Mary Anne Coxe*: Mary Anne Coxe (née Cumming, 1811–*c.* 1875), a daughter of Lesley Baillie and Robert Cumming of Logie, County Moray. Her mother was the subject of Robert Burns's 'Bonnie Lesley', as noted in the novelist Charles Gibbon's dedication of *In Honour Bound*, vol. 1 (London: Richard Bentley and Son, 1874) to Lady Coxe. In 1841 she married the Scottish physician James Coxe (1811–78), her second marriage, and became Lady Coxe after he was knighted in 1863. Her first marriage, in 1828 to William James Fraser (d. 1838), resulted in four children. Tyndall claimed to have toasted Lady Coxe on the summit of Mont Blanc in 1857.
2. *your welcome and pleasant proof of remembrance*: likely referring to a letter from Tyndall, which is missing.
3. *how thoroughly... Atlantic*: presumably Tyndall had described in the letter how his American lecture tour was proceeding (see Introduction, n. 8).
4. *"sent... vein"*: quote from the Scottish poet William Edmondstoune Aytoun's poem 'The Island of the Scots' (*Lays of The Scottish Cavaliers, and other Poems* (Edinburgh: William Blackwood and Sons, 1849), p. 142, lines 137–49).
5. *Mrs Rose Innes*: Grace Louisa Rose-Innes (née Fraser, d. 1910), a daughter of Mary Anne Coxe from her first marriage to William James Fraser (see n. 1). Grace married Thomas Gilzean Rose-Innes (b. 1831) in 1853. Their three children were Mary Grace Edith Rose-Innes (b. 1854), Elizabeth Frances Eugénie Rose-Innes (*c.* 1857–1925), and Georgina Everilda May Rose-Innes (1858–1928).
6. *Mrs Stuart*: Elvorinda Eliza Maria Stuart (née Fraser, *c.* 1837–1903), a daughter of Mary Anne Coxe from her first marriage to William James Fraser (see n. 1). Elvorinda married John Glass Gordon Stuart in 1857.
7. *Dinan*: a coastal town in northwestern France.
8. *Rose Innes's*: see n. 5.
9. *Lord Arch. Campbell*: Archibald Campbell (1846–1913), second son of George Campbell.
10. *D[uke]. of Argyll*: George Campbell (see letter 3747, n. 1).
11. *Mercantile House*: an import-export business.
12. *Coutts's—House*: Coutts & Co., a private bank founded in 1692.
13. *Lord Walter*: Walter Campbell (1848–89), third son of George Campbell. George Campbell and his first wife, Elizabeth Georgiana Campbell (née Leveson-Gower, 1824–78), had twelve children in all, five sons and seven daughters.
14. *your letter*: see n. 2.

From Henry Bence Jones 3–4 January 1873 3954

Jan[uar]^{y} 3^{d} Friday 1873

Dear friend

I am soon writing to you again partly because I have got some information about Gramme's Magneto Electric Light machine[1] & partly because I have a sad thing to tell you which however will happily not I think affect us.

Our architect Mr Harwood six weeks ago got an attack like the one you had in your leg.[2] The leg swelled &c & he is surveyor for the admiralty & went to the Portsmouth? works.[3] The veins blocked up & the clots stopped his circulation & he has died last Tuesday. He looked the healthiest of us all & was only 49. His wife was dead & his children I hear are well provided for but the suddenness of the attack & the rate of progress of the bad symptoms which only existed a few days has affected us all. Unless any trouble arises regarding changes of the Contractors &c I do not see that we shall directly suffer,[4] but indirectly his loss may be great for he knew better than any one where, what & how every thing should be & we may want many small changes.

However he is gone & I envy him

With regard to the Electric Light

A small company has been formed here; Col Stuart Wortley[5] & a German civil Engineer R. Werdermann[6] who is the brains of the company came here yesterday & they have brought Grammes patent for England & are making machines for the post office telegraphs &c.

I shewed him a drawing of the size of the furnace room & he says that the space is more than enough & that any thing the company can do for you they will do. The people who made L[or][d] Lindsays magnet[7] are making the machine here for the company. I could not get any thing definite as to cost, but as to size that is settled and I almost think the light machine could be hired if the purchasing price proves to great & Mr Werdermann wishes you to have one on trial. He is a scientific German & has worked with Magnus & was driven out of Paris by the French[8] & now means to remain here. I gave him an admission to the Friday evenings[9] so I hope you will see him when you return. Every body is delighted to hear that we have an early chance of your return. We all feared you might be tempted by success to remain until after Easter.

I was very glad to get all the New York papers you sent me. I passed them on to Sir. H. H.[10] who liked them very much but fears you will be spoiled by your great success.

Your salutation to the young lady who gave you the flowers amused my grown up children.[11] We have had nothing new a dull Christmas Odlings lectures[12] have done very well. £151 against £121 in 70–71. £100 less than your last Xmas course.[13]

His experiments have done well though he had no laboratory.

Saturday. Lecky[14] came in last night & he tells me he has written a very strong article on Froude's new book[15] in Macmillan for this month.[16] He has sent a note to Froude to say that he fears he has offended Froude for life but that he was compelled to speak out for Ireland. Lecky says the feeling in Ireland against it is excessive. That he will never go there again.

& now every good wish for the new year & a quick & happy return home. I expect you about the 10th of Feb[17] & count the days. I go on badly

Ever yours | H Bence Jones

RI MS JT/1/J/76
RI MS JT/1/TYP/2/706–7

---

1. *Gramme's Magneto Electric Light machine*: see letter 3944, n. 5.
2. *an attack like the one you had in your leg*: in the Alps in the summer of 1869, Tyndall slipped while bathing in a stream, injuring his leg. The ensuing wound and infection kept him bed-ridden for weeks and had a serious impact on his health and productivity in the following months. See the Introduction in *Tyndall Correspondence*, vol. 11.
3. *Portsmouth? works*: in 1872 the Royal Navy began an expansion of Portsmouth Dockyard.
4. *change... suffer*: a reference to the ongoing renovations of the RI's laboratories (see letter 3776, n. 14).
5. *Col Stuart Wortley*: Henry Stuart Wortley (1832–90), an English army officer with an interest in photography and optical devices. See K. DiGiulio, *Natural Variations: Photographs by Colonel Stuart Wortley: An Exhibition at the Henry E. Huntington Library and Art Gallery* (San Marino, CA: Huntington Library Press, 1994).
6. *Mr R Werdermann*: Karl Sigismund Richard Werdermann (1837–82), a Prussian-born engineer and associate of the inventor and electrical engineer Zénobe Gramme (1826–1901), who secured the British and American patents rights for the Gramme magneto. In 1873 he led a project to illuminate the House of Commons clocktower using electric lighting ('Karl Sigismund Richard Werdermann', *The Electrician*, 2:11 (November 1883), p. 337).
7. *L[or]^d^ Lindsay's magnet*: a reference to James Ludovic Lindsay (1847–1913), a British politician and amateur astronomer, who in 1872 constructed what was at the time the largest electromagnet in the world (*ODNB*).
8. *driven out of Paris by the French*: probably referring to the Franco-Prussian War of 1870–17.
9. *Friday evenings*: the upcoming season of Friday Evening Discourses at the RI (see letter 3880, n. 15).
10. *H. H.*: Henry Holland.
11. *Your salutation... children*: a reference to a passage in 'Capital Bon Ton', *New York Herald*, 22 December 1872, p. 5: 'Professor Tyndall, a day or two ago, paid a visit to the Franklin school, and, getting among the "sweet girl graduates"—that are to be—of the grammar class, was presented with a handsome bouquet of flowers by a pretty miss of Georgetown. Taken unprepared for an appropriate speech, he at once pleaded the privilege of age, and, bending his head, heartily kissed the young lady amid the plaudits of all present'.
12. *Odlings lectures*: see letter 3782, n. 4.
13. *your last Xmas course*: Tyndall delivered the annual Christmas lectures at the RI on 'Ice, Water, Vapour, and Air' on 28 and 31 December 1871, and 2, 4, 6 and 9 January 1872; they are described as 'adapted to a Juvenile Auditory' (*Roy. Inst. Proc.*, 6 (1870–72), p. 384). The Christmas lectures began in 1827 and still continue. Tyndall delivered them a dozen times over his time with the RI, his first in 1861 (see F. A. J. L. James (ed), *Christmas at the Royal Institution: An Anthology of Lectures* (Hackensack, NJ: World Scientific, 2007)).
14. *Lecky*: William Edward Hartpole Lecky (see letter 3922, n. 9).

15. *Froude's new book*: see letter 3922, n. 11.
16. *Macmillan for this month*: see letter 3922, n. 12.
17. *10th of Feb*: see letter 3928, n. 18.

From Emily Peel 4 January 1873 3955

[VILLA LAMMERMOOR, | GENÈVE][1] | Jan[uary]: 4/73

Dear Professor Tyndall,

It gave me great pleasure to receive your letter[2] & I am glad to see by it that you are all well.

We came here on our way, as we had thought, to Nice, but the climate here has been Italian & we have lingered on at house, (it is always preferable to an Hotel,) until it is too late to go South. We shall therefore soon be in London where I hope to have the pleasure of seeing you. In the meanwhile pray accept all my best wishes, &

Believe me yours | sincerely | Emily Peel

RI MS JT/1/P/62

1. *VILLA LAMMERMOOR, | GENÈVE*: see letter 3844, n. 15.
2. *your letter*: a previous extant letter from Tyndall to Peel is letter 3844, but as it is from September 1872 and this letter from Peel is from January 1873, it is more probable that Peel was referring to a different letter that is missing.

From W. S. Tisdale[1] 4 January 1873 3956

[EDITORIAL ROOMS | OF | THE COAL AND IRON RECORD. | Western & Company, Publishers, | No. 37 PARK ROW, | P.O. 5969, | *New York,*] Jan 4th [*187*]3

Professor John Tyndall

Dear Sir

You have said at the conclusion of your recently delivered lecture that:—"To keep society, as regards science, in healthy play, three class classes of workers are necessary"—++++ "These three classes ought to co-exist and interact upon each other." ++++ "In no case is the adage truer." "Other men labour but ye are entered into their labour"—"That in the case of the discoverer and the applier of natural truth"—++++ While I say that practical men are not the men to make the necessary antecedent discoveries, the cases are rare in which the discoverer knows how to turn his labour to practical account" ++++ They are the necessary supplements to each other" &c &c

Now my Dear Sir—following out this train of thought it appears to me to be eminently proper and just that the discoverer should enjoy a very large proportion of the practical wealth of his labor—or, in other words the laborer should enter into the enjoyment of the fruits of his labor—And with this object in view in reference to yourself and myself as representing the discoverers and the practical men—I propose to reduce a conception or invention growing out of a careful reading of your lectures as reported in the "Tribune"—to practice for our natural good and profit. I must premise by saying that I have not had the pleasure of witnessing your illustrations of light refraction, &c and have therefore depended upon the reports and a vivid imagination in working out the idea which I will now submit. Cut out of any suitable opaque substance having flat and even surfaces any desired lettering configuration, or emblem and place the part from which it was cut between two sheets of clear, white, fluid glass. Seal the edges all around excepting a small opening and fill the space between the plates of glass (through the small opening) with spirits, glycerine, water, or any other liquid. Then place a light at the side opposite yourself and a beautifully illuminated sign presents itself which may be seen in a store window from across the street. Make now in imagination the configuration of a druggist's emblematic jar (such as they have in their windows) and fill it in the manner just described with colored spirits and we have the emblem in such a convenient forum as will enable us (by using four of them) to make a lamp-post-sign which may be seen from a very great distance. Let us now fancy the figure of a cross cut out of properly prepared wood, and the wood from which the cross is cut (intaglio[2]-type) placed upon one of the sheets of glass—We fill the open figure of the cross with prisms, properly secured and placed, and placing the other sheet of glass over all proceed as before to seal the edges and fill in with spirits. This, I think, would make a very beautiful altar piece for an Episcopal church or a Romish Cathedral. Letters figures or emblems might be filled in with little globules or other forms of glass containing colored water or spirits. Indeed there appears to be no end to the brilliant striking and fanciful designs and exquisite conceptions which may be carried out on this principle. It would also be applicable to other purposes which I could name. Here is something practical—based upon your discoveries and the discoveries of others elucidated by yourself to whom I am indebted for the invention, something which appeals directly to the interests of men who seek by advertisement and display to draw custom and "their name is legion".[3] This is an every day and every night business appliance in which there is a great deal of money, and this consideration brings me right to the point of interest to ourselves. Conjointly we can perfect this invention as I could reduce the idea to a practical basis and you could suggest valuable additions and correct scientific rules and views by which I would be governed. It would afford me great pleasure to have an interview with you if you think this matter of sufficient importance. Please consider the subject seriously for a few days as there is more in it than may strike you casually—I have thus exposed to you an invention which is susceptable of being made exceedingly valuable when carried out in detail and I request of you perfect secrecy as I desire to secure the English Patent first in consequence of the now internationality of the Patent laws in Great Britain and the United States. My desire is to make a business proposition to you, in view of having you participate equally with myself in all the benefits which may accrue from the invention. I sincerely believe that the English Patent alone would realize for us ten thousand pounds. There would be a large demand for signs (lettered, &c) and emblems of such novelty and beauty—and the Patents would be of the nature of those easily managed and returning large profits—

Please be careful of this and respond at your earliest convenience—

With sentiments of great respect for your brilliant talents and record I subscribe myself

Very truly | Your obedient serv[ant] | W. S. Tisdale | PO Box 5969 | New York | City

RI MS JT/1/T/35
RI MS JT/1/HTYP/614–5

1. *W. S. Tisdale*: one of the editors of *The Coal and Iron Record*, a short-lived industry periodical established in 1872.
2. *intaglio*: a figure or design incised or engraved (*OED*).
3. *"their name is legion"*: Mark 5:9.

From Henry Bence Jones 7 January 1873 3957

Jan[uar]y 7 1873

My Dear Friend

I have just got your letter of the 22d of Dec.[1] We read your loss of cash in the papers here & we were sorry for you. But as long as you are well & get back safely that is all I care about. I want you sadly home. I cannot meddle with your laboratory[2] & you shall have the lower one which Mr Wills seems to be fitting up regardless of expense until your own is to your liking. But it wants you & Cotterell[3] to say how the place is to be fitted up. I am rarely able to go to the Institution & I shall put every thing wrong & I cannot keep appointments with other people who ought help me to get the rooms nice. So do come home. Every thing & every one is in want of you. I hope the gales will blow themselves out this month & have the Atlantic like glass for you in Feb[ruar]y.[4] The workmen will all clear out in a fortnight but as soon as you come back people will be set to work at your rooms. The course of American lectures[5] will do capitally after Easter & I hope a month or six weeks will fix up your laboratory but pray remember that the laboratories are for future professors far on. That we may not be monstrously inferior to other places: Your name & fame is of more consequence to us than any new discovery just now can be & except as adding to <u>your</u> fame no new discovery will be of great use to the Institution at present. It will be very different when you leave us by health or death[6] then indeed the Institution will have to depend on what its professors <u>then</u> do for it but if you do not do an atom more work & of nothing at all comes out of the Institution for the next ten years I believe that your name & repute will make it flourish. This is the substance of a talk I have had very lately with Sir H. H.[7] who was anxious about Odling's successor.[8] It really matters but little whilst you reign.

I fear when you come back we must fix on my successor.[9] I have gone all wrong again & water rises fast & my drugs fail me alas tho' I do not spare them & I am longing for my last *[rests]*.

The Times yesterday!! published your Washington prayer[10] at last & in the Evening it was in the Echo[11] the piece you copied out & sent me. Many thanks for the papers. I like to see your muses greatly & it reads here or will do so, but here they do not hear enough of it. Send me word what day I am to expect you. I have made an appointment for you to see the magneto Electric machine[12] which is to be at Wheatstone & Cooke;[13] 190 Westminster bridge R[oa]d; Cooke went to look at the furnace house there is plenty of room. The company Col S Wortley[14] Edward[15] &c have offered you the machine at cost price but that price is very great. I fear machine & engine would be £700 however hopefully some arrangement can be made when you return & when we see what our bills come to. Mr Harwood, Berkley, & Co. have done the work too grandly & expensively & Wills & Odling seem to think money is dirt for them to scatter. Send Cotterell home if you do not come yourself but I long to see you. When oh when.

I am so glad you have had such success & wishing you every warm & good wish.

I am yours ever | H Bence Jones | no news of any kind

RI MS JT/1/J/77

1. *your letter... 22d Dec*: letter 3933.
2. *your laboratory*: a reference to the ongoing renovations of the RI's laboratories (see letter 3776, n. 14).
3. *Cotterell*: John Cottrell.
4. *in Feb[ruar]y*: see letter 3928, n. 18.
5. *course of American lectures*: see letter 3871, n. 10.
6. *when you leave us by health or death*: Tyndall resigned from the RI in 1887.
7. *Sir H. H.*: Henry Holland.
8. *Odling's successor*: John Hall Gladstone was Fullerian Professor of Chemistry at the RI from 1874–7.
9. *my successor*: in mid-1873, William Spottiswoode succeeded Bence Jones as Secretary of the RI.
10. *The Times... Washington prayer*: 'Superfluous Profession of Faith', *Times*, 6 January 1873, p. 4.
11. *Echo*: it is not clear to which of multiple newspapers titled *The Echo* this refers.
12. *magneto Electric machine*: see letter 3944, n. 5.
13. *Cooke*: William Fothergill Cooke (1806–79), an English inventor who began his career in medicine and served in the British Army in India. He shares the invention of the electric telegraph with Charles Wheatstone which they patented in 1837. Cooke also co-founded the world's first public telegraph company (Electric Telegraph Company) in 1846 (*ODNB*).
14. *Wortley*: Henry Stuart Wortley (see letter 3954, n. 5).
15. *Edward*: not identified.

From Joseph Henry[1] 7 January 1873 3958

Smithsonian Institution, | Washington, Jan. 7. '73.

My dear Sir,

I regret very much that I have not an opportunity of discussing the subject of your lecture orally, and at leisure. Were it not that the annual meeting of the Board of Regents[2] takes place on the 16th. Inst[ant].,[3] and that in the meantime that I am obliged to prepare an account of the expenditures and operations of the Institution for the past year, I would visit New York for the purpose in question. I must however content myself with exchanging views with you by letter.

It is my own opinion as well as that of others with whom I have conversed on the subject, that you are not called upon by any consideration usually thought valid by the world to make any other disposition of the proceeds of your lectures, which have been of so much labor to yourself and of profit to your hearers, than that of furnishing more enlarged means for the advancement of science through your own investigations. But I presume your mind is made up on this point and that nothing I could say would change your determination.[4] I shall therefore proceed to give you such views as a somewhat hasty consideration of the subject suggests.

Your proposition to expend your money for the advancement of science in our country is a noble example of the liberalizing tendency of scientific pursuits on a mind of generous sympathies and ought to be one which can permanently be referred to as a prominent illustration of the truth in question. For this purpose as well as for more certain and lasting effects I would suggest that instead of expending it in one hazard the money be permanently invested and the annual proceed from it only be applied perpetually to assist in supporting a student in a German, or other foreign, university while being trained with reference to the prosecution of original research in the line of Physics? This fund, which, if judiciously invested, would yield six or seven hundred dollars a year, should be put in charge of a Board of Trustees consisting, if you please, of myself, Prof. Youmans and your cousin,[5] than whom no better man could be selected, this Board to perpetuate itself by selecting new members to fill any vacancies that may occur by death or otherwise. I suggest this plan because there are at present, according to a recently published statement, upwards of a thousand American students in different schools in Europe and among these as one of my assistants Dr. Endlich,[6] (who has lately returned from a ten years residence at Freiburg), informs me there are at least two hundred in Germany alone devoted to the acquisition of science.* By addressing letters to the Professors of Physics in the German Universities information might be had as to whether there are any young men from the United States pursuing their studies who would be worthy recipients of aid from the fund in question.

There being such a demand for practical applications of science in this country, and the remuneration of professors being so small, few of the young men educated abroad devote themselves to pure science on their return home. Another reason for the scarcity of men devoted to original science in this country is owing to the fact that there are no honorary distinctions of any kind which can be conferred by the government and therefore the tendency of talent is to seek social position and influence through wealth and politics. There are at the present time two first class positions vacant in the United States, viz, the chair of Physics in the University of Pennsylvania, at Philadelphia, and the "Joseph Henry Chair of Physics" in the College of New Jersey at Princeton—a chair lately endowed by Mr. John C. Green[7] of New York.

What is wanted in this country is well endowed professorships which will give those occupying them a liberal support and afford them sufficient time and ample apparatus for pursuing original research while the drudgery of teaching is done by assistants under the professors direction. To this should be added a higher appreciation on the part of the public of the importance of pure science. I think however we are making progress in the latter and that your lectures will have done good service in this direction.

Of course what I have said to you in this letter is merely in the way of suggestion to be adopted by you in part, or in whole, or to be entirely disregarded as may suit your views. Whatever plan you may decide will, I am sure, result in good and in keeping you in grateful remembrance by the more intelligent portion of this country.

Since commencing this letter I have learned through the newspapers that you are not well, and I fear you have overworked yourself. Your labors have been immense, but I trust, most sincerely, that the invigorating sea-air and the entire rest of the voyage will restore you to your wanted vigor.

Do not forget before leaving for England[8] to give General Woodruff[9] a visit at the Light House Depot, Tompkinsville, Staten Island. Professor Youmans has been there and will no doubt accompany you.

Your lectures in Washington were a great success. From the President of the United States to the lowest Employee of the Government—all were delighted with your course.

With kind regards from Mrs. Henry and my daughters,

I am | Very truly | your friend and servant | Joseph Henry

John Tyndall | Brevoort House[10] | N.Y.

* Freiburg 30; Berlin 20; Stuttgart 10; Tübingen 6, or 8; Göttingen 4; Erlangen 2; Heidelberg 4, or 6.—

RI MS JT/1/H/78

1. This letter has previously been published as document 204 in *Papers of Joseph Henry*, pp. 435–8.
2. *Board of Regents*: of the Smithsonian Institution, of which Henry was secretary.
3. *Inst[ant]*: said of the current calendar month (*OED*).
4. *your determination*: for Tyndall's donation of his lecture tour profits, see Introduction, n. 15.
5. *your cousin*: Hector Tyndale.
6. *Endlich*: Frederic Miller Endlich (1851–99), an American mineralogist and chemist at the Smithsonian Institution (1872–80) and assistant on the US Geological and Geographical Surveys under Ferdinand V. Hayden (1873–9) (W. D. Miles (ed), *American Chemists & Chemical Engineers*, 2 vols (Washington: American Chemical Society, 1976–94), vol. 2, pp. 88–9; *Papers of Joseph Henry*, p. 437).
7. *Mr. John C. Green*: John Cleve Green (1800–75), an American merchant and major donor to Princeton University (*ANB*).

8. *leaving for England*: see letter 3928, n. 18.
9. *Woodruff*: Israel Carl Woodruff (1815–78), a lieutenant colonel in the US Army Corps of Engineers, and engineer officer in charge of the Light-House Board's third district from 1870–78 (*Papers of Joseph Henry*, p. 437).
10. *Brevoort House*: see letter 3843, n. 3.

To Henry Bence Jones 10 January 1873 3959

Brooklyn,[1] 10th. Jan. 1873.

Dear Friend

I cannot resist the temptation of sending you "Hooker's prayer"[2] copied for me by the Rev Dr Ellis of Boston. It ought to be published, together with the preceding account of it. It is excellent.

The other letter[3] is from Dr Moorhouse,[4] one of Miller's[5] Physicians. It may comfort his father and mother.

I have already sent you a telegram saying that I sail (in the Cuba) on the 5th of February.[6] I am terribly pressed with the publication of my lectures: but I begin to see my way through the work.[7] Every thing has been as successful as it possibly could be—Only think of £2000 being taken in New York for a course of six lectures![8]

Good bye: keep strong till I see you.

Yours ever | John Tyndall

RI MS JT/1/TYP/3/822
LT Typescript Only

1. *Brooklyn*: Tyndall delivered six lectures in Brooklyn from 4–20 January 1873 at the Academy of Music.
2. *"Hooker's prayer"*: referring to Richard Hooker (1554–1600), an Anglican theologian and philosopher (*ODNB*), and the fifth book of his *Of the Lawes of Ecclesiastical Politie* (1597), a defense of the Common Prayer Book, which was later published on its own in the twentieth century as J. S. Marshall (ed), *Hooker's Theology of Common Prayer: The Fifth Book of the Polity Paraphrased and Expanded into a Commentary on the Prayer Book* (Sewanee, TN: University Press at the University of the South, 1956).
3. *other letter*: if this is a letter to Tyndall from Moorhouse, it is missing.
4. *Dr Moorhouse*: not identified, but presumably a physician who cared for Miller. In letter 3945 to Tyndall, Hector Tyndale names two physicians who cared for Miller as 'Doctors Murchison & McClellan'.
5. *Miller's*: see letter 3896, n. 10.
6. *I sail (in the Cuba) on the 5th of February*: see letter 3928, n. 18.

7. *through the work*: *Six Lectures on Light*, which was published in July 1873.
8. *New York for a course of six lectures*: Tyndall delivered six lectures in New York from 17–31 December 1872 at the Cooper Institute.

To Joseph Henry 11 January 1873 3960

11th. Jan[uar]y. 1873.

My dear Prof. Henry.

I fully accept your views.[1]

The surplus will be larger than the sum I mentioned.[2]

My cousin[3] will go to Washington to arrange details.

Pray put the matter[4] in a practical form! When I know the precise form I will write you a letter asking you to carry the thing out.

faithfully yours | John Tyndall

I shall be at the Brevoort House[5] till Tuesday morning next, afterwards for the week at the Montague St. Brooklyn.

Rec. | The audience in Brooklyn[6] is very large & influential.

John Tyndall to Joseph Henry, 11 January 1873, mssRH 3978, box 43, William Jones Rhees papers, The Huntington Library, San Marino, California

1. *your views*: see letter 3958.
2. *The surplus... I mentioned*: see Introduction, n. 15.
3. *my cousin*: Hector Tyndale.
4. *the matter*: see n. 2, and letter 3958.
5. *Brevoort House*: see letter 3843, n. 3.
6. *Brooklyn*: Tyndall delivered six lectures in Brooklyn from 4–20 January 1873 at the Academy of Music.

To Thomas Archer Hirst 11 January 1873 3961

Brooklyn[1] 11th. Jan[uar]y. 1873.

My dear Tom

When I last wrote to you[2] I was suffering from the effects of a cold brought on by the terrible weather in New York;[3] I came soon afterwards to Brooklyn, and remained somewhat low for a few days. But the weather brightened though the snow did not disappear, and the smooth ice on the pavements has already caused numberless broken legs. The weather I say brightened and enabled me to get out from time to time. Happily there is a park, vocal with the sound of sleigh-bells, about a mile & a half from here. Thither I betook myself for an hour's walk almost daily and in this way have in great part recovered my lost strength. Besides the approaching end of my labours has something reviving in it. I have already given three lectures in Brooklyn, and the remaining three lectures of the course will be given next week. On the 22nd & 23rd I lecture in Newhaven,[4] and there my labour of lecturing ends.[5]

I think I must have come to this Country at a critical period, when a universal hunger for scientific instruction is prevalent. At New York I had wonderful audiences; and they held on to me through all obstacles, of the most terrible weather known here for a score of years.[6] At Brooklyn the audiences are still more imposing, for the people better massed together in the Academy of Music than they could be in the Cooper Institute. I thought the first audience a large one the second was still larger, and last night an amazing crowd gathered to hear me. And the wonder is that they are as loyal to the spoken words, to the expositions, as to the experiments. A gentlemen came to me last night & said "I am a teacher of science, and you are entirely altering the character of experimental teaching in these cities." "How so?" I said. "Why because in our lectures hither to the experiments have been the principal of object of the lecturer's care, and he has neglected exposition—that is now changing fast." I am perfectly astonished at the interest the people take in my work and me. The New York Tribune has sown a report of the lectures broadcast over the country.[7] I am told the circulation has been enormous. Taking the circulation of the paper itself into account I am assured that hundreds of thousands of copies have been purchased in the East alone. The West are now beginning to send in their orders; and there is no calculating how many will be thus sent abroad.

Last night after the lecture I went, on the invitation of the members, to the Brooklyn Club.[8] The kindness and appreciation of the people are simply indescribable. And there is a genuine warmth about them which gives a tone of intense reality to what they say and do. Were it my wish, or did I take pleasure in such things, I might be fêted in the most extraordinary fashion. So it is they will not hear of my going away without a public dinner to be given on the 4th of February, the evening before my departure from New York.[9] Last night at the Club, they quenched all speechifying with the exception of the proposal of my health, which was given with concentrated warmth by the President of the Club, Mr. Silliman[10]—a cousin of the man of Science[11]—and responded to with the same warmth. This morning I find here a splendid vase of roses and other flowers sent to me with the compliments of the Club.

Up to the present time, however, I have incurred the risk of being deemed unsociable. I have hardly dined with anybody, but sternly kept my mind upon my work, and now I see that it will be successfully done. Matters have moved as I best like them to move: beginning in a small way, and growing more and more successful: Certainly, successful as all the other lectures were they did not come up to the effect produced in New York and Brooklyn.

The sleighing goes on merrily, and the sun shines from an unclouded heaven. In the park a few days ago the trees enamelled with ice, and with icicles pendent from their twigs presented the most extraordinary appearance in the sunshine that I ever witnessed. In certain aspects of the light, they were simply of dazzling beauty. But I must stop my gossip, for I return to New York today to eat one or two friendly dinners. I return to my work here on Tuesday, and finish on the Saturday.

Your letter about your appointment has just come[12]—need I say that it delighted me? Tenderness is the word to apply to Lady Mary![13] Hers is a rare nature.

Ever affectionately | John.

RI MS JT/1/T/699
RI MS JT/1/HTYP/617–18

---

1. *Brooklyn*: Tyndall delivered six lectures in Brooklyn from 4–20 January 1873 at the Academy of Music.
2. *When I last wrote to you*: letter 3951.
3. *New York*: Tyndall delivered six lectures in New York from 17–31 December 1872 at the Cooper Institute.
4. *in Newhaven*: Tyndall's last stop in his American lecture tour was in New Haven, CT, for two lectures on 22 and 23 January 1873 at Yale College.
5. *labour of lecturing ends*: for Tyndall's American lecture tour in general, see Introduction, n. 8.
6. *most terrible weather known here for a score of years*: a storm deposited eighteen inches of snow in New York on 26 December 1872.
7. *New York Tribune... country*: see letter 3962, n. 2.
8. *Brooklyn Club*: a private social club established in 1865 with a building at the corner of Pierrepont and Clinton Streets, Brooklyn, New York.
9. *public dinner... my departure from New York*: see Introduction, n. 23.
10. *Silliman*: Benjamin D. Silliman (1805–1901), an American lawyer in New York *(Obituary Record of Graduates of Yale University, Deceased during the Academical year ending in June, 1901*, pp. 3–5).
11. *the man of Science*: Benjamin Silliman (1779–1864), an American chemist and Yale University professor noted for establishing the *American Journal of Science* (*ANB*; *CDSB*).
12. *Your letter... appointment*: letter missing, but Tyndall was referring to Hirst's appointment as Director of Studies at the new Royal Naval College, Greenwich.
13. *Lady Mary*: Mary Egerton.

From Mary Adair 11 January 1873 3962

Heatherton[1] | Jan[ua]ry 11th 1873.

Dear Dr. Tyndall,

The "New York Tribune"[2] duly reached me a few days ago; very many thanks for sending it. The American accounts are very interesting, and one cannot help feeling amused at their frank avowal of surprise at finding that an English philosopher can discourse of the deep things of science without rhetoric or "bunkum"! I often covet the power of explaining clearly, when trying to solve the puzzles of a terrible questioner—a bright little nephew of mine aged seven.

We heard with regret of the death of Mr. Miller[3]—

Sir Charles[4] says that you valued him, and that he was devoted to you, so we fear the loss is severe. We trust that this may be the only drawback, and that you will really enjoy this brilliantly successful Tour.[5]

Your's always sincerely | H. M. Adair.

RI MS JT/1/A/6

---

1. *Heatherton*: see letter 3887, n. 1.
2. *"New York Tribune"*: the *New York Tribune* ran detailed reports of Tyndall's New York lecture course on its front page, reproducing large portions of the lectures. For example, 'Prof. Tyndall on Light: Opening Lecture of the Course at the Cooper Institute', *New York Tribune*, 18 December, 1872, pp. 1–2; 'Prof. Tyndall on Light: Second Lecture of the Course', *New York Tribune*, 20 December 1872, pp. 1–2; 'Crystals and Light: Third Lecture of Prof. Tyndall's Course', *New York Tribune*, 23 December 1872, pp. 1–2. The newspaper also published Tyndall's New York lectures, verbatim with illustrations, separately as a supplement which cost 3 cents ('Tyndall's Six Lectures on Light', *New York Tribune*, suppl., Lecture Extra 1 (1873)). The newspaper later reported about the successful sales of the supplement ('Cheap Literature', *New York Tribune*, 26 February 1873, p. 4).
3. *death of Mr. Miller*: see letter 3896, n. 10.
4. *Charles*: probably Charles Hamilton (see letter 3887, n. 6).
5. *brilliantly successful Tour*: for Tyndall's American lecture tour in general, see Introduction, n. 8.

From Mrs. J. S. Sherlock[1,2] 11 January [1873][3] 3963

Cincinnati Jan[uar]y. 11

Mr Tyndall | Sir

Excuse me for troubling you, my only excuse is curiosity for I have read so much of you and your history in the papers that I thought you must be a relative. I am the only daughter of the late Sarah Bruin[4] née Tyndall, of Dublin, who emigrated to this country between fifty years ago with her brothers William and Henry Tyndall[5] now of Philadelphia. And as I have never heard much of my mother's family I thought it would be pleasant to hear from you if a relative.

I was in hopes that you would extend your visit to this country to a longer period, and that you would visit Cincinnati. But I hear it is your intention to leave for home this month.[6]

I am sorry that you met with such a misfortune in the loss of your assistant[7] as it must have been.

Hoping this may find you enjoying good health, and that I have not occupied too much of your time.

I remain | Mrs. J. S. Sherlock.

If you see fit to reply, direct to P. O. Cin[cinna]'ti. Ohio

RI MS JT/1/T/1482
RI MS JT/1/TYP/5/1717

---

1. This letter appeared as an enclosure with letter 3974.
2. *Mrs. J. S. Sherlock*: not identified.
3. *[1873]*: the year is given based on the dates of Tyndall's American lecture tour (see Introduction, n. 8).
4. *Sarah Bruin*: not identified.
5. *William and Henry Tyndall*: not identified.
6. *intention to leave for home this month*: see letter 3928, n. 18.
7. *loss of your assistant*: George Miller (see letter 3896, n. 10).

From Carel Brutel de la Rivière[1,2] 12 January 1873 3964

Leiden the 12th January 1873.

Dear Sir!

I beg you to accept an atlas for mathematics and physics[3] with explanatory text; which I have forwarded to you by the post at the same time as this letter.

I wish you to consider it as a proof I never will forget the kindness with which you met some time ago my request for assistance to remove a difficulty, I tried in vain to overcome when translating your lecture on "the scientific use of the imagination.[4]

I would feel happy if my unworthy production might remember you kindly of

Your most devoted | C. J. E. Brutel de la Rivière

RI MS JT/H/254

1. This letter appeared as an enclosure with letter 3970 (in fact, it appears that Thomas Hirst wrote his own letter on the backside of this letter).
2. *Carel Brutel de la Rivière*: Carel Johan Emerentius Brutel de la Rivière (1835–1912), a Dutch mathematician and educator.
3. *atlas for mathematics and physics*: C. J. E. Brutel de la Rivière, *Atlas voor de meet- en natuurkunde* (Leiden: Sijthoff, 1872).
4. *translating your lecture on "the scientific use of the imagination*: for Tyndall's lecture, see letter 3818, n. 26. A Dutch translation has not been identified.

From Mariano Bichi[1] 13 January 1873 3965

Fermo (Marches) 13 Janvier 1873.

Très-illustre Professeur,

J'ai lu votre Traité sur le son traduit en français par M. l'Abbé Moigno, Paris 1869—Gauthier-Villars et je l'ai trouvé excellent soit par rapport aux idées neuves que vous y exposez, soit par rapport à sa remarquable simplicité qui le mettant à la portée des intelligences les moins éclairées dispense de bien des connaissances mathématiques—Me proposant de faire quelques expériences sur cette matière, je viens vous prier de m'éclaircir quelques doutes qui me sont nés.—Dans la leçon Vme pag. 204, je trouve que pour déterminer la vélocité du son dans les liquides vous remplissez des tuyaux d'orgue convenablement construits, et d'après la tonalité qu'on en obtient vous jugez de la puissance qu'auraient ces mêmes liquides de propager le son. Il me faudrait connaitre d'abord de quelle matière ces tuyaux doivent être construits, et quelle serait leur conformation la plus convenable ; savoir ensuite par quel moyen on pourrait les exciter au son étant remplis de liquide—Ma précise intention étant d'expérimenter de quelle manière les fluides animaux sont parcourus par le son suivant leur densité, leur quantité ou leur composition, je voudrais m'en valoir dans les maladies surtout de poitrine, et au moyen de l'auscultation arriver à connaître si entre le poumon et le thorax se trouve un liquide séreux, ou sanguin, ou purulent ou d'une autre nature—car je crois que parmi les différentes causes qui peuvent faciliter ou retarder la propagation des bruits qui des voies respiratoires vont au thorax on puisse aussi reconnaître quel rôle y joue la qualité et la quantité des liquides ramassés dans la plèvre. Il serait aussi mon idée d'arriver à des expériences directes en mesurant l'intensité d'une sonnerie de montre à réveil qui serait renfermée dans une boite sphérique et qu'on immergerait dans divers fluides animaux, pour, en l'expérimentant sur des malades et sur des cadavres, pouvoir en conclure sur les rapports qui existent entre les liquides ramassés en dedans et les bruits ou les sons entendus à la surface pendant la vie. Mais attendu qu'il n'est pas bien facile d'arriver à un jugement bien précis même par ces expériences là comme il le serait en s'enquerrant sur la vélocité du son par sa tonalité, (chose facilement déterminable au moyen de la Syrène) et comme on n'a pas toujours dans les Conduites Médicales ni de cadavres, ni même de malades à sa disposition, et en considération surtout que les effets observés dans l'organisme animal ne sont pas toujours attribuables à de causes simples, mais plutôt multiples, j'ai cru, Monsieur, ne pouvoir mieux faire que de m'adresser à vous pour des éclaircissements dont je vous parle plus haut. J'aurais aussi une autre prière à vous faire, ce serait de me dire, si par le mot Resonnance vous entendez un simple accroissement de son qui nait de ce qu'un corps se met à l'unisson avec un autre qui sonne comme on le définit à la page 181–182, ou bien tout redoublement de son engendré par une cause quelconque. Car d'après les faits rapportés dans l'Appendice p. 227 vous sembleriez l'attribuer également au réfléchissement des ondes sonores (tel que le retentissement des cavernes) et aussi à la propagation des tremblemens harmoniques rendue plus facile, comme il arrive des pincettes à feu suspendues par des cordons et mises en contact avec l'ouïe d'une personne.

De toutes ces explications que je vous demande, aussi bien que de l'ennui que je vous donne veuillez m'en excuser M[r] le Professeur grâce à votre incomparable mérite et à la valeur de votre merveilleux ouvrage sur le Son qui en sont les seules causes. Croyez surtout, Monsieur, qu'à part le vif souvenir qui sera constamment gravé dans mon esprit par les sentimens d'estime et de reconnaissance pour une semblable faveur, je me regarderai toujours heureux de ne reconnaitre que votre seule autorité, si dans les faits que je pourrai observer il s'en trouve de telle nature qui me déterminent à les communiquer au public comme une interessante vérité.

En attendant votre gracieuse réponse j'ai l'honneur de me dire

Monsieur le Professeur | Votre très-humble serviteur | Mariano Bichi | Médecin en Chef au service du Municipe de Fermo (Italie)

Fermo (Marche) 13 January 1873.

Most renowned Professor,

I read your Treatise on sound,[2] translated into French by Father Moigno, Paris 1869—Gauthier-Villars,[3] and I found it excellent, both for the new ideas that you develop in it and for the remarkable simplicity that makes mathematical knowledge accessible even to the least educated minds. As I am thinking of doing some experiments on this subject, I would like to ask you to clarify some of the doubts I had. In the Vth lesson on p. 204, I find that to determine the velocity of sound in liquid you fill properly constructed organ pipes, and you judge the capacity of the liquid to diffuse sound by the ensuing tonality. Firstly, I need to know of what material the organ pipes should be made and what their most suitable shape is; then to know how we could stimulate them with sound while they are filled with liquid. My specific intent is to do an experiment on how animal fluids are driven by sound according to their density, their quantity or their composition; I would like to apply it to diseases of the chest, and to know via auscultation[4] if there is a serous,[5] bloody, or purulent[6] liquid, or liquid of another kind between the lung and the thorax—as I believe that among the different causes that could facilitate or delay the diffusion of noises that go to the thorax via the respiratory tracts, it is worth recognizing the role played by the quality and the quantity of liquids collected in the pleura. It is also my intention to create direct experiments by measuring the intensity of a ringing alarm clock that will be enclosed in a spherical box and that we submerge in various animal fluids to be able, by trying it on the sick and the dead, to draw conclusions about the relationship between liquids collected inside and the noises or sounds heard at the surface over a lifetime. But since it is not very easy to arrive at a precise verdict, even by these experiments, as it would be to investigate the velocity of sound by its tonality, (which is easy to determine using the Siren method[7]) and as we do not always have cadavers or sick people at our disposal in Medical */Activities/*, and above all considering that the effects observed in the animal body are not always due to simple causes, but multiple, I thought, Sir, that the best thing was to ask you for clarification on what I mentioned above. I also have another request for you—to tell me if by the word Resonance you mean a simple increase in sound made by a body brought into unison with another that sounds as it is defined on page 181–182, or rather any repetition of sound generated by any cause. Since according to the data reported in the Appendix p. 227, you seem to apply the term also to the reflection of sound waves (such as the resonance of caves) as well as the propagation of harmonic tremors facilitated by fireplace tongs suspended by cords and brought into contact with human hearing.

Please excuse me, Professor, for all the explanations I ask of you and the trouble I bring you, thanks to your incomparable merit and value of your great work on sound, which are the only reasons why I do so. Believe above all, Sir, that aside from the vivid memory that will be etched in my mind by feelings of esteem and gratitude for such a favour, I will always be glad to recognize your sole authority if the nature of the facts that I observe is such that I will be spurred to communicate them to the public as an interesting truth.

While waiting for your gracious reply, I have the honour of being

Professor | Your very humble servant | Mariano Bichi | Chief Medical Officer of the Fermo Municipality (Italy)

RI MS JT/1/B/99

1. *Mariano Bichi*: not identified any further than what is given in the signature of the letter.
2. *Treatise on sound*: *Sound.*
3. *translated... Gauthier-Villars*: J. Tyndall, *Le son: cours expérimental fait à l'Institution royale*, trans. Abbé Moigno (Paris: Chez Gauthier-Villars, 1869).
4. *auscultation*: the action of listening, with ear or stethoscope, to the sound of the movement of heart, lungs, or other organs, in order to judge their condition of health or disease (*OED*).
5. *serous*: of or pertaining to serum (*OED*).
6. *purulent*: of the nature of or resembling pus (*OED*).
7. *Siren method*: a technique developed by Charles Cagniard de la Tour (1777–59), a French engineer and physicist (*CDSB*), which used a siren he invented to correlate pitch with discrete number of vibrations of the siren's noise-making element.

Albert Jackson[1] 13 January 1873 3966

Brooklyn. Jan 13, 1873 | Prof. John Tyndall Esq

Dear Sir—

Will you permit me to express the hope, in common I am sure, with many of those who are listening to your final course of lectures[2] in this country, that you will not conclude them without making some reference to your views on the subject of Prayer in its scientific aspects; and especially regarding the proposed test as to its objective value which you recently seconded.[3] You are doubtless aware how largely your opinions are misunderstood and misstated, perhaps quite as much in our country as your own; and I cannot but think that a word at this time would be of great value in inducing a more correct appreciation of your position as a scientist upon this subject.—

The same platform from which your lectures are delivered[4] is occupied each Sunday by one of the most prominent of the Brooklyn Clergy[5] (another Presbyterian, by the way)[6] who a few weeks since delivered a widely advertised sermon on what he styled—"Tyndall's Prayer Gauge" enlarging especially on the infidelity of science in general. At that time he spoke from his own pulpit, in one of the largest and latest built of our churches;—and doubtless little imagined that within a month it would burn to the ground, in open day, on a Sunday morning, from a fire which the merest chance should have discovered,—which science might have prevented, but which prayer certainly did not.[7] I am sure the world cannot be too thoroughly enlightened as to the attitude of scientific men upon this subject of prayer; to know that they do not deny or disregard its subjective value and influence for good; and would question simply its power objectively; to learn, in time that greatest of truth, that "the acquisition of knowledge must need be preceded by a love of inquiry, and therefore by spirit of doubt; since without doubt there will be no inquiry, and without inquiry there will be no knowledge."[8]—

I am sir, | Very Resp[ectfull]y yours | Alb[er]t Jackson

RI MS JT/2/12, page unknown

1. *Albert Jackson*: not identified.
2. *your final course of lectures*: Tyndall delivered six lectures in Brooklyn from 4–20 January 1873 at the Academy of Music. Jackson may have been unaware that Tyndall planned to end his lecture tour at Yale College in New Haven, CT.
3. *the subject of Prayer... recently seconded*: a reference to the prayer gauge controversy (see letter 3780, n. 2).
4. *platform from which your lectures are delivered*: at the Academy of Music in Brooklyn, New York.
5. *one of the most prominent of the Brooklyn Clergy*: Thomas De Witt Talmage (1832–1902), a Brooklyn-based Presbyterian minister with a large following (*ANB*). His congregation met at the Academy of Music after its tabernacle burned down in December of 1872.
6. *another Presbyterian, by the way*: probably a reference to the Presbyterian minister John Hall in New York (see letter 3943).
7. *burn... did not*: the Brooklyn Tabernacle, on Shermerhorn Street, in Brooklyn, New York, built in 1870 and at the time the nation's largest church, was destroyed by a fire on the morning of Sunday, 22 December 1872 ('The Brooklyn Tabernacle', *New York Times*, 23 December 1872, p. 8).
8. *"the acquisition... no knowledge"*: a quote taken, loosely, from T. H. Buckle, *History of Civilization in England*, 2 vols (London: John W. Parker and Son, 1857), vol. 1, p. 307.

To Joseph Henry 14 January 1873 3967

14th. Jan[uar]y. 1873

My dear sir,

At the request of Prof. Youmans and General Tyndale[1] I sent you yesterday the telegram[2] which you were good enough to answer so promptly.

My work done, and as far as I can see, not altogether ineffectual, the course most agreeable to myself would be to return at once to England, and resume my quiet occupation there, for I am not given to banquetting or festivities of any kind. But many excellent men here think that in the present state of the public mind; if men of weight in science could be collected together, and would but express their views, something of permanent value to the science of the country might accrue. It is solely on this public ground that I presumed to meddle in the matter of the request of my two friends; and it is purely on this ground, and not on any personal one that I should like to see you taking any part in the proposed gathering.[3]

I depart on the 5th of February,[4] and as my visit to America becomes a thing of the past I can look back upon it as a memory without a single stain.[5]

Faithfully yours | John Tyndall.

John Tyndall to Joseph Henry, 14 January 1873, mssRH 3979, box 43, William Jones Rhees papers, The Huntington Library, San Marino, California

1. *General Tyndale*: Hector Tyndale.
2. *telegram*: telegram missing.
3. *proposed gathering*: see Introduction, n. 23.
4. *I depart on the 5th of February*: see letter 3928, n. 18.
5. *without a single stain*: excepting the death of his younger assistant George Miller (see letter 3896, n. 10).

From George J. Brush[1] and Othniel C. Marsh[2] 15 January 1873 3968

[Sheffield Scientific School[3] of Yale College, | *New Haven, Connecticut.*] | January 15, 1873

Prof. John Tyndall | New York

Sir:

The Governing Board of the Sheffield Scientific School desire the pleasure of your company at an informal supper at the New Haven House on

Thursday, Jan[uar]^{y} 23^{d}. after the close of your lecture.[4] We think that this will afford an agreeable opportunity for you to meet our men of science, and it will not interfere with the general reception to be given by our colleague Professor Lyman[5] on Friday. Trusting that (you will kindly) favor us with an early reply, we are

very resp[ectfull]y yours, | Geo[rge]. J. Brush | O. C. Marsh | Committee | of the | Governing Board

RI MS JT/1/S/72

1. *George J. Brush*: George Jarvis Brush (1831–1912), an American mineralogist and the first director of Yale's Sheffield Scientific School (*ANB*).
2. *Othniel C. Marsh*: Othniel Charles Marsh (1831–99), an American paleontologist who was educated at the Sheffield Scientific School and gained notoriety through his public rivalry with Edward Drinker Cope (1840–77) in who could discover and name more new dinosaur species in the American West, known as 'The Bone Wars' (*ANB*; *CDSB*).
3. *Sheffield Scientific School*: a school of Yale College, focusing on science and engineering education, named in honor of the American railroad magnate Joseph E. Sheffield (1793–1882).
4. *your lecture*: Tyndall's last stop in his American lecture tour was in New Haven, CT, for two lectures on 22 and 23 January 1873 at Yale College.
5. *Lyman*: Chester Smith Lyman (1814–90) was educated at Yale and was professor of Industrial Mechanics and Physics at Yale's Sheffield Scientific School ('Prof. Chester S. Lyman,' *New York Times*, 30 January 1890, p. 5).

From Henry Bence Jones 17 January 1873 3969

Friday | Jan 17 1873

My dear Friend

It is very pleasant indeed to read of your success.[1] I am heartily glad that you have had it. It is worth all you have gone through & when you are here in the quiet of the Institution resting this year & when you get nearer your long rest it will be pleasant to think of what you have done & of the effect which your mental power has had on others.

Moreover it will reach here & make your attractive power at the Institution still more & no less work will be wanted to keep it up. So I am heartily glad you went for I shall drag on until I see you though my fear is I can find no drug to help me. The devil used to fail his protegé at last & take him off & my medicines will serve me the same—I only wish they would take me off quickly & set my matter & force free for some better action then they now perform. Debus gave his first lecture yesterday.[2] He did not speak above a loud whisper I am told so that there has been no good judgement on his matter. His experiments did but I fear Wills was too officious. However I shall hear more from Col York[3] & Odling & Frankland who were there. Spottiswoode has got a capital lecture for to night[4] & the laboratories[5] will be lighted up afterwards for the members &c. When you return you will take possession of the lower room if you wish it & I hope it will be in tolerable order. I mentioned your wish[6] to the managers but no one will hear of your refusal of the little we give & I hope next meeting even more may be offered you but our accounts are not made up & I see 6000 will not pay our bills. I also mentioned Grammes Electric light[7] I have all the papers & the managers will get it when you have expressed a wish for it of that I am sure from what was said by Mr Pole.[8] I have drawn up a curious return of the attendance on Friday evening Xmas & afternoon lectures of Frankland & Odling. There is nothing great in favour of Odling. Four fewer than Frankland on Friday evenings. Thirty more than 7 in Juveniles[9] & only fifteen more in courses in the afternoons I have not been able to crawl about the Institution this week indeed I have not been out & I do not see when I shall get out. I wish I had something better to talk about. I must finish my *[pages]* tomorrow hoping to have Spottiswoode to talk about Thursday 18. Sir H H[10] is greatly pleased with your grand success & sends you his "affectionate regards". He has twice come home in the Cuba[11] & it is a first rate ship. He wished me to tell you. He wants you to bring home a wife & from his tone I was very near asking him whether it was true that he was going to be married but I forebore. Spottiswoode was first rate but his audience small 323 still the thing is made known. I send you a report from the Standard[12] which you may like to see. I count the days till you are back; but why for when you are back I have only to congratulate you & nothing to say. But I am in an unstable state of nerve I suppose & long for all sorts of things & without reason but your return will be a great pleasure to me. Goodbye dear friend all health & happiness to you

Ever yours most truly | H B J

Many many thanks for the telegram[13] if I have not before thanked you. Thanks for the papers from Mr H[14] as well for myself. Your <u>six</u> lectures for 3 cents!!![15]

RI MS JT/1/J/78

---

1. *your success*: referring to Tyndall's American lecture tour (see Introduction, n. 8). At the time Bence Jones wrote this letter, Tyndall was nearing the end of his lectures in Brooklyn (six lectures from 4–20 January 1873), with two final lectures at Yale College in New Haven, CT, on 22 and 23 January 1873.

2. *his first lecture yesterday*: Debus delivered a course of three lectures at the RI on the subject of oxidation on 16, 23, and 30 January 1873 (*Roy. Inst. Proc.*, 6 (1870–72), pp. 559).
3. *Col York*: probably Philip James Yorke (1799–1874), a British Army officer, chemist, and FRS who served several times as a manager of the RI. Yorke was a founder of the Chemical Society (*ODNB*).
4. *capital lecture for to night*: Spottiswoode delivered a Friday Evening Discourse at the RI on 17 January 1873, 'On the Old and New Laboratories at the Royal Institution' (*Roy. Inst. Proc.*, 7 (1873–75), pp. 1–10).
5. *laboratories*: the RI's laboratories underwent renovation while Tyndall was on his lecture tour (see letter 3776, n. 14).
6. *your wish*: for Tyndall's donation of his lecture tour profits, see Introduction, n. 15.
7. *Grammes Electric light*: see letter 3944, n. 5.
8. *Mr Pole*: William Pole was at this time among the managers of the RI.
9. *Juveniles*: the RI's annual juvenile Christmas Lectures. For the series that Odling had just delivered, see letter 3782, n. 4.
10. *Sir H H*: Henry Holland.
11. *Cuba*: see letter 3928, n. 18.
12. *the Standard*: the *London Evening Standard*; the exact clipping Bence Jones sent is not identified.
13. *telegram*: telegram missing.
14. *H*: perhaps a reference to Joseph Henry.
15. *Your six lectures for 3 cents!!!*: see letter 3962, n. 2.

From Thomas Archer Hirst 17 January 1873 3970

Jan[uar]y 17th /73

My dear John.

I was <u>very</u> glad to receive your New York letter;[1] although it told of a new ailment for you (Tooth ache) the tone of it indicated good spirits and a most wholesome desire to come back again to us who want you so much. I was going to write you a long letter but my "chief"[2] has just sent me some work to do I must therefore send this or loose a post[3] which I should not like to do for the enclosed letters[4] may have some interest for you.

Ever yours | Tom—

Jan 17th 1873

Last night I sat in the chair of Newton[5] and presided at the meeting of the Royal Society, Airy having very gracefully made me one of the vice presidents of the Year![6] It was a commonplace thing enough to do in the estimation of those who know what slow affairs meetings of the Royal Society have become, but I could not help, as I sat there, going back in memory to the time when you taught me to make "pot hooks",[7] and saying to myself who would have ventured then to predict that this could come out of the "Pot Hooks".

RI MS JT/1/H/254

1. *your New York letter*: letter 3951.
2. *my "chief"*: probably a reference to Astley Cooper Key (1821–88), the first president of the Royal Naval College, Greenwich (*ODNB*), and Hirst's new boss (see letter 3911, n. 12).
3. *a post*: outgoing mail pick up.
4. *enclosed letters*: Hirst appears to have written his letter on that from Carel Brutel de la Rivière (see letter 3964). Other enclosed letters not identified.
5. *chair of Newton*: George Biddell Airy was President of the RS from 1871–3, and here Hirst is commenting on Airy having previously been the Lucasian professor of Mathematics at the University of Cambridge (from 1826–8), which Isaac Newton held from 1669–1702. So, in presiding over the meeting of the RS on 16 January 1873, Hirst was sitting in the 'chair of Newton'. He did so again on 23 January (see 'Proceedings of Learned Societies', *Phil. Mag.*, 45 (1873), p. 306).
6. *made me one of the vice presidents of the Year*: at the RS's Anniversary Meeting of 30 November 1872. The other Vice-Presidents for the year were William Spottiswoode, George Busk, William Sharpey (1802–80), and Francis Sibson (1814–76).
7. *"pot hooks"*: a curved or hooked stroke made with the pen, esp. as a component of an unfamiliar or unintelligible script or when learning to write (*OED*).

From Elizabeth Rutherford — 17 January 1873 — 3971

13. Shandwick Place—| Edinburgh | Scotland, | 17th January 1873.

My Dear Friend,

I need not tell you, how much I felt and sympathised in your sorrow & distress, in losing far away from England your friend and helper.[1] It must have required all your philosophy to bear so great a grief, and it must have sadly damped the pleasure of your success and triumph.—

I thank you much, very much, for sending me so many tidings of your progress in the States.[2] The journals have interested me much with regard to your Lectures, and also amused me excessively in their remarks on your person and manner. Now, I imagine your thoughts are beginning to be turned homewards and that you will not be sorry to embark for a more genial clime. We have absolutely had no winter as yet, here, altho' the excess of rain has been very unpleasant.

It was stated that you were to be home in February,[3] and therefore as I think it probable that you may not get this letter, I shall make it a very short one, as I merely write now, that you may not think me ungrateful for your kind remembrance of your old friend.

I hope in the early part of summer to have the pleasure of seeing & shaking hands with you once again—and I <u>hope</u> I shall not find you spoiled by all the praises and admiration you have met with. I shall desire for you a prosperous and pleasant voyage across the great Herring Pond.[4] You will at all events have the satisfactory remembrance of having entirely fulfilled your object and the delightful feeling that you leave many friends behind you. Perhaps also, there will be some regrets at parting ... Eh?—for "Joy whose hand is ever at his lips bidding Adieu"[5]—is never far from veiled sadness—Thus the world is constituted, and 'twill ever be the same old story. And now, Farewell—

Believe me Always | Very Sincerely & Affectionately Yours, | <u>Elizabeth Rutherford</u>

RI MS JT/1/R/89

---

1. *your friend and helper*: George Miller (see letter 3896, n. 10).
2. *your progress in the States*: for Tyndall's American lecture tour, see Introduction, n. 8.
3. *home in February*: see letter 3928, n. 18.
4. *Herring Pond*: the sea or ocean, esp. the North Atlantic Ocean (*OED*).
5. *"Joy... Adieu"*: from John Keats' poem 'Ode on Melancholy', published in *Lamia, Isabella, the Eve of St Agnes, and Other Poems* (London: Taylor and Hessey, 1820), p. 141, lines 140–2.

From Mary H. Birkbeck[1] 19 January [1873][2] 3972

20. Berkeley Sq[ua]r[e] 19th. Jan[ua]ry

Dear Mr. Tyndall

I should have written before to thank you so much for sending the papers, but I heard a report you were coming back the middle of this month, & I thought we should have the pleasure of thanking you by <u>word</u>. What a great work you are doing among the Americans[3] & how deeply they seem to be appreciating it. You will have kindled a fire which they will not allow to expire.

It will be very pleasant indeed to see you back again here, & if this warm weather continues I fancy <u>you</u> will be glad to leave the sharp cold of America & to return to these warmer parts.[4] I am so thankful it <u>is</u> mild here, it is better for my mother,[5] who lies in her bed without gaining strength, but losing it, I fear—Nevile[6] & his wife[7] started yesterday for the West Indies, & we hope in a few days to have my mother carried into their house. The change will I hope be good for her.

We were so sorry to miss M^r^. Spottiswoode's lecture yesterday,[8] but Robert[9] only returned from Suffolk close upon the time. His train being late, just lost us the lecture. We very much enjoyed D^r^. Odling's lectures.[10] I cannot tell you how deeply we felt for your sorrow in losing your assistant.[11] I am thankful that M^r^. Cottrell[12] keeps well. I hope so much this letter may just catch you, as Sir Henry Holland speaks of y[ou]r sailing on the 4th You will be most warmly welcomed back again—

Always Sincerely y[ou]^rs^ | Mary H Birkbeck

RI MS JT/1/B/100

---

1. *Mary H. Birkbeck*: Mary Birkbeck (née Lubbock, 1835–1910), a sister of John Lubbock who married Robert Birkbeck in 1857 (see *Ascent of John Tyndall*, p. 323; R. Birkbeck, *Notes on the History and Genealogy of the Family of Lubbock* (London: Mitchell and Hughes, 1891), pp. 42).
2. *[1873]*: the year is given by reference to Tyndall's American lecture tour.
3. *great work you are doing among the Americans*: for Tyndall's American lecture tour, see Introduction, n. 8.
4. *return to these warmer parts*: see letter 3928, n. 18.
5. *my mother*: Harriet Lubbock (née Hotham, d. 1873). Her and John Lubbock (1803–65) were married in 1833 (R. Birkbeck, *Notes on the History and Genealogy of the Family of Lubbock*, p. 35).
6. *Nevile*: most probably Nevile Lubbock (1839–1914), Mary's brother who made his career in the sugar trade in the West Indies (R. Birkbeck, *Notes on the History and Genealogy of the Family of Lubbock*, p. 38).
7. *his wife*: Harriet Charlotte Lubbock (née Wood, d. 1878) (R. Birkbeck, *Notes on the History and Genealogy of the Family of Lubbock*, p. 38–9).
8. *lecture yesterday*: see letter 3969, n. 4.
9. *Robert*: Robert Birkbeck, Mary's husband (m. 1857). He was the compiler of *Notes on the History and Genealogy of the Family of Lubbock*.
10. *lectures*: see letter 3782, n. 4.
11. *losing your assistant*: George Miller (see letter 3896, n. 10).
12. *Cottrell*: John Cottrell.

To Isaac B. Choate[1] 19 January 1873 3973

19th. Jan[uar]y 1873

Dear Sir,

I do not think the effect to which you refer arises from the shadow of the wire.[2]

faithfully yours | John Tyndall

[verso] Isaac B. Choate Esq | 434 Franklin Avenue | Brooklyn

HM 75992–75998. The Huntington Library, San Marino, CA

1. *Isaac B. Choate*: probably Isaac Bassett Choate (1833–1917), from Maine, who graduated from Bowdoin College in 1862, then admitted to the Cumberland Bar in 1865. He taught at several schools, and then was professor of Greek and Latin at Buchtell College in Akron, Ohio. Choate was an avid writer contributing many poems and prose to numerous publications. He later established himself in Boston ('Border State Ballads', *The Republican Journal*, 30 January 1913, p. 4; N. Cleaveland and A. S. Packard (eds), *History of Bowdoin College: With Biographical Sketches of Its Graduates from 1806 to 1879, Inclusive* (Boston: James Ripley Osgood & Company, 1882), pp. 770–1; and T. W. Herringshaw (ed), *Local and National Poets of America* (Chicago: American Publishers' Association, 1890), p. 704).
2. *effect to which... shadow of the wire*: probably a reference to something in Tyndall's lectures on light delivered in Brooklyn (six lectures from 4–18 January 1873). Presumably Choate traveled to New York for the lectures, or for another reason but partook in the lectures. The letter from Choate to which Tyndall replied is missing. For Tyndall's lecture tour in America, see Introduction, n. 8.

To Hector Tyndale [19 January 1873][1] 3974

Sunday

My dear Hector

I finished last night amid thunders of applause—but oh! how weary & sleepless I am.

I send you a note[2] that has just reached me: what curious things turn up.

Best love to your wife[3] and kindest regard to the young ladies.[4]

ever affection[atel]y| John.

I wish I could now vanish.

RI MS JT/1/T/1482
RI MS JT/1/TYP/5/1716

1. *[19 January 1873]*: the date is given by the reference to Tyndall's final lecture in Brooklyn, on 18 January 1873.
2. *a note*: Tyndall enclosed letter 3963 to Tyndale.
3. *your wife*: Julia Tyndale.
4. *young ladies*: presumably their daughters (not identified).

To Henry Bence Jones 21 January 1873 3975

21st. January. 1873.

My dear Friend

I am on the point of packing up for New Haven where I lecture on Wednesday and Thursday,[1] but before starting must send you a line. I do not go joyfully for your two last letter notes[2] have rendered me gloomy. I hope however that the toughness of constitution which has carried you victoriously over so many assaults will carry you over this one also. The death of poor Harwood[3] is very shocking: how foolish men will be through ignorance of their dangers; had I not lain by with my leg[4] something similar might have happened to me.

I came over from Brooklyn—where my lectures[5] ended on Saturday night—on the Sunday. The Brooklyn audience was the most impressive that I have seen. The Hall[6] was an exceedingly fine one. First there was a broad sea of faces spreading far forward and to the right and left. Then were the stage boxes. Then the first *[deep]* tier of seats running round the Hall Then an upper tier equally deep. These tiers corresponded to our dress circles, and were packed with people—In fact for the last three lectures the house was crammed as full as it could hold. And when I finished on Saturday the storm of applause was deafening, and it did not and would not end until I had made a second farewell bow. I afterwards joined a few gentlemen whom I might take as representatives of the general feeling. Certainly nothing could be more hearty, nor I believe more genuine.

Thus my work in America virtually, and assuredly though tried by unexpected private grief[7] I have had no reason to repent of my visit. It has been to me a succession of pleasures, and the greatest pleasure of all was contemporaneous with my arrival in Washington. Dare I as a student withdraw from the arena of politics—a cultivation of that science which is of no country and no nationality—Dare I, in passing, express to you the profound gratification with which I read this morning the words of the President of the United States,[8] that not a shadow now rests on the friendly relations of the governments of America and England, and the expression of his sincere hope that these relations may remain for ever unclouded as they are now. I am not, as the proprietors of this beautiful hall (the Young men's Christian Association)[9] well know, in the ordinary sense of the term, a praying man.[10] But if prayer be the heart's sincere desire then I may claim to pray, that the hopes and wishes of General Grant may receive ratification and fulfillment in the coming time and if I except the death of poor Miller[11] it is without a drawback of any kind. From Boston to Brooklyn[12] I have been welcomed with warmth, and my parting with each audience has been enthusiastic.

But how glad I am that the thing is drawing to a close! I shall have about 3000 pounds to dispose of. I think I shall put it into the hands of trustees for the encouragement of original research.[13]

Sir Henry[14] is perhaps right judging from the data in his possession[15]—but the data may be wrong. Instead of feeling elated at all this I have gone through it as a duty, at times a very anxious one—The most that I feel is that clearness and relief of mind which accompanies the successful execution of an arduous task.

I return here on Saturday, and remain here until the following Tuesday when I go and steep myself in quiet for three or four days in Philadelphia. I should willingly retreat in quiet now that my work is done; but certain excellent people here say that they owe it to themselves to give me a banquet before I go.[16] This is to come off on Tuesday the 4th and on Wednesday—doubtless with a heavy head—I start in the Cuba.[17]

The weather here is wonderful. We stand between a vast continent and a vast ocean, the wind from the former keen and clear and dry: cold and sharp as razors; but it has only to veer to send over us the vapours of the Gulfstream. Hence a climate of extremes—embracing the most beautiful and the most abominable weather. On Saturday it was muggy; on Sunday clear and keen; with lustres of ice hanging from every twig and telegraph wire. To day it is again muggy: and thus the weather shifts and changes. Take care of yourself till I come back. You are bound to let me find you in fair condition. That this may be the case is assuredly the earnest wish of

Your friend | John Tyndall

RI MS JT/1/TYP/3/823–4
LT Typescript Only

---

1. *New Haven where I lecture on Wednesday and Thursday*: Tyndall's last stop in his American lecture tour was in New Haven, CT, for two lectures on 22 and 23 January 1873 at Yale College.
2. *your last two letter notes*: letters 3954 and 3957.
3. *death of poor Harwood*: on the death of Hamilton Edward Harwood, see letter 3954.

4. *lain by with my leg*: see letter 3954, n. 2.
5. *my lectures*: Tyndall delivered six lectures in Brooklyn from 4–20 January 1873 at the Academy of Music.
6. *The Hall*: in the Academy of Music in Brooklyn.
7. *unexpected private grief*: referring to the death of his assistant George Miller (see letter 3896, n. 10).
8. *I read this morning the words of the President of the United States*: the source of Ulysses S. Grant's words, here paraphrased by Tyndall, is not identified.
9. *Young men's Christian Association*: the Washington, DC, chapter of the YMCA was based in Lincoln Hall. Tyndall referred to the YMCA again in letter 4125. Tyndall gifted $1,500 to the YMCA (*Yale Literary Magazine*, 38 (February 1873), p. 214).
10. *I am not... a praying man*: a reference to the prayer gauge controversy (see letter 3780, n. 2).
11. *Miller*: see n. 7.
12. *From Boston to Brooklyn*: for Tyndall's American lecture tour in general, see Introduction, n. 8.
13. *into the hands... of original research*: see Introduction, n. 15.
14. *Sir Henry*: Henry Holland.
15. *judging from the data in his possession*: see letter 3954, in which Bence Jones reports Holland's impression that Tyndall would be spoiled by his success, as reported in the American newspapers.
16. *a banquet before I go*: see Introduction, n. 23.
17. *I start in the Cuba*: see letter 3928, n. 18.

From Henry Bence Jones — 21 January [1873][1] — 3976

Tuesday Jan 21

Dear friend

I write a last line on the chance of its reaching. Tho I have no news. I have sent the letter to the Millers[2] & will see about the prayer[3]

I have nothing to say & yet I long for you to be back to say how pleased I am w/ y[ou]r success £2000 for six lectures![4] I give £60 & think I do well but the sheet of your lectures for 3 cents[5] as a bit of a daily paper is to me a greater success even /& marks/ what interest those who do not hear you take in Scientific clearness. Fancy one of our dailys publishing such a supplement. Some trial or scandal might sell but science never. There must be a great future for such wits & /words/ as exist on the other side the Atlantic.

I cannot get on but I shall see you. I hope the Westerly gales may blow you here in 9½ days & when I have seen you I will go to sleep.[6]

Good bye dear friend. How can we welcome you home I have not been to the Institution & hear little about it.

I have already got one legacy of £1000 perhaps it may be £2000.
If I could I would do something but I have my chicks[7] to feed.
Kindest regards from all. | Ever yours most truly | H Bence Jones

RI MS JT/1/J/79

1. *[1873]*: the year is given by the dates of Tyndall's American lecture tour.
2. *the letter to the Millers*: a letter from a Philadelphia doctor, which Tyndall asked Bence Jones to forward to the family of George Miller, Tyndall's deceased assistant. See letter 3959.
3. *the prayer*: Tyndall sent Bence Jones a piece of writing, since lost, entitled 'Hooker's Prayer'. See letter 3959.
4. *£2000 for six lectures!*: in letter 3959, Tyndall reported this as his profit for his lecture series in New York.
5. *the sheet of your lectures for 3 cents*: see letter 3962, n. 2.
6. *go to sleep*: in previous letters, Bence Jones had commented to Tyndall that he wouldn't see him in heaven, and that he was nearing his end. Indeed, Bence Jones died on 20 April 1873, fortunately having time to visit with Tyndall.
7. *my chicks*: see letter 3860, n. 6.

To Henry Bence Jones 24 January 1873 3977

New Haven,[1] 24th. Jan. 1873.

My dear Friend

I came down here from New York on Wednesday and lectured the same evening. As in all other cases the audience was, I am told, altogether out of the ordinary run of things in New Haven. The Hall was smaller than those at New York[2] and Brooklyn,[3] still there cannot have been much less than 1500 people present. My first lecture was ill prepared and ill digested. In fact I had taken my type to pieces and had not time to reset it—I was in better case at the second lecture which therefore went off pleasantly—The parting was enthusiastic in the extreme.

After the lecture there was a kind of banquet at the principal Hotel.[4] The Governor of the State[5] was there, also President Porter[6] of Yale College: expresident Wolsey,[7] the Governing body and a number of Professors—about 50 in all. After the banquet speeches were started by President Porter, and several of the guests spoke. In fact when I came to America I had no notion how well the people like me. It has been hammered into me in town after town, and nowhere with more decision and emphasis than among the Professors and students of Yale College. They gave me 1000 dollars for my two lectures and I made a present of 250 to the Yale scientific Club! a society of young fellows who take a lively interest in scientific matters, and who arrange the scientific lectures.

After the turmoil of New York the quiet of New Haven is soothing. Still I have been heavily drawn upon. I have been out all this day seeing men, women, and things,—the men cordial—the women handsome and interesting and the things relating to science and education. I have got away for a couple of hours rest; to night there is to be a "reception," at the house of one of the Professors[8]—about 150 people are invited—tomorrow morning I return to New York. There, for my sins, I have three dinner parties on three successive evenings. My only way of escape from ruin is to dine on something plain and wholesome, at my hotel, to drink half a bottle of good wine, and then to abstain in toto at the dinner. This I have already done and found it to answer.

This New Haven in summer must be a pretty place. The streets for the most part are planted by graceful elms; Yale College stands very prettily and when surrounded by the summer foliage must appear beautiful. At present dread winter clasps us: the snow began again to fall yesterday, so that I was obliged to have a carriage and pair[9] to drive me to the lecture Hall last night. It was quite touching to see the people, men and women, trudging through the snow to hear the lecture. In fact the weather, though terrible, hardly perceptibly thinned the audience.

What a relief it is to have more lectures reeled off! and what an additional relief it will be when the banquet of the 4th[10] is over! Agassiz is to preside, and I believe it is going to be a grand affair. I shall hardly write to you again before I start. My most earnest wish now is that I may find you with this last assault repelled when I return.

Ever and ever Yours | John Tyndall

RI MS JT/1/TYP/3/825
LT Typescript Only

---

1. *New Haven*: Tyndall's last stop in his American lecture tour was in New Haven, CT, for two lectures on 22 and 23 January 1873 at Yale College.
2. *New York*: Tyndall delivered six lectures in New York from 17–31 December 1872 at the Cooper Institute.
3. *Brooklyn*: Tyndall delivered six lectures in Brooklyn from 4–20 January 1873 at the Academy of Music.
4. *principal hotel*: New Haven House (see letter 3968).
5. *Governor of the State*: Marshall Jewell (1825–83), Governor of Connecticut from 1869–70 and 1871–3 (*ANB*).
6. *President Porter*: Noah Porter (1811–92), an American philosopher and 11th president of Yale, serving from 1871–86 (*ANB*).

7. *Wolsey*: Theodore Dwight Woolsey (1801–89), the 10th president of Yale, serving from 1846–71, who also founded the Sheffield Scientific School (see letter 3968, n. 3) (*ANB*).
8. *"reception," at the house of one of the Professors*: professor not identified.
9. *pair*: of horses.
10. *banquet of the 4th*: see Introduction, n. 23.

To Hector Tyndale [27 January 1873][1] 3978

Monday

My dear Hector.

I hope the Road will not be blocked up tomorrow—Today matters look ominous.

If the powers of the air permit I will start at 1 oC[lock] and reach Philadelphia at 4 P.M.

I have had an exceedingly interesting letter from Miller's father[2] in which he speaks with boundless gratitude of Gen. Tyndale[3] & his "loving lady"[4]—give my love to the said lady & kind regards to all others.

John

RI MS JT/1/T/1474
RI MS JT/1/TYP/5/1718

1. *[27 January 1873]*: the date is given by the timing of Tyndall's departure for Philadelphia.
2. *letter from Miller's father*: from W. Miller, the father of George Miller, Tyndall's younger assistant who died of typhoid fever on 30 November 1872 (see letter 3896, n. 10). The letter is missing.
3. *Tyndale*: Hector Tyndale.
4. *"loving lady"*: Julia Tyndale.

To Thomas Archer Hirst [29][1] January 1873 3979

2<9>th. Jan[uar]y 1873

My dear Tom.

Here I am once more with Hector[2] with a terribly bad cold when I ought to be particularly strong. I came down to put my thoughts in order for the "banquet"[3] which takes place on the 4th. But as yet my thoughts have been directed to getting rid of a sore throat and headache. Still I suppose you will have sufficient belief in my powers to get through the duties of the banquet. Agassiz is to preside, and many great guns[4] are to fire a shot on that occasion.

I think I told you of the successful issue of the Brooklyn lectures.[5] Since then I have been down to Newhaven lecturing besides to the general public to the President Professors & Students of Yale College.[6] I cannot describe to you the warmth of these people. They also had their banquet Reception and said the kindliest things.

On the 5th. I step on board the Cuba,[7] and if the powers of the air should prove propitious shall, I hope, be near you on the 15th

Until then goodbye | Affectionately yours | John Tyndall

RI MS JT/1/T/700
RI MS JT/1/HTYP/618

1. *[29]*: there is an ink blot over the second numeral in the day, but we believe it to be '29' as Tyndall wrote to Tyndale on 27 January that he was going to Philadelphia the next day (letter 3978).
2. *Here I am once more with Hector*: in Philadelphia with his cousin Hector Tyndale.
3. *banquet*: see Introduction, n. 23.
4. *a great many guns*: numerous New York luminaries, including several of Tyndall's critics from the religious community, were invited to speak at his farewell banquet. See Introduction, n. 23.
5. *Brooklyn lectures*: Tyndall delivered six lectures in Brooklyn from 4–20 January 1873 at the Academy of Music.
6. *President... Yale College*: Tyndall's last stop in his American lecture tour was in New Haven, CT, for two lectures on 22 and 23 January 1873 at Yale College. The President of Yale College at the time was Noah Porter (see letter 3977, n. 6).
7. *Cuba*: see letter 3928, n. 18.

From Richard Sears Mac-Culloch[1] 30 January 1873 3980

Lexington Virginia | Jan 30. 1873

To | Prof. Tyndall

Many thanks, my dear Sir, for your kind & much esteemed response just rec[ieve]d.[2]

I now wish to say that I send another copy of the book[3] to make sure of your getting one. Should both fail to reach you, please speak of it to my friend Prof[esso]r. Mayer[4] of Hoboken.

Let me add that being only lectures delivered to a class of young men for their instruction, they pretend to nothing more than an elementary presentation of the labours of Rumford, Davy, Carnot, Joule, Mayer, Thomson, Rankine, & Clausius,[5]—yet the setting of their diamonds you will, I think, find after a fashion somewhat different from any familiar to you—and it is this feature which I think will be interesting to you.

To save you the trouble & time of looking for it, let me tell in few words what is the peculiarity of my mode of composition of the subject:

The equation, which Rankine, in his Applied Mechanics,[6] sometimes uses & calls the law of "conservation of energy," but which I would rather call the "general formula for all energy & work"

$$\Sigma\int Pdp - \Sigma\int m \frac{d^2s}{dt^2} ds = 0$$

is used by me (after Prof. Bartlett[7] of the U.S. Military School at West Point, who thus used it in 1854[8]) as the only necessary & one single fundamental law of Analytical Mechanics—an expression for all work & all energy (potential or kinetic), & embracing alike Newton's action & reaction, D'Alembert's[9] principle, & Lagrange's[10] princ[iple]. of virtual velocities.

You will see at a glance that the integration of its first term $\Sigma\int Pdp$ is either potential energy or work done, the "travail moteur" & "travail resistant" of French writers—and that the second term is kinetic energy, half vis-viva

$$\Sigma m \frac{v^2}{2},$$

or "work of inertia". This briefly is what you will find set forth in Lectures II & III more particularly.

All mechanical science (including Heat &c) is I believe, destined to soon crystallize into some such very single & general form. My poor attempt for my class may lack transparency and be but a confused mass of crystalline particles—but others more able will obtain a beautiful regular clear & perfect solid.

The death of Prof. Rankine[11] is a loss to the world, deeply to be regretted. I was indebted to him for kindness, as I am to Sir C. Wheatstone, Prof[esso][r]. Stokes, Clifton[12] & Tait, Sir W. Thomson & to yourself, not only for yours, but also for theirs, to whom you did me the honor to make me known & who were all very kind to me

And now renewing to you the assurances of my most grateful recollection & highest esteem, with my best wishes for your safe return home, my regret at not having been able to do anything to contribute to the pleasure of your visit, & my hope that we may meet again,—I am very truly yours

R. S. Mac-Culloch

RI MS JT/1/M/1

---

1. *R. S. Mac-Culloch*: Richard Sears Mac-Culloch (1818–94) was a professor of natural

and experimental philosophy and chemistry at Columbia College, New York, before the American Civil War. In 1863 he resigned his chair and fled to Richmond, Virginia, to join the Confederacy. After the war he was appointed to a professorship in natural philosophy at Washington College (which later became Washington and Lee University) in Lexington, Virginia ('Richard Sears McCulloch', in W. D. Miles (ed), *American Chemists & Chemical Engineers*, 2 vols (Washington: American Chemical Society, 1976–94), vol. 1, pp. 325–6). His surname was inconsistently spelled.

2. *response just rec[ieve]*[d]: letter missing.
3. *the book*: likely a draft manuscript of R. S. McCulloch, *Treatise on the Mechanical Theory of Heat and Its Applications to the Steam-engine, etc.* (New York: D. Van Nostrand, 1876).
4. *Prof[esso]*[r]*. Mayer*: Alfred Mayer.
5. *Rumford... Clausius*: all scientists who made contributions to thermodynamics: Benjamin Thompson, Humphry Davy, Sadi Carnot (1796–1832), James Joule, Julius Mayer, William Thomson, William Rankine, and Rudolf Clausius.
6. *Applied Mechanics*: W. J. M. Rankine, *A Manual of Applied Mechanics* (London: Richard Griffin, 1858).
7. *Prof. Bartlett*: William Holms Chambers Bartlett (1804–93), an American engineer and professor of natural and experimental philosophy at West Point (*ANB*).
8. *thus used it in 1854*: see W. H. C. Bartlett, *Elements of Analytical Mechanics* (New York: A. S. Barnes and Company, 1853), p. 345, where the equation above appears, absent the integral signs.
9. *D'Alembert*: Jean le Rond D'Alembert (1717–83), a French mathematician and physicist (*CDSB*).
10. *Lagrange*: Joseph-Louis Lagrange (1736–1813), an Italian mathematician and astronomer who was later naturalized French (*CDSB*).
11. *death of Prof. Rankine*: William Rankine died on 24 December 1872.
12. *Clifton*: Robert Bellamy Clifton (1836–1921), a British physicist who studied under George Gabriel Stokes and previously was Professor of Natural Philosophy at Owens College, Manchester (*ODNB*).

To Frederick Barnard [January 1873][1] 3981

Wednesday

Dear D[r]. Barnard.

I have not yet heard any thing from you regarding Saturday[2] would you kindly send me a line to say whether that day will answer & if so kindly naming the time?

faithfully yours | <u>John Tyndall</u>

I am staying at M[r]. Dickenson's[3] | 84 Montague Street | Brooklyn

Columbia University Manuscripts; Box 2 Page 95; University Archives, Rare Book & Manuscript Library, Columbia University Libraries, Rare Book & Manuscript Library, Columbia University in the City of New York

---

1. *[January 1873]*: the date is given based on when Tyndall was in Brooklyn during his American lecture tour (4–21 January 1873).
2. *regarding Saturday*: previous letters from Tyndall to Barnard (3842 and 3867) indicate that Barnard was attempting to schedule a time to meet Tyndall while he was in the United States. At this time Barnard was President of Columbia College in New York (now Columbia University).
3. *M^r^. Dickenson's*: H. H. Dickinson (dates unknown), whose establishment Brooklyn Heights Pharmacy was located at 84 Montague Street, presumably with his own residence and extra lodging upstairs (*Proceedings of the Medical Society of the County of Kings* (Brooklyn, 1877), p. 388).

Governing Board of the Sheffield Scientific School [January 1873][1] 3982

[Sheffield Scientific School[2] of Yale College, | *New Haven, Connecticut.*]

Sir:

A word of explanation is due from us in regard to your reply to an invitation[3] received this morning through Prof Barker.[4] It is evident that you misapprehend the source of the letter. The Governing Board of this department of college consists of the President of the Univ[ersit]y and fourteen professors in Science. We desired as a body to extend to you the courtesy of a supper where we could meet you socially, and regret that any misunderstanding should have prevented the realization of what, to us, would have been so agreeable an occasion.

Very resp[ectful]ly | The Governing Board of the Sheffield Scientific[5]

George Jarvis Brush Family Papers (MS 108). Manuscripts and Archives, Yale University Library

---

1. *[January 1873]*: the date is given by when Tyndall visited New Haven, CT.
2. *Sheffield Scientific School*: see letter 3968, n. 3.
3. *an invitation*: see letter 3968.
4. *Barker*: George Frederick Barker (see letter 3926, n. 1).
5. *Sheffield Scientific*: the word 'School' was not written.

From Asa Gray 3 February 1873 3983

[BOTANIC GARDEN, | CAMBRIDGE, MASS.] | Feb. 3, 1873

My dear Mr. Tyndall

I cannot go to the New York dinner,[1] and say "good-bye" in person. I have been ill myself, and my wife[2] is still very much so, tho. not in a way to give me much anxiety.

Pray come soon again, and then we will see more of you, if you will.

I hear by rumors—without particulars—that you have been doing <u>liberal things</u>—which is too much after all the labor you have bestowed upon us.

You will be glad to hear that my memorial to Gladstone I saw not a disclaimer of any intention on the part of Government to interfere with Kew[3]—which has much pleased Hooker—all the acknowledgements of Memorials in England having been strictly non-commital.

Mrs. Gray joins me in heartiest good wishes for a pleasant voyage[4] and a happy return to your home.

Sincerely Yours | Asa Gray

RI MS JT/1/G/15

1. *the New York dinner*: see Introduction, n. 23.
2. *my wife*: Jane Loring Gray (1821–1909).
3. *interfere with Kew*: a reference to Hooker's dispute with Acton Ayrton (see letter 3755, n. 1).
4. *pleasant voyage*: see letter 3928, n. 18.

From Joseph Henry 3 February 1873 3984

Feb[uar]y 3rd 1873

My Dear Dr. Tyndall:

I cannot express to you in words the disappointment I feel in being obliged at the last moment to give up the idea of seeing you again before you leave our shores.[1] I have not had an opportunity of saying as much to you as I wished, nor of telling you how indelibly and pleasurably you are impressed on the minds of my family[2] and myself.

[I] [3–4 words illeg] for your native land your departure will be attended with a shade of sadness in having to bid adieu to so many *[persons]* in whom you have awakened feelings of the warmest kind. I fear you will be much oppressed hereafter with correspondence from this country and that many, through thoughtlessness, will occupy your valuable time and thus deprive the world of a most precious article.

I shall send you through *[3 lines illeg]* good health, but my throat is very sensitive to changes of the weather and the setting-in of what threatens to be a severe snow-storm prevents my venturing on a journey northward.

I shall hope, however, during the remainder of my life to hear from you frequently.

The management of the fund you have created for the advancement of science in this country[3] will be, I trust, a pleasant bond of friendly union. It will require careful management and we can scarcely hope that in all cases of its application it will produce important fruits. But if in a single instance it serves to call into scientific existence a Davy[4] or a Faraday its reputation will be forever established.

You must feel as if an immense load were removed from your mind in having finished your lectures;[5] yet *[I]* *[6–7 words illeg]* one for Mrs Faraday,[6] to whom I would be pleased to have you kindly remember me.

Please give my regards and those of my daughter to Mr. & Mrs. Huxley[7] and *[say]* to the former that we would be glad to have a course of lectures from him,[8] and that if I can do anything through the Smithsonian Institution to facilitate his investigations in the way of supplying specimens he can command my service.

I need scarcely say that it will always give me pleasure to be the medium of communication between yourself and your friends in this Country. You *[would]* have no *[reason to]* fear of burdening me in this way since through the assistance I have in this Institution[9] our business matters can be readily transacted without trouble to myself.

I must not forget to ask you to give to Sir Frederick Arrow[10] and Capt. Webb[11] the assurance of my high regard for them. Their visit to this country forms an era in the operations of the Light House System of the United States and we shall not fail to persist by the remarks in the published journal of their tour.

Permit me, in closing, to wish you a pleasant voyage across the ocean, a warm welcome from your friends and the members of the Royal Institution, long continuance of health and, above all, the prolonged power of enlarging the domain of science—

Truly your friend—Joseph Henry

Prof. *[John]* <Tyndall>

Smithsonian Institution Archives. RU 7001, Box 4A, Folder 1, Pages 725–8[12]

---

1. *before you leave our shores*: Henry was unable to attend the farewell banquet in New York on 4 February (see Introduction, n. 23). He sent a letter to the committee that organized the banquet, and it was read there and published in *Proceedings at the Farewell Banquet* (pp. 11–24).
2. *my family*: see letter 3869, n. 7.

3. *the fund... in this country*: see Introduction, n. 15.
4. *Davy*: Humphry Davy.
5. *your lectures*: for Tyndall's American lecture tour, see Introduction, n. 8.
6. *Mrs Faraday*: Sarah Faraday.
7. *Mr. & Mrs. Huxley*: Thomas and Henrietta Huxley.
8. *course of lectures from him*: Huxley completed a lecture tour in the United States in 1876 (he arrived in in New York on 6 August; the tour did not include any lectures for the Smithsonian Institution). See Finnegan, *Voice of Science*, ch. 2; L. Huxley (ed), *Life and Letters of Thomas Henry Huxley*, 2 vols (London: MacMillan and Co., 1900), vol. 1, pp. 459–70; T. H. Huxley, *American Addresses, with a Lecture on the Study of Biology* (London: MacMillan and Co., 1877), also published by Appleton and Co. in New York; and J. V. Jensen, 'Thomas Henry Huxley's lecture tour of the United States, 1876', *Notes and Records of the Royal Society of London*, 42 (1988), pp. 181–95.
9. *this Institution*: Smithsonian Institution.
10. *Sir Frederick Arrow*: Fredrick Arrow was on a trip to the United States at the same time as Tyndall, during which time he met Henry.
11. *Capt. Webb*: John Sydney Webb (1816–98), an Elder Brother of Trinity House ('Death of Sir Sydney Webb', *Flintshire Observer*, 3 November 1898, p. 3), made a visit to the United States in 1872, during which time he met Henry.
12. The Smithsonian Institution Archives has provided us with the best possible images of this letter available; unfortunately, there remain numerous illegible words and lines. This letter is not among those about Tyndall's American lecture tour that were published in *Papers of Joseph Henry*.

From Alfred Mayer 3 February 1873 3985

[Stevens Institute of Technology, | HOBOKEN, NEW JERSEY,] | February 3[d] [*187*]3.

To | Professor John Tyndall F.R.S.

Respected Sir;

With this note I have, at the Brevoort House,[1] a letter for you from Professor Agassiz,[2] which came enclosed in one to me; also the book[3] which my relative McCulloch[4] requested me to present to you.

After ten days of solid work I have completed the agreeable duties which fell to me as Secretary of the "Tyndall Dinner;"[5] and I can say, from the character of the men of science who will be present that you will meet an appreciative and sympathetic assemblage. From those who could not come I have received letters[6] expressive of full appreciation of you as a man of Science; and of hearty well-wishes for your success and happiness, coupled with thanks for the great treat you have offered the American people in your clear and fascinating lectures.

It is needless for me to speak of the pleasure and privilege I have experienced in taking you by the hand and in looking you in the face—that you know full well. But I cannot but regret that I have seen so little of you and that the hope I had of your spending a day with me in my laboratory could not be realized. I will, however, look forward to a day with you in the Royal Institution[7] where I can see and touch the very tools with which Young, Davy, Faraday and you[8] have worked out such great results. That will indeed be the greatest pleasure that can befall me.

I regret that I cannot send you some of the soft Norway-iron filings before you leave, but my assistant lost my bar of annealed iron the very day I proposed making the filings for you, and I have, as yet, found no time to select and prepare another bar.[9] I will bring them with me if I reach London next May.

Wishing you a safe voyage, and trusting that pleasant memories of your American friends may long linger with you,

I remain | Yours respectfully & cordially | Alfred M. Mayer

RI MS JT/1/M/74

1. *Brevoort House:* see letter 3843, n. 3.
2. *letter for you from Agassiz*: letter from Louis Agassiz is missing.
3. *the book*: see letter 3980, n. 3.
4. *McCulloch*: Richard Sears McCulloch (see letter 3980, n. 1).
5. *"Tyndall Dinner"*: see Introduction, n. 15.
6. *I have received letters*: the published proceedings of the farewell banquet includes full letters from Joseph Henry, Louis Agassiz, Oliver Wendell Holmes, John Lawrence Smith, Asa Gray, Chester S. Lyman (see letter 3968, n. 5), and William Henry Brewer (1828–1910), an American explorer and botanist who worked on the first California Geological Survey and was the first Chair of Agriculture at Yale University's Sheffield Scientific School (*ANB*); and extracts of letters from James Dwight Dana (see letter 4001, n. 1); Joseph Leidy (see letter 3902, n. 6); Charles A. Young (see letter 3920, n. 3); John LeConte (1818–91), an American scientist and resident of the University of California from 1869–70 and 1875–81 (*ANB*); Isaac Lea (1792–1886), an American naturalist and publisher (*ANB*); John Wells Foster (1815–73), an American geologist and archaeologist ('John Wells Foster', *American Historical Record*, 2 (1873), pp. 382–3); William Barton Rogers (1804–82), an American geologist and educator who founded the Massachusetts Institute of Technology in 1861 (*ANB*); Truman Henry Safford (1836–1901), an American astronomer, mathematician, and first director of the Dearborn Observatory at the University of Chicago (*ANB*); Jeffries Wyman (1814–74), an American naturalist and anatomist (*ANB*); Joseph le Conte (1823–1901), an American geologist and educator (*ANB*); John Torrey (1796–1873), an American botanist (*ANB*); Reverdy Johnson (1796–1876), an American lawyer, U.S. attorney general, and U.S. senator (*ANB*); Cleveland Abbe (1838–1916), an American meteorologist and astronomer (*ANB*); James Craig Watson (1838–80), a

Canadian-American astronomer and director of the University of Michigan's Detroit Observatory in Ann Arbor (*ANB*); Daniel Coit Gilman (1831–1908), an American educator and second president of the University of California, Berkeley (*ANB*); Arnold Henri Guyot (1807–84), a Swiss-born geographer and geologist who emigrated to the United States in 1848 and was appointed professor of geography at Princeton University (*ANB*; *CDSB*); John William Dawson (1820–99), a Canadian geologist and university administrator (*CDSB*); and Edward S. Morse (see letter 3920, n. 4). See *Proceedings at the Farewell Banquet*, pp. 11–39.

7. *look forward to a day with you in the Royal Institution*: Mayer visited London in May 1873 and presumably also traveled to Manchester. See letters 4003, 4019, 4084, 4085, 4099, 4100, 4101, 4107, 4108, 4123, and 4124.
8. *Young... and you:* Thomas Young, Humphry Davy, Michael Faraday, and Tyndall were all professors at the RI.
9. *I regret... another bar*: Mayer developed a method for photographing magnetic spectra that relied on annealed Norway iron filings. See 'Magnetic Spectra', in *Documents of the Assembly of the State of New York*, vol. 9 (Albany: The Argus Company, 1871), pp. 1092–3.

From Nathaniel H. Morison[1] — 3 February 1873 — 3986

Peabody Institute | Baltimore, Feb. 3, 1873

My dear Sir,

My engagements are such that I cannot, without serious inconvenience, be at the banquet to-morrow night;[2] but I am unwilling that you should leave the country without expressing to you my sense of the great obligation which the country owes to you for coming to America, and giving so complete an illustration of the power of public lectures to interest and instruct a popular audience.[3] I think that your few lectures in this place[4] have done great service to our people and to science here; and your exhortation to the audience, at the close of your course, to provide liberally for the study of pure science without reference to results, will long be remembered by our practical people. They needed such a reminder of the value of scientific investigation.

I trust you may have a pleasant voyage across the Atlantic,[5] and that you will bear with you pleasant recollections of our people and of our efforts to rise in the intellectual scale.

Personally, I am only sorry not to have seen more of you while in Baltimore: but the fear of intruding upon the quiet which your letters informed me that you desired, and an undefined feeling that you might have suffered in the other cities from over attentions, prevented my seeing so much of you as I desired.

Hoping that your visit, though laborious, has, on the whole, been an agreeable one, and that your expectations in coming here have not been disappointed.

I am, my dear Sir | Yours very truly, | N. H. Morison

Prof. John Tyndall LL.D. F.R.S.

RI MS JT/1/M/144

---

1. *Nathaniel H. Morison*: Nathaniel Holmes Morison (1815–90), first provost of the Peabody Institute from 1867–90 ([A. S. Morison], *Memorial of Nathaniel Holmes Morison (1815–1890)* (Baltimore: Privately Printed, 1892)).
2. *banquet to-morrow night*: see Introduction, n. 23.
3. *so complete... popular audience*: for Tyndall's American lecture tour, see Introduction, n. 8.
4. *in this place*: Tyndall delivered three lectures in Baltimore from 26–9 November 1872 at the Peabody Institute.
5. *pleasant voyage across the Atlantic*: see letter 3928, n. 18.

To Joseph Henry 7 February 1873 3987

7th. Feb[ruar]y. 1873.

My dear Professor Henry

I have made my "will"[1] in due form and signed it in the presence of witnesses.

My desire and intention on accepting the invitation of my friend[2] were, as you know, to hand over the proceeds to Chicago. But the recovery from calamity is quick in this country, so that Chicago, not only does not need my feeble aid,[3] but would be willing of her abundance to add to my wealth.

My disbursements, as I told you are heavy. Living I have found to be exceedingly expensive in the United States; hence the balance which I am able to hand over to the Board of Trustees is not so large as I could wish it to be. It however accounts to a little more than thirteen thousand dollars.

I have bestowed some care on the accounts, and do not think I carry home with me a single cent of American money. But I carry home what is to me incomparably more precious, and that is the assured good will of the American people.

The instruments that I take home with me I intend to present to the Royal Institution,[4] where they will be turned to good account. My hands will be then entirely clean, and no foreign element will mingle with the bright memory of the time I spent here.

My best regards to your wife and daughters.[5]

Goodbye. | Ever yours faithfully | John Tyndall | Prof. Joseph Henry | &c &c

[Enclosure][6]

1. I, John Tyndall, Professor of Natural Philosophy in Royal Institution of Great Britain, having been solicited by my friend[7] to lecture in the United States and the lectures having been favourably received and liberally patronized, desire as an evidence of my appreciation of the kind feeling that has been manifested towards me and the interest that has been shown in my lectures to devote the proceeds of said lectures that many remain after deducting my expenses (which, I regret to say here have been unusually heavy) to the advancement of abstract science—especially in the line of Physics—, in the United States.

2. To accomplish the object in question I hereby appoint Professor Joseph Henry, Director of the Smithsonian Institution, D[r] E. L. Youmans of New York and General Hector Tyndale of Philadelphia, my Kinsman, to act as a Board of Trustees to take charge of the proceeds of said lectures, to carefully invest the same in permanent securities and to expend the interest therefrom in the manner to be hereafter mentioned.

3. I further direct that vacancies that may occur in said Board of Trustees by death or otherwise be filled by the President of the National Academy of Sciences.

4. I further direct that said Board of Trustees shall appropriate the interest of the fund for the present in supporting, or in assisting to support, at such European Universities as may hereafter be denigrated two American pupils who many evince decided talents in the line of original research in Physics and who may express a determination to devote their lives to this work.

5. If, however, in the progress of science in the United States it should at any time appear to the said Board that the end herein proposed would be better subserved by granting aid to students for some special researches in this Country the Board is authorized to make appropriations from the income of the fund for such purposes.

6. The said Board shall make an annual report to the National Academy of Sciences of the receipts, expenditures and condition of the fund and of the results which have been derived from it.

7. If in the course of any year the whole amount of the interest which accrues from the fund be not expended in the manner before mentioned the surplus may be added to the principal or may be expended in addition to the annual interest of another year.

8. If at any time an organization shall be estab[lishe][d]. & monies provided by other parties for the promotion of original research such as I have in view, I authorize the said Board of Trustees to exercise their discretion as to cooperating in such a work from the income of this fund.

[in verso] Tyndall | foundation. | Tyndall Trust Fund. |—1873—

John Tyndall to Joseph Henry, 7 February 1873, mssRH 3977, box 43, William Jones Rhees papers, The Huntington Library, San Marino, California

1. *my "will"*: for Tyndall's donation of his lecture tour profits, see Introduction, n. 15.
2. *invitation of my friend*: perhaps a reference to Henry himself, who led the effort to invite Tyndall to America (see letters 3504 and 3510, *Tyndall Correspondence*, vol. 12).
3. *But the recovery... feeble aid*: a reference to the Great Chicago Fire of October 1871 and the subsequent rebuilding of the city.
4. *The instruments that... present to the Royal Institution*: at the 3 March 1873 meeting of the RI, Tyndall was congratulated on his successful American lecture tour, welcomed home to 'what they are proud to be able to designate as his own scientific home', and thanked for his donation of the scientific apparatus he employed in his American lectures (*Roy. Inst. Proc.*, 7 (1873–75), pp. 55–6).
5. *your wife and daughters*: see letter 3869, n. 7.
6. This is a draft (SIAUNB, 000032, Box 01, Folder 04, Document 01) of 'The Tyndall Trust for the Promotion of Science in the United States', *Annual report of the Board of Regents of the Smithsonian Institution, showing the operations, expenditures, and condition of the Institution for the year 1872*. Serial Set Vol. No. 1573, Session Vol. No. 3, 42nd Congress, 3rd Session (Washington: Government Printing Office, 1873), pp. 104–6, which includes monetary statements and the names of two witnesses, C. Burritt Watte and L. E. Fuller. This was also published in *Popular Science Monthly*, 3 (3 May 1873), pp. 100–01. In the published versions, the following is included: 'In witness whereof I have hereunto set my hand and seal this 7th of February, 1873, in the city of New York'.
7. *my friend*: Joseph Henry.

To William Rhees [before 8 February 1873][1] 3988

Monday

Dear Mr. Rhees.

I will do what you request me to do[2] on Wednesday—earlier I cannot.

Kindest regards to Prof Henry & his family.[3]

faithfully yours | John Tyndall

John Tyndall to William Jones Rhees, 1872, mssRH 3975, box 43, William Jones Rhees papers, The Huntington Library, San Marino, California

1. *[before 8 February 1873]*: the date is given as Tyndall was in communication with Rhees regarding his American lecture tour from October 1872 to February 1873. Without knowing what Rhees had requested Tyndall do, we place this letter as '[before 8 February 1873]', when Tyndall left the United States.
2. *what you request me to do*: letter from Rhees missing.
3. *his family*: see letter 3869, n. 7.

To unidentified 8 February 1873 3989

BREVOORT HOUSE,[1] | Fifth Avenue. | 8th. Feb[ruar]y. 1873

My dear Sir

I cannot leave Miss Pryor[2] under the impression that I found fault with her method of tuition. It was far from my intention to do so.

In all the schools that I have visited in the United States I have noticed the culture of the memory: and were I to undertake a thorough inspection of your mode of teaching it is to this point that I would devote my chief attention. For years memory was cultivated in my own case, but this was not sufficient to develop such original power as I possessed, & this is the point regarding which I should like to assure myself if I had the time.

But pray request Miss Pryor on my part to dismiss from her mind the thought that I intended to find fault with her. Her note[3] is full of the earnestness of a true teacher.

I am at the point of leaving the country,[4] and wish to assure Miss Pryor's peace of mind before I go.

Faithfully yours | John Tyndall

Historical Society of Philadelphia, Gratz Collection

1. *Brevoort House*: see letter 3843, n. 3.
2. *Miss Pryor*: not identified.
3. *Her note*: letter missing.
4. *at the point of leaving the country*: Tyndall departed New York on the SS *Cuba* on 8 February 1873, the same day he wrote this letter (see letter 3928, n. 18).

From Mary F. Egerton 12 February [1873][1] 3990

[Mountfield Court, | Robertsbridge, |
Hawkhurst.][2] | Febr[ua]ry. 12th

Welcome—Thrice welcome back[3] to home & friends! I must and will be among the first to greet you, dear friend, but it shall only be by a single line, for I see a vision of the piles of letters and papers that will be awaiting you!

Your letter[4] was most welcome. I hope you got one from me addressed to New York,[5] which crossed it, thanking you for the delightful lectures. I trust you have left all colds far behind you, & that the teeth have returned to their condition of perfection, notwithstanding the cold you must have experienced on the voyage. When I hear the bitter East wind howling round the house, I cannot help thinking with somewhat of a shudder, of the great Atlantic and—but why speak of that now! When you receive this, all perils will be over, & all hardships forgotten! I have several more papers to thank you for. The article in "the Liberal Christian"[6] was after my own heart. I might not venture to draw perhaps all its inferences, but in spirit, it is what I have always felt and said. When you come back, I shall quite miss reading pleasant things about you every few weeks! But I hope we shall meet ere very long. We think of going to town next week, (on the 20th,) and I hope to hear Mr. Clarke Maxwell's lecture on the Friday.[7] But when shall I get such another comfortable chat as we had that evening when the girls[8] had the mumps! To me it makes all the difference, talking to one person or for the benefit of half a dozen, especially when they are one's own criticising children! But it is quite different for you & I dare say you are not "de mon avis."[9]

Once again I will say, "auf wiedersehen."[10] Mind you don't get tired of telling your adventures before I hear them!

Y[ou]rs ever | M F Egerton

The young ones would send you a greeting if they knew I was writing

RI MS JT/1/E/17

---

1. *[1873]*: the year is given by the timing of Tyndall's return from his American lecture tour (see Introduction, n. 8).
2. *Mountfield Court, | Robertsbridge, | Hawkhurst*: see letter 3852, n. 2.
3. *welcome back*: Tyndall reached England, returning from his American lecture tour, on 19 February 1873, arriving in London later the same day.
4. *your letter*: letter missing.
5. *the one... New York*: letter missing.
6. *The article in "The Liberal Christian"*: the article referenced is not identified, but *The Liberal Christian: A Monthly Miscellany* was a periodical founded in 1823 to provide a forum for free religious inquiry.
7. *lecture on the Friday*: James Clerk Maxwell delivered a Friday Evening Discourse at the RI, 'On Action at a Distance', on 21 February 1873 (*Roy. Inst. Proc.*, 7 (1873–75), pp. 44–54).
8. *girls*: see letter 3805, n. 14.
9. *"de mon avis"*: in my opinion (French).
10. *"auf wiedersehen"*: farewell (German).

From Henry Bence Jones [before 19 February 1873][1] 3991

Dearest Friend

I send you all I have of some Grand Vin | Chateau Laffite[2] | 1844

Drink a bottle the night you arrive[3] to my health & another the night after I go to sleep for good[4] to my better health. You dear friend how glad I am you are back safely.[5] I should like to send you six dozen but such stuff is not be bought.

Ever dear friend | Your well wisher | H Bence Jones

RI MS JT/1/J/111
RI MS JT/1/TYP/2/708

---

1. *[before 19 February 1873]:* the date is given by reference to Tyndall's return to England from his lecture tour in the United States, on 19 February 1873 (see n. 3).
2. *Chateau Laffite*: Château Lafite is a highly regarded wine producer in the Bordeaux region of France.
3. *the night you arrive:* Tyndall reached England, returning from his American lecture tour, on 19 February 1873, arriving in London later the same day.
4. *after I go to sleep for good*: in numerous letters to Tyndall while he was in the United States, Bence Jones wrote about nearing his end, and worried that he would not see Tyndall. Fortunately, he did, as he did not pass on until 20 April 1873.
5. *back safely*: see n. 3.

From Henry Bence Jones [19 February 1873][1] 3992

Dear friend. Welcome home.[2] How glad I am you are here

After your dinner & a good dose of some claret you will find upstairs come in for ten minutes. I do not go to bed till Eleven & do not go to sleep till one

Ever Yours | H Bence Jones

Thanks for all the telegrams—| Wednesday evening | Prof. Tyndall

RI MS JT/1/J/110
RI MS JT/1/TYP/2/709

---

1. *[19 February 1873]*: the date is given by reference to Tyndall's return to England from his lecture tour in the United States, on 19 February 1873 (see n. 2).

2. *Welcome home*: Tyndall reached England, returning from his American lecture tour, on 19 February 1873, arriving in London later the same day.

From Sarah Faraday 22 February 1873 3993

Barnsbury Villa,[1] | 320, Liverpool Road, N. | 22nd Feb[ruar]y. 1873.

Dear Dr. Tyndall,

My Niece[2] reminds me that I intended writing a few words to welcome you home,[3] and I do it most heartily, but as I sit in a good deal of pain I must ask you, my kind and sympathizing friend, to excuse all my deficiencies. After a time, for you to subside, I shall hope to see you face to face, but not to add to your burdens just as present. I am very glad for Dr. Bence Jones' sake, of your return, for I know how he longed to see you. Farewell, and believe me, dear Dr. Tyndall,

Your very sincere friend, | S. Faraday.

RI MS JT/1/TYP/12/4189
LT Typescript Only

---

1. *Barnsbury Villa*: Sarah Faraday and Jane Barnard moved from Hampton Court to Barnsbury Villa shortly after Michael Faraday's death in August 1867 (Frank A. J. L. James, personal communication).
2. *My Niece*: Jane Barnard.
3. *welcome you home*: Tyndall reached England, returning from his American lecture tour, on 19 February 1873, arriving in London later the same day.

From Thomas Henry Huxley 23 February 1873 3994

4 Marlborough Place | Abbey Road N.W. | Feb. 23rd 1873.

My dear Tyndall,

The enclosed[1] was due on the 1st January but I could not pay it into your bankers not knowing who they were.

The wife[2] was very vexed not to be at home when you called yesterday, but I hear that Leonard[3] did the honours. I hope you liked the house.[4] I think it may be very jolly when we have once settled down in it, but that will not be for months.

What about our having a run to the Isle of Wight?[5] I cannot get away this week before Friday morning and I should have to be back by Monday evening as I have an examination at South Kensington[6] on Monday and Wednesday next week. Will that suit you?

Ever yours faithfully | T. H. Huxley

Don't trouble to send any receipt as the cheque is payable to order and you can't do me if you try.

RI MS JT/1/TYP/9/2970
IC HP 8.133
Typed Transcript Only

---

1. *The enclosed*: enclosure missing.
2. *wife*: Henrietta Huxley.
3. *Leonard*: Leonard Huxley.
4. *the house*: see letter 3947, n. 1.
5. *a run to the Isle of Wight?*: it appears weather prevented Tyndall and Huxley from taking a quick trip to the Isle of Wight (see letters 3997 and 3998).
6. *South Kensington*: see letter 3711, n. 4 and n. 5.

To Joseph Dalton Hooker — 24 February 1873 — 3995

Royal Institution of Great Britain | 24th Feb[ruar]y 1873.

My dear Hooker,

I looked forward to the pleasure of going out to you yesterday, but when my work was complete the day was too far spent.

How are you? How is Mrs Hooker?[1] How are the children?[2] I hope all well. Need I say how rejoiced I am to find the Royal Society behaving thus towards you.[3] What a comment on the conduct of that Bear's cub![4]

Yours affect[ionate]ly | John Tyndall

Dr. Ellis of Boston was very civil to me when I was there. His daughter in law Mrs. Ellis[5] and her brother Mr. Little[6]—a business man, but devoted to Botany—wish much to know you. I do not wish to take up your time; but I shall give Mrs. Ellis a note of introduction to you.

Of course I am charged with all manner of kind messages from Gray[7]

Wellcome Library MS. 7777/15

---

1. *Mrs Hooker*: Frances Hooker.
2. *children*: see letter 3841, n. 5.
3. *Royal Society... towards you*: Hooker was elected president of the RS at its Anniversary Meeting on 1 December 1873, but it was known long before that it would be him. He served in this position until 1877 (*Roy. Soc. Proc.*, 22 (1874), p. 12).

4. *that Bear's cub*: probably a reference to Hooker's dispute with Acton Ayrton (see letter 3755, n. 1).
5. *Mrs. Ellis*: Grace Atkinson Ellis (née Little, later Oliver, 1844–99), an American writer and women's rights activist ('Mrs Grace A. Oliver Dead. Woman of Rare Executive Ability and Noted for Her Charities', *The Boston Globe*, 22 May 1899, p. 5). Her first husband was John Harvard Ellis (1841–70), the only child of the American minister and historian George Edward Ellis (1814–94) and his first wife Elizabeth Bruce Ellis (née Eager, m. 1840, d. 1842).
6. *Mr. Little*: it is not clear to which of Grace's five brothers this refers to.
7. *Gray*: Asa Gray.

To Thomas Henry Huxley — 25 February [1873][1] — 3996

Royal Institution | 25th. Feb.

My dear Huxley

If I remain a bachelor I will circumvent you—If not—not.

It cleaves to me like dirt—And that is why you wish to get rid of it.

I am game for Friday. When will you be ready to start and when must you return?

Yours ever | John Tyndall

I mean what <u>hour</u>. You naturalists speak of the spectrum as if it had no subdivisions.

RI MS JT/1/TYP/9/2971
IC HP 8.134
Typed Transcript Only

1. *[1873]*: the year is given by relation to letters 3994 and 3997 from Huxley regarding a quick trip to the Isle of Wight.

From Thomas Henry Huxley — [26 February 1873][1] — 3997

ATHENAEUM CLUB,[2] PALL MALL, S. W.

My dear Tyndall,

I called at the Royal Institution just now, and had the ill-luck to miss you.

What do you think about the weather? The papers say that the last two days' fall of snow has been general over England, and if this rain is as widespread the Isle of Wight will be as knee deep in mud as the Isle of Dogs,[3] to say nothing of the possibilities and probabilities of a continued downfall of rain.

Don't you think under these circumstances it might be wise to defer our jaunt for a week.

Let me have your mind on the subject by note to 4 Marlborough Place[4] this evening.

Ever yours very sincerely | T. H. Huxley.

P.S. I am not to be deterred by any amount of bribery and corruption from bringing you under the yoke of a "rare and radiant"[5],—whenever I discover one competent to undertake the ticklish business of governing you. I hope she will be "radiant",—uncommonly "rare" she certainly will be!

RI MS JT/1/TYP/9/2972
IC HP 8.135
Typed Transcript Only

1. *[26 February 1873]*: the date is given by LT's notes on the letters, and relation to letters 3994, 3996, and 3998.
2. *ATHENAEUM CLUB*: see letter 3710, n. 5.
3. *Isle of Dogs*: a marshy peninsula into the River Thames, known in the Victorian era for its docks.
4. *4 Marlborough Place*: see letter 3947, n. 1.
5. *"rare and radiant"*: a quote from Edgar Allan Poe's 'The Raven', *Evening Mirror*, 29 January 1845: 'For the rare and radiant maiden whom the angels name Lenore'. Huxley was referring to finding Tyndall a wife.

From Henrietta Huxley 26 February [1873][1] 3998

4 Marlborough Place Abbey Road[2] | 26th Feb

Dear Brother John

I was so annoyed that I was out when you called.[3]

Can you not come and dine with us soon? Tell me one or two days on either of which you could come—This bad weather will prevent you and Hal[4] from going Southwards.[5] It would be doing him a real kindness to take him away for a few days.

Leonard[6] told me that he had shown you some of the rooms in my new house.[7] It is very pleasant to have so much room but it has given me an infinity of trouble—and will yet—I cannot get the drawingroom finished for 2 months, and the mantle piece for the dining room has yet to come.

1872—was a horrid year[8] and I am glad it is gone Such an anxious time about Hal it was—besides other troubles. But the sun seems to be thinking of shining again for Hal is so much better thanks to Dr Andrew Clark's treatment. How I wish he (Hal) could give up lecturing, but I don't see any way to that—So much hangs by his lectures at S. K.[9]

Jess[10] is delighted at having completed her 15th—birthday—not so I nor her Father. We all want to see you so, say when you can come and do battle with

Yours very affectionately | Nettie Huxley.

RI MS JT/1/TYP/9/2973
LT Typescript Only

---

1. *[1873]*: the year is given relation to letter 3994, and by reference to Jessie Huxley's 15th birthday. She was born in 1858, making her fifteen in 1873. Also, Henrietta's comment that 1872 'was a horrid year' implies that she wrote this the following year.
2. *4 Marlborough Place Abbey Road*: see letter 3947, n. 1.
3. *when you called*: in letter 3994, Thomas Huxley wrote to Tyndall, 'The wife was very vexed not to be at home when you called yesterday, but I hear that Leonard did the honours'.
4. *Hal*: a nickname for Thomas Huxley.
5. *going Southwards*: to the Isle of Wight (see letters 3994, 3996, and 3997).
6. *Leonard*: Leonard Huxley.
7. *my new house*: see n. 2.
8. *a horrid year*: in early 1872 Thomas Huxley traveled in the Mediterranean and Egypt to treat his illness (see *Tyndall Correspondence*, vol. 12). The move to their new residence in 1872 was likely also stressful (see letter 3947).
9. *his lectures at S. K.*: Huxley's third summer course of biology lectures for science teachers at the Royal School of Mines at South Kensington (see letter 3711, n. 7).
10. *Jess*: Jessie Oriana Huxley.

From Osmund Airy[1] 27 February 1873 3999

[Wellington College, | Wokingham.] | 1873. Feb. 27.

My dear Sir,

Pray pardon the unnecessary trouble[2] I have given you. The only excuse I can offer is my ignorance up till now of the law of testimonials and of the line at which 'acquaintance' begins. I sinned however in good company.

Faithfully yours | Osmund Airy.

Prof. Tyndall.

RI MS JT/1/A/64
RI MS JT/1/TYP/1/50

1. *Osmund Airy*: Osmund Airy (1845–1928), the youngest son of George Biddell Airy, who later became an historian. At this time he was Assistant Master at Wellington College (1868–76) ('Mr. Osmund Airy', *Times*, 4 December 1928, p. 21).
2. *trouble*: the difficulty Airy might have caused Tyndall is unclear.

From Joseph Dalton Hooker — 27 February 1873 — 4000

Kew. | February 27th 1873.

Dear old Tyndall

I am heartily glad that you are back[1] and I long to shake you by the hand.

Pray let me know where Dr Ellis is,[2] I should like to pay his party some attention and would write or call or both.

I have just read the fragment of your spirited and inspiriting farewell address in "Nature"[3] and could fancy I heard your words ring in my ears. I felt secure of your pleasure in hearing of my nomination to the Presidentship.[4] Well I am hoisted up on your and other good men's shoulders. Thank God you are not windbags, but what is it that you have all hoisted so high? That neither am I a windbag is, I hope, not a too flattering estimate of myself, but I can say little more of my fitness. There is no fear of my either soaring beyond your reach or being carried away by a gust of self-sufficiency; but I do fear that I shall prove a heavy weight upon your shoulders, and in some respects a disappointment to the Society. I am not ready, I have to beat about not only for words but for expressions, thoughts, and ideas, and on the spur of the moment for matter too. Need I say I should have declined, did I not feel it to be my clear duty to accept:—the council's wish was so decided and well considered; my scientific friends were unanimous, Botanists regarded my acceptance as a duty to their science, and my relations as demanded in respect of my father's[5] name. So I have accepted, determined to do with all my might that which I have set my face to do.

I am supposed to be at S. Leonards[6] where I left Harriette[7] with Decimus Burton[8] on Tuesday. I return on Saturday, thither. Tomorrow we dine at Colvile's, perhaps we shall meet there. The newspapers you sent were capital; it was good of you to send them.

Ever aff[ectiona]tely yours | J. D. Hooker.

RI MS JT/1/TYP/8/2717
LT Typescript Only

1. *you are back*: Tyndall reached England, returning from his American lecture tour, on 19 February 1873, arriving in London later the same day.
2. *Pray let me know where Dr Ellis is*: see letter 3995.

3. *farewell address in "Nature"*: *Nature* printed Tyndall's speech at the banquet held for him in New York before his departure. See 'Parting Banquet to Prof. Tyndall', *Nature*, 7 (20 February 1873), pp. 303–5. See also Introduction, n. 15.
4. *the Presidentship*: see letter 3995, n. 3.
5. *my father's name*: William Jackson Hooker.
6. *S. Leonards*: likely the Sussex town of St. Leonards-on-Sea.
7. *Harriette*: Hooker's wife, Frances Harriet Hooker.
8. *Decimus Burton*: the architect and urban planner Decimus Burton (1800–81) who designed much of the Sussex town of St. Leonards-on-Sea and maintained a residence there (*ODNB*).

From James Dwight Dana[1] — 28 February 1873 — 4001

New Haven, Feb[ruar]y 28. 1873—| Prof. Tyndall

My dear Sir:

I have just addressed to you a copy of a new paper on the Old Glacier over New England[2] which may interest you.—In connection I wish to thank you for your strong words while on this side of the ocean on behalf of high science,[3] and also for your generous gift to our Yale Scientific School[4] & to the cause of scientific education in this land.[5]

I regret much that the state of my health prevented my participating in the pleasures which were so much enjoyed by others during your visit to our city—

With high esteem, yours truly | James D. Dana—

RI MS JT/1/D/9

1. *James Dwight Dana*: James Dwight Dana (1813–95), a prominent American geologist, was the Silliman Professor of Natural History and Geology at Yale College (*ANB*; *CDSB*).
2. *new paper on the Old Glacier over New England*: probably J. D. Dana, 'On the Glacial and Champlain eras in New England', *American Journal of Science*, 5, no. 27 (March 1873), pp. 198–211.
3. *your strong words... behalf of high science*: referring to Tyndall's American lecture tour (see Introduction, n. 8).
4. *generous gift to our Yale Scientific School*: Tyndall's last stop in his American lecture tour was in New Haven, CT, for two lectures on 22 and 23 January 1873 at Yale College. Tyndall gifted $500 'to the Yale Scientific Club to found a prize in the Scientific School, to be called the Tyndall prize' (*Yale Literary Magazine*, 38 (February 1873), p. 214).
5. *to the cause of scientific education in this land*: this probably refers to both Tyndall's donation of his lecture tour profits (see Introduction, n. 15), and to Tyndall's words encouraging the pursuit of pure science in the United States.

From Jane Barnard [February 1873][1] 4002

My dear Dr. Tyndall,

My brother[2] was too tired to go up to your rooms with me, and much as I should have enjoyed it, I could not urge him. It is pleasant to have you home again.[3]

Your sincere friend. | Jane Barnard.

RI MS JT/1/TYP/12/4189
LT Typescript Only

1. *[February 1873]*: the date is given by reference to Tyndall's return to England from his lecture tour in the United States, on 19 February 1873 (see n. 3).
2. *My brother*: very likely Jane's brother Frank Barnard (1828–95), son of John and Margaret Barnard (née Faraday) and nephew of Sarah Faraday ('Deaths', *Morning Post*, 21 August 1895, p. 1). She had another brother as well, John Barnard (1835–1915).
3. *have you home again*: Tyndall reached England, returning from his American lecture tour, on 19 February 1873, arriving in London later the same day.

From Alfred Mayer 1 March 1873 4003

[Stevens Institute of Technology, | HOBOKEN, NEW JERSEY,]
U.S.A. | March 1st [*187*]3

My dear Professor Tyndall;

As you requested me, I write to tell you that I will sail for England on the 23rd of next April. I look forward with exceeding great pleasure to the date of my prepared departure; for I am not well; I feel the effects of too much work, and the action of this severe winter in a body developed in a southern clime and not yet used to the rigours of our northern winters.

I will come for <u>rest</u>, recreation and the quiet observation of the work and workshops of my scientific brethren in England. Specially am I interested in every thing that pertains to the planning and organization of laboratories in England devoted to research and to practical instruction in Physics; for I am convinced that the only way open, for us in America, to create a corps of <u>working</u> young men is to give them love and ardour for science before the love and admiration of the power of money (so deeply seated and so early developed with <u>so called</u> "Anglo-Saxon") obtains its devilish possession of them. This demon can only be kept off by letting them taste early in life the exquisite delight following the successful solution of experimental problems, and by developing at an early age that imaginative insight into molecular actions, which alone comes from the constant dynamic explanation of the phenomena of matter.

I am therefore possessed with the desire to visit all the laboratories of England and to become acquainted with the plans of instruction adopted by their respective directors.

Indeed, I have little desire or time for continental travel and will probably remain in England until September, and pass much of my time in quiet rural places.

I had not the opportunity see you after the "Tyndall-dinner."[1] I hardly know what to say of it. A success as our minimal assemblage of nearly all the workers in science in America. But let me assure you, Professor Tyndall, that we did not anticipate that the parsons would demonize[2] us; and that Mr Godwin[3] would depart from his "toast" and give us the *[*Bohemian*]* balderdash he did. To say the least of it—and the least said by me, probably the better—some of the speakers certainly were wanting in good taste.

All this would have been prevented had Henry and Agassiz not failed us,[4] and had Mr Evarts[5] called out men of science (as he was requested by Youmans and by me) present who were able and exceedingly eager to make the parsons feel that science could take care of herself without clerical patronage. But probably I had better keep my pen dry when these ideas come into my mind and defer my own personal feelings and opinions until I see you—However, Dr Barnard[6] and I have taken a decided stand against the publication these homilies.[7] Of course Youmans agrees with us and he will do as we wish, and thus he will control the Appletons[8] in the matter. I have obtained several letters from prominent scientific men dated on this acceptance of the invitation to the dinner. These letters will give proper balance and character to the celebration and will tend to further the objects of your visit among us instead of frustrating it as was the tendency of the speeches of the dominies.[9]

But the dinner was a decided success and gave pleasure to all who were invited and all who were invited and present wished to do all they could to express to you their considerations of respect and their good feelings.

with highest regard and well wishes I remain | Yours faithfully | Alfred M. Mayer

RI MS JT/1/M/75

1. *"Tyndall-dinner"*: the farewell banquet in New York held for Tyndall prior to his return to England (see Introduction, n. 23).
2. *parsons would demonize*: a reference to speeches at the banquet by the American clergyman Roswell Dwight Hitchcock (1817–87), who stated that when pitted against each

other, 'science, and not religion, must go down the cliff', since 'mankind at large have managed to get on very well without much science, but they have never managed to get on very well without religion'; and American journalist Parke Godwin (1816–1904), who was critical of Tyndall and others' perceived stepping over the boundaries of science to make claims about natural phenomena that went beyond factual knowledge (*Proceedings at the Farewell Banquet*, pp. 65–73; Finnegan, *Voice of Science*, pp. 59–61).

3. *Godwin*: Parke Godwin (see n. 2).
4. *Henry... failed us*: both Henry and Agassiz were approached to preside over the dinner, but neither could attend, and sent letters to be read instead (*Proceedings at the Farewell Banquet*, pp. 11–25).
5. *Evarts*: William M. Evarts presided over the banquet (*Proceedings at the Farewell Banquet*, pp. 11–25).
6. *Barnard*: Frederick Barnard.
7. *decided stand against the publication these homilies*: the speeches from the farewell banquet were published in *Proceedings at the Farewell Banquet*.
8. *the Appletons*: see letter 3823, n. 9.
9. *Dominie*: in the United States, the title of a pastor of the Dutch Reformed Church; whence in New York, New Jersey, etc., extended colloquially to ministers or parsons of other churches (*OED*).

From Elizabeth Emsley[1] 3 March 1873 4004

4 Nelson Terrace | City Road N. | March 3rd 1873.

Sir,

I beg to remind you that at an interview I had with you prior to your leaving England "respecting a bill left unpaid by Mr Henry Tyndall"[2] you desired me to write to you upon your return, if I did not hear from you before you left.

I enclose the bill[3] and beg to apologize for troubling you in such a matter, for nothing could have induced me to do so, but the necessitous situation I am in, making the loss of this small sum, of much greater importance to me, than can be supposed by those in a more affluent station than myself.

If you will kindly use your influence with Mr H Tyndall on my behalf I shall always feel Your most grateful servant | Elizabeth Emsley

RI MS JT/1/E/105

1. *Elizabeth Emsley*: not identified.
2. *Henry Tyndall*: see letter 3795, n. 6.
3. *I enclose the bill*: enclosure missing.

From Margaret Ginty 3 March 1873 4005

4 Hemingford Terrace | Birkenhead | March 3rd. 73.

Dear Dr. Tyndall,

Before I left London I saw Mr James[1] and had several conversations with him—I told him everything and what I so much wished to do, but, thinking of your words and his also I have come to the conclusion that you are all very wise and I am terribly foolish—I can't help it—but I do wish that I could see things at once clearly and act for myself—but like a stupid woman I only see and understand when it is pointed out and then it is clear enough. Mr James thinks that in my state of health I would not be able to take charge of the poor little darling[2]—Ah! I am so sorry. Sometimes I hate Holman[3] with all my heart and soul, for I think he did not take the best care of my poor dear girl.—Then again I think I may do him a great injustice for he is only weak not wicked, and I never can forget how good and tender he was to my husband[4] at the last—Oh I wish I had no memory at all for it brings such cruel grief! Mr James told me that he was going to call on you one of these days. He gave me such kind good advice, you don't know how good he is (I am only finding it out by degrees) and his sweet wife too! Pray forgive this long note—I wish you were not looking so tired and so very very thin!

I remain | dear Dr. Tyndall | Yours truly | M. Ginty.

Some day when you have time you might write to me,[5] but let it be a very kind letter because I am so sad,

Just think again about the helpless babies? I set my heart on having at least the little tiny ones.[6] It is so hard to be wise and prudent—I hate it, and I do not like to have to abandon my proposal.

It was so good and kind of you dear Dr. Tyndall to come to see me that day, in the cold and snow.

RI MS JT/1/TYP/11/3671
LT Typescript Only

1. *Mr James*: Bartlett James (dates unknown), was manager, or 'Gerente', of the Imperial Gas Works Company in Rio de Janeiro, Brazil, established by Baron de Maua, where he worked with Tyndall's friend William Ginty, who was the engineer, or 'Engenheiro' (*Almanak Administrativo, Mercantil e Industrial da Corte e Provincia do Rio de Janeiro para o anno de 1865* (Rio de Janeiro: Eduardo & Henrique Laemmert, 1865), pp. 414–15). According to a 5 December 1882 letter from Ginty's daughter Fanny Unwin (née Ginty) to Tyndall, James had taken on the role of executor to her late father's will (Ginty died in 1866) (RI MS JT/1/TYP/11/3682).

2. *poor little darling*: not identified, but she refers to the same person as 'my poor dear girl' in the next sentence. The children of Margaret and William Ginty were sons William Robert, Georgie (b. 1851), and Frederic (b. *c.* 1854); and daughters Kathleen and Fanny (1851–1937). William Robert, Kathleen, and Fanny were all listed on the 1851 census; given this, all of Margaret's children were all at least twenty years of age in 1873. Thus, it is not clear who 'poor little darling' and 'my poor dear girl' refers to.
3. *Holman*: not identified, though possibly the future husband of the Gintys' daughter Kathleen. See letter 3505, *Tyndall Correspondence*, vol. 12.
4. *my husband*: William Ginty, who died in Brazil in 1866.
5. *you might write to me*: see letter 4006, Tyndall's reply to this letter.
6. *Just think... tiny ones*: it is not clear to whom this refers.

To Margaret Ginty 4 March 1873 4006

Royal Institution | 4th. March 1873.

My dear Mrs Ginty,

If you do not always clearly see the way to be pursued in this complicated world, you have a rich endowment that atones for the want of worldly wisdom—you have a nature full of motherly affections—and these assuredly though they bring you grief sometimes belong to a woman's highest endowments. I can understand your feelings regarding the children,[1] and I shall be willing to promote your views as far as I can when your health has been re-established, and its permanence secured. Nobody however can contravent the wishes of Mr Holman.[2] The utmost that can be done is to submit to him a statement of <u>opinion</u>—and this must come from those who have the data on which to found an opinion.

Yours faithfully | John Tyndall.

RI MS JT/1/TYP/11/3672
LT Typescript Only

---

1. *your feelings regarding the children*: see letter 4005.
2. *Holman*: see letter 4005, n. 3.

From Arthur Russell[1] 7 March 1873 4007

7 March | 1873

Hochgeehrter Herr Professor,[2]

I cannot say how much pleasure your approbation[3] has given me. I wish to show you the enclosed letter[4] because it illustrates an odd state of mind and is quite honest evidently—

y[ou]rs truly | Arthur Russell

RI MS JT/1/R/62

1. *Arthur Russell*: probably Arthur John Edward Russell (1825–92), a British Liberal Party politician who became Lord Russell in 1872 (*ODNB*).
2. *Hochgeehrter Herr Professor*: Esteemed Professor (German).
3. *your approbation*: while Tyndall's letter to Russell is missing, it is possible that in it Tyndall praised Russell for a paper he presented to the Metaphysical Society on 14 January 1873 in which he was supportive of Darwin's theory of evolution. See I. Hesketh, 'Evolution, Ethics, and the Metaphyiscal Society, 1869–1875', in C. Marshall, B. Lightman, and R. England (eds), *The Metaphysical Society (1869–1880): Intellectual Life in Mid-Victorian England* (Oxford: Oxford University Press, 2019), pp. 185–203, on p. 186; and A. Russell, 'Darwinians and Idealists', in *Papers Read at the Meetings of the Metaphysical Society* (Privately Printed, 1896), pp. 12–14 (this is also Paper no. 31 in C. Marshall, B. Lightman, and R. England (eds), *The Papers of the Metaphysical Society, 1869–1880. A Critical Edition*, 3 vols (Oxford: Oxford University Press, 2015), vol. 2).
4. *enclosed letter*: letter missing.

From Thomas Henry Huxley 8 March 1873 4008

4 Marlborough Place | Abbey Road N.W.[1] | March 8th. 1873.

My dear Tyndall,

I am ready for a start on Saturday 15th, but I must be back on Tuesday evening at latest[2] as the Royal Commission[3] meets on Wednesday 19th. I am ready to start whenever you like on Saturday and leave it to you to fix the hour.

Shall we make old Hooker come?

Ever yours faithfully | T. H. Huxley.

I am very glad to see that the Royal Institution people have welcomed you back so prettily.[4]

RI MS JT/1/TYP/9/2976
IC HP 8.136
Typed Transcript Only

---

1. *4 Marlborough Place | Abbey Road N.W.*: see letter 3947, n. 1.
2. *I am ready... evening at latest*: while Tyndall and Huxley had wanted to take a short trip to the Isle of Wight in February, for which the weather kept them home (see letters 3994, 3996, 3997, and 3998), this mid-March trip was presumably to Bath, as Tyndall wrote a letter from there to the editor of *Nature* (letter 4015).
3. *Royal Commission*: see letter 3712, n. 2.
4. *welcomed you back so prettily*: at the 3 March 1873 meeting of the RI, Tyndall was congratulated on his successful American lecture tour (see Introduction, n. 8), welcomed back to 'what they are proud to be able to designate as his own scientific home', and thanked for the donation of the scientific apparatus he employed in his American lectures (*Roy. Inst. Proc.*, 7 (1873–75), pp. 55–6).

From Mary Agatha Russell — 8 March [1873][1] — 4009

[41, South Street | Grosvenor Square. W.] | March 8.

Dear Professor Tyndall

I have long meant to thank you for your kindness in sending me your lecture from America;[2] & in writing me such a very interesting letter, but I did not know where to direct to you in America & since you have been in London[3] I have always hoped to see you at the lectures at the Royal Institution.

Papa[4] & Mama[5] would have asked you to dinner before this, as they hoped so much to see you on your return to England but they have both been unwell; Papa with a cold & cough & Mama with a bad sore throat which has made her very unwell indeed so that we have been obliged to give up any dinners for the present as there is no chance of her coming downstairs for some time.—But I hope before very long they will be well enough to see you if you will be good enough to call some Sunday afternoon when they are always at home.

Yours very sincerely | Agatha Russell.

RI MS JT/1/R/59

---

1. *[1873]*: the year is given by reference to Tyndall's return from his American lecture tour (see n. 3).
2. *your lecture from America*: as the book version of Tyndall's American lectures (*Six Lectures on Light*) had not yet been published, this was probably the *New York Tribune*'s

supplemental printing of his New York lectures, which sold tremendously (see letter 3962, n. 2).

3. *since you have been in London*: Tyndall reached England, returning from his American lecture tour, on 19 February 1873, arriving in London later the same day.
4. *Papa*: John Russell.
5. *Mama*: Frances Russell.

From A. F. F.[1] 10 March 1873 4010

U.S. America, March 10th, 1873.

J. TYNDALL,—How it ought to 'heap coals of fire on your head,'[2] that, in return for your *insults to their Religion*,[3] in your various works, the American people treated you with distinguished consideration. You have repeatedly raised your puny arm against God and His Christ! You have endeavoured to deprive mankind of its only consolation in life, and its only hope in death (*vide* 'Fragments of Science,'[4] &c.), *without offering anything instead*, but the 'dry-light' of your molecules and atoms. Shall we praise you for this? We praise you not!

'Do not I hate them, O Lord, that hate Thee?'[5]

Every *suicide* in our land (and they are a daily occurrence) is indirectly the effect of *the bestial* doctrines of yourself, Darwin, Spencer, Huxley, *et id omne genus*.[6]

'The pit is digged up for you all?'

'Woe unto you that laugh now, for ye shall mourn and lament,'[7]

With the supremist contempt, I remain, | A. F. F—.

H. Spencer, *The Study of Sociology*, 8th edn, (London: C. Kegan Paul & Co., 1880), p. 419

---

1. *A. F. F.*: the author's identity is not identified. Herbert Spencer reproduced this letter as a note in his *The Study of Sociology* (London: Henry S. King, 1873), p. 419, to demonstrate the proposition that 'frequent repetition of Christian doctrines does not conduce to growth of Christian feelings' (p. 418). Tyndall presumably forwarded this letter to Spencer (letter missing); for Spencer's acknowledgment of receiving it, see letter 4114.
2. *'heap coals of fire on your head'*: Proverbs 25:22.
3. *insults to their Religion*: a reference to the prayer gauge controversy. See letter 3780, n. 2.
4. *'Fragments of Science'*: *Fragments of Science*.
5. *'Do not I hate them, O Lord, that hate Thee?'*: Psalm 139:21.
6. *ed id omne genus*: and all that sort (Latin).
7. *'Woe unto you that laugh now, for ye shall mourn and lament'*: Luke 6:25.

From Charles Lyell 10 March 1873 4011

[73, Harley St. | London. W.] | March 10. 1873

Dear Tyndall

I am very much obliged to you for looking through the abstracts.[1]

Please do not take any more trouble about the matter.[2] I have found some passages in Huxley's Anatomy of Vertebrated Animals[3] from which I can take what I require and though I should have preferred that lecture[4] as being specially to the point & popularly expressed yet I think now I can manage without it.

With many thanks | y[ou]rs very truly | Cha Lyell

RI MS JT/1/L/56
RI MS JT/1/TYP/3/857

1. *abstracts*: not identified.
2. *obliged… the matter*: previous correspondence missing. This letter appears to relate to Lyell's working on the most recent edition of his book *The Geological Evidences of the Antiquity of Man: with an Outline of Glacial and Post-Tertiary Geology and Remarks in the Origin of Species, with Special Reference to Man's First Appearance on the Earth*, 4th edn (London: John Murray, 1873), first published in 1863. In a section about ancestral horses in America, Lyell quotes from Huxley's *A Manual of the Anatomy of Vertebrated Animals* (pp. 491–2).
3. *Huxley's Anatomy of Vertebrated Animals*: T. H. Huxley, *A Manual of the Anatomy of Vertebrated Animals* (London: J. A. Churchill, 1871).
4. *that lecture*: not identified.

To Edward L. Youmans 11 March 1873 4012

ROYAL INSTITUTION, LONDON, *March 11, 1873.*

Many thanks to you, my dear Youmans, and many thanks to the Tribune[1] for the cordial expression of good will contained in the number which you have just sent me.[2]

Two hundred thousand copies![3] It is certainly a most extraordinary phenomenon, and one which the English public will probably take to heart. Nothing could be more gratifying.

I am throwing my experiences at Niagara[4] into a readable shape, intending to make a Friday evening lecture out of them on the 4th of April. As soon as ever the paper is ready I shall send it to you.

I have not yet got properly into harness; indeed, this is always a difficulty with me. When I get into a rut I tend to persist in it.

Had a letter yesterday from Hector.[5] He tells me that he has forwarded the deed of trust[6] to Prof. Henry. I did not keep a copy of it, and should like some time to have one, but there is no hurry....

Yours ever, | JOHN TYNDALL.

E. A. Youmans, 'Tyndall's American Visit', *Popular Science Monthly*, 44 (February 1894), pp. 502–14, on pp. 513–14

---

1. *Tribune*: the *New York Tribune*, which published transcripts of Tyndall's New York lectures as a supplement (see letter 3962, n. 2).
2. *the number... sent me*: an article about the success of the supplement, 'Cheap Literature', *New York Tribune*, 26 February 1873, p. 4.
3. *Two hundred thousand copies*: the newspaper noted that fifty thousand copies of the supplement had been sold, while they expected 'eventually to sell more than two hundred thousand copies'.
4. *my experiences at Niagara... 4th of April*: Tyndall paid a visit to Niagara Falls early in his American tour, and delivered a lecture about it at the RI in on 4 April 1873 (see letter 3738, n. 3).
5. *letter yesterday from Hector*: letter from Hector Tyndale missing.
6. *deed of trust*: establishing a scholarship for American students to study in Europe (see Introduction, n. 15; and letters 3949 and 3958.

From Thomas Henry Huxley 14 March 1873 4013

Council on Education, Kensington Museum | March 14 1873.

My dear Tyndall,

I will be with you to-morrow at Waterloo at 11.30 or ten minutes' before. Your route is capital.[1]

Ever yours very sincerely | T. H. Huxley.

P.S. | Please don't go and decide in a hurry about the British Association.[2] I thought it was much better to let other people talk the other day, but I was greatly disappointed at your non-acceptance.

It is physic[3] you will have to swallow one day or another. Be a good boy and take it at once.

RI MS JT/1/TYP/9/2977
IC HP 8.137
Typed Transcript Only

1. *Your route is capital*: see letter 4008, n. 2.
2. *decide in a hurry about the British Association*: later in September 1873, Tyndall was officially elected president of the BAAS for 1874, but showed some reluctance to accept the responsibility (see letters 4033, 4171, 4175, and 4177). In this role he delivered his notorious 'Belfast Address', widely seen as scandalously materialist (see *Ascent of John Tyndall*, chapter 20; see also the forthcoming 14th volume of *The Correspondence of John Tyndall*).
3. *physic*: a medicinal substance (*OED*).

From Mary Egerton 16 March [1873][1] 4014

[45 Eaton Place, | S.W.][2] | March 16th

My dear Mr. Tyndall,

How good of you to give me that beautiful Photograph. It is really a splendid specimen as well as being an excellent likeness. At first I thought it too grave, almost sad—a sort of—"Oh! That the desert &c"—expression;[3] but sitting for one's picture is apt to produce a certain solemnity of countenance, and certainly it is you down even to the eye glass![4] Very many thanks for it.

You have never been to see us yet! The armchair has been wheeled round to a cozy position by the fire all ready for you!

May[5] & I shall have the house to ourselves most of this next week, as the others are going out of town on a visit. We are mainly endeavouring to think of some "lark"[6] we can have on our own account!

Do you—? No you don't, remember what a talk we had about Prayer before you went away?

A thought seemed to help me, the other day to see why some prayers are answered & not others, & it connected itself so naturally with the "Prayer test",[7] that I am tempted to send it you, though there is nothing new in it. (You see five months absence have not quite cured my old propensity!)

How they did preach about that in America![8] I used to get so provoked at their insisting on calling it the "Tyndall prayer test", because I shall always believe that had you exercised your own (not scientific but) sympathetic imagination, you would never yourself have suggested what involved such an entire misconception of the very nature of the thing!

Pardon my bringing back a subject of which you must have had more than enough![9] "I won't do it again" as the children say!—

Bye the bye today is my birthday, so your present was almost a birthday one! For the good of posterity you should have written your own name & not mine under the picture.

Y[ou]rs ever gratefully | M. F. Egerton

RI MS JT/1/E/24

1. *[1873]*: the year is given based on mention of the 'Prayer test', proposed in July 1872 (see n. 7); the mention of Tyndall's absence of five months, the length of his recently concluded American lecture tour (see Introduction, n. 8); and the discussion of the response of American preachers to Tyndall on prayer (see n. 8).
2. *45 Eaton Place, | S.W.*: in London.
3. *"Oh! That the desert &c"—expression*: a reference to a much-quoted line from near the conclusion of Lord Byron's early nineteenth-century poem 'Childe Harold's Pilgrimage' (Canto IV, stanza CLXXVII): 'Oh! that the Desert were my dwelling-place, | With one fair Spirit for my minister, | That I might all forget the human race, | And, hating no one, love but only her!' (Lord Byron, *The Works of the Right Honourable Lord Byron*, vol. 7, *Childe Harold's Pilgrimage, Cantos III and IV* (London: John Murray, 1819), p. 179).
4. *certainly it is you down even to the eye glass!*: this photograph of a seated Tyndall, eyeglass in hand, taken in January 1873 by Frederick Gutekunst, serves as the frontispiece to this volume. An albumen print mounted on card survives in the National Portrait Gallery, London. Frederick Gutekunst (1831–1917) was a Philadelphia photographer (J. Hannavy (ed), *Encyclopedia of Nineteenth-Century Photography*, 2 vols (New York: Routledge, 2008), vol. 1, p. 629). The photograph is referred to also in letter 4073 from Joseph Henry.
5. *May*: Mary Alice Egerton.
6. *"lark"*: a frolicsome adventure, a spree (*OED*).
7. *"Prayer test"*: a proposal introduced and then defended by Tyndall in summer 1872 for testing the power of prayer by seeing whether prayed-for hospital patients show lower mortality rates compared with patients suffering from the same diseases in similarly well-managed hospitals. On Tyndall and prayer, see letter 3780, n. 2.
8. *How they did preach about that in America!*: Tyndall's views on putting prayer to empirical test generated a great deal of commentary from American pulpits during his American lecture tour. See Finnegan, *Voice of Science*, chapter 1.
9. *a subject of which you must have had more than enough!*: see letter 3780, n. 2.

To the Editor of *Nature*[1] 17 March 1873 4015

[Perception in the Lower Animals]

THE theory of taking olfactory notes by the way, as suggested by Mr. Wallace[2] in explanation of the faculty possessed by animals of finding their way home, seems to meet with general acceptance amongst your correspondents; yet it totally fails to account for those instances in which the animal finds its way back by quite a different route to that by which it was taken away.

A good example is given by "F. R. G. S.," in the last number of NATURE; the anecdote of his riding-horse, by Mr. Darwin, also seems to illustrate this point.[3] In an article on the "Consciousness of Dogs," in the *Quarterly Review*, of last October,[4] the following remarkable instance, amongst others, is men tioned on indisputable authority. A hound "was sent by Charles Cobbe, Esq., from Newbridge, county Dublin, to Moynalty, county Meath, and thence, long afterwards, conveyed to Dublin. The hound broke loose in Dublin, and the same morning made his way back to his old kennel at Newbridge, thus completing the third side of a triangle by a road he had never travelled in his life."

Now as Mr. Wallace's theory does not explain these and similar instances, it clearly cannot be received as a solution of the question. Moreover, not only does the faculty exist in other animals not remarkable for their sense of smell, but we find it in cases where this sense has nothing to do with it. Take, for example, the direct homeward flight of the carrier pigeon. Under the same head may be brought the migrations of birds and fishes, and the habits of the turtle, as mentioned by Mr. Darwin.

The writer in the *Quarterly* suggests a sense of the magnetic currents of the earth—a sort of internal mariner's compass in fact.[5] But it is difficult to see how this could have helped the dog to find its way from Dublin to Newbridge, for instance, unless it was also able to consult a map so as to ascertain the relative position of the two places.

It seems then that the problem still remains unsolved. Either we must extend almost indefinitely the range of smell and sight; or, we must suppose the existence of some peculiar sense of the nature of which we are ignorant, which enables its possessor to retain, as F. R. G. S. expresses it, "a constant perception of the bearing of its old home."[6]

J. T. | Bath,[7] March 17

Nature, 7 (20 March 1873), p. 384

1. *Editor of Nature*: Norman Lockyer.
2. *theory of taking olfactory notes by the way, as suggested by Mr. Wallace*: Alfred Russel Wallace had suggested that dogs and other animals find their way home not by instinct but by smell (A. R. Wallace, 'Inherited Feeling', *Nature*, 7 (20 February 1873), p. 303).
3. *A good example... illustrate this point*: see F. R. G. S., 'External Perception in Dogs', *Nature*, 7 (13 March 1873), p. 361, about a dog in Brazil; and C. Darwin, 'Perception in the Lower Animals', *Nature*, 7 (13 March 1873), p. 360, about a blind horse who knew different public-houses on a familiar route by smell.
4. *an article on... of last October:* F. P. Cobbe, 'The Consciousness of Dogs', *Quarterly Review*, 133 (1872), pp. 419–51, on p. 425.
5. *a sense of the magnetic currents*: Cobbe, 'Consciousness of Dogs', pp. 425–6. Darwin also expressed doubt about animals having such a sense (see n. 3).
6. *as F.R.G.S. expresses it*: see note 3.
7. *Bath*: see letter 4008, n. 2.

From Emily Peel 18 March 1873 4016

[Drayton Manor, | Tamworth.][1] | March 18/73

Dear Professor Tyndall,

I wonder if you ever received my letter[2] written to you during your absence in the States, & addressed to the Royal Institution? I hope that it was not lost, for although uninteresting in itself, it however had one merit of affording you proof how fully I appreciated your goodness in sending me the newspaper containing reports of your Lectures,[3] & also for a charming little work,[4] which I found *[at]* my address on my return from Geneva last autumn.

I hope you have returned in good health, & not overfatigued by your Trans Atlantic exertions.

I shall probably be in London very shortly, where I look forward to the pleasure of seeing you & hearing some more about your travels.

Your sincerely | Emily Peel

RI MS JT/1/P/63

1. *Drayton Manor, | Tamworth*: a Peel family estate in Drayton Bassett, Staffordshire, England.
2. *my letter*: the only surviving letter from Emily Peel to Tyndall from the period of his American tour is letter 3955, sent in January 1873 when she was back in Geneva (en route to Greece); there is no mention there, however, of receiving the Americana discussed in the present letter.
3. *your Lectures*: for Tyndall's American lecture tour, see Introduction, n. 8. Tyndall may have sent her the *New York Tribune* supplement that contained the text of his New York lectures and illustrations (see letter 3962, n. 2).
4. *charming little work*: *Forms of Water*.

To Charles Lyell 19 March 1873 4017

Royal Institution 19th. March. 1873

My dear Sir Charles.

Huxley's view of the pedigree of the Horse have been published in one of his addresses to the Geological Survey.[1] He wrote no abstract of his lecture here.[2]

I have been trying to understand the data on which you base your conclusions that after a recession of two miles the height of the fall of Niagara will have diminished from fd to ki (Travels in America p. 45)[3] To me it appears more likely that they should deepen for a time through the further excavation of the shale bed now at the bottom of the Fall. Is there any thing that you would recommend me to read in connexion with this subject? I am to talk about Niagara on the 4th of April.[4]

ever faithfully yours | John Tyndall

I am rejoiced to find that notwithstanding his petulant outburst in the House of Commons we have given his quietus to the First Commissioner. He has ceased to trouble our friend.[5]

RI MS JT/1/T/1051
RI MS JT/1/TYP/3/841

---

1. *Huxley's view of... the Geological Survey*: T. H. Huxley, 'On the Progress of Palaeontology: Anniversary Address Delivered before the Geological Society', *Nature*, 1 (24 February 1870), pp. 437–43, 439–40. Tyndall was responding to a request from Lyell in letter 4011.
2. *his lecture here*: Huxley delivered a Friday Evening Discourse at the RI, 'On the Pedigree of the Horse', on 8 April 1870 (*Roy. Inst. Proc.*, 6 (1870–72), p. 129.
3. *your conclusions... (Travels in America p. 45)*: see C. Lyell, *Travels in North America, in the Years 1841–2: With Geological Observations on the United States, Canada, and Nova Scotia*, 2 vols (London: John Murray, 1845), vol. 1, p. 45 ('fd to ki' refers to labelled lines on an illustration of a cross-section of the strata along the Niagara River).
4. *I am to talk about Niagara on the 4th of April*: Tyndall delivered a Friday evening discourse on Niagara at the RI on 4 April 1873 (see letter 3738, n. 3).
5. *He has ceased to trouble our friend*: a reference to Hooker's dispute with Acton Ayrton (see letter 3755, esp. n. 1).

To Emily Peel 19 March 1873 4018

Royal Institution. | 19th March, /73

My dear Lady Emily,

Before I opened your note[1] this morning a pang of reproach passed through my mind. For I felt that I ought to have answered earlier your most friendly note.[2] But I have been hard pressed since I came here. First of all we have been committed to new buildings: the sacred walls of our old laboratories have been pulled down and new laboratories have been erected.[3] This has thrown the place into a chaotic condition. Then I had to deal with about six cubic feet of papers and correspondence which had accumulated during my absence.[4] Think of all this pray and be lenient with me.

I can hardly account for the magnitude of my good fortune in the United States. The scrap I now send[5] you will give you some idea of the thirst of the people for scientific knowledge. In fact I think if our government have any further arrangements to make with that of the United States they had better send over a Commission of Men of Science.

Kind regards to Sir Robert.[6]

Most faithfully yours, | John Tyndall.

RI MS JT/1/TYP/3/968
LT Typescript Only

---

1. *your note*: letter 4016.
2. *earlier your most friendly note*: perhaps a reference to letter 3955; see also letter 4016, n. 2.
3. *new laboratories have been erected*: the RI's laboratories underwent renovation while Tyndall was on his lecture tour (see letter 3776, n. 14).
4. *during my absence*: for Tyndall's American lecture tour, see Introduction, n. 8.
5. *The scrap I now send*: not identified; perhaps *Nature*'s editorial noting that the overwhelming response of Americans to his lectures—with newspapers carrying them in full, and demand for a published version so high that it could be sold incredibly cheaply—was 'one among many signs that the untrammelled Americans are rapidly outstripping us in the love for and the spread of scientific knowledge' ('Prof. Tyndall in America', *Nature*, 7 (23 January 1873), pp. 224–5). Perhaps also 'Cheap Literature', *New York Tribune*, 26 February 1873, p. 4.
6. *Sir Robert*: Robert Peel (see letter 3844, n. 15).

From Sir Charles Lyell 21 March 1873 4019

73 Harley Street[1] | March 21. 1873.

Dear Tyndall,

The passages I can best recommend you to read on Niagara are Chap XV of my Principles of Geology Vol I (p 354 to 358—11th edition)[2] and parts of the accompanying work by James Hall[3] in which he treats fully of the geology of Niagara.[4] At p 399 I have marked a passage in which he says that the Falls will diminish in height as they recede.

Do you know that all geologists agree in inferring that the Falls have diminished in height in receding from Kingston or Lewiston to their present position?[5]

I fear you were too sanguine about the immediate exodus of the First Commissioner.[6] ever truly y[ou]rs | Cha Lyell

Could you send me three tickets for your lecture on Niagara[7]

RI MS JT/1/L/57
RI MS JT/1/TYP/3/858

1. *73 Harley Street*: Lyell's residence in London.
2. *Chap XV of my Principles of Geology Vol I (p 354 to 358—11th edition)*: C. Lyell, *Principles of Geology*, 11th edn, 2 vols (London: John Murray, 1872), vol. 1, pp. 354–8.
3. *James Hall*: James Hall (1811–98), an American geologist and paleontologist employed by the state of New York, had accompanied Lyell when he visited the Falls in 1841 (*ANB*). See G. M. Friedman, 'Charles Lyell in New York State', *Lyell: the Past is the Key to the Present* (Geological Society, London, Special Publications), 143 (1988), pp. 71–81; L. G. Wilson, *Lyell in America: Transatlantic Geology, 1841–1853* (Baltimore: John Hopkins University Press, 1988), pp. 22–5.
4. *work by James Hall in which he treats fully of the geology of Niagara*: J. Hall, *The Geology of New York, Part IV. Comprising the Survey of the Fourth Geological District* (Albany: Carroll and Cook, 1843), pp. 383–404.
5. *Do you know... their present position*: Lyell was responding to Tyndall's request for more information on the evidence for Lyell's view that as Niagara Falls receded the height of the Falls had diminished (see letter 4017).
6. *the immediate exodus of the First Commissioner*: a reference to Hooker's dispute with Acton Ayrton (see letter 3755, esp. n. 1; and letter 4017).
7. *your lecture on Niagara*: Tyndall delivered a Friday evening discourse on Niagara at the RI on 4 April 1873 (see letter 3738, n. 3).

To Alfred Mayer 22 March 1873 4020

Royal Institution of Great Britain | 22nd March 1873

My dear Professor Mayer

Your letter[1] has just reached me. I shall be glad to welcome you to London. Your invitation to the Athenaeum[2] shall be duly arranged, and I am sure you will find the Club agreeable.

It concerns me much to hear you complain of your health; this however is a trial through which most earnest workers have to pass—I wish you will restore yourself in the air and the rest of England.

The science of England, as regards its introduction into education is still in a chaotic condition. It is perhaps well that it should continue so for a time. It is better to go through the ordeal of trial and error, to endure the turbulence of testing, selecting and rejecting, than harden too soon into an imperfect system.

The line taken by our friends the theologians at the dinner[3] interested me profoundly. Their position, I fear, is not a comfortable one.

If I knew where you intended to lodge I would try and have the invitation of the Athenaeum sent there so to meet you on your arrival.

Presenting kind regards to President Morton,[4] and with best wishes for your health

believe me | faithfully yours | John Tyndall

Tyndall, John (1820–1893); Hyatt and Mayer Collection, C0076, Manuscripts Division, Department of Special Collections, Princeton University Library

1. *Your letter*: letter 4003.
2. *Athenaeum*: the Athenaeum Club (see letter 3710, n. 5).
3. *line taken by our friends the theologians at the dinner*: see letter 4003, n. 2.
4. *President Morton*: Henry Morton (see letter 3881, n. 6).

From Joseph Henry 24 March 1873 4021

Smithsonian Institution | Washington, M[ar]ch 24th. 1873

My dear Professor Tyndall:

Accompanying this I send you a photograph at the request of Prof[esso]r. S. H. Lockett,[1] of the Louisiana State University of which the following is his explanation; "In my drawing room I kept a wash basin in which to rinse out the color from my water color brushes. This color gradually formed a uniform sediment of an indefinite tint over the bottom of the basin. On the night of the 26th of December last, which was an unusually cold one for this climate, the water in the basin froze. On the melting of the ice the next day the beautiful figure you see on the photographs[2] was left in the sediment. I carefully poured the water from the basin, let the sediment dry and thus perfectly preserved the figure. It has been accurately photographed by an artist in this city The negative is preserved and if you would like to have any more copies they can readily be obtained.

We are not much accustomed in this warm warm country of ours to the beautiful 'Forms of Water', and this has struck me as a little remarkable and worthy of being kept."

The fact that the results have been produced by colored sediment indicates a method of exhibiting the effects of crystallization in an interesting manner.

I am, | Very truly yours, | Joseph Henry | Sec[retar]y. S.I.[3]

Prof. John Tyndall, | Royal Institution, | London, E.

Smithsonian Institution Archives. RU 33, Vol. 33, Pages 210–13

PLUMES PRODUCED BY THE CRYSTALLIZATION OF WATER.

Photographed by Professor Lockett.

1. *Prof[esso]r. S. H. Lockett*: Samuel Henry Lockett (1837–91), Professor of Mechanics and Engineering at Louisiana State University from 1867–73 (L. C. Post, 'Samuel Henry Lockett (1837–1891): A Sketch of His Life and Work', *Louisiana History: The Journal of the Louisiana Historical Association*, 5 (1964), pp. 421–41).
2. *beautiful figure you see on the photographs*: the photograph he sent to Tyndall via Joseph Henry—of a phenomenon with the scantiest link to Tyndall's 1872 popular bestseller *The*

Forms of Water—was published as the frontispiece to the English edition of Tyndall's *Six Lectures on Light*, and is here reproduced following this letter.

3. *S.I.*: Smithsonian Institution.

To George Frederic Watts[1] 24 March 1873 4022

Royal Institution | 24th March 1873

My dear Sir

Mr. Spottiswoode has informed me of your friendly and flattering desire to paint my portrait.[2] Were I less doubtful of my own claims on the time of a man of genius I should have promptly written to express to you the high value which I attach even to the unrealized intention, not to speak of its fulfillment in fact.

I can only say that should your wish survive reflection I shall place myself at your disposal after the fifth of April.

I am my dear Sir | Most faithfully yours | John Tyndall

G. F. Watts Esq[ui]re | R.A.[3] &c. &c. &c.

Letter from John Tyndall, Royal Institution to G. F. Watts, 24 March 1873, p. 118, GFW/1/4/98, Heinz Archive, National Portrait Gallery, London

1. *George Frederic Watts*: George Frederic Watts (1817–1904), a widely esteemed sculptor and painter, especially of portraits (*ODNB*).
2. *paint my portrait*: there appears to be no record of a painting of Tyndall by Watts.
3. *R.A.*: Watts was a member of the Royal Academy of Arts.

To Charles Lyell 25 March [1873][1] 4023

25th March.

My dear Sir Charles

I send you the three tickets with very great pleasure.[2]

That the Fall is lower than it has been is I think obvious, but considering the present position and dip of the bed of soft shale I should not like to predict that the bed of the river will rise in the immediate future.[3] ever faithfully yours | John Tyndall

RI MS JT/1/T/1052
RI MS JT/1/TYP/3/841

1. *[1873]*: the year is written next to the month and day in another hand; it makes sense as this letter relates to letters 4017 and 4019.
2. *I send you the three tickets*: Tyndall delivered a Friday evening discourse on Niagara at the RI on 4 April 1873 (see letter 3738, n. 3). Lyell requested tickets from Tyndall in letter 4019.
3. *I should not... in the immediate future*: Tyndall here challenged not merely Lyell's claim about the regular diminution of the height of Niagara Falls (see letters 4017 and 4019), but a basic tenet of Lyellian geology: that change witnessed in the present can be straightforwardly extrapolated into the future as well as into the past. By the time of the lecture, however, Tyndall had come around to Lyell's view, arguing in conclusion that, in five thousand years, what was now the Falls would be an unspectacular bend in a gorge-coursing river.

To Hermann Helmholtz 26 March [1873][1] 4024

Royal Institution. | 26th. March

My dear Friend

The head and front of my offending in "Forms of Water"[2] is that I have done injustice to the late Principal Forbes.[3] When the time comes I shall know how to deal with this charge; but it will be ungrateful to me to fight across a grave; and it is such a deplorable loss of time.

Since his death[4] I have never entertained a thought hostile to Forbes, nor uttered a word inconsistent with tenderness to his memory. But if I am attacked I shall know how to defend myself.[5]

The matter is still in the state of rumour, but I have the best reasons for regarding the rumour as well founded. Meanwhile let the matter rest: I do not wish to deflect my assailants from their course.

Yours affectionately | John Tyndall.

Years ago I had a sharp passage of arms with Professor Tait[6] of Edinburgh, but we had, as I thought, become friends: he however seems to have only skinned his wounds while I have healed mine. This between us.

RI MS LT/53/19
LT Typescript Only

1. *[1873]*: the year is given by reference to the attack on Tyndall regarding Forbes in his *Forms of Water* (see letters 4033 and 4038).
2. *"Forms of Water"*: *Forms of Water*
3. *late Principal Forbes*: James David Forbes.
4. *his death*: Forbes died on 31 December 1868.
5. *how to defend myself*: Tyndall's eventual published response was: J. Tyndall, 'Principal Forbes and His Biographers', *Contemporary Review*, 22 (August 1873), pp. 484–508.
6. *sharp passage of arms with Professor Tait*: Tyndall and Peter Guthrie Tait had been in

dispute before. During the 1860s, the Scottish physicists Tait and William Thomson took issue with Tyndall over his claims about Julius Mayer's priority over a fellow North British man of science, James Joule, in developing the mechanical equivalent of heat. This heated dispute spilled over into popular journals before they resolved their differences at the Dundee meeting of the BAAS in 1867, where, Tyndall wrote they 'seemed like brothers together' (see letter 2646, *Tyndall Correspondence*, vol. 10). On this 'passage of arms', see *Ascent of John Tyndall*, pp. 169–78. See also C. Smith, 'North Britain versus Metropolis: Territorial Controversy in the History of Energy', in *The Science of Energy: A Cultural History of Energy Physics in Victorian Britain* (Chicago: University of Chicago Press, 1998), pp. 170–91.

To George Grove[1] 29 March [1873][2] 4025

Royal Institution of Great Britain | 29th. March

My dear Grove.

Next week I will send you a proof of my paper on Niagara:[3] you can then judge of its fitness or unfitness for the magazine.

The paper is partly composed of narrative and partly of science which every body can understand.

In fact I have seen more of Niagara than falls to the lot of most people—and I have endeavoured to write down my impressions and explanation in the clearest possible language.

ever faithfully yours | John Tyndall

Wellcome Library MS. 7777/49

1. *George Grove*: George Grove (1820–1900), then editor of *Macmillan's Magazine* (*ODNB*).
2. *[1873]*: the year is given based on the discussion of Tyndall's article 'Niagara' (see n. 3).
3. *my paper on Niagara*: Tyndall's paper based on his 4 April 1873 Friday Evening Discourse at the RI about his visit to Niagara Falls was published in several places (see letter 3738, n. 3).

To Moncure Daniel Conway[1] 30 March [1873][2] 4026

Royal Institution of Great Britain | 30th March

My dear Mr. Conway

Thank you much for your note and its enclosure.[3] Thanks also with a little blood heat[4] added for your friendly congratulations.

There are curious "clouds" of organisms sometimes in the air, but the one described in the article you send me[5] looks a little overdrawn.

Yours ever | John Tyndall.

If I ever again cross the Atlantic I must make it a point to visit my friends at Columbus.[6]

Invaluable Auctions or Goldberg Coins

---

1. *Moncure Daniel Conway*: Moncure Daniel Conway (1832–1907), a Virginia-born, London-based minister, abolitionist, and freethinker (*ANB*; *ODNB*). He was a part of Tyndall's circle from the 1860s; see M. D. Conway, 'Tyndall', *Open Court*, 8 (18 January 1894), pp. 3943–4.
2. *[1873 or later]*: the year is given based on Tyndall's reference at the end of the letter to his recently concluded American trip (which was from late September 1872 to early February 1873). We recognize that this letter could be from 1873 on, but place it in this volume.
3. *your note and its enclosure*: letter and enclosure missing.
4. *blood heat*: used figuratively, the temperature of the blood in the periphery of the healthy human body, usually taken as 37°C (98.6°F) (*OED*).
5. *article you sent me*: article not identified, but the reference to 'clouds' of micro-organisms suggests it was related to the debate over spontaneous generation. For Conway's role in publicizing Tyndall's contribution to that debate, see 'Force not the Equivalent of Personality', *Universalist Quarterly and General Review*, 34 (October 1877), pp. 439–52, on p. 444.
6. *Columbus*: Tyndall must have meant Cincinnati, Ohio, where Conway lived as a minister from 1856–63 prior to moving to England.

To the Editor of *Nature*[1] March 1873 4027

[Effect of Resistance in modifying Spectra]

In a review of M. Guillemin's[2] work "The Forces of Nature"[3] which appeared in last week's *Athenaeum*,[4] the following reference, by M. Guillemin, to the experiments of M. Mitscherlich[5] is quoted: "Suivant ce physicien il arrive que le presence de certaines substances dans une flame a pour effet d'empecher de se produire les spectres des autres substances, d'enteindre leur raies principales." The English editor adds that the effect "may probably be explained by the observations of Frankland and Lockyer."

In relation to this subject of the extinction of the bands of one metal by another, you will perhaps permit me to quote a paragraph from one of the lectures which I have recently had the honour of delivering in the United States.[6] The arcs of thallium and silver had just been compared, and their similarity of colour pointed out. The power of prismatic analysis to show that, notwithstanding the apparent identity of colour, the arcs really belonged to two different metals, was then demonstrated. The metals were afterwards subjected *together* to the action of the Voltaic current, and it was shown that the band of thallium fell midway between the two bands of silver. Hence the similarity of colour. The lecture then proceeds thus:—

"But you observe here another interesting fact. The thallium band is at first far brighter than the silver bands; indeed the latter have wonderfully degenerated since the bit of thallium was put in. The reason of this is worth knowing. It is the *resistance* offered to the passage of the electric current from carbon to carbon that calls forth the heating power of the current. If the resistance were materially lessened, the heat would be materially lessened; and if *all* resistance were abolished there would be no heat at all. Now thallium is a much more fusible and vaporisable metal than silver, and its presence facilitates the passage of the current to such a degree as to render it almost incompetent to vaporise the more retractory silver. But the thallium is a gradually consumed; its vapour diminishes, the resistance consequently rises, until finally the silver bands are rendered as brilliant as at first."[7]

In the spectra of mixed substances derived from the electric spark the action here referred to must come frequently into play. If neither the fact, nor its proposed explanation, be new, I would thank you to commit this document to your waste-paper basket.

JOHN TYNDALL | Royal Institution, March 1873

Nature, 7 (20 March 1873), p. 384

---

1. *Editor of Nature*: Norman Lockyer.
2. *M. Guillemin's*: Amédée Guillemin (1826–93), a French scientific writer ('Amédée Guillemin', *Nature*, 48 (25 May 1893), p. 82).
3. *"The Forces of Nature"*: A. Guillemin, *The Forces of Nature*, trans. W. J. Lockyer, ed. J. N. Lockyer (London: Macmillan, 1872).
4. *in last week's Athenaeum*: 'The Forces of Nature', *Athenaeum*, 8 March 1873, pp. 313–5; the quoted text that follows is from p. 315.
5. *M. Mitscherlich*: Eilhard Mitscherlich (1794–1863), a German chemist (*ADB*).
6. *delivering in the United States*: for Tyndall's American lecture tour, see Introduction, n. 8.
7. *"But you observe... brilliant at first"*: the passage is from Tyndall's sixth lecture, published in slightly modified form in *Six Lectures on Light*, p. 192.

From Thomas Henry Huxley — 1 April 1873 — 4028

4 Marlborough Place | Abbey Road N.W.[1] | April 1st /73.

My dear Tyndall,

The printers (deuce take them) have only just sent me the accompanying proof of my preface[2] which I begged you to look at.

I have made such corrections as have struck me to be needful, but I shall be very glad of any suggestions. And you know that I am particularly anxious not to annoy Spencer in any way.[3]

Considering how bad I have been for weeks past[4] (I have a lucid interval to-day) I really think the preface is not so bad. There is life in the old dog yet.

Ever yours very sincerely | T. H. Huxley.

RI MS JT/1/TYP/9/2978
IC HP 8.138
Typed Transcript Only

---

1. *4 Marlborough Place | Abbey Road N.W.*: see letter 3947, n. 1.
2. *my preface*: the proof that Huxley sent must have been for the preface to his then-forthcoming volume of collected essays, *Critiques and Addresses* (London: Macmillan, 1873). The preface is dated April 1873 (p. xii).
3. *anxious not to annoy Spencer in any way*: the volume republished Huxley's attack on his friend Herbert Spencer's *laissez-faire* political philosophy, 'Administrative Nihilism' (a lecture delivered to members of the Midland Institute on 9 October 1871), even though Spencer had already complained that it had caricatured his views ('Specialized Administration', *Fortnightly Review*, 60 (December 1871), pp. 627–54). Huxley in his preface at once acknowledged Spencer's concerns, expressed great admiration and affection for him, and gently but firmly defended the line taken in the 1871 essay; T. H. Huxley, 'Preface', in *Critiques and Addresses* (London: Macmillan, 1873), pp. v–xii, on pp. v–vii.
4. *Considering how bad I have been for weeks past*: between habitual overwork, money troubles, and the pressure of a recently concluded lawsuit, Huxley was in such a poor state by early April 1873 that Tyndall, Spencer, and other friends (including Darwin) went on to contribute funds to help Huxley financially and for a health-restoring holiday (see *Ascent of John Tyndall*, p. 179; and Desmond, *Huxley*, pp. 426–30). Tyndall, Darwin, Hooker, and Spencer met on 8 April to discuss this effort, which raised over £2,000 from eighteen subscribers (these were William Armstrong; William Bowman (1816–92), a British surgeon and anatomist (*ODNB*); George Busk; Charles Darwin; Erasmus Darwin (see letter 4036, n. 7); John Evans (1823–1908), a British archaeologist, numismatist, and paper manufacturer (*ODNB*); Thomas Henry Farrer; William Henry Flower (1831–99), a British zoologist and museum curator (*ODNB*); Thomas Hirst; Joseph Hooker; Henry Bence Jones; John Lubbock; Charles Lyell; James Paget; Herbert Spencer; William

Spottiswoode; John Tyndall; and Joseph Whitworth (1803–87), a British mechanical engineer best remembered for the Whitworth rifle, considered the most accurate rifle of its time (*ODNB*)). Katherine Lyell, sister-in-law of the geologist Charles Lyell, also made a contribution, but Darwin and Tyndall were hesitant to make this known as they did not wish others to think that they solicited donations beyond the smaller group of scientific friends (see letters 4043 and 4054). For the donation fund, see also letters 4032, 4034, 4035, 4036, 4037, 4042, 4060, 4061, 4067, 4069, 4071, 4072, and 4079.

To Thomas Henry Huxley 2 April [1873][1] 4029

Royal Institution. | 2nd. April.

My dear Huxley,

I consider this preface admirable.[2]

I do not think after your warm introductory words that Spencer can possibly quarrel with what comes afterwards. Nevertheless your retort is perfectly characteristic—with a good spice of the sabre hack in it.[3]

Your reply to Miss Helen Taylor[4] is also very good: I think it will tell well upon the public. The strength of her position would be that millions of people hold a different view from yours; and that where this is the case doubt must exist.[5]

Mivart[6] you have handled most admirably. The remarks about his "guidance," are excellent, and will give satisfaction to all good men.[7]

And finally your hit at the Bastianites is also very delightful. Nothing could be better than the figure employed.[8]

I wish I could take you and myself away somewhere for three months. The life would return as a spring tide.[9]

Yours ever | John Tyndall.

IC HP 8.139
Typed Transcript Only

1. *[1873]*: the year is given as this letter is a response to letter 4028.
2. *this preface admirable*: in letter 4028, Huxley asked Tyndall to look over the proof of the preface to a new volume of essays, *Critiques and Addresses*. Tyndall's remarks here follow the topic ordering in Huxley's preface. T. H. Huxley, 'Preface', in *Critiques and Addresses* (London: Macmillan, 1873), pp. v–xii.
3. *I do not think... sabre hack in it*: Huxley sought reassurance that he had managed to criticize Herbert Spencer without ruffling his feathers unduly (see letter 4028, n. 3).
4. *Miss Helen Taylor*: Helen Taylor (1831–1907), an English political writer and supporter of women's rights (*ODNB*).
5. *strength of her position... doubt must exist*: Huxley capped his prefatory discussion of

Spencer with a swipe at Helen Taylor, who likewise attacked Huxley's 1871 lecture 'Administrative Nihilism', and in the same issue of the *Fortnightly Review* which carried Spencer's attack (see letter 4028, n. 3). Taylor, the stepdaughter of the liberal philosopher John Stuart Mill, had objected to Huxley's view that the State should actively suppress religious doctrines judged to be contrary to the general welfare (H. Taylor, 'The New Attack on Toleration', *Fortnightly Review*, 10 (December 1871), pp. 718–27). See Huxley, 'Preface', pp. viii–ix.

6. *Mivart*: St. George Jackson Mivart (1827–1900), a London anatomist and Catholic critic of Darwin (*ODNB*).
7. *The remarks about... all good men*: Mivart was featured in Huxley's essay 'Mr. Darwin's Critics' (1871), republished in the new volume. What drew Huxley's ire in his preface (pp. ix–xi) was not Mivart's views on evolution but a recent double allegation Mivart had made against Huxley's religious learning: that he misconstrued a major Jesuit authority as a Biblical literalist, and had done so because lacking proper 'guidance' on how to interpret his writings (see St. G. J. Mivart, 'Evolution and its Consequences: A Reply to Mr. Huxley', *Contemporary Review*, 19 (January 1872), pp. 68–72). On both counts, Huxley begged to differ.
8. *your hit at the Bastianites... figure employed*: Tyndall here saluted Huxley's witty put-down of the life-from-nonlife or 'abiogenesis' brigade, led in Britain by the Darwinian London physician and neurologist Henry Charlton Bastian (see letter 3782, n. 12), and dealt with at length in Huxley's 'Biogenesis and Abiogenesis' (1870), republished in the volume. Although 'the public may have been led to believe that so much noise must indicate rapid progress, one way or the other', wrote Huxley in his preface (p. xii), 'an impartial critic will admit, with sorrow, that the question has been "marking time" rather than marching. In mere sound, these two processes are not so very different'.
9. *The life would return*: Huxley had been unwell, to the alarm of his friends (see letter 4028, n. 4).

To Eliza Spottiswoode — 4 April [1873][1] — 4030

Royal Institution of Great Britain | 4th. April

My dear Mrs. Spottiswoode

Tell your kind husband,[2] please, that I mortally dread dissatisfaction tonight,[3] and that this and other things will *[compel]* me to have my solitary chop at the Athenaeum[4]

Yours ever | John Tyndall

RI MS JT/1/T/1350

---

1. *[1873]*: the year is given by the hint of a Friday Evening Discourse occurring on the date of the letter (see n. 3), the 'dissatisfaction' Tyndall feared being that of the audience at the RI.

2. *your kind husband*: William Spottiswoode.
3. *tonight*: Tyndall delivered a Friday Evening Discourse, 'Some Observations on Niagara', at the RI on 4 April 1873 (see letter 3738, n. 3).
4. *Athenaeum*: the Athenaeum Club (see letter 3710, n. 5).

From Louisa Baring [4 April 1873][1] 4031

I was so sorry not to see you after the Lecture[2] this evening to thank you for the delicious treat you had given us—No voyage into fairy Land[3] could reveal such glories, as you showed us tonight—1000 thanks for your kindness in giving me the tickets. I did not know how much I was asking, or I sh[oul]d. not have asked—I am very much obliged—Next time I shall know better—I hope you are fast asleep, and not over done after such an effort as it must have been to crowd so much into so short a time.

Always yours most truly | L[ad]y. Ashburton | Friday E[ven]ing | 1 oc[loc]k—

Elton Hall, Ashburton Collection
RI MS JT/1/TYP/1/125

1. *[4 April 1873]*: the date is given by reference to a lecture at the RI (see n. 2).
2. *Lecture*: Tyndall delivered a Friday Evening Discourse at the RI on 4 April 1873 (see letter 3738, n. 3).
3. *fairy Land*: a nickname for Niagara Falls.

From Herbert Spencer 5 April 1873 4032

5 April 73

My dear Tyndall

I saw the manager to-day. He said that a sum might be paid[1] in to any account anonymously, and that they could keep counsel. He agreed however in the suggestion that if a payment was made into their bank by another banker, it would put an additional obstacle in the way of tracing its source.

Would not Farrer be the best man to effect the transaction,[2] as being least likely to be taxed by Huxley. The banker who paid in the sum might be one with whom we none of us bank, so as to avoid giving a clue.

Or if a lawyer who could be trusted, would undertake to make the payment through his banker to the Lon & West, and his reply, if traced, were to be that the sum was handed to him for the purpose by a client of his, the source would be pretty well hidden.

I leave town on Thursday, & will put my contribution in your hands if you think anything can be done at once.

Sincerely yours | Herbert Spencer[3]

RI MS JT/1/S/100
RI MS JT/1/TYP/3/1117

1. *a sum might be paid*: the sum in question—nearly £2,000—was raised by Huxley's friends to pay for a health-restoring holiday (see letter 4028, n. 4).
2. *to effect the transaction*: Farrer was successfully persuaded to transfer the funds to Huxley, on the grounds that 'it would be untraceable to the obvious suspects' (*Ascent of John Tyndall*, p. 179).
3. On the back of the letter is a list of names with numbers, presumably some of those who contributed to the fund for Huxley and how much (£); they are: Charles Darwin (200; later increased to 300, see letter 4036), Joseph Hooker (50), Erasmus Darwin (200, see letter 4036, n. 5), George Busk (200), John Lubbock (100), Herbert Spencer (100), William Spottiswoode (100), Thomas Hirst (50), Andrew Clark (20), Thomas Farrer (100), and Tyndall (200). Frank (Francis) Galton is listed but no amount given. For the names of all the friends who donated, see letter 4028, n. 4.

To Joseph Henry 6 April 1873 4033

Prof Tyndall | 6th. April 1873.

My dear Professor Henry

I thank you most heartily for this beautiful example of the "forms of water". It is exquisite. I must try and have it reproduced. Pray be good enough to convey my best thanks to Professor Lockett.[1]

By the way that little book,[2] which I imagined to be without offence, is likely to bring me into trouble. You remember I dare say some years ago a passage of versus between Professor Thomson & Tait and myself regarding Mayer[3] and the Mechanical Theory of Heat.[4] I thought all ill feeling had ended for we met each other as friends in Dundee;[5] and I was delighted to be assured that no further quarrels could arise between us—I was too sanguine. For while it was my earnest wish in the "Forms of Water" that the pupil should *[emerge]* from its perusal with no other feeling than one of respect for all of those who had wrought at the glaciers, Prof. Tait I am told has taken deep offence at my references to Forbes, and intends to call me to account in a Life of Forbes on which he is now engaged.[6] It will be an ungracious thing to quarrel over a grave, but I may be freed to that which will be exceedingly disagreeable to me.

In a brief review the editor of Nature has given expression to Mr. Tait's feeling: certainly no change had ever less justification than that I wished to "rake up" a buried controversy.[7]

Your excellent letter written before I quitted New York[8] reached London before I did. I thank you most heartily for it. I asked General Mayer[9] to say to you that I had seen your friends at Trinity House[10] & to make known to you how cordially they spoke of yourself and Mrs. Henry.[11]

I am hard pressed by the Council of the British Association to accept the Presidency at Belfast next year.[12] I declined emphatically, but they urge it upon me with such earnestness, and kindness that I am in a strait regarding it. It would be the third refusal on my part to accept the post, and I do not like to show myself *[unmindful]* of such kindness. But the acceptance of the office will entail loss of time & a great deal of labour.

I have been just writing an article on Niagara[13] where I had reasons to refer to what you said a great many years ago in the Transactions of the Albany Institute regarding the cutting of the gorge.[14] I intend sending the paper to Youmans when it is finished, for publication in the Popular Science Monthly.[15] I showed the last number of the Monthly, containing your letter & Draper's & White's speeches,[16] to some of my scientific friends. They were perfectly delighted and predict a great scientific future for a country in which such a Spirit can be shown.

Give my very best regards to Mrs. Henry and the young ladies[17]
and believe me always | your faithful friend | John Tyndall
Kind remembrances also to Mr. Rhees.

[Archive source unknown][18]

---

1. *convey my best thanks to Professor Lockett*: Tyndall was replying to Joseph Henry's letter dated 24 March 1873, in which Henry passed along a photograph plus explanation from Samuel Henry Lockett, of Louisiana State University, showing a striking pattern that formed after the freezing of water in a basin full of residue from paintbrushes used in watercolors (see letter 4021, n. 1 and n. 2).
2. *"that little book"*: *Forms of Water*.
3. *Mayer*: Julius Mayer.
4. *passage of versus... Mechanical Theory of Heat*: see letter 4024, n. 6.
5. *we met each other as friends in Dundee*: see letter 4024, n. 6. On the friendly reconciliations with Thomson and Tait at the BAAS meeting in Dundee in 1867, see *Ascent of John Tyndall*, p. 205.
6. *a Life of Forbes on which he is now engaged*: on Tyndall's learning that Tait was to use his role as co-editor of the imminent *Life and Letters of Forbes*, co-edited with John Campbell Shairp and A. Adams-Reilly, to hit back at the perceived slighting of Forbes in *Forms of Water*, see *Ascent of John Tyndall*, p. 317.

7. *In a brief review... buried controversy*: 'Tyndall's Forms of Water', *Nature*, 7 (27 March 1873), pp. 400–01, quotation on p. 401. The editor of *Nature* at this time was Norman Lockyer.
8. *Your excellent letter written before I quitted New York*: letter 3984. Henry wrote to Tyndall on 3 February 1873, five days before Tyndall left New York to return home to England from his American lecture tour, arriving in London on 19 February (see Introduction, n. 8).
9. *General Mayer*: Alfred Mayer.
10. *your friends at Trinity House*: in his letter, Henry expressed gratitude for the visit of British experts on lighthouses from Trinity House, Fredrick Arrow and John Sydney Webb (see letter 3984, n. 11). On Tyndall as scientific adviser to Trinity House, see letter 3869, n. 4.
11. *M^rs^. Henry*: see letter 3869, n. 7.
12. *the Presidency at Belfast next year*: see letter 4013, n. 2.
13. *article on Niagara*: Tyndall quoted Henry (at secondhand) on p. 59 of 'Niagara', *Macmillan's Magazine*, 28 (May 1873), pp. 49–62; see letter 3738, n. 3.
14. *Transactions of the Albany Institute regarding the cutting of the gorge*: in a paper read to the Albany Institute on 28 October 1829 and published in the first volume of its *Transactions*; republished as J. Henry, 'Topographical Sketch of the State of New-York, designed chiefly to show the General Elevations and Depressions of its Surfaces', in *Scientific Writings of Joseph Henry*, 2 vols (Washington: Smithsonian Institution, 1886), vol. 1, pp. 8–36.
15. *for publication in the Popular Science Monthly*: see letter 3738, n. 3.
16. *the last number of the Monthly, containing your letter & Draper's & White's speeches*: the April 1873 issue of *Popular Science Monthly* included the speeches that John William Draper and Andrew Dickson White (1832–1918), an American historian and educator who was President of Cornell University (*ANB*), delivered at Tyndall's farewell banquet at Delmonico's in New York on 4 February 1873: J. W. Draper, 'English and American Science', *Popular Science Monthly*, 2 (April 1873), pp. 734–6; and A. D. White, 'Science and Public Affairs', *Popular Science Monthly*, 2 (April 1873), pp. 736–9. The 'Tyndall number' of *Popular Science Monthly*, as he noted it was called in letter 4046, also included, F. A. P. Barnard, 'Science and Our Educational System', pp. 695–8; W. H. Brewer, 'Discovery of Mount Tyndall', pp. 739–41; and J. Henry, 'On the Importance of the Cultivation of Science', *Popular Science Monthly*, 2 (April 1873), pp. 641–50. All of these were soon published in *Proceedings at the Farewell Banquet* as well. For the banquet itself, see Introduction, n. 15.
17. *young ladies*: see letter 3869, n. 7.
18. The provenance of this letter is unclear, and its archive source unknown. It does not come from the collections at the Huntington Library, where many of the Henry letters in this volume are held.

From George H. Darwin[1] [7 April 1873][2] 4034

16 Montague St. | Portman Sq.[3] | Monday morning.

Dear Professor Tyndall

My father begs me to write you a note to say that he particularly wishes to see you on business not connected with himself.[4] He will therefore call on you at about 4.15 this evening; if you cannot see him, will you have the great kindness to call here this evening at 5 p.m. or any later hour which may suit you.

Yours very sincerely | G. H. Darwin

P.S. | He is very much fatigued with a journey he has had to take to the city this morning, and I therefore venture to say on my own responsibility that it would be a very great kindness to him if you could call here from 3.30 to 3.45 so as just to stop him starting. If you do come please not to mention that I have asked you to come as I'm sure he would think I had taken an unwarrantable liberty, tho' I trust you will not think so too.

RI MS JT/1/TYP/9/2819
LT Typescript Only

1. *George H. Darwin*: George Howard Darwin (1845–1912), mathematically trained son of Charles and Emma Darwin, who collaborated with Darwin on various scientific projects (*ODNB*).
2. *[7 April 1873]*: the date was established by the DCP in light of a letter to Joseph Hooker dated 6 April 1873. See *Darwin Correspondence,* vol. 21, p. 161; this is letter 8843 in the online edition of the Darwin correspondence.
3. *16 Montague St. | Portman Sq.*: see letter 4036, n. 2.
4. *business not connected with himself*: Darwin wished to discuss further arrangements for transferring the money that he, Tyndall, and other friends had collected to donate to Thomas Huxley for a health-restoring holiday (see letter 4028, n. 4).

To Charles Darwin 8 April [1873][1] 4035

Royal Institution of Great Britain | 8th April

My dear Darwin.

I drove to Spottiswoode's yesterday but he was not at home: then I drove to the Athenaeum[2] and caught Spencer whom I knew to be a dissipated individual and a billiard player—during the afternoon. I thought it well to secure his presence today;[3] and he therefore like myself will be with you at about half past one oClock.

Yours ever | John Tyndall

GBR/0012/MS DAR 106: C10[4]
RI MS JT/1/TYP/9/2820

---

1. *[1873]*: the year was established by the DCP by relation to other letters (see letters 4028, n. 4). Also, the DCP notes that Darwin's wife Emma recorded the referred to meeting in her diary (see n. 4).
2. *Athenaeum*: the Athenaeum Club (see letter 3710, n. 5).
3. *to secure his presence today*: Spencer was among the group due to gather at Down House that afternoon to discuss an anonymous donation to Thomas Huxley for a health-restoring holiday (see letter 4028, n. 4).
4. This letter has been published in *Darwin Correspondence*, vol. 21, p. 167; this is letter 8849 in the online edition of the Darwin correspondence.

From Charles Darwin 8 April [1873][1] 4036

Private | 16 Montague St. | Montague Sq. W.[2] | Ap. 8

My dear Tyndall,

After you left I thought of the extreme interest and kindness which Spottiswoode showed on our subject,[3] and it appeared to me proper that I should write to him which I have done,[4] and I mention it to you to save your taking the trouble. As he would have known if present what was done I gave him an outline of our decision and thought it fair to tell him what had been subscribed. I presume he will communicate directly with Lubbock. As you will naturally desire to know how the subscription stands I may mention that my brother[5] will give £100 and that I have changed mine to £300[6] These will be despatched to-morrow. I really think we have done a good days work and I remain

Yours most cordially | Ch. Darwin.

Our meeting this morning will give me a higher opinion of human nature for the rest of my life than I ever had before, though I am not one of those who think lowly of mankind.

RI MS JT/1/TYP/9/2821[7]
LT Typescript Only

---

1. *[1873]*: the year is given by reference to the effort to raise funds for a health-restoring holiday for Thomas Huxley (see n. 3); also, the DCP notes that Darwin's wife Emma recorded the referred to meeting in her diary (see n. 7).
2. *16 Montague St. | Montague Sq. W.*: a rented house in London, where Darwin stayed from 15 March to 10 April 1873 (P. van Helvert and J. van Wyhe (eds), *Darwin: A Companion.*

With Iconographies by John van Wyhe (London: World Scientific, 2021), p. 197). Some Darwin letters from this time give Montague Square, while others give Portman Square.

3. *our subject*: the meeting concerned the funds that Thomas Huxley's friends were collecting for anonymous donation to him to pay for a health-restoring holiday (see letter 4028, n. 4).
4. *which I have done*: for Darwin's letter to William Spottiswoode of the same day, see *Darwin Correspondence*, vol. 21, pp. 167–8; this is letter 8851 in the online edition of the Darwin correspondence.
5. *my brother*: Erasmus ('Ras') Alvey Darwin (1804–81), the older and only brother of Charles Darwin.
6. *give £100 and that I have changed mine to £300*: see the list of donations included in letter 4032.
7. This letter has been published in *Darwin Correspondence*, vol. 21, pp. 168–9; this is letter 8850 in the online edition of the Darwin correspondence.

From Walter B. Woodbury[1] 08 April 1873 4037

Cliff House, Greenhithe,[2] April 8

A CURIOUS thing in connection with colour having come under my notice, and never having before seen it remarked in any scientific journal, I take the liberty of bringing it before your notice; if it is new to you it will interest you, if not I must ask you to excuse me for troubling you. On looking through a piece of blue glass (of which I forward you a sample) at a planter tree lit up by the sun, those leaves that are lit from behind, or rather by transmitted light, appear of a rich crimson, the other leaves, seen only by reflected light, merely like the colour given by the glass. In the case of a geranium plant, those leaves become almost the same colour as the flowers. As a scientific fact, if new, it is curious.[3]

WALTER B. WOODBURY | To Prof. Tyndall

Nature, 7 (24 April 1873), p. 481

1. *Walter B. Woodbury*: Walter Bentley Woodbury (1834–85), a Manchester-born photographer and inventor (*ODNB*).
2. *Cliff House, Greenhithe*: Woodbury's residence in Greenhithe, a town in Kent, England, on the River Thames.
3. *it is curious*: Tyndall's reply is letter 4040. Both this letter and Tyndall's reply were sent by Tyndall to the journal *Nature*, and were published with a note from Tyndall (see letter 4052).

To Louis Agassiz 9 April 1873 4038

Royal Institution of Great Britain | 9th. April 1873

My dear Friend,

Soon after my return from the United States[1] I learned that two distinguished Scotch professors[2] were engaged upon the life of the late James D. Forbes,[3] the author of "Travels in the Alps".[4]

To my amazement I was informed that great outrage had been taken with the manner in which I had spoken of you and Forbes in my little volume the "Forms of Water",[5] and I learn that in the life of Forbes I am to be attacked and punished for what I have written in the little volume referred to.

This I say astonishes me—for my aim and desire were to cause the young pupil to rise from the perusal of that little book with a feeling of respect for all those who have worked at the glaciers.[6]

You can bear testimony as to how tenderly I spoke of Forbes's memory when I had the pleasure of meeting you at Cambridge.[7]

But if I am attacked I must defend myself,[8] and I am inclined to think that my assailants will gain but little from their assault.

I write to ask you to jot down such memoranda as may occur to you relating to the behaviour of Forbes, as regards the glacier question. For example. I wish particularly to know whether he was aware when he published his books and memoirs that you, in 1841, had fixed those stakes in the glacier of the Aar,[9] and that in 1842 you had measured their motions and communicated the results of your measurements to the Academy of Sciences.[10] Is it not the fact that he ignored these measurements of yours, and that the people of England were left in ignorance of them until public attention was directed to them by myself?

I wish you would describe in simple, temperate language Forbes's visit to you in 1841, and whether he then made known to you his intention of reparing to the Mer des Glace the following year.

It is in the very highest degree repugnant to me to quarrel over a grave. But I am assured that there will be no escape from it.

I am told that Forbes's son,[11] who has been recently elected to a professorship in Scotland is very hot upon this subject. On him, therefore, and on his friends will rest the responsibility of any damage that may be done to his father's memory.

It is, I am informed, my attitude toward <u>you</u> that has given most offence. I verily thought in writing the book that you, rather than Forbes, would be the dissatisfied person. As regards your relationship to him I wish you would kindly see whether I have committed any historical mistakes. I do not think so: at least none from which Forbes suffers.

I should be thankful to you if at the same time you would tell me where the account of your labours, which have been criticized be Geikie, is to be found.[12]

I am very sorry to take up your time with this subject; and God knows I grudge the expenditure of my own.

Kind regards to M^rs^. Agassiz.[13]

Believe me always | Yours faithfully | John Tyndall

One of the gentlemen engaged upon the Life of Forbes wrote to Huxley, charging me with unfairly making up the controversy between you & Forbes. Huxley told him promptly that he saw no making up, but that he by no means approved of the manner in which Forbes had treated you. Another long letter came demanding reasons—Huxley cut the matter short by declining to enter upon the discussion.

Huxley, I am sorry to say, is by no means well.[14]

Darwin is staying in town for a little time: he looks, at times, very fresh & well, but at times very tired.[15]

I have written this rapidly for I am pressed by many labours—I should like to be able to publish anything you may send me, if I thought it advisable to do so.

Tyndall, John, 1820–1893. A.L.s. (John Tyndall) to [Louis Agassiz]. [n.p.], 9 Apr 1873. Louis Agassiz correspondence and other papers, MS Am 1419, (687). Houghton Library, Harvard College Library

---

1. *my return from the United States*: for Tyndall's American lecture tour, see Introduction, n. 8.
2. *two distinguished Scotch professors*: Peter Guthrie Tait and John Campbell Shairp.
3. *the life of the late James D. Forbes*: *Life and Letters of Forbes*.
4. *"Travels in the Alps"*: J. D. Forbes, *Travels through the Alps of Savoy* (Edinburgh: Adam and Charles Black; London: Longman, Brown, Green, and Longmans, 1843).
5. *you and Forbes in my little volume*: the impression conveyed in Tyndall's *Forms of Water* was that Agassiz was the pioneering investigator and Forbes the consolidator and populariser (see esp. pp. 161–2, 187).
6. *the young pupil to rise from the perusal*: Tyndall here echoes closely what he wrote to Joseph Henry on 6 April (see letter 4033).
7. *meeting you at Cambridge*: along with Ralph Waldo Emerson, Oliver Wendell Holmes, Henry Wadsworth Longfellow, the Harvard-based Agassiz was among the 'perfect galaxy of genius', as Tyndall wrote to Hirst, that he met with when his recent American lecture

tour took him to Boston (see letter 3873; see also *Ascent of John Tyndall*, p. 307; and DeYoung, *Vision of Modern Science*, pp. 70–1).

8. *I must defend myself*: for Tyndall's eventual published response, see letter 4024, n. 5.
9. *you, in 1841, had fixed those stakes in the glacier of the Aar*: Tyndall described this in *Glaciers of the Alps*, pp. 271–2.
10. *communicated the results of your measurements to the Academy of Sciences*: Tyndall noted in *Glaciers of the Alps* that Agassiz's results, which included observations made in 1842 as well, 'were laid before the [French] Academy of Sciences on the 29th of August 1842, and are published in the "Comptes Rendus" of the same date' (p. 273). See 'Lettre de M. Agassiz à M. Arago, sur les glaciers', *Comptes Rendus Hebdomadaires des Séances de l'Académie des Sciences*, 15 (1842), pp. 435–46.
11. *Forbes's son*: George Forbes.
12. *criticized by Geikie*: neither Agassiz's account of his labours nor its criticism by the Scottish geologist Archibald Geikie (1835–1924) have been identified. As Geikie was a great supporter of Agassiz's in his debate with Charles Lyell over the ice-age theory, however, any criticisms would probably have been mild.
13. *M^rs^. Agassiz*: Elizabeth Cabot Cary Agassiz (1822–1907), married in 1850. She was an educator, naturalist, and writer, and very involved in her husband's scientific work. She was widowed not long after this letter when Louis died on 14 December 1873. She served as the first president of Radcliffe College in Cambridge, Massachusetts from 1882–1903 (*ODNB*).
14. *Huxley... is by no means well*: on efforts to support Thomas Huxley at this time, see 4028, n. 4.
15. *Darwin is staying in town*: Charles Darwin and some of his family were staying at 16 Montague Street in London (see letters 4034 from George Darwin and 4036 from Charles Darwin).

To Charles Darwin — 9 April [1873][1] — 4039

Royal Institution of Great Britain | 9th April

My dear Darwin,

You have stolen a march upon me but I am content. Our excellent friend Spottiswoode & his sweet wife[2] were here yesterday. I told him all about our meeting & that we had made a shot at him for £100[3]—He was delighted. It was a capital stroke of work yesterday: and I feel more and more confident, considering the perfectly brotherly spirit of the transaction that Huxley will raise no permanent objection to it.

I wrote to Armstrong yesterday.[4]

I also saw Bence Jones[5] He gave me (without asking) a cheque for fifty pounds

Yours ever | John Tyndall.

GBR/0012/MS DAR 106: C11[6]
RI MS JT/1/TYP/9/2822

1. *[1873]*: the year was established by the DCP in light of letter 4036.
2. *his sweet wife*: Eliza Spottiswoode.
3. *a shot at him for £100*: in Spottiswoode's absence from the gathering of Thomas Huxley's concerned friends the day before, it was suggested that Spottiswoode would donate £100 to a fund for a health-restoring holiday for Huxley (see letter 4028, n. 4).
4. *wrote to Armstrong yesterday*: letter missing.
5. *Bence Jones*: Henry Bence Jones died twelve days after Tyndall saw him.
6. This letter has been published in *Darwin Correspondence*, vol. 21, pp. 169–70; this is letter 8852 in the online edition of the Darwin correspondence.

To Walter B. Woodbury[1] 9 April 1873 4040

THE observation you describe[2] is interesting, and if you have taken care to exclude subjective colouring, is, I should think, to be explained in this way.

The light from your leaves contains a quantity of red; it appears as a yellowish green, I suppose, and contains little or no blue.

Your glass is of a kind which transmits the two ends of the spectrum while cutting out its centre. It is very hostile to the yellow, hence on placing it before the eye, and receiving through it light which has already been deprived of its blue, the glass quenches the yellow, and red alone remains.

JOHN TYNDALL | April 9

Nature, 7 (24 April 1873), p. 481

1. *Walter B. Woodbury*: see letter 4037, n. 1.
2. *THE observation you describe*: Tyndall was replying to letter 4037 from Woodbury. Both that letter and Tyndall's reply were sent by Tyndall to the journal *Nature*, and were published with a note from Tyndall (see letter 4052).

From Henrietta Huxley 9 April [1873][1] 4041

4 Marlborough Place | Abbey Road |
London N.W.[2] | 9th April (1873)[3]

Dear Brother John

It was very kind of you to offer me a reserved seat at your lecture[4]—but in the first place I did not like to bother you and in the second place I had a young friend[5] with me. The evening was an untoward affair for us—for my little friend nearly fainted and had to be lifted out—and indeed was so ill that I had to drive to an hotel on my way home with her. So I only heard the first ten minutes discourse of yours, to my great disappointment.

Hal[6] goes to Aberdeen[7] on Saturday. I never more regretted any journey than this. He will have too much talking by far—perhaps too much hospitality. I cannot think that it is right of him to go by sea just now It is so cold and he will get nothing right to eat for two days.

How disappointing it is to me to see him no better[8] I cannot tell you.

Ever Yours affectionately | Nettie Huxley.

RI MS JT/1/TYP/9/2979
LT Typescript Only

---

1. *[1873]*: the year is given by a note on the typescript by LT (see n. 3). This year is appropriate as the date of 9 April follows a Friday Evening Discourse that Tyndall delivered on 4 April 1873 (see n. 4), and that Huxley became Lord Rector of Aberdeen University in 1873 (through 1880), having been elected by the students there on 14 December 1872. This trip to Aberdeen University must have been an administrative one, as he did not deliver his Inaugural Address to the university, titled 'Universities: Actual and Ideal', until 27 February 1874, citing at the start of the address having had to postpone it until his health bettered (P. J. Anderson (ed), *Rectorial Addresses delivered in the Universities of Aberdeen, 1835–1900* (Aberdeen: Aberdeen University, 1902), pp. 170–98, 376–7; Desmond, *Huxley*, pp. 439–40). The address was published as T. H. Huxley, 'Universities: Actual and Ideal', *Contemporary Review*, 23 (1873–4), pp. 657–79; and T. H. Huxley, 'Universities: Actual and Ideal', in *Collected Essays*, 9 vols (London: Macmillan and Co., 1893–4), vol. 3 (1893), pp. 189–233). See letter 3947, n. 28 and n. 29.
2. *4 Marlborough Place | Abbey Road | London N.W.*: see letter 3947, n. 1.
3. *(1873)*: this date annotation on LT's typescript is accompanied by a note reading 'prob[ab]ly'.
4. *your lecture*: Tyndall delivered a Friday Evening Discourse, 'Some Observations on Niagara', at the RI on 4 April 1873 (see letter 3738, n. 3).
5. *young friend*: not identified.
6. *Hal*: a nickname for Thomas Huxley.
7. *Aberdeen*: see n. 1.
8. *to see him no better*: see letter 3998, n. 8; and letter 4028, n. 4.

To Charles Darwin 10 April 1873 4042

Royal Institution of Great Britain | 10th April 1873

My dear Darwin.

I have got £100 with a sweet good letter from Armstrong[1]—He desires me to call upon him again if more be needed.

We shall, however, have quite sufficient for our purpose. I was also sure of Farrer,[2] and he has justified my confidence in him.

Ever heartily yours | John Tyndall

GBR/0012/MS DAR 106: C12[3]
RI MS JT/1/TYP/9/2823

1. *£100 with a sweet good letter from Armstrong*: William Armstrong was among the eighteen friends of Thomas Huxley who contributed to anonymous fund to give Huxley a health-restoring holiday. Tyndall wrote to Darwin that he had written to Armstrong in letter 4039. See also letter 4028, n. 4. The letters to and from Armstrong are missing.
2. *I was also sure of Farrer*: Herbert Spencer suggested Thomas Farrer to Tyndall as best placed to transfer the funds to Huxley (see letter 4032, n. 2).
3. This letter has been published in *Darwin Correspondence*, vol. 21, p. 170; this is letter 8855 in the online edition of the Darwin correspondence.

From Charles Darwin 11 April 1873 4043

Down, | Beckenham, Kent. | Ap. 11. 1873.

My dear Tyndall,

You have had magnificent success.[1] I enclose a list[2] which I think includes all the names; if it does not please to inform me.

I heard from Lubbock this morning and the figures opposite the names shew the sums actually received by him.

We are certain of 1700 and shall probably get 1800. Is not this grand?

I rather expect a subscription from Dr. A. Clark, who was delighted at my news under a medical point of view. I wish Mrs L.[3] had not subscribed at all; though she has so good a heart—we had better keep her amount secret.

I further much wish that Hooker had not given so much. As soon as I hear from Lubbock that all have subscribed (or we know that some will not) Lubbock will place the amount to H[uxley]'s acc[oun]t. and I will write to him. I have made up my mind, if H[uxley]. refuses, to write to him and to say that I will not accept his answer until he has taken a few more days for consideration, and will supplicate him, and use every argument that I can think of, to make him change his mind. I wish you had undertaken this part of the job, for I believe you would have done it better. Hooker, however, saw and approved of a rough copy of my intended letter.[4] I will of course tell you everything as it happens.

My dear Tyndall | Yours most sincerely | Charles Darwin.

RI MS JT/1/TYP/9/2824[5]
LT Typescript Only

1. *You have had magnificent success*: Darwin was congratulating Tyndall in response to his letters of 9 and 10 April (4039 and 4042), where Tyndall reported good news in executing the plan—agreed with Darwin and others on 8 April—to collect and donate funds to Huxley for a health-restoring holiday (see letter 4028, n. 4).
2. *enclose a list*: enclosure missing; presumably the list included with letter 4032.
3. *Mrs. L.*: identified by the DCP as Charles Lyell's sister-in-law. Katharine Murray Lyell (1817–1915), an English botanist and an editor of important volumes on the correspondence of several notable scientists, including Charles Lyell, Leonard Horner, and Charles Bunbury (R. Desmond, *Dictionary of British and Irish Botanists and Horticulturalists* (London: BMNH, 1994); M. Ogilvie and J. Harvey (eds), *Biographical Dictionary of Women in Science*, 2 vols (London: Routledge, 2000), vol. 2, p. 813).
4. *my intended letter*: Darwin's letter to Thomas Huxley announcing the donated funds, dated 23 April 1873, is published in *Darwin Correspondence*, vol. 21: pp. 183–4; this is letter 8872 in the online edition of the Darwin correspondence.
5. This letter has been published in *Darwin Correspondence*, vol. 21, pp. 170–1; this is letter 8856 in the online edition of the Darwin correspondence.

From Henrietta Huxley [11 April 1873][1] 4044

4 Marlborough Place[2] | Good Friday

Dear Brother John

Your kind note[3] is comforting—"A word in season, how good it is"![4]

Hal[5] is very downcast "Melancholy mad" as he himself describes it—I am ashamed to confess it but last night, after having pent up all my anxiety about him for two long years, my fortitude gave way and I, who have never shed a tear all that time, sobbed for an hour. I have got a horrid sore throat and stiff neck—but I shall still endeavour to get away tomorrow with the 3 elder children[6] to Stratford on Avon,[7] where the elder Flowers[8] have very kindly invited us for Easter. For I think if I can get out of the ring of association for a time I shall get brighter, and this is so essential for Hal—Some well meaning but ill-doing friend told him I was looking ill—and it made him miserable.

This is a doleful screed but you must pay the penalty of having sympathized—You know how a dog will whine when he is pitied.

With love | Yours affectionately | Nettie Huxley.

RI MS JT/1/TYP/9/2979a
LT Typescript Only

1. *[11 April 1873]*: the date is given by reference to Thomas Huxley being ill. He fell ill in 1871, took a trip to the Mediterranean and Egypt in 1872 to alleviate this, but in 1873 his

ill health continued. This matches Henrietta's words in this letter, 'having pent up all my anxiety about him for two long years'. Good Friday in 1873 fell on 11 April.

2. *4 Marlborough Place*: see letter 3947, n. 1.
3. *Your kind note*: letter missing; Tyndall may have written to Henrietta about Thomas Huxley accepting the funds collected for him to take a much-needed health-restoring holiday (see letter 4028, n. 4).
4. *"A word in season, how good it is"*: Proverbs 15:23.
5. *Hal*: a nickname for Thomas Huxley.
6. *3 elder children*: Jessie Oriana Huxley (1858–1927), Marian Huxley (1859–87), and Leonard Huxley (1860–1933).
7. *Stratford on Avon*: Stratford-upon-Avon, a town in the county of Warwickshire, England, birthplace of William Shakespeare.
8. *elder Flowers*: the parents of William Henry Flower (1831–99), a British zoologist, museum curator, and friend and supporter of Huxley's whose parents lived in Stratford-upon-Avon (*ODNB*).

To C. Cecil Trevor 12 April 1873 4045

Royal Institution 12th April 1873.

Sir,

I have proposed to the Elder Brethren[1] in accordance with your wish expressed in H. 1791[2] the inspection of the lights at Haisbro' on the night of Monday the 21st.[3]

I have appended to this proposition the following remark:—"It will aid me, and render my excursions more prompt and satisfactory, to have Mr. John Wigham, the inventor of the Burner employed at Haisbro', at my side at the time. Questions will probably arise which he can best answer, and which it will be necessary to answer, prior to the writing of my report. I should therefore feel much obliged if the Elder Brethren would take upon themselves the trouble of inviting Mr. Wigham to join me on the 21st."

This arrangement will, I trust, be agreeable to the Board of Trade,[4] and I am sure it will be conducive to the clearing up and final settlement of the question of the employment of gas in Lighthouses.

I have the honour to be Sir, | Your obedient servant | John Tyndall | Thomas Gray Esq[ui]re | &c &c &c

National Archives MT 10/220. *Illumination of lighthouses by gas*. H2075 'Further Papers Relative to a Proposal to Substitute Gas for Oil as An Illuminating Power in Lighthouses', C 282, Parliamentary Papers (1871), pp. 27–30

1. *Elder Brethren*: see letter 3736, n. 3 and n. 5.
2. *H. 1791*: presumably this is a letter from Tyndall to Trinity House and the Board of Trade, but it is missing.
3. *I have proposed... the 21st*: prompted by a report concluding that gas was better for fuelling lighthouse burners than paraffin or other alternatives, Tyndall had proposed an inspection of the gas-burning system in use at the notoriously treacherous sandbanks at Haisborough (or Haisboro or Haisbro') off of the Norfolk coast of England (*Ascent of John Tyndall*, p. 316).
4. *Board of Trade*: see letter 3736, n. 3.

To Edward L. Youmans 12 April 1873 4046

April 12, 1873.

MY DEAR YOUMANS: The "Tyndall number," as the World calls it, of The Popular Science Monthly duly reached me.[1] I wish you had sent over a dozen of them. I took the number to Bence Jones (who to my great grief is dying) and to others. They were mightily struck by its tone, and Bence Jones predicts all manner of great things for a nation which can evoke the spirit manifested in the address of President White.[2]

. . . I send you by this post a proof of my little paper on Niagara;[3] it may be printed as it stands if time be an element of importance, otherwise I am having a little map of the Falls prepared which will add to the clearness of the paper. . . .

Faithfully yours, | JOHN TYNDALL.

E. A. Youmans, 'Tyndall's American Visit', *Popular Science Monthly*, 44 (February 1894), pp. 502–14, on p. 514

1. *The "Tyndall number"... duly reached me*: the April 1873 issue of *Popular Science Monthly* included several of the Tyndall-honoring reflections given at his farewell banquet at Delmonico's in New York on 4 February 1873 (see letter 4033, n. 16).
2. *address of President White*: A. D. White, 'Science and Public Affairs', *Popular Science Monthly*, 2 (April 1873), pp. 736–9. On Andrew Dickson White, see letter 4033, n. 16.
3. *proof of my little paper on Niagara*: 'Niagara', *Macmillan's Magazine*, 28 (May 1873), pp. 49–62; see letter 3738, n. 3.

To Thomas Archer Hirst [13 April][1] 1873 4047

Easter 1873

My dear Tom

Considering the amount of work in front of me, and that the Trinity House & the Board of Trade[2] have both come down upon me I do not think I can afford time to go to the Isle of Wight.[3] I must keep myself up by shorter excursions made nearer home. That I am sorry for—but I hardly think I should enjoy quietude of mind if I acted otherwise.

Y[ou]rs affection[ate]ly | John
Sunday.

RI MS JT/1/T/986

---

1. *[13 April]*: Easter Sunday in 1873 fell on 13 April.
2. *Trinity House & the Board of Trade*: see letter 3736, n. 3 and n. 5. On Tyndall as scientific adviser to Trinity House, see letter 3869, n. 4.
3. *Considering the amount... Isle of Wight*: some of Tyndall's commitments at this time were preparation for a lecture course at the RI (see letter 3872, n. 10), preparing *Six Lectures on Light* for publication, and lighthouse work for the Board of Trade (see letter 4045).

To Ellen Tucker Emerson 14 April [1873][1] 4048

Royal Institution of Great Britain | 14th April

My dear Miss Emerson

I hope before you leave London that your father[2] and yourself will be able to take luncheon with me here some day. I am a miserable bachelor and dine at my club.[3]

I wish one or two of his old disciples to meet him. Huxley is away in Aberdeen,[4] but will return next week. My friend Hirst would also be charmed to meet your father. I hope this can be arranged.

I shall make it a point to see him tomorrow evening at Lady Amberley's.[5]

I hope he still finds that "Spring makes spring in the mind when sixty years are told".[6]

Yours ever faithfully | John Tyndall

[n.p.] 14 Apr [1873], 1873. Emerson family correspondence, MS Am 1280.226, (4123). Houghton Library, Harvard College Library

1. *[1873]*: the year is established since the proposed luncheon took place on 23 April 1873; see the entry for Ralph Waldo Emerson in the Biographical Register for this volume.
2. *your father*: Ralph Waldo Emerson, who was on his third and last European tour this time in the company of his daughter Ellen, between October 1872 and April 1873. See 'Ralph Waldo Emerson: The Reluctant Traveler', in W. W. Stowe, *Going Abroad: European Travel in Nineteenth-Century American Culture* (Princeton, NJ: Princeton University Press, 1994), pp. 74–101.
3. *my club*: the Athenaeum Club (see letter 3710, n. 5).
4. *away in Aberdeen*: see letter 4041, n. 1.
5. *Lady Amberley's*: Katherine Amberley, Viscountess Amberley (1842–74), known at the time for her socially progressive views, is perhaps best remembered now as the mother of the philosopher Bertrand Russell (*ODNB*).
6. *"Spring makes... are told"*: a (near) quotation of the first two lines of the final stanza of Ralph Waldo Emerson's poem 'The World-Soul' (1843): 'Spring still makes spring in the mind, | When sixty years are told; | Love wakes anew this throbbing heart, | And we are never old. | Over the winter glaciers, | I see the summer glow, | And, through the wild-piled snowdrift, | The warm rosebuds below' (R. W. Emerson, *Poems* (Boston: James Munroe and Company, 1847), pp. 30–5, on p. 35). On Tyndall as Emersonian, see DeYoung, *Vision of Modern Science*, pp. 69–78; and R. A. Haugrud, 'Tyndall's Interest in Emerson', *American Literature*, 41 (January 1970), pp. 507–17.

To Heinrich Rudolf Vieweg 14 April [1873][1] 4049

Royal Institution of Great Britain | 14th April

Dear Sir,

I propose adding to the next edition of "Heat a mode of motion"[2] a sixteenth chapter containing a resumé of my latest researches on aqueous vapour, and on the points of difference between the late Professor Magnus & myself.

Will you add this to the forthcoming edition in Germany? It will be ready in a week or two.

I also propose publishing my lectures on Light delivered in America[3] I do not know whethe[r] you would be disposed to undertake this publication.

faithfully Yours | John Tyndall

TU Braunschweig/Universitätsbibliothek/UABS V 1 T : 63

1. *[1873]*: the year is given by mention of Tyndall's American lecture tour (see Introduction, n. 8), and by relation to letter 4111 (dated 13 June 1873), in which Vieweg referred to this letter.
2. *next edition of "Heat a mode of motion"*: the next English edition was J. Tyndall, *Heat: A Mode of Motion*, 5th edn (London: Longmans: Green, and Co., 1875), and its German translation was J. Tyndall, *Die Wärme, betrachtet als eine Art der Bewegung*, 3rd edn, trans. H. Helmholtz and G. Wiedemann (Braunschweig: F. Vieweg und Sohn, 1875).
3. *my lectures on Light delivered in America*: *Six Lectures on Light* was published in German as J. Tyndall, *Das Licht: Sechs Vorlesungen, gehalten in Amerika im Winter 1872–1873*, trans. G. Wiedemann (Braunschweig: F. Vieweg und Sohn, 1876). As was typical, while the translation was done by Clara Wiedemann, her husband Gustav's name was given on the publication.

From François-Napoléon-Marie Moigno 14 April 1873 4050

[LES MONDES | Revue Hebdomadaire | DES SCIENCES | *et de leurs applications* | aux Arts et à l'Industrie | *PAR* | M. L'ABBÉ MOIGNO | Abonnement: | Paris un an 25^{f} | Départements 30^{f} | BUREAUX | 32, rue du Dragon, 32

Paris, le] 14 avril [187]3

Mon cher Mr Tyndall

Permettez moi de vous donner un petit signe de vie et de vous féliciter du triomphe que vous avez remporté en Amérique. J'ai rendu compte fidèlement de vos succès dans Les Mondes, j'avais même inséré l'article dans lequel le Scientific American vous faisait gagner des sommes fabuleuses; je ne pensais pas qu'il *[*pût*]* vous nuire, par le principe qu'il vaut mieux inspirer l'envie que la pitié. Cependant je me suis empressé d'insérer la rectification de Mr Bence Jones.

Pendant votre absence je n'ai pas cessé de presser Mr Gauthier-Villars de publier le plus promptement possible la nouvelle édition de la Chaleur. Il a *[*marché*]* mais beaucoup trop lentement à mon gré, j'aurais voulu qu'il vous *[*fut*]* une agréable surprise en terminant l'impression avant votre retour. Je n'ai pas réussi à mon grand regret. Nous n'en sommes encore pour le tirage qu'à la page 309 un peu plus de la moitié. Mais l'édition sera très jolie, *[*format*]* en 18 comme la première.

Combien je regrette de plus en plus la trahison de Giraud, nous aurions continué à faire des affaires ensemble; la librairie des Mondes et les actualités scientifiques vont très bien, j'aurais certainement écoulé déjà deux éditions de la Chaleur.

Avez vous reçu ma dernière actualité L'optique Moléculaire qui est pleine de mémoires de vous.

Mr Gauthier-Villars est disposé à traiter avec vos éditions pour la traduction française de *[*ceux*]* de vos ouvrages que vous voudrez. Mais peut-être avez vous des engagements avec *[2 words illeg]*.

J'ai fait aussi ma grande *[*campagne*]* des Salles du Progrès, mais je suis loin d'avoir *[*eu*]* le *[1 word illeg]* *[*qui*]* *[*a*]* *[*suivi*]* *[*partout*]* *[1 word illeg]* pas. *[1 word illeg]* si comme vous j'avais été en Amérique? Mais j'ai ouvert mes conférences de science illustrées de tous les soirs, dans un lieu trop en dehors de la circulation et dans un milieu politique trop agité. Après quatre mois d'essai et de succès relatif, succès d'estime, non succès d'argent; j'ai cru qu'il était nécessaire de *[*replier*]* ma *[*tente*]* de *[*suspendre*]* en attendant des jours meilleurs, et je suis rentré dans ma chère solitude où je travaille plus que jamais.

Si vous n'avez pas encore disposé de tous vous bénéfices, j'en *[*réserve*]* une petite partie pour quand je rouvrirai de nouveau ma Salle du Progrès.

Une pensée m'a préoccupé pendant votre longue absence. Vous avez moissonné gloire et argent; avez-vous en outre rencontré une fidèle et douce compagne de votre noble *[*vie*]* ou êtes vous toujours Old Bachelor. Pardonnez moi cette question indiscrète. Je serais heureux d'apprendre que vous êtes à tous les points de vue au comble de vos *[*voeux*]*.

J'ai l'honneur d'être dans les sentiments du respect le plus affectueux et le plus dévoué

Votre humble serviteur et ami l'abbé F. Moigno

Vous savez cher ami que je suis toujours à votre disposition pour tout ce que vous voudrez publier. Les Mondes ont *[*maintenant*]* une assez grande circulation.

[LES MONDES[1] | Weekly Review | of the sciences | *and their applications* | to the Arts and the Industry | *BY* | FATHER MOIGNO | Yearly Subscriptions: | Paris for a year 25[f] | Departments 30[f] | OFFICES | 32 Dragon street, 32

Paris,] 14 April [187]3

My dear M[r] Tyndall,

Allow me to give you a small sign of life and to congratulate you on the accomplishments that you achieved in America.[2] I faithfully accounted for your success in Les Mondes,[3] and I also added the article in which the Scientific American turned you into a maker of fabulous sums of money.[4] I did not think that it could hurt you, because of the principle that it is better to inspire envy in others rather than pity. However, I hastened to add M[r] Bence Jones' correction.[5]

During your absence, I did not stop pressuring M[r] Gauthier-Villars to start publishing the new edition of Heat[6] as soon as possible. He worked on it, but it was too slow to my liking—I would have wanted him to give you a pleasant surprise of finishing the printing before your return. To my great disappointment, I did not succeed. We are just printing page 309, a little more than half. But the edition will be beautiful, *[*format*]* in 18 like the first one.

How I regret more and more the betrayal of Giraud. We would have continued to do business together.[7] The bookselling of Les Mondes and the scientific news is going very well—I would certainly have sold off two editions of the Heat already.

Have you received my last news on Molecular Optics,[8] which is full of your reports?

Mr Gauthier-Villars[9] is willing to deal with your edits for the French translation of the books that you would like. But perhaps you have commitments with *[2 words illeg]*.

I also did my large campaign of Salles du Progrès,[10] but I am far from having had the *[1 word illeg]* that followed *[*everywhere*]* *[1 word illeg]*. *[1 word illeg]* if like you I was in America? But I opened my illustrated scientific conferences every night in a place too far from traffic and in a very agitated political environment. After four months of trials and of relative success—a success in terms of reputation, not of money—I thought that it was necessary to *[*fold my tent*]* and suspend until better days,[11] and I returned to my cherished solitude, where I work more than ever.

If you have not yet disposed of all of your *[*earnings*]*, I *[*save*]* a small part for when I will reopen my Salle du Progrès.

I was preoccupied with a thought during your absence. You have *[*earned*]* glory and wealth; have you also met a loyal and gentle companion for your noble life, or are you still an Old Bachelor? Forgive me for this bold question. I would have been happy to know that you have, in all possible ways, fulfilled all of your *[*wishes*]*.

I have the honour of expressing my most affectionate and devoted respect.

Your humble servant and friend | Father F. Moigno

You know, dear friend, that I am always at your disposal for everything that you would like to publish. Les Mondes *[*now*]* has a big enough circulation.

RI MS JT/1/M/122

---

1. *LES MONDES*: see letter 3829, n. 1.
2. *accomplishments that you achieved in America*: for Tyndall's American lecture tour, see Introduction, n. 8.
3. *accounted for your success in Les Mondes*: 'Le professeur Tyndall à New-York', *Les Mondes*, 30 (13 February 1873), pp. 265–9.

4. *the Scientific American... sums of money*: under the title 'Professor Tyndall as a Money Maker', *Scientific American* had reported that although money had been stolen from Tyndall's hotel while he was giving his first lecture on light in New York, he was earning so much from his exceptionally well-prepared and well-delivered lectures—'the Professor is bagging ten thousand dollars a week'—that he would hardly notice the loss ('Tyndall as Money Maker', *Scientific American*, 28 (4 January 1873), p. 7). For Moigno's notice of this, see 'Le professeur Tyndall et sa mine d'or', *Les Mondes*, 30 (16 January 1873), pp. 91–2.
5. *Bence Jones' correction*: Henry Bence Jones sent a notice to Moigno correcting the claim of *Scientific American* that Tyndall was making as much money as he did ('La mine d'or de M. Tyndall', *Les Mondes*, 30 (30 January 1873, p. 152). He shared what Tyndall had written to him regarding his expenses in America in a letter from New York of 3 January 1873: 'As it is a balance will doubtless remain to me. My expenses, including those attendant on poor Miller's death, have been terribly high. Still two thousand pounds will remain as a surplus. I wrote yesterday to Henry of Washington regarding the application of this money. Doubtless we shall come to some arrangement about it soon' (letter 3950). Bence Jones later learned that Tyndall planned to donate the surplus of his lecture earnings to a fund for pure science education in America (see Introduction, n. 15).
6. *new edition of Heat*: see letter 3829, n. 3.
7. *the betrayal.... business together*: Étienne Giraud (fl. 1850–60s) was a French publisher and bookseller whose firm was located at 20 rue Saint Sulpice in Paris. He published Abbé Moigno's *Les Mondes*, as well as the first French edition of Tyndall's *Heat* (1863) which appeared as *La chaleur* (1864). Letters in *Tyndall Correspondence*, vol. 10, suggest that Giraud went bankrupt before paying Tyndall for the sales of *La chaleur*.
8. *Molecular Optics*: F.-N.-M. Moigno, *Optique Moléculaire: Effets de Précipitation, de Décomposition, d'Illumination, Produits par la Lumière* (Paris: Gauthier-Villars, 1873). This collection of scientific articles includes six from Tyndall; two from Alexandre Lallemand (1816–86); two from Jacques-Louis Soret (1827–90), a Swiss physicist and chemist who taught at the University of Geneva from 1866 (*HLS*); and one from Jean François Auguste Morren (1805–70).
9. *Mr Gauthier-Villars*: Jean-Albert Gauthier-Villars.
10. *Salles du Progrès*: Salles du Progrès, or Halls of Progress, was a short-lived program initiated by Moigno in 1872 to educate the French public in scientific topics: 'to give popular instruction, varied by recreative musical and physico-chemical, optical, and other experiments and exhibitions, for the purpose, in the first place, of aiding the education of a large class of people in Paris who now spend too much of their time in frequenting cafés, chantants, and other places of amusement of a questionable character; and also of affording instruction to young men in science in its widest sense, inclusive of art and applied science' ('Salles du Progrès, Soirées et Matinées de Science Illustrée', *Chemical News*, 25 October 1872, p. 203).
11. *necessary to [fold my tent] and suspend until better days*: the journal *Nature* reported the

end of Moigno's program thus: 'The Abbé Moigno, after all, has brought his *Salles du Progrès* to a termination; not, however, we are glad to learn, because his praiseworthy scheme of popular education has turned out a failure, but because he has found another means of accomplishing his "great object of intellectual regeneration by true science, good and beautiful." The founders of the Catholic Circle of Workmen… have proposed to make use of the Abbé's services, his educational staff and *matériel*, to carry out their educational plans both in Paris and in the provinces' (*Nature*, 7 (13 February 1873), p. 290). Moigno's notice of ending his program is: 'Salles du Progrès', *Les Mondes*, 30 (6 February 1873), p. 193–4.

From Herbert Spencer 14 April 1873 4051

38 Queen's Gardens | Bayswater, W.[1] | 14 April 1873.

My dear Tyndall,

For security I send you a written reminder that you are engaged to me on the 22d at half-past seven.[2]

Dress interdicted & whoever comes late takes the consequences.

Sincerely yours, | Herbert Spencer

RI MS JT/1/S/101
RI MS JT/1/TYP/3/1117

---

1. *37 Queen's Gardens | Bayswater, W.*: Spencer's London residence since September 1866.
2. *you are engaged… half-past seven*: a reminder of an upcoming X Club meeting. On the X Club, see 3757, n. 2.

To the Editor of *Nature*[1] 15 April 1873 4052

[Reflected and Transmitted Light]

[The following correspondence has been sent us by Prof. Tyndall]:—

[…][2]

[…][3]

I detained this note until the arrival of sunshine, which enables me to say that the explanation here given is correct. Employing a blue medium, which does not transmit red you get no effect of the kind you describe. The tender leaves of spring are best suited for the experiment: the hard green leaves of ivy, for example, do not produce the effect.

It is not necessary, nor indeed desirable, to have the leaves between the eye and the sun.

April 15.

Nature, 7 (24 April 1873), p. 481

1. *Editor of Nature*: Norman Lockyer.
2. Tyndall copied letter 4037 here.
3. Tyndall copied letter 4040 here.

From Millicent M. Bence Jones[1] [mid-April 1873][2] 4053

84 Brook St[3] | Monday

My dear Dr. Tyndall

Papa[4] asked me to write & tell you that he is suffering such horrible pain from rheumatism in his ancle that he feels he would be quite unable to talk today. He is so very sorry to postpone your visit for there is nothing he enjoys so much as a talk with you, & he begged me to tell you how it vexed him to put off the pleasure of seeing you; he is in such bad pain that he could not talk.

He has enjoyed his talks with you so much since you returned,[5] nothing has pleased him so much since his illness.

Believe me dear Dr Tyndall | Yours very sincerely | Millicent M Bence Jones

RI MS JT/1/J/125
RI MS JT/1/TYP/2/718

1. *Millicent M. Bence Jones*: Millicent Mary Bence Jones (1843–1932), eldest child of Henry and Millicent Bence Jones.
2. *[mid-April 1873]*: the date is given since this letter falls between Tyndall's return to England (see n. 5 below) and the death of Henry Bence Jones on 20 April 1873. We place this letter in mid-April 1873, although it possibly could by late February or March.
3. *84 Brook St*: the Bence Jones's London residence.
4. *Papa*: Henry Bence Jones.
5. *since you returned*: Tyndall returned to London from his American lecture tour on 19 February 1873 (see Introduction, n. 8).

To Charles Darwin 16 April 1873 4054

Royal Institution of Great Britain | 16th April 1873

My dear Darwin.

Your excellent letter[1] reached me on Saturday last. You will doubtless set the matter before Huxley effectually in your own gentle way;[2] but for my part the idea of "entreaty" would not have occurred to me. It is his clear and bounden duty to do what we wish him to do: his duty to himself: his duty to his wife and children:[3] his duty to us and to the world.

Compare his case with that of Faraday, the most delicate minded man that ever lived, the most sensitive to the touch of anything questionable in the way of money. He accepted a pension of £300 a year.[4] Now I take it that we can measure Huxley's merits as well as the Prime Minister of the day could measure those of Faraday; and I also take it that what is done by his most intimate friends ought to raise less scruple in his mind than anything that could be done by the taxpayers of England.

I wish with you that Mrs. L.[5] had not subscribed—It suggests the idea of an effort, which ought to be entirely absent from the movement

Heartily yours | John Tyndall

GBR/0012/MS DAR 106: C13–C14[6]
RI MS JT/1/TYP/9/2825

---

1. *Your excellent letter*: letter 4043.
2. *the matter before Huxley effectually in your own gentle way*: on how to communicate to Thomas Huxley of the anonymous donation of funds that were pulled together by friends in order to give him a health-restoring holiday (see letter 4028, n. 4).
3. *his wife and children*: Huxley and his wife Henrietta had seven surviving children (see letter 3834, n. 12).
4. *pension of £300 a year*: Michael Faraday was awarded a civil list pension of £300 a year in 1836, and as the DCP notes, it 'was the subject of some political embarrassment because Faraday had at first refused the offer from the prime minister of the day, William Lamb (Lord Melbourne), who reportedly referred to such pensions as "humbug"' (*ODNB*; see also letter 861, William IV to The Commissioners of the Treasury, 9 December 1835, in F. A. J. L. James (ed), *The Correspondence of Michael Faraday*, vol. 2, 1832–1840 (London: Institution of Engineering and Technology, 1993), p. 315).
5. *Mrs. L.*: see letter 4043, n. 3.
6. This letter has been published in *Darwin Correspondence*, vol. 21, pp. 176–7; this is letter 8860 in the online edition of the Darwin correspondence.

From Thomas Gray[1] 16 April 1873 4055

Professor Tyndall F.R.S | Royal Institution | Albemarle Street

Sir

With reference to your letter on the 12th Instant[2] stating that you have suggested to the E.B. of the T.H.[3] that Mr. John Wigham sh[oul]d. accompany you on your proposed inspection of the Lights at Haisbro' on the 21st Inst[ant], I answer to acquaint you that this arrangement will be in every way agreeable to them

I am *[sir]*, | *[1 word illeg]* Ap[ril]. 16 | Signed by Mr Gray | 16.4.73[4]

National Archives, MT 10/220. *Illumination of lighthouses by gas*. H2075

1. *Thomas Gray*: Thomas Gray (1832–90), head of the Maritime Department of the Board of Trade from 1869 until his death ('Memoir of Mr. Thomas Gray, C.B.', *Transactions of the institution of Naval Architects*, 31 (1890), pp. 289–90).
2. *your letter on the 12th Instant*: see letter 4045, proposing the shipping-lights inspection and associated arrangements which Gray, as head of the relevant department at the Board of Trade, here approved.
3. *E.B. of the T.H.*: Elder Brethren of Trinity House (see letter 3736, n. 3 and n. 5). On Tyndall as scientific adviser to Trinity House, see letter 3869, n. 4.
4. *Signed by Mr Gray | 16.4.73*: in a different hand.

From Juliet Pollock 16 April [1873][1] 4056

59 Montague Square W.[2] | April 16th

Dear good wicked hard unbelieving true friendly rough tender Boreas[3] after this address comes my son Fred,[4] in whom you believe and all whose blossoms and leaves you honour.

It is all quite true, and in particular Clerk Maxwell's theory of electrical action was done into English (for the first time I believe) so

<Handwritten Letter begins>

as to be intelligible without high mathematics. I really think the lecture[5] was so contrived that any one with moderate training in elementary mathematics and tolerably clear mechanical ideas could follow it and get solid profit from it. I don't say that some people may not have gone away with a confused notion of electricity being like a bag of india rubber balls and small shot, and light like a row of iron bars with somebody hitting them at the end—but such people will exist and rejoice in invincible un-geometry to the end of time, and it can't be helped. If any thing could overcome invincible insensibility to scientific ideas it w[oul][d]. be such teaching as Clifford's. F. P.[6]

Henry Pollock[7] has writ[ten]. a note to Sir F[rederick].[8] expressing great admiration of this discourse of Cliff[ord]'s especially of the last half the Maxwell electricity half.—

RI MS JT/1/P/154
RI MS JT/1/TYP/6/2121

1. *[1873]*: the year is given based on the mention of a lecture of Clifford's (see n. 5).
2. *59 Montagu Square | W.*: the Pollocks' residence in Marylebone, London.
3. *Boreas*: Tyndall's and Juliet Pollock's nicknames for each other were 'Boreas' and 'Eolian'.
4. *my son Fred*: Frederick Pollock, who appears to have dictated his words to his mother Juliet in this letter.
5. *lecture*: William Kingdon Clifford delivered a Friday Evening Discourse at the RI on 28 March 1873, 'The Meaning of Force and Energy'. As Frederick Pollock noted some years later, 'By some accident no trace of it, not even the date or title, appears in the printed *Proceedings*' (see F. Pollock, 'Energy and Force', *Nature*, 22 (10 June 1880), pp. 122–4). James Joseph Sylvester also gave a discourse at the RI on 28 March 1873, 'On the Meaning of Force and Energy', and likewise this is not recorded in *Roy. Inst. Proc.*
6. *F. P.*: Frederick Pollock.
7. *Henry Pollock*: not identified.
8. *Sir F[rederick].*: William Frederick Pollock.

To Henry Bence Jones 17 April [1873][1] 4057

Royal Institution, 17th. April

My dear Friend

The earliest communication from you that I can find in my journal reached me on the 16th. of October 1852.[2] You ask me there to undertake the translation of 12 to 14 sheets of Liebig's Report.

On Sunday the 17th of October a letter from you[3] reached me stating that you had mentioned me to Barlow[4] as likely to give a good course of lectures,[5] and asking me to give a lecture on some Friday evening after Easter.[6]

A letter from Mr Barlow[7] reached me on the 24th asking me to fix some day in February for the Friday evening lecture.

On the 6th of January I called on you, but you were not at home.

You had asked me to dine with yourself: but on the 21st a letter[8] reached me saying that you had asked Dr Hofmann[9] to join us, but that Sir James Clarke[10] had asked you to go to him and talk with Playfair. You told him how you were circumstanced and then he asked you to bring Hofmann and myself with you. You mention also that you had asked Du Bois[11] to write about Animal Electricity for Churchill,[12] that he had declined, and then you asked whether I would undertake it.

I was not I find uncomfortable at Sir James Clarke's: I say of Lady Clarke[13] that she must have been a beautiful woman I say that I sat between Playfair and Sir James and that talk flowed on amicably and freely. Upstairs afterwards I signed Franklands certificate for the Royal Society.[14] One kind word of Hofmann's I have noted down. He said "We must have Tyndall here along with us—He is too far away from us." It is pleasant to meet these little touches of good will.

Your ever affectionate friend | John Tyndall

RI MS JT/1/TYP/3/796
LT Typescript Only

---

1. *[1873]*: the year is given since just before Henry Bence Jones died on 20 April 1873, he wrote some autobiographical notes in March and April which his family published much later in 1929. See *Henry Bence-Jones, M.D., F.R.S, 1813–1873. An Autobiography with elucidations at later dates* (Privately printed, 1929). Tyndall here writes of their earliest interactions, and one item made it into the autobiography, about John Barlow setting Tyndall up with a Friday Evening Discourse at the RI in February 1853 (p. 29).
2. *in my journal reached me on the 16th. of October 1852*: the letter from Bence Jones is dated 12 October 1853. Tyndall wrote it out into his journal (RI MS JT/2/6/159–60), and it was published as letter 0671 in *Tyndall Correspondence*, vol. 3.
3. *On Sunday the 17th of October a letter from you*: Tyndall may have been mistaken, as the letter he refers to was written, according to the RI typescript, on 19 October (it is possible, however, that the typescript is in error and should have been 17 October). Either way, the letter was published as letter 0674 in *Tyndall Correspondence*, vol. 3.
4. *Barlow*: John Barlow (1799–1869), an Anglican clergyman and secretary of the RI. Barlow studied at Trinity College, Cambridge, and was ordained in 1823. From 1842–60 he was the secretary of the RI, and together with Michael Faraday he undertook the day-to-day running of the institution. Barlow lived near the RI and entertained many of its wealthy and aristocratic supporters as well as fellow scientists, including Tyndall, who was a frequent guest (*Alum. Cantab.*).
5. *good course of lectures*: Tyndall delivered 'Four Lectures on Air and Water' at the RI in May and June 1853 (*Roy. Inst. Proc.*, 1 (1851–54), p. 276)
6. *give a lecture on some Friday evening after Easter*: Tyndall delivered a Friday Evening Discourse at the RI on 11 February 1853, 'On the influence of Material Aggregation upon the manifestations of Force' (*Roy. Inst. Proc.*, 1 (1851–54), pp. 254–9).

7. *A letter from Mr Barlow*: Barlow's letter is dated 21 October 1852, and was published as letter 0675 in *Tyndall Correspondence*, vol. 3.
8. *letter*: this letter is dated 17 January 1853, and was published as letter 0704, in I. Hesketh and E. Sera-Shriar (eds), *The Correspondence of John Tyndall, Volume 4: The Correspondence, January 1853–December 1854* (Pittsburgh: University of Pittsburgh Press, 2018).
9. *Dr. Hofmann*: August Wilhelm Hofmann (see letter 3776, n. 15).
10. *Sir James Clarke*: James Clark (1788–1870), a British physician and Physician-in-Ordinary to Queen Victoria from 1837–60 (*ODNB*).
11. *Du Bois*: Emil du Bois-Reymond.
12. *write about Animal Electricity for Churchill*: while du Bois-Reymond declined this offer, Bence Jones had previously published a selection of du Bois-Reymond's works. See H. Bence Jones (ed), *On Animal Electricity: Being an Abstract of the Discoveries of Emil du Bois-Reymond* (London: John Churchill, 1852).
13. *Lady Clarke*: Barbara Clark (née Stephen, d. 1862), married in 1820.
14. *Franklands certificate for the Royal Society*: Edward Frankland was elected FRS on 25 January 1853, and his certificate bears the signatures of James Clark, Michael Faraday, August Wilhelm Hofmann, Henry Bence Jones, Lyon Playfair, and John Tyndall, among others (RS, Certificates of Election and Candidature, EC/1853/06).

To unidentified 17 April 1873 4058

Royal Institution of Great Britain | 17th April 1873

My dear sir

Most willingly would I comply with any request of yours—but I find as I grow older the multiplication of small demands upon my time tending to take all dignity & wholeness of purpose out of life, reducing it in fact to driblets.

This is especially my case at present—surrounded as I am by new labor alone which need filling up.[1] I am forced therefore to sternly limit my out goings—and though applied to from time to time by friends whom it would be a pleasure to me to oblige I am forced to ask them to extend their tolerance to me when I decline their invitations.

I am sorry to say that I must now request this tolerance on your part.

Faithfully yours, | John Tyndall

Wellcome Library MS. 7777/16

1. *new labor alone which need filling up*: some of Tyndall's commitments at this time were preparation for a lecture course at the RI (see letter 3782, n. 10), preparing *Six Lectures on Light* for publication, and lighthouse work for the Board of Trade (see letter 4045).

To Thomas Archer Hirst 18 April 1873 4059

Royal Institution of Great Britain | 18th April '73

My dear Tom.

I want you & Huxley here to meet Emerson at luncheon[1] some day next week. What days are you at liberty?

Yours aff[ectionate]ly | John Tyndall

RI MS JT/1/T/701
RI MS JT/1/HTYP/620

1. *at luncheon*: writing to Emerson's daughter Ellen four days earlier, Tyndall had expressed his hope of hosting a luncheon that would bring together the visiting poet-philosopher with two 'old disciples', Thomas Hirst and Thomas Huxley (see letter 4048).

From Charles Darwin 18 April [1873][1] 4060

Down, | Beckenham, Kent. | April 18.

My dear Tyndall,

I write a line to say that the total amounts to £1755,[2] Lubbock having given £200. This will be paid in to-morrow and I will write to-morrow.[3] Huxley does not return till Saturday or Monday night. I tremble about his answer[4]— when I definitely hear, I will send a "private circular" to the 18 subscribers,[5] telling the total amount sent, with no other details, and H[uxley]'s answer, either in extenso or in abstract according to its nature. Whatever H[uxley]. may think (and I quite agree with what you say in your note)[6], it is a pleasure to think what a relief it will be to Mrs H.[7]

My dear Tyndall | Ever yours most truly | Ch. Darwin.

RI MS JT/1/TYP/9/2826
LT Typescript Only[8]

1. *[1873]*: the year has been established by the DCP editors in light of the letter's relationship to Tyndall's of 16 April (letter 4054).
2. *total amounts to £1755*: this was not the final total (see n. 5).
3. *I will write to-morrow*: see letter 4043, n. 4.
4. *I tremble about his answer*: Darwin was unsure whether Thomas Huxley would accept the donated funds that a circle of friends—eighteen in all—had pulled together to enable him to take a health-restoring holiday (see letter 4028, n. 4). Huxley's answer to Darwin, dated 24 April 1873, began: 'Your most kind and affectionate letter, with its wholly unexpected,

I may say unimaginable, announcement fairly overwhelmed me when it arrived last night (along with a notification from my banker) and I have spent many sleepless hours in wondering what I have done to make my friends care so tenderly & thoughtfully for my welfare'. Referring to the purpose of the donated funds, he wrote: 'I shall go & take a long holiday in the summer now without feeling that I am potentially guilty of fraud—and when my familiar blue-devil dances about me (as I daresay he will for some time yet) I shall shy the cheque for £2100 at his head as Luther did the inkstand' (*Darwin Correspondence*, vol. 21, pp. 184–5; this is letter 8876 in the online edition of the Darwin correspondence).

5. *"private circular" to the 18 subscribers*: this is dated [25 April 1873] and published in *Darwin Correspondence*, vol. 21, p. 189; this is letter 8876 in the online edition of the Darwin correspondence. It read, in part: 'The sum of £2150, subscribed by 18 of Prof. Huxley's friends, was paid into his account on April 23rd. & 25th. On April 23rd., Mr. Darwin wrote to inform Prof Huxley of the fact, & he has this morning received an answer, of which the enclosed is a copy'. Charles Lyell had given an additional £50 (see letter 4067).
6. *your note*: letter 4054.
7. *Mrs. H*: Henrietta Huxley.
8. This letter has been published in *Darwin Correspondence*, vol. 21, p. 179; this is letter 8873 in the online edition of the Darwin correspondence.

To Charles Darwin 21 April 1873 4061

Royal Institution of Great Britain | 21st April 1873.

My dear Darwin.

Whitworth has sent us £150[1]—So that we have got clear over the 2000—I have sent the money on to Lubbock who will do what is right with it.

Don't be fainthearted—no purer aim or object ever animated a man than that which animates you.

Kind regards to Mrs. Darwin[2]

ever & always yours | John Tyndall

Huxley has returned.[3] I wish I could persuade you to join Emerson and Huxley at luncheon[4] here on Wednesday

GBR/0012/MS DAR 106: C15[5]
RI MS JT/1/TYP/9/2828

1. *Whitworth has sent us £150*: this contributor to the fund to give Thomas Huxley a health-restoring holiday has been identified by the DCP as Joseph Whitworth. See letter 4028, n. 4.
2. *Mrs. Darwin*: Emma Darwin (1808–96), wife and first cousin of Charles Darwin, married in 1839. Her grandfather was Josiah Wedgwood of the famous Wedgwood pottery company (*ODNB*).

3. *Huxley has returned*: from Aberdeen (see letter 4041, n. 1).
4. *luncheon*: writing to Emerson's daughter Ellen a week earlier, Tyndall had expressed his hope of hosting a luncheon that would bring together the visiting poet-philosopher with two 'old disciples', Thomas Hirst and Thomas Huxley (see letter 4048).
5. This letter has been published in *Darwin Correspondence*, vol. 21, p. 182; this is letter 8870 in the online edition of the Darwin correspondence.

To Ellen Tucker Emerson — 21 April [1873][1] — 4062

Royal Institution of Great Britain | Monday 21st. Ap[ri]l.

My dear Miss Emerson

I hope that you and your excellent father[2] will be able to come here to the Royal Institution to take luncheon with me on Wednesday next at one o'clock precisely.

Huxley has just returned from Aberdeen[3]—He will be delighted to meet you.

I have just lost one of my most intimate friends; Dr. Bence Jones,[4] the Honorary Secretary of the Royal Institution. Such occurrences make the world appear a phantasm.

Yours ever faithfully | John Tyndall

[n.p.] 21 Apr [1873], 1873. Emerson family correspondence, MS Am 1280.226, (4124). Houghton Library, Harvard College Library

---

1. *[1873]*: the year is established in light of the previous correspondence about the upcoming luncheon that Tyndall was hosting for Ralph Waldo Emerson, in the company of Thomas Huxley and Thomas Hirst (see letters 4048, 4059, and 4061).
2. *your excellent father*: Ralph Waldo Emerson.
3. *has just returned from Aberdeen*: see letter 4041, n. 1.
4. *I have just lost... Dr. Bence Jones*: Henry Bence Jones died on 20 April 1873. Tyndall saw the dying man on 8 April (see letter 4039), showing him the 'Tyndall number' of the *Popular Science Monthly* (see letter 4033, n. 16). Bence Jones had been well enough not only to share his prediction of great things for America scientifically but, unbidden, to contribute £50 to the Huxley holiday fund (see letters 4028, n. 4).

From Millicent Bence Jones 21 [April 1873][1] 4063

84 Brook St[2] | Monday 21st

My dear Dr Tyndall

My first feeling on reading your note[3] was that of regret that he could not see it[4]—but that bitterness must now be mingled with everything—and vainly one longs that all the expressions of sympathy and sorrow for his loss could reach him

I thank you most truly for your kindness in writing and the sympathy of a friend of his so truly valued and so long—is gratefully fẹlt by me—

Your brother's feeling towards him was fully returned by him and none of the visits of his many kind friends gave him I think so much pleasure as your's—when he could see no one else, he always excepted you—The end has come so suddenly I cannot yet realize that all we have of late been living for is gone and it is difficult to face the sad future

Thank you again for your words of comfort and also for your kind wish to be of some use to us at this sad time—

I shall indeed feel that if I should need help or advice I cannot apply to a truer friend than the one who has so loved him—

Ever yours sincerely | & gratefully | M Bence Jones.

RI MS JT/1/J/127
RI MS JT/1/TYP/2/736

1. *[April 1873]*: the month and year are given based on when Henry Bence Jones died.
2. *84 Brook St*: residence in London.
3. *your note*: letter missing.
4. *he could not see it*: Henry Bence Jones died the day before, 20 April 1873.

To Thomas Archer Hirst 23 April 1873 4064

April 23rd/1873

Dear Tom: you have doubtless had my telegram.[1] Be here a little before me tomorrow—Emerson & his daughter[2] will be here.[3]

Yours aff[ectionate]ly | John Tyndall

RI MS JT/1/T/998

1. *my telegram*: telegram missing.
2. *his daughter*: Ellen Tucker Emerson.
3. *will be here*: for a luncheon in which Thomas Huxley was also invited (see letters 4048, 4059, 4061, and 4062).

From Millicent Bence Jones — 23 April [1873][1] — 4065

21 Up[per]: Berkeley St[2] | April 23—

My dear Dr Tyndall

I found your note[3] in Wimpole Street[4] this afternoon & write at once to say that either this week or the next—& any time after as long as this continues to be the house of his children[5] you will find a welcome here—I have shrunk indeed from seeing friends in general—even if the several moves we have had to make from Wimpole St to Lodgings & then finally to this house—had not prevented my doing so by giving the girls & myself more to do than we had time or strength for—but we are getting settled—& it will be a pleasure however saddened to see you again. Any hour or day that you would name you would be certain to find us—but we are always at home by five o'clock—

I should be glad to speak to you alone for a few minutes on the subject of your note—

It was a suggestion[6] in a letter from Dr du Bois Reymond[7] that made me mention the subject in the beginning of the winter to you more as consulting you about it than as expressing a wish of my own—My object was that his wishes should be carried out & I thought you of all his friends the most likely to know what they were. If I knew for certain that he had no desire for it—I should myself prefer his living in the memories & hearts of those who knew him best—rather than to have him talked over by those who could not judge or understand him—but as you are kindly coming some day—I need not trouble you with a longer note now—Believe me always—| Your's very sincerely | M Bence Jones

RI MS JT/1/J/120
RI MS JT/1/TYP/3/741

1. *[1873]*: the year is given by reference to the death of Henry Bence Jones, on 20 April 1873.
2. *21 Up[per]: Berkeley St*: one of the Bence Jones's London residences.
3. *your note*: letter missing.

4. *Wimpole Street*: one of the Bence Jones's London residences.
5. *house of his children*: for the Bence Jones children, see letter 3860, n. 6.
6. *suggestion*: not identified. Possibly related to the autobiographical notes that Henry Bence Jones worked on in the months before his death (see letter 4057, n. 1). Tyndall wrote a letter providing some details of their earliest interactions (4057), and perhaps du Bois-Reymond had done the same or at least was asked. Or, this may have to do with efforts to have du Bois-Reymond made ForMemRS (see letter 4090).
7. *Dr du Bois Reymond*: Emil du Bois-Reymond.

From Margaret Ginty 23 April 1873 4066

220 Marylebone Road | April 23rd—73

Dear Dr. Tyndall,

I have just finished burning old letters (which were treasures to me) and it now only remains for me to send a few lines to you. It may be to ask a last favour. If it is in your power at any time to say a kind word for my boys,[1] will you say it for my sake, you know, a kind word from you goes a long way. Do dear Dr. Tyndall speak kindly for my gentle Willie[2]—poor stormy George[3] and even poor little Fred[4]—Oh, I have forgiven him his mistakes long ago and taken him back quiet close to my heart, you will despise me for this weakness, but I cannot help it, what is the use of love if we turn away from our friends in trouble—and Freddy is my own boy. He left me to go to South America six months ago—feeling deeply repentant for all the pain he gave me. I am sure he will now succeed—A few words for myself—Ever since I came to England I have been very troublesome to you, I often went to see you when I think it would have been wise to stay away, but I had many perplexities and my state of health often made me look at things in a wrong light. But a few words from you always gave me strength and courage. Before I left Rio,[5] I had yellow fever, which left a disease behind, steadily increasing ever since. It has lately developed to such an extent that I am obliged to undergo an operation at once (to-morrow) to save my life. If the Drs. succeed they promise me perfect health. Then I will feel stronger than ever, and so independent—perhaps, that I might never give you the least trouble. If I die—and I may—I will be where there is no more tears or sorrow—and that terrible struggling and striving. I can tell my dear husband[6] how kind and good you were to me. Oh! if I was only certain of meeting him how little I would care for all the doctors, but nobody seems to know any thing you will forgive all the trouble I have given you, won't you?

and believe | dear Dr. Tyndall | that I am very very grateful | M. Ginty.

RI MS JT/1/TYP/11/3673
LT Typescript Only

---

1. *my boys*: her adult sons Frederic, George, and William Roberts Ginty, whose prospects in life she was surely concerned about.
2. *my gentle Willie*: William Roberts Ginty.
3. *poor stormy George*: George Ginty.
4. *poor little Fred*: Frederic Ginty.
5. *left Rio*: Ginty and her children returned to England following the death in 1866 of her husband William Ginty, a surveying-colleague friend of Tyndall's. Along with Margaret and their growing family—eventually comprising three sons and two daughters—Ginty had moved in 1854 to Rio de Janeiro, where he died in 1866. On Tyndall serving as informal 'Trustee' for Margaret and her children long after William's death, see *Ascent of John Tyndall*, p. 198. See also letters 3766, 3849, 3858, 4005, and 4006.
6. *my dear husband*: William Ginty.

From Charles Darwin 25 April [1873][1] 4067

Private | April 25

My dear Tyndall,

Is not this a charming letter?[2] Ought the circular and copy to be sent to Lady Millicent Jones?[3] I should doubt it under present circumstances; but you will know by far best. If it is sent perhaps it had better go to her last.

I am so happy about the whole affair

In haste | Ever yours | C. Darwin.

Poor dear Sir C. Lyell sent in this morning an additional £50.[4]

RI MS JT/1/TYP/9/2827[5]
LT Typescript Only

---

1. *[1873]*: the year has been established by the DCP in light of the reference to the circular concerning Thomas Huxley's accepting, as he wrote to Darwin on 24 April, 'the splendid gift you & my other friends offer, frankly and in the spirit of the givers', adding that by this acceptance Huxley pledged himself 'to make the best I can of this cranky frame of mine and get it in order for the best and highest kind of work of which I am capable'. For the circular and Huxley's reply to Darwin, see n. 2. See also letter 4028, n. 4.
2. *a charming letter*: enclosed with the 25 April circular (see letter 4060, n. 5) was a copy of Huxley's 24 April letter to Darwin (see letter 4060, n. 4).
3. *Lady Millicent Jones*: Millicent Bence Jones, wife of the recently deceased Henry Bence Jones who pledged money to Huxley.
4. *Poor dear Sir C. Lyell... an additional £50*: Charles Lyell's wife had just died, and his

contribution arrived too late to be included in the sum gifted to Huxley. Mary Elizabeth Lyell (née Horner, 1808–73) was a geologist, conchologist, and wife and scientific partner to Charles Lyell (see M. Ogilvie and J. Harvey (eds), *Biographical Dictionary of Women in Science*, 2 vols (London: Routledge, 2000), vol. 2, pp. 813–14).

5. This letter has been published in *Darwin Correspondence*, vol. 21, p. 189; this is letter 8877 in the online edition of the Darwin correspondence.

From Henrietta Huxley 25 April [1873][1] 4068

4 Marlborough Place | Abbey Road, N.W.[2] | 25th April

Dear Brother John

You are one of the kind conspirators who have plotted that Hal[3] should have the rest he so much wants,[4] and I cannot resist writing a line to tell that my heart is very full of grateful love to you. Shakespeare's birthday[5] will not easily be forgotten by us. I felt as if my treasure had been given back to me—by reason of the rest, that is now placed within his reach—whether he will or not.

We wanted so much to have seen your dear old face at dinner here on the 5th by way of keeping Hal's birthday of 4th[6] but Leonard[7] is laid up with the chicken pox, and at any moment I may have 2 or 3 more ill[8]—added to the confusion of the house—because I cannot get certain workpeople to complete their work. So that I have put off keeping the birthday for a time.

Ever—with love | Your true sister | Nettie Huxley.

RI MS JT/1/TYP/9/2983
LT Typescript Only

---

1. *[1873]*: the year is given by a note from LT on the typescript, and that in 1873 the Huxley children came down with chicken pox (see n. 8).
2. *4 Marlborough Place | Abbey Road, N.W.*: see letter 3947, n. 1.
3. *Hal*: a nickname for Thomas Huxley.
4. *one of the kind conspirators... rest he so much wants*: Tyndall helped to organize a donation fund—from eighteen friends of Huxley's in all—to enable him to take a health-restoring holiday. See letter 4028, n. 4.
5. *Shakespeare's birthday*: English playwright William Shakespeare was born in April 1564, and while the actual day is not recorded, it is traditionally celebrated on 23 April.
6. *Hal's birthday of 4th*: Thomas Huxley was born on 4 May 1825.
7. *Leonard*: Leonard Huxley.
8. *may have 2 or 3 more ill*: in an 8 May 1873 letter to his doctor Andrew Clark, Thomas Huxley shared that his wife and six children (see letter 3834, n. 12) had all come down with chicken pox (Thomas Henry Huxley Papers, American Philosophical Society).

From Charles Darwin 25 April 1873 4069

April 26th 1873 | Down Beckenham.

My dear Tyndall,

I am very sorry to give you trouble; but it has just occurred to me that I ought not to leave a copy of H[uxley]'s confidential letter to me[1] in the hands of any one. As I do not know to whom the copies were sent last I cannot write myself which I would otherwise have done. Will you therefore be so kind as to write a line to the two men who will ultimately receive the two copies which were sent to you and ask them to be so good as to return them to "C. Darwin Down Beckenham. Kent"

Believe me | My dear Tyndall | Yours very sincerely | Ch. Darwin.

RI MS JT/1/TYP/9/2830[2]
LT Typescript Only

---

1. *copy of H[uxley]'s confidential letter to me*: for Huxley's 24 April reply to Darwin's letter regarding the donation of funds from friends for him to be able to take a health-restoring holiday, see letter 4060, n. 4. On the effort to collect these donations, see letter 4028, n. 4.
2. This letter has been published in *Darwin Correspondence*, vol. 21, p. 190; this is letter 8879 in the online edition of the Darwin correspondence.

From Charles Darwin 25 April [1873][1] 4070

Ap[ril]. 25 | Evening

My dear Tyndall,

Here is another copy which you can send off to any one of your 3 names

Yours—| C. Darwin

RI MS JT/1/TYP/9/2829[2]
LT Typescript Only

---

1. *[1873]*: the year has been established by the DCP in light of the previous correspondence on the acceptance letter circulating among the subscribers to the donation fund for giving Thomas Huxley a health-restoring holiday (see letter 4067, n. 1). See also letter 4028, n. 4.
2. This letter has been published in *Darwin Correspondence*, vol. 21, p. 190; this is letter 8878 in the online edition of the Darwin correspondence.

To Thomas Archer Hirst 26 April 1873 4071

26th. April. /73

My dear Tom

Huxley has accepted in the most charming manner.[1] I ought to have forwarded you his letter to Darwin,[2] but take the liberty of deflecting it to Armstrong & Whitworth.

I shall have it back & then you can read it.

Yours aff[ectionate]ly | John Tyndall

RI MS JT/1/T/702
RI MS JT/1/HTYP/620

1. *Huxley has accepted in the most charming manner*: Huxley accepted the donation of funds collected by eighteen of his friends to give him a health-restoring holiday (see letter 4028, n. 4).
2. *his letter to Darwin*: for Huxley's 24 April reply to Darwin's letter regarding the donation of funds, see letter 4060, n. 4. Although Darwin subsequently changed his mind (see letter 4069), he sent Tyndall two copies of Huxley's letter for circulation to the subscribers of the fund, which included, among others, William Armstrong, Thomas Hirst, and Joseph Whitworth (for the full list of subscribers, see letter 4028, n. 4).

To Henrietta Huxley 26 April 1873 4072

Royal Institution. | 26th April 1873.

Dear True Sister

Your love is to me more precious than gold and the gratification Hal's behaviour in this matter has given me is not to be described.[1] I knew he would act so—and I said he would act so, but the actual verification of what I thought and said is profoundly pleasant to me. He must now take rest.

The thing was simmering in many minds before it took definite shape, and dear old Spencer was one of the very first to put it into form.[2] But it <wo>uld have done your soul good to see the true brotherly heartiness with which the thing was done. Dear old Darwin said it made him think better of human nature than he had previously done.[3] The man is half perverted by it, being on the point of abandoning the Ascidian theory.[4]

Your letter[5] is delightful to me; for I like to see your mind bright, and it is evidently now in this condition.

And now goodbye | and never doubt me to be | Your affectionate Brother | John.

RI MS JT/1/TYP/9/2983A
LT Typescript Only

1. *the gratification... to be described*: Tyndall was expressing his pleasure at the grace with which Henrietta's husband Thomas Huxley ('Hal') had accepted his friends' gift of funds so that he could take a much-needed health-restoring holiday (see letter 4028, n. 4).
2. *Spencer was one of the very first to put it into form*: see letter 4032.
3. *Dear old Darwin... previously done*: after the meeting at Darwin's London residence on the morning of 8 April where the holiday-donation plan was agreed, Darwin wrote to Tyndall that the meeting 'will give me a higher opinion of human nature for the rest of my life than I ever had before, though I am not one of those who think lowly of mankind' (see letter 4036).
4. *The man is... Ascidian theory*: exaggerating for comic effect, Tyndall here suggested that the magnanimity of Huxley's friends was so great that it had made Darwin reconsider his notorious claim (in *The Descent of Man, and Selection in Relation to Sex*, 2 vols (London: John Murray, 1871), vol. 1, pp. 204–6) that humankind traces back evolutionarily to a worm-like ancestor. On the Amphioxus-ascidian theory of vertebrate origins, see J. Maienschein, '"It's a long way from Amphioxus": Anton Dohrn and late 19th-century debates about vertebrate origins', *History and Philosophy of the Life Sciences, 16* (1994), pp. 465–78.
5. *Your letter*: letter missing.

From Joseph Henry[1] 26 April 1873 4073

Smithsonian Institution, | Washington, April 26. 1873.

My dear Professor Tyndall

Your letter of April 6th[2] has just been received and to save time and labor I dictate the following to an assistant.

The money of your fund[3] has been paid over to the Trustees by Mr Appleton and through the kind agency of Mr Joseph Patterson of Philadelphia[4] (with whom you dined[5]) has been temporarily invested; with the exception of between seven and eight hundred dollars now in charge of Gen[era]l. Tyndale;[6] in bonds yielding 7. per cent in currency. The money in Gen[era]l. Tyndale's hands will with the interest on the bonds be sufficient by the first of September to purchase another bond making, in all, thirteen one thousand dollar bonds which are a little above par and therefore represent something more than $13.000.

General Tyndale is very anxious to have the fund increased. This might have been readily done at the time the money was left by you had the matter been brought before the public at the Banquet at Delmonicos;[7] but I doubt whether much could be effected at present in the way of enlarging the fund without great exertion, since most persons who have the power of giving desire the results to appear in their own names. The income of the fund as it is, however, will do much good. We have already had several applicants for assistance from it and one of these I think is of the right character, as the young man has evinced originality in the laboratory at the scientific school at Yale[8] and is warmly recommended by the Professors.

I am sorry to learn that it is probable your time will be absorbed and your thoughts distracted by controversy. This ought not to be; yet in defence of yourself and of truth all other considerations must be disregarded.[9]

To be called on to preside at the meeting of the British Association[10] is an honor which you cannot well decline; yet it is to be regretted that anything should occupy your thoughts even to the temporary exclusion of the subject of your investigations. It is only by continued and uninterrupted thought that the mind assumes that condition in which it *[teems]* with new suggestions and is *[active] [in]* way of invention for their verification.

I have just been dipping into Maxwells work on Electricity and Magnetism[11] but as yet have not been able to see in what manner the idea of an intervening medium[12] can assist in the way of a working hypothesis unless clear conceptions are given of the nature and mode of operation of such medium. Thus far I have never been able to get a conception of any medium which shall be an advance on the principle of action at a distance which as in the case of gravitation is instantaneous.

In the phenomena of light in which transmission in time through space is exhibited a medium consisting of matter, that is, something obedient to the laws of force and motion including inertia, is required. But even the constitution of such a medium demands the existence of action at a distance between its particles in order to explain the high elasticity. But this may be on account of the difficulty I experience at my age in entirely changing views long since adopted.

Mr. Childs of Philadelphia,[13] has presented me with a portrait of yourself[14] which is an excellent specimen of the art of photography and the artist, Mr. Gutekunst,[15] has presented another to the Institution.[16] The latter occupies a prominent place on the walls of my office; the former will be displayed in the parlor of my house.

I leave Washington and the cares of the Smithsonian Institution next week to devote myself for some days to an inspection of the annual supply of oil and the various projects that have been presented to the Light House Board[17] during the past year. I look forward to the visit as a pleasant change in the way of breaking up for a while the monotony of Smithsonian duties.

Major Elliot[18] of the Light House Board leaves Washington on the 30th of this month on a tour of examination of Light Houses in Europe. The object of his mission is to collect such facts as may tend to improve the Light House system of the United States.

Mrs. Henry[19] has lately received a very kind letter from Sir Frederic Arrow[20] who has added to our pleasure by sending us a copy of his photograph. I intend giving him an account of the changes which have taken place in the Light House Board since his visit and to keep alive his interest in our affairs.

I am, | Very truly yours, | Joseph Henry

Prof[esso]r John Tyndall, | Royal Institution, | London, | Eng[lan]d.

Smithsonian Institution Archives. RU 33, Vol. 33, Pages 722–6

---

1. This letter has previously been published as document 212 in *Papers of Joseph Henry*, pp. 454–7.
2. *your letter of April 6th*: letter 4033.
3. *money of your fund*: see Introduction, n. 15.
4. *Mr Joseph Patterson of Philadelphia*: see letter 3888, n. 3.
5. *with whom you dined*: see letter 3888, n. 4.
6. *Gen[era]l. Tyndale*: Hector Tyndale.
7. *Banquet at Delmonicos:* the farewell banquet in New York on 4 February 1873 (see Introduction, n. 23).
8. *scientific school at Yale*: the Sheffield Scientific School (see letter 3968, n. 3).
9. *I am sorry... must be disregarded*: Henry was here responding to Tyndall's report in his 6 April letter about how his recent book *Forms of Water* had rekindled the old dispute with Peter Guthrie Tait (see letters 4024, 4033, and 4038).
10. *To be called on... British Association*: Tyndall was hesitant to accept the Presidency of the BAAS for 1874 (see letter 4013, n. 2).
11. *Maxwells work on Electricity and Magnetism*: J. C. Maxwell, *Treatise on Electricity and Magnetism*, 2 vols (Cambridge: Cambridge University Press, 1873).
12. *intervening medium*: Maxwell famously posited that light waves were electromagnetic waves travelling in a medium, the ether.
13. *Mr. Childs of Philadelphia*: not identified; possibly the Philadelphia publisher George W. Childs (1829–94) (*ANB*).
14. *portrait of yourself*: see letter 4014, n. 4.
15. *The artist, Mr Gutekunst*: Frederick Gutekunst (see letter 4014, n. 4).
16. *Institution*: Smithsonian Institution.
17. *devote myself for... Light House Board*: Henry served on the American Light-House Board since 1852, and became chair in 1871.
18. *Major Elliott*: George Elliott (1831–1900), an American engineer with the American Light-House Board, and author of *European Light-House Systems: Being a Report of a Tour of Inspection Made in 1873* (London; Lockwood, 1875); see *Ascent of John Tyndall*, p. 318.

19. *Mrs. Henry*: see letter 3869, n. 7.
20. *Sir Frederic Arrow*: Frederick Arrow was on a trip to the United States at the same time as Tyndall.

To George Back[1] 27 April 1873 4074

27th April 1873

My dear Sir George

Count on me for Tuesday the 20th of May[2]—It is very kind of you indeed to look so long in advance and to give me a choice of days.

Yours ever | John Tyndall

Letter, John Tyndall to George Back, 27 April 1873. (SGB/1/79. Royal Geographical Society-Institute of British Geographers)

1. *George Back*: George Back (1796–1878), a British naval officer and Arctic explorer; he was a member of the Council of the Royal Geographical Society, having previously served as Vice President (*ODNB*).
2. *Count on me for Tuesday the 20th of May*: while this is not identified for certain, it most likely was to meet for lunch or dinner.

To Thomas Archer Hirst 29 April 1873 4075

Royal Institution | 29th April 1873

My Dear Tom.

I want you to do me a favour and a service. Huxley has been looking at the Life of James Forbes by Messrs Sharp, Tait, and "Reilly".[1] He says they make an "onslaught" upon me.[2]

I am very busy preparing my American lectures for publication,[3] & would not willingly swerve just now from this good work—I think it will be good work when completed—What I want you to do for me is to purchase and look over the portion of this volume[4] which relates to me,[5] and ascertain whether there is any part of it which calls for immediate attention. The mischievous remark Huxley is going to take in hand—it seems they make out him & me to be fellow conspirators against the peace & fame of Forbes. He will dispose of this charge.[6]

The volume is at the Athenaeum[7]—but I shall need a copy to send to Agassiz[8] who I doubt not will deal with a portion of it in a manner that my assailants little anticipate. Therefore I would ask you to purchase the volume & look over it at home.

It will not I trust take you more than a couple of hours. I grudge even this invasion upon time.

Yours ever | John Tyndall

RI MS JT/1/T/703
RI MS JT/1/HTYP/620

1. *Life of James Forbes by Messrs Sharp, Tait, and Reilly*: this much-anticipated volume—*Life and Letters of Forbes*—had at last been published. See letters 4033 and 4038.
2. *He says... upon me*: writing to Charles Darwin on 26 April, Huxley had indeed written that in the new volume on Forbes 'there is a regular onslaught on Tyndall' (see *Darwin Correspondence*, vol. 21, pp. 190–2; this is letter 8882 in the online Darwin correspondence).
3. *my American lectures for publication*: Tyndall sent the English edition of his lectures on light to the press in June; see 'Preface to the English Edition' in *Six Lectures on Light*, p. v.
4. *this volume*: *Life and Letters of Forbes*.
5. *which relates to me*: in letter 4033 to Joseph Henry, Tyndall wrote: 'Prof. Tait I am told has taken deep offence at my references to Forbes, and intends to call me to account in a Life of Forbes on which he is now engaged'. See also letter 4038.
6. *He will dispose of this charge*: Huxley defended Tyndall in T. H. Huxley, 'Forbes and Tyndall', *Nature*, 8 (22 May 1873), p. 64.
7. *Athenaeum*: the Athenaeum Club (see letter 3710, n. 5).
8. *Agassiz*: see letter 4038 for Tyndall's putting Louis Agassiz on notice about the looming controversy.

From Mary Egerton [April 1873][1] 4076

[45, Eaton Place, | S.W.][2] | Saturday

My dear Mr Tyndall,

I dreamt last night that you were explaining to me the sewing Machine!! and that you said you were going to start for Vienna next Monday!

But, (except it seems in my dreams!) it is just as I said it w[oul]^d^ be, I know less about you now, than I did when you were in America.[3] It was so pleasant in following your progress by the papers you sent me, (do you know you sent 18, if not more; but I kept 18 *[1 word illeg]* to count; and I think that was very good of you!) as well as hearing of you from Mr Hirst & his papers. I learnt to know him too, & the deep feeling & tenderness that lies under his calm exterior. I shall not soon forget his anxiety about you at the time of poor young Miller's death,[4] nor his longing for your return. How much you must miss his constant vicinity! Is he able to be much in London?

I have always wondered what made him write to tell me of your arrival in America? Did you ask him to do it?

If you are not staying the Evening at Kew, would you be disposed to look in tomorrow Evening as soon after 1/2 past 8 as you like? You would find only ourselves[5] & Hugh,[6] & a Cousin[7] who is coming to see him. I despair of a real good chat! but this w[oul]d be better than nothing. No need to answer about this.

We are going to the Vaudeville, (School for Scandal)[8] tonight. Suppose you come! Our stalls are 8 to 11 row B; it w[ould]d be fun if you got behind or before!

We go down to my brother's at Thoresby[9] next Saturday for a week or so. Let me hear if you go anywhere—I like to know where to place my friends when I think of them! Niagara[10] was delightful and that sand process most curious. one c[oul]d not have believed it w[oul]d have such power! Does any common sand do for it?

I hope to take back y[ou]r. picture today. I c[oul]d. not fetch it on Thursday.

Ever y[ou]r attached friend | M F Egerton

RI MS JT/1/E/86

---

1. *[April 1873]*: the month and day are given by reference to Tyndall's Friday Evening Discourse about Niagara Falls at the RI on 4 April 1873 (see n. 10). We recognize that it could also be May or June 1873, but do not believe it would be any further than that based on the lack of discussion regarding summer travels.
2. *45, Eaton Place, | N.W.*: in London.
3. *when you were in America*: for Tyndall's American lecture tour, see Introduction, n. 8.
4. *poor young Miller's death*: see letter 3896, n. 10.
5. *ourselves*: not her husband as he died in 1869, so it is unclear as to who she means by 'ourselves'.
6. *Hugh*: Hugh Edward Egerton (1855–1927), the second son of Mary Egerton and Edward Christopher Egerton. Educated at Oxford, Hugh became a civil servant and historian, and was the first Beit Professor of Colonial History at Oxford (1905–20) (*ODNB*). Their other son was Charles Augustus Egerton (see letter 4089, n. 4).
7. *Cousin*: not identified.
8. *Vaudeville, (School for Scandal)*: at London's Vaudeville Theatre, Richard Brinsley Sheridan's comedy *The School for Scandal* opened on Thursday, 18 July 1872, and played continuously up to Friday, 31 October 1873.
9. *my brother's at Thoresby*: Thoresby Park, Ollerton, a small town in Nottinghamshire, England, was the estate of Egerton's brother Sydney William Herbert Pierrepont (1826–1900), 3rd Earl Manvers, a British nobleman and politician.
10. *Niagara*: Tyndall delivered a Friday Evening Discourse, 'Some Observations on Niagara', at the RI on 4 April 1873 (see letter 3738, n. 3).

From Mary Egerton [April 1873][1] 4077

[45, EATON PLACE, | S.W.][2] Tuesday

Dear Mr Tyndall,

I hope you won't be very angry but I want to ask you a great favour. Do you think you could let Rose Kingsley[3] have an order for your lectures on Light,[4] for I know that she is extremely anxious to go to them but as she will only be able to manage some, I don't know whether she would afford herself the luxury of a ticket for the set. Of course I have said nothing about it to her, only I thought I would just ask you & please don't take any notice of my cool[5] request if it is against the rules or bothers you in any way. I assure you I am writing in fear & trembling for I do so hate asking favours, only I know you will be good & not mind, whether you grant it or no.[6]

Ever yours very sincerely | Mary Egerton

RI MS JT/1/E/69

1. *[April 1873]*: the date is given by relation to letter 4089, dated to [14 May 1873]. The lecture course referred to by Egerton began on 24 April (see n. 4), so we place this letter in April as it seems Egerton wrote to Tyndall before the course began.
2. *45, EATON PLACE, | S.W.*: in London.
3. *Rose Kingsley*: Rose Georgina Kingsley (1845–1925), a daughter of the English clergyman Charles Kingsley (1819–75) (*New York Times*, 19 August 1925, p. 19), who was a friend of Egerton's daughter Mary Alice Egerton.
4. *your lectures on Light*: see letter 3782, n. 10.
5. *cool*: of a person, an action, or a person's behaviour: assured and unabashed where diffidence and hesitation would be expected; composedly and deliberately audacious or impudent in making a proposal, demand, or assumption (*OED*).
6. *grant it or no*: in letter 4089, Egerton wrote to Tyndall: 'Thanks for the ticket for Rose Kingsley'.

To [Thomas Archer Hirst][1] 2 May 1873[2] 4078

Royal Institution | May 2nd 1873

It appears to me that my friend Tom and Mrs. Tunstal[3] take the matter of luncheon far more to heart than I do, or have done. Tait I can assure Tom troubles me in no way whatever—I am a little worn by my work and that is all.[4]

Tom's last letter[5] read more like an official homily than any thing I had previously received from him.

J. T.

RI MS JT/1/T/997
RI MS JT/1/HTYP/621

1. *[Thomas Archer Hirst]*: the letter refers rather archly to Hirst; if Hirst was the addressee, the archness was thus even more pronounced.
2. *2 May 1873*: the date on the letter was written by a different hand.
3. *M^rs^. Tunstal*: Jane Tunstall.
4. *It appears... that is all*: Tyndall seems to have cancelled on a luncheon date at Tyndall's rooms at the RI, leading Hirst to wonder if Tyndall was in distress over the looming dispute with Peter Guthrie Tait over the reputation of James David Forbes (see letters 4033, 4038, and 4075).
5. *Tom's last letter*: letter missing, as Hirst's last letter to Tyndall in this volume is from January (3970).

From Charles Darwin 2 [May][1] 1873 4079

Down, | Beckenham, Kent. | Ap[ril]. 2. 1873.

My dear Tyndall,

I am very much obliged for all the trouble you have taken, and I hope you do not think that I have been over-cautious in asking for the copies to be returned.[2]

I think I told you that H[uxley]. had asked for a list of the names, which I have sent in alphabetical order.[3] And now our work is completed. I return the 2 enclosed letters, but you will see that I have spitted one by mistake.[4]

My dear Tyndall | Yours very sincerely | Ch. Darwin.

RI MS JT/1/TYP/9/2831[5]
LT Typescript Only

1. *[May]*: the month was established by the DCP in light of the letter's relationship with letter 4069, sent towards the end of April. Also, LT left a note next to the date on the typescript: 'prob[ab]ly May'.
2. *the copies to be returned*: after sending Tyndall two copies of Thomas Huxley's gracious letter accepting funds from his friends to pay for a much-needed health-restoring holiday, Darwin had second thoughts about the wider circulation of the letter and asked Tyndall to return the copies; see letters 4067, 4069, and 4070. See also letter 4028, n. 4.
3. *list of the names, which I have sent in alphabetical order*: enclosure missing; for the names of the eighteen friends who donated, see letter 4028, n. 4.
4. *I return... by mistake*: the DCP notes that Darwin put letters that he was not actively engaged with on spits, though exactly which letters Darwin was here referring to as 'enclosed' have not been identified.

5. This letter has been published in *Darwin Correspondence*, vol. 21, p. 199; this is letter 8894 in the online edition of the Darwin correspondence.

From François-Napoléon-Marie Moigno 5 May 1873 4080

[LES MONDES | Revue Hebdomadaire | DES SCIENCES | *et de leurs applications* | aux Arts et à l'Industrie | par | M. l'Abbé MOIGNO | RÉDACTION: | 2, Rue d'Erfurth, 2

Paris, le] 5 mai [187]3

Mon cher Mr Tyndall,

Ma dernière lettre est restée sans réponse, et j'en suis tout */attristé/*.

Nous touchons à la fin de la nouvelle édition de la chaleur, et je */tiens/* absolument puisque vous me l'avez presque proposé à */l'enrichir/* du */septième/* chapitre. Envoyez moi la copie, je vous en prie.

Je */tiens/* aussi beaucoup à faire le premier et bientôt la traduction de vos conférences d'Amérique, donc Mr Gauthier-Villars fera les */frais/* dans les mêmes conditions que pour la Chaleur, ou dans des conditions nouvelles si vous le voulez; je vous en prie */confiez/* moi bientôt le texte anglais.

J'ai l'honneur d'être dans les sentiments du dévouement le plus affectueux.

Votre humble serviteur | L'abbé F. Moigno

Pardon de tant [d'] */impertinences/* mais il s'agit des progrès des */sciences/* et de votre gloire.

[LES MONDES[1] | Weekly Review | OF THE SCIENCES | *and their applications* | to the Arts and the Industry | by | Father MOIGNO | EDITORIAL OFFICES: | 2 Erfurth street, 2

Paris,] 5 May [187]3

My dear Mr Tyndall,

My last letter[2] remained unanswered and it saddens me.

We are reaching the end of the new edition of Heat,[3] and I definitely want, since you nearly suggested it, to enrich it with the */seventh/* chapter. Please do send me a copy.

I also wish just as much to do the first and soon the translation of your American lectures.[4] Therefore, Mr Gauthier-Villars[5] will do the */pricing/* in the same conditions as for Heat, or within new conditions if you wish. Please entrust me with the English text soon.

I have the honour of expressing my most affectionate devotion.

Your humble servant | Father F. Moigno

Forgive such *[impertinence]* but it's about the progress of science and your glory.

RI MS JT/1/M/123

1. *LES MONDES*: see letter 3829, n. 1.
2. *My last letter*: letter 4050.
3. *new edition of Heat*: Moigno reported on progress on publication of the new French edition of *La Chaleur* in his previous letter (4050). See also letter 3829, n. 3.
4. *translation of your American lectures*: Moigno here offered to translate the forthcoming *Six Lectures on Light*. This would appear two years later as J. Tyndall, *La Lumière: six leçons faites en Amérique pendant l'hiver de 1872–1873*, trans. F.-N.-M. Moigno (Paris: Gauthier-Villars, 1875).
5. *Mr Gauthier-Villars*: Jean-Albert Gauthier-Villars.

From Thomas Henry Huxley 6 May 1873 4081

4 Marlborough Place,[1] | May 6th 1873.

My dear Tyndall,

There is something refreshing about Agassiz's letter[2] (which I return) and I hope he will be as good as his word

I must take some notice of the passage in Forbes's life[3] of which I spoke to you the other day,[4] and I am thinking of printing in "Nature" the document I submitted to the Council of the R. S. and of which I sent you a copy some time back.[5] Can you let me have that copy again as it has Hopkins' notes.[6]

I should have sent it sooner but I have been very weary these few days.

Ever yours very sincerely | T. H. Huxley.

RI MS JT/1/TYP/9/2984
IC HP 8.140
Typed Transcript Only

1. *4 Marlborough Place*: see letter 3947, n. 1.
2. *Agassiz's letter*: letter missing. In letter 4038, Tyndall asked Louis Agassiz for his memories 'relating to the behaviour of Forbes, as regards the glacier question'. The response may have been from Agassiz's son Alexander Agassiz (1835–1910), the Harvard-based naturalist-zoologist (*CDSB*), since Alexander subsequently published a letter to *Nature* on the matter, dated 15 April: 'Originators of Glacial Theories', *Nature*, 8 (8 May 1873), pp. 24–5.
3. *Forbes's life*: *Life and Letters of Forbes*.
4. *I must take notice... the other day*: Tyndall referred to this conversation with Thomas Huxley in a 29 April letter to Thomas Hirst (see letter 4075). The passage in question

is presumably the one Huxley quoted at the beginning of his letter to *Nature* (see n. 5) 'I believe that the effect of the struggle—though unsuccessful in its immediate object—will be to render Tyndall and Huxley and their friends more cautious in their further proceedings. For instance, Tyndall's book, again withdrawn from Murray's "immediate" list, will probably be infinitely more carefully worded relative to Rendu than he at first intended' (*Life and Letters of Forbes*, p. 387).

5. *I am thinking... some time back*: Huxley published a notice in *Nature* reporting (though not quoting from) his 1859 testimonial to the RS's Council on why James David Forbes should not be awarded the Copley Medal: 'Forbes and Tyndall', *Nature*, 8 (22 May 1873), p. 64.
6. *Hopkins' notes*: William Hopkins (1793–1866), the Cambridge mathematician and geologist, had also crossed swords with Forbes over glaciers (*ODNB*). Tyndall saluted Hopkins' contribution in 'Principal Forbes and His Biographers', *Contemporary Review*, 22 (August 1873), pp. 484–508, esp. p. 500.

From Mary Egerton [7 May 1873][1] 4082

[45, Eaton Place, | S.W.][2] | Wed[nesda][y]

Dear Friend,

We never accomplished seeing those laboratories![3] It is only a piece of curiosity very different to the feeling one had in seeing the old one! But I sh[oul][d] like to see the improvements. Would it be convenient if we stopped after the lecture tomorrow?[4] or on Saturday? And if so, will you leave word with the porter for there is always such a buzzing round you after the lecture one can hardly get a word.

When will that book[5] be done? Take care you do not become "an infinitely attenuated bookworm"![6] or come to realize M[r] Rodwell's doleful picture in Nature (p. 9)[7] of the imaginary being who by perpetual stooping more & more (over his writing of course!), at length "grows so"; and "gets thinner and thinner and thinner", until at last he becomes infinitely flat, and can only perceive space of two dimensions!! Behold your fate! (but I confess even that is preferable to the whirling of multiform motion by which he endeavours to enlarge ones conceptions up to four dimensions!) The girls[8] say "the Professor has quite given us up—he never comes down in the evening now!" You ought not to work in the evening, & you might rest here sometimes, just as well as at the club![9] (This, by the way!)

Y[ou][rs] ever | M F Egerton

I did not see your name at the Academy dinner?[10]

RI MS JT/1/E/75

1. *[7 May 1873]*: the year is given by mention of the new laboratory at the RI, and the date by reference to a *Nature* article of 1 May 1873 (see n. 7), the Wednesday following being 7 May.
2. *45, Eaton Place, | S.W.*: in London.
3. *those laboratories*: a reference to the renovations of the RI's laboratories while Tyndall was in the United States (see letter 3776, n. 14).
4. *lecture tomorrow*: see letter 3782, n. 10.
5. *book*: probably *Six Lectures on Light*.
6. *"an infinitely attenuated bookworm"*: quoted in Rodwell's article (see n. 7 below) as 'like infinitely attenuated book-worms in an infinitely thin sheet of paper', these words come from James Joseph Sylvester (1814–97), a British mathematician, in 'A Plea for the Mathematician', *Nature*, 1 (30 December 1869), pp. 237–8, on p. 238.
7. *M[r] Rodwell's doleful picture in Nature (p. 9)*: G. F. Rodwell, 'On Space of Four Dimensions', *Nature*, 8 (1 May 1873), pp. 8–9.
8. *girls*: one of these girls would probably be Egerton's daughter Mary Alice Egerton.
9. *the club*: probably the Athenaeum Club (see letter 3710, n. 5).
10. *Academy dinner*: annual dinner of the Royal Academy of Arts on 3 May 1873 (*Medical Times and Gazette*, 10 May 1873, p. 500).

To Charles Mathews[1] 9 May [1873][2] 4083

Royal Institution of Great Britain | 9th May

Dear Sir

A tradition has prevailed among the Tyndalls in Ireland that they were descended from the family of the martyr William Tyndall;[3] but none of them possess the documentary evidence that could prove it.

This however is certain. Some members of the martyr's family emigrated from Strickcombe in Gloucestershire to Ireland in the 17th century. The very year of their emigration is known. It is probable that my origin is to be traced to one of these emigrants—very probable I think; but there is no proof of the matter that would satisfy a scientific man.

Yours &c | John Tyndall

RI MS DPM/1H

1. *Charles Mathews*: Charles Elkin Mathews (1851–1921), who at the beginning of his career was a London bookseller and publisher (*ODNB*).
2. *[1873]*: the year is established by the preservation of the date-stamped envelope which carried the letter.
3. *William Tyndall*: William Tyndall (1494–1536), 'Old Martyr Tyndall', was a religious

reformer during the Protestant Reformation, and translator of the Bible. A Protestant martyr, he was executed by strangulation and then burned at the stake (*ODNB*). It was a family tale that Tyndall was descended from him, and he wrote in his journal in 1860 that he 'was struck by the remarkable reproduction of his temper amongst us;—give us the same circumstances and I am sure either Hector [Tyndale] or myself would act as he did tomorrow' (RI MS JT/2/13c/1187). In 1869 Tyndall visited the supposed birthplace of William Tyndale, and the monument to him erected at Nibley Knoll in Gloucestershire (see letter 3083, *Tyndall Correspondence*, vol. 11).

To Alfred Mayer 9 May [1873][1] 4084

Royal Institution of Great Britain | 9th May

My dear Prof. Mayer.

If you would care to hear a lecture in our theatre on the site of ancient Troy,[2] pray come here this evening at 8.45. Say please to the porter that I requested you to go in to the library—where you saw our librarian[3] this morning—and I shall join you there & take you in to the theatre.

Yours faithfully | John Tyndall

Mr Cyrus Field[4] would be glad to see you—He stays at the Buckingham Palace Hotel.

Tyndall, John (1820–1893); Hyatt and Mayer Collection, C0076, Manuscripts Division, Department of Special Collections, Princeton University Library

1. *[1873]*: the year is given by reference to a lecture at the RI (see n. 2), and that Alfred Mayer was in London in 1873. A letter discussing Mayer's visit, with the year given, is letter 4108.
2. *lecture in our theatre on the site of ancient Troy*: Mountstuart Elphistone Grant-Duff (see letter 3947, n. 31), delivered a Friday Evening Discourse at the RI, 'A Fortnight in Asia Minor', on 9 May 1873 (*Roy. Inst. Proc.*, 7 (1873–75), pp. 117–33).
3. *our librarian*: Benjamin Vincent (1818–99), a British translator and editor. He served as Assistant Secretary and Librarian of the RI for most of the second half of the nineteenth century (F. A. J. L. James (ed), *The Correspondence of Michael Faraday*, vol. 5, 1855–1860 (London: Institution of Engineering and Technology, 2008), p. lvii).
4. *Mr Cyrus Field*: Cyrus West Field (1819–92), an American businessman and financier who, with others, created the Atlantic Telegraph Company and laid the first telegraph cable across the Atlantic Ocean in 1858 (K. E. Hendrickson Jr., 'Field, Cyrus West (1819–1892)', in K. E. Hendrikson, III (ed), *The Encyclopedia of the Industrial Revolution in World History* (Lanham, MD: Rowman & Littlefield, 2015), p. 326). Field was certainly in London in 1873, as he organized a banquet on 10 March 1873 at the Buckingham Palace Hotel, in 'Commemoration of the Signature of the Agreement on the 10th of March, 1854, for the

Establishment of a Telegraph across the Atlantic'; Tyndall was a guest and gave a speech (C. W. Field, *Report of the Proceedings at an Anniversary Banquet given by Mr. Cyrus W. Field* (London, 1868)).

From Alfred Mayer 9 May [1873][1] 4085

My dear Prof[esso]r. Tyndall;

I will be most happy to attend the lecture this evening.[2]

Y[ou]rs truly | Alfred Marshall Mayer.[3]

[RI Envelope][4] Professor Alfred Mayer | (of New York) | Tavistock Hotel | Covent Garden | To be delivered immediately

Tyndall, John (1820–1893); Hyatt and Mayer Collection, C0076, Manuscripts Division, Department of Special Collections, Princeton University Library

1. *[1873]*: the year is given by relation to letter 4084.
2. *lecture this evening*: see letter 4084, n. 2.
3. The text of this note is written on a printed signature card bearing the name 'Alfred Marshall Mayer'.
4. This envelope attached to the letter is the one Tyndall used to send his letter to Mayer (letter 4084), indicating that Mayer may have sent his reply back in the same envelope.

To Thomas Henry Huxley 10 May [1873][1] 4086

Royal Institution. | 10th. May.

My dear Huxley,

The document[2] duly labelled with the date on which it came into my possession is here, and I will send it to you whenever you wish. I would not I think publish it at the outset. Let the fight go on for a time, and then you will see your way more clearly.[3]

You see the fight has already begun. In the next number of Nature we shall probably have a rejoinder from Tait.[4]

You might, if you liked, contradict point blank that stupid assertion that there was any concert between you and me as regards the Copley Medal.[5] There is an inherent meanness in the minds that harbour such notions.

I have been longing to get up to see you all, but have been defeated by all manner of obstacles.

Yours ever | John Tyndall.

Hirst says I may wait till my book is finished before I join in the fray.[6] It will, I trust be finished within a month from the present time; and then I shall discourse like Jeremy Taylor on justice, mercy, and judgment to come.[7]

RI MS JT/1/TYP/9/2980
IC HP 8.141
Typed Transcript Only

---

1. *[1873]*: the year is established on the basis of the relationship with letter 4081, where Thomas Huxley asked Tyndall to return a copy of Huxley's 1859 testimonial to the Council of the RS against the awarding of the Copley Medal to James David Forbes.
2. *document*: Huxley's anti-Forbes testimonial from 1859 (see n. 1).
3. *I would not... more clearly*: Huxley seems to have taken the advice, not sending his letter to *Nature* until 17 May, and then only referring to his testimonial rather than printing it (see letter 4081, n. 5).
4. *You see the fight... a rejoinder from Tait*: Alexander Agassiz's pro-Tyndall letter had just appeared in the 8 May issue of *Nature* (see letter 4081, n. 2); the expected rejoinder came not, however, from Peter Guthrie Tait, co-editor of *Life and Letters of Forbes* that had antagonized Tyndall and Huxley, but from Forbes' son George. See G. Forbes, 'Forbes and Agassiz', *Nature*, 8 (22 May 1873), pp. 64–5.
5. *You might... Copley Medal*: Huxley did as requested, writing in his subsequent letter to *Nature*: 'in justice to Prof. Tyndall, I am bound to say that he knew nothing about the battle until after it was over' ('Forbes and Tyndall', *Nature*, 8 (22 May 1873), p. 64).
6. *Hirst says... in the fray*: in Tyndall's letter to Thomas Hirst of 29 April, he indicated his preference for concentrating on finishing the book of his American lectures, *Six Lectures on Light* (see letter 4075); it seems that Hirst affirmed that preference in a missing letter, but it is described in a letter from Tyndall to Hirst of 2 May as 'more like an official homily than any thing I had previously received from him' (see letter 4078).
7. *I shall discourse... judgment to come*: Tyndall here compared the intervention he was then contemplating against Tait to the religious discourses of the Anglican cleric Jeremy Taylor (1613–67), who enjoyed a high reputation (subsequently fallen) for the poetry of his prose (*ODNB*).

From Frances Russell 12 May 1873 4087

[Pembroke Lodge.[1] | Richmond Park.] | May 12/73

Dear Mr Tyndall

I fear you are too busy for it to be of much use to tell you that we are to be found all Friday & Sunday afternoons—but if there is <u>any</u> day on which you will propose y[ou]rself to dine & sleep here come as early as you can, it is almost sure to suit us—

Y[ou]rs very sincerely | F. Russell

RI MS JT/1/R/64

1. *Pembroke Lodge*: during John Russell's first time as Prime Minister (1846–52), he was granted use of Pembroke Lodge by Queen Victoria and it remained the Russell family home until Frances died in 1898.

From Mary Egerton [13 May 1873][1] 4088

[45, Eaton Place, | S.W.][2]

One line in May's[3] note.[4] It pained me you should think I do not understand you! I do. But it was thoughtless to talk of rest! I forgot we were so many. Were I alone, & could put you in an armchair and play dreamy tunes to you, & not speak a word, it would be different; but I see it all, so don't think me unreasonable.

I would have the book[5] as perfect as work can make it. But is time more consequence than the wear & tear of brain?

Y[ou]rs Ever | MFE | Tuesday
John Tyndall Esq[ui]r[e].

RI MS JT/1/E/99a

1. *[13 May 1873]*: as this letter follows that from Egerton dated to [7 May 1873] (letter 4082), a Wednesday, this Tuesday letter would be 13 May 1873.
2. *45, Eaton Place, | S.W.*: in London.
3. *May's*: Mary Alice Egerton.
4. *note*: if this refers to a letter from Mary Alice Egerton to Tyndall, it is missing.
5. *book*: probably *Six Lectures on Light*.

From Mary Egerton 14 May [1873][1] 4089

[45, Eaton Place, | S.W.][2] | May 14th

No, dear Friend—you must not, shall not come till your hardest work is over. I sh[oul]d only feel I had teased you into it & sh[oul]d feel guilty if I let you talk, & yet we could hardly realize the dream of "gentle dulness"[3] by which I tried to prove to you that I could form some idea of "Rest"!

I hope you will get your Country Excursion, and then if the weather becomes less starving, you will drink in new life from the springing life of leaves & flowers. But at any rate do not come to us on Saturday; Charley[4] comes to town that day & his first evening will be a whirl of chatter, & business talk! Later on when you are more free, if you are disposed to name any evening, I will tell you if we shall be at home, & than shall sing to you.

When the weather is warm again & the "Rhodoras" are out, could we not have another drive to Kew some day? I want to call on Mrs Hooker,[5] they were so kind to us last year.

We have been hearing Mendelsohn's glorious music at St Paul's today.[6] Perhaps we meet at the Spottiswoodes[7] on the 23rd? Thanks for the ticket for Rose Kingsley.[8]

Ever yours | M F Egerton

RI MS JT/1/E/35

---

1. *[1873]*: the year is established by the reference to a performance of Mendelsohn that day at St. Paul's for the Festival of the Sons of the Clergy; for a notice of that performance, see *Harper's Weekly*, 17 (28 June 1873), p. 554.
2. *45, Eaton Place, | S.W.*: in London.
3. *"gentle dulness"*: from Alexander Pope's *Dunciad* (1728–43): 'Gentle dullness ever loves a joke' (the misspelling of 'dullness' is Egerton's).
4. *Charley*: Charles Augustus Egerton (1846–1912), the first son of Mary Egerton and Edward Christopher Egerton. Their second son was Hugh Edward Egerton (see letter 4076, n. 6).
5. *Mrs Hooker*: Frances Hooker.
6. *Mendelsohn's glorious music at St Paul's today*: see n. 1.
7. *Spottiswoodes*: William and Eliza Spottiswoode.
8. *ticket for Rose Kingsley*: see letter 4077.

To Thomas Henry Huxley — 15 May 1873 — 4090

Royal Institution. | 15th. May, 1873.

My dear Huxley,

The circumstance that on one or two former occasions I have been asked to express an opinion as to the merits of foreign candidates proposed for the Copley Medal,[1] or for the foreign fellowships of the Royal Society, induces me to hope that I may not be out of order in saying one word with regard to the choice of candidates to day.

I will confine my remarks to one of them, and these remarks shall be brief. I do not know a man of broader scientific culture than DuBois Reymond,[2] the Perpetual Secretary of the Academy of Sciences in Berlin. I hardly know a man to compare with him in skill and power as an experimenter. In the domain of Animal Electricity he has no rival. He has never had the slightest recognition from the Royal Society, though I question whether another man of his rank and merit exists who has failed to receive such recognition. Were I on the Council I would earnestly advocate his claims as a measure of justice both to him and to the Society, whose fellowship he would unquestionably adorn, and in whose ranks he ought certainly to find a place.

I am my dear Huxley | ever Yours faithfully | John Tyndall.

T. H. Huxley Esq[ui]re. | Sec[retary]. of the R.S.

RI MS JT/1/TYP/9/2981
IC HP 8.142
Typed Transcript Only

---

1. *asked to express... Copley Medal*: Tyndall wrote to Thomas Huxley in the latter's capacity as Secretary of the RS, in relation to a then-current application being considered by the society's Council; but they had just been corresponding about an earlier case in which Huxley had provided a testimonial (see letters 4081 and 4086).
2. *DuBois Reymond*: Emil du Bois-Reymond, famed for his studies of neuro-electricity, was a good friend of Tyndall's from the time of his studies in Germany, an association that led ultimately to Tyndall's appointment as Faraday's successor at the RI (S. Forgan, 'Tyndall at the Royal Institution', W. H. Brock, N. D. McMillan, and R. C. Mollan (eds), *John Tyndall: Essays on a Natural Philosopher* (Dublin: Royal Dublin Society, 1981), pp. 49–59, on p. 49). Despite Tyndall's intervention, du Bois-Reymond was not elected ForMemRS until 1877 (*Roy. Soc. Proc.*, 26 (1877), p. 504); for his election certificate in the RS's archive, see RS EC/1877/22.

From Mary Agatha Russell — 15 May [1873][1] — 4091

[Pembroke Lodge.[2] | Richmond Park.] | May 15.

Dear Professor Tyndall

As I am unfortunately unable to go up to the Royal Institution for your lecture today,[3] I must write to thank you for so very kindly sending me the Popular Science Monthly in which I was much interested in reading the account by Prof. Brewer of the mountain you mention.[4]—Mama[5] & I have just been reading those fine passages (I think in a speech by Pres. White)[6] which you quoted, if I remember rightly, in your Friday evening lecture on Niagara.[7]—It seems to be a most delightful journal[8]—& we are extremely obliged to you for sending it.

Believe me, very sincerely yours | Agatha Russell.

RI MS JT/1/R/60

1. *[1873]*: the year is given by reference to Tyndall's course of lectures on light at the RI (see n. 3), an issue of *Popular Science Monthly* (see n. 4), and Tyndall's Friday Evening Discourse on Niagara (see n. 7).
2. *Pembroke Lodge*: see letter 4087, n. 1.
3. *your lecture today*: Tyndall delivered a course of Thursday lectures at the RI, 'On Light' (see letter 3782, n. 10). The date 15 May 1873 was indeed a Thursday.
4. *account by Prof. Brewer of the mountain you mention*: W. H. Brewer, 'Discovery of Mount Tyndall', *Popular Science Monthly*, 2 (April 1873), pp. 739–41.
5. *Mama*: Frances Russell.
6. *speech by Pres. White*: in the published version of Tyndall's discourse on Niagara (see n. 7), there are no quotes from Andrew Dickson White (1832–1918), an American historian and educator who was President of Cornell University (*ANB*), who gave a speech at Tyndall's farewell banquet (see Introduction, n. 23), which was published as A. D. White, 'Science and Public Affairs', *Popular Science Monthly*, 2 (April 1873), pp. 736–9.
7. *in your Friday evening lecture on Niagara*: Tyndall delivered a Friday Evening Discourse, 'Some Observations on Niagara', at the RI on 4 April 1873 (see letter 3738, n. 3).
8. *a most delightful journal*: see letter 4033, n. 16.

To Thomas Archer Hirst 16 May 1873 4092

Royal Institution | 16th May /73

My dear Tom.

Could you at your leisure, and from time to time as I send them to you look over these proofs[1] for me? I have tried to make them so plain that you will lose little time in hunting for my meaning. If you cannot look at these earlier ones then perhaps you will look at those where the composition & resolution of vibrations are treated.[2]

Yours aff[ectionate]ly | John Tyndall

RI MS JT/1/T/704
RI MS JT/1/HTYP/621

1. *these proofs*: probably a reference to the English edition of Tyndall's *Six Lectures on Light*, about which he was corresponding with Thomas Hirst (see letters 4075, n. 3, and 4086, n. 6).
2. *those where the composition & resolution of vibrations are treated*: a fair description of the final chapters/lectures in *Six Lectures on Light*.

To Joseph Henry 17 May 1873 4093

17th May 1873.

My dear Professor Henry

Many thanks to you for your long and interesting letter.[1] I do not wish to trammel the action of the Trustees in any way: but my private desire is not to increase the fund[2]—At all events should it be increased it must not bear my name. My sole object is to do good in a quiet way, and it will be a far deeper satisfaction to me should the little trust fund bear fruit as it now stands, than to have it swelled with a view of rendering my name more conspicuous. In fact my happiest and best hours are spent in quiet noiseless labour, not in hearing myself spoken of in the newspapers.

This Forbes controversy is a bad and foolish business[3]—His memory cannot profit by it. I believe it will be conducted more between Forbes's friends and Agassiz than between them and me. But I am in duty bound to make good every point that I have asserted in Agassiz's behalf and that I shall assuredly do.

Maxwell's theory[4] is still very difficult to grasp. I think by long pondering over his style, he might have diminished this difficulty. Most certainly hitherto we have been accustomed to regard the ether as composed of particles, and to see the action at insensible distances has always presented the same difficulty as action at sensible ones.

Today is Saturday, and on Monday morning I go down to Dover with the view of observing the performance of Fog Signals at the South Foreland. A letter just received from the Trinity House informs me that Major Elliott will go down in the same train with me.[5] This will be a great pleasure.

Depend upon it Arrow brought nothing but affectionate memories regarding you and your family from the United States.[6]

What you say regarding the photographs makes me very proud—It was very thoughtful in Mr. Child to send them to you,[7] but had I thought you would have cared for the photographs I should have sent you copies myself.

I am expanding my American Lectures & hope to have them out here in a few weeks. I intend to introduce that lovely specimen of water crystallization which you were good enough to send me, among the illustrations.[8]

Professor Mayer of Hoboken[9] is now here. By the way the printing of a new edition of my lecture notes on Electricity caused me to look up your old memoirs, and to give them due mention.[10] This ought to have been done long ago; but it is not easy to read every thing. Give my kindest regards to Mrs. Henry & the young ladies

ever yours faithfully | John Tyndall

Smithsonian RU 26, Vol 131, p. 183

1. *your long and interesting letter*: letter 4073.
2. *I do not wish... the fund*: Henry in his letter had written about how desirable it would be to increase the trust fund created from the profits from Tyndall's American lecture tour (see letter 4073, and Introduction, n. 15). The trustees of the fund were Henry, Hector Tyndale, and Edward L. Youmans.
3. *This Forbes controversy... foolish business*: Tyndall expended much energy in the spring of 1873 responding to attacks from the partisans of the late Scottish glaciologist James David Forbes—mainly Peter Guthrie Tait—for what they perceived as Tyndall's unjust slighting of Forbes (and elevating of Louis Agassiz) in *Forms of Water* (see letters 4033, 4038, 4075, 4081, and 4086).
4. *Maxwell's theory*: Henry expressed his puzzlement over the conception of the ether in James Clerk Maxwell's recently published *Treatise on Electricity and Magnetism* (see letter 4073, esp. n. 11). For context, see G. N. Cantor and M. J. S. Hodge (eds), *Conceptions of Ether: Studies in the History of Ether Theories, 1740–1900* (Cambridge: Cambridge University Press, 1981).
5. *On Monday morning... same train as me*: on Monday, 19 May, Tyndall indeed went, in the company of the American Light-House Board engineer George Elliott (see letter 4073, n. 18), to inspect the acoustic fog-signal system in use at the electrified South Foreland lighthouse in Dover (see *Ascent of John Tyndall*, p. 318). The letter from Trinity House that Tyndall mentions is missing. On Tyndall as scientific adviser to Trinity House, see letter 3896, n. 4.
6. *Arrow brought... United States*: Henry in his letter reported that his wife received a friendly letter from Frederick Arrow, who, while Tyndall was in the United States, had traveled around North America inspecting the fog-signalling systems there (see *Ascent of John Tyndall*, p. 318).
7. *regarding the photographs... send them to you*: see letter 4073, nn. 13–15.
8. *I intend... the illustrations*: Tyndall published the water-crystallization image sent via Henry from Samuel Lockett of Louisiana State University (see letters 4021 and 4033) as the frontispiece in *Six Lectures on Light*.
9. *Professor Mayer of Hoboken*: Alfred Mayer.
10. *a new edition... due mention*: in the 1874 American edition of Tyndall's *Light and Electricity* he added a short final paragraph acknowledging Henry's role in the study of the induced currents of the Leyden Battery (see J. Tyndall, *Light and Electricity: Notes of Two Courses of Lectures Before the Royal Institution of Great Britain* (New York: D. Appleton, 1874), p. 194).

To Thomas Archer Hirst 17 May 1873 4094

Royal Institution | 17th May /73

My dear Tom.

I hardly think I can get down to you before luncheon tomorrow—I have two examination papers to set.[1]

I will run down after luncheon if not before.

Yours ever | John Tyndall

RI MS JT/1/T/705
RI MS JT/1/HTYP/621

1. *two examination papers to set*: it is not clear what papers these were, or for which institution. On Tyndall's service as examiner at various times for the University of London and the Council for Military Education, see DeYoung, *Vision of Modern Science*, p. 134.

To Thomas Henry Huxley 17 May 1873 4095

Royal Institution | 17th, May, 1873.

My dear Huxley

I could hardly have believed that a year has elapsed since you placed this document in my hands.[1] But like a falling stone life moves with accelerated velocity as we make our final approach to the earth.

I have not yet looked at the life of Forbes; but I hope to make my deliverance upon the subject a final one.[2]

Yours ever | John Tyndall

Have you made up your mind about going anywhere in particular during the summer?[3] We might soften each other's ills by sympathy for my head is beginning to be shaky. I must however dispose of Mr Tait,[4] before I move, and the best of it is that my case is so good that I feel in no way anxious about the matter.

RI MS JT/1/TYP/9/2984
IC HP 8.143
Typed Transcript Only

---

1. *since you placed this document in my hands*: Tyndall is referring to a copy Huxley's 1859 testimonial to the Council of the RS setting out the case against James David Forbes as recipient of the Copley Medal; Huxley had recently asked to see the copy again, for purposes of writing a letter to *Nature* challenging the account of the Copley Medal opposition in the recently published *Life and Letters of Forbes*. See letters 4081 and 4086.
2. *I have not yet looked… final one*: for Tyndall's eventual published response to *Life and Letters of Forbes*, see letter 4024, n. 5.
3. *Have you… the summer?*: Huxley and Tyndall went on to holiday separately that summer, with Huxley traveling to the Auvergne with Joseph Hooker and Tyndall to the Swiss Alps via Paris with Thomas Hirst (see, respectively, L. Huxley, *Life and Letters of Thomas Henry Huxley*, 2 vols (London: Macmillan and Co., 1900), vol. 1, pp. 390-7; and *Ascent of John Tyndall*, pp. 322–4).
4. *I must… Mr Tait*: although the Forbes volume had three co-editors, Tyndall early on formed the impression that Peter Guthrie Tait was the source of the animus expressed in the volume (see letters 4033, 4078, and 4086).

To William Spottiswoode — 17 May [1873][1] — 4096

17th May

My dear Spottiswoode

Let no man dare to say that there is anything absurd in what you propose.[2]

Yours ever | John Tyndall

A pair of perforated poles placed between the polarizer & analyzer of a polariscope[3]—say with colourless selenite—Make circuit & see whether any colour appears—By no means absurd.

RI MS JT/1/T/1357
RI MS JT/TYP/3/1245

---

1. *[1873]*: the year is given since on 23 May 1873, Spottiswoode delivered a Friday Evening Discourse at the RI, 'On Spectra of Polarized Light' (*Roy. Inst. Proc.*, 7 (1873–75), pp. 134–8). It seems reasonable that Spottiswoode would be seeking advice from Tyndall a week before the lecture.
2. *what you propose*: letter from Spottiswoode missing.
3. *polariscope*: an instrument for showing the polarization of light, or for viewing objects in polarized light (*OED*).

From Thomas Henry Huxley — 18 May 1873 — 4097

4 Marlborough Place | Abbey Road, N.W.[1] | May 18 1873.

My dear Tyndall,

Another illustration of the difficulty of writing history![2] The document you have returned to me[3] is after all not that which was read to the Council! I have discovered this by collating it with Frankland's copy, which exhibits sundry deviations and additions. Now I recall the whole circumstance to mind, the original draft which I sent to you was somewhat altered and expanded before it went to the Council.

I have written a letter to "Nature"[4] putting the facts of the case clearly and as moderate in tone as I can make it. They shall have the statement laid before the Council, and Hopkins's letter[5] to boot if they ask for them, but I shall volunteer neither.

I have been thinking of taking myself off to the Auvergne country, and it would be extremely jolly if we could go together. Hooker has promised he will come if he does not go to America. I cannot get away before the very end of June. Would that suit you?[6]

Ever yours | T. H. Huxley.

RI MS JT/1/TYP/9/2985–7
IC HP 8.144
Typed Transcript Only

---

1. *4 Marlborough Place | Abbey Road, N.W.*: see letter 3947, n. 1.
2. *Another illustration... writing history!*: Huxley is referring to Tyndall's expression of surprise at realizing that a document he believed that he received very recently was in fact sent to him a year before (see letter 4095, n. 1).
3. *The document you have returned to me*: at Huxley's request, Tyndall had returned a draft copy of Huxley's 1859 testimonial against the award of the RS's Copley Medal to James David Forbes (see letter 4095, n. 1).
4. *letter to "Nature"*: on Huxley's letter, published in the 22 May issue, see letter 4081, n. 5.
5. *Hopkins's letter*: on the role of William Hopkins in the Forbes controversies, see letter 4081, n. 6.
6. *I have... suit you?*: in the end Tyndall and Huxley holidayed separately that summer (see letter 4095, n. 3).

From C. Cecil Trevor 19 May 1873 4098

Board of Trade (Harbour Department), | Whitehall Gardens, S.W., 19th May 1873.

SIR,

WITH reference to previous correspondence on the subject of the use of gas for lighthouse illumination at Haisbro',[1] I am directed by the Board of Trade[2] to state that the Elder Brethren[3] have called the attention of this Board to an arrangement which was come to when this experimental trial of gas at Haisbro' was undertaken, to the effect that the gas referees should be judges upon all points connected with the materials, manufacture, and cost of the gas, and to the fact that down to the present time the gas referees have had every detail of the plant and working under their cognizance and approval.[4]

The Elder Brethren in drawing attention to this arrangement have requested that upon the present occasion they may be furnished by you, as a gas referee, with a report upon the materials, manufacture, and cost of the gas used at Haisbro', and they propose that upon the receipt of such a report you should accompany them, in your capacity of scientific adviser to the Corporation,[5] in a series of careful demonstrations which they intend making afloat, with the view of determining the relative value of the gas as a lighthouse illuminant.

The Board of Trade request therefore that you will favour them at your earliest convenience with a report embracing the various points upon which the Elder Brethren desire information.

I am, &c. | (Signed) C. Cecil Trevor.

Professor Tyndall, F.R.S., | Royal Institution, Albemarle Street.

'Further Papers Relative to a Proposal to Substitute Gas for Oil as an Illuminating Power in Lighthouses, Part III', C 1511, Parliamentary Papers (1875), p. 34

1. *Haisbro'*: see letter 4045, n. 3.
2. *Board of Trade*: see letter 3736, n. 3.
3. *Elder Brethren*: see letter 3736, n. 3 and n. 5.
4. *With reference... cognizance and approval*: in April Tyndall had arranged a visit to the gas-burning lighting system in use at Haisborough to check whether gas really was superior to alternative lighthouse fuels, and was now being chased for his report (see letters 4045 and 4055).
5. *your capacity of scientific adviser to the Corporation*: see letter 3869, n. 4. The previous year, Tyndall became a Gas Referee (see letter 3830, n. 17).

To William Cooke[1] 23 May [1873][2] 4099

Royal Institution of Great Britain | 23rd May

My dear Sir

Would you allow me to introduce to you my friend Prof. Mayer[3] of New York: he wishes to see Gramme's Machine,[4] and I have ventured to promise that you will show it to him.

Very faithfully yours | John Tyndall

Cooke Esq[ui]r[e] | &c &c

[Envelope] Cooke Esq[ui]re | Westminster Road.

Tyndall, John (1820–1893); Hyatt and Mayer Collection, C0076, Manuscripts Division, Department of Special Collections, Princeton University Library

1. *William Cooke*: William Fothergill Cooke (see letter 3957, n. 13).
2. *[1873]*: the year is given by reference to Alfred Mayer's visit to London in 1873. A similar letter to another correspondent, with the year given, is letter 4108.
3. *Prof. Mayer*: Alfred Mayer, who visited London in 1873.
4. *Gramme's Machine*: see letter 3944, n. 5.

To Edward Frankland 23 May [1873][1] 4100

Royal Institution of Great Britain | 23rd May

My dear Frankland.

My good friend Prof. Mayer[2] of New York wishes to make your acquaintance and it gives me great pleasure to give him this note of introduction to you.

Yours faith[full]y | John Tyndall

[Envelope] Professor Frankland. | Chemical Laboratory | Dept of Science & Art | South Kensington

Tyndall, John (1820–1893); Hyatt and Mayer Collection, C0076, Manuscripts Division, Department of Special Collections, Princeton University Library

1. *[1873]*: the year is given by reference to Alfred Mayer's visit to London in 1873. A similar letter to another correspondent, with the year given, is letter 4108.
2. *Prof. Mayer*: Alfred Mayer, who visited London in 1873.

To Samuel Jeffery[1] 23 May [1873][2] 4101

Royal Institution of Great Britain | 23rd May

My dear Sir

Would you allow me to introduce to you my excellent friend Prof. Mayer[3] of New York? He wishes much to see the Kew Observatory[4] & what you are now doing there.

faithfully yours | John Tyndall

[Envelope] To the Director of the | Kew Observatory | Richmond.

Tyndall, John (1820–1893); Hyatt and Mayer Collection, C0076, Manuscripts Division, Department of Special Collections, Princeton University Library

1. *Samuel Jeffery*: Samuel Jeffery (*c.* 1822–1903) was Director (Superintendent) of the Kew Observatory from August 1871 to February 1876 (L. Macdonald, *Kew Observatory and the Evolution of Victorian Science, 1840–1910* (Pittsburgh: University of Pittsburgh Press, 2018), pp. 147–52; M. Walker, *History of the Meteorological Office* (Cambridge: Cambridge University Press, 2012), p. 103).
2. *[1873]*: the year is given by reference to Alfred Mayer's visit to London in 1873. A similar letter to another correspondent, with the year given, is letter 4108.
3. *Prof. Mayer*: Alfred Mayer, who visited London in 1873.
4. *Kew Observatory*: Kew Observatory, in Richmond, London. See L. Macdonald, *Kew Observatory and the Evolution of Victorian Science, 1840–1910.*

To Edward L. Youmans 26 May 1873 4102

London, *May 26, 1873.*

MY DEAR YOUMANS: I was somewhat concerned yesterday to gather from a letter of yours to Spencer that the religious world were bearing hard upon you[1]—seeking to damage Appleton, and in so far damaging you; but your letter of this morning,[2] which is all brightness, dissipates my concern. Surely, surely, a man like you must have labour room in the wide area of the States. The men on this side whom you work with are not scoffers; and if they did not lead the movements of the age, trust me, these movements would fall into more irreverent hands.

Those that have seen The Popular Science Monthly here call it "a delightful magazine."[3]

Yours ever, | JOHN TYNDALL.

J. Fiske (ed), *Edward Livinsgton Youmans: Interpreter of Science for the People* (New York: D. Appleton and Company, 1894), p. 319

1. *I was somewhat... hard upon you*: the letter to Herbert Spencer, which might have been Tyndall's forwarding letter 4010 to him, is missing. That year, Youmans defended the religious credentials of the sort of naturalistic science he promoted in the *Popular Science Monthly*—science à la Tyndall, Huxley, and Spencer—in an address, 'The Religious Work of Science'. For the address, see J. Fiske (ed), *Edward Livingstone Youmans: Interpreter of Science for the People* (New York: D. Appleton and Company, 1894), pp. 491–501; for context, see pp. 320–1. See also J. C. Ungureanu, 'Edward L. Youmans and the "Peacemakers" in the *Popular Science Monthly*', *Fides et Historia*, 51 (2019), pp. 13–32.
2. *your letter of this morning*: letter missing.
3. *"a delightful magazine"*: see letter 4033, n. 16.

From Charles Lyell 27 May 1873 4103

73 Harley Street[1] | May 27/73

Dear Tyndall

If you have quite done with James Hall's State Report[2] which you borrowed from me[3] when you were preparing for your Niagara lecture[4] I would be obliged to you if you would return it to me

Ever most truly yours, | Cha Lyell

RI MS JT/1/L/58
RI MS JT/1/TYP/3/859

1. *73 Harley Street*: Lyell's residence in London.
2. *James Hall's State Report*: see letter 4019, n. 4.
3. *borrowed from me*: see letter 4019.
4. *your Niagara lecture*: see letter 3738, n. 3.

To Thomas Archer Hirst [29 May 1873][1] 4104

I should like to speak to you about a very kind proposal from oxford. They wish to make me DCL, but I have some scruples.[2]

J. T.

RI MS JT/1/T/996
RI MS JT/1/HTYP/599

1. *[29 May 1873]*: the date is written on the handwritten letter in another hand, '29-5-73'.
2. *a very kind... some scruples*: on 18 June, Oxford awarded Tyndall with an honorary doctorate (Doctor of Civil Law), notwithstanding lingering concerns at the university about Tyndall's views on miracles and prayer. Tyndall accepted (*Ascent of John Tyndall*, p. 321; 'List of Honorary Degrees Conferred from 1870', in *The Historical Register of the University of Oxford* (Oxford: The Clarendon Press, 1900), p. 518). On Tyndall and prayer, see letter 3780, n. 2.

To Thomas Henry Huxley — 30 May [1873][1] — 4105

Royal Institution. | 30th. May.

My dear Huxley,

You have probably by this time read Shairp. He seems perfectly unconscious that he is exhibiting his friend stripped of all inner dignity.[2]

Do not, at all events just yet, let him cause you to publish that document.[3]

Yours ever | John Tyndall.

IC HP 8.147
Typed Transcript Only

1. *[1873]*: the year is established by the relation to letter 4106.
2. *You have probably... inner dignity*: Huxley took away much the same impression about the letter that John Campbell Shairp published in *Nature* ostensibly in defense of James David Forbes; see, for details, the notes for letter 4106.
3. *Do not... that document*: Huxley had been thinking about publishing his 1859 testimonial to the Council of the RS against the award of the Copley Medal to James David Forbes, but changed his mind; Tyndall here was advising Huxley to stick with the decision not to publish. For context see letters 4081, 4086, 4095, and 4097.

From Thomas Henry Huxley — 30 May 1873 — 4106

Marlborough Place | Abbey Road, N.W.[1] | May 30th 1873.

My dear Tyndall,

I have read Shairp's letter[2] and find it droll. I show that Forbes[3] has made statements which are incorrect, and his defender tells me there are much worse things in his letters.[4]

I do not think it necessary for me to take any further steps in the matter.

When you are hammering them,[5] just put a little power into your elbow for me!

Ever yours faithfully | T. H. Huxley.

RI MS JT/1/TYP/9/3014
IC HP 8.148
Typed Transcript Only

---

1. *Marlborough Place | Abbey Road, N.W.*: see letter 3947, n. 1.
2. *Shairp's letter*: John Campbell Shairp responded to Huxley's letter to *Nature* in the 22 May issue, where Huxley had given his views in relation to a passage in a letter from the late James David Forbes, included in the recently published *Life and Letters of Forbes*, of which Shairp was a co-editor, commenting on what he perceived to be the roles of Huxley and Tyndall in the decision of the RS's Council not to award the 1859 Copley Medal to Forbes (see letters 4081, n. 4, and 4095, n. 1). See J. C. Shairp, 'Forbes and Tyndall', *Nature*, 8 (29 May 1873), p. 84. A Scottish man of letters, not science, Shairp had succeeded his friend Forbes to the principalship of University College, and had 'masterminded' the volume's publication (*ODNB*).
3. *Forbes*: James David Forbes.
4. *his defender... his letters*: in Shairp's *Nature* letter (see n. 2), he insisted that, far from seeking to provoke, by selection he kept to a minimum the publication of letters on a topic about which Forbes had been much exercised.
5. *When you are hammering them*: for Tyndall's eventual published response, see letter 4024, n. 5.

To Alfred Mayer [May 1873][1] 4107

Monday m[ornin]g

Dear Prof. Mayer

Will you give me the pleasure of your company at the Athenaeum[2] on Thursday next at 6 precisely?

You will have an opportunity of meeting Mr. Herbert Spencer & some other big men.

Yours faith[ful]ly | John Tyndall

[Envelope, front]

By hand—| Wait answer.

Professor Mayer | 22 Jermyn St.[3]

J. Tyndall

[Envelope, verso]

Could you tell me Mr. Craik's[4] address?

Tyndall, John (1820–1893); Hyatt and Mayer Collection, C0076, Manuscripts Division, Department of Special Collections, Princeton University Library

---

1. *[May 1873]*: the year and month are given based on numerous other letters from the same time during Mayer's visit to London.
2. *Athenaeum*: the Athenaeum Club (see letter 3710, n. 5).
3. *22 Jermyn St.*: presumably the address of where Mayer was that day.
4. *M^{r}. Craik's*: probably the American clergyman and lawyer James Craik (1806–82).

To Latimer Clark[1] 5 June 1873 4108

ATHENAEUM | PALL MALL[2] | 5th. June 1873

My dear Sir,

Would you permit me to introduce to you my friend Prof. Mayer[3] of New York. I shall feel personally obliged to you for any attention that you may be kind enough to show him.

Yours faithfully | John Tyndall

Latimer Clark. Esq[ui]^{r}[e] | &c &c &c.

[Envelope] Latimer Clark Esq[ui]r[e]. | &c &c &c

Tyndall, John (1820–1893); Hyatt and Mayer Collection, C0076, Manuscripts Division, Department of Special Collections, Princeton University Library

---

1. *Latimer Clark*: Josiah Latimer Clark (1822–98), an English civil engineer (*ODNB*).
2. *ATHENAEUM | PALL MALL*: the Athenaeum Club (see letter 3710, n. 5).
3. *Prof. Mayer*: Alfred Mayer, who visited London in 1873.

From Jean-Albert Gauthier-Villars 5 June 1873 4109

[IMPRIMERIE ET LIBRAIRIE | *Pour les* Mathématiques, *les* Sciences *et les* Arts, | QUAI DES GRANDS-AUGUSTINS, 55. | GAUTHIER-VILLARS, SUCCESSEUR DE MALLET-BACHELIER, | IMPRIMEUR-LIBRAIRE | Du Bureau des Longitudes; de l'École Polytechnique; de l'Observatoire Impérial de Paris; de l'École Centrale des Arts et Manufactures.—Éditeur des *Comptes Rendus Hebdomadaires des Séances de l'Académie des Sciences*; du *Journal de Mathématiques*; par M. LIOUVILLE; du *Bulletin des Sciences mathématiques et astronomiques*, par MM. DARBOUX et HOÜEL; des *Nouvelles Annales de Mathématiques*, par MM. GERONO et BOURGET; des *Annales de l'Observatoire de Paris*; des *OEuvres de Lagrange*, etc.

Paris, le] 5 juin [187]3.

Monsieur,

C'est avec un vif regret que j'ai appris par votre lettre du 26 mai que l'édition française de la Chaleur, actuellement sous presse, n'était pas exactement conforme à l'édition anglaise.

Lorsque M[r] l'abbé Moigno m'a remis la copie à composer, je lui ai demandé si elle était bien conforme à la 4e édition anglaise; Il m'a répondu affirmativement, et a ajouté qu'il collationnerait avec le plus grand soin les épreuves avec le texte anglais. Comme c'est vous-même qui aviez manifesté le désir de voir confier la traduction à l'abbé Moigno, comme, de plus, vous aviez paru satisfait des traductions faites précédemment par lui, de la Chaleur et du Son; j'ai marché avec confiance, et j'ai cru que vous seriez content de la traduction et des soins matériels apportés à l'impression de votre livre.

Je savais cependant, par une longue expérience, que l'œil du maître ne se remplace pas. Aussi, je vous rappelle que dans le principe, je vous avais demandé de vouloir bien examiner le manuscrit de la traduction ou les épreuves; mais vous m'avez déclaré ne pouvoir en aucune façon y consacrer le moindre temps! Que pouvais-je donc faire de plus.

Vous ajoutez dans votre lettre que les éditions françaises de vos ouvrages vous ont occasionné beaucoup d'ennuis, sans vous rapporter un penny; et que la France est le seul pays dont vous ayez eu ainsi à vous plaindre. Notre malheureux pays est tellement attaqué de toutes parts, que ce reproche m'a été pénible. Permettez-moi d'y répondre brièvement.

En ce qui concerne la 1ère édition de la Chaleur, votre éditeur a fait faillite, et vraiment le pays entier ne peut en être responsable. Quant au Son, que j'ai publié, je ne croyais pas *[1 word illeg]* avoir donné quelque désagrément, mais si je me trompe, je vous serai grandement obligé de préciser vos reproches. Cet ouvrage n'a pas eu le prompt écoulement qu'il méritait, par ce qu'il y a eu une édition allemande, et que maintenant les Français n'étant plus jugés bons à rien, on achète, même en Italie, l'édition allemande plutôt que l'édition française. Du reste, vous assez bien reconnu le tort que l'édition allemande devait faire à l'édition française, car, à cette époque, vous m'aviez proposé, avec une excessive délicatesse, de partager la perte. Je *[m]*'y suis refusé, bien entendu.

Excusez-moi, Monsieur, d'être entré dans tous ces détails; mais je n'ai pas l'honneur d'être connu de vous, et il est pénible de paraître n'avoir pas rempli entièrement ses devoirs d'éditeur, quand au contraire on consacre ses efforts à produire des publications correctes et soignées.

Dans le cas actuel c'est pour moi un regret particulièrement vif, car personne au monde n'est un plus sincère admirateur de votre œuvre.

Laissant le passé, voyons le moyen de réparer le mal. M^r^. l'abbé Moigno, auquel j'ai communiqué votre lettre me répond ceci,

«Je ne puis pas m'imaginer quelles ont pu être les différences qui ont tant *[froissé]* M^r^. Tyndall, ce dont je suis désolé, je tiens ma traduction pour parfaite, plus parfaite que l'édition allemande.

«M'importe pour me rendre compte des différences qui inquiètent Mr Tyndall, je vais faire faire par un homme très compétent le travail de comparaison qui est demandé, on va relever avec le plus grand soin toutes les différences, on les lui communiquera, et il se convaincra par lui-même qu'elles ne *[motivent]* en aucune manière la mesure rigoureuse dont il vous parle.»

A la suite de cette révision, s'il y a des passages importants à modifier, je recomposerai et retirerai les pages à remplacer; s'il y a des notes à ajouter en appendice, je les ajouterai; en un mot, je ferai mon possible pour vous donner satisfaction.

J'ai arrêté le tirage de la feuille qui était sous presse lorsque votre lettre m'est parvenue; je fais disposer des épreuves de cette feuille *[en] [1 word illeg]*, et je me permets de vous les envoyer. Afin de ne pas perdre votre temps, veuillez confier la révision de cette feuille à un de vos assistants, ou à toute autre personne vous inspirant confiance; il en sera de même pour les autres feuilles.

Vous n'aurez qu'à me faire connaître les dépenses nécessitées par cette révision et je vous en enverrai immédiatement le montant. Je ne suis pas *[homme]*, et c'est en cela que je trouve être peu connu de vous, à accepter votre trop généreuse proposition de prendre ces frais à votre charge.

Je tiens si particulièrement à publier une bonne édition française, supérieure aux autres traductions, que si vous pouviez me faire donner, soit des additions de texte, soit de nouvelles et nombreuses figures à graver, j'en serais heureux.

Peut-être, avez-vous remarqué déjà que certaines figures mal gravées de la première édition française ont été refaites avec soin.

Je vous prie d'agréer, Monsieur, l'assurance de ma considération très distinguée. | Gauthier-Villars

P.S. Je pourrais joindre mes plaintes aux vôtres, au sujet des traductions. Plus de 80 ouvrages importants dont je suis propriétaire ont été traduits en Allemagne, en Italie et même en Angleterre. Pour un seul (Théorie Mécanique de la Chaleur par M. Briot), on a jugé à propos de demander l'autorisation à M. Briot et à moi; cette cession pour toutes les éditions, nous a rapporté à chacun 300f. Un ouvrage important (Géométrie Descriptive de Leroy), dont j'ai acheté il y a trois ans la propriété aux héritiers de l'auteur moyennant 36,000 francs, vient d'être traduit et mis en vente par l'État-Major de l'Armée Italienne, tout cela sans autorisation.

Ma maison n'a pas, au contraire, le reproche à se faire d'avoir traduit un seul ouvrage, sans y avoir été préalablement autorisé par l'auteur et par l'éditeur.

Mr. J. Tyndall, à Royal Institution à Londres.

[Printing and Books | For Mathematics, Science and the Arts | Quai des Grands-Augustins, 55, | Gauthier-Villars, successor of Mallet-Bachelier | Printer-Seller | Of the Bureau of Longitude; of the Polytechnic school; of the Imperial Observatory of Paris; of the Central School for the Arts and Manufacturing. – Publisher of the *Proceedings of the Weekly meetings of the Academy of Sciences*; of the *Journal of Mathematics*, by Mr Liouville; of the *[Bulletin] of Mathematical and Astronomical Science*, by Mr Darboux and Mr Houël; of the *New Annals of Mathematics*, by Mr Gerono and Mr Bourget; of the *Annals of the Observatory of Paris*; of the *Works of Lagrange*, etc.

Paris,] 5 June [187]3.

Sir,

It is with deep regret that I learned, through your letter of 26 May,[1] that the French edition of Heat,[2] which is currently being printed, did not entirely conform to the English edition.

When Father Moigno gave me the text[3] to type up, I asked him if it was in line with the 4th English edition. He replied in the affirmative and also added that he carefully compared it to the copies of the text in English. As it was you who expressed the desire to entrust the translation to Father Moigno, and also since you seemed satisfied with his earlier translations of Heat and Sound,[4] I was confident to continue and I thought that you will be happy with the translation and the necessary material for the printing of your book.

However, due to an extensive experience, I knew that the eye of the master is irreplaceable. Also, I will remind you that in the beginning, I asked you to kindly review the manuscript of the translation or the samples, but you told me that you could not spend any time on it! What more could I do?

In your letter you add that the French editions of your books have caused you a lot of trouble, without making a penny, and that France is the only country that you had to complain about. Our unfortunate country is so heavily attacked from all directions that this criticism was painful for me. Allow me to respond briefly.

With regard to the first edition of Heat, your publisher went bankrupt, and really the whole country cannot be held responsible. As for Sound, which I published, I did not believe that it gave you any inconvenience, but if I am wrong, please do specify your criticism. This book did not have the swift sales that it deserved because there was a German edition[5] and now that the French are no longer considered to excel in anything, we buy, even in Italy, the German edition rather than the French edition. Besides, you already knew well enough the harm that the German edition would cause to the French one because at that time you asked me, with extraordinary care, to share in the losses.[6] I refused, of course.

I apologize, Sir, for going into all these details, but I have not had the honour of meeting you, and it is painful to seem unable to fulfil my duties as a publisher, when on the contrary we dedicate all our effort into producing publications that are both meticulous and accurate.

In the present case, I am especially disappointed because there is no one in the world that is a more sincere admirer of your work.

Leaving the past behind, let us see how we may repair the damage. Father Moigno, with whom I shared your letter, answered me:

"I cannot imagine what could be the differences that have so offended M^r^. Tyndall, and about which I am sorry, as I consider my translation being perfect, more perfect than the German edition.

"*[No matter]* which differences worried M^r^. Tyndall, I will have a very competent man do the job of comparison that is required. We will take care of all the differences with the utmost care, we will tell him about it, and he himself will be convinced that they do not require in any way the rigorous measure about which he speaks to you."

Following this revision, if there are any important passages to modify, I will retype and reprint the pages that need to be replaced. If there are notes to add to the appendix, I will do so, in one word, I will do my best to make you satisfied.

I stopped the printing of the sheet that was being produced as soon as your letter reached me. I will *[make use]* of the samples of this sheet or *[1 word illeg]* and I take the liberty of sending them to you. In order to not waste your time, please entrust the revision of this paper to one of your assistants, or to any other person that inspires confidence in you; it will be the same for the other sheets.

You could just let me know of the expenses necessary for this revision and I will immediately send you that sum. I am not [a man], and this is why I find to be [little known to you], to accept the too generous offer to pay for these expenses yourself.

I have such a strong desire to publish a good French edition, better than any other translations, that if you could give me either additional text or many new figures to etch, I would be happy for it.

Perhaps you already noticed that some badly [printed] figures in the first French edition have been remade with great care.

Please accept, Sir, my kindest regards. | Gauthier-Villars

P.S. I could add my own complaints to yours about the translations. More than 80 major books that I own have been translated in Germany, Italy and even in England. For only a single one (Mechanical Theory of Heat by Mr. Briot[7]), it was seen fit to ask for the authorization of Mr Briot and myself. This [transfer] for all the editions brought us in 300[f] each. An important book (Descriptive Geometry by Leroy[8]), which I bought the property from the heirs of the author rights three years ago, which cost 36,000 francs, has just been translated and put up for sale by the Italian Army Staff, all that without authorization.

On the contrary, my publishing house does not have to blame itself for doing any translations without first receiving authorization from the author and the publisher.

M[r]. J. Tyndall, at the Royal Institution in London.

RI MS JT/1/V/20

1. *your letter of 26 May*: letter missing.
2. *French edition of Heat*: see letter 3829, n. 3.
3. *When Father Moigno gave me the text*: François-Napoléon-Marie Moigno wrote to Tyndall about the new French edition of *Heat* then in progress in letters 4050 and 4080.
4. *earlier translations of Heat and Sound*: Tyndall's *Heat* (1863) was first published in French as J. Tyndall, *La Chaleur considérée come un mode de mouvement*, trans. F. Moigno (Paris: Étienne Giraud, 1864); and his *Sound* (1867) as J. Tyndall, *Le Son, Cours Expérimental fait à l'Institution Royale*, trans. F. Moigno (Paris: Gauthier-Villars, 1869).
5. *German edition*: J. Tyndall, *Der Schall: acht Vorlesungen gehalten in der Royal Institution von Grossbritannien*, trans. H. Helmholtz and G. Wiedemann (Braunschweig: Friedrich Vieweg, 1869). In practice, the book was not translated by Hermann Helmholtz and Gustav Wiedemann, but by their wives Anna Helmholtz and Clara Wiedemann.
6. *at that time you asked me, with extraordinary care, to share in the losses*: letter missing.
7. *Mechanical Theory of Heat by Mr. Briot*: C. Briot, *Théorie Mécanique de la Chaleur* (Paris: Gauthier-Villars, 1869. Charles Auguste Briot (1817–82), a French mathematician.
8. *Descriptive Geometry by Leroy*: C. F. A. Leroy, *Traité de Géométrie Descriptive*, 2 vols (Paris: Gauthier-Villars, 1867).

From Jean-Albert Gauthier-Villars — 12 June 1873 — 4110

[IMPRIMERIE ET LIBRAIRIE | *Pour les* Mathématiques, *les* Sciences *et les* Arts, | QUAI DES GRANDS-AUGUSTINS, 55. | GAUTHIER-VILLARS, SUCCESSEUR DE MALLET-BACHELIER, | IMPRIMEUR-LIBRAIRE | Du Bureau des Longitudes; de l'École Polytechnique; de l'Observatoire Impérial de Paris; de l'École Centrale des Arts et Manufactures.—Éditeur des *Comptes Rendus Hebdomadaires des Séances de l'Académie des Sciences*; du *Journal de Mathématiques*; par M. LIOUVILLE; du *Bulletin des Sciences mathématiques et astronomiques*, par MM. DARBOUX et HOÜEL; des *Nouvelles Annales de Mathématiques*, par MM. GERONO et BOURGET; des *Annales de l'Observatoire de Paris*; des *OEuvres de Lagrange*, etc.

Paris, le] 12 juin 1873 | N° 8472

Monsieur,

Avant tout, je vous remercie très sincèrement de la lettre si bienveillante que vous m'avez adressée, et dont j'ai été très heureux.

Je partage entièrement votre manière de voir au sujet de la traduction et ferai mes efforts pour réparer, dans la limite du possible, les omissions et erreurs commises.

La faute de la première page est très regrettable; je vais faire recomposer en conséquence les 2 pages 1,2 et les pages correspondantes ou plutôt *[*attenantes*]* 11,12. Le *[1 word illeg]* sera substitué aux pages *[1 word illeg]*.

Afin de ne pas jeter au *[1 word illeg]*trop de feuilles, j'ai l'honneur de vous proposer de reproduire à la fin les notes omises. La traduction aura soin d'en avertir le lecteur dans un petit avertissement. Les épreuves de cet avertissement et des notes vous seront envoyées.

Vous *[*ne*]* me répondez pas au sujet de la révision des épreuves, à mes frais, par une personne de confiance choisie par vous. Ce sera, je vous assure, une grande satisfaction de me charger de cette dépense, et d'avoir ainsi la certitude d'une traduction bien conforme à l'original.

Si vous pouvez me faire retourner bientôt les épreuves que vous m'avez envoyées, je ferai exécuter les corrections avec le plus grand soin.—J'aurai des placards à vous envoyer dans quelques jours; les premières épreuves de ces placards ont été complètement bouleversées par l'abbé Moigno, qui maintenant se *[*pique*]* *[*d'honneur*]*. C'est un peu tard malheureusement.

J'accepte avec grand plaisir le soin de publier une traduction de vos lectures américaines; et ce travail se fera, j'en suis sûr, à notre mutuelle satisfaction.

Vous pouvez compter sur tous mes soins.

L'ouvrage est-il publié déjà en anglais? Pouvez-vous me faire envoyer les feuilles déjà parues? Combien de pages aura l'ouvrage?

Excusez, je vous prie, toutes ces questions.

Pour la traduction, je prendrai, ou non, l'abbé Moigno, suivant votre désir, et aussi suivant les convenances de l'abbé qui a déjà beaucoup de travail.

Voici mon avis à ce sujet:

Personne ne peut mieux que lui rendre le charme de votre style dans une forme parfaitement française; mais il faudra que la traduction *[donnée]* par lui soit revue, en Angleterre ou en France, par un homme exact, consciencieux, comparant la traduction avec l'original phrase à phrase. C'est le procédé que nous suivrons, si cela vous convient.

Veuillez agréer, Monsieur, avec mes nouveaux remerciements, l'assurance de ma haute considération

Gauthier-Villars

P.S.—Je viens de relire les feuilles tirées de la Chaleur; je puis vous assurer, et c'est aussi l'avis de gens bien compétents, qu'abstraction faite des omissions et de la non-conformité de quelques passages avec la dernière édition, la traduction a une forme d'une grâce incomparable et a su vraiment conserver la fleur de votre style.

Mr Tyndall *[Prof]* à Royal Institution, London

[Printing and Books | For Mathematics, Science and the Arts | Quai des Grands-Augustins, 55, | Gauthier-Villars, successor of Mallet-Bachelier | Printer-Seller | Of the Bureau of Longitude; of the Polytechnic school; of the Imperial Observatory of Paris; of the Central School for the Arts and Manufacturing. – Publisher of the *Proceedings of the Weekly meetings of the Academy of Sciences*; of the *Journal of Mathematics*, by Mr Liouville; of the *[Bulletin] of Mathematical and Astronomical Science*, by Mr Darboux and Mr Houël; of the *New Annals of Mathematics*, by Mr Gerono and Mr Bourget; of the *Annals of the Observatory of Paris*; of the *Works of Lagrange*, etc.

Paris,] 12 June 1873 | N° 8472

Sir,

First of all, I am sincerely grateful for the kind letter that you sent me,[1] and which made me very happy.

I entirely agree with your point of view on the subject of translation, and I will do my best to fix, as much as possible, the omissions and the errors that were made.

The mistake on the first page is very unfortunate; consequently, I will rewrite the 2 pages, 1, 2 and the corresponding pages, or rather the *[*adjoining*]* pages 11, 12. The *[1 word illeg]*will be substituted by *[1 word illeg]* pages.

In order to not throw *[*away*]* too many pages, I have the honour of proposing to you the reproduction of the omitted notes at the end. The translation will take care of notifying the reader in a small notice. The drafts of this notice and of the notes themselves will be sent to you.

You do not reply to me in regards to the proofreading of the drafts, at my expense, using a trusted person chosen by you. I assure you that it would be to my pleasure to take care of those expenses, and thus to have the assurance of a good translation conforming to the original.

If you could soon return to me the drafts that you sent me, I will make the corrections with the utmost care.—I will have *[*posters*]* to send to you in a few days; the first drafts of those posters were completely *[*reworked*]* by Father Moigno,[2] who now *[*prides himself*]* in it. It is a little late, unfortunately.

I accept with great pleasure the task of publishing the translation of your American lectures; and this work will be, I am sure, to our mutual satisfaction. You can count on my utmost care.

Is the book already published in English?[3] Could you send me the papers that were already published? How many pages will the book have?

I beg of you to excuse me for all these questions.

For the translation, I will either employ Father Moigno or not, depending on your wishes, and also depending on his availability because he already has a lot of work to do.

Here is my opinion on this matter:

No one can render the charm of your style in a perfectly French manner better than him. But the translation done by him should be proofread, either in England or in France by a meticulous and thorough man, in order to compare the translation with the original text, sentence by sentence. This is the procedure that we will follow, if you agree with it.

Please accept, Sir, with renewed gratefulness, the assurance of my highest regards Gauthier-Villars

P.S.—I just reread the pages taken from Heat. I can assure you, and that is also the opinion of other experts, that apart from the omissions and some non-conforming passages in accordance with the latest edition, the translation was done with incomparable grace and was able to truly keep the *[*essence*]* of your style.

Mr Tyndall *[*Prof*]* at the Royal Institution, London

RI MS JT/1/V/21

1. *the kind letter that you sent me*: letter missing, but presumably a response from Tyndall to Gauthier-Villars' letter of 5 June (letter 4109), addressing Tyndall's complaints about the new French edition of *Heat*.
2. *Father Moigno*: François-Napoléon-Marie Moigno.
3. *Is the book already published in English?*: Tyndall had finished work on the English edition of his American lectures on light—*Six Lectures on Light*—earlier that month (see letter 4075, n. 3). For the translation, see letter 4080, n. 4.

From Heinrich Rudolf Vieweg[1] 13 June 1873 4111

Braunschweig d. 13ⁿ Juni 1873. | Herrn Professor J. Tyndall | London.

Am 12 Aug 1872 erlaubten wir uns einen Brief an Sie nach Bel-Alp, Wallis zu adreßiren, in dem wir Ihnen anzeigten daß die Uebersetzung Ihres Werkes „Hours of exercise" zur Ausgabe gekommen und anfragten, wo wir Ihnen das fällige Honorar in Betrage von 15£ zahlen sollen.

Wir haben keine Antwort erhalten. Wahrscheinlich ist der Brief nicht in Ihre Hände gelangt. Sie haben wohl gelegentlich die Güte sich des einliegenden Wechsels auf G & A Worms in London zu bedienen, und die Honorarforderung zu erledigen.

Am 14ⁿ April schrieben Sie uns, daß Sie uns zu der neuen Auflage Ihres Buches „Heat a mode of motion" ein 16ⁿ Capitel senden wollten. Es würde uns angenehm sein, wie wir schon die Ehre hatten in unserm Brief vom 16ⁿ April Ihnen zu schreiben, wenn Sie uns diesen Abschnitt einschicken wollten, damit *[H]* Wiedemann Zeit behält nach Bequemlichkeit die Uebersetzung zu veranlaßen und zu revidiren.

Wir hatten Sie in demselben Briefe gebeten uns Ihre in Amerika gehaltenen Vorlesungen „on Light" zu senden. Wir haben dieselben nicht durch den Buchhandel erhalten können und uns zur Herausgabe der Uebersetzung autorisiren zu wollen. Wir hoffen daß H. Wiedemann *[& H.]* Helmholz auch diese Arbeit wieder unter Beifugung ihrer Namen und von ihnen revidirt erscheinen laßen werden. Wenigstens hat H. Wiedemann uns in diesem Sinne geschrieben. Wir erlauben uns diese Mittheilungen zu wiederholen weil wir seit zwei Monat nicht die Antwort erhalten und fürchten daß der Brief Sie nicht erreicht hat.

Mit größter Hochachtung | ergebenst | Friedr Vieweg & Sohn

Braunschweig 13th June 1873. |
Professor J. Tyndall | London.

On 12 Aug 1872, we took the liberty to send you a letter[2] to Bel-Alp, Valais,[3] where we informed you that the translation of your work "Hours of exercise" had been published[4] and enquired where to send the payable royalties of £15.

We have not received an answer. The letter has probably not reached you. Please be so kind as to make use of the enclosed promissory note drawn on G & A Worms in London to finalise the payment of your royalties.

You wrote to us on 14th April that you wanted to send us a 16th chapter for the new edition of your book "Heat a mode of motion".[5] It would be agreeable for us, as we had the honour to mention in our letter from 16th April,[6] if you sent us the passage so that *[Mr]* Wiedemann has sufficient time to arrange the translation and revise it at his convenience.

In the same letter, we had requested that you send us the lectures "on light" that you held in America.[7] We were not able to obtain them through the book trade, and to authorise us to publish the translation. We hope that Mr Wiedemann *[& Mr]* Helmholtz will allow the publication also of this work under mention of their names and with their revisions. At least, Mr Wiedemann has written to us to this effect. We allow ourselves to repeat these communications, as we have not received a response in two months, and we fear that the letter may not have reached you.

Respectfully | your humble | Friedr Vieweg & Son.

RI MS JT/1/V/14
TU Braunschweig/Universitätsbibliothek/UABS V 1 T : 63
Vieweg-Archive der Universitatsbibliothek Braunschweig 933

1. *Heinrich Rudolf Vieweg*: while the letter is signed 'Friedr Vieweg & Sohn', we believe the sender is Heinrich Rudolf Vieweg, who took over the family publishing firm, Friedrich Vieweg und Sohn, from his father Eduard Vieweg (1797–1869).
2. *letter*: letter 3820.
3. *Bel-Alp, Valais*: see letter 3771, n. 5.
4. *your work "Hours of exercise" had been published*: Tyndall's *Hours of Exercise in the Alps* was published in German as J. Tyndall, *In den Alpen* (Braunschweig: Friedrich Vieweg und Sohn, 1872).
5. *You wrote to us... "Heat a mode of Motion"*: the letter mentioned is missing, though around the same time Tyndall had asked that a new French edition of the book likewise include a new chapter from the most recent English edition; see, e.g., letter 4080. The new German edition was J. Tyndall, *Die Wärme, betrachtet als eine Art der Bewegung*, trans. H. Helmholtz and G. Wiedemann (Braunschweig: Friedrich Vieweg und Sohn, 1875). It is not clear if Helmholtz's and Wiedemann's wives Anna and Clara, respectively, actually did the

translations while their husbands received the credit, as had happened for the first German translation of *Heat* and other of Tyndall's books.

6. *our letter from 16th April*: letter missing.
7. *we had requested... held in America*: for the same wish to translate and publish the *Six Lectures on Light* volume expressed at the same time from Tyndall's French publisher, see letter 4110. The German edition was J. Tyndall, *Das Licht. Sechs Vorlesungen gehalten in Amerika im Winter 1872–1873*, trans. G. Wiedemann (Braunschweig: Friedrich Vieweg und Sohn, 1876).

To Heinrich Rudolf Vieweg 16 June 1873 4112

Royal Institution of Great Britain | 16th. June 1873

Gentlemen.[1]

I thank you much for the draft for £15 which you have sent me.[2]

You will probably before the week is ended have the whole of the proofs of my American Lectures[3] in your hands. It will be especially agreeable to me if my eminent friends[4] who have already placed my popular works in so noble a fashion before the German people, can undertake the superintendence of this one.

I shall very soon send you the chapter to which you refer.[5] Just now I am excessively busy.

I have not received any copies of the Fragments of Science—Is it published?[6]

Yours faithfully | John Tyndall

TU Braunschweig/Universitätsbibliothek/UABS V 1 T : 63

1. *Gentlemen*: while this letter is addressed to 'Gentlemen', we believe the recipient is Heinrich Rudolf Vieweg, who took over the family publishing firm, Friedrich Vieweg und Sohn, from his father Eduard Vieweg (1797–1869).
2. *draft for £15 which you have sent me*: in his previous letter, Vieweg informed Tyndall that the translation of his *Hours of Exercise in the Alps* (1871) had been published and asked where to send the royalties of £15 (see letter 4111, n. 4).
3. *proofs of my American Lectures*: in his previous letter, Vieweg asked Tyndall to send a copy of his published American lectures (*Six Lectures on Light*), as he was unable to obtain a copy through the book trade (see letter 4111, n. 7)
4. *my eminent friends*: Hermann Helmholtz and Gustav Wiedemann.
5. *chapter to which you refer*: in his previous letter, Vieweg asked Tyndall to 'send us a 16th chapter for the new edition of your book "Heat a mode of motion"' (see letter 4111, n. 5).
6. *Is it published?*: see letter 3735, n. 3.

From Thomas Henry Huxley 16 June 1873 4113

4 Marlborough Place[1] | June 16th 1873

My dear Tyndall,

I return Agassiz's letter[2] herewith. It is a nuisance having to belabour these fellows,—but as the work has to be done, much power to your elbow![3]

Ever yours very sincerely | T. H. Huxley.

RI MS JT/1/TYP/9/3015
IC HP 8.149
Typed Transcript Only

1. *4 Marlborough Place*: see letter 3947, n. 1.
2. *I return Agassiz's letter*: probably the letter from Alexander Agassiz which Thomas Huxley had intended to return to Tyndall as an enclosure on 6 May; see letter 4081, n. 2.
3. *It is a nuisance... to your elbow!*: Tyndall was then at work on his defense of what he wrote about the priority claims of Louis Agassiz versus James David Forbes, resulting in 'Principal Forbes and His Biographers' (see letter 4024, n. 5). For an excellent summary, see *Ascent of John Tyndall*, pp. 317–8.

From Herbert Spencer 17 June [1873][1] 4114

The Athenaeum[2] | 17 June

My dear Tyndall

All right. I feared that I had offended <u>you</u> and am glad it is not so.

Thanks for the letter.[3] If you are not going to use it, and do not object to publication of it, I should like to give it as an illustration of Christian charity.[4]

Ever yours | Herbert Spencer

RI MS JT/1/S/183
RI MS JT/1/TYP/3/1174

1. *[1873]*: the year is given based on the letter being discussed.
2. *The Athenaeum*: see letter 3710, n. 5.
3. *letter*: Tyndall presumably sent to Spencer the letter he received from an American clergyman in March 1873 (letter 4010).
4. *If you are not... illustration of Christian charity*: Spencer published the letter from the American clergyman in his *The Study of Sociology* (London: Henry S. King, 1873), p. 419, to demonstrate the proposition that 'frequent repetition of Christian doctrines does not conduce to growth of Christian feelings' (p. 418).

From Frances Russell 19 June 1873 4115

[Pembroke Lodge.[1] | Richmond Park.] | June 19/73

Dear Mr Tyndall

Now that y[ou]r lectures are over[2] I hope you can come to dine & sleep here—When can it be? Tomorrow? the Friday following, 27th? Pray name some day—*[At]* any except the next two Saturdays w[oul]d suit us,

Y[ou]rs sincerely | F Russell

I think I told you that we are to be found all Friday & Sunday afternoons—

RI MS JT/1/R/65

1. *Pembroke Lodge*: see letter 4087, n. 1.
2. *y[ou]r lectures are over*: see letter 3782, n. 10.

From Herbert Spencer 19 June 1873 4116

38 Queen's Gardens, | Bayswater, W.[1] | 19 June 1873

Dear Tyndall,

The balance due per head, to defray the X expenditure this year, is £3.2.7.[2]

Truly yours, | Herbert Spencer

RI MS JT/1/S/102
RI MS JT/1/TYP/3/1117

1. *37 Queen's Gardens | Bayswater, W.*: Spencer's London residence since September 1866.
2. *The balance due... £3.2.7*: it fell to Herbert Spencer, as Treasurer of the X Club, to calculate and collect the annual dues. On the X Club, see letter 3757, n. 2.

From Gustav Wiedemann 19 June 1873 4117

Leipzig [d.] 19 Juni 73

Mein lieber Tyndall!

Zuerst meinen herzlichen Glückwunsch zu den schönen Erfolgen Ihrer amerikanischen Reise, der Sie ja Selbst noch nachträglich den schönsten Abschluß gegeben haben. Ich wollte, ich hätte Sie begleiten können; schon lange möchte ich *[einmal]* nach Amerika hinüber und suche nur einen zwingenden Grund zur Reise. Vielleicht kann ich einmal eine Kabellegung dorthin mitmachen.

Ihre ersten 5 Bogen On light habe ich eben von Vieweg erhalten und will morgen in Berlin mit Helmholtz sprechen, ob die Uebersetzung von unseren Frauen übernommen werden kann. Das Leben in den größeren Städten macht *[mehr]* Ansprüche an sie, so daß sie für eine wissenschaftliche Arbeit wenig Zeit behalten. Jedenfalls sorge ich für eine tüchtige Uebersetzung, wenn die Damen nicht Zeit gewinnen können, was ich aber nicht hoffen und wünschen möchte. Auf wieviel Bogen haben Sie das Buch etwa *[berechnet]*?

Meine Frau reist vom 13 Juli–13 August nach Scheveningen, um dann mit den Kindern der Schule wegen hierzubleiben. Ich bin nachher ganz frei und dachte an ein englisches Seebad, wo ich vielleicht mit einigen Bekannten von hier zusammentreffen könnte. Können Sie mir über die *[Bäder]* nicht einige Auskunft geben, über Ventnor *[&]* Ryde oder die beiden an der Küste des englischen Festlandes? Gewiß können Sie, was Sie nicht aus eigener Erfahrung kennen, durch Ihre verheiratheten *[&]* unverheiratheten Freunde erfahren; vielleicht auch Einiges über die Reise. Für alle Notizen wäre ich Ihnen sehr verbunden.

Was denken sie zu unternehmen? Es würde mich doch recht sehr erfreuen, wenn ich nach England käme, Sie dort wiederzusehen, wenn Sie nicht vielleicht vorher auf *[dem]* Wege nach Wien durch Leipzig passiren.

Wo ist die British assoc. in diesem Jahr und zu welcher Zeit?

Mit den herzlichsten Grüßen von meiner Frau

Ihr | stets treu ergebener | G. Wiedemann

Leizpig 19 June 73

My dear Tyndall!

First of all, my warm congratulations for the fine successes of your American trip,[1] to which you yourself have of course given still later the finest conclusion.[2] I wished I could have accompanied you; for a long time now, I have been wanting to go over to America *[sometime]*, and am only looking for a compelling reason to make the trip. Perhaps I can participate in the laying of a cable there sometime.

I have just received your first 5 sheets of On light[3] from Vieweg and shall speak with Helmholtz in Berlin tomorrow as to whether the translation[4] can be undertaken by our wives.[5] Life in the larger cities makes *[more]* demands on them, with the result that they are left with little time for a scientific work. At any rate, I shall provide a competent translation if the ladies cannot find time, which, however, I would not want to hope or wish for. Approximately how many sheets have you *[calculated]* for the book?

My wife will be travelling to Scheveningen from 13 July–13 August, in order to stay here then with the children[6] because of school. I shall be quite free afterwards, and was thinking of an English seaside resort where I could perhaps meet with some acquaintances from here. Can you not give me some information about the *[resorts]*, about Ventnor[7] *[&]* Ryde,[8] or the two on the coast of the English mainland? I'm sure you can find out what you do not know from your own experience from your married [and] unmarried friends; perhaps also something about the trip. I would be very obliged to you for any quick note.

What are you thinking of doing? I would be most delighted to see you again if I came to England, unless perhaps you are passing through Leipzig on [the] way to Vienna before then.

Where is the British assoc. this year and at what time?[9]

With the warmest regards from my wife

Your | ever loyally devoted | G. Wiedemann

RI MS JT/1/W/48–48TRANS

1. *your American trip*: for Tyndall's American lecture tour, see Introduction, n. 8.
2. *finest conclusion*: see Introduction, n. 15.
3. *On light*: *Six Lectures on Light*.
4. *translation*: see letter 4111 for the attempt by Tyndall's German publisher Vieweg to secure rights to publish a German edition of *Six Lectures on Light*, with the translation overseen by Wiedemann (though, plainly, the main work was to be done by his wife Clara Wiedemann). The translation appeared as J. Tyndall, *Das Licht: Sechs Vorlesungen, gehalten in Amerika im Winter 1872–1873*, trans. G. Wiedemann (Braunschweig: Friedrich Vieweg und Sohn, 1876).
5. *our wives*: Clara Wiedemann and Anna Helmholtz. Only Clara would work on the translation of *Six Lectures on Light*.
6. *children*: Eilhard Ernst Gustav Wiedemann (1852–1928), who became professor of physics at Erlangen in 1886; Carl Alfred Wiedemann (1856–1936), who was appointed to the extraordinary professorship of Egyptology at Bonn in 1892; and Margaretha Laura Clara von Domarus (née Wiedemann, 1861–1927).
7. *Ventnor*: an English seaside resort town located on the southeast coast of the Isle of Wight.
8. *Ryde*: an English seaside resort town located on the northwest coast of the Isle of Wight.
9. *British assoc. this year and at what time*: the forty-third annual meeting of the BAAS was held in Bradford from 17–24 September 1873 (*Brit. Assoc. Rep. 1873*).

To unidentified[1] 24 June [1873][2] 4118

Royal Institution | 24th June.

My dear Friend,

It was only yesterday that I heard from Pollock[3] that you had been ill, and right sorry I was to receive the intelligence. I should be very near the limit of my labours here were it not for the "Life of Forbes"[4] which must have some of my attention, although twelve years ago I let the assault which it contains pass without rejoinder. But it is now republished with a challenge for an answer, the quarrelsome man[5] who is responsible for its publication, backed I fear by

Mrs Forbes[6] and her son,[7] imagining that because I did not answer at the time I could not answer. I will say what I have to say as mildly as I can though the temptation is a sore one to say things strongly.[8]

On the 2nd I grieve to say the Trinity Corporation, of which I am the scientific adviser,[9] have their annual dinner, and I must be there. The rest of the week also I fear I shall be away, steaming between the South Foreland and the coast of France, and making observations on guns and trumpets and other means of warning the mariners in fogs.[10] Of course if I ever come near to Edinburgh I will call to see you, and it is a great cause of regret to me that I am unlikely to see you now.

Good bye my dear friend | believe me always | affectionately yours | John Tyndall

Give my kindest regards to Miss Herries;[11] and tell Charles Herries[12] that I miss his face at the Athenaeum,[13] and that I hope he will soon be well.

RI MS JT/1/TYP/3/1021
LT Typescript Only

---

1. The recipient is possibly the Scottish physiologist William Rutherford (1839–99), who served as professor of physiology at the RI from 1871–4, before taking up a professorship in Edinburgh (*ODNB*).
2. *[1873]*: the year is established by reference to the controversy regarding the volume *Life and Letters of Forbes*.
3. *Pollock*: probably William Frederick Pollock. Letter missing.
4. *"Life of Forbes"*: *Life and Letters of Forbes*.
5. *quarrelsome man*: Peter Guthrie Tait.
6. *Mrs Forbes*: Alicia Forbes (née Wauchope, 1824–85), married in 1843 (*ODNB*).
7. *her son*: George Forbes.
8. *I should be... say things strongly*: Tyndall's eventual published response, in which he answered the charge that he plotted against Forbes in order to rob him of due scientific recognition, was his 'Principal Forbes and His Biographers' (see letter 4024, n. 5).
9. *Trinity Corporation, of which I am the scientific adviser*: see letter 3869, n. 4.
10. *The rest of... mariners in fogs*: for Tyndall's work in this period evaluating different warning systems for ships in fog, see *Ascent of John Tyndall*, p. 318.
11. *Miss Herries*: probably a sister of Charles Herries, as he had no children (see n. 12).
12. *Charles Herries*: Charles John Herries (1815–83), a well-known London financier (*ODNB*).
13. *Athenaeum*: the Athenaeum Club (see letter 3710, n. 5). Herries was elected a member in 1849 (*Athenaeum. Rules & Regulations and List of Members. 1874* (London, 1874), p. 62).

From Rudolf Clausius 25 June 1873 4119

BONN, 25 Juni 73.

Lieber Tyndall,

Ich lese mit grosser Freude in einer Deutschen Zeitung, dass Du in Oxford zum Ehren-Doctor vorgeschlagen bist, und ich wünsche Dir von Herzen Glück dazu. Schon früher habe ich mit grossem Interesse in verschiedenen Zeitschriften davon gelesen, wie grosse Erfolge Du in America durch Deine Vorlesungen erzielt hast. Zugleich habe ich mit Bewunderung gehört, dass Du die bedeutenden Einnahmen welche Du Dir dabei in der ehrenvollsten Weise verdient hast, und welche Dir persönlich für Dein künftiges Leben viele Annehmlichkeiten bereiten könnten, in so uneigennütziger Weise einem gemeinnützigen wissenschaftlichen Zwecke geopfert hast. Dein Name der schon in so ruhmreicher Weise verbreitet ist, wird dadurch mit um so mehr Verehrung in allen Ländern genannt werden.

Ich denke mir, dass die Reise nach America und der dortige Aufenthalt für Dich auch sehr genussreich gewesen sein muss. Obwohl die americanischen Verhältnisse aus den englischen hervorgegangen sind, so hat sich dort doch auch manches Neue und Eigenthümliche entwickelt, und der Aufschwung America's hat so grossartige Dimensionen angenommen, dass man dort gewiss viel Gelegenheit zu interessanten Beobachtungen hat.

Ich erhalte soeben das Buch von Maxwell über Electricität und Magnetismus zugesandt, und ich ersehe daraus, dass Maxwell mir das Buch sendet, zu meiner Freude, dass er mir meine Reclamation, zu der ich mich früher genöthigt gesehen habe, nicht zu übel genommen hat. Er scheint darin sehr verschieden von Tait zu sein. Es ist mir das um so angenehmer, als ich ihn wirklich als Gelehrten sehr hoch achte. Nun lebe wohl und nimm noch einmal die herzlichsten Glückwünsche von meiner Frau und mir zu den so wohl verdienten Ehrenbezeugungen welche Dir dargebracht werden, und freundliche Grüsse an Dich und Hirst von | Deinem | Clausius

BONN, 25 June 73.

Dear Tyndall,

I read with great joy in a German newspaper that you have been proposed as an honorary doctor at Oxford,[1] and I sincerely wish you good luck for this. I read quite a while ago now in different journals and with great interest about how you achieved great success in America through your lectures. I heard with admiration at the same time that you gave up in such an unselfish way the considerable income which you earned yourself in the process in the most honourable way, and which could provide you personally with many comforts for your future life, for a charitable scientific purpose.[2] Your name, which is already widespread in such a glorious way, will through this be mentioned with all the more veneration in all countries.

I imagine that the voyage to America and the stay there must have been very enjoyable for you also. Although conditions in America grew out of those in England, a good many new and curious things have nevertheless developed there too, and the rise of America has assumed such magnificent dimensions that one certainly has much opportunity there for some interesting observations.

I have just this minute received the book by Maxwell on electricity and magnetism,[3] and I see from it that Maxwell is sending me the book and, to my delight, that he did not take the complaint to which I saw myself compelled earlier[4] too badly. He seems in that respect to be very different from Tait.[5] That is all the more agreeable to me considering I really hold him in very high regard as a scholar. Fare well now, and accept once again the warmest congratulations from my wife and me on the such well-deserved tributes which are being offered to you, and kind regards to you and Hirst from | Your | Clausius

RI MS JT/1/TYP/7/2329
LT Typescript Only

1. *honorary doctor at Oxford*: see letter 4104, n. 2.
2. *I heard with admiration... a charitable scientific purpose*: for Tyndall's donation of his lecture tour profits, see Introduction, n. 15. On his American lecture tour in general, see Introduction, n. 8.
3. *book by Maxwell on electricity and magnetism*: for Tyndall's less-than-admiring comments on James Clerk Maxwell's *Treatise on Electricity and Magnetism*, see letter 4093.
4. *the complaint... compelled earlier*: over a decade before, Clausius had criticized Maxwell's kinetic theory of gases; see E. W. Garber, 'Clausius and Maxwell's Kinetic Theory of Gases', *Historical Studies in the Physical Sciences*, 2 (1970), pp. 299–319.
5. *very different from Tait*: Clausius was then in the midst of controversy with Peter Guthrie Tait over Clausius' claim to priority in formulating the second law of thermodynamics; see, e.g., P. G. Tait, 'On the History of the Second Law of Thermodynamics, in reply to Professor Clausius', *Phil. Mag.*, 43 (1872), p. 516–18.

To Rudolf Clausius 27 June 1873 4120

27th June 1873.

My dear Friend Clausius

It gave me the very greatest pleasure to receive your most friendly letter.[1] I had often thought of writing to you, and feared when I received your letter that I should find you upbraiding me for not doing so, but instead of this I find it filled by your old kindness.

The people at Oxford treated me very kindly. It was a great stretch of liberality on their part to accept me. I was proposed to them three or four years ago, but they would not hear of it on religious grounds—I[t] shows the advance of liberal sentiments in the university. The undergraduates received me in the most cordial manner.[2]

There has been a little correspondence in the newspapers about it; but nothing unkind has been spoken.

The American people also treated me with wonderful cordiality. They had been sending me invitations for five or six years, and in the end I resolved to go to them, but at the same time I resolved to let no foreign element, or motive, mingle with what I regarded as a purely friendly visit.[3]

I am very glad to hear what you say about Maxwell. Depend upon it he is a noble fellow, and as different from Tait as day is from night.[4] Tait may be very able fellow, but he is a quarrelsome, vulgar man. He and two others have been recently writing the Life of James Forbes, and have taken occasion to publish an attack upon me.[5] Tait also goes out of his way to say something unpleasant about you. I am now engaged in writing a reply to this attack. They never yet gained anything by their attacks on me, and they will gain just as little now.[6]

It is very disgusting having to occupy time in this way, and I shall take occasion to tell the public that I do not feel called upon to notice Tait for the future.

Give my kindest regards to your wife[7] & my love to Johnny[8] & the children[9]—I wish I could see them again.

Yours always faithfully | John Tyndall

RI MS JT/1/T/205

1. *your most friendly letter*: see letter 4119.
2. *The people at Oxford... the most cordial manner*: on the awarding of an honorary doctorate by Oxford (which Clausius had read about in the newspapers), see letter 4104, n. 2.
3. *The American people... purely friendly visit*: for Tyndall's American lecture tour, see Introduction, n. 8.
4. *I am very glad... day is from night*: Tyndall was agreeing with Clausius' evaluation in his letter (4119), where he compared Peter Guthrie Tait unfavorably to James Clerk Maxwell, who, despite criticisms of earlier work from Clausius, had sent him a copy of his new book, *Treatise on Electricity and Magnetism.*
5. *He and two others... an attack upon me*: *Life and Letters of Forbes*. See letters 4033 and 4038.
6. *Tait may be... as little now*: Tyndall's reply to the attack by Tait came in August, 'Principal Forbes and His Biographers' (see letter 4024, n. 5).

7. *your wife*: Adelheid Clausius (née Rimpau, 1833–75), the wife of Rudolf Clausius. They married in Zürich in 1859 and had six children.
8. *Johnny*: Rudolf John 'Johnny' Clausius (b. 1864–*c.* 1915), a son of Adelheid and Rudolf Clausius, who was named after Tyndall, who was also his godfather. He completed a doctorate of law in 1890 at Friedrich Wilhelms University Bonn.
9. *children*: the Clausius's five other children were Mathilde (1861–1907), Helene (1863–1919), Hedwig (1866–1946), Alfred (1868–1950), and Else (1872–1954).

To Hector Tyndale 27 June 1873 4121

Royal Institution | 27th June 1873.

My dear Hector.

You will have long ago concluded that I have entirely forgotten you,[1] but I have not ceased to think of you through my tangle of heavy work. I had my lectures to give[2] when I returned. I had the English edition of them to prepare, and this took me a long time. I wished to bring them out on a better [dress] than I was able to put upon them in New York, and they will now constitute the first English edition and the second American edition of the lectures.[3] Then we had workmen about in the laboratories—indeed they are still there.[4] Then I had experiments on Fog Signals for the Trinity House and the Board of Trade,[5] which took me several times down to Dover, and which will take me there again next week. And finally to my supreme disgust I found on my return, that that harmless little book the Forms of Water, which I intended to be a model of fairness had raised the ire of the friends and relatives of the late Principal Forbes. A Life of this philosopher has been lately published under the joint auspices of three different authors, and in it an attack upon me is revived, which I am called upon to answer.[6] It is not the first assault made upon me from the north.[7] Hitherto they have gained little by their attacks, and when you see my rejoinder[8] you will say that they gain but little <u>now</u>.

Some years ago when the Marquis of Salisbury became Chancellor of the University of Oxford a number of scientific men were proposed for the degree of DCL. Darwin and myself were amongst them. Froude was also proposed.[9] But it was threatened that if we were brought forward hot work would ensue. We were withdrawn. I believe they agreed to take one of us, namely Darwin, but he was ill and did not accept the degree. I had no hope that Oxford would depart from this attitude in my lifetime—In fact I never thought about the matter. Some time ago one of the best men in the University wrote to me[10] saying that I was to receive the degree, and after various skirmishes, in which my friends behaved towards me in the most admirable way the degree was decided upon. I went down and doffed the red gown upon the 18th and was cordially received by the undergraduates. To my great gratification I met

Lowell there and we were 'doctored' together—Lowell I had been sorry to remark in his 'Study Window'[11] had been rather querulous about the 'airs' of Englishmen—but the frank and hearty reception which he experienced at Oxford will have gone far to dissipate all notions of this kind.

I shall be going to Ireland to look after the Lighthouses in a few days,[12] and shall probably run down to Gorey. By the way they sent me some time ago a photograph of Emma,[13] and said that they had sent you one. I wish you would burn it—It is not fair to her to have her rendered permanent while in infirm health.

Give my love to your wife,[14] and my kind regards to the other ladies.[15] You did not let me know your wishes regarding the headstone. It must now I fear stand over till my return from the Alps whither I go in a few days. Give my kindest regards to Lesley, his wife and two daughters.[16] I bear them all in aromatic remembrance; I have the photographs of the two girls which are very charming. Kind regards also to my numerous other friends in Philadelphia. It would fill a big sheet to name them all.

Goodbye dear Hector | Your affe[ctionate] Cousin | John

Give my best thanks to General *[Tilghman]*[17] to whom I intended to write months ago.

RI MS JT/1/T/1456
RI MS JT/1/TYP/5/1721–2

1. *You will have... forgotten you*: Tyndall had not written to his cousin Hector, who helped to organize Tyndall's American lecture tour (see Introduction, n. 8), since returning to London in mid-February.
2. *my lectures to give*: re-delivering his American lectures on light at the RI (see letter 3782, n. 10).
3. *first English edition and the second American edition of the lectures*: Tyndall's published American lectures appeared in the United States as *Lectures on Light delivered in the United States in 1872–'73* (New York: D. Appleton and Company, 1873), and in England as *Six Lectures on Light delivered in America in 1872–1873* (London: Longmans, Green, and Co., 1873), abbreviated to *Six Lectures on Light* elsewhere in this volume. By 'second American edition', Tyndall was referring to the printing of the full lectures in New York by the *New York Tribune* as a '3 cent' supplement (see letter 3962, n. 2).
4. *we had workmen... still there*: on the renovation of the RI's laboratories, see letter 3776, n. 14.
5. *experiments on Fog Signals... Board of Trade*: on these experiments, see *Ascent of John Tyndall*, pp. 318–21.
6. *to my supreme disgust... called upon to answer*: for a summary of the events that drew Tyndall into a debate about James David Forbes and his claims as an innovative glaciologist, see *Ascent of John Tyndall*, pp. 317–8.
7. *It is not... from the north*: on Tyndall's skirmishing with William Thomson and Peter

Guthrie Tait over whom to credit for the calculation of the mechanical equivalent of heat, see letter 4024, n. 6.

8. *my rejoinder*: for Tyndall's eventual published response, see letter 4024, n. 5.
9. *Froude was also proposed*: the English engineer William Froude (1810–79) was celebrated for his work on the science of ship design (*ODNB*). On the background at Oxford to the eventual awarding of an honorary doctorate to Tyndall in June 1873, see *Ascent of John Tyndall*, p. 321.
10. *one of the best men in the University wrote to me*: letter missing and thus the writer not identified.
11. *I met Lowell... in his 'Study Window'*: James Russell Lowell (1819–91), of Harvard in the United States, was awarded an honorary degree (Doctor of Civil Law) at Oxford alongside Tyndall. Lowell published the essay collection *My Study Windows* in 1871 (*ANB*).
12. *I shall be going to Ireland to look after the Lighthouses in a few days*: Tyndall sent a report to Trinity House regarding observing on 8 July 1873 a new light at the lighthouse at Howth Baily (see letter 4141).
13. *Emma*: Emily Tyndall.
14. *your wife*: Julia Tyndale.
15. *other ladies*: presumably their daughters (not identified).
16. *his wife and two daughters*: Lesley's wife was Susan Inches Lesley (née Lyman, 1823–1904), married in 1849. Their daughters were Mary Lesley (1854–1929), who later married Charles Wilberforce Ames (1855–1921) in 1883; and Margaret White Lesley (1857–1944), who became a painter and married the American sculptor Henry Kirke Bush-Brown (1857–1935) in 1886.
17. *General [Tilghman]*: not identified.

From Edward Frankland 27 June 1873 4122

Royal College of Chemistry | June 27/73

My dear Tyndall

I have no documentary evidence of the kind you mention;[1] but I may mention that on the part of the late Master of Trinity[2] & others there was undoubtedly a general belief that to Principal Forbes alone was due the credit[3] of establishing the analogy with liquid motion furnished by the greater central motion of a glacier. Neither the rough but intelligent work of Rendu, nor the actual measurements of Agassiz[4] were at that time generally known in this country, and to the best of my recollection they were never once alluded to by the proposer & seconder of Forbes for the Copley Medal.[5]

Yours sincerely | E. Frankland.

RI MS JT/1/F/54

1. *evidence of the kind you mention*: the letter to which Frankland was responding is missing.
2. *late Master of Trinity*: William Whewell (1794–1866), the Cambridge historian and philosopher of the sciences, had served as Master of Trinity College from 1841 until his death (*ODNB*).
3. *to Principal Forbes alone was due the credit*: on the dispute then roiling over the priority of the late James David Forbes in understanding the mechanics of glacial movement, see *Ascent of John Tyndall*, pp. 317–18.
4. *Neither the rough... measurements of Agassiz*: for Tyndall's own adjudication of the rival claims of Forbes and the Swiss students of glaciers Louis Rendu (1789–1859) and Louis Agassiz, see 'Principal Forbes and His Biographers' (see letter 4024, n. 5).
5. *to the best... Copley Medal*: in 1859 Forbes had been proposed by Roderick Impey Murchison and William Whewell for the Copley Medal of the RS, unsuccessfully. This disappointment rankled a great deal, for which he blamed the plotting of Tyndall and Thomas Huxley (see, e.g., letter 4106). In a letter to Thomas Huxley of 26 April 1873, Charles Darwin identified Forbes's proposers as Murchison and Whewell (F. Burkhardt, et al. (eds), *Darwin Correspondence*, 24 vols (New York: Cambridge University Press, 2005), vol. 21, pp. 190–2; this is letter 8882 in the online edition of the Darwin correspondence). See also N. K. L. Kaalund, 'A frosty disagreement: John Tyndall, James David Forbes, and the early formation of the X-Club', *Annals of Science*, 74:4 (2017), pp. 282–98.

To Henry Enfield Roscoe[1] 30 June [1873][2] 4123

Royal Institution of Great Britain | 30th. June.

My dear Roscoe

Permit me to introduce to you my friend Professor Mayer[3] of New York. He wishes to see Owen's College[4] & you will do a good work in obliging so good a man.

Yours ever | John Tyndall

[Envelope] Professor Roscoe | Owen's College | Manchester

Tyndall, John (1820–1893); Hyatt and Mayer Collection, C0076, Manuscripts Division, Department of Special Collections, Princeton University Library

1. *Henry Fairfield Roscoe*: Henry Enfield Roscoe (1833–1915), an English chemist who studied at University College, London, before earning his PhD under Robert Bunsen at Heidelberg (for Bunsen, see letter 4135, n. 3). After Edward Frankland left for London, in 1857 Roscoe replaced him as professor of chemistry at Owens College, Manchester, where he remained until 1886. Roscoe's research included collaborations with Bunsen on photochemistry between 1855 and 1862 as well as his own work on vanadium, of which he

was the first to prepare a pure sample. He wrote a number of popular works on chemistry, including *Lessons in Elementary Chemistry* (1866) and *Spectrum Analysis* (1869) (*CDSB*; *ODNB*).

2. *[1873]*: the year is given by reference to Alfred Mayer's visit to England in 1873. A similar letter to another correspondent, with the year given, is letter 4108.
3. *Professor Mayer*: Alfred Mayer, who visited London in 1873, and presumably also traveled north to Manchester.
4. *Owen's College*: founded in 1851, later the University of Manchester.

To Balfour Stewart[1] 30 June [1873][2] 4124

Royal Institution of Great Britain | 30th June

My dear Stewart

Doubtless you already know Professor Mayer[3] of New York: but through this note will probably be what the Catholics would call a work of supererogation[4] I write it to introduce to your kind attention a very good man.

Yours faithfully | John Tyndall

[Envelope] Professor. Balfour Stewart | Owen's College | Manchester[5]

Tyndall, John (1820–1893); Hyatt and Mayer Collection, C0076, Manuscripts Division, Department of Special Collections, Princeton University Library

---

1. *Balfour Stewart*: Balfour Stewart (1828–87), a Scottish physicist, meteorologist, devoted churchman, director of Kew Observatory (1859–71), and later professor of natural philosophy at Owens College, Manchester. Stewart was co-author with Peter Guthrie Tait of *The Unseen Universe* (London: Macmillan and Co., 1875) (*ODNB*).
2. *[1873]*: the year is given by reference to Alfred Mayer's visit to England in 1873. A similar letter to another correspondent, with the year given, is letter 4108.
3. *Professor Mayer*: Alfred Mayer, who visited London in 1873, and presumably also traveled north to Manchester.
4. *supererogation*: in Catholic theology, the performance of good works beyond what God commands or requires, held to contribute to a store of merit which the Church may dispense to others to make up for their deficiencies (*OED*).
5. *Owen's College | Manchester*: founded in 1851, later the University of Manchester.

To [Edward L. Youmans][1] [June 1873] 4125

I confess to reading with some amazement the article on the 'Tyndall Banquet,' in the *Intelligencer*. I am there charged with attacking, in my lectures,[2] both the Christian faith and one large class of its professors. If the telling of the truth be a necessary entry on the passport to 'the better land,' then assuming the maker of this charge to be *not* in a state of invincible ignorance, I would not exchange my chances on the frontier of immortality for his. The fact is that, though solicited to do so, I steadily refused to quit the neutral ground on the intellect during my visit to the United States. My audiences in Boston, Philadelphia, Baltimore, Washington, Brooklyn, and New Haven, can testify whether a single word relating to religion was heard in any lecture of mine delivered in those cities. New York can answer whether, in five out of the six lectures there delivered, a syllable was uttered, *pro* or *con*, regarding religion. And I confidently appeal to that heroic audience which paid me the memorable compliment of coming to hear me on the inclement night[3] when the words were spoken on which this charge is hung, whether, as regards its substance or its tone, what I then said[4] could, with fairness, be construed into an attack 'upon religious faith, and one large class of its professors.'[5] Put my words and manner before them, and I would fearlessly trust to the manhood of any Young Men's Christian Association[6] in the Union for a verdict in this matter. The writer in the *Intelligencer*, moreover, fails to see on conclusion to which his assertions inevitably lead; for, were they true, the perfectly unmistakable manner in which the 'attack' was received by the audience would prove the state of 'religious faith' in New York to be the reverse of creditable to him and others who have the care of it.

The head and front of my offending hath this extent: At the conclusion of one of my lectures, I referred, for two minutes, in mild language, to the reported words—*reported, I would add by a Presbyterian*—of the intemperate occupant of a single Presbyterian pulpit,[7] and this is wilfully twisted by that occupant into an attack upon the Presbyterian body. The charge, as originally made, and as now echoed by the *Intelligencer*, is so silly that I did not think it worth public refutation. Why should I care about refuting it, when the sympathetic kindness of the very men I was reported to have assailed assured me that they did not believe a word of the indictment? I carried no more pleasant memory with me from the United States than that of my reception at the Presbyterian College of Yale.[8] The high-minded youths and cultured gentlemen whom I met there, as indeed the Presbyterian body generally, a few hot-headed fanatics excepted, knew how to rate at its proper worth the statement of Dr. Hall,[9] and they will, I am persuaded, assign to its echo in the *Intelligencer* the self-same arithmetical value.

Should you deem this letter, or any part of it, necessary to public enlightenment, you are at liberty to make public use of it,

Ever yours faithfully, | JOHN TYNDALL

'A Correction.—Letter From Prof. Tyndall', *Popular Science Monthly*, 3 (June 1873), pp. 241–3, on pp. 242–3

1. *[Edward L. Youmans]*: this letter is known only in the form published in the June 1873 issue of the *Popular Science Monthly*, as part of an item in the 'Editor's Table' entitled 'Correction—A Letter from Prof. Tyndall'. The first part of the article quotes at length from an uncomplimentary report about Tyndall's farewell banquet from the 13 February 1873 issue of the *Christian Intelligencer*, under the title 'The Tyndall Banquet', stressing the irreligious character of the occasion. After the extract is written: 'Upon which, Prof. Tyndall remarks as follows, in a letter to a friend'. Although the 'friend' is not identified, Youmans, as editor of the magazine, is a plausible candidate.
2. *in my lectures*: for Tyndall's American lecture tour, see Introduction, n. 8.
3. *inclement night*: a 'terrible snowstorm', as he wrote to Hirst (see letter 3951) on 26 December 1872, the day of his fourth lecture in New York.
4. *what I then said*: Diarmid Finnegan notes that the *New York Tribune* and *New York Times* reports of Tyndall's fourth lecture in New York on 26 December 1872 differ slightly. At the close of the lecture, Tyndall indirectly referred to the prayer gauge controversy (see letter 3780, n. 2), stating that he regretted that he aroused bad feelings in some Presbyterians, with the *New York Tribune* reporting that Tyndall referred to the Bible and stated that he could only 'pray for them that despitefully use you and persecute you', and the *New York Times* reporting that he resolved to not respond in kind but to offer his blessing (Finnegan, *Voice of Science*, p. 56).
5. *'upon religious faith, and one large class of its professors'*: quoted from the *Christian Intelligencer* report (see n. 1).
6. *Young Men's Christian Association*: see letter 3975, n. 9.
7. *the intemperate occupant of a single Presbyterian pulpit*: given the remainder of the paragraph, the pulpit occupant who provoked Tyndall's comments appears to be the 'Dr. Hall' referred to (see n. 9); but the *Intelligencer* extract provides no further clues.
8. *Presbyterian College of Yale*: Yale College in New Haven, CT.
9. *Dr. Hall*: see letter 3943, n. 2.

From Thomas Mellard Reade 1 July 1873 4126

Heath House | Blundellsands | W. Liverpool | July 1st 1873

Dear Sir—

I desire to bring under your notice a physical fact so far as I am aware first discovered by Mr Robertson of Glasgow[1] when engaged in washing clays for the purpose of extracting foraminifera & ostracoda.[2] These glacial clays consist of a very large proportion (—80 per cent sometimes—) of very fine mud & Mr Robertson observed that in these washed in sea water on being allowed to settle the mud was precipitated much more rapidly than in fresh water—I have tried the experiment myself with fine clay dissolved in salt, fresh & brackish water—The sea water certainly precipitates the finer particles much more rapidly than the fresh but the addition of 1/4 of salt water to a phial containing clay dissolved in fresh water makes a striking difference in its precipitating powers—with 3 phials all shaken up at the same moment and allowed to settle a child can sure readily distinguish which is the salt or brackish water from this superior clearness & purity—

This all seems a very simple matter but now for the application of the knowledge—I am engaged investigating the drift depurity & have been trying to interpret within what maximum depth the mass of the clay has been laid down—There are several ways of approaching the subject one only of which I need mention here—viz. the precipitating power of salt water will limit the distance to which the fine materials derived (in our case) mostly from the washings of local materials on sea margins & within certain limits of depth can be carried out to sea by currents—It would also be interesting to know what velocity is required to keep the fine particles in suspension.

My main object however in writing to you is to point out that the extreme blackness of the sea in its greater depths is due as you have so clearly shewn[3] to the absence of solid particles in suspension must to a considerable extent arise from the great precipitating power of salt water as well as depth & distance from land—

It would be an interesting experiment to try if the addition of sea water to water from the lake of Geneva would precipitate the finer particles that give the beautiful turquoise tint to the latter—Of course the sea water would first have to be certainly freed from impurities itself—

If merely gravity were concerned in the sinking of the particles the fresh water ought to become clarified first—You will no doubt be able to suggest an explanation of the phenomena.[4]

Yours faithfully | T. Mellard Reade

Liverpool University Library Special Collections and Archive (TMR2.T.1.1)
Transcript Only

---

1. *Mr Robertson of Glasgow*: David Robertson (1806–96), a Scottish businessman and naturalist (*ODNB*).
2. *when engaged in washing clays for the purpose of extracting foraminifera & ostracoda*: see H. W. Crosskey and D. Robertson, 'The Post-Tertiary Fossiliferous Beds of Scotland', *Transactions of the Geological Society of Glasgow*, 2 (1867), pp. 267–82; 3 (1868), pp. 113–29; and 3 (1869), pp. 321–41. Robertson also published, a year after this letter, 'Note on the Precipitation of Clay in Fresh and Salt Water', *Transactions of the Geological Society of Glasgow*, 4 (1874), pp. 257–9.
3. *as you have so clearly shewn*: see J. Tyndall, 'On the Colour of the Lake of Geneva and the Mediterranean Sea', *Nature*, 2 (20 October 1870), pp. 489–90; and *Ascent of John Tyndall*, p. 271.
4. *suggest an explanation of the phenomena*: Reade wrote an additional short letter on the same day (letter 4127), and Tyndall replied on 4 July (letter 4129).

From Thomas Mellard Reade — 1 July 1873 — 4127

Professor Tyndall—

It is I think quite worth while making some quantitative experiments but also qualitative to find out if the action[1] is merely mechanical dependent upon the fineness of the particles & not on their constitution—

I shall be happy to send you clays from this locality as well as from the Clyde[2] and as I am preparing a paper on the drift for the London Geological Society[3] in which I shall embody the information sent you [know] I should be very glad to hear what you have to say on the subject—

I am dear Sir | Yours faithfully | T. Millard Reade

[in verso] Copy of letter to | Professor Tyndall | July 1st. 1873[4]

Liverpool University Library Special Collections and Archive (TMR2.T.1.1)

---

1. *to find out if the action*: Reade is referring to the greater power of salty water to precipitate out solid particles, as observed when washing clay from fossil specimens (see letter 4126).
2. *Clyde*: in Scotland, either the Firth of Clyde (the mouth of the River Clyde), or more generally the Clyde clay formation.
3. *paper on the drift for the London Geological Society*: T. M. Reade, 'The drift-beds of the north-west of England.—Part I. Shells of the Lancashire and Cheshire low-level boulder-clay and sands', *Quarterly Journal of the Geological Society*, 30 (1874), pp. 27–37.
4. This letter is a second letter to Tyndall of the same date; the first is letter 4126, and Tyndall's reply is letter 4129.

From Thomas Henry Huxley 2 July 1873 4128

4 Marlborough Place[1] | July 2d 1873.

My dear Tyndall,

The governors of the London Hospital have asked Dr Clark (my doctor of doctors) to petition you to make them an address at the Hospital (probably on the efficacy of prayer) next October.[2] I gave Clarke my card by way of introduction and promised him I would let you know what he wanted, but I do not think he is sanguine of success. However entreat him kindly for my sake.

I hope to be off to-night.[3]

Ever yours faithfully | T. H. Huxley.

Six months rent enclosed.[4] Punctuality [is the] soul of business!

RI MS JT/1/TYP/9/3016
IC HP 8.150
Typed Transcript Only

1. *4 Marlborough Place*: see letter 3947, n. 1.
2. *make them an address at the Hospital (probably on the efficacy of prayer) next October*: it does not appear that Tyndall delivered such an address.
3. *I hope to be off to-night*: Huxley was due to leave on holiday for the Auvergne with Joseph Hooker (see letter 4095, n. 3).
4. *Six months rent enclosed*: why Huxley was paying Tyndall rent has not been identified; it is possible that 'rent' is a metaphor for something else.

To Thomas Mellard Reade 4 July [1873][1] 4129

4th. July

Dear Sir,

Your facts[2] are interesting—very—I agree with you that the subject would be a promising one for investigation, and I trust that you will follow it out.

Yours faithfully | John Tyndall

Liverpool University Library Special Collections and Archive (TMR1.T.7.1)

1. *[1873]*: the year is established by the relation with letters 4126 and 4127.
2. *Your facts*: Reade had reported striking observations and reflections about the greater precipitating power of salt versus fresh water (see letter 4126).

From Millicent M. Bence Jones[1] 7 July [1873][2] 4130

84 Brook St.[3] | July 7th

My dear Dr Tyndall

It is so very kind of you to think of sending me your new book on Light, & I value the kind thought even more than the book itself.

I hope when we return from Reigate early in the autumn that we shall see you & thank you for all your kindness.

I must add my thanks for part of the preface[4] to your book. I cannot write what I feel about it, but I am sure you will understand

With very kind regards from all of us | Believe me | Very sincerely yours | M. M. Bence Jones

RI MS JT/1/J/121
RI MS JT/1/TYP/2/737

1. *Millicent M. Bence Jones*: see letter 4053, n. 1.
2. *[1873]*: the year is established by the publication of the English edition of Tyndall's American lecture tour—*Six Lectures on Light*—in early summer 1873.
3. *84 Brook St.*: the Bence Jones' London residence.
4. *part of the Preface*: the English edition reprinted the American edition's (see letter 4121, n. 3) preface, written in February 1873 when Tyndall was preparing to return to England. He expressed his hopes that America would one day enjoy a scientific institution as capably and selflessly managed as Bence Jones had managed the RI, and that readers in the American interior and west would forgive and understand Tyndall's need to go home before visiting them, owing in part to Bence Jones' grave illness (see *Six Lectures on Light*, pp. viii–ix).

From Mary Adair 9 July 1873 4131

Heatherton Park[1] | July 9th 1873

My dear Dr. Tyndall,

Your kind note[2] and charming book[3] have duly reached me. Accept my best Thanks for both. I anticipate great pleasure in reading again the Lectures on Light[4] and with, at last, coming across Dr. Young's masterly Reply to Lord Brougham.[5] How glad you will be to be free from London in this heat! and I hope after Ireland you will enjoy Switzerland[6] as of old. I was very glad to see the Class Lists at the Oxford Commemoration,[7] though you can afford to think very lightly of Honorary Degrees! &c.

With our kindest regards, believe me, always sincerely your's | H. M. Adair

RI MS JT/1/A/7

1. *Heatherton Park*: see letter 3887, n. 1.
2. *your kind note*: letter missing.
3. *charming book*: *Six Lectures on Light*.
4. *reading again the Lectures on Light*: Adair may be referring to the printing of the full lectures in New York by the *New York Tribune* as a '3 cent' supplement (see letter 3962, n. 2), which Tyndall possibly sent to her as he did others.
5. *D^r. Young's masterly Reply to Lord Brougham*: in *Six Lectures on Light*, Tyndall included, as a lengthy appendix, extracts from Lord Brougham's attack (in the *Edinburgh Review*) on Thomas Young's wave theory of light as well as Young's reply (pp. 227–55).
6. *after Ireland you will enjoy Switzerland*: Tyndall and Thomas Hirst left for the Alps on 1 August 1873 after Tyndall made a visit to Ireland (see letter 4121, n. 12), and then a few days in Folkestone beginning on 28 July (see letter 4145). See *Ascent of John Tyndall*, p. 322. On the way to Switzerland, Tyndall spent a few days in Paris visiting with friends, and he returned to London on 10 September, as he wrote to Elizabeth Dawson Steuart (see letter 4159).
7. *Class Lists at the Oxford Commemoration*: Tyndall received an honorary doctorate from Oxford on 18 June 1873 (see letter 4104, n. 2).

From Agatha Russell — 9 July 1873 — 4132

[Pembroke Lodge.[1] | Richmond Park.] | July 9. /73.

Dear Professor Tyndall

I do not know how to thank you enough for your kind presents to me. This one is particularly delightful as I shall be able to read over & study those parts of your lectures,[2] which, owing to my ignorance of the subject, I was unable to understand at the time.[3] I look forward with immense pleasure to reading it with my brother[4] who always helps me through any difficulties. I hope you are not gone to Switzerland yet[5] & that we may see you again before you leave London. Have you bought your little lake in the mountains?[6] Thanking you once more for sending me such a very interesting book,

believe me, most sincerely yours | Agatha Russell

RI MS JT/1/R/48

1. *Pembroke Lodge*: see letter 4087, n. 1.
2. *read over & study those parts of your lectures*: referring to the recently published English edition of Tyndall's American lecture tour, *Six Lectures on Light*.
3. *unable to understand at the time*: this indicates that Russell attended Tyndall's re-delivery of his American lectures on light at the RI (see letter 3782, n. 10).
4. *my brother*: probably Rollo Russell.

5. *not gone to Switzerland yet*: see letter 4131, n. 6.
6. *Have you bought your little lake in the mountains?*: Tyndall had previously thought about purchasing land in Switzerland near the Bel Alp hotel where he often stayed during alpine excursions. Still three years later he wrote to Joseph Peter Lesley that he was 'thinking of buying a scrap of land & building a chalet for myself among my beloved mountains' (Mss.B.L56. American Philosophical Society). This came to frutition in 1877 when he and Louisa Tyndall (married in February 1876) built their summer cottage, Alp Lusgen, at Bel Alp (now Belalp), on the northern side of Valais, above Brieg.

To Thomas Archer Hirst — 10 July [1873][1] — 4133

10th. July

Dear Tom.

I shall not probably see you until Monday; though I return on Saturday evening. Meanwhile you perhaps might have time to glance at the Scotsman.[2] It may influence you in your judgment of <u>my</u> work.

You will doubtless find the paper at Albemarle Street.[3]

Yours aff[ectionate]ly | John Tyndall

RI MS JT/1/T/706
RI MS JT/1/HTYP/622

---

1. *[1873]*: the year is given by the year added onto the letter (in a different hand), and given as such by LT on her typescript.
2. *glance at the Scotsman*: the newspaper had published a review of *Life and Letters of Forbes*. In his published reply to the charge that he plotted against Forbes in order to rob him of due scientific recognition, Tyndall quoted two passages from the review in the *Scotsman*, one disparaging Louis Agassiz, the other disparaging Tyndall himself ('[t]he most persistent opponent of Forbes's reputation... represent[ing] a certain *coterie* in the Royal Society, who from their metropolitan eminence affect to look down on everything provincial'). See 'Literature', *The Scotsman*, 8 July 1873, p. 2; and J. Tyndall, 'Principal Forbes and His Biographers', *Contemporary Review*, 22 (August 1873), pp. 484–508, on p. 486 (emphasis in original).
3. *Albemarle Street*: the RI was located at 21 Albemarle Street.

From Frances Russell 12 July 1873 4134

[Pembroke Lodge.[1] | Richmond Park.] | July 12/73

Dear Mr Tyndall

Since we saw you I have had some talk with Rollo[2]—& I find that he wishes as strongly as we that it were possible for him to make Physics the pursuit of his life—how can you suggest a beginning? can you help him by advice? I will not trouble you with a longer note—indeed I feel that it is wrong to trouble you at all, especially as all I have said is so vague—but you will excuse it I am sure—

I remain y[ou]rs ever sincerely | F Russell

RI MS JT/1/R/66

1. *Pembroke Lodge*: see letter 4087, n. 1.
2. *Rollo*: Rollo Russell.

To Gustav Kirchhoff[1] 15 July 1873 4135

My dear Kirchhoff

I have been taking your name in vain in America and therefore send you a copy of book which will inform you of what I have said regarding you.[2] A little haze has surrounded your discoveries which I thought it my duty to clear away.

I hope you are well, and I hope Bunsen[3] is well. It is not usual in England for one man to send his love to another, but I ask you to give Bunsen my love.

Ever yours | John Tyndall

Staatsbibliothek zu Berlin (Stabi) Slg Darmstaedter 1855 Tyndall Be. 1–58, 38

1. *Gustav Kirchhoff*: Gustav Robert Kirchhoff (1824–87), a German physicist who made fundamental contributions in several fields. In Königsberg, Kirchhoff studied physics under Franz Neumann and mathematics under Friedrich Julius Richelot. His formulation of the laws of electrical circuits was part of his doctoral dissertation, which he finished at Königsberg in 1847. During an associate professorship at the University of Breslau in 1850, he became acquainted with Robert Bunsen, who made sure that Kirchhoff followed his move to Heidelberg in 1854. Together with Bunsen, he worked on thermal radiation and

spectrum analysis. Regarding his law of thermal radiation, he fought a long struggle for recognition of priority with the Scottish physicist Balfour Stewart (1828–87). Kirchhoff and Tyndall were close friends and colleagues. Tyndall had recommended him for the RS's Rumford Medal, which he received in 1862; Kirchhoff was unable to attend the ceremony and Tyndall accepted it in his place. He was perhaps best known for the invention of the spectroscope and subsequent discovery of the elements found in the Sun, work done in collaboration with the chemist Robert Bunsen (see n. 3) (*ADB*; *CDSB*).

2. *what I have said regarding you*: for Tyndall's remarks in his American lectures on Kirchhoff's research in spectroscopy and its significance, see *Six Lectures on Light*, pp. 194–204.
3. *Bunsen*: already associated with the laboratory burner that still bears his name—Tyndall in his *Six Lectures* refers to 'a Bunsen's burner' (p. 201)—the German chemist Robert Bunsen (1811–99), who mentored both Tyndall and Edward Frankland at Marburg, became celebrated for his collaborations with Gustav Kirchhoff on the spectroscopy of heated elements (see n. 1) (*CDSB*).

To Frederick Barnard 16 July 1873 4136

16th. July 1873

My dear Dr. Barnard.

It would be a very great happiness to me to meet you in Switzerland; but I do not yet know my destinations. I have never seen the Dolomites,[1] and a friend[2] who proposes being my companion this year is much taken by the idea of going to them. If I go to Switzerland[3] I am sure to be heard of at the AEggischhorn,[4] and the Bel Alp,[5] about four hours distant from the AEggischhorn. These are two grand stations, and for some years past the Bel Alp has been a favourite spot with me.

I am still entangled in my work here. My American tour[6] caused many things to be postponed which now come down upon me. I am already more than a fortnight beyond my usual time of departure, and I fear I shall be here nearly a fortnight longer.

If Mrs. Barnard[7] should be with you pray give her my very kindest regards.

Yours ever faithfully | John Tyndall

A trip to Ireland on Light house business[8] has rendered this reply more tardy than it would otherwise have been.

Should you wish to communicate with Mr. Ball[9]—the author among other things of Ball's Alpine Guide[10]—his address is

24 St. Georges Road | Eccleston Square | London

Columbia University Manuscripts; X973 C72 V01.2; Box 2 Folder 96; University Archives, Rare Book & Manuscript Library, Columbia University Libraries, Rare Book & Manuscript Library, Columbia University in the City of New York

1. *Dolomites*: mountain range located in northeastern Italy. They form a part of the Southern Limestone Alps and extend from the River Adige in the west to the Piave Valley in the east.
2. *friend*: Thomas Hirst.
3. *If I go to Switzerland*: he did indeed (see letter 4131, n. 6).
4. *AEggischhorn*: Eggishorn (see letter 3828, n. 15).
5. *Bel Alp*: see letter 3771, n. 5.
6. *My American tour*: see Introduction, n. 8.
7. *M[rs]. Barnard*: see letter 3842, n. 2.
8. *trip to Ireland on Light house business*: in letter 4121 to Hector Tyndale, Tyndall wrote that he 'shall be going to Ireland to look after the Lighthouses in a few days'. Tyndall sent a report to Trinity House regarding observing on 8 July 1873 a new light at the lighthouse at Howth Baily (see letter 4141).
9. *M[r]. Ball*: John Ball.
10. *Ball's Alpine Guide*: Ball authored popular guides to the Western, Central, and Eastern Alps (first published in 1863, 1864, and 1868), as well as titles for more specific alpine locations, published by Longmans, Green, and Co. in London.

From William Frederick Pollock — 19 July 1873 — 4137

59 Montagu Square[1] | 19th July 1873

Dear Tyndall,

You are very good in sending me your latest book[2]—It is a Light & a Delight to me—and you know how much I am interested in the vindication of Thomas Young. He will sell, I take it, more than one copy of his pamphlet now[3]—& in whatever sphere he is now rejoicing in company with other philosophers, I trust he may know what has been done for him by his successor at the Royal Institution.

Ever Yours Truly | W. F. Pollock

RI MS JT/1/P/240
RI MS JT/1/TYP/6/2143

1. *59 Montagu Square | W.*: the Pollocks' residence in Marylebone, London.
2. *your latest book*: *Six Lectures on Light*.
3. *vindication of Thomas Young... his pamphlet now*: in *Six Lectures on Light*, Tyndall included an appendix quoting extensively from Young's magisterial defense of his wave theory of light, in a pamphlet written in response to an attack on the theory by Lord Brougham, also quoted extensively (see letter 4131, n. 5).

From François-Napoléon-Marie Moigno 20 July 1873 4138

[LES MONDES | Revue Hebdomadaire | DES SCIENCES | *et de leurs applications* | aux Arts et à l'Industrie | par | M. l'Abbé MOIGNO | RÉDACTION: | 2, rue d'Erfurth, 2

Paris le] 20 Juillet [187]3.

Mon cher Monsieur,

Merci de votre bonne petite lettre d'hier. Je ne m'étais pas effrayé des planches que vous avez adressées à Mr Gauthier-Villars, parce que je savais que ma traduction était excellente. Seulement elle avait été faite il y a trois ans sur la troisième édition, la quatrième n'existait pas alors, et quoique je l'eusse revue sur la quatrième édition, quelques notes ou quelques variations de mots avaient pu m'échapper. Je n'avais rien à me reprocher, bien loin de là. Mr Gauthier-Villars avait égaré dans le déménagement de son imprimerie mon manuscrit tout entier, et j' *[avais]* eu le courage de le refaire. Sans la mauvaise foi de Mr Giraud, mon cher Mr Tyndall, mauvaise foi qui m'a couté trente mille francs, j'aurais continué à éditer votre chaleur et nous aurions aujourd'hui trois ou quatre éditions françaises. Je *[vais]* d'un bon gré d'abord compter sur moi pour l'optique et la traduire moi-même aussi parfaitement que possible et je pousserai l'impression avec la plus grande énergie.

Il n'est resté dans mon esprit aucun nuage du douloureux incident; je vous suis plus dévoué que jamais, et je vous verrai avec un très grand bonheur à votre passage par Paris.

Puisque vous le voulez bien, votre petite note sera prête, mais j'étais tout prêt à vous servir des Mondes gratuitement. Ils sont à vous tout entier et leur circulation est devenue relativement très grande; *[et donc]* *[partout]*

A vous de coeur | Votre admirateur et votre ami | L'Abbé F. Moigno

[LES MONDES[1] | Weekly Review | OF THE SCIENCES | *and their applications* | to the Arts and the Industry | by | Father MOIGNO | EDITORIAL OFFICES: | 2 Erfurth street, 2

Paris] 20 July [187]3

My dear Sir,

Thank you for your good short letter from yesterday.[2] I had not been worried by the plates you had sent Mr Gauthier-Villars[3] because I knew my translation was excellent. However, it was done three years ago on the third edition, the fourth did not exist then, and, while I have reviewed it on the fourth edition, a few notes or a few variations in words might have slipped from my grasp. I had nothing to feel ashamed of, far from that. M[r] Gauthier-Villars had lost my whole manuscript when moving his printing works, and I */had/* the courage to do it again. Without the dishonesty of M[r] Giraud,[4] my dear M[r] Tyndall, dishonesty that cost me thirty thousand francs, I would have continued to edit your Chaleur and we would have today three or four French editions. I will gladly first count on myself for the Optique[5] and translate it myself as perfectly as possible, and I will push for its publication with all my strength.

There is no cloud left in my mind of the disturbing incident; I am ever more devoted to you, and will see you with great delight when you come by Paris.[6]

Since you accept it, your little bill will be ready, but I was ready to graciously serve you some Mondes.[7] They are yours fully and their circulation has become relatively large; */and hence everywhere/*.

Yours with my heart | Your admirer and your friend | Father Moigno

RI MS JT/1/M/124

1. *LES MONDES*: see letter 3829, n. 1.
2. *your good short letter from yesterday*: letter missing, but from context (see, in addition to the rest of this letter, letters 4050, 4080, 4109, and 4110), it concerned the new French edition then in progress of Tyndall's *Heat* (*Chaleur*).
3. *Mr Gauthier-Villars*: Jean-Albert Gauthier-Villars.
4. *dishonesty of Mr Giraud*: on problems with Étienne Giraud, see letter 4050, esp. n. 7.
5. *Optique*: see letter 4050, n. 8.
6. *when you come to Paris*: Tyndall and Thomas Hirst visited Moigno in Paris, en route to Switzerland, in early August (see *Ascent of John Tyndall*, p. 322; though Jackson misdates the visit as in early July).
7. *graciously serve you some Mondes*: Moigno was here offering to give Tyndall copies of his journal *Les Mondes* (see letter 3829, n. 1).

To William Luson Thomas[1] 21 July 1873 4139

21st. July 1873.

It is not perhaps a subject in which I ought to speak but I trust the proprietors of the 'Graphic' will allow me to express my high appreciation of the rendering by their excellent artist of my most difficult face.[2]

<u>John Tyndall</u>

Staatsbibliothek zu Berlin (Stabi) Slg Darmstaedter Be.3

THE GRAPHIC [July

JOHN TYNDALL, LL.D., F.R.S.

---

1. *William Luson Thomas*: the founder and, in Tyndall's term, 'proprietor' of the *Graphic* at the time of Tyndall's writing, Thomas (1830–1900) started the newspaper with his brother in 1869 (*ODNB*).
2. *the rendering... difficult face*: a wood-engraving portrait of Tyndall appeared in the 19 July 1873 issue of the *Graphic*, accompanying a biographical article about him.

From John B. Monckton[1,2] 23 July 1873 4140

Guildhall, E.C., July 23, 1873.

Gentlemen,—I have to inform you that the circular issued by the Gas Referees,[3] with respect to the maximum of sulphur as proposed to be fixed by you for the winter months at gas-works other than Bow and of Beckton, has been laid before the Gas and Water Committee of the City of London, and they have directed me to state that they feel much disappointed at the change proposed, as they had been led to expect that the maximum would have been diminished at all the stations rather than increased at any; the more especially as in their opinion ample time has been given to the companies for the necessary alteration in their apparatus for purification. I am further to call attention to the great desirability of diminution in the quantity of sulphur in gas from the deleterious effects it produces, not only upon substances exposed to its influences, but also to the health of those inhabiting the rooms in which gas charged with it is burned.

I am, &c., | JOHN B. MONCKTON, Town-Clerk.

Sulphur Compounds in Gas. Report of the Proceedings before a Committee of the House of Commons on the Crystal Palace District Gas and The Gaslight & Coke Company Bills. Session 1877 (London: William B. King, 1877), p. 140

1. This letter is from John B. Monckton to Tyndall, Augustus Vernon Harcourt, and William Pole, the three together comprising the Metropolitan Gas Referees appointed by the Board of Works (see letter 3830, n. 17), to the engineers of the Chartered Gas Light and Coke Company, one of several companies supplying London with coal gas.
2. *John B. Monckton*: John Braddick Monckton (1832–1902), a British lawyer and civil servant. He served as Town Clerk of London from 1873–1902 ('Obituary', *The Law Journal*, 37 (8 February 1902), p. 86).
3. *circular issued by the Gas Referees*: the notice to the Chartered Gas Light and Coke Company regarding the gas-works at Bow and Beckton, dated 30 June 1873, stated: 'The Gas Referees have to make the following announcement as to the "sulphur maximum." For gas made at Beckton and Bow the maximum shall remain fixed at 2 grains. But the Referees have determined, in consideration of various difficulties which the gas companies have encountered in attempting to reduce the amount of sulphur below 20 grains at works less favourably situated, to allow at these works, from this date until the 30th of September, a maximum of 25 grains. During the winter half year it is the intention of the Referees, in further consideration of the increased quantity of gas which the companies have to make, to raise the maximum to 30 grains. But the Referees wish it to be understood that they

sanction these maxima for the periods names only because it appears that at several gas-works the purifying apparatus is on an insufficient scale, and in the hope that the time thus granted will be used by the companies in making such additions and improvements as will secure to the public a higher standard of purity for the future' (*Sulphur Compounds in Gas. Report of the Proceedings before a Committee of the House of Commons on the Crystal Palace District Gas and The Gaslight & Coke Company Bills. Session 1877* (London: William B. King, 1877), p. 123).

To C. Cecil Trevor 25 July 1873 4141

Royal Institution | 25th July, 1873.

Sir,

A little more than a year ago, I ventured to recommend to the Board of Trade the purchase of an annular lens with the view of enabling Mr. Wigham to execute experiments on a new form of light of his invention, called by him the Triform Light.[1]

I have now the honour to report to the Board that on Tuesday the 8th of this month I proceeded to Ireland to inspect the performance of this light; and also to instruct myself as to the merits of the gas gun proposed by Mr Wigham for fog signalling.[2]

A few preliminary remarks will render the character and performance of the Triform Light generally intelligible.

A luminous point placed naked in the air radiates its light in all directions—to the horizon, to the zenith, to the earth.

When such a point is surrounded by a fixed dioptric[3] apparatus the skyward and earthward rays are so deflected that the whole of the light impinging on the apparatus, making deduction of what is reflected at the surface of the glass, and absorbed within it, is sent forth as an illuminating sheet, or layer which grazes the horizon.

In a revolving apparatus, the light, instead of being diffused in a continuous layer, is gathered up into distinct beams, which sweep in succession the space between the apparatus and the horizon.

Thus, in the fixed apparatus, the light is economised by preventing its upward and downward waste; and it is intended to present to the mariner a constant source of illumination. In the revolving apparatus, on the contrary, the mariner at certain intervals, is deprived wholly of the light, in order that he may enjoy it at other intervals with greatly augmented intensity.

The fixed apparatus consists of three parts namely:—

1. A glass belt which surrounds the light and acts upon its rays by refraction, so as to send them forth in the required direction.

2. A series of prisms below this belt which receive the rays directed obliquely downwards, and by totally reflecting them send them forth in the same direction.

3. A series of prisms forming a cupola, which receive the rays directed upwards, and after totally reflecting them sends them forth in the same direction

In the revolving apparatus, for the continuous refracting belt is substituted a series of annular lenses, generally eight in number, and therefore forming an octagon. As in the fixed apparatus each annular lens is aided above and below by a series of totally reflecting prisms.

The quantity of light issuing from the refracting belt in the one case, and from the annular lenses in the other, is greatly in excess of that issuing from the top and bottom prisms taken together. Probably three fourths of the light which reaches the mariner is in most cases sent to him by the refracting portions of each apparatus.

In the Triform Light Mr. Wigham substitutes for the bottom prisms a refracting belt or lens, and for the top prisms a second refracting belt or lens; thus forming, in the case of a fixed light, a round tower, and in case of a revolving light, a polygonal tower of lenses.

In the focus of each of the three refracting belts is placed a burner; each "tower" therefore contains three burners placed along the same vertical line. By an arrangement of annular spaces and tubes the products of combustion of each burner are prevented from acting injuriously upon the others.

The apparatus being thus placed distinctly before the mind its performance will be readily understood. Its range of power may in the first instance be referred to.

The light-keeper can confine himself to a single 28-jet burner; and a single refracting belt. By lighting two such burners he raises his power to 56, and by lighting three of them he raises it to 84.

The 48, 68, 88 and 108-jet burners can be treated in the same manner. The light keeper can, in each case employ one, two, or three burners, according to the state of the atmosphere. Three 48-jet burners would give a light of 144; three 68-jets, a light of 204; three 88-jets, a light of 264, while three 108-jets would give you a light of 324. The range of the apparatus stretches therefore from 26 jets to 324 jets, the latter shedding forth nearly twelve times the luminous power of the former.

On the evening of the 8th of July I had an opportunity of witnessing the performance of the Triform Light erected at Howth Baily,[4] from a point near the Salthill Hotel between five and six miles distant from the lighthouse. The triform arrangement was placed in the experimental shed referred to in former reports, which stands a little below the base of the lighthouse tower. In each case the triform light was compared with the ordinary Baily light. That is to say the triform of 28 jets below was compared with the single 28 jets above, and so of the other burners; ending with the 108 jet burner in the upper light-house, and three burners of the same size in the lower one. The difference, as might be expected, was in each case marked in favour of the triform, though not quite so marked as I expected.

When I next witnessed the performance of the triform I had the advantage of being accompanied by an independent and most sagacious observer.[5] My visit to Ireland being meant to embrace experiments on the gas-gun, and these experiments connecting themselves immediately with those now in prosecution at the South Foreland,[6] I suggested to the Trinity House that it might be desirable for one of the Elder Brethren[7] to accompany me to Ireland. On the morning of the 9th accordingly I had the gratification of being joined by Admiral Collinson,[8] and on the evening of the 10th I had an opportunity of witnessing in his company the performance of the Triform Light.

The observations began near Kingstown, our last place of observation being the deck of the gunboat "Hawk" in Kingstown Harbour.

When the 28-jet burner was ignited above and the three 28's below, the preponderance of the latter was, as might be expected, manifest. Still, though 84 jets were alight in the triform I thought a 68-jet burner in the upper house would have been fully equal to the 84 jets in the lower. This I am informed by Mr. Wigham he subsequently found to be the case. It arises from the fact that the burner in the upper house is aided by the top and bottom prisms, while each burner in the lower house radiates through the refracting belt alone.

Experiments of the same kind were made with the other burners. The weather was clear and the single burner in each case showed a very fine light; but in each case it was distinctly transcended by the Triform.

The concluding observation was, as I have said, made from Kingstown Harbour. The 108-jet burner was in the upper house and showed a splendid light; three 108 jet burners were in the lower house. The distance of the place of observation from the lighthouse was a little over five miles and at this distance the flames of the three burners seemed blended together to a single ball of light of extraordinary intensity.

To my eye the difference of <u>size</u> between the light in the lantern and that in the experimental shed did not appear remarkable, but the latter appeared as if into it had been compressed a distinctly greater amount of light. I hardly think that in my experience of lights, I ever saw anything so powerful; and I have reason to believe that my judgment in this respect is entirely corroborated by that of Admiral Collinson.

When the burner in the upper house was changed from 108 jets to 28 jets, it became instantly a diminutive satellite of the large light, and could hardly be imagined to be the same pure, strong light which it showed itself to be when burning without the overmastering presence of the triform.

The record of my observations on the triform fixed light is here concluded. I shall add a few general remarks regarding it further on.

The Board[9] will probably remember my recommending that the gas apparatus erected for experimental purposes at Rockabill[10] should be permitted to remain there permanently; and I accompanied my recommendation by the remark that it was desirable to have in the neighbourhood of Dublin two establishments where experiments on the two classes of lights might be conducted.

Since that time the gas has continued to act at Rockabill in a highly efficient manner, and its existence there enabled us to extend our experiments on the triform light to the revolving apparatus. Accompanied by Admiral Collinson I went to Rockabill on the 9th. inspected the arrangement, and drew out a programme of experiments. Three panels of the upper and lower prisms had been removed, and instead of them we found two annular lenses, which, with the lens of the apparatus itself, made three. In the focus of each of these lenses was placed a burner, the arrangement of the three burners being the same as that already referred to in the case of the fixed triform.

Thus five panels of the apparatus were in their ordinary condition; two of its annular lenses were without the usual aid of top and bottom prisms while one side of the octagon consisted of the three annular lenses so placed that their three centres were in the same vertical line.

The night of the 9th was thick, and heavy rain fell at intervals. We began with the 28-jet burner, and advanced to the 48 and 68 jet burners. We did not try a higher power. The time of an entire revolution of the Rockabill apparatus is 96 seconds; and as there are eight panels the time from flash to flash, counting from their centres, is 12 seconds.[11] During a complete revolution we had five beams of the ordinary kind where lenses and prisms acted in conjunction; two beams where the lenses alone acted, and one beam from the triform. In all cases the flash from the triform proved itself most luminous.

On the night of the 10th these experiments were repeated, with the same substantial results. Admiral Collinson expressed the opinion that the superiority of the triform was more marked in thick than in clear weather. In this I agree with him.

These experiments leave no doubt upon the mind that the triform arrangement places in our hands a practical and powerful means of illumination, and that it is equally applicable to fixed and to revolving lights. By it we are enabled to bring an amount of luminous power into play which would be quite unattainable with a single burner.

This must be understood as the great feature of the triform—the possibility of applying power in case of need. In it indeed economy is to some extent sacrificed to power. For the top and bottom prisms being abandoned the light derivable from these sources is lost. It is I believe to some extent made good by the heating of the air which feeds the two upper burners, but to what extent I am not prepared to say.

I think it in the highest degree important that clear notions should be entertained of the relation of this question of light power to 'thick' and 'foggy' weather. In the first place it is hopeless to aim at penetrating a really dense fog to any considerable extent with any light whatever. Wishing for some definite data here, on Saturday the 19th of July my assistant at my request went to Haggerston Station, Kingsland Road,[12] and made the following observations on a Steam-cloud issuing from a slowly moving locomotive, and crossing the disk of the unclouded sun.

1st Observation. Time 3.39 P.M. Total obscuration of sun's disk when steam cloud passed over it.

2nd Observation. Time 3.47 p.m. Total obscuration as before.

3rd Observation. Time 3.52 p.m. Total obscuration.

4th Observation. Time 4.2 p.m. Total obscuration followed every gush of steam-cloud.

Other observations were made; but these suffice to show that the sun itself, even when very powerful, is unable to drive its rays through a few feet of a steam cloud.

I noticed a similar effect produced upon the powerful Baily light a few days ago in Ireland.

But the sun which, as a defined object, ceases to be seen through the artificial steam-cloud; or through a natural atmospheric cloud, may notwithstanding do important service by illuminating the cloud which hides it. We transact the business of the world on cloudy days when the sun is unseen, by the light imparted to the clouds and scattered by them upon the earth. And so a powerful light in one of our lighthouses, may though unseen as a defined object, do good service by illuminating the fog which intercepts its direct rays.

This is not only a theoretic deduction but a practical fact. I took occasion recently to question separately the Captain, and the Second officer of the mail boat "Leinster" plying between Holyhead and Kingstown, and they were both perfectly agreed that the illumination of the fog by the South Stack Light[13] is frequently seen, and is of the highest utility even when the light itself has ceased to be visible. It is needless to say that the stronger the light the greater the distance which this fog-gleam penetrates the atmosphere, and the greater the chance of its being useful to the mariner.

But fogs vary in density, and between our densest fog and our clearest air, there are infinite gradations. It is not sound reasoning to conclude that because in a dense fog weak lights and strong are alike extinguished, the lights are therefore as regards fog penetration, equally valueless. For there are states of the atmosphere when it is of the highest importance that the mariner should see his light, and where if a low power be used he will not see it, while if a high power be used he will. Fogs so dense as to reduce all lights to equal helplessness are happily not frequent, and it is in the commonest forms of "thick" weather that the superiority of a strong light would be most beneficially felt.

I have already stated that the triform flash, which occurs at Rockabill once every revolution, is superior to all the others. Reasoning alone would lead us to conclude that with a gradually thickening atmosphere the feebler flashes would disappear first; the strong beam of the triform proving itself to be useful after the others had ceased to be so. This is corroborated by observations made from the Hill of Howth[14] and from the "Kish" Lightship.[15] At all times when the light is seen, one flash in eight is paramount; but on so many occasions the seven minor flashes disappeared altogether, and the eighth flash alone was visible.

In the experiments hitherto recorded with the triform arrangements, the lenses were so placed that their centres were in the same vertical line. This, for shortness' sake, may be called the vertical arrangement. By displacing the lenses so that their centres are no longer in the same vertical line we have what may be called the oblique arrangement. Here the advantage gained by the superior brilliancy of the triform beam disappears, being exchanged for an advantage of another kind. By this arrangement the number of beams reaching the mariner during an entire revolution would be trebled without any change in their duration. But each of them coming from a single refracting belt, its intensity would be proportionally reduced.

On a former occasion I was honoured by the company of Sir Leopold McClintock[16] in observing the lights of Howth Baily and Rockabill, and in a note written at the time he expressed the opinion that the applicability of gas both to fixed and revolving lights had been reduced to demonstration. This I have reason to know is also the conviction of Admiral Collinson. Into the question of cost of course these distinguished officers have not entered; and on this score we may hope for additional information from the investigation to be commenced I trust in a few days at Haisboro.[17] In one important particular, which I have already dwelt upon, no other mode of illumination can compete with gas, and that is the readiness with which it lends itself to individualization. By its wise employment those disastrous shipwrecks which arise from the mistaking of one light for another may be materially lessened, if not rendered impossible.

The expense of the triform apparatus would probably be not greater than that of an ordinary apparatus with top and bottom prisms.

There is no proposition at present before me with regard to the intro duction of this light, and I limit myself therefore to describing its character and performance. But I think that the skill and energy, which without extravagance, and proceeding by the sure way of tentative experiment, have achieved so much in Ireland, are well worthy of the favourable consideration of the Board of Trade, not only on their own account, but through the wholesome stimulus which they are calculated to impart to lighthouse illumination generally.*

Gas-gun Experiments

A mixture of coal gas and atmospheric air, and still more a mixture of coal gas and oxygen, is known on ignition to explode with loud detonation. Taking advantage of the existence of gas at Rockabill the Board of Irish Lights have authorized Mr. Wigham to erect in the gas-house yard an apparatus for the generation of oxygen. They have also authorized the construction of a gas-gun, an instrument of the most simple description. It consists of a piece of an iron gas-main from 12 to 18 feet long and from a foot to 18 inches in diameter.

A gas holder of measured capacity receives from the oxygen apparatus a quantity of oxygen sufficient to half fill it: coal gas is then added until it is full. The coal gas being the lighter of the two ascends through the oxygen and thoroughly mixes with it. From the gas-holder the mixture is transferred to the gun. It enters at the breech which is slightly elevated above the muzzle, and the gas being lighter than atmospheric air has no tendency to flow down by its own gravity. It is fired by a friction tube

We experimented with two gas guns, twelve feet long & 12 inches in diameter, and a longer one 18 inches in diameter. With these were compared a brass gun & a carronade, each firing a charge of 3lbs of gunpowder. The two 12 inch gas guns could be fired together as in quick succession, thus giving a peculiar character to the sound.

On the evening of my arrival the guns were all fired at Howth Baily, while I stood beside the railway near the Salthill Hotel. The wind was light, and blowing from the place of observation towards the Baily. No sound whatever was heard either from the gas or the gunpowder guns. The flashes were seen those of the gas-guns being brightest.

I was not surprised to learn that the sound was usually heard with distinctness at our place of observation; for previous experiments at the South Foreland had revealed the fact that a day's interval can so change the conditions of the atmosphere that a sound capable on one day of being heard to a distance of 10 miles, on the day following ceases to be heard at a distance of 3 miles.

Various experiments were subsequently made with these guns: it is needless to describe these in detail. The result may be generally stated to be that as regards the loudness of the sound the carronade had the advantage of the gas-guns; while as regards the vividness of the flash the gas-gun had the advantage of the carronade.[18] This is also a point on which I questioned the intelligent and courteous Captain of the 'Leinster',[19] and he informed me that only on two occasions of impenetrable fog has he failed to see, and to derive great advantage from seeing, the flash of the gun at the North Stack[20] near Holyhead. The flash therefore is an important element in this problem. Experiments in hazy weather would be necessary to decide the comparative penetrative powers of the gas flash and the gunpowder flash.

Pending the decision of this point all that can be said is that the gas-gun, as a means of fog-signalling, is well worthy of further investigation. The cost of the powder is, I am informed from $2^s/7'/n$ to $2^s/9$ per shot, whereas the cost of the gas for the larger gun is $1^s/3'/n$ and for the smaller gun $6\frac{3}{4}^d$ per explosion. Before recommending the gas-gun at gas stations these figures would need a searching examination.

I may be permitted to express my recognition of the prompt, earnest, and thorough manner in which my suggestions have hitherto been carried out by Capt. Hawes[21] and Mr. Wigham, and also of the highly efficient manner in which all my wishes on this last occasion were seconded by Mr. Legget[22] who commanded the gun-boat "Hawk".

I have the honour to be Sir, | Your obedient servant, | John <u>Tyndall</u> | C. Cecil Trevor Esq[ui]re | &c &c &c

* I think it a fortunate circumstance when the interests of a really intelligent and skilful man coincide, as I believe they do in the case of Mr. Wigham, with the interests of the public. Hence my desire that he should have fair-play and encouragement. He has done much, and is destined I think to do much more, towards the improvement of lighthouse illumination. I should, however, exceedingly regret any attempt on his part to make his truly meritorious work appear an iota better than it is in reality.

National Archives MT 10/165. H3908 2481–504
Transcript Only

---

1. *I ventured to recommend... the Triform Light*: see letter 3736. On Tyndall as scientific adviser to Trinity House, see letter 3869, n. 4.
2. *I have now... for fog signalling*: for an excellent summary of Tyndall's work at this time in evaluating the use of gas to power systems for warnings ships, see *Ascent of John Tyndall*, pp. 316–21.
3. *dioptric*: relating to the refraction of light; pertaining to dioptrics; *esp.* (of a telescope, etc.), refractive, refracting (*OED*).
4. *Howth Baily*: see letter 3736, n. 2.
5. *independent and most sagacious observer*: see n. 7.
6. *South Foreland*: see letter 4093, n. 5.

7. *one of the Elder Brethren*: Richard Collinson (1811–83), a British naval officer, was a member of the Elder Brethren of Trinity House since 1862 (*ODNB*). On the Elder Brethren, see letter 3736, n. 3 and n. 5.
8. *Admiral Collinson*: see n. 7.
9. *Board*: Board of Trade (see letter 3736, n. 3).
10. *Rockabill*: see letter 3736, n. 1.
11. At this point there is an addition in a different hand—'*Note at back'—indicating the following note written on the back of the sheet of paper: '*I do not know why this rapid succession was adopted: it seems to me to be simply tantalizing to the mariner'.
12. *Haggerston Station, Kingsland Road*: a train station in London.
13. *South Stack Light*: a lighthouse on a small island off the northwest coast of Holy Island, Anglesey, Wales, built in 1809.
14. *Hill of Howth*: a hill on Howth Head, a peninsula northeast of Dublin, Ireland.
15. *"Kish" Lightship*: a lightship moored at the sandy shallows of the Kish bank off Dublin, for the purpose of aiding mariners in and out of Dublin Bay.
16. *Sir Leopold McClintock*: (Francis) Leopold McClintock (1819–1907) made his reputation as an Arctic explorer investigating the fate of the Franklin expedition; in 1873 he was admiral-superintendent of Portsmouth dockyard (*ODNB*).
17. *Haisboro*: see letter 4045, n. 3.
18. *carronade*: a short piece of ordnance, usually of large calibre, having a chamber for the powder like a mortar: chiefly used on shipboard (*OED*).
19. *Captain of the 'Leinster'*: not identified.
20. *North Stack*: a small island just off Holy Island on the northwest coast of Anglesey, Wales.
21. *Capt. Hawes*: Elliot H. Hawes, RN, was the Assistant Inspector of Lights for the Commissioners of the Irish Lights.
22. *Mr. Legget*: not identified.

From W. Jackson[1] 26 July [1873][2] 4142

St. Giles | Oxford. | July 26

My Dear Dr. Tyndall

You may like to hear that Mrs[3] & Miss Adair[4] leave England for Switzerland on the 13th of August.

After travelling there they proceed to Italy.

The illness of the little grandchild[5] has prevented their plans from being settled earlier. I shall probably hear more soon.

In haste to catch the midday post I am always

Yours very faithfully | W. Jackson | Saturday morning

RI MS JT/1/J/2

1. *W. Jackson*: not identified.
2. *[1873]*: 26 July fell on a Saturday in the years 1862, 1873, 1879, and 1884. Given that Tyndall barely knew Mary Adair in 1862, and that he was married as of 1876, we place this letter in the year 1873.
3. *M^rs*: Harriet Eliza Adair.
4. *Miss Adair*: Mary Adair.
5. *little grandchild*: not identified.

From Alfred Vialay[1] 26 July 1873 4143

Paris, le 26 Juillet 1873.

Monsieur.

Je me présente à vous en inconnu: je n'ai pour faire excuser mon audace de vous écrire que l'admiration profonde que j'éprouve, pour un homme qui a exposé d'une façon si *[entrainante]* *[ses]* grandes recherches dans le monde naturel. J'ai lu et relu vos divers ouvrages "Forms of Water": "On Sound": "Heat a mode of motion". Avec quel talent vous avez *[pénétré]* dans la nature intime des corps et y avez découvert les mouvements moléculaires par le moyen si original du rayonnement de la chaleur! Avec quel art vous faites passer dans l'esprit de celui qui vous lit la notion de ces phénomènes si compliqués qu'il semble au premier abord que *[les]* mathématiques seules puissent en débrouiller le *[1 word illeg]* et que l'on est *[tant]* étonné et *[tant]* heureux de voir exprimés dans un langage qui est celui de la vie ordinaire. Tout ce qui est signé de votre nom *[m'attire]*, *[aussi]* attendais-je impatiemment votre nouvel ouvrage "On Light", qui était annoncé depuis quelques mois. J'ai été probablement un des premiers à le lire en France et l'impression que j'en ai ressentie est telle que le désir m'est venu de faire connaître à mes concitoyens ces leçons que vous avez professées aux Etats Unis, au milieu de gens si avides de vous voir et vous entendre. Je ne sais comment je puis oser vous le dire, à vous *[dont]* les ouvrages ont été traduits dans les langues étrangères par des hommes de l'éminence de Helmholtz et Wiedemann.

Pour mener à bonne fin une tâche si difficile je n'ai pour moi que ma bonne volonté et mon grand désir d'être utile, car mes titres scientifiques sont complètement nuls.

Sorti, il y a quelques années, de l'Ecole Centrale des Arts et Manufactures, je suis entré à la Compagnie Parisienne des ingénieurs, chargé au service de la canalisation. Presque tous les loisirs que cette *[fonction]* me donne, je les consacre à l'étude des langues étrangères vers lesquelles je me sens porté naturellement. De la langue allemande *[que]* je possède d'une façon assez complète, je suis passé à la langue anglaise; désireux de connaître dans leur forme originale les travaux scientifiques de la Grande Bretagne. Du moment où j'ai pu la comprendre assez facilement je me suis empressé de chercher dans vos ouvrages ces préceptes que vous exposez d'une façon si magistrale. J'étais d'autant plus avide de lire vos leçons que j'y voyais *[1 word illeg]* applications immédiates, notamment celles dont peut bénéficier la météorologie cette science presque nouvelle: Devenu par mon mariage le plus proche parent d'un homme qui s'est adonné depuis de longues années à l'étude de la météorologie, j'y consacre mes faibles moyens *[tant]* pour m'instruire que pour entrer dans une plus grande communauté de pensée avec lui. Je *[veux]* parler de Monsieur Belgrand, membre de l'Institut, Inspecteur général des *[Ponts]* et Chaussées, et auquel ses grands travaux des *[eaux]* et des égouts de Paris ont fait un nom Européen. Sa grande expérience et ses hauts conseils ne me font jamais défaut.

En vous demandant l'autorisation de traduire votre nouvel ouvrage j'ai la crainte d'arriver trop tard; peut-être avez-vous déjà accordé cette faveur à un homme que ses titres scientifiques vous recommandent. Quoiqu'il en soit, alors que je ne devrais pas réussir, je m'estimerais heureux, la lettre que je prends la liberté de vous écrire vous dit la joie avec laquelle j'accueille vos travaux.

Dans le cas où vous voudriez vous donner la peine de m'honorer d'une réponse et me faire savoir si vous daignez me choisir pour traduire votre ouvrage "On Light" et à quelles conditions vous voulez qu'il paraisse en français j'ose vous donner mon adresse.

Monsieur Al Vialay ingénieur de la Compagne Parisienne du Gaz, chez Monsieur Belgrand, inspecteur général des *[Ponts]* et Chaussées à Joinville-le-Pont (Seine).

Veuillez agréer, Monsieur, l'expression des sentiments très distingués avec lesquels j'ai l'honneur d'être votre dévoué serviteur

Al Vialay

Paris, 26 July 1873.

Sir.

I am presenting myself without being known to you: I have nothing to excuse my audacity of writing to you except for the deep admiration that I feel for a man who presented in such a stirring manner his great research of the natural world. I read and reread your different books "Forms of Water"; "On Sound"; "Heat a Mode of Motion".[2] With such talent you have ventured into the intimate nature of bodies and discovered there the molecular movements through such an original method as heat radiation! With such art you give into the mind of the one that reads you the notion of such complicated phenomena, that it seems at first that only mathematics could unravel the *[1 word illeg]* and that we are so astonished and so happy to see expressed in the language of ordinary life. All that is signed by your name attracts my attention, and I also looked forward to your new book "On Light",[3] which was announced a few months ago. I was probably one of the first to read it in France, and the impression that I got was of a desire to spread amongst my fellow citizens the lessons that you professed in the United States; among such people eager to see and hear you. I do not know how I could dare tell you this, to you whose books were translated into foreign languages by eminent men such as Helmholtz and Wiedemann.

To successfully complete such a difficult task, I have nothing to stand for me except for my good will and my great desire to be useful, since my scientific titles are nonexistent.

Graduated a few years ago from the Central School of Arts and Manufactures, I entered the Parisian Company of Engineers, responsible for the [maintenance of pipes]. I dedicate almost all of the leisure that this [job] gives me to the study of foreign languages towards which I feel naturally inclined. From German, which I mustered well enough, I moved onto English, wishing to know the scientific works of Great Britain in their original form. From the moment when I could understand it easily enough, I hastened to look in your books for the precepts that you present in such a masterly way. I was even more eager to read your lessons when I saw *[1 word illeg]* immediate applications, especially those that could benefit meteorology; this relatively new science. Having become, through marriage, the next of kin to a man who gave many years to the study of meteorology, I spent my limited resources as much for self-study as for entering into a larger [sphere] of thought with him. I am speaking of Mr Belgrand,[4] member of the Institute, General Inspector of Bridges and Roads, and to whom his great works on [the waters] and sewers of Paris have made him a name in Europe. His vast experience and his vital advice never fail me.

By asking you for permission to translate your new book, I am afraid to have come too late; perhaps you have already given this favour to a man[5] whose scientific titles recommend him to you. In either case, if I should not succeed, I will consider myself fortunate for the letter that I took the liberty of writing to you in order to tell you how joyfully I welcome your work.

In the case that you would like to trouble yourself with honouring me with a reply, and to let me know if you deign to choose me for the translation of your book "On Light" and under which conditions you would like to see it appear in French, I dare give you my address.

Mr Al Vialay, Engineer of the Parisian Gas Company, at Mr Belgrand, General Inspector of Bridges and Roads in Joinville-le-Pont (Seine).

Please accept, Sir, the expression of my highest regards, with which I have the honour of being your devoted servant

Al Vialay

RI MS JT/1/V/7

1. *Alfred Vialay*: Alfred Vialay (dates unknown), a French engineer.
2. *"Forms of Water"; "On Sound"; "Heat a Mode of Motion"*: as Vialay noted later in this letter, he learned English in order to read 'the scientific works of Great Britain in their original form'. Thus, he was referring to *Forms of Water* (1872), *Sound* (1867), and *Heat: A Mode of Motion* (London: Longmans, Green, and Co., 1870).
3. *"On Light"*: *Six Lectures on Light.*
4. *Mr Belgrand*: Marie-François-Eugène Belgrand (1810–78), a French engineer who made significant contributions to the modernization of the Parisian sewer system (J. Bertrand, 'Éloge Historique: d'Eugène Belgrand', *Mémoires de l'Académie des sciences de l'Institut de France*, 42 (1883), pp. ix–xxxii).
5. *perhaps you... to a man*: François-Napoléon-Marie Moigno was established as Tyndall's French translator, and he already asked Tyndall to translate his American lectures on light, and eventually did so (see letter 4138).

To [William] Walton[1] 27 July [1873][2] 4144

MY DEAR SIR:—Your idea[3] appears to me to be a very ingenious one. I can say no more, as my thoughts have never been turned to this subject.

Faithfully yours, | JOHN TYNDALL. | July 27.

W. Walton, 'Deep Sea Soundings', *Scientific American*, 29 (13 September 1873), p. 164

1. *[William] Walton*: the 'W. Walton' who authored this letter published in *Scientific American* is possibly William Walton (dates unknown), author of a number of mid-nineteenth-century mathematics textbooks published in England. In the letter, he wrote of himself that he was 'late of the Science and Art Department, South Kensington, England', and currently residing in Williamsburg, New York.
2. *[1873]*: the year is given based on the publication year of the letter in *Scientific American.*
3. *Your idea*: Tyndall's endorsing letter was published—and now only exists—as part of a longer letter sent by Walton to the *Scientific American* outlining a new scheme for the taking of deep sea soundings (*Scientific American*, 29 (13 September 1873), p. 164).

To Elizabeth Hamilton 28 July [1873][1] 4145

28th. July

My dear Lady Claud H.

I go to Folkestone[2] today—remain there two or three days to get a little work finished, for London had broken me down, then pass over on Thursday or Friday to Paris. There I must remain for one or two days to see some friends—thence to the mountains.[3]

My hotel in Paris is

Grand Hotel de St. James | Rue Honoré St.

I go to the Bel Alp[4] near Brieg, Canton de Valais Switzerland.

I do not know your arrangements—Is there to be any gentlemen in your party? I never took charge of a lady or a party of ladies in my life and I fear you would find me a poor hand at looking after the small needs & comforts of a ladies party—But I can show you all about the Bel-Alp—The Sparrenhorn,[5] the Nessel[6]—the Ober Aletsch Glacier,[7] the Great Aletsch,[8] the Massa Gorge.[9] If you are coming alone I will do every thing I can for you—I only doubt my own power.

I would also go with you to the Aeggischhorn,[10] and show you all that is to be seen there. These two places might well occupy you a fortnight—I do not know what I shall do afterwards.

My friend Hirst, the Director of Studies in the Greenwich Naval College[11] accompanies me—<u>You will like him</u>. and <u>he will like you</u>. In fact he will cut me out.

Write to me please to the post office Folkestone—I think it would be well to be in Paris on Friday evening—you would not object to halting a day there? or even two.

Count on my desire to make the Alps lovely & lovable to you; but I am a wild savage and fear my ability to look after the smaller odds & ends.

There is "honesty."

most faithfully Yours | John <u>Tyndall</u>

I am writing to the Bel Alp—would you like a small sitting room? They have <u>one</u> & I believe only one.

BL Add MS 63092, ff. 55–6

---

1. *[1873]*: the year is established by the relation to Tyndall's upcoming holiday with Thomas Hirst to the Alps via Paris—a holiday spent in part with Elizabeth Hamilton and her children Emma Hamilton, Douglas Hamilton, and Tyndall's future wife, Louisa Hamilton. On Tyndall's connections with the Hamiltons, see *Ascent of John Tyndall*, p. 323.
2. *Folkestone*: Tyndall often stayed at a hotel in Folkestone in order to focus on getting 'a little work finished'; he stayed at the Westcliff Hotel and worked on lighthouse business, writing letters to C. Cecil Trevor and Robin Allen (4149 and 4152).
3. *thence to the mountains*: see letter 4131, n. 6.
4. *Bel Alp*: see letter 3771, n. 5.
5. *Sparrenhorn*: now Sparrhorn, a peak above Bel Alp, with a wonderful view.
6. *Nessel*: a village about a two hour walk from the Bel Alp.

7. *Ober Aletsch Glacier*: the Oberaletsch (Upper Aletsch) Glacier is located on the south side of the Bernese Alps, in the Swiss canton of Valais.
8. *Great Aletsch*: the longest glacier in the Alps, which flows past the Eggishorn.
9. *Massa Gorge*: a gorge that was slowly carved out over time by masses of ice from the Aletsch glacier cutting through the rock.
10. *Aeggischhorn*: Eggishorn (see letter 3828, n. 15).
11. *Director of Studies in the Greenwich Naval College*: see letter 3911, n. 12.

To Thomas Archer Hirst 29 July [1873][1] 4146

29th July

My dear Tom:

Calais is best;[2] therefore I choose Calais.

Charing X at 7.40 Friday: reach Paris at 6. PM.

The Folkestone route is a few shillings (very few) cheaper—but it does not reach Paris till 8.15.

I meet you at Dover.

Yours aff[ectionate]ly | John

RI MS JT/1/T/912
RI MS JT/1/HTYP/622

---

1. *[1873]*: the year is written on the letter in a different hand, and makes sense for the trip to Switzerland with Hirst that occurred that year (see letter 4131, n. 6).
2. *Calais is best*: with their holiday on the Continent coming up, Tyndall here expressed his preference to travel the faster but more expensive Dover-to-Calais route, with a morning start from Charing Cross rail station in London and then a final trip by rail to Paris.

From Emily Peel 29 July 1873 4147

[VILLA LAMMERMOOR | SÉCHERON | GENÈVE][1] | July 29/73

Dear Professor Tyndall,

Altho' this letter will probably be returned to you in Switzerland, I cannot delay until y[ou]r return to England[2] before thanking you, for having so kindly remembered me & sent me your most interesting Volume on "Light",[3] recently published. I shall read your Lectures over again[4] & try to understand them. How I wish you would come & spend a day or two here on y[ou]r way to England; please remember us & if you have a few spare hours, & devote them to our blue lake.[5]

With renewed thanks for your kind remembrance I remain your very sincerely | Emily Peel

RI MS JT/1/P/64

1. *VILLA LAMMERMOOR | SÉCHERON | GENÈVE*: see letter 3844, n. 15.
2. *y[ou]ʳ return to England*: from his trip to Switzerland via Paris with Thomas Hirst (see letter 4131, n. 6).
3. *your most interesting Volume on "Light"*: *Six Lectures on Light*.
4. *read your Lectures over again*: Peel may be referring to the printing of the full lectures in New York by the *New York Tribune* as a '3 cent' supplement (see letter 3962, n. 2).
5. *our blue lake*: Lake Geneva.

To Juliet Pollock 30 July [1873][1] 4148

30th. July

My dear Lady Pollock

I ran down here for two days through sheer incapacity to finish my work in London.[2] I had a terrible spell of sleeplessness.

Today I was exceedingly taken by an article in the Contemporary[3]—of its kind are of the very best I ever read—judge of my surprise & pleasure when I found your name at the end of it.

Goodbye, I had no head, no heart, no time to call to see you before quitting London.[4]

Yours ever | John Tyndall

RI MS JT/1/T/1170
RI MS JT/1/TYP/6/2063

1. *[1873]*: the year is established by the relation with the article written by Pollock, published in the June 1873 issue of *Contemporary Review* (see n. 3).
2. *I ran down... work in London*: Tyndall was in Folkestone at the Westcliff Hotel, having arrived on 28 July aiming to 'remain there two or three days to get a little work finished, for London had broken me down', as he wrote to Elizabeth Hamilton (see letter 4145). He worked on lighthouse business, writing letters to C. Cecil Trevor (4149) and Robin Allen (4152).
3. *an article in the Contemporary*: J. Pollock, 'The Imaginative Literature of America', *Contemporary Review*, 22 (June 1873), pp. 347–71.
4. *before quitting London*: Tyndall returned to London only to then head off to Switzerland via Paris with Thomas Hirst (see letter 4131, n. 6).

To C. Cecil Trevor [30 July 1873][1] 4149

Royal Institution of Great Britain[2] |
Westcliff Hotel | Folkestone | Wednesday

My dear M^r^. Trevor,

You have just caught me. I am here till tomorrow afternoon: then off to the Alps.[3]

It gave me pleasure to correct the proof.[4]

I think it highly desirable that copies of the correspondence regarding Haisboro[5] should be placed in M^r^. Valentin's hands

Faithfully yours, | John Tyndall

I have been examining the cliff round the S. Foreland[6,7]

National Archives, MT 10/165. *Fog signals - gun flashes*. H3908

---

1. *[30 July 1873]*: the date is given by reference to the South Foreland (see n. 6 below), and that Tyndall left London on 1 August 1873 with Thomas Hirst to start making their way toward Switzerland (see letter 4131, n. 6).
2. This letter was written on RI letterhead, but written from Folkestone.
3. *I am here till tomorrow afternoon: then off to the Alps*: between lighthouse business in Ireland and then to Switzerland, Tyndall stayed in Folkestone to do some work (he wrote letters on lighthouse business, this one and letter 4152 to Robin Allen).
4. *proof*: probably the 'report upon the materials, manufacture, and cost of the gas used' at Haisbro lighthouse (see letter 4098 from Trevor).
5. *Haisboro*: see letter 4045, n. 3.
6. *I have been examining the cliff round the S. Foreland*: on 19 May 1873, Tyndall went 'down to Dover with the view of observing the performance of Fog Signals at the South Foreland' (see letter 4093). On 24 June, he wrote to William Rutherford that for 'The rest of the week also I fear I shall be away, steaming between the South Foreland and the coast of France, and making observations on guns and trumpets and other means of warning the mariners in fogs' (see letter 4118). In a 25 July 1873 letter to Trevor, Tyndall wrote that there were experiments 'now in prosecution at the South Foreland' (see letter 4141). Finally, a 31 July 1873 letter to Robin Allen refers to the South Foreland experiments as 'recently conducted' (see letter 4152).
7. A note on the letter from Trevor reads: 'Done | 9 Aug | CCT'.

From Henri Étienne Sainte-Claire Deville[1] 30 July 1873 4150

[*Laboratoire de Chimie* | École Normale Supérieure | *Paris, le*] 30 Juillet [*187*]3

Cher ami,

Soyez assez bon pour bien recevoir un savant & professeur distingué, le Dr de Saint Germain, qui va visiter l'Angleterre & qui voudrait bien visiter le laboratoire de M. Davy, de Faraday & de Tyndall. Je vous serai bien reconnaissant si vous pouvez lui faciliter sa /tournée/ scientifique dans votre pays. /Nous aurons vous/ à Schaffhausen ?

Votre ami dévoué, | M. Sainteclaire Deville

[*Chemistry Laboratory* | École Normale Supérieure | *Paris,*] 30 July [*187*]3

Dear friend,

Would you be so kind as to welcome a scientist and distinguished professor, the Dr. de Saint Germain,[2] who will be visiting England & wishes to visit the laboratories of Mr Davy, of Faraday & of Tyndall?[3] I would be grateful if you could facilitate his scientific /tour/ in your country. /Will we see you/ at Schaffhausen?[4]

Yours truly, M. Sainteclaire Deville

RI MS JT/1/D/133

---

1. *Henri Étienne Sainte-Claire Deville*: Henri Étienne Sainte-Claire Deville (1818–81), a French chemist who worked at the Chemistry Laboratory, Ecole Normale Supérior, Paris. He developed a process for producing aluminum (*CDSB*).
2. *Dr. de Saint Germain*: not identified.
3. *laboratories of Mr Davy, of Faraday & of Tyndall*: the laboratories of the RI, where Humphry Davy, then Michael Faraday, and then Tyndall himself conducted their scientific work.
4. *Schaffhausen*: a town in northern Switzerland.

From Thomas Henry Huxley 30 July 1873 4151

Hotel de France, | Baden Baden.[1] | July 30, 1873.

My dear Tyndall,

We find ourselves here after a very successful cruise in the Auvergne and Ardèche,[2] successful at least so far as beauty and geological interest go. The heat was killing and obliged us to give up all notion of going to Ursines[3] as we had at first intended to do. So we turned our faces north and made for Grenoble[4] hoping for a breath of cool air from the mountains of Dauphiny.[5] But Grenoble was hotter even than Clermont[6] (which by the way quite deserves its reputation as a competitor with hell), a neighbour's drains were adrift close to the hotel, and we got poisoned before we could escape. Luckily we got off with nothing worse than a day or two's diarrhea—After this the best thing seemed to be to rush northward to Gernsbach[7] which had been described to me as a sort of earthly paradise. We reached the place late last Saturday night, and found ourselves in a big rambling hotel crammed full of people and planted in the bottom of a narrow valley, all hot and steaming. A large pigstye "convenient" to the house mingled its vapours with those of the seventy or eighty people who eat, drank and excreted without any other earthly occupation that we could discern during the three days we were bound by stress of letters and dirty linen to stop. On Monday we made an excursion over here, prospecting—and the air was so fresh and good, and things in general looked so promising that I made up my mind to put up in Baden-Baden until the wife joins me.[8] She writes that you talk of leaving England on Friday and I may remark that Baden is on the high road to Switzerland. Verbum sap.[9]

I am wonderfully better and really feel ashamed of loafing about when I might very well be at work. But I have promised to make holiday and make holiday I will.[10]

No proof of your answer to Forbes' biographer[11] reached me before I left, so I suppose you had not received one in time. I am dying to see it out.

Poor dear Sammy![12] His end has been all too tragic for his life. For once reality and his brains came into contact and the result was fatal.[13] I suppose that Lord Westbury and he have made up their little differences about eternal punishment in some warm corner or other.[14]

Hooker is down below but I take upon myself to send his love. He is in great force now that he has got rid of his Grenoble mulligrubs.

Ever yours | T. H. Huxley.

RI MS JT/1/TYP/9/3017–8
IC HP 8.151
Typed Typescript Only

---

1. *Baden Baden*: Baden-Baden is a resort town in southwestern Germany's Black Forest, near the border with France.
2. *in the Auvergne and Ardèche*: see letter 4095, n. 3.
3. *Ursines*: not identified; this may be a mis-transcription by LT for another place name in France.
4. *Grenoble*: a city in the Auvergne-Rhône-Alpes region of southeastern France, south of Geneva, Switzerland.
5. *mountains of Dauphiny*: the Dauphiné Alps are a group of mountain ranges in southeastern France.
6. *Clermont*: Clermont-Ferrand, a city in central France.
7. *Gernsbach*: a town in the Black Forest of Germany, just to the east of Baden-Baden.
8. *wife joins me*: Henrietta Huxley and their son Leonard Huxley joined Thomas in Cologne, France on 11 August (see letter 4154).
9. *Verbum sap*: a Latin phrase used in place of making or concluding a full statement or explanation of something (*OED*). Here Huxley may have been hinting to Tyndall to detour to Baden Baden in Germany while on his way to Switzerland. On Tyndall's trip to Switzerland via Paris with Thomas Hirst, see letter 4131, n. 6.
10. *make holiday I will*: see letter 4028, n. 4.
11. *your answer to Forbes' biographer*: 'Principal Forbes and His Biographers' (see letter 4024, n. 5).
12. *Poor dear Sammy!*: Samuel Wilberforce (1805–73), Bishop of Oxford, died on 19 July after he was thrown from a horse. He and Huxley had famously crossed swords over the Darwinian theory at the July 1860 meeting of the BAAS in Oxford (*ODNB*). See I. Hesketh, *Of Apes and Ancestors: Evolution, Christianity and The Oxford Debate* (Toronto: University of Toronto Press, 2009); and N. K. L. Kaalund, 'Oxford Serialized: Revisiting the Huxley–Wilberforce Debate through the Periodical Press', *History of Science*, 52:4 (2014), pp. 428–53.
13. *For once reality... result was fatal*: this quip has become one of the many immortal one-liners from Huxley.
14. *Lord Westbury and he... warm corner or other*: Richard Bethnell, 1st Baron Westbury (1800–73), who served as the Lord High Chancellor of Great Britain in the early 1860s, had died on 20 July, the day after Bishop Wilberforce (see n. 12). In the public controversy over the supposedly heretical character of *Essays and Reviews* (1860), a collection of seven essays on religion by Anglican clergymen, Wilberforce and Westbury took opposite sides (*ODNB*).

To Robin Allen 31 July 1873 4152

Copy | Folkestone | 31st July 1873

Sir

Pending the detailed Report of the observations recently conducted at the South Foreland,[1] I beg to lay before the Elder Brethren some suggestions arising out of the observations thus far made.

At a distance of nearly 90 yards from the upper fog-horns there is an indentation in the chalk cliff, and for a further distance of 70 yards the cliff falls almost vertically. On this cliff face here exposed the sound waves proceeding direct from the fog horns impinge, and from it they are are[2] reflected seaward. A line drawn from the fog horns to the edge of this projecting cliff, and produced, strikes the Admiralty Pier at Dover at some distance from its end. The space between this line and the land lies in what is sometimes called the sound shadow. But immediately after passing the cliff referred to the sound waves diverge—technically are diffracted—into the shaded space. These diffracted waves also meet the cliff and are reflected from it.

Between Dover Pier and the fog signal station we have repeatedly noticed a remarkable subsidence of the sound. This is not accounted for by the direction given to the mouths of the horns, for the stifling of the sound has also been observed in the case of guns. It is moreover not to be referred to the state of the wind, for it is always observed. It is not therefore due to the operation of a variable but a permanent cause. This cause I take to be the interference of the reflected with the direct waves of sound.

The results suggest the desirability, whenever a fog signal is established, of determining the local circumstances most favorable to the free propagation of the sound. It is highly desirable, in the present instance, to compare the top and bottom horns with reference to this special point. The cliff referred to does not present itself in the same manner to the waves issuing from the top and bottom horns, the interference therefore will probably be found to be different in the two cases.

The usual effect of the wind in stifling sound has been noticed in our observations. On this point I think we need more certain information than we now possess. It is one of the darkest points in the science of sound. It is easy to see that the velocity may be augmented by a wind blowing in the same direction and diminished by an opposing wind, but why the <u>intensity</u> or loudness should suffer, as it is known to suffer is, I have stated, an enigma. The only quantitative experiments known to me on the influence of the wind are those of De la Roche executed (I believe) in 1816; and the only attempt made to explain the results of De la Roche we owe to Professor Stokes. His explanation will be found in the Report of the British Association from 1857.[3]

It is in my opinion highly desirable to confirm or confute the conclusions of De la Roche. This could be done by experimenting, as he did, on two sources of sound of the same pitch and intensity—say two bells of the same size in which two equal hammers are permitted to fall from the same height. Placed in the open air, with a wind blowing, an observer between the bells would be able to determine with considerable precision the point at which the two sounds appear equally loud. De la Roche found that for short distances—say, of 20 or 30 yards—the point of equal intensity, was always sensibly midway between the two bells; but for long distances it way always nearest to the bell towards which the wind was blowing at the time.

He also found the sound to be propagated with less loss of intensity at right angles to the wind than in the direction of the wind.

In my detailed report I shall if it be deemed necessary endeavour by means of a diagram to render the explanation of Professor Stokes simpler than it is as it now stands. I may remark that it was in reference to it that I recommended the stationing of two men, the one on Dover Pier, and the other on the summit of Shakespeare's Cliff. A captive balloon would furnish the best means of clearing up the questions which here present themselves.

But the greatest obstruction to fog-signalling by sound remains yet to be mentioned, it is the want of homogeneity in the air traversed by the sound. Whenever sound changes its medium say from dense to rare air, or from rare air to dense—a portion of the sound-wave is always reflected; and if the atmosphere be so mottled with spaces of different densities as to render the reflections very frequent and numerous, the sound is wasted in echoes before it can penetrate the air to any considerable distance.

I was not prepared—nor do I think that men of science in general are prepared—for the extraordinary dissipation of sound which may be thus brought about over the sea, in days of perfect calmness. The influence of the wind is small in comparison. As bearing upon this point the observations made at the South Foreland are of high scientific interest; and they will also if I mistake not, enable us to explain many historic occurrences, the causes of which have been only guessed at hitherto.

Here, I say, is our greatest enemy, and inasmuch as we cannot manipulate the atmosphere and render its density uniform, our only resource is to make our sounds so forceful as to be able to bear great loss without absolute extinction.

It gave me therefore great pleasure to be informed that the Elder Brethren[4] contemplate trying the Canadian Steam-Whistle and also the steam syren, a fine specimen of which I had the opportunity of listening to in the courteous company of General Woodruffe,[5] at Sandyhook. The observations could not be considered complete if they did not embrace these instruments.

A very useful series of experiments might be carried out with the view of determining the best form of Gun for fog signals. It is certain that the 18 pounder generally in use is not the best form. In our experiments a howitzer of one half of the length of the 18-pounder had a distinct advantage over it.

The "Syren" receives its name from its power of sounding under water. The instrument thus sounded is well known to scientific men. It would be a problem worthy of the practical skill of Mr Holmes[6] to determine whether by means of a water syren, sound cannot be transmitted to great distances through the sea. This if successful would liberate us from the caprices of the atmosphere. I hope on my return from Switzerland I have the time and health necessary to bring these observations to a conclusion

I have &c. | (signed) John Tyndall | To Robin Allen Esq. | &c &c

P.S. It is of course highly desirable to determine the influence of a fog on the propagation of sound. Derham[7] and others state it to be the very great. But so much looseness of statement has been indulged in with reference to this subject that I fear we must verify every point independently.

National Archives MT 10/195. *For Signals, Report of Experiments on Lighthouses*. H4140
Transcript Only

---

1. *Report of the observations recently conducted at the South Foreland*: for Tyndall's work in this period evaluating different warning systems for ships in fog, see *Ascent of John Tyndall*, p. 318. On Tyndall as scientific adviser to Trinity House, see letter 3869, n. 4. Tyndall's report on fog signalling was sent to Trinity House in May 1874; see 'Copy of Report by Professor Tyndall to the Trinity House, upon recent Experiments with regard to Fog Signals', 188, Parliamentary Papers (1874), pp. 1–77. This report was republished in abbreviated form as J. Tyndall, 'On the Atmosphere as a Vehicle of Sound', *Phil. Trans.*, 164 (1874), pp. 183–244; and again, in two parts, as J. Tyndall, 'On the Atmosphere in Relation to Fog-Signalling', *Contemporary Review*, 24 (November 1874), pp. 819–41, and 25 (December 1874), pp. 148–68.
2. *are*: in the transcript, the word 'are' is repeated.
3. *The only quantitative... from 1857*: see G. G. Stokes, 'On the Effect of Wind on the Intensity of Sound', *Brit. Assoc. Rep. 1857*, pp. 22–3, in which he discussed F. Delaroche, 'Observations sur l'influence que le vent exerce dans la propagation du son, sous le rapport de son intensité', *Annales de chimie et de physique*, 1 (1816), pp. 176–96. François-Étienne de la Roche (1780–1813) was a Geneva-based physician, chemist, and botanist; the paper cited by Stokes was published posthumously (*CDSB*).
4. *Elder Brethren*: see letter 3736, n. 3 and n. 5.
5. *General Woodruffe*: Woodruff (dates unknown), a lighthouse district engineer in the United States, accompanied Tyndall on his visit to the lighthouses at Staten Island and Sandy Hook to see the steam 'syren' (see Tyndall, 'On the Atmosphere as a Vehicle of Sound', p. 186).

6. *Mr Holmes*: Frederick Hale Holmes (b. 1812), a Professor of Chemistry at the Royal Panopticon of Science and Art, and pioneer of the use of electric lighting in lighthouses.
7. *Derham*: William Derham (1657–1735), an Anglican clergyman and natural philosopher who produced the earliest, reasonably accurate estimate of the speed of sound in 1709 (*ODNB*).

From Mary Egerton 4 August [1873][1] 4153

Mountfield Court[2] | August 4th

My dear Mr Tyndall,

This day last year we were with you on the glacier, and I must celebrate the anniversary by a little chat. The day lives in my memory not only for the beauty of the blue ice clefts and black coves among which we rambled, but by the pleasant talk we had pacing up & down the terrace in the quiet twilight afterwards. Half an hour of real conversation, in earnest, as friend to friend, is worth all the drawing-room chit-chat & "badinage"[3] of the year put together, to my mind;

I suppose you are now in your Eyrie.[4] Each day as I looked on the cloudless blue of the skies we have often had lately, I have longed for a ridge of sparkling snow peaks with their pearly shadows to shine out against them. Oh dear! I don't know when I shall be rich enough to go & see them again! Every thing is so dreadfully expensive now-a-days: You may well sing "Vive les garçons!"[5] they have not 20 mouths to feed every day, and can do what they like!

We spent a very pleasant afternoon with the Spottiswoodes[6] on our way down here. What a charming place it is! And will be still more so, for there is much left to do, in the way of opening out views, and letting in the air, &c; but with those splendid trees it must always be delightful. I don't know whether it is owing to having been born and bred in Sherwood Forest, that fine trees always seem so homelike to me; my love for the mountains is a passion, deep—intense; but like all love set on an object beyond its reach, full of excitement, and unsatisfied yearnings! But the influence of forest scenery is to me like a calm trusting friendship; I could sit and dream away the day under the shadow of a glorious old tree, while its giant arms bending round, seem to enfold me with a loving protection. It is quite different with you;—you are the favoured lover of the queenly mountains!—For you the proud cold brow relaxes into smiles—and you are admitted to the inmost recesses of the noble heart whose sternness tempers with awe the admiration of humbler worshippers!—There is a romantic paragraph for you!!

We saw Hughie's[7] Caves, but the Magnesium Wire[8] being unfortunately not forthcoming, humble candlesticks had to do duty for the illumination.

I wonder whether the little book[9] has even been opened? I do not want you to agree with me intellectually, only to observe how intellects on a par with your own, have thought and felt spiritually. And I know you can appreciate beautiful thoughts, even though you may not agree in the views they imply.

I see your reply about Forbes is in the Contemporary;[10] I shall get it at once, for I am very anxious to see what you say upon the one point that puzzles me. Not that you will suspect me of thinking "Galahad"[11] capable of anything but what is chivalrously fair and honorable in intention, but the very warmth of generous indignation may make even "Sir Galahad" hasty & perhaps prevent his fully estimating the effect of his own manner of stating the case. At any rate after looking the whole subject carefully over, it certainly does appear on the face of it, to be an unfortunate mistake, when you are accusing Forbes of not doing justice to Rendu, to have omitted that very part of the quotation from the former, in which he seems distinctly to acknowledge Rendu having attributed the different rates of glacier motion to the true cause. In fact Forbes does not appear himself to claim originality for the idea of river motions, so much as his "Reviewer" whom you quote; (& who would seem also to have overlooked that passage!) does for him. I can well understand your annoyance at the pertinacity with which Forbes tries to make out that nothing done before him was of any real value, but one cannot think he meant to be otherwise than truthful as to facts, though viewing them in a somewhat depreciative light.[12]

For my own part, dear friend, you know I "trust you," both on the ice, & about the ice, or anything else; but I shall be heartily glad to see those "four dots,"[13] either explained, or frankly acknowledged, as an omission, of which the full effect had been overlooked, for the sake of many who may read Forbes's life,[14] who do not know you as I do.

I am sorry to see Mr Joule has to give up the Presidentship from health. Will this make any difference about your going to Bradford? Are you really already beginning preparations for Belfast?[15] It seems a long time beforehand! but I can well conceive you would have no rest in your next year's holiday if it was left till the last. Did Mr Hirst's foreign expedition[16] take place? And has he—or will he—be able to join you for awhile in the course of it? Do write a line when you can, to waft me a breath of mountain air, and tell me what you are doing, or,—better still—what you are thinking. "Nature" has been very dull of late; and the only scientific excitement that has reached as far as here, has been about this Transit of Venus Question.[17] Surely Sir G. Airy must see he was mistaken! how much higher he w[oul]^d^ stand had he frankly confessed it! I don't want them to sacrifice precious lives on that wretched "Possession Island" or some such unattainable spot, but it seems a pity some southern station should not be occupied by us!

I wonder whether you are in your old gite[18] or elsewhere? When do you return? We are going to my sister's[19] in Wales the end of this month, and after that to Thoresby[20] and then, I hope you will come & see us.

The limes have been delicious, but we have not once read Tennyson[21] under them! But it is more than time to end.

Ever Y[ou]rs | M.F. Egerton

RI MS JT/1/E/43

1. *[1873]*: the year is established by the discussion of Tyndall's 'Principal Forbes and His Biographers', which was published that August in the *Contemporary Review* (see n. 10).
2. *Mountfield Court*: see letter 3852, n. 2.
3. *"badinage"*: humorous, witty, or trifling discourse; banter; frivolous or light-hearted raillery (*OED*).
4. *Eyrie*: a human residence or retreat occupying an elevated position; a high vantage point (*OED*); Egerton is here referring to the Bel Alp in Switzerland, Tyndall's usual homebase in the Alps.
5. *"Vive les garçons!"*: Long live the boys! (French).
6. *Spottiswoodes*: William and Eliza Spottiswoode.
7. *Hughie's*: Hugh Egerton (see letter 4076, n. 6).
8. *Magnesium Wire*: when burned, magnesium wire produces a dazzling light—something perhaps witnessed by the author at one of Tyndall's RI lectures.
9. *little book*: not identified.
10. *Your reply... the Contemporary*: Tyndall's 'Principal Forbes and His Biographers' (see letter 4024, n. 5), in which Tyndall answered the charge from Peter Guthrie Tait that he plotted against James David Forbes in order to rob him of due scientific recognition.
11. *"Galahad"*: a knight of King Arthur's Round Table in Arthurian legend who embodies gallantry and purity (A. Lupack, *The Oxford Guide to Arthurian Literature and Legend* (Oxford: Oxford University Press, 2005), pp. 444–5). Egerton may be referencing Alfred Tennyson's poem 'Sir Galahad' (1834; published in 1842).
12. *I am very anxious... a depreciative light*: in contrast to Thomas Huxley, who congratulated Tyndall on his rebuttal of the charge that by selective quotation he had made the Swiss geologist Louis Rendu look better than he deserved (see letter 4154, n. 5), Egerton, responding to the same pages, found that it was not Rendu but his rival James David Forbes who suffered the truly egregious selective quotation at Tyndall's hands, with Forbes made to look worse than he deserved.
13. *"four dots"*: the phrase is Forbes', quoted in 'Principal Forbes and His Biographers' (p. 502), complaining about the ellipses that Tyndall had used in lieu, as Forbes saw it, of exculpatory direct quotation.
14. *Forbes's life*: *Life and Letters of Forbes*.
15. *I am sorry... preparations for Belfast*: James Joule was unwell and so unable to take up the

Presidency of the 43rd annual meeting of the BAAS at Bradford from 17–24 September 1873 (*Brit. Assoc. Rep. 1873*). While Tyndall had been asked to take up the role, and declined, Joule's replacement was Alexander Williamson (1824–1904), a University College, London chemist best known for his work on the formation of ethers (*ODNB*). See *Ascent of John Tyndall*, p. 324. Tyndall did attend the meeting, staying at the Victoria Hotel (*Bradford Observer*, 18 September 1873, p. 7; *PMG*, 18 September 1873, p. 5; *Spiritualist*, 15 September 1873, p. 9). On Tyndall as President of the BAAS for its meeting the following year, in Belfast, see letter 4013, n. 2.

16. *Mr Hirst's foreign expedition*: Thomas Hirst traveled with Tyndall to Paris and then Switzerland before heading to Italy on his own (see *Ascent of John Tyndall*, p. 323).
17. *this Transit of Venus Question*: on the controversies surrounding preparations in England for expeditions to observe the transit of Venus across the Sun in late 1874, and the role of the Astronomer Royal George Biddell Airy, see J. Ratcliff, *The Transit of Venus Enterprise in Victorian Britain* (London: Routledge, 2008).
18. *gite*: a stopping-place, lodging (*OED*).
19. *my sister's*: Egerton's younger sister was Annora Charlotte Williams-Wynn (née Pierrepont, *c.* 1822–88), who in 1853 married Charles Watkin Williams-Wynn (1822–96), a Welsh Conservative politician who sat in the House of Commons from 1868–80.
20. *Thoresby*: see letter 4076, n. 9.
21. *Tennyson*: Alfred Tennyson (1809–92), 1st Baron Tennyson from 1883, a major Victorian poet. Born in Lincolnshire, he was educated at Trinity College, Cambridge. In 1850 he published *In Memoriam* (London: Edward Moxon). In 1853 he moved to Farringford on the Isle of Wight, where he published *Maud* (London: Edward Moxon, 1855). Tyndall greatly admired Tennyson's poetry and met him in 1858. He subsequently developed a close friendship with the Tennysons and stayed with them at their home in Farringford. Queen Victoria appointed him Poet Laureate in 1850, succeeding William Wordsworth, and, after he declined honors or titles four times between 1865 and 1880, made him Baron Tennyson in 1883 (*ODNB*).

From Thomas Henry Huxley 24 August 1873 4154

Axenstein, | Luzerne. | Aug. 24 1873.

My dear Tyndall,

The copies of your booklet[1] intended for Hooker[2] and me reached me just as I left Baden[3] last Tuesday. Hooker had left me for home a fortnight before and I hardly know whether to send his to Kew or keep them for him until I return. I have read mine twice and I think that nothing could be better than the tone you have adopted. I did not suspect that you had such a shot in your locker as the answer to Forbes[4] about the direction of the "crevasses" referred to by Rendu.[5] It is a deadly thrust; and I shall be curious to see what sort of parry the other side will attempt. For of course they will attempt something. Scotland is I believe the only country in the world in which you can bring an action for "putting to silence" an adversary who will go on with an obviously hopeless suit. The lawgivers knew the genius of the people; and it is to be regretted they could not establish a process of the same sort in scientific matters.

I wrote to you a month ago[6] to tell you how we had been getting on in France. Hooker and I were very jolly, notwithstanding the heat, and I think that the Vivarrais[7] is the most instructive country in the world for seeing what water can do in cutting down the hardest rocks. Scrope's book[8] is very good on the whole though the pictures are a little overdone.

My wife[9] and Leonard[10] met me at Cologne[11] on the 11th. Then we went on to Baden and rested till last Tuesday when we journeyed to Luzern[12] and, getting out of that hot and unsavoury hole as fast as we could, came here last Thursday.

We find ourselves very well off. The hotel is perched up 1800 feet above the lake,[13] with a beautiful view of Pilatus[14] on the west and of the Urmer See[15] on the south. On the north we have the Schwys valley so that we are not shut in, and the air is very good and fresh. There are plenty of long walks to be had without much fatigue which suits the wife. Leonard promises to have very good legs of his own with plenty of staying power. I have given him one or two sharp walks and I find he has plenty of vigour and endurance. But he is not thirteen yet and I do not mean to let him do overmuch, though we are bent on a visit to a glacier. I began to tell him something about the glaciers the other day but I was promptly shut up with, "Oh yes! I know all about that. It's in Dr Tyndall's book"[16]—which said book he seems to me to have got by heart. He is the sweetest little fellow imaginable; and either he has developed immensely in the course of the last year, or I have never been so much thrown together with him alone, and have not had the opportunity of making him out. I am beginning to form great hopes of his future.

You are a fatherly old bachelor and will not think me a particularly great donkey for prattling on in this way about my swan who probably to unprejudiced eyes has a power of goose about him.

I suppose you know that in company with yourself and Hooker, the paternal gander (T.H.H.) has been honoured by the King of Sweden and made into a Polar Goose by the Order of the North Star.[17] Hooker has explained to the Swedish Ambassador that English officials are prohibited by order in Council from accepting foreign orders, and I believe keeps the cross and ribbon on these conditions. If it were an ordinary decoration I should decline with thanks, but I am told it is a purely scientific and literary affair like the Prussian "pour le merite"; so when I get back I shall follow Hooker's line.

I met Langel[18] on board the Luzern steamboat the other day and he told me that you were at the Belalp[19]—gallivanting as usual and likely to remain there for some time. So I send this on the chance of finding you. With best love from us all,

Ever yours | T. H. Huxley.

I am as well as I ever was in my life—regularly set up—in token whereof I have shaved off my beard.

RI MS JT/1/TYP/9/3019–20
IC HP 8.152
Typed Transcript Only

---

1. *your booklet*: Tyndall's 'Principal Forbes and His Biographers' was first published in the *Contemporary Review* (see letter 4024, n. 5) but then as a standalone pamphlet by his publisher Longmans, Green, and Co.
2. *Hooker*: Joseph Hooker traveled to France with Huxley (see letter 4131, n. 6).
3. *Baden*: Baden-Baden in Germany (see letter 4151, n. 1).
4. *Forbes*: James David Forbes.
5. *the direction... by Rendu*: Huxley here was admiring Tyndall's line of argument in defense of both himself and Louis Rendu in 'Principal Forbes and His Biographers' (see letter 4024, n. 5), pp. 500–02. Following Forbes, his biographers (see letter 4033, n. 6) had accused Tyndall of selectively quoting from Rendu's work to disguise the fact that Rendu had arrived at something like the correct (Forbesian) theory of glacial movement only by good luck or bad reasoning, on the basis of a wholly incorrect idea about crevasses and the direction in which they break. But what Forbes and his biographers had failed to grasp, according to Tyndall, adducing further quotations, was that Rendu had meant by 'crevasses' something quite different from what they took him to mean; and once Rendu's meaning was understood, it became clear that the accusation against him was baseless.
6. *I wrote to you a month ago*: letter 4151.
7. *Vivarrais*: a region in the southeastern France.
8. *Scrope's book*: either G. Scrope's *The Geology and Extinct Volcanos of Central France* (1858) or an earlier book covering much the same territory, G. Scrope, *Memoir on the Geology of Central France, including the Volcanic formations of Auvergne, the Velay and the Vivarais* (1827). A meticulous geological observer as well as a Member of Parliament, George Julius Poulett Scrope (1797–1876) was, with Charles Lyell, part of the generation of English geologists intent on improving its standing as a science (*CDSB*; *ODNB*).
9. *My wife*: Henrietta Huxley.
10. *Leonard*: Leonard Huxley.
11. *Cologne*: a city in Germany, to the northwest of Baden-Baden.
12. *Luzern*: or Lucerne, a city in central Switzerland.
13. *lake*: Lake Lucerne.

14. *Pilatus*: a mountain massif in the Swiss Alps overlooking Lucerne.
15. *Urmer See*: a lake, but not further identified.
16. *Dr Tyndall's book*: either *Forms of Water*, *Glaciers of the Alps*, or *Hours of Exercise in the Alps*.
17. *in company with yourself... order of the North Star*: at his coronation in Stockholm on 18 July 1873, King Oscar II of Sweden made Tyndall, Huxley, and Joseph Hooker each a Knight of the Order of the Polar Star. News of this was published in *Nature*, 8 (7 August 1873), p. 292. Hooker wrote to Huxley on 8 August about whether or not he should accept the honor (see L. Huxley, *Life and Letters of Sir Joseph Dalton Hooker*, 2 vols (London: John Murray, 1918), vol. 2, pp. 187–8). See *Ascent of John Tyndall*, p. 324.
18. *Langel*: not identified; Tyndall describes their time together in Bel Alp in letter 4158.
19. *Belalp*: the Bel Alp (see letter 3771, n. 5).

From Sydney Pierrepont[1] 31 August 1873 4155

Thoresby Park | Ollerton, | Newark.[2] | August 31st 1873.

Dear Sir,

I have heard from my sister Lady Mary Egerton[3] that you have expressed a desire to see our Forest Country, and as she and two of her daughters will be here between the 8th and 20th of September I beg to say that it would give Lady Manvers[4] and myself much pleasure if you could pay us a visit here during that time. You may possibly be at the meeting at Leeds[5] on the 17th in which case this would be on your way or not much out of it from London, but having heard of you last in Switzerland I shall not be surprised if I do not receive a very speedy reply. Tuxford would be our best station from London by 12.10. train from Kings Cross.

I am | faithfully yours | Manvers

RI MS JT/1/TYP/6/2144
LT Typescript Only

---

1. *Manvers*: Sydney William Herbert Pierrepont, 3rd Earl Manvers (1825–1900), brother of Mary Egerton.
2. *Thoresby Park | Ollerton, | Newark*: see letter 4076, n. 9.
3. *I have heard... Lady Mary Egerton*: see letter 4156.
4. *Lady Manvers*: Georgine Jane Elizabeth Fanny Pierrepont (née Franquetot, 1826–1910), married in 1852.
5. *meeting at Leeds*: Manvers probably meant to refer to the 43rd annual meeting of the BAAS at Bradford, next to Leeds (see letter 4153, n. 15).

From Mary Egerton — 31 August–1 September [1873][1] — 4156

Coed y Maen | Welshpool[2] | August 31st

My dear Mr. Tyndall,

The pamphlet[3] duly arrived many thanks for sending it. I thoroughly like your rejoinder;[4] it is so temperate & fair, though full of spirit, & makes all as clear as the day. It is exactly as I expected with regard to the omitted sentence;[5] you were looking at the spirit of the passage, which certainly was anything but that of giving credit to others, & you left out words which did not alter that, not adverting to the interpretation the omission might be made to bear. It is a pity they should have had even this shadow of an excuse for accusing you of unfairness, but when people are so looking out for offence, they will make an excuse if they cannot find one! With regard to the other omission, it is rather amusing their being convicted of such a complete misunderstanding of Rendu[6] certainly without the previous explanation, no one would have guessed that by "crevasses" he only meant the wrinkles of the dirt bands! but Forbes[7] who professes to have studied his book,[8] was bound to know it! Rendu's Nomenclature does not seem to have come into it at all generally? I hardly needed to refer to the passage in "Mountaineering"[9] I remembered it well; it is indeed sad that a spirit so different from yours, should ever prevail with reference to subjects that ought to exalt the mind above all petty jealousies & self seeking! I am glad you "gave it them" about that mean insinuation! It made my blood boil, I confess!

We had driven over to Heathfield[10] the very day L[or]d Claud & his remaining daughter[11] had heard of the arrival of the party at Bel Alp.[12] You may guess if my heart gave a jump when I heard where they were, and what doing! I think L[ad]y Claud is very attractive, & so handsome! And I daresay she w[oul]d be still more charming when free from all home cares, which give her sometimes rather an oppressed look, I fancy. I am so glad you have had fine weather to show the dear Mountains to appreciative observers, though I cannot help a little pang of envy as I say so! Not so much for what we lost, for we got much enjoyment out of our week, notwithstanding all, but for your disappointment after so kindly devoting yourself to us. And you have had Mr Hirst too![13] How he would enjoy being with you again on the glaciers! Altogether this year will have been so much brighter than the last, and our visit will only be associated with clouds and darkness! Your visit at Paris must have been very interesting I wonder why I fancied I had heard of old M. Chasle's death last year![14]

By the bye I must congratulate you on your foreign honours![15] "Sir

Galahad[16] of the Northern Star!" How well it sounds! Worthy to be the title of another Idyll!

Perhaps I ought not to bother you with another letter, but I wanted to talk about the Pamphlet,[17] & I always feel a desire to remark upon the subjects contained in any letter I receive that interests me & thus prove that I have cared about them.

I trust you are getting so thoroughly "wound up" that you will not "run down" again in a hurry, but may be "warranted to go," without repair for the rest of the year at least! When you feel like that, cannot you fancy a time when this curious human clockwork shall be replaced by some perfectly elastic material, that may work without friction & without fatigue, and be supplied from the ethereal substances around it? You will laugh at my favorite fancy—never mind!

I have been much interested with Helmholtz's Popular lectures,[18] as much of them at least as I can understand. He has such a grand philosophic way of viewing things. The young party send kindest remembrances. I suppose Mr Hirst will have gone on to Italy ere this, or I should send the same to him. Farewell! You once called me "a true & constant lady" for remembering you from one London season to the next;—now—after seven years—let me adopt your words with a slight variation, and subscribe myself

Your "true & constant" friend, | M. F. Egerton

Excuse my having taken the paper the wrong way.[19]

I had never till yesterday read Emerson's beautiful essay on "Nature"[20] I thought of you!

Sept[embe]^r 1st

Since writing this, (finished too late for yesterday's post) I have received a letter from my brother, telling me he has written[21] to try to persuade you to come to Thoresby.[22] I knew he wished to ask you sometime so I told him you might probably be going to Bradford[23] which would be a good opportunity. Do think about it! It would be so pleasant if you would come while we are there; I should so like to show you the old forest![24]

Only you must not expect too much, for the poor old trees keep losing their "stag heads," & thus growing shorter every few years, so that but few are now fine in point of height, thought[25] very picturesque; neither is the extent great, for much of it was cut up into plantations by the late D[uke]. of Portland![26] We stay here (at my sister's) till Monday the 8th when we go to Thoresby till probably the 18th or 19th. We are among the pretty Welsh hills! I went up a "mountain" on Saturday, only "stopping to take breath" about a dozen times!

RI MS JT/1/E/45

1. *[1873]*: the year is established by its relation to letter 4153 (see n. 4 below).
2. *Coed y Maen | Welshpool*: the residence of her sister and brother-in-law, as Egerton shared with Tyndall in letter 4153 that she was 'going to my sister's in Wales the end of this month'.
3. *The pamphlet*: see letter 4154, n. 1.
4. *I thoroughly like your rejoinder*: in what followed Egerton withdrew the mild anticipatory criticisms she ventured in her previous letter (see letter 4153, n. 12; and also letter 4154, n. 5).
5. *omitted sentence*: see letter 4153, n. 13.
6. *Rendu*: Louis Rendu.
7. *Forbes*: James David Forbes.
8. *his book*: L. Rendu, *Theorie des Glaciers de la Savoie* (Chambéry: Chez Puthod, Imprimeur-Libraire, 1840). An English translation, edited by George Forbes, was published by Macmillan in 1874, and includes a chapter by Peter Guthrie Tait that attacks Tyndall.
9. *passage in "Mountaineering"*: in his *Mountaineering in 1861*, Tyndall discussed Rendu's contributions to glacial movement theory (pp. 68–71).
10. *Heathfield*: Heathfield Park, the Hamilton family estate in Hawkhurst.
11. *Lord Claud & his remaining daughter*: Claud Hamilton and his youngest daughter Mary Hamilton.
12. *arrival of the party at Bel Alp*: Tyndall spent time in the Alps in August 1873 with Elizabeth Hamilton and her children Emma Hamilton, Douglas Hamilton, and Tyndall's future wife, Louisa Hamilton. On Tyndall's connections with the Hamilton family, see *Ascent of John Tyndall*, p. 323.
13. *you have had Mr Hirst too!*: Hirst accompanied Tyndall to Paris and then Switzerland, later departing for Italy on his own.
14. *why I fancied... death last year*: the French mathematician Michel Chasles, born in 1793, in fact lived until 1880 (*CDSB*).
15. *your foreign honours!*: on the King of Sweden making Tyndall a Knight of the Order of the Polar Star (see letter 4154, n. 17).
16. *Sir Galahad*: see letter 4153, n. 11.
17. *Pamphlet*: see n. 3.
18. *Helmholtz's Popular lectures*: H. Helmholtz, *Popular Lectures on Scientific Subjects*, trans. E. Atkinson (New York: D. Appleton and Company, 1873); this included an introduction by Tyndall (pp. xv–xvi). At this time no British edition of this book had been published.
19. *Excuse my... wrong way*: a *mea culpa* for the criticisms of her earlier letter (see n. 4).
20. *Emerson's beautiful essay on "Nature"*: R. W. Emerson, *Nature* (Boston: James Munroe and Co., 1836).
21. *my brother... has written*: Sydney Pierrepoint's letter to Tyndall is 4155.
22. *Thoresby*: see letter 4076, n. 9.
23. *to Bradford*: for the annual meeting of the BAAS (see letter 4153, n. 15).
24. *old forest*: Sherwood Forest. In letter 4153, Egerton wrote that she had 'been born and bred in Sherwood Forest'.

25. *thought*: Egerton misspelled 'though'.
26. *late D[uke]. of Portland*: William Henry Cavendish-Scott-Bentinck, 4th Duke of Portland (1768–1854).

To Martha Somerville[1] [August 1873][2] 4157

Saturday

My dear Miss Somerville.

According to an arrangement made in England we were joined here by Lady Claud Hamilton her two daughters & her son.[3] I think they will stay here for three or four days & afterwards I propose going with them for a few days to the AEggischhorn.[4] You & your sister[5] would I think take pleasure in going on little excursions: and I shall be delighted to have you with us.

Ever faithfully yours | John Tyndall

Heartiest thanks for the photographs.

Oxford, Bodleian Library, Dep c.372 (MST-1)

---

1. *Martha Somerville*: Martha Charters Somerville (1815–79), the oldest of two surviving children of Mary Somerville (1780–1872), a Scottish science writer who studied mathematics and astronomy, with her second husband William Somerville (1771–1860). Their first two children died young; the fourth child was Mary Charlotte Somerville (1817–75) (*ODNB*).
2. *[August 1873]*: the month and year are given by reference in other letters to Tyndall spending time with the Hamiltons while in Switzerland. See letters 4145, 4156, and 4164.
3. *joined here by... & her son*: the children of Claud and Elizabeth Hamilton were Louisa Hamilton, Tyndall's future wife; Emma Hamilton; Mary Hamilton; and Douglas Hamilton. On Tyndall's connections with the Hamilton family, see *Ascent of John Tyndall*, p. 323.
4. *AEggischhorn*: Eggishorn (see letter 3828, n. 15).
5. *your sister*: Mary Charlotte Somerville (1817–75).

To Thomas Henry Huxley 2 September 1873 4158

Eggischhorn[1] | 2nd. Sept. 1873.

My dear Huxley

I am right glad to hear that the reply to Forbes pleases you.[2] Imagine my self denial in permitting that thing to remain unanswered for a dozen years. I doubt whether I was right in doing so, for on the part of an outsider it would take a good deal of rummaging to find the proper answers, and I might have died with the stigma of foulplay attached to my name. An analysis of Forbes would be in the highest degree instructive as a piece of psychology and if they force me to it they may yet have it.

I tried when writing the Reply to dislodge all mere irritation from my mind; and having printed it I ran it through Hirst's brain where the last traces of bitterness were extracted from it.

The terms in which you speak of Leonard[3] are exactly those which my own observation of the boy would lead me to apply to him. He is a sweet little fellow, and will I doubt not grow up into a true gentleman. I wish heartily I could have given him a lesson on the glaciers.

After my arrival I spent some time at the Bel Alp,[4] where I had the good fortune to see Langel.[5] He however disappeared with extraordinary suddenness perhaps because his quarters were not good ones. About a week ago I came over here, and to day I depart for Zermatt.[6] I shall try and return in time for Bradford[7] as I should like to support Williamson.[8] Thanks to my being one of a pleasant little party I have taken more exercise than usual, and though to day I feel slumberous, I am on the whole very strong.

Your postscript was cheering in the highest degree. The molecules I doubt not have been shaken into their proper places. Give them time to set.[9]

Give my love to your wife and Leonard. I would write you a longer letter were my brain less sleepy than it is.

Yours ever | John Tyndall

RI MS JT/1/TYP/9/3021
IC HP 8.154
Typed Transcript Only

---

1. *Eggischhorn*: Eggishorn (see letter 3828, n. 15).
2. *I am right glad... pleases you*: on 23 August, Huxley wrote to congratulate Tyndall on his 'Principal Forbes and His Biographers' (see letter 4154). Tyndall's letter as a whole is a point-by-point response to Huxley's.
3. *The terms in which you speak of Leonard*: Huxley had written in letter 4154 about how impressed he was with his twelve-year-old son Leonard, not least in his enthusiasm for Tyndall's work.
4. *Bel Alp*: see letter 3771, n. 4.
5. *Langel*: not identified; Huxley describes meeting him and learning about the time in Bel Alp with Tyndall in letter 4154.
6. *Zermatt*: a Swiss town in the Matter Valley, located in the canton of Valais. It is surrounded by the Pennine Alps including Switzerland's highest peak, the Monte Rosa, which Tyndall climbed for the first time in 1858.
7. *Bradford*: for the annual meeting of the BAAS (see letter 4153, n. 15).

8. *support Williamson*: at the annual meeting of the BAAS in Bradford (see letter 4153, n. 15).
9. *Your postscript... time to set:* at the end of his letter (4154), Huxley had reported: 'I am as well as I ever was in my life—regularly set up'. Tyndall had been instrumental in organizing Huxley's friends in not only forcing him to take a health-restoring holiday—desperately needed, they reckoned—but paying for it (see letter 4028, n. 4).

To Elizabeth Dawson Steuart 11 September 1873 4159

Royal Institution | 11th Sep. 1873

My dear Mrs Steuart,

Your note[1] followed me to Switzerland, whence I returned last night.[2] A mountain of correspondence is now before me.

Of course I would gladly give my best attention to any request of yours, but I know nothing about the Blue Coat School[3] nor its Governors, nor its modes of election. Of course I am aware of its celebrity as a historic school, but beyond this of nothing. Nor do I know the lad,[4] to whom reference is made, and I am rather stern in demanding from myself a strict reason for any responsibility of this kind that I may undertake. In short my devotion to my own pursuits has rendered me helpless and inoperative as regards the world around me.

I wished to write to you at once regarding this matter—when my present work has been to some extent diminished I shall be able to write you a longer letter.

Yours ever faithfully | John Tyndall

If you thought a note to old Mr Bewley[5] whom I know would help in Dublin I would write it to the effect that it would give me pleasure if he could help the lad. But then I must state that it is at the instance of a friend that I interest myself in this case.

RI MS JT/1/TYP/10/3409
LT Typescript Only

1. *Your note*: letter missing.
2. *whence I returned last night*: see letter 4131, n. 6.
3. *Blue Coat School*: the term 'Blue Coat School' came to be used for a range of Christian schools throughout England; but given the reference in the letter to helping out a boy in Dublin, the particular Blue Coat School under discussion was probably the one founded in 1669 in Dublin.
4. *lad*: not identified.
5. *Mr Bewley*: not identified.

From Mary Egerton [13–14 September 1873][1] 4160

Thoresby Park, Ollerton[2] | Sat[urda][y] Ev[enin][g]

How tiresome! I had been planning where we would go, & now to show you the best of the forest in the shortest time. And how, even if you could only stay one night, the 9 o'clock train from King's X,[3] would bring you to Retford[4] at 12.25, & here in time for luncheon, and the 3–49 from Retford would get you quite early to Bradford[5] each day (5.36), so we should have the morning for another ramble. Could it not be? You were to come before Bradford, mind, so it w[oul][d] not have prevented your getting back to town as soon as you liked. The country is in such beauty just now, and it is a chance our being here another year just at the fine season. Well! It might rain! Your "Excellent Friend"[6] must take that for a consolation if there is no better at hand!

And I wanted so much to hear about your adventures with the Claud Hamiltons.[7] I have heard something about a storm on the glacier[8] and a night in the hut, but no particulars yet, but I gather that they all behaved gallantly, and enjoyed it.

How about <u>your heart</u> after these achievements coming at the end of such a pleasant fortnight? The only safety is, you could not well fall in love with <u>both</u> the young ladies,[9] & the Mother[10] is <u>better</u> looking than either of them! I only hope the <u>Heath</u>[11] will not quite supersede the <u>Mount</u>[12] in your <u>field</u> of vision, (Do you see?)

<u>Sunday.</u>

Just got your note.[13] You terribly busy people, are very unsatisfactory as friends! Not a syllable to say whether you ever received <u>my</u> letter![14] Why not come here <u>instead</u> of Bradford? it w[oul][d] do you much more good! Tuesday till Thursday or Friday. We do not go till <u>Friday</u> now. The same post brings me "Nature" forwarded from home. You would be amused if you c[oul][d] see May's[15] "ohs!!!" and pencil annotations on the Tait article.[16] In one place she has put, "<u>Green-Eyed Jealousy</u>"! As to myself I am too angry to talk about it! As if you meant any disrespect to men like Wheatstone,[17] by selling original research above money making application of science!

I earnestly wish you may find it right not to answer it, for I see an interminable vista of bickerings opening out! but I feel <u>I cannot</u> judge. I am curious about that rainbow reflection,[18] because someone in "Nature" asserted the other day they had <u>seen it</u>![19] You would be quite different to my thoughts of you, were you not willing to acknowledge a mathematical oversight, <u>should</u> you by chance have made one, in a matter which only came curiously under your attention, so that I feel no hesitation or scruple in alluding to it, much as I detest the tone in which it is brought forward on this occasion.

If you cannot come here, will there be any chance of seeing you in London when we go through on the 30th or 1st? I shall write & let you know if we stop there.

Y[ou]rs Ever | M F Egerton

RI MS JT/1/E/99g
RI MS JT/1/E/91

1. *[13–14 September 1873]*: the date is given since this letter, in the RI archive in two parts but connected here, was written after Tyndall returned from his trip to Switzerland where he spent time with the Hamiltons (10 September 1873), but before the annual meeting of the BAAS in Bradford (17–24 September). The only Saturday and Sunday that fell between these dates were 13 and 14 September 1873.
2. *Thoresby Park, Ollerton*: see letter 4076, n. 9.
3. *King's X*: King's Cross railway station in London.
4. *Retford*: a town in England near Ollerton, considerably north from London.
5. *Bradford*: for the annual meeting of the BAAS (see letter 4153, n. 15).
6. *"Excellent Friend"*: referring to herself.
7. *adventures with the Claud Hamiltons*: Tyndall wrote on 15 September to Elizabeth Dawson Steuart (letter 4164): 'I had a very pleasant expedition in the Alps. Lady Claud Hamilton, two of her charming daughters, her son and myself formed a little party—and it gave me additional pleasure to observe their delight'. See also letter 4145. On Tyndall's connections with the Hamilton family, see *Ascent of John Tyndall*, p. 323.
8. *the glacier*: possibly Aletsch glacier, below the Eggishorn (see letter 3828, n. 15).
9. *both the young ladies*: Louisa Hamilton, Tyndall's future wife; and Emma Hamilton. The youngest sister, Mary Hamilton, remained at home. Unbeknownst to Egerton, Tyndall had previously been in love with two sisters simultaneously (of the Drummond family; see *Tyndall Correspondence*, vol. 6).
10. *Mother*: Elizabeth Hamilton.
11. *the Heath*: see letter 4156, n. 10.
12. *the Mount*: see letter 3852, n. 2.
13. *your note*: letter missing.
14. *my letter*: letter 4156.
15. *May's*: Mary Alice Egerton.
16. *Tait article*: P. G. Tait, 'Tyndall and Forbes', *Nature*, 8 (11 September 1873), pp. 381–2.
17. *Wheatstone*: in his letter, Tait called 'attention to the way in which Sir Charles Wheatstone, and (by implication) Sir William Thomson, and others, some of whose splendid scientific labours have had the misfortune to become profitable in a pecuniary sense, are treated in Dr. Tyndall's "Lectures on Light," just published. The contrast between the utter contempt for money shown by their censor, and the (implied) opposite which is condemned as unworthy of scientific men, is brought out with all the flow of word-ainting and

righteous indignation which Dr. Tyndall so abundantly possess. Besides, the monstrous doctrine is inculcated that men who devote themselves to practical applications are men incapable of original research' (p. 382).

18. *rainbow reflection*: in his letter, Tait quotes a passage from Tyndall's American lectures on light about the lack of a reflection of a rainbow viewed over the tranquil sheet of water, and asks, 'What confidence can one have in the accuracy of any statement on a scientific matter made by the author of it?' (p. 382). Tyndall replied to Tait's letter in general the following week (J. Tyndall, 'Tyndall and Tait', *Nature*, 8 (18 September 1873), p. 399; this is letter 4167 in this volume, and see also letter 4161, n. 5), and to Tait's remark about rainbow reflections the next week ('Reflection of the Rainbow', *Nature*, 8 (25 September 1873), p. 432–3; this is letter 4169 in this volume).
19. *someone in "Nature" asserted the other day they had seen it!*: see 'Reflected Rainbows', *Nature*, 8 (4 September 1873), pp. 361–2 (signed from 'Z. X. Y.').

From Mary Egerton [after 14 September 1873][1] 4161

Monday

Dear friend,

I wrote to you to the Athenaeum,[2] but concluded you were out of town.[3] It was to tell you we should be in town for a night today, and should be so glad to see you if you had time to look in upon us any time this evening from 6 o'clock upwards; only between ½ p7 & 8 the room might smell of mutton chops! So if you don't come before, you had better give it till ½ p8 to air.

I should have liked a chat about Switzerland,[4] but I am glad if you are getting a little fresh air for I know you must have been bothered and vexed. I am glad "Nature" took your advice, but put in the paragraph which I felt sure you would send on cooler thoughts.[5] Don't think me impertinent, for believe me, I feel with you in the whole matter!

Ever yours | M F Egerton

RI MS JT/1/E/68

1. *[after September 1873]*: this letter follows Egerton's letter of [13–14 September 1873] (see letter 4160), and as it was written on a Monday it could be 15 or 22 September.
2. *Athenaeum*: the Athenaeum Club (see letter 3710, n. 5).
3. *out of town*: for the forty-third annual meeting of the BAAS which Tyndall did not attend (see letter 4153, n. 15).
4. *about Switzerland*: Tyndall wrote on 15 September to Elizabeth Dawson Steuart (letter 4164): 'I had a very pleasant expedition in the Alps. Lady Claud Hamilton, two of her charming daughters, her son and myself formed a little party—and it gave me additional

pleasure to observe their delight'. See also letter 4145. On Tyndall's connections with the Hamilton family, see *Ascent of John Tyndall*, p. 323.

5. *"Nature" took your advice... send on cooler thoughts*: J. Tyndall, 'Tyndall and Tait', *Nature*, 8 (18 September 1873), p. 399. This is letter 4167 in this volume. An editorial note (from Norman Lockyer) accompanied Tyndall's letter, which stated that while words of a personal nature were removed from Tait's letter attacking Tyndall before its publication (P. G. Tait, 'Tyndall and Forbes', *Nature*, 8 (11 September 1873), pp. 381–2), that it is 'under strong protest that he allows to pass in Prof. Tyndall's letter... personalities' that, in due course, might lead the journal to 'easily fall from the position of absolute justice and impartiality in all scientific matters which it now occupies and become the mere mouthpiece of a clique' (p. 399). In the following issue of *Nature*, Lockyer issued a statement that he would cease publishing anything further from Tait and Tyndall regarding the subjects of dispute ('Tait and Tyndall', *Nature*, 8 (25 September 1873), p. 431).

From Joseph Dalton Hooker — 14 September 1873 — 4162

Kew, Sept. 14th 1873.

Dear old Tyndall

I have just read Professor Tait's letter in "Nature" with contempt, and cannot resist writing a few lines to you feeling as I do a recurrence of that pain which communications similar in regard of tone and spirit inflicted on me 12 months ago, pain felt not at all on my own account, for I knew the attacks would no more sully me in the opinions of those whom I loved and respected than they did in my own opinion,—but pain for the wounded honour of science, and for the outraged dignity of scientific controversy, coupled with indignation at the association of my name with such an exhibition of dishonourable motives and dishonouring epithets.[1]

Not that I was insensible to the virulence of the hatred or the bitterness of the jealousy: such traits of corrupted morals, like physical distortions, distressed me—as I fancy this exhibition of them must you:—they preyed on me by night and I could no more away with them than I could with the unsubstantial horrors of a wicked dream, the recollection of which cleaves like a sin to the sinless soul, long after it has waked to the light reflected from its own purity.

Well, dear Tyndall, I can't help feeling as I did then, but if your soul is stern enough to throw off indignation and its concomitant disquietings, then I rejoice, and far more abundantly.

As to the matter of the attack, it is not worth notice, it is well that when malice is so black it should be as naked as black; and that when motives for traduction are so evil, they should be as undisguised as evil; and in these respects the letter leaves nothing to be desired.

I often wonder where you are and what about; with Huxley I hope, and keeping him as idle as I did, and as well as he knows how to keep himself. We are all prepared for Bradford on Wednesday.[2] We have taken a lodging for ourselves, Strachey,[3] Dyer[4] and Lawson of <O>xford.[5] We go to Lord Houghton's from Sat. to Monday, and to Sir J. K. Shuttleworth's[6] for a day or two after the meeting.

G. Henslowe[7] is much better and now with us, carried from room to room, but can stand upright a little. I was at Down a fortnight ago. Darwin was wonderfully well, but was terribly ill after it, as bad as he ever was: he is now better again. He is doing splendid work. George[8] is still wretchedly ill and Horace[9] not at all well, the rest are flourishing. I lunched at the Lubbocks' where everything blossomed like the rose. Amy[10] is so handsome and graceful. Prof. <L>aurence Smith[11] and wife come here tomorrow and I expect to hear from them of your doings in the U.S.A. A thousand thanks for your book,[12] I have not read much of it yet. With much love from us all,

Ever your aff[ectiona]te | J. D. Hooker.

Though not much to the purpose I must add that I think it very wrong of "Nature" to print letters containing personal traduction. It is a sad thought that future generations may light on some of its columns and ask what sort of gentlemen were their scientific forefathers considered to be in their day and generation? What can such men as Tyndall and Hooker have done to call up such animosity however unmerited and innocuous.

RI MS JT/1/TYP/8/2718–9
LT Typescript Only

1. *I have just read... dishonouring epithets*: Hooker was responding to Peter Guthrie Tait's letter to *Nature* (see P. G. Tait, 'Tyndall and Forbes', *Nature*, 8 (11 September 1873), pp. 381–2). In his reply to *Nature* (letter 4167), Tyndall quoted from this paragraph, crediting an 'illustrious person'. Hooker's reference to a painful public controversy a year earlier was to the Ayrton affair (see letter 3755, esp. n. 1).
2. *for Bradford on Wednesday*: for the annual meeting of the BAAS (see letter 4153, n. 15).
3. *Strachey*: Richard Strachey (1817–1908), a government administrator in India who was esteemed for his contributions to physical and botanical geography ('Sir Richard Strachey, G.C.S.I., F.R.S.', *Nature*, 77 (27 February 1908), pp. 395–7).
4. *Dyer*: William Turner Thiselton-Dyer (1843–1928), a British botanist and support of Darwin who would go on to succeed Hooker as Director of the Royal Botanic Gardens, Kew (*CDSB).*
5. *Lawson of <O>xford*: Marmaduke Alexander Lawson (1840–96), Sherardian Professor of Botany and Rural Economy at Oxford from 1868 (*ODNB*).

6. *Sir J. K. Shuttleworth's*: James Phillips Kay-Shuttleworth (1804–77), an English physician and civil servant (*ODNB*).
7. *G. Henslowe*: George Henslow (1835–1925), an English botanist and Anglican curate, son of Darwin's botany teacher John Stevens Henslow (*ODNB*).
8. *George*: George Darwin (1845–1912), a barrister and an astronomer, son of Charles Darwin (*ODNB*).
9. *Horace*: Horace Darwin (1851–1928), an engineer and scientific-instrument manufacturer, son of Charles Darwin (*ODNB*).
10. *Amy*: Amy Harriet Lubbock (1857/8–1929), daughter of John Lubbock and his wife Ellen Frances Lubbock (see letter 3783, n. 22). The Lubbocks were neighbors of the Darwins in the village of Downe in Kent; their estate High Elms was close to Down House.
11. *Prof. <L>aurence Smith*: probably the American chemist John Lawrence Smith (1818–83), who served as President of the AAAS in 1872 (*ANB*).
12. *your book*: *Six Lectures on Light*.

To the Editor of *Nature* 15 September [1873][1] 4163

[Reflection of the Rainbow]

Draw a circle to represent a rain-drop, or rather a section of it, by a plane passing through its centre, the sun, and the eye. Draw a straight line through the centre to represent a solar ray of mean refrangibility. At the front and back of the drop reflection occurs, and the incidence being normal, the incident and reflected beams will coincide after the emergence of the latter from the drop. Now suppose the ray through the centre to move parallel to itself, the incidence grows more and more oblique, refraction occurs at entrance and at emergence, the ray finally becoming a tangent to the drop. Let the incident and the twice refracted and once reflected rays be produced backwards till they intersect behind the drop: the angle enclosed between them augments with the obliquity, reaches a maximum, and then diminishes. The ray corresponding in obliquity with this maximum angular value, and those in its immediate vicinity, quit the drop sensibly parallel, and these are the rays which are effectual in the rainbow. This angle being for red light 42°, and for violet light 40°, for light of mean refrangibility it is 41°.

If those parallel rays before reaching the observer's eye impinge upon a surface of calm water, they are, in part, reflected according to the usual law, and *a* rainbow is then seen by reflection. But the absolute position of the bow changes with every change in the position of the observer's eye; hence the bow seen mirrored in the pool is *not* the reflection of that seen at the same time directly in the heavens. Suppose the shower to be fixed in space, then the drops which produce the bow seen directly, would not be those which produce the bow as seen by reflection.

In the paragraph to which your correspondent "Z.X.Y." has called attention,[2] I meant to combat the notion, entertained by many, that the rainbow is reflected after the fashion of an ordinary floating cloud which emits light in all directions, and which, by the light thus emitted, paints its image in the water. A few additional words might have made my meaning clearer; but as I was dealing at the time more with historic statement than with scientific exposition, I desired to be brief. I can hardly think, however, that your correspondent will be angry with me for giving him what must have been agreeable as well as successful occupation at the Falls of the Rhine.

JOHN TYNDALL | Royal Institution, Sept. 15

Nature, 8 (25 September 1873), pp. 432–3

1. *[1873]*: the year is established by the publication of this letter in the 25 September 1873 issue of *Nature*.
2. *In the paragraph... has drawn attention*: Tyndall was replying to both a letter in *Nature* from 'Z. X. Y.', in Schaffhausen, who tested—and found wanting—Tyndall's claim in his *Six Lectures on Light* (on pp. 25–6) that a rainbow's reflection cannot be seen in the surface of a still lake, and Peter Guthrie Tait's criticism of Tyndall regarding this; see 'Reflected Rainbows', *Nature*, 8 (4 September 1873), pp. 361–2 (signed from 'Z. X. Y.'). See also letter 4161, n. 5.

To Elizabeth Dawson Steuart — 15 September 1873 — 4164

Royal Institution | 15th Sep. 1873

My dear Mrs Steuart,

Mr Conwill[1] informs me that Edward Heydon,[2] whom I knew when a boy—is in great distress. Would you kindly act the part of almoner and hand over to him the cash represented by this cheque?

Yours ever | John Tyndall

I had a very pleasant expedition in the Alps. Lady Claud Hamilton, two of her charming daughters, her son[3] and myself formed a little party – and it gave me additional pleasure to observe their delight.[4]

RI MS JT/1/TYP/10/3410
LT Typescript Only

1. *Mr Conwill*: John Conwill (1802–80) was Tyndall's teacher back in Leighlinbridge, County Carlow, Ireland.
2. *Edward Heydon*: not identified.
3. *Lady Claud Hamilton, two of her charming daughters, her son*: see letter 4157, n. 3.
4. *I had... their delight*: on Tyndall's time in the Alps that summer with Elizabeth Hamilton and her children, see *Ascent of John Tyndall*, pp. 323–4.

From Mary Egerton [between 15 and 25 September 1873][1] 4165

[2]you as suspicious as Mr Tait, if you make me mean things I do not say!! I sent your message to May.[3] I hope the Rainbow will be in the next Nature.[4] Now I must end, or you will "increase & multiply" this note in your imagination, as you did the other day when you said "all your letters reached me,"[5] & I had only written one! I believe you don't like one to care the least bit about you or your concerns, (which (N.B.) is very ungrateful of you!) but I fear it is "too late to mend" in my case, so you must then submit to your fate with as good a grace as you can!

I should not have written now, only I wanted to ask about Monday.[6] Send me one line here, (we stay till Monday morning,) to say whether if you can call at all, it would suit you best to come soon after 6, or in the evening early, that I may arrange accordingly

Y[ou]rs ever | M F Egerton

RI MS JT/1/E/99s

1. *[between 15 and 25 September 1873]*: the date range is established by reference to Mary Egerton's expressed hope that Tyndall's letter on rainbows would be published imminently in *Nature*. Tyndall's 'Reflection of the Rainbow', dated 15 September, appeared in *Nature*, 8 (25 September 1873), pp. 432–3 (this is letter 4163 in this volume).
2. Letter incomplete; fragment only.
3. *May*: Mary Alice Egerton.
4. *the Rainbow will be in the next Nature*: see n. 1 above. Tyndall's letter to *Nature* on rainbows was a response to a letter published in the 4 September issue querying his apparent claim, in his *Six Lectures on Light*, that a rainbow above a still body of water is not reflected in its surface. For further details, see letter 4163, n. 2. Tait's letter in the 11 September issue (see letter 4160, n. 16) ridiculed Tyndall for the same claim—hence Tyndall's remark to Thomas Huxley, in a letter of 24 September (letter 4171): 'By the way I sent to Nature a little communication about the rainbow, regarding which Tait has found a mare's nest' (that is, in Tyndall's view, Tait had mistaken an unremarkable passage in Tyndall's book for a reputation-scupperingly huge mistake).

5. *when you said "all your letters reached me"*: any letters from Tyndall to Egerton are missing.
6. *Monday*: since Tyndall wrote his letter to *Nature* on 15 September 1873, the following Monday was 22 September, during the annual meeting of the BAAS in Bradford (see letter 4153, n. 15). Egerton had tried to get Tyndall to visit her at Thoresby Park (see letter 4076, n. 9) while on his way to Bradford (see letter 4160), and this may be her attempt for a visit on his return trip to London.

From Elie-François Wartmann[1] 15 September 1873 4166

Les Colombières sur Versoix,
Canton de Genève. | 15 septembre 1873

Mon cher ami,

A mon retour de Vienne j'ai trouvé votre brochure sur Forbes et votre ouvrage sur l'Optique, que j'ai lus avec un grand plaisir. Vous avez le secret précieux d'exprimer en un langage simple et accessible à tout auditeur attentif les vérités les plus sublimes de la science.

Votre définition de la vie (page 260) est la même que celle que je professe depuis nombre d'années. Quant au principe de la moindre action, mentionné à la page 23, il y avait lieu, dans un but de haute philosophie, d'en généraliser l'application à d'autres phénomènes que celui de la réfraction simple. Vous avez sans doute remarqué qu'à la page 110, l'imprimeur vous fait dire: The velocity in air is to its velocity in glass as 2:3 au lieu de as 3:2. Il y aura lieu à une correction dans notre future édition, et, puisque je suis sur ce sujet, permettez-moi de vous en signaler une autre à laquelle j'attache une certaine importance. Vous attribuer (page 186) à De la Povostaye et Desains la découverte de la magnétisation d'un rayon ou d'un faisceau de chaleur. Cette découverte m'appartient, et, pour vous en assurer, veuillez ouvrir les Archives des Sciences physiques et naturelles,

Tome I, page 417, Genève 1846

Tome IV, page 50, note; 1847

ou relire la note des deux physiciens français dans les Comptes rendus de 1849 tome 29, page 352. Je tiens à cette rectification, parce que mes expériences de 1846 furent faites dans des conditions singulièrement difficiles et pénibles. J'avas dû construire moi-même mes électro-aimants, mes piles de mica, etc. Je professais à Lausanne et n'avais aucun secours instrumental ou pécuniaire. La révolution venait d'éclater: elle eut pour effet de chasser de cette ville, avec moi-même, tous mes collègues de l'Académie: j'y perdis mes appareils. A Genève où la révolution grondait, M. de la Rive résignait ses fonctions et me désignait pour lui succéder mais je n'avais ni les ressources de laboratoire ni la liberté d'esprit nécessaires pour poursuivre mes expériences. Pendant ce temps De la Povostaye les répétait, les confirmait et y introduisait l'importante modification de placer sous un angle de 45° les sections principales du polarisateur et de l'analyseur. Au surplus, les traités français de physique m'ont toujours cité comme ayant été l'auteur de la découverte et il me serait pénible de laisser une erreur se propager dans un ouvrage tel que le vôtre, appelé à une grande circulation et à de nouvelles éditions.

Sir John Herschel cite, page 392 de ses Familiar Lectures, un liquide trouvé par M. Jellett de Trinity College, Dublin, qui serait right-handed pour une extrémité du spectre et left-handed pour l'autre. Savez-vous quel est ce liquide?

Believe me, my dear Tyndall, very truly yours | E. Wartmann

Les Colombières sur Versoix,
Canton of Geneva. | 15 September 1873

My dear friend,

When I came back from Vienna, I found your booklet on Forbes,[2] and your book on Optics[3] that I read with great pleasure. You have the precious ability to express in a simple language accessible to every attentive listener the most sublime truths of science.

Your definition of life (page 260) is the same that I have been teaching for several years. As for the principle of the slightest action, mentioned page 23, it would have been desirable, with the purpose of high philosophy, to generalize the application to other phenomena than the one of the simple refraction. You have probably noticed that on page 110, the editor makes you say: The velocity in air is to its velocity in glass as 2:3 instead of 3:2. There will be a correction in our future edition, and since I am on this subject, please let me signal another mistake which is important to me. You attribute (page 186) to De la Povostaye and Desains[4] the discovery of the magnetism of a ray or beam of heat. This discovery belongs to me, and to prove it to you, please open the Archives des Sciences physiques et naturelles,

Volume I, page 417, Geneva 1846[5]

Volume IV, page 50, note, 1847[6]

or read the note of the two French physicists in the Reports of 1849, volume 29, page 352.[7] This correction is important to me because my experiments of 1846 were made in particularly difficult conditions. I had to build myself my electro-magnets, my mica batteries etc. I was teaching in Lausanne and did not have any funding and instrumental help. The revolution had just broken out, it had for effect to chase away from the city myself and all my colleagues from the Academia: I lost my instruments. In Geneva where the revolution was rumbling, M. De la Rive resigned and designated me to take over his position but I did not have the laboratory resources nor the freedom of spirit to continue my experiments. During this time, De la Povostaye repeated them, confirmed them, and introduced the important modification to place under an angle of 45 degree the main sections of the polarizer and the analyzer. Moreover, the French treaties of physics have always cited me as the author of this discovery, and it would be painful to me to let this mistake spread in a book like yours, meant for a great distribution and new editions.

Sir John Herschel[8] cites, page 392 of his Familiar Lectures,[9] a liquid found by Mr. Jellet of Trinity College, Dublin,[10] which would be right-handed for one extremity of the spectrum and left-handed for the other. Do you know what is this liquid?

Believe me, my dear Tyndall, very truly yours | E. Wartmann

RI MS JT/1/W/9

1. *Elie Wartmann*: Elie Wartmann (1817–86), a Swiss physicist and Professor of Natural Philosophy at the Academy of Geneva, and author of *Mémoire Sur La Diathermansie Électrique Des Couples Métalliques* (1841), among other works (*HLS*).
2. *your booklet on Forbes*: see letter 4154, n. 1.
3. *your book on Optics*: *Six Lectures on Light* (the page references that follow are all to this book).
4. *De la Povostaye and Desains*: F. De la Povostaye (dates unknown) and Paul Desains (1817–85), French physicists and collaborators.
5. *Volume I, page 417, Geneva 1846*: E. Wartmann, 'Mémoire sur de nouvelles liaisons entre la chaleur, l'électricité, et le magnétisme', *Bibl. Univ. Archives*, 1 (1846), pp. 417–18.
6. *Volume IV, page 50, note, 1847*: E. Wartmann, 'Troisième mémoire sur divers phénomènes d'induction', *Bibl. Univ. Archives*, 4 (1847), pp. 49–63.
7. *note of the two French physicists in the Reports of 1849, volume 29, page 352*: F. De la Provostaye and P. Desains, 'Rotation di plan de polarization de la chaleur produite par le magnétisme', *Comptes Rendus Hebdomadaires des Séances de l'Académie des Sciences*, 29 (1849), pp. 352–5.
8. *Sir John Herschel*: John Frederick William Herschel (1792–1871), an English mathematician and astronomer. He was the son of the famous astronomer William Herschel (1738–1822) and himself became one of the most eminent scientists in Victorian Britain. He graduated from University of Cambridge in 1813 as senior wrangler, received two RS Copley Medals (1821 and 1847), and was awarded a knighthood (1831) followed by a baronetcy (1838). His *Preliminary Discourse on the Study of Natural Philosophy* (London: Longmans, Rees, Orme, Brown & Green, 1830) was widely read and greatly admired. In 1849 he published *Outlines of Astronomy* (London: Longman, Brown, Green, and Longmans), a popular presentation of astronomy that went through eleven editions by 1871. He made many contributions to the science of photography and also edited the admiralty's

Manual of Scientific Enquiry (London: John Murray, 1849), a science field guide for the use of the Royal Navy. From 1850 to 1855 he served as master of the mint, and on his death was buried next to Isaac Newton in Westminster Abbey (*CDSB*; *ODNB*).

9. *Familiar Lectures*: J. F. W. Herschel, *Familiar Lectures on Scientific Subjects* (London: Alexander Strahan, 1872).
10. *Mr. Jellet of Trinity College, Dublin*: John Hewitt Jellett (1817–88), an Irish mathematician and university administrator who was best known for his work in pure and applied mathematics (*ODNB*).

To the Editor of *Nature*[1] 16 September [1873][2] 4167

[Tyndall and Tait]

I HAVE hitherto refrained from intruding upon your space with reference to this deplorable Forbes' controversy,[3] but now that the occasion has come when a brief deliverance on my part seems called for, I trust to your courtesy, if not to your justice, to allow me room for it.

In the first place I would ask permission to inform such of your readers as may feel an interest in the subject, that if they wish to form a correct opinion of the tone and logic of my rejoinder to Principal Forbes and his biographers, they will consult the rejoinder itself, as published by Longmans,[4] and not the extracts and inferences of Professor Tait.

They will thus learn, among other things, that what Professor Tait calls "plausible," is simply unanswerable.

With regard to the taking up of the various points in Principal Forbes's reply, item by item, that may be done some day should I deem it a worthy occupation. In my rejoinder I converged attention on the two points which Principal Forbes himself considered the really serious ones, and having broken the neck of the argument in both these cases I cared little about prolonging the controversy. Nevertheless if circumstances show it to be necessary it may be prolonged.

Professor Tait invariably writes on the hypothesis that what is not contradicted cannot be contradicted, and must therefore be accepted as true—a natural, if not inevitable, assumption on his part. For example, Forbes's argument regarding the crevasses of Rendu[5] was left unanswered by me, hence the conclusion that it was unanswerable. That argument, however, is now in shreds, as it might have been, had I so willed, any time during the last dozen years. Again, Principal Forbes makes an assertion regarding his tutelage of Agassiz; the assertion is left uncontradicted; it must therefore be accepted as true, and I am unjust because I do not so accept it. Thirteen years ago, however, I was in possession of a diametrically opposite assertion from M. Agassiz. Quite as distinctly, though not so specifically, he writes thus within the present year. "When Forbes came to visit me upon the glacier of the Aar, he knew not only everything that I had done, but also my plans for the future. When he left he positively declined to express any opinion concerning glacier phenomena, under the plea that he only came to gratify his curiosity, and had no intention of following up the subject, as he had no desire to be involved in the controversy then raging regarding the former extension of glaciers.* When he showed his hand I did not enter into a protracted discussion, but simply made a statement of facts and let the matter rest.... When I look," adds M. Agassiz, "on the whole transaction it seems incredible. There is in it no vestige either of the gentleman or the honest investigator."[6]

With statements of this character confronting the assertions of Principal Forbes, the proper course for me was to ignore assertions on both sides, and to confine myself to demonstrable facts. This I accordingly did.

With regard to Mr. Tait's criticism of my "popular" writings it has, of course, nothing to do with his defence of Forbes, but is the product of mere ignoble spite. He asks me to reply to him not according to the letter, but according to the spirit of his attack. If I might use the expression I would say, "God forbid!" for how could I do so without lowering myself to some extent to his level. The antecedents of Mr. Tait with reference to me are pretty well known. When I sought to raise from the dust a meritorious man whose name is now a household word in science, who has been elected by acclamation a member of the French Academy, and who has received the crowning honour of the Royal Society—when I sought to place Dr. Mayer in the position which he now holds, and from which no detraction can remove him, it was Mr. Tait who, in *Good Words*, charged me with misleading the public; who followed up his attack in the "Philosophical Magazine," and who when publicly hoisted by his own petard, retired to void his venom against me in the anonymous pages of the "North British Review."[7] It is this man whose blunders and whose injustice have been so often reduced to nakedness, without ever once showing that he possessed the manhood to acknowledge a committed wrong, who now puts himself forward as the corrector of my errors and the definer of my scientific position. That position is happily not dependent upon him, and his opinion regarding it, is to me, as it will be to most others, a trifle light as air. But graver considerations than mere personal ones here arise. Might I venture, Mr. Editor, to express a doubt as to the wisdom of permitting discussions of this kind to appear in your invaluable journal. Having opened your columns to attack you are, of course, in duty bound to open them to reply, but if I might venture a suggestion, you would wisely use your undoubted editorial rights, and consult the interest of science, by putting a stop to proceedings which dishonor it. An illustrious person[8] writes to me thus:—"I have just read Professor Tait's letters in NATURE, and feel a recurrence of that pain which similar communications once inflicted on myself—pain felt, not on my own account, for I knew that the attacks would no more sully me in the opinion of those whom I loved and respected, than they did in my own opinion; but pain for the wounded honour of science and the outraged dignity of scientific controversy."[9]

JOHN TYNDALL | Athenaeum Club,[10] Sept. 16

* This tallies with Forbes's own account (Travels,[11] page 38). "Far from being ready to admit, as my sanguine companions wished me to do in 1841, that the theory of glaciers was complete, and the cause of their motion certain, after patiently hearing all they had to say, and reserving my opinion, &c." This reservation of opinion is probably the reticence referred to by Agassiz.

Nature, 8 (18 September 1873), p. 399

---

1. *Editor of Nature*: Norman Lockyer.
2. *[1873]*: the year is established by the letter's publication in *Nature* in 1873.
3. *this deplorable Forbes controversy*: Tyndall's letter was a response to one in the previous issue of *Nature* from Peter Guthrie Tait, one of the co-editors of *Life and Letters of Forbes* and a stalward defender of the Scottish geologist's reputation against perceived slights from Tyndall. See P. G. Tait, 'Tyndall and Forbes', *Nature*, 8 (11 September 1873), pp. 381–2.
4. *the rejoinder itself, as published by Longmans*: see letter 4154, n. 1.
5. *Forbes' argument... Rendu*: see letter 4154, n. 5.
6. *Thirteen years ago... honest investigator"*: these testimonials from Louis Agassiz have not been identified.
7. *When I sought... "North British Review"*: on the earlier controversy with Tait over the Tyndall-backed Julius Robert Mayer, who received the RS's Copley Medal in 1871, see letter 4024, n. 6.
8. *An illustrious person*: Joseph Hooker.
9. *"I have just read... of scientific controversy"*: see letter 4162.
10. *Athenaeum Club*: see letter 3710, n. 5.
11. *Travels*: see letter 4038, n. 4.

To Joseph Dalton Hooker — 16 September [1873][1] — 4168

16th. September

Dear, good, loving Hooker,

A million thanks for your note, which I have already turned to good account.[2] It breathes the spirit for which I always gave you credit—pure and high and warm with affection. I am strong my boy, and the next number of Nature[3] will prove this to you,

Ever affectionately yours | John Tyndall

Love to all.

RI MS JT/1/TYP/8/2720
LT Typescript Only

1. *[1873]*: the year is established by the relation to letter 4162, to which this letter is a reply.
2. *A million thanks... to good account*: Tyndall quoted from Hooker's letter (4162) in a letter to *Nature* (see letter 4167).
3. *the next number of Nature*: Tyndall's letter, printed under the title 'Tyndall and Tait', appeared in the 18 September issue, with Hooker identified only as an 'illustrious person' (see letter 4167).

To the Editor of *Nature*[1] — [*c.* 19 September 1873][2] — 4169

My letter[3] was rapidly written, and the proof of it reached me, not on the Tuesday evening as I expected, but on the Wednesday morning when I was in the midst of my preparations for Bradford.[4] I had therefore little time to give it the calm thought which it ought to have received. On re-reading it I find two passages in it which I think it desirable to cancel. The first is that in which I speak of lowering myself to the level of Prof. Tait; the second that in which I reflect upon his manhood. These passages I wish to retract.

'Tait and Tyndall', *Nature*, 8 (25 September 1873), p. 431

1. *Editor of Nature*: Norman Lockyer.
2. *[c. 19 September 1873]*: the date is established by relation to letter 4167, published in the 18 September 1873 issue of *Nature*, and the publication of this extract in the 25 September issue, in an editor's note at the head of the 'Letters' column explaining that further

correspondence on the Forbes controversy would not be printed. See 'Tait and Tyndall', *Nature*, 8 (25 September 1873), p. 431.

3. *My letter*: letter 4167.
4. *for Bradford*: for the annual meeting of the BAAS (see letter 4153, n. 15).

From Heinrich Rudolf Vieweg[1] [16 September][2] 1873 4170

Braunschweig */*d. 16 Septemb.*/* 1873 | Herrn Professor Tyndall | London. Royal Institution | Albemarle Street.

Hochgeehrter Herr!

In Erwiderung Ihres Schreibens vom 14^{tn} dieses Monats danken wir Ihnen sehr, daß Sie zu der neuen Auflage Ihres Werkes über die „Wärme", welche wir in einigen Wochen zu drucken beginnen werden, Zusätze senden wollen. Sobald wir dieselben erhalten werden wir dieselben Herrn Wiedemann & Helmholtz einschicken und eine Uebertragung in's Deutsche ersuchen. Es wird der neuen Auflage zur Zierde gereichen, wenn sie durch diese neue Mittheilung vermehrt werden kann.

Was Ihre in Amerika gehaltenen Vorlesungen betrifft so werden wir Sie bitten müßen, uns ein Exemplar derselben zur Ansicht senden zu wollen. Wir haben diese */*Arbeiten*/* nicht gesehen, kennen den Inhalt nicht und finden dieselbe<n> nicht in den buchhändlerischen Anzeigen.

Wir würden uns sehr freuen, wieder in deutscher Uebertragung ein Werk von Ihnen den deutschen */*Leserkreisen*/* zugänglich machen zu können.

Mit größter Hochachtung ergebenst
Friedr Vieweg & Sohn

Braunschweig */*16 September*/* 1873 | Mr Professor Tyndall | London. Royal Institution | Albemarle Street.[3]

Esteemed Sir!

In reply to your letter of the 14th of this month[4] we thank you very much for wanting to send addenda to the new edition of your work on "Heat",[5] which we shall begin printing in a few weeks. As soon as we receive them, we shall submit them to Herr Wiedemann & Herr Helmholtz and request a translation into German.[6] It will be an adornment to the new edition, if it can be enlarged through this new information.

As far as the lectures you held in America[7] are concerned, we shall have to ask you if you would send us a copy of them for us to view, we have not seen these */*papers*/*, do not know their contents, and do not find them in the booksellers' advertisements.

We would be very happy to be able to make a work by you accessible to German *[readers]* in German translation again.[8]

With the greatest esteem your humble servants

Friedrich. Vieweg & S[on]

TU Braunschweig/Universitätsbibliothek/UABS V 1 T : 63

---

1. *Heinrich Rudolf Vieweg*: while the letter is signed 'Friedr Vieweg & Sohn', we believe the sender is Heinrich Rudolf Vieweg, who took over the family publishing firm, Friedrich Vieweg und Sohn, from his father Eduard Vieweg (1797–1869).
2. *[16 September]*: the date is faded on the letter, but we believe it to be 16 September.
3. *Royal Institution | Albemarle Street.*: this text is in a different hand than that of the rest of the letter.
4. *your letter of the 14th of this month*: letter missing.
5. *new edition of your work on "Heat"*: the translation of the new chapter to be added to the new German edition of Tyndall's book on heat was a topic of correspondence in April and June 1873 (see letter 4111).
6. *a translation into German*: see letter 4111, n. 5.
7. *the lectures you held in America*: *Six Lectures on Light*.
8. *a work by you... German translation again*: the book was published as *Das Licht: sechs Vorlesungen gehalten in Amerika im Winter 1872–1873*, trans. G. Wiedemann (Braunschweig: Friedrich Vieweg und Sohn, 1876). While Gustav Wiedemann was recognized as the book's translator, it was actually translated by his wife Clara, who had previously translated some of Tyndall's books into German with Anna Helmholtz without due credit (see letter 4117, n. 4).

To Thomas Henry Huxley — 24 September [1873] — 4171

West Cliff Hotel. | Folkestone 24th. Sept.

My dear Huxley

When I returned from Switzerland[1] I found the Clarendon Gardens, immediately under my bedroom[2] possessed by a devil in the shape of a steam engine and roller employed for the purpose of crushing bricks. They began their racket sometimes as I began my doze—So I had to scamper off for the sake of quiet.

I am staying here at the West Cliff Hotel where Spencer once deprived me of a night's sleep by asking me just as I was retiring to bed whether I believed in the externality of matter.

I am exceedingly glad to hear that you have got back all right.[3] I hope God will give you grace to keep yourself right.

He forsook me for a moment when I allowed that burst of anger to overtake me as regards Tait:[4] But

"Wer nicht liebt und wer nicht irrt | Der lasse sich begraben."[5]

So saith Goethe:[6] I have tried to set matters right in "Nature."[7] But with the little man who has hitherto ruled it at the head of affairs[8] I am never safe. My notions regarding him have been too undisguised. However I will henceforth possess my soul in patience.

I have read the mad letter[9] which you sent me It is a sample of a multitude—America is specially prolific in such things and I get the benefit.

I am trying to get a little work done here; but for the most part thus far have dissipated my time and demoralized my religious character over Spencer's First Principles.[10] But as a set off I have been comforted by a number of Scribner's monthly in which they pitch into him.[11]

I wish to Heaven you had not persuaded me to accept that Belfast duty.[12] They do not want me.[13] Well in return I may be less tender in talking to them than I otherwise should have been. But Spottiswoode assures me that no individual offered the slightest support to the two unscientific persons who showed opposition.[14]

So I suppose I am in for it—and so are you you know and Hooker too.

By the way I sent to Nature a little communication about the rainbow,[15] regarding which Tait has found a mare's nest. The proof of it reached Mr L[ockyer]. before the manuscript of the letter regarding Tait quitted my hands. I told him I was willing to postpone the latter. But he wanted to have his fling; so he published it and withheld the other.

I have spread abroad the intelligence that you are perfectly restored—do not belie me by your impudence.

Love to all | Ever Yours | John Tyndall

RI MS JT/1/TYP/9/3022
IC HP 8.155
LT Typescript Only

1. *returned from Switzerland*: Tyndall returned to London on 10 September, as he wrote to Elizabeth Dawson Steuart (see letter 4159).
2. *Clarendon Gardens, immediately under my bedroom*: the Clarendon Hotel at 169 New Bond Street in London was directly behind the RI, where Tyndall lived.
3. *I am exceedingly glad... got back all right*: letter missing, but presumably Huxley had informed Tyndall of his return to London from his summer holiday abroad, and enclosed a letter from America referred to below.
4. *that burst of anger... as regards Tait*: Tyndall came to regard as intemperate his response in the 18 September issue of *Nature* (see letter 4167) to Peter Guthrie Tait's letter in the previous issue on the Forbes controversy.

5. *"Wer nicht... begreben"*: 'A man who no longer loves and no longer errs | should have himself buried straight away' (German), from a longer stanza in Goethe's *Faust* (D. Luke (ed), *Goethe. With Plain Prose Translations of Each Poem* (New York: Penguin Books, 1964), p. 287).
6. *Goethe*: Johann Wolfgang von Goethe (1749–1832), the outstanding German literary figure of the late eighteenth and early nineteenth centuries. His extensive writings included his theory of colours and a study of the structure and function of plants. During his time in Marburg, Tyndall became familiar with some of Goethe's works (J. Williams, *The Life of Goethe: A Critical Biography* (Oxford: Blackwell, 1998).
7. *I have tried... in "Nature"*: see letter 4169.
8. *the little man... head of affairs*: Norman Lockyer.
9. *mad letter*: enclosure missing.
10. *Spencer's First Principles*: H. Spencer, *First Principles* (London: Williams & Norgate, 1862); this was the opening book of Spencer's 'Synthetic Philosophy'.
11. *a number of Scribner's... pitch into him*: A. Blauvelt, 'Modern Skepticism', *Scribner's Monthly*, 6 (1873), pp. 424–32, 582–96, 725–39.
12. *that Belfast duty*: Tyndall had agreed to serve as President at the next BAAS meeting, due to take place in Belfast in August 1874 (see letter 4013, n. 2).
13. *They do not want me*: at the Bradford meeting of the BAAS (see letter 4153, n. 15), when Tyndall was formally proposed as President of the Belfast meeting, the mayor of Belfast James Henderson (1846–1914) objected, suggesting that the honor should go to a man of science based in Belfast, the chemist Thomas Andrews (1813–85); but he was judged too ill for the role. See the report in *Nature*, 8 (25 September 1873), p. 448, as well as *Times*, 23 September 1873, p. 5, and *Times*, 27 September 1873, p. 12.
14. *the two unscientific persons who showed opposition*: while one of the opponents to Tyndall's Presidency was the mayor of Belfast (see n. 13), the other opponent has not been identified.
15. *a little communication about the rainbow*: see letter 4163.

To Hermann von Helmholtz 25 September 1873 4172

Royal Institution | 25th. September 1873.

My dear Helmholtz,

Vieweg has written to me several times[1] regarding a XVIth chapter to my book on Heat, in which I unfold, and endeavour to solve, the differences between Magnus and myself as regards the action of aqueous vapour upon Radiant Heat. I am almost ashamed to occupy a mind which must be already overladen, with this subject. But I thought you would allow me to say that nothing would grieve me more than to appear captious or critical as regards the work of Magnus; and that if this chapter should appear to you to bear this character I would suppress it. I have confined myself to pure facts and reasonings, and have not omitted to speak with due appreciation of Magnus, still notwithstanding all my efforts the chapter may appear ungenerous, and if so, as I have said, I will abandon the idea of publishing it.[2]

It gave me pleasure to hear of the recognition which you have received at the hands of the King of Prussia. But you knew beforehand that the unanimous verdict of those who knew your labours would be that this distinction is your right: I have strong reason for hoping, though this is private, that the Council of the Royal Society will add its unanimous verdict this year to that of the advisers of the King of Prussia.[3]

Huxley's health was, as you probably know, exceedingly critical for more than a year: he went to Egypt but it did him no good. His friends took strong measures to induce him to take a good holiday: So he went to Axenstein in Switzerland, and I am happy to say has returned, if I may trust his own account of matters, perfectly restored.[4] I have not seen him, for I am staying at Folkestone. But it is a great comfort to be assured that he has surmounted an attack which once threatened to cripple him altogether. His main enemy was a fearful dyspepsia.[5] Darwin oscillates—sometimes very well, and sometimes much broken: I am told by Hooker that he is doing splendid work.[6]

I have not written to you since the death of Bence Jones.[7] It was a sore loss to me: for as the years passed we grew more and more closely intimate, until before he died I had become his closest English friend. He wrote to me a series of most interesting letters to America,[8] through which I could trace the advance of his disease, until finally I began to fear that I should not find him alive. His illness hastened my departure from America.[9]

I dare say you will see from Nature that I have again got into difficulties with Tait.[10] I regret much that I permitted myself to speak so angrily about him. But in the forth-coming number I retract those passages which I deemed inappropriate to scientific discussion.[11] This will do no good to Tait, but it will set me right in the eyes of men of science. The attack upon me was absolutely unexpected, but his temper is incalculable. He is aided by M^r^. Lockyer whom I have had to cut once or twice against the grain in personal intercourse, and who now, as Editor of Nature, seems disposed to pay me off. What Tait has to say about my work I know not: if he has nothing of greater gravity to object to than what I said about the Rainbow,[12] it will not be difficult to meet him. When Zöllner's book came out Tait wrote to me expressing the "consolation" he experienced that we were to have "a row!": he must be very happy now. It is also a "consolation" to me to think that my work cannot be so atrociously bad, for if it were you would not have had any thing to do with it.[13]

I met one of your assistants this year at the Bel Alp[14] and was sorry to learn from him that they were not so prompt as you wished them to be in the construction of your laboratory.

Give my kind regards to Du Bois: I ventured to tell the Council of the Royal Society many months ago, that no man approaching him in ability is at present outside the pale of the Society.[15] They will do their duty, I doubt not, in due time. The delay arises in part from the inability of the Council to fully enter into his labours.

With kindest regards to M^rs^. Helmholtz[16] and to your daughter[17]

Believe me | ever yours faithfully | John Tyndall

I should have been rejoiced to send a scientific contribution to the volume[18] intended for presentation to Poggendorff,[19] but the request reached me as I was starting for Switzerland in a state of utter exhaustion. I wished much however to connect myself with the movement, and offered to undertake part of the printing expenses. Indeed I forwarded a cheque to Dove[20] but have heard nothing about it since.

BBAW NL Helmholtz, Nr. 477, BI_20–1
RI MS JT/1/TYP/2/490

1. *Vieweg has written to me several times*: letters 4111 and 4170.
2. *abandon the idea of publishing it*: the chapter described was published in Tyndall's *Die Wärme* (1875), and it does not appear to have been published in later English editions.
3. *It gave me... King of Prussia*: in 1873 Helmholtz received both the Order of Merit for the Sciences and Arts from the King of Prussia and, by year's end, the Copley Medal from the RS (D. Cahan, *Helmholtz: A Life in Science* (Chicago: University of Chicago Press), pp. 434–5, 489; *CDSB*).
4. *Huxley's health... perfectly restored*: Tyndall was among a large circle of Huxley's friends who collected funds to pay for the health-restoring holiday to the Continent that they reckoned he needed very much (see letter 4028, n. 4, and letters 4151 and 4154). For letters from Huxley during his travels in the Mediterranean and Egypt in 1872 when ill, see *Tyndall Correspondence*, vol. 12.
5. *dyspepsia*: see letter 3839, n. 3.
6. *doing splendid work*: see letter 4162. In 1873, Charles Darwin's active scientific investigation included the study of carnivorous plants, which may be what Hooker was referring to (see *Darwin Correspondence*, vol. 21, which covers the year 1873).
7. *not written to you since the death of Bence Jones*: Henry Bence Jones died on 20 April 1873.
8. *series of most interesting letters to America*: for letters from Bence Jones to Tyndall while he was on his American lecture tour, see letters 3880, 3889, 3897, 3912, 3922, 3932, 3944, 3954, 3957, 3969, and 3976.
9. *my departure from America*: see letter 3928, n. 18.
10. *difficulties with Tait*: see letter 4165.

11. *I regret much... scientific discussion*: see letter 4169.
12. *what I said about the Rainbow*: see letter 4163.
13. *When Zöllner's book... do with it*: for Tait's delight at the attack on Tyndall in J. C. F. Zöllner's 1872 book *Ueber die Natur der Cometen, Beiträge zur Geschichte und Theorie der Erkenntniss*, see P. G. Tait, 'Sensation and Science. II.', *Nature*, 6 (4 July 1872), pp. 177–8. For the reception of the book generally, see D. Cahan, *Helmholtz: A Life in Science* (Chicago: University of Chicago Press, 2018), pp. 494–7. For the letter from Tait referred to by Tyndall, dated 18 May 1872, see letter 3695, *Tyndall Correspondence*, vol. 12.
14. *Bel Alp*: see letter 3771, n. 5.
15. *Give my kind regards... pale of the Society*: for Tyndall's testimonial in support of Emil du Bois-Reymond to be elected ForMemRS, see letter 4090.
16. *M^rs^. Helmholtz*: Anna Helmholtz.
17. *your daughter*: Ellen Ida Elisabeth Helmholtz (1864–1941), who later married Arnold von Siemens (1853–1918), a German telecommunications industrialist.
18. *volume*: scientific friends of Poggendorff edited a volume of *Poggendorff's Annalen* for its fiftieth anniversary in 1874 (*Nature*, 9 (16 April 1874), p. 469; *Nature*, 10 (28 May 1874), p. 75–6). Referred to as the 'jubilee volume', or 'Jubelband', the volume consisted of sixty-two contributions, the final one being from Tyndall, 'Vorläufige Mittheilung über eine Untersuchung der Fortpflanzung des Schalles durch die Atmosphäre', or 'Preliminary Communication on an Investigation of the Propagation of Sound through the Atmosphere' (pp. 668–82).
19. *Poggendorf*: Johann Christian Poggendorff (1796–1877), a German physicist at the University of Berlin. His major contribution to science was as editor for fifty-two years (1824–76) of *Annalen der Physik und Chemie*, which was known as *Poggendorff's Annalen* during his editorship. He used his broad knowledge of science and scientific men in producing a *Biographisch-literarisches Handworterbuch zur Geschicte der exacten Wissenschaften* (Leipzig: Johann Ambrosius Barth), an approximate German version of the RS's *Catalogue of Scientific Papers* (*CDSB*; *NDB*).
20. *Dove*: probably Heinrich Wilhelm Dove (1803–79), a German physicist (*CDSB*).

From Mary Egerton 25 September [1873][1] 4173

Haselbech Hall | Northampton[2] | Sept[embe]^r^ 25th

My dear Mr Tyndall,

We are on our way home, I shall sleep in Eaton Place[3] on Monday. Would there be any chance of seeing you? It would be so pleasant to have a little talk about Switzerland. You never told me the Hamiltons had gone on with you to Zermatt![4] I had a letter from L[ad]^y^ Claud[5] before I left Thoresby,[6] full of delight at the whole expedition and all its circumstances, & of gratitude to you. I had gone out alone, and I sat down on the root of a tree; to read her letter, and try to rejoice in their enjoyment, but somehow the stream of sympathy would not flow quite clear, and I am afraid there was a taint of less pure waters in the drops that blinded my eyes! Not that for a moment I grudge them their enjoyment, but I cannot help thinking how very very great it would have been to me! Though of course, I c[oul][d] not have done half what they did! As one grows older, each year takes a larger proportion out of the remaining chances of realizing ones ideals of happiness, and there are some things in this solid material world, so delightful that one cannot bring oneself to think that anything in another can quite do instead of them!

I am more angry than ever with M[r] Tait! Not only has he attacked you most unjustly and unfairly,[7] (by that he would have done himself much more harm than you) but he has provoked you to come down from the vantage ground of calm dignity where you were so proudly entrenched before,[8] & whence his blows glanced off harmless, and to join in a hand to hand combat of hard words! I know it would have been almost superhuman to resist, but I look for heroic deeds from "Galahad"![9]

By the bye I did not "tremble" about the Rainbow![10] It was *[1 word illeg]* because I had "faith" in your candour, that I did not fear to allude to the possibility of an oversight. I shall begin to think investigations; but I have nothing of interest to offer you in return, & therefore I sometimes feel I might slip out of your memory altogether, and any expression of friendliness, or anything you tell me which seems to accept or expect my sympathy is very pleasant to me.

If the thoughts of Switzerland & its associations made me a little sentimental, forgive it; you ought, for you love it as much as I do, I have put it all aside now! And you shall see what a sober sedate old friend I will be, if you will let me! There—"liberavi animam meam"![11] (is that good Latin?) and there's an end of it!

You will have to write to us for civility's sake! So give me a general notion of what you expect your plans to be, and whether you could not snatch a day or two of rest here, in the course of your various flights up & down?

Ever & always yours | M F Egerton

RI MS JT/1/E/51

1. *[1873]*: the year is given by reference to Tyndall's dispute with Tait.
2. *Haselbech Hall | Northampton*: the country estate of Selina Wentworth-Fitzwilliam (1812–83), Viscountess Milton, wife of William Charles (1812–35), Viscount Milton. They were married in 1833 (see M. Leyland, *Patronage and the Architectural Profession: The Country House in Nineteenth-Century Northamptonshire* (PhD dissertation, University of Leicester, 2016), p. 45).
3. *Eaton Place*: 45 Eaton Place, Egerton's London residence.

4. *Hamiltons had gone on with you to Zermatt!*: Tyndall wrote on 15 September to Elizabeth Dawson Steuart (letter 4164): 'I had a very pleasant expedition in the Alps. Lady Claud Hamilton, two of her charming daughters, her son and myself formed a little party—and it gave me additional pleasure to observe their delight'. See also letter 4145. On Tyndall's connections with the Hamilton family, see *Ascent of John Tyndall*, p. 323.
5. *L[ad]y Claud*: Elizabeth Hamilton.
6. *Thoresby*: see letter 4076, n. 9.
7. *attacked you most unjustly and unfairly*: P. G. Tait, 'Tyndall and Forbes', *Nature*, 8 (11 September 1873), pp. 381–2.
8. *provoked you to... proudly entrenched before*: J. Tyndall, 'Tyndall and Tait', *Nature*, 8 (18 September 1873), p. 399. This is letter 4167 in this volume.
9. *"Galahad"*: see letter 4153, n. 11.
10. *Rainbow*: see letter 4161, n. 5; and letter 4163.
11. *"liberavi animam meam"*: free my soul (Latin).

From Thomas Henry Huxley — 25 September 1873 — 4174

4 Marlborough Place[1] | Sept. 25 1873.

My dear Tyndall,

I am sure you are mistaken about the Belfast people. That blundering idiot of a Mayor wanted to make himself important and get up a sort of 'Home Rule' agitation in the Association, but nobody backed him and he collapsed.[2] I am at your disposition for whatever you want me to do, as you know, and I am sure that Hooker is of the same mind. We shall not be ashamed when we meet our enemies in the gate.[3]

The grace of God cannot entirely have deserted you since you are aware of the temperature of that ferocious epistle.[4] Reeks, whom I saw yesterday, was luxuriating in it, and said (confound his impudence) that it was quite my style.[5] I forgot to tell him by the bye, that I had resigned in your favour ever since the famous letter to Carpenter.[6] Well so long as you are better after it there is no great harm done.

Somebody has sent me the two numbers of Scribner with Blanvelt's articles on "Modern Skepticism".[7] They seem to be very well done, and he has a better appreciation of the toughness of the job before him than any of the writers of his school with whom I have met. But it is rather cool of you to talk of his pitching into Spencer when you are chief target yourself. I come in only par parenthèse,[8] and I am glad to see that people are beginning to understand my real position and to separate me from such raging infidels as you and Spencer.[9]

Ever thine | T. H. Huxley.

RI MS JT/1/TYP/9/3023
IC HP 8.158
Typed Transcript Only

---

1. *4 Marlborough Place*: see letter 3947, n. 1.
2. *I am sure... and he collapsed*: Huxley was responding to Tyndall's remarks, in letter 4171, about the mayor of Belfast's preferring a Belfast man of science to Tyndall as President of the upcoming BAAS meeting in Belfast in August 1874.
3. *I am at your disposition... enemies at the gate*: in letter 4171, Tyndall warned of trouble ahead in Belfast for Huxley and Joseph Hooker, too.
4. *The grace of God... ferocious epistle*: Huxley here echoed Tyndall's religious language when, in letter 4171, he rued what Huxley acknowledged was indeed a very angry letter to *Nature* (see letter 4167).
5. *Reeks... my style*: Huxley here reported being teased by a London colleague who loved Tyndall's 'ferocious epistle': Trenham Reeks (1823–79), a mineralogist and registrar of the Royal School of Mines ('Obituary', *Geological Magazine*, 6 (1879), p. 288).
6. *famous letter to Carpenter*: letter to William Carpenter not identified.
7. *Somebody has sent... on "Modern Skepticism"*: see letter 4171, n. 11.
8. *par parenthèse*: by way of parenthesis, incidentally, by the way (*OED*).
9. *people are beginning... you and Spencer*: see B. Lightman, 'Huxley and Scientific Agnosticism: The Strange History of a Failed Rhetorical Strategy', *British Journal for the History of Science*, 35 (2002), pp. 271–89, on p. 275.

To Thomas Archer Hirst — 27 September 1873 — 4175

27th Sept. 1873.

My dear Tom.

The weather is so fine here, and my sitting room so tranquil that I am trying to get a little work done: and I believe I shall be more successful than I should be if I returned to London.[1] I shall remain therefore for a day or two longer than I at first intended.

I thought from the Report in the Times[2] that the people in Belfast wished to have nothing to do with me. They made it out or rather an intemperate minister made it out in New York that I had attacked the Presbyterians,[3] so that I was not surprised at a demurrer to my appearance among the Presbyterians of the North of Ireland.[4] However Spottiswoode & Huxley both assure me[5] that it was only an asinine attempt to establish the principle of Home Rule in the affairs of the Association.

I am rather at a loss what to do when I return. The possible racket[6] will I fear drive me from the Institution, and where I am to lodge I hardly know.

I have not seen Nature since I came here. Doubtless that coarse and reckless man[7] will have his say—You will let me know should anything require immediate reply. My notion is never to look at his lucubrations till he has ended them and then to deal with them in a dignified and conclusive manner.[8]

I have been reading Spencer's First Principles: the first part is unspeakably good; but if I am a competent judge there is a tremendous falling off where he comes to deal with forces and motions, and the transformation & equivalence of forces.[9] I must have a talk with him about it.

It is exceedingly pleasant to receive such letters from Huxley as I have recently received. I believe he is really & thoroughly well. As regards Belfast he is ready for anything.[10]

I wish you & Lilly[11] could come drive here for tomorrow. It would do you a world of good. My love to her

Yours affe[ctionate]ly | John Tyndall

RI MS JT/1/T/707

---

1. *my sitting room so tranquil... returned to London*: Tyndall wrote from the West Cliff Hotel in Folkestone, where he went to escape the noise of the construction very near the RI (see letter 4171).
2. *the Report in the Times*: 'The British Association', *Times*, 23 September 1873, p. 5; the report was quite clear that the objection to Tyndall as President of the BAAS for the next meeting in Belfast in August 1874 was not to him personally but 'to the choice of an Englishman as president of an Irish meeting'. The mayor of Belfast (see letter 4171, n. 13) countered this claim against him in a short letter, for which see *Times*, 27 September 1873, p. 12.
3. *They made it out... attacked the Presbyterians*: for Tyndall's account of how he came to refer, at the end of one of his New York lectures on light, to a critic who was, Tyndall put it, 'the intemperate occupant of a single Presbyterian pulpit', and how that defense in turn got reported as an attack by Tyndall upon Presbyterians everywhere, see letter 4125.
4. *demurrer to my appearance among the Presbyterians of the North of Ireland*: the mayor of Belfast (see letter 4171, n. 13).
5. *Spottiswoode & Huxley both assure me*: Tyndall wrote of Spottiswoode's assurance in letter 4171; for Huxley's assurance, see letter 4174.
6. *possible racket*: see n. 1.
7. *that course and reckless man*: Peter Guthrie Tait, then in full flow in attacking Tyndall over his treatment of James David Forbes (see letter 4167).
8. *My notion is never... dignified and conclusive manner*: Tyndall was regretful about having

dashed off what he later judged to be an undignified response to Tait's latest attack (see letters 4169 and 4171).

9. *Spencer's First Principles... equivalence of forces*: a few days earlier, in a letter to Thomas Huxley, Tyndall had joked that he 'dissipated his time and demoralized his religious character' in reading Herbert Spencer's book (see letter 4171, n. 10).
10. *letters from Huxley... ready for anything*: for Huxley's pungently expressed attitude of readiness for any fighting to be done at the 1874 BAAS meeting in Belfast, see letter 4174.
11. *Lilly*: Emily Hirst.

To Heinrich Rudolf Vieweg — 27 September 1873 — 4176

27th Sept. 1873

Dear Sir,

It gives me great pleasure to learn that Madame Wiedemann has undertaken the translation of the Lectures on Light.[1] The sum which circumstances would warrant you in sending me for permission to translate the work I beg of you to forward to Madame Wiedemann.

And also with respect to the translation of Fragments of Science,[2] be good enough to hand over my share in that book to Madame Helmholtz. She must have had infinite trouble in translating the work. Be good enough also in your edition to omit the words "for unscientific people" in the title of the book, and make it simply "Fragments of Science".

The last chapter of the book on Heat[3] is already written out in manuscript and I shall have the pleasure of sending it to you in a day or two. I am now staying at Folkestone,[4] but hope to return to London in a few days. It is necessary to compare some lettering in the manuscript with the plate which is in London. When this has been done the manuscript shall be posted to you.

Faithfully Yours | John Tyndall

TU Braunschweig/Universitätsbibliothek/UABS V 1 T : 63

---

1. *It gives me great pleasure... Lectures on Light*: Tyndall had learned in June that Clara Wiedemann was to translate his *Six Lectures on Light* into German (see letter 4117, n. 4). The more recent letter from Vieweg apparently confirming this arrangement for Wiedemann is missing.
2. *the translation of Fragments of Science*: J. Tyndall, *Fragmente aus den Naturwissenschaften. Vorlesungen und Aufsätze*, trans. A. Helmholtz (Braunschweig: Freidrich Vieweg und Sohn, 1874).
3. *The last chapter of the book on Heat*: on the new final chapter which appeared in the German edition, see letter 4172.

4. *I am now staying at Folkestone*: on Tyndall's decision to stay in a hotel there rather than endure the noise of construction near the RI, see letter 4171.

To Gustav Wiedemann 27 September 1873 4177

27th. Sept. 1873.

My dear Wiedemann

I am again indebted to you for Lieferung[1] of your admirable work.[2] By your steadfast carrying out of this great task you are doing an incalculable good to scientific workers in this age and in the age to come.

It gave me immense pleasure to learn from Vieweg that Madame Wiedemann[3] had undertaken the translation of the Lectures on Light.[4] She will not I think fall into the mistake of some who have read my book and consider that I denied that a rainbow could be seen by reflection.

What I did seek to correct was the common error that the bow thus seen is the optical image of the bow seen at the same time directly in the heavens: that in other words the same raindrops were conserved in the production of the direct and reflected bow.[5]

You will see from Nature that my old enemy Tait has recommenced his attacks upon me.[6] I must regret having permitted him to make me so angry: he shall do so no more.[7] He lacks both the Candour of a philosopher and the good manners of a gentleman. For some time to come I shall pay no attention whatever to him; when he has exhausted himself I shall know how to deal with him.

He is now supported by Mr. Lockyer, whose conceit has been such as to disgust his best friends.[8] I never disguised my opinion of him and he is now disposed to avenge himself. It is sorry work for men of science. It might have been well had nature thrown a little more clay on the fire of my character: warm as it is, however, it will never betray me into anything unjust.

I hope you have had a refreshing holiday. Give my kindest regards to Mrs. Wiedemann.

Always Yours | John Tyndall

I wish you could be induced to come to the meeting of the British Association next year in Belfast.[9] I declined the presidency three times; but at length the council have pressed it upon me.[10]

RI MS JT/1/T/1490

1. *Lieferung*: delivery (German).
2. *your admirable work*: not identified, but presumably translation work.
3. *Madame Wiedemann*: Clara Wiedemann.

4. *Lectures on Light*: Tyndall remarked on his pleasure at his German publisher's confirmation that Clara Wiedemann would serve as translator of *Six Lectures on Light* into German in letter 4176. For the German edition, see letter 4170, n. 8.
5. *She will not... reflected bow*: to Tyndall's dismay, *Six Lectures on Light* had attracted a complaint in *Nature* about a passage where Tyndall had seemed to claim, falsely, that a rainbow's reflection could not be seen on the flat surface of a lake (see letter 4163).
6. *You will see... attacks upon me*: Peter Guthrie Tait's latest attacks on Tyndall were in defense of the late James David Forbes (see letter 4167, n. 3).
7. *I must regret... do so no more*: for Tyndall's remorseful retraction of the lowest blows in his response to Tait in *Nature*, see letter 4169.
8. *He is now... his best friends*: Tyndall reckoned that Lockyer had deliberately rushed Tyndall's intemperate letter about Tait into print (see letter 4171).
9. *meeting of the British Association next year in Belfast*: the 44th annual meeting of the BAAS was held in Belfast in August 1874 (*Brit. Assoc. Rep. 1874*).
10. *pressed it upon me*: see letter 4013, n. 2.

From Sarah Faraday 27 September [1873][1] 4178

Barnsbury Villa,[2] | 320, Liverpool Road, N. | Sep. 27th.

Dear Dr. Tyndall

I have seen Mr. Vincent[3] and he tells me that you have gone to Folkestone out of health and out of spirits[4] which fills me with compunction and sorrow for I know that I wrote my little note[5] in a very selfish tone bemoaning over my own troubles and not knowing or thinking of what yours might be. I wish I could say or do any thing to soothe your troubled spirit, I do not know what to say but I can imagine that after all your glorification in the United States[6] it is hard to come down to common life—but I shall hope to see you. Till then

Believe me very truly your friend | S. Faraday | Dr. Tyndall

RI MS JT/1/TYP/12/4186
LT Typescript Only

---

1. *[1873]*: the year is given by reference to going to Folkestone and his recent lecture tour in the United States (see n. 6).
2. *Barnsbury Villa*: see letter 3993, n. 1.
3. *Mr. Vincent*: Benjamin Vincent (see letter 4084, n. 3).
4. *gone to Folkestone out of health and out of spirits*: Tyndall stayed at the West Cliff Hotel in Folkestone.
5. *my little note*: letter missing.
6. *all your glorification in the United States*: referring to Tyndall's American lecture tour (see Introduction, n. 8). He returned to London on 19 February 1873.

From Thomas Archer Hirst 28 September 1873 4179

Sep[tembe]r 28th 1873

My dear John.

I received your letter[1] last night,—too late to start to Folkstone[2] or we might have been with you to day. Lilly has come up to town to hear Stopford Brook[3] and I have been taking a two hours walk. We have come in here to beg a bit of luncheon from Mrs Tunstall.[4]

I saw an account in Nature of the meeting of the General Committee at Bradford where the Mayor of Belfast made an ass of himself[5]—I agree with Huxley and Spottiswoode.[6] I do not see that you can do anything but ignore it. The council of the British Association has deliberately selected you and you have accepted their invitation after drawing their attention to the claims of Andrews.[7] It would never do to reopen the question because a Mayor has been foolish enough to express a preference for Andrews Besides it should not be forgotten that next August he will no longer be Mayor.

I am glad to see that Lockyer has decided to publish a letter of Taits in reply to your last and has closed the controversy so far as the pages of Nature are concerned by simply quoting from your last letter the passage wherein you recall the two expressions you employed in the previous letter.[8] Of course Tait will break out in some other place on some future day, but thanks to your withdrawal of the two expressions this phase of the controversy has ended honourably for you and but by no means satisfactorily so far as he is concerned.

Lockyer also printed your letter to X.Y.Z. about the Rainbow.[9] The College opens on Wednesday.[10] I shall be terribly hard worked this week and unfortunately I am not in first rate condition at the commencement of it. Lilly[11] sends her love with mine.

Ever affectionately yours | T Archer Hirst

RI MS JT/H/255

---

1. *your letter*: see letter 4175.
2. *Folkestone*: Tyndall was staying at the West Cliff Hotel in Folkestone, as the noise of construction near the RI made it difficult for him to work.
3. *Stopford Brook*: the Irish chaplain Stopford Brooke (1832–1916) was at that time a popular preacher in London, whose admirers included Queen Victoria (*ODNB*).
4. *Mrs Tunstall*: Jane Tunstall.
5. *I saw an account... an ass of himself*: see letter 4171, n. 13.
6. *agree with Huxley and Spottiswoode*: Tyndall wrote of Spottiswoode's assurance of his

accepting the Presidency of the BAAS for the 1874 meeting in Belfast in letter 4171; for Huxley's assurance, see letter 4174.

7. *Andrews*: Thomas Andrews (see letter 4171, n. 13). See also letter 4013, n. 2.
8. *I am glad to see... in the previous letter*: see letters 4167 and 4169.
9. *your letter... the Rainbow*: see letter 4163.
10. *College opens on Wednesday*: Hirst was hired as the first Director of Studies of the new Royal Naval College, Greenwich, which opened in October 1873 (see letter 3911, n. 12).
11. *Lilly*: Emily Hirst.

To Louisa Baring [29 September 1873][1] 4180

Folkestone Monday

I, certainly, have not forgotten you but have thought of you many a time. I did not think that you had "forgotten" me, but I certainly did picture myself as a portion of that lumber[2] of indifferent memories which every human mind is forced to carry through life. And I thought at times I might have displeased you, but could not remember how. However your little note[3] gave me the joy of thinking that I am not the mere lumber that I thought I was, and that if I have displeased you, you have forgiven me.

I will ask permission to come to see you soon after your return to town. Would that I could go to Seaton,[4] but from next Monday I am to be rivetted in London.

I did not know that "May"[5] attended those lectures.[6] Had I known it I should assuredly have sought her out.

I return to town tomorrow—Lady Claud[7] has been suffering from a severe cold all during her stay here. But today it seems to be giving way, and I hope when she returns she will feel the benefit which has failed to manifest itself here.

I have not seen our grand old friend[8] for a long time, the reason in part being that I am frequently so tired at night that I shrink from afflicting him with my dullness. But I must see him soon after my return.

Deeply concerned was I to hear from Lord Haughton a day or two ago that you had been ill. But Lady Claud assures me that you have conquered your illness, and your note assures me that your heart is still full of life and kindness. May it long continue thus to add to the happiness of

Your friend | John Tyndall

If I thought it would not offend her I would send to "May" so much love.

She ought to know how bravely Douglas Hamilton struggled to rescue her goblet on the top of the Aeggischhorn.[9]

National Library of Scotland

1. *[29 September 1873]*: the year is established by the reference to being recently with the Hamiltons on the Eggishorn (see n. 9); and the date is established by the relationship to letter 4171, where Tyndall reported his arrival at Folkestone on 24 September, and that on 2 October he was at an X Club meeting in London, and on 5 October visiting with the Lubbocks. The Monday between his arrival in Folkestone and his return to London was 29 September.
2. *lumber*: useless or cumbrous material (*OED*).
3. *your little note*: letter missing.
4. *Seaton*: a seaside town in Devon.
5. *"May"*: Mary 'May' Florence Baring (1860–1902), the daughter of Louisa Baring, Lady Ashburton and Bingham Baring, 2nd Baron Ashburton. She married William Compton, 5th Marquess of Northampton in 1884, and held the title Marchioness of Northampton from 1897.
6. *those lectures*: probably Tyndall's redelivery at the RI of his American lectures on light (see letter 3782, n. 10).
7. *Lady Claud*: Elizabeth Hamilton.
8. *our grand old friend*: Thomas Carlyle.
9. *top of the Aeggischhorn*: the Eggishorn (see letter 3828, n. 15). Tyndall spent time with the Hamiltons in the Alps in summer 1873 (see *Ascent of John Tyndall*, p. 323), writing in a 28 July 1873 letter to Elizabeth Hamilton beforehand that 'I would also go with you to the Aeggischhorn' (see letter 4145).

From Mary Egerton [September 1873][1] 4181

[Heathfield Park | Hawkhurst][2] | Wed[nesday]y. Morning Early

I don't wonder you were charmed! They[3] are charming one & all! Lady Claud asked us over on purpose to have a talk about Switzerland,[4] and we did talk! From one o'clock when we arrived till nearly 1/2 past 7, (utterly ignoring the clicking bell) with only the interruption of a little singing, for a walk (chiefly in the rain) among their beautiful trees, was no interruption; and then from dinner till past 11 o'clock. They are all so agreeable, so amusing, on all subjects; but somehow, whatever subjects deep or light we got upon, we were sure to be back in Switzerland or its neighbourhood before long. Did "your ears tingle"[5] much? Dear friend, it was very pleasant to hear you talked of & to talk of you in the way I like! I did not know "Mary"[6] before; how pretty & bright she is! We had "the Bridge"[7] & the "Shattered Lamp",[8] &c &c; & heard many little episodes of your joint adventures. I will not say it was not a little trying, it must all have been so very very pleasant! the Great Expeditions were beyond my ambitions, but those sits by moonlight and the easier rambles, & the interesting talk and discussions, all that, I could have enjoyed!

I suppose thinking so much of the mountains made me wake at nearly mountain hours, namely 6 o'clock; and being tired of thinking I have got up to awake myself at your expense, by telling you how much I like your friends, whom I knew but slightly till this mutual interest of Switz[erlan][d] brought us together.

Had I known you were going to be away,[9] I would not have directed my letter to the Athenaeum,[10] where perhaps they do not forward them; for I would not willingly have lost a post in replying to your last[11] which, as you may suppose, will be a safe secret between you & me. Perhaps you will have been amused at the plain spoken way in which I took it up. But what is the use of "beating about the bush!"

I was so glad to see that your old friend Fichte[12] had been again one of your travelling companions. Another of them,[13] from what I have read of him in the "Contemporary"[14] I should like to plunge in the depths of the sea!

We go home today. The house here is nicer than I thought, & the drawing room, though dull from having no view, is most comfortable.

When am I ever going to see you again, & hear your side of the story?

Y[ou][rs] ever, | M F Egerton

RI MS JT/1/E/71

---

1. *[September 1873]*: the year and month are given since Tyndall spent time with the Hamiltons in Switzerland in August 1873, and the referred to article in the *Contemporary Review* was published in August 1873. Also by relation to letter 4173, in which Egerton has jealously heard about Tyndall's time with the Hamiltons but not yet met with them to talk about the trip.
2. *Heathfield Park | Hawkhurst*: the Hamilton family estate.
3. *They*: the Hamiltons, a family that Tyndall became acquainted and friendly with in 1873, were Claud Hamilton, Elizabeth Hamilton, and their children Louisa Hamilton, Tyndall's future wife; Emma Hamilton; Mary Hamilton; and Douglas Hamilton. On Tyndall's connections with the Hamilton family, see *Ascent of John Tyndall*, p. 323.
4. *talk about Switzerland*: Tyndall spent time with the Hamiltons (the mother, son, and Louisa and Emma) while in Switzerland in August.
5. *"your ears tingle"*: Egerton was suggesting that Tyndall may have sensed he was being spoken of.
6. *"Mary"*: Mary Hamilton.
7. *"the Bridge"*: not identified, but probably referring to something that happened in Switzerland that the Hamiltons shared with Egerton.
8. *"Shattered Lamp"*: not identified, but probably referring to something that happened in Switzerland that the Hamiltons shared with Egerton.
9. *to be away*: for the annual meeting of the BAAS at Bradford (see letter 4153, n. 15).

10. *Athenaeum*: the Athenaeum Club (see letter 3710, n. 5).
11. *your last*: letter missing.
12. *Fichte*: book by Johann Gottlieb Fichte not identified.
13. *Another of them*: referring to another friend, not companion, by which Egerton was sarcastically referring to Peter Guthrie Tait whom Tyndall was in dispute regarding James David Forbes.
14. *read of him in the "Contemporary"*: see letter 4024, n. 5.

BIOGRAPHICAL REGISTER

This register contains the names and biographical details of people who are mentioned three or more times in the letters included in this volume. Biographical information about people mentioned only once or twice is contained in the notes appended to the appropriate letter(s).

Conventions

The sources for the biographical entries are standard biographical dictionaries, such as the *Complete Dictionary of Scientific Biography*, the *Oxford Dictionary of National Biography*, Poggendorff's *Biographisch-literarisches Handwörterbuch der exacten Wissenschaften, Historisches Lexikon der Schweiz, Neue Deutsche Biographie*, and the *Dictionary of Nineteenth-Century British Scientists.* These have not been specified in the entries unless necessary. Additionally, the editors have found material from the Tyndall correspondence and from the journals of Tyndall and T. A. Hirst very useful. Where additional sources add important material, they are given in parentheses at the end of the entry.

Adair, Harriet Eliza (née Atkinson, 1811–78), was the mother of Mary Adair. In 1828 she married Alexander Adair (1791–1863) (J. B. Adair, *Adair History and Genealogy* [Los Angeles: Published by the author, 1924], 23).

Adair, Henrietta Mary (1843–1921), was the youngest daughter of Alexander Adair (1791–1863) and Harriet Eliza Atkinson (1811–78). She attended Tyndall's lectures at the RI and they conversed about scientific matters. Tyndall fell in love with Mary and eventually proposed to her in late 1869; she declined. In 1879 she was admitted as a member to the RI. She never married (*Ascent of John Tyndall*, 257).

Agassiz, Louis (1807–73), was a Swiss naturalist. Agassiz studied medicine, botany, geology, and zoology in Germany and France before taking a position at the University of Neuchâtel. He was appointed professor of zoology and geology at Harvard in the United States in 1847 and founded the Museum of Comparative Zoology in 1859. He was a key figure in developing the study of the earth's history, proposed the theory of ice ages, and made a close study of glaciers in the Alps. In 1840 he published his important *Études sur les glaciers*, 2 vols. (Neuchâtel: Gent et Gassmann), which helped establish the scientific study of glaciers. A believer in the fixity of species and a leading advocate of a polygenist form of old earth creationism, Agassiz was a leading opponent of Darwinism in the United States.

Airy, George Biddell (1801–92), was educated at University of Cambridge, graduating senior wrangler in 1823. In 1826 he was appointed Lucasian Professor of Mathematics, and in 1828 to the Plumian Chair of Astronomy and directorship of the University of Cambridge observatory. He married Richarda Smith (1804–75) in 1830; they had nine children. In 1835 he was appointed Astronomer Royal, a position he held until 1881, and set about improving the instrumentation and research activities of the Royal Observatory, Greenwich. He pursued research on a wide range of topics, including optics, sound, electricity, calculating the earth's mean density, and analyzing the inequalities in the motions of Earth and Venus. From 1872 to 1873, Airy was president of the RS.

Allen, Robin (1820–99), was a British poet who was appointed as a clerk in the Secretary's Office at Trinity House in 1837. From 1867 to 1881 he was the secretary of Trinity House. Some of his correspondence is held at the University of Cambridge Library (MS Add.9225; "The Trinity House," *Times*, 25 April 1881, 9; "Deaths," *Times*, 4 April 1899, 1).

Appleton, William Henry (1814–99), was an American publisher who joined the family business at the age of sixteen. Appleton clerked with his father, Daniel Appleton (1785–1849), and became a buyer for the firm in 1835, when he made his first trip to London. He became a partner in the firm in 1838, and the head of the firm in 1848. William Appleton was the firm's London representative by 1853 and was a prominent figure in American and international scientific publishing. After the death of his father, Appleton expanded the interests of D. Appleton and Company to include the fields of travel, poetry, and science, becoming the first American publisher to release editions of the works on evolution by Darwin and Spencer. Appleton published Tyndall's books in the United States, and also published the magazine *Popular Science Monthly*, edited by Edward L. Youmans. Letters in this volume indicate that Appleton and Company printed two thousand copies of Tyndall's *Forms of Water* (see letter 3933). In 1844 he married Mary Moody Worthen (1824–84). At their Wave Hill estate in Hudson Hill, New York, they hosted guests such as Thomas Huxley.

Armstrong, William George (1810–1900), was a British armaments manufacturer and industrialist, who also worked on hydraulics and electrostatics. Armstrong performed extensive service for the BAAS as well as a number of engineering institutions. He was knighted in 1859 for giving the patent for his design for the Armstrong gun to the Crown. In 1887 he was elevated to the peerage as Baron Armstrong of Cragside.

Arrow, Frederick (1818–75), was captain and deputy master of Trinity House (1865–75), a corporation reporting to the Crown on the safety and progress of navigation. Born at Fort William, Kolkata, he married Harriet Stileman (later Lady Arrow) in 1850. At the request of the hereditary khedive and viceroy of Egypt, Ismail, Arrow was invited to the opening of the Suez Canal in 1869. In *Fortnight in Egypt*, Arrow detailed his impressions of the event and his extraordinary trip. He died suddenly at the age of fifty-six at his residence, Pilgrim's Hall, in Brentwood, Essex ("The Late Sir Frederick Arrow," *Illustrated London News* 67 [31 July 1875]: 100; see also the India Office Family History from the British Library).

Ayrton, Acton Smee (1816–86), was a politician and first commissioner of the Office of Works in the Gladstone administration. Competent and knowledgeable, Ayrton was also known for his brusque demeanor, which brought him into conflict with several prominent people during the course of his duties. Among them was Joseph Hooker, the director of Royal Botanic Gardens, Kew; their dispute over the budget, management, and collections at Kew is referenced in many letters in this volume. A petition on Hooker's behalf was organized and signed by Tyndall, Charles Lyell, Charles Darwin, Thomas Huxley, and others and publicly aired in the House of Commons in 1872. The following year, Ayrton was transferred from the commissioner's office to that of judge-advocate-general, and he retired from government in March 1874.

Ball, John (1818–89), was an Irish politician and glaciologist and first president of the Alpine Club (1858–60). There, he edited *Peaks, Passes, and Glaciers*, the club's earliest publication. Other publications include three volumes of *Ball's Alpine Guide* (1863, 1864, 1868), popular and comprehensive mountaineering handbooks. As undersecretary of state for the colonies from 1855 to 1857, he assisted in efforts to publish colonial floras. A trip to Morocco and the Atlas Mountains in 1871 with Joseph Hooker resulted in a published travelogue of their excursion. Ball was elected member of the Royal Irish Academy (1840), fellow of the Linnean Society (1856), and FRS (1868). He was married twice, first to Eliza Parolini (d. 1867) and then to Julia O'Beirne in 1869. He died after a sudden illness in 1889.

Baring, Louisa Caroline (née Stewart-Mackenzie, 1827–1903), Lady Ashburton, was a philanthropist and art collector. In 1858 she married the extremely wealthy Bingham Baring (1799–1864), 2nd Baron Ashburton. After her husband's death, Baring amassed a significant collection of paintings and sketches, which filled her numerous residences where she entertained Britain's social and cultural elites, including Tyndall.

Barnard, Frederick Augustus Porter (1809–89), was a scientist and university president. In the family tradition, Barnard studied at Yale and began his own career in education following his graduation in 1828. He held several academic positions over the years: professor of mathematics and natural philosophy at the University of Alabama (1837–55); professor of physics, astronomy, and civil engineering (as well as chancellor) at the University of Mississippi; and president of Columbia College (1864–89). It was at Columbia that Barnard was most active in education reform, which included support for coeducation. Barnard College, named in his honor, was a compromise solution to the dilemma of coeducation, a separate but affiliated women's college, across the street from Columbia.

Bence Jones, Henry (1813–73), was a wealthy English physician and medical chemist. He was educated at Harrow School and Trinity College, Cambridge, before studying medicine at Saint George's Hospital, London. Attracted by the potential applications of chemistry to clinical medicine, he studied organic chemistry with Justus Liebig at Giessen (1841–43). He then went on to a career as a successful private physician and physician at Saint George's Hospital. In May 1842 he married his cousin Lady Millicent Acheson, with whom he had three sons and four daughters. He published important research on protein chemistry and made an abstract in English of Emil Heinrich du Bois-Reymond's work on electrophysiology (1852) that annoyed Carlo Matteucci, the Italian pioneer of electrophysiology. In 1853 he became a manager (trustee) of the RI and in 1860 he was elected secretary, a position he held until 1873. His *The Life and Letters of Faraday* was published in two volumes in 1870. He frequently entertained Tyndall both at his London home and at Folkestone, where he kept a summer house (in 1869 Tyndall convalesced there after injuring his leg in the Alps). Bence Jones suffered from ill health for several years. As letters from this volume show, his condition worsened while Tyndall was in America, which hastened Tyndall's return trip to England. Bence Jones died in April 1873.

Bence Jones, Millicent (née Acheson, 1812–87), was a daughter of Archibald Acheson (1776–1849), 2nd Earl of Gosford, and Mary Sparrow (1777–1841). She married her cousin Henry Bence Jones in 1842 and they had seven children: Henry Robert (1844–1902), Ralph Noel (1845–66), Archibald Bence (1856–1937), Millicent Mary (1843–1923), Olivia Mary (1845–1919), Anabella Mary (1846–51), and Edith Mary (1853–1919). They lived first at 30 Lower Grosvenor Street and later at 84 Brook Street, Grosvenor Square, London. As letters in this volume indicate, Bence Jones was moved by Tyndall's tribute to her late husband, included in the preface to his published lectures from America, *Six Lectures on Light* (1873) (*Henry Bence-Jones, M.D., F.R.S. 1813–1873: Autobiography with Elucidations at Later Dates* [London: Crusha and Son, 1929]).

Bentham, George (1800–84), was an English botanist fluent in several languages. For five years he worked on translating an appendix to *Chrestomathia*, a text by his uncle, the philosopher and reformer Jeremy Bentham (1748–1832). Bentham had a lifelong association with Joseph Hooker and the Royal Botanical Gardens, Kew, where he donated his herbarium and subsequently researched. Hooker also mentored Bentham in his botanical work, and together they cowrote *Genera plantarum* (1862–83), three volumes still in regular use today. Bentham's other publications include *Catalogue des plantes indigènes des Pyrénées et du Bas Languedoc* (1825) and *Handbook of British Flora* (1858). He was the recipient of the RS's Royal Medal in 1859 and the longest-serving president of the Linnean Society (1862–74).

Berkley, George (1821–93), was a civil engineer. At age fourteen he began an apprenticeship with marine engineers the Samuda brothers. In 1840 Berkley started working in the office of Robert Stephenson, who sent him to Dublin to experiment with atmospheric traction for railways in 1844. The following year he presented his findings from these experiments to the RI. From 1849 onward Berkley worked as principal and consulting engineer in several railway projects in England, India, and South Africa. For many years he was involved with the RI, first as an associate (1845), then member (1854), coun cil member (1872), and one of the board of managers at the time of the renovation of the laboratory. He died suddenly during an asthmatic coughing spell in 1893 (*Minutes of Proceedings of the Institution of Civil Engineers* 115 [1894]: 382–85).

Burns, John (1829–1901), 1st Baron Inverclyde, was a Scottish shipowner who joined the family firm, G. and J. Burns, in 1850. After his father, George Burns (1795–1890), retired in 1860, he took over the management and development of both G. and J. Burns and his father's share in Cunard Line, a partnership formed with shipping magnate Samuel Cunard and David MacIver. Cunard specialized in transatlantic steamer services based in Glasgow and Liverpool. During the 1860s, Cunard's original wooden paddle steamers were gradually replaced with iron ships, eventually incorporating those with screw propellers. Tyndall traveled to America on the SS *Russia* liner, which was launched by Cunard Line in 1867.

Burrows, George (1801–87), was an English physician who at the time of the Ayrton dispute was president of the Royal College of Physicians (1871–76).

Busk, George (1807–86), was a British naturalist and surgeon. Busk was educated at Dr. Hartley's school, in Bingley, Yorkshire, and then at the College of Surgeons. He served as editor of a number of noted scientific publications, including *Natural History Review* (1861–65), and served in leadership roles in a number of professional organizations, including as president of the Microscopical Society, which he founded. Busk was elected FRS in 1850, was a founding member of the exclusive X Club (1864), and served for many years as treasurer of the RI. He was honored with the RS's Royal Medal (1871) and the Geological Society of London's Lyell Medal (1878) and Wollaston Medal (1885). In 1843, he married his cousin Ellen Busk (1816–90). Two discoveries were named after him, the parasite *Distoma buskii* and a genus of Bryozoa called *Buskia*.

Cardwell, Edward (1813–86), 1st Viscount Cardwell, was a British Liberal politician educated at Winchester College and at Balliol College, Oxford. He married Annie Parker in 1838. They had no children, but her nephew Charles (1829–1910) was like a surrogate son to them. Cardwell served as president of the Board of Trade (1852–55). In the Gladstone administration, he served as the secretary of state for war (1868–74). During his tenure in the War Office he reduced the military budget and introduced reforms such as the War Office Act (1870), the Army Enlistment Bill (1870), and the Army Regulation Bill (1871), which abolished the purchase system. From 1872 to 1873 Cardwell introduced the linking in pairs of the infantry battalions, his most significant program of reform. For these advancements, Cardwell may be remembered as the most important reformer of the British army in the nineteenth century.

Carlyle, Thomas (1795–1881), was a Scottish essayist, novelist, and historian, and one of the most notable intellectual figures in the Victorian era. His marriage to Jane Baillie Welsh (1801–66) in 1826 was not a happy one, as their letters indicate, but they were inimitable hosts who entertained an impressive circle of friends, particularly during their time in London. Carlyle was a major literary influence on Tyndall, who read his publications avidly. After reading *Past and Present* (1843), Tyndall commented that Carlyle "must be a true hero. My feelings towards him are those of worship 'transcendental wonder' as he defines it" (Journal, 18 July 1847, RI MS JT/2/13a/231). During the 1860s Tyndall became a close friend of Carlyle's, supporting him in the Governor Eyre controversy and accompanying him to Louisa Baring's home in Mentone, France, in December 1866.

Clark, Andrew (1826–93), was a Scottish physician who treated many prominent people, including William Gladstone, Thomas Huxley, and Charles Darwin. Considered one of the leading society physicians, Clark had the largest London consulting practice of his day. He was married twice, first in 1851 to Seton Mary Percy Forster (d. 1858) and then in 1862 to Helen Annette Doxat (d. 1922). Clark was created baronet (1883), made FRS (1885), and elected president of the Royal College of Physicians (1888). He died in 1893 after suffering from a cerebral hemorrhage and was buried near his country home in Hatfield, Hertfordshire.

Clausius, Rudolf Julius Emanuel (1822–88), was a German mathematician and physicist and one of the founders of modern thermodynamics. His famous paper on the theory of heat, "Ueber die bewegende Kraft der Wärme," was published in the *Annalen der Physik* in 1850. He held a professorship in mathematical physics at the Polytechnikum in Zurich from 1855, then moved to the University of Würzburg in 1867 and finally to the University of Bonn in 1869. Tyndall translated many of his major works into English for publication in the *Phil. Mag.* and *Scientific Memoirs* in the early 1850s. He was very close to Tyndall, as was his wife, Adelheid, and they named their son Rudolf John after Tyndall.

Clifford, William Kingdon (1845–79), was an English mathematician and philosopher. He attained second wrangler at the University of Cambridge in 1867, and in 1868 he was elected fellow at Trinity College. In the same year Clifford gave a Friday Evening Discourse at the RI, applying Herbert Spencer's philosophy of evolution to the development of mind. He also delivered a Friday Evening Discourse at the RI on 28 March 1873, "The Meaning of Force and Energy," which, as Frederick Pollock noted years later, was never recorded in the RI's proceedings (see letter 4054, n. 5). Clifford was appointed professor of mathematics and mechanics at University College, London, in 1871, a post he held until his death in 1879. He made contributions to a variety of subjects, including ethics, the philosophy of science, and algebra.

Colville [Colvile], James William (1810–80), was a lawyer and civil servant. Educated at Eton College and at Trinity College, Colville was called to the bar in 1835. He became advocate general of Bengal (1845), and was appointed puisne judge of the supreme court of Bengal (1848), knighted (1848), and appointed chief justice (1855). In 1857 he married Lady Frances Elinor Grant (1838–1919) and retired to London, residing at 8 Rutland Gate, in 1859. Colville was elected FRS in April 1875.

Cottrell, John (1843–90), was Tyndall's senior laboratory assistant at the RI from ca. 1866 until 1885. In 1872 he traveled to the United States with Tyndall to assist in his American lecture tour, and in 1875 he assisted with observations at the Aletsch glacier.

Darwin, Charles (1809–82), was an English naturalist and geologist, best known for his book *On the Origin of Species* (London: John Murray, 1859). In it, Darwin introduced his theory of natural selection, born out of research stemming from his five-year research trip aboard HMS *Beagle* (1831–36). In 1838 he married his first cousin Emma Wedgwood (1808–96). Though a devout Christian, she supported her husband's materialist research. Together they had ten children (seven of whom survived to adulthood) and made their family home at Down in Kent. Earlier in Tyndall's career, Darwin encouraged his research on glaciers and supported him in his disagreement with James David Forbes. During the 1860s Darwin and his supporters had vociferously promoted the *Origin*, which was in its fifth edition by 1869 and had been translated into French, German, Italian, Dutch, Swedish, and Russian. In 1864 he was awarded the RS's Copley Medal but only after being nominated three years in a row. By the time *The Descent of Man* was published in 1871, *Darwinism* had become a catchword among scholars. His evolutionary writing cycle was completed in 1872 with the publication of *The Expression of the Emotions in Man and Animals*. Letters in this volume discuss the "Huxley fund," a collective donation among scientific friends organized by Darwin, Tyndall, and others to enable Thomas Huxley to take a health-restoring holiday.

Davy, Humphry (1778–1829), was an English chemist and inventor born in Cornwall, England. From 1801 until 1825 he was director of the RI laboratory, where he was also professor of chemistry (1802–12) and honorary professor (1813–23). He achieved fame as a lecturer and for his chemical research that isolated such elements as sodium, potassium, and chlorine. He is often credited with the invention of the miner's safety lamp. He served as president of the RS from 1820 to 1827.

Debus, Heinrich (1824–1915), was a German chemist who studied at Marburg (1845–48), where he met Tyndall, Edward Frankland, and Thomas Hirst. Debus moved to England in 1851 and taught chemistry at Queenwood

College until 1867, when he began teaching at Clifton College, Bristol. Later Debus lectured at Guy's Hospital Medical School (1870–88) and became professor of chemistry at the Royal Naval College, Greenwich (1873–88).

de la Rive, Auguste Arthur (1801–73), was a Swiss physicist and one of the founders of the electrochemical theory of batteries. He served as professor of natural philosophy at the Academy of Geneva and worked principally on electricity, although he also investigated the specific heats of gases and calculated the temperature of the earth's crust. His first publication, in 1822, was about Earth's magnetic field. He also published on electrical discharges through rarefied gases and on the aurora borealis. His principal book was *Traité d'électricité théorique et appliquée*, 3 vols. (Paris: J.-B. Baillière et H. Baillière, 1854–58). De la Rive was a member of the Society of Physics and Natural History in Geneva, of which Tyndall was elected an honorary member in 1859.

Delane, John Thadeus (1817–79), was editor of the *Times* in London from 1841 to 1877.

Derby, Lord: see Stanley, Edward Henry.

Dewar, James (1842–1923), was a Scottish chemist and physicist who succeeded John Hall Gladstone as Fullerian Professor of Chemistry at the RI in 1877.

Draper, John William (1811–82), was an English-born American chemist and historian known for contributions to early photography, photochemistry, and intellectual history. Born at St Helens, near Liverpool, England, Draper was the son of a scientifically inclined Methodist minister. He went to the University of London, which had recently been inaugurated in 1829, to study chemistry. Draper emigrated to America in 1832 after his father's death, and thereafter studied medicine at the University of Pennsylvania. In 1836 he was appointed professor of chemistry, physiology, and natural philosophy at Hampden-Sydney College, followed by a call to the University of the City of New York in 1839 as professor of chemistry and natural philosophy. There he founded the university's School of Medicine in 1840. In 1850 he became the school's president, a position he held until 1873. In 1837 he began experimenting with photography and between 1839 and 1840 took the first known photographs of the moon. Draper had a strong interest in all types of radiant energy and in 1847 published experimental results showing that all solid bodies become incandescent at the same temperature. Draper also spoke at the BAAS meeting at Oxford in 1860, which ignited the infamous exchange between Bishop Samuel Wilberforce and Thomas Huxley. In

1869 Draper also became founding president of the American Union Academy of Literature, Science and Art, and in 1876 founding president of the American Chemical Society. In 1875 he was awarded the Rumford Medal by the American Academy of Arts and Sciences for his research on radiant heat. Draper authored the massive *A History of the Intellectual Development of Europe* (1863), and is probably best known for his *History of the Conflict between Religion and Science* (1874), which many historians of science claim is an exemplar of the "conflict thesis," the idea that science and religion are fundamentally and irrevocably in conflict.

du Bois-Reymond, Emil Heinrich (1818–96), was a German physician and physiologist who studied with Johannes Müller at the University of Berlin. Under Müller's guidance, he developed a lifelong interest in animal electricity. Upon Müller's death in 1858, du Bois-Reymond was offered the position of professor of physiology at Berlin. He wrote extensively on animal electricity, publishing the first volume of his *Investigations of Animal Electricity* in 1848 and the second in 1884. In 1877, after his intensive lobbying, the Prussian government finally funded the provision of a modern physiological laboratory. Du Bois-Reymond and his colleagues Carl Ludwig, Hermann Helmholtz, and Ernst Brücke were instrumental in shaping the methods and direction of German physiology in the second half of the nineteenth century, integrating approaches from physics and engineering into the analysis of organic systems.

Egerton, Mary (May) Frances (née Pierrepont, 1819–1905), was the daughter of Charles Pierrepont (1778–1860), 2nd Earl Manvers. In 1845 she married Edward Christopher Egerton (1816–69), with whom she had five children: Mary Alice, Georgina, Charlotte, Hugh Edward, and Charles Augustus. Egerton was a frequent correspondent of Tyndall's and attended his lectures.

Egerton, Mary (May) Alice (1848–1924), was the eldest daughter of Edward and Mary Egerton. She married Beauchamp Tower in 1902 ("Births," *Times*, 19 February 1848, 8; "Wills and Bequests," *Times*, 11 June 1924, 15).

Ellis, George Edward (1814–94), was an American clergyman and historian, born in Boston, Massachusetts. Following his graduation from Harvard, Ellis traveled abroad, including a visit to the Vatican, where he had a personal audience with Pope Gregory XVI. He married twice, first in 1840, to Elizabeth Bruce Eager, who died shortly after the birth of their son, and second in 1859, to Lucretia Goddard Gould. Ellis was on the Board of Overseers at Harvard College (1850–79), and the first professor of systematic theology at the Divinity School. Following the deaths, within a year, of his second wife and only son, he devoted his time to literary and historical interests, with frequent contributions to the *New York Review*, *Atlantic Monthly*, and *North American Review*, along with several other publications. He died at home, in his Boston townhouse, in 1894.

Emerson, Ellen Tucker (1839–1909), was the eldest of three children of Ralph Waldo Emerson and Lidian Jackson Emerson (née Lydia Jackson, 1802–92), presumably named after her father's beloved first wife, Ellen Louisa Tucker, who died of tuberculosis in 1831, two years into their marriage. Emerson devoted her life to helping her parents and teaching Sunday school at the Emerson family home. From October 1872 to May 1873 Emerson traveled with her father to Europe and Egypt, and her letters home detail their social events abroad. When her father's health began to deteriorate in the early 1870s, he relied on his children and his close friend James Elliott Cabot for assistance with editing and publishing. Emerson helped with the publication of a final volume of essays, *Letters and Social Aims* (1876), and, with Cabot, published *Fortune of the Republic* (1878), and "The Preacher" (1880) (E. E. W. Gregg, ed., *The Letters of Ellen Tucker Emerson*, 2 vols. [Kent, OH: Kent State University Press, 1982]).

Emerson, Ralph Waldo (1803–82), was an American author, poet, and lecturer, educated at Harvard. He was a leading figure in the transcendentalism movement, which he heavily influenced through such essays as "Self-Reliance"‘ (1841), "Nature" (1836), and "The American Scholar" (1837). During the 1860s he continued a productive output, editing posthumous works of Henry David Thoreau (1817–62), publishing the collection of poems *May-Day and Other Pieces* (1867), and delivering hundreds of lectures across the United States. Emerson's trip to Europe and Egypt (1872–73) with his daughter Ellen included many social engagements with London's literary and scientific elite, such as the Froudes, Max Müller, Lord and Lady Amberly, Lady Augusta Stanley of Alderly, Henry Holland, and William Frederick Pol lock. Emerson had lunch with Tyndall, Thomas Huxley, and Thomas Hirst on 23 April 1873 (E. E. W. Gregg, ed., *The Letters of Ellen Tucker Emerson*, 2 vols. [Kent, OH: Kent State University Press, 1982]).

Evarts, William M. (1818–1901), was a prominent New York lawyer and politician who served as attorney general under President Andrew Johnson and would later serve as a senator and as secretary of state.

Faraday, Michael (1791–1867), was a natural philosopher, Fullerian Professor of Chemistry at the RI, and a member of the Sandemanian Church. His extensive research focused on electricity and encompassed electrochemistry, electromagnetism, diamagnetism, and magneto-optics (the Faraday effect).

Faraday also held the positions of director of the laboratory (1825–67) and superintendent of the house (1852–67) at the RI. He was Tyndall's patron and helped him to obtain the position of professor of natural philosophy at the RI in 1853. Despite significant differences in their views on religion and on the nature of magnetism, they held the highest regard for each other. Faraday died in 1867, and Tyndall succeeded him as superintendent of the house and director of the laboratory at the RI. Tyndall's *Faraday as a Discoverer*, published the following year, was his tribute to his mentor. Tyndall took over from Faraday as scientific adviser to Trinity House on matters regarding lighthouses and, later, fog-signaling.

Faraday, Sarah (née Barnard, 1800–79), was the daughter of Mary Boosey and Edward Barnard, a successful London silversmith and Sandemanian. She likely met Michael Faraday at the Sandemanian meeting house in Paul's Alley. After they married in 1821 she moved into his rooms at the RI. They had no children, but two nieces, Margery Ann Reid (1815–88) and Jane Barnard (1832–1922), lived with them at different times for extended periods. In later years Faraday would take on many of her husband's administrative responsibilities at the RI. From 1858 onward they lived at Hampton Court Palace, provided for them by Queen Victoria. Following the death of her husband in August 1867, Faraday lived at Barnsbury Villa with Jane Barnard.

Farrer, Thomas Henry (1819–99), 1st Baron Farrer, was an English civil servant. On the recommendation of his brother-in-law Stafford Northcote (1818–87), he started work at the Board of Trade as an assistant (1849), soon moving up the ranks and eventually becoming the board's first sole permanent secretary (1867–86). During his time there, he had significant influence over policy. A short retirement was followed by a brief career in local politics as a Liberal alderman. Farrer was married twice, first to Frances Erskine, who died in 1870, and second to Katherine Euphemia Wedgwood (d. 1931), Charles Darwin's niece.

Forbes, George (1849–1936), was the son of James David Forbes, who as of 1872 was professor of natural philosophy at Anderson's University, Glasgow.

Forbes, James David (1809–68), was a Scottish physicist and geologist. He was elected FRS (1832) and made professor of natural philosophy at Edinburgh following the death of his predecessor, John Leslie, in 1833. Forbes received the RS's prestigious Rumford Medal in 1838 for his experimental investigations on the polarization of heat, which he was the first researcher to demonstrate in November 1834. In 1860 he became principal of the college of

Saint Salvator and Saint Leonard (the United College), at University of Saint Andrews in 1860. He was also a pioneering mountaineer, and published *Travels through the Alps of Savoy* (Edinburgh: A. and C. Black) in 1843. While in the Alps, Forbes studied the motion of glaciers and theorized that glaciers were viscous bodies, which brought him into controversy with Tyndall and Huxley in the late 1850s. Tyndall was particularly critical of Forbes's misuse of terminology, and for taking credit for work that Tyndall thought belonged to Louis Rendu (1789–1859). The dispute created two camps of supporters and played out in the pages of *Nature*, but was put to rest with the reprinting of Rendu's *Memoir* in 1874, by Forbes's son George.

Frankland, Edward (1825–99), was an English chemist and one of Tyndall's closest friends. He worked as an assistant to the chemist Lyon Playfair in 1845 before teaching at Queenwood College in 1847, where he met Tyndall. He and Tyndall studied together at the University of Marburg in 1848. After completing his degree in 1849, Frankland worked briefly in Justus Liebig's laboratory at Giessen and taught at the Putney College of Engineering, near London, before assuming the chair of chemistry at Owens College, Manchester. Frankland moved to London in 1857 and held a number of positions, including at Saint Bartholomew's Hospital and the East India Company's military college. He was awarded the RS's Royal Medal after Tyndall nominated him. Frankland was professor of chemistry at the RI (1863–68) and afterward at the Royal College of Chemistry (1868–85). In 1864 Frankland joined Tyndall, Thomas Hirst, Joseph Hooker, Thomas Huxley, and four others as members of the X Club, a monthly dining club whose members promoted science in British society.

Franklin, Benjamin (1706–90), was an American natural philosopher, inventor, writer, printer, publisher, statesman, and diplomat. Born and raised in Boston, Franklin moved to Philadelphia at age seventeen to look for work in the printing business. There he met and settled down with Deborah Read (ca. 1704–74), cofounded his own printing office (1728), and was elected clerk for the Pennsylvania assembly (1736). By 1747 Franklin began experimenting with electricity, leading to his famous kite demonstration (1752), which proved that electricity was a basic element of nature. In addition to his scientific and meteorological experiments, Franklin was also the inventor of several devices that improved quality of life. As one of the "founding fathers" of the United States, Franklin was best known for helping to draft the Declaration of Independence (1776) and the US Constitution (1787). He also assisted in negotiating the Treaty of Paris (1783) which ended the Revolutionary War. A slave owner for many years, Franklin became an outspoken abolitionist in later life. He was also an early environmentalist and vegetarian. Franklin suffered from multiple health problems throughout his life, especially gout, which worsened with age. He died of pleurisy in 1790 and was buried at Christ Church in Philadelphia.

Froude, James Anthony (1818–94), was an English historian and editor, educated at Oxford University. His *The Nemesis of Faith* (1849) was highly controversial at Oxford and a copy was set on fire by subrector William Sewell. A pivotal meeting with Thomas Carlyle in 1849 developed into a close lifelong friendship between the two men, with Froude not only editing Carlyle's posthumus *Reminiscences* but also writing a multivolume biography. In 1856 he began work on his twelve-volume *History of England from the Fall of Wolsey to the Death of Elizabeth*, and in 1860, after the death of his first wife, he moved to London and took on the editorship at *Fraser's Magazine*. Froude was a member of the Metaphysical Society, established in 1869, where he would have encountered fellow member Tyndall. In the 1870s he presented controversial views on the Irish Question and imperial policy in South Africa. Froude traveled to the United States in 1872 for a lecture tour, and took passage on the same ship Tyndall took to New York, the SS *Russia*.

Gascoyne-Cecil, Robert Arthur Talbot (1830–1903), Lord Salisbury, was an English statesman, the fifth of six children born to James Brownlow William Cecil (1791–1868), 2nd Marquess of Salisbury, and his first wife, Frances Mary Gascoyne. He was a pious man and a devoted husband to his wife, Georgina Alderson, with whom he had eight children. Considered one of the greatest British foreign secretaries, Salisbury was a Conservative politician who served as prime minister three times (1885–86, 1886–92, 1895–1902), acting as his own foreign minister while in office. Following his death in 1903 his daughter Lady Gwendolen Gascoyne-Cecil authored a well-researched and well-regarded biography, *Life of Robert, Marquis of Salisbury*, published in several volumes dating until 1892.

Gauthier-Villars, Jean-Albert (1828–98), was a French engineer and publisher. Born to a family of printers in Lons-le-Saunier, he studied at the École Polytechnique (1848–50), became a telegraph engineer, and served in the military. In 1864 he purchased the Mallet-Bachelier printing house, which had published the *Comptes rendus hebdomadaires des séances de l'Académie des sciences*. Under his direction the firm, renamed Gauthier-Villars, specialized in mathematical and scientific books and magazines, including French translations of Tyndall's books.

Ginty, Margaret (née Roberts, b. 1826), was the widow of Tyndall's friend William Ginty. Margaret married William in Ireland in 1846, after which they relocated to Manchester. In 1854 the family moved to Rio de Janeiro, Brazil, where William died in 1866. After her husband's death, Tyndall served as Ginty's trustee and supervised the education and employment of her children. Letters in this volume show how Ginty continued to seek out Tyndall's assistance in her affairs.

Gladstone, John Hall (1827–1902), was an English chemist who trained at University College, London, before completing his PhD at the University of Giessen under Justus Liebig. Gladstone was FRS, and from 1874 was the Fullerian Professor of Chemistry at the RI.

Gladstone, William Ewart (1809–98), was an author and prime minister of the United Kingdom. Born into a wealthy evangelical family and then educated at Eton and Christ Church, Oxford, Gladstone was destined and encouraged by his family and connections, for a future in politics. In 1839 he married Catherine Glynne (1812–1900), who was from a minor aristocratic family. Tyndall and others expected Gladstone to show more support for Joseph Hooker in his dispute with Acton Ayrton over the management of the Royal Botanic Gardens, Kew, but that proved not to be the case. Tyndall and Gladstone would clash in the 1880s and early 1890s in the periodical press over the question of Irish Home Rule, which Tyndall was against. Gladstone died of cancer of the mouth in 1898; he was honored with a state funeral and burial in Westminster Abbey.

Grant, Ulysses S. (1822–85), was the eighteenth president of the United States (1869–77). He graduated from West Point in 1843, and served in the Mexican–American War. Grant resigned from the army in 1854, but re-enlisted at the start of the American Civil War (1861–65). He rose through the ranks to become commanding general of the US Army from 1864 to 1869, leading the Union to victory over the commander of the Confederate States Army, Robert E. Lee, who surrendered to Grant on 9 April 1865 in Appomattox, Virginia. Grant served as secretary of war from 1868 to 1869 before his two-term presidency, during which he was noted for his progressive views and actions, especially with regard to racial issues.

Granville, Lord: see Levenson-Gower, Granville George.

Gray, Asa (1810–88), was an American botanist born in upstate New York. After completing his medical degree at the College of Physicians and Surgeons of the Western District of New York in 1831, Gray practiced medicine for a few years before abandoning it to work in botanical research. At Harvard University, he became the Fisher Professor of Natural History in 1842.

Gray's notable contributions to botany—in which Gray was the first to make a living—include his extensive correspondence with Charles Darwin, which helped inform Darwin's theory of evolution, and his enduring publication *Memoirs of the American Academy of Arts and Sciences* (1859). Letters in this volume show Gray putting his support behind friend and fellow botanist Joseph Hooker in the Ayrton affair.

Hamilton, Claud (1813–84), was a British Conservative politician. The second son of James Hamilton (1786–1814) and Harriet Douglas, he was educated at Harrow and Trinity College, Cambridge. He served as a member of Parliament for County Tyrone (1835–37, 1839–74), treasurer of the Household (1852, 1858–59), and vice-chamberlain of the Household (1866–68). Hamilton married Elizabeth Proby in 1844, with whom he had four children, including Tyndall's future wife, Louisa.

Hamilton, Douglas James (1856–1931), was the only son of Claud and Elizabeth Hamilton. He became a soldier and politician.

Hamilton, Elizabeth Emma (née Proby, 1821–1900), was the second daughter of the Honourable Granville Leveson Proby (1781–1868) and Isabella Howard (d. 1836). In 1844 she married Claud Hamilton (1813–84), a British Conservative politician. They had one son, Douglas, and three daughters, Louisa, Emma, and Mary. Louisa married Tyndall in 1876. In 1885 Hamilton translated *Louis Pasteur: His Life and Labours* from French into English and her son-in-law, Tyndall, provided the introduction. She died on 24 June 1900 at the Tyndalls' Hindhead home after suffering from a lengthy illness.

Hamilton, Emma Frances (1847–1924), was the second daughter of Claud and Elizabeth Hamilton.

Hamilton, Louisa Charlotte (1845–1940), was the eldest daughter of Claud and Elizabeth Hamilton. She married John Tyndall in 1876 and supported him in his work. Following his death in 1893 due to an accidental overdose of his medication that she administered, she compiled his journals, papers, and letters with the intention of producing a biography of her late husband. That task proved monumental, and was incomplete upon her death in 1940; it was taken up by others and published five years later (A. S. Eve and C. H. Creasey, *Life and Work of John Tyndall* [London: Macmillan and Company, 1945]).

Hamilton, Mary Stuart (1851–1939), was the youngest daughter of Claud and Elizabeth Hamilton. In 1878 she married Wilbraham Frederic Tollemache

(1832–1904), 2nd Baron Tollemache, the son of John Jervis Tollemache (1805–90), 1st Baron Tollemache, a British Conservative politician.

Harcourt, Augustus George Vernon (1834–1919), was a British chemist at Oxford University noted for his research in chemical kinetics. He was the nephew of William Vernon Harcourt (1789–1871), who founded the BAAS in 1831. Harcourt was educated at Harrow School and then at Balliol College, Oxford, where he earned a degree in natural science (1858). With Tyndall and William Pole, Harcourt was one of the metropolitan gas referees, who were appointed by the Board of Trade under the City of London Gas Act (1868) to work with the city's suppliers of gas to maintain safety. He was made FRS in 1863, and was secretary of the Chemical Society from 1865 to 1873 and its president in 1895.

Harwood, Hamilton Edward (1824–72), was an architect and surveyor. Born in London, Harwood articled with architect Francis Thompson (1840), where he assisted on railway stations at North Midland, South Eastern, Chester, and Holyhead. In 1850, he joined the office of engineer Charles Heard Wild, where he worked on projects such as the Great Exhibition and the Crystal Palace. He was appointed surveyor to the Board of Admiralty for the Naval Yard (1855) as well as running his private practice as architect and surveyor. As letters in this volume indicate, Harwood designed the new laboratory at the RI, but died unexpectedly during its construction (F. A. J. L. James and A. Peers, "Constructing Space for Science at the Royal Institution of Great Britain," *Physics in Perspective* 9, no. 2 [2007]: 162–63; *Minutes of Proceedings of the Institution of Civil Engineers* 38 [1873–74]: 313–14).

Helmholtz, Anna (née Mohl, 1834–99), was a Berlin salonnière and translator of scientific works, and the wife of Hermann Helmholtz, whom she married in 1861. With Clara Wiedemann, she translated Tyndall's *Heat* (1863) and *Sound* (1867) for their German editions in 1867 and 1869, respectively. Although the two women translated the work, only their husbands were credited (*HLS*; B. Meijer et. al., eds., *Nordisk familjebok*, 2nd ed. [Stockholm: Nordisk familjeboks, 1904–26]).

Helmholtz, Hermann Ludwig Ferdinand (1821–94), was a German physicist and physician. Educated in military medicine at the Friedrich-Wilhelms Institute in Berlin, he was first a teacher of anatomy at the Berlin Academy of Arts in 1848. He became professor of Physiology at the University of Königsberg (1849), and then professor for Anatomy and Physiology at the University of Bonn (1855), where he turned his attention to physics. He was professor of physiology at Heidelberg from 1858 until he became professor of physics at Humboldt University, Berlin. Helmholtz made many contributions to numerous fields, including thermodynamics and electrodynamics, as well as physiological studies of sound and optics.

Helps, Arthur (1813–75), was a public servant and author. In 1860 he was made clerk of the privy council on the recommendation of Granville George Levenson-Gower, a position that included him in the Gladstone and Disraeli administrations and the management of the Cattle Disease Prevention Act (1866). The privy council also connected Helps to the royal family, which led to a royal assignment as editor of *Leaves from the Journal of Our Life in the Highlands* (1868), a publication of extracts from Queen Victoria's journal. Of his numerous publications, *Friends in Council* (1847–59) was his most successful. *The Life of Hernando Cortes* (1871) and *Conversations on War and General Culture* (1871), dedicated to Thomas Carlyle and the British art and social critic John Ruskin (1819–1900), respectively, were formatted on the model of a Cambridge conversazione. As letters in this volume indicate, Helps was made KCB in 1872.

Henry, Harriet Alexander (1808–82), was the wife and cousin of Joseph Henry. They married in 1830 and had four children, William Alexander (1832–62), Mary Anna (1834–1903), Helen Louisa (1836–1912), and Caroline (1839–1920).

Henry, Joseph (1797–1878), was an American physicist, who attended Albany Academy, a well-regarded academic high school, from 1819 to 1822. Following his studies, he held various roles there such as laboratory assistant and teacher, eventually becoming professor of mathematics and natural philosophy in 1826. His early research focused on electromagnetism. Henry was appointed to the chair of natural philosophy at the College of New Jersey in 1832 and elected to the American Philosophical Society in 1835. An 1837 trip to Europe gave him the opportunity to meet many leading scientists, including Michael Faraday. Henry is probably best remembered for his appointment in 1846 as the first secretary of the new Smithsonian Institution in Washington, DC. In 1870 he took a second trip to Europe, visiting England, Scotland, Ireland, Belgium, Switzerland, France, and Germany, and attended the international conference on the metric standard in Paris. Letters in this volume discuss plans for Tyndall's American lecture tour and updates during the tour. Along with Hector Tyndale and Edward L. Youmans, Henry was named a trustee of the fund established with Tyndall's lecture profits for pure science research in America.

Hirst, Emily (Lily/Lilly) Anna (1853–83), was a niece of Thomas Hirst who lived with him for stretches in the 1870s. Her parents were William and

Rachel Hirst (Brock and MacLeod, *Hirst Journals*, index, 50). Thomas Hirst's journals of the time indicate their closeness and the paternal interest he took in securing her future (Brock and MacLeod, *Hirst Journals*, chapter 6). She studied at Bedford College.

Hirst, Thomas Archer (1830–92), was an English mathematician, educator, and Tyndall's closest friend and correspondent. His friendship with Tyndall began in Halifax, Yorkshire, in 1845 when they were both working for the surveyor Richard Carter. Encouraged by Tyndall—who acted as a sort of mentor to him—Hirst studied at the Universities of Marburg, Göttingen, and Berlin between 1849 and 1852. In 1853 he replaced Tyndall as mathematics teacher at Queenwood College; he resigned in 1856 to look after his sick wife, Anna, who died in Paris in 1857. During the 1860s Hirst established himself as an active and respected member of the London scientific community. He taught mathematics at University College, London (1860–64), and subsequently held professorships of physics and mathematics (1865, 1867–70). He joined the RS council (1864); was a founder of the X Club, a monthly dining club the members of which promoted science in British society; and served as general secretary of the BAAS (1866–70). In 1873, Hirst was appointed as the first Director of Studies of the Royal Naval College at Greenwich, serving in that position until 1883.

Holland, Henry (1788–1873), was an English physician and travel writer educated at the University of Edinburgh. He accompanied George MacKenzie and Richard Bright on an expedition to Iceland in 1810, and his work on Icelandic diseases formed the basis of his doctoral thesis. He became a licentiate of the Royal College of Physicians in 1816 and a fellow in 1828. In 1816 he opened a practice that served members of London's wealthy and fashionable society. He became physician-extraordinary to Queen Victoria (1837) and physician-in-ordinary to Prince Albert (1840). In 1853 Holland was made 1st baronet, after declining the offer in 1841. In his role as president of the RI (1865–73), he promoted lectures as a way of popularizing science. He was also a supporter of Faraday and Tyndall in their efforts to improve the RI and established the Holland Fund for the promotion of scientific research.

Holmes, Oliver Wendell (1809–94), was an American physician, teacher of anatomy, and writer. Holmes graduated from Harvard (1829), and after a brief period in law school he returned to Harvard to study medicine (1831–36). In 1840 he married Amelia Lee Jackson. His greatest contribution to medicine was his publication "The Contagiousness of Puerperal Fever" (1843), but in addition to his several medical essays he also published on homeopathy. Holmes returned to Harvard as dean of the medical school and Parkman Professor of Anatomy and Physiology in 1847. A founding contributor to *Atlantic Monthly*, Holmes wrote several conversational essays and serials for the publication. Considered Boston's unofficial poet laureate, Holmes's best-known poems include "The Living Temple" (1858) and "The Last Leaf" (1833). He was a member of the Saturday Club, along with friends Louis Agassiz, poet James Russell Lowell (1819–91), and Ralph Waldo Emerson.

Hooker, Frances Harriet (née Henslow, 1825–74), was a daughter of Harriet Jenyns and the Cambridge botanist John Stevens Henslow (Charles Darwin's mentor). In 1851 she married Joseph Hooker. She was fluent in French and German, and in addition to making private translations for her husband she published a translation of Emmanuel le Maout, *A General System of Botany* (London: Longman, 1873).

Hooker, Joseph Dalton (1817–1911), was a British botanist who specialized in taxonomy and the geographical distribution of plants. He arguably became the most famous British botanist of the Victorian era. He was the younger son of William Jackson Hooker, the director of the Royal Botanic Gardens, Kew. After studying medicine at the University of Glasgow he became a naval surgeon and spent the years 1839–43 on James Clark Ross's geomagnetic survey of the southern oceans, where he collected plants as well as serving as surgeon. Subsequently he explored and collected plants in India (1847–51), achieving considerable mountaineering success, before joining his father as assistant director at Kew in 1855. In 1864 Hooker joined Thomas Huxley, Tyndall, and others in founding the X Club, a monthly dining club in which members promoted science in British society. He succeeded his father as director of Kew in 1865. He served as president of the BAAS (1868) and the RS (1873–77), and was knighted several times and received many scientific awards, including the RS's Royal Medal (1854) and Copley Medal (1887).

Hooker, William Jackson (1785–1865), was a British botanist and director of the Royal Botanic Gardens, Kew from 1841 to 1865. He was the father of Joseph Hooker.

Houghton, Lord: see Milnes, Richard Monckton.

Huxley, Henrietta (Nettie) Anne (née Heathorn, 1825–1915), was born in the West Indies to Henry Heathorn and Sarah Henrietta Richardson (née Harris), and was educated in Neuwied, Germany, for two years as a youth. In 1843 she, her mother, and her half sister moved to Australia to join her father. In 1847 she was keeping house for her sister Oriana and brother-in-law William Fanning in Sydney, Australia, when she met Thomas Huxley, who was then a surgeon and zoologist on HMS *Rattlesnake*. They were married in London in 1855, after an eight-year engagement. Henrietta Huxley assisted her husband in his work by translating German and drawing diagrams for his lectures. She was also a poet and at the age of eighty-six published a volume of collected poems, *Poems of Henrietta A. Huxley with Three of Thomas Henry Huxley* (London: Duckworth, 1913). As with many of the wives of this small group of scientific friends, Huxley was very much a part of their community; she corresponded with the Darwins as well as with Tyndall ("Death of Mrs. Huxley," *Times*, 6 April 1915, 12).

Huxley, Jessie Oriana (1856–1927), was the eldest daughter of Thomas and Henrietta Huxley. In 1877 she married the British architect Frederick Sandham Waller (1822–1905).

Huxley, Leonard (1860–1933), was a biographer and writer and the son of Thomas and Henrietta Huxley. After graduating from Balliol College, Oxford, he took a teaching post at Charterhouse School and joined the publishing firm of Smith, Elder and Company in 1884. He and his first wife, schoolmistress Julia Frances Arnold (1862–1908), had four children, the youngest of whom was novelist Aldous Huxley (1894–63). Leonard Huxley's publications include biographies of his father, Thomas Huxley (1900), and Joseph Hooker (1918), a volume of poems (1920), and an account of the publishing firm where he had worked (1923). He died in 1933 and was survived by his second wife.

Huxley, Thomas Henry (1825–95), was a British zoologist, educator, science popularizer, and one of Tyndall's closest friends. Following medical training at Charing Cross Hospital in London, he was a ship's surgeon aboard the HMS *Rattlesnake* expedition to Australia (1846–50) under Captain Owen Stanley. While in Australia, he met Henrietta Anne Heathorn, whom he married in July 1855 after an eight-year engagement. After eking a living by writing for Chapman's *Westminster Review*, he became professor of natural history at the Royal School of Mines, London (1854–71) and moved with the school to South Kensington, where he taught in a new laboratory until 1885. He joined Tyndall twice at the RI as Fullerian Professor of Natural History (1855–58, 1866–69). In the summer of 1856, and again in 1857, he and Tyndall studied the movement of glaciers and climbed in the Alps together (they published their research in January 1857 as "On the Structure and Motion of Glaciers," *Phil. Trans.* 147 [1857]: 327–46). In 1864 Huxley joined Tyndall, Edward Frankland, Thomas Hirst, Joseph Hooker, and four others as members of the X Club, a monthly dining club the members of which promoted science in British society. In this volume, a group of eighteen friends, including Tyndall and Charles Darwin, donate funds to enable Huxley to take a much-needed health-restoring holiday to France with Hooker.

Jamin, Jules Célestin (1818–86), was a French physicist, educated at the École Normale Superieur. He was professor of physics at the École Polytechnique (1858–81), and from 1863 he was professor of physics at the Faculté des sciences, Paris. He worked in many branches of physics, including optics, acoustics, electricity, magnetism, and radiation, and in 1858 was awarded the RS's Rumford Medal.

Joule, James Prescott (1818–89), was an English physicist and mathematician. The son of a brewer, he studied for a brief time under John Dalton before focusing his own work on electromagnetism. In 1838 Joule published his first paper on electromagnetic engines. This research led to his formulation of the mechanical equivalent of heat, which he presented to the BAAS in 1843, and for which the RS awarded him a Royal Medal (1852). However, Tyndall had come to believe that the German physicist Julius Mayer also deserved credit for the discovery rather than Joule alone. Tyndall appealed to his friends and correspondents in Germany to send him copies of Mayer's works, which convinced him of the truth of his position, and he then set about publishing and promoting Mayer's work in England, a move that upset Joule and his colleagues.

Kennedy, Thomas Stuart (1841–94), of Leeds, was an engineer and alpinist (*Proceedings of the Institution of Mechanical Engineers* 48 [1894]: 598).

King, Clarence Rivers (1842–1901), was an American geologist and the first director of the US Geological Survey. Influenced by Tyndall's description of the Alps, he claimed and named Mount Tyndall in 1864. His fieldwork between the Sierra Nevada and the Great Plains (1867–72) included topographical mapping and a survey of the geological and natural resources of the fortieth parallel. In 1872 he became famous for uncovering the Great Diamond Hoax in northwest Colorado, one of biggest frauds ever committed. Under the alias James Todd, King met and married Ada Copeland (ca. 1860–1964), a formerly enslaved woman working as a nursemaid in New York City. Together they had five children. Their marriage was kept secret from King's family and friends, and his true identity was kept from Ada until his death from tuberculosis in 1901.

Lee, Thomas (1818–92), was a Scottish mathematics teacher and amateur astronomer. Lee was born in the parish of Galston, but moved to Kilmarnock, Ayreshire, at age fourteen. For several years he taught mathematics at the Kilmarnock Academy. For astronomy instruction he used the facilities of the Morton Place Observatory in Kilmarnock, Scotland, built by Bailie Thomas Morton. Lee married Morton's daughter Agnes Morton (b. 1820) in 1841. In 1856 he was made fellow of the Royal Astronomical Society. He became director of the observatory in 1862, eventually taking ownership of the facility upon the death of his father-in-law. Lee was listed as an RI visitor in 1872–73. According to the RI's 1800 prospectus, visitors elected from the membership were involved in the governance of the institution, giving Tyndall and Henry Bence Jones reason to take issue with how certain of those visitors dispensed with that responsibility ("Obituary: List of Fellows and Associates Deceased during the Year: Thomas Lee," *Monthly Notices of the Royal Astronomical Society* 53 [February 1893]: 220).

Lesley, Joseph Peter (1819–1903), was an American geologist and director of the second Geological Survey of Pennsylvania. At the time of Tyndall's American lecture tour, Lesley was editor of the *United States Railroad and Mining Register* and professor of geology, mining, and engineering, as well as dean of the science faculty at the University of Pennsylvania. The Second Pennsylvania Survey (1874–89), which expanded the work of the first and detailed the economic resources of the state, is considered his most significant contribution to science. Indefatigable in his work ethic, Lesley suffered from recurrent illnesses and ultimately a breakdown in 1893, after which he became a semi-invalid and was cared for by his wife, Susan until his death in 1903.

Levenson-Gower, Granville George (1815–91), 2nd Earl Granville, was a Whig politician. He was the eldest son of Granville Leveson-Gower (1773–1846), 1st Earl Granville and Lady Henrietta Elizabeth Cavendish (1785–1862). Educated at Eton and Christ Church, Oxford (1832–36), he later became a member of Parliament for Morpeth (1837–40). He married twice, first to Marie-Louise Pellina (Lady Acton, 1812–60) and second to Castalia Rosaline (1847–1938). Under the Gladstone administration, Granville held several positions, including leader of the Liberal Party in the Lords (1866), colonial secretary (1868–70), and foreign secretary (1852–54, 1870–74, 1880–85). Letters in this volume show Tyndall soliciting Granville's support for Joseph Hooker in the Ayrton affair.

Lockyer, Joseph Norman (1836–1920), was an English astronomer and the first editor of *Nature*, a position he retained for more than fifty years. Privately educated, he wrote his first scientific paper for the Royal Astronomical Society in 1863. Lockyer was an early astrophysicist and used a spectroscope for solar observations. Lockyer was the first to suggest that a yellow spectral line in sunlight observed during a solar eclipse in 1868 was due to an as yet unknown element, which he labeled "helium." This discovery was not verified until the 1890s. As editor, Lockyer appears to have encouraged controversy in the pages of *Nature*, as evidenced in the letters in this volume concerning the bitter exchange between Tyndall and the physicist Peter Guthrie Tait. Lockyer also used the journal to support causes he championed. At the RS, he was elected fellow (1869), was awarded the Rumford Medal (1874), and gave the Bakerian Lecture (1888).

Longfellow, Henry Wadsworth (1807–82), was an American poet and professor of modern languages. After graduating from Bowdoin College in 1825, Longfellow was offered a professorship, contingent on further study in Europe. Harvard offered him a professorship of French and Spanish in 1834, which once again required an overseas trip to study, this time with his wife Mary Storer Potter (1812–35). She died during this trip, after suffering a miscarriage. After a rocky seven-year courtship, he married Fanny Appleton (1817–61) in 1843, and together they had six children. Beloved in his lifetime, Longfellow's most successful works include the long poem *Evangeline: A Tale of Acadie* (1847), *The Song of Hiawatha* (1855), *The Courtship of Miles Standish* (1858), the collection of narrative poems *Tales of a Wayside Inn* (1863), and a translation of Dante's *Divine Comedy*, published in three volumes (1865–67).

Lowe, Robert (1811–92), was a British statesman who was Gladstone's chancellor of the Exchequer from 1868 to 1873.

Lowell, John Amory (1798–1881), was an American businessman and philanthropist. The Lowells were a prominent Boston family. Following family tradition, at age thirteen Lowell entered Harvard, where he was later made a fellow. In 1822 he became director at the Suffolk Bank, a position he held for fifty-nine years. He married twice, first in 1822 to his cousin Susan Cabot Lowell (1801–27), and after her death in 1827 to her cousin Elizabeth Cabot Putnam (1807–81). Lowell was made the sole trustee of the Lowell Institute in Boston, founded by his cousin John Lowell Jr. (1799–1836). Under his direction, the institute flourished into an impressive educational facility where lectures were given by esteemed figures in literature, politics, philosophy, and science, Charles Lyell and Louis Agassiz among them (F. Greenslet, *Lowells and Their Seven Worlds* [Boston: Houghton Mifflin, 1946]).

Lubbock, John (1834–1913), 1st Baron Avebury, was an English banker, anthropologist, and Liberal politician. In his youth he befriended Charles Darwin and John Tyndall and later became part of Darwin's inner circle. In 1865 he published the highly influential *Pre-Historic Times: As Illustrated by Ancient Remains*, in which he coined the terms *paleolithic* and *neolithic*. He was also an amateur biologist and in the 1870s published works on insects and plants. In 1856 he married Ellen Frances Horden (1834–79), and the couple were among Tyndall's closest friends. After Ellen's death, John married Alice

Fox-Pitt (ca. 1862–1947), daughter of archaeologist Augustus Pitt-Rivers (1827–1900). Lubbock was the youngest member of the X Club and joined Tyndall on his climbing adventures in the Alps on a number of occasions. He was also a member of the RI and the Geological Society of London, and was made FRS in 1858. When his father died in 1865, he became Sir John Lubbock, 4th baronet. Starting in 1870, he served numerous times in Parliament and was created the 1st Baron Avebury in 1900.

Lyell, Charles (1797–1875), was a highly influential Scottish geologist. Lyell was at first trained as a lawyer; he was educated at Exeter College, Oxford, and studied law at Lincoln's Inn, London. He was briefly professor of geology at King's College, London (1831–33), after which he lived on private means. He married Mary Elizabeth Horner in July 1832; she assisted him in his scientific work. Lyell was most noted for his controversial three-volume work *Principles of Geology* (1830–33), in which he argued that the only legitimate type of causal explanation for the past had to be of the kind seen in operation in the present ("uniformitarianism"); natural laws were nothing if not consistent. In consequence of this philosophical commitment, Lyell was forced to argue that the earth was in fact many millions of years older than even the most generous of his contemporaries had hitherto been willing to admit. The "catastrophism" of Georges Cuvier was then the dominant philosophy among British geologists. Though controversial, Lyell was highly regarded, both as a philosopher of science and a geologist. A prominent figure in the Geological Society of London, Lyell was twice elected as its president (1835–37; 1849–51). He was knighted in 1848. Lyell mentored Charles Darwin from his return to England from his voyage on HMS *Beagle* in 1836, and in 1858, with Joseph Hooker he had pressed Darwin to publish an abstract of his big species book, which became *On the Origin of Species* (1859). By the 1860s, Lyell was one of the elder statesmen of science. The discovery in 1859 that humans had lived alongside the mammoths of the last ice age led to Lyell's last major publication, *Geological Evidences of the Antiquity of Man* (London: John Murray, 1863).

Magnus, Heinrich Gustav (1802–70), was a German chemist, physicist, and professor of technology and physics at the University of Berlin from 1833 until his death. He established a physics laboratory at his home in Berlin, to which he invited young physicists to work, including Tyndall in 1851. Although his early research was primarily in chemistry, by the 1850s he had begun to investigate a range of physical problems, including the absorption of heat by gases. Magnus was one of the founding members of the German Chemical Society (1868). He is perhaps best remembered for demonstrating how a spinning cylinder or sphere curves as it moves through a fluid, such as air—what is referred to as the Magnus effect. Tyndall's series of collected research *Contributions to Molecular Physics in the Domain of Radiant Heat* (1873) discusses in detail his academic debates with Magnus.

Maxwell, James Clerk (1831–79), was a Scottish physicist and mathematician. He graduated in the Cambridge Mathematical Tripos as second wrangler in 1854 after which he held professorships at Marischal College, Aberdeen (1856–60), and King's College, London (1860–65), and from 1871 held the Cavendish professorship at the University of Cambridge. Maxwell is best known for his contributions to the study of electromagnetic radiation. He developed Faraday's field theory and demonstrated that electricity, magnetism, and light were instances of the same phenomenon. Maxwell's work culminated in his 1873 publication, *A Treatise on Electricity and Magnetism.*

Mayer, Alfred Marshall (1836–97), was an American physicist. He became professor of physics and astronomy at the newly established Lehigh University in 1867, where he worked extensively on electromagnetism and astronomy. His groundbreaking study of the 1869 solar eclipse, funded by the US Nautical Almanac Office, used photographs he had taken of the phenomenon. In 1871 he became professor of physics at Stevens Institute of Technology in Hoboken, New Jersey, known for its state-of-the-art facilities. In his first decade there, Mayer published thirty-five articles and three books and became the leading authority on sound research in addition to his continued work in electromagnetism and heat conduction. As letters in this volume demonstrate, Mayer was involved in the organization of Tyndall's farewell dinner in New York on 4 February 1873, and he traveled to England later that same year, where he was received by Tyndall and other eminent men of science. Mayer's first marriage was to Katherine Duckett Goldsborough (d. 1868); they had one son. His second marriage was to Maria Louisa Snowdon and they had two sons, both of whom died young.

Mayer, Julius Robert (1814–78), was a German physicist, chemist, and physician. In 1842 he published a short text that contained one of the first statements on the conservation of energy and the mechanical equivalent of heat. Mayer was a scientific outsider; his arguments in this paper were based on philosophical principles and thought experiments, and his papers on heat and the conservation of force received little recognition in the 1840s. In the late 1850s and early 1860s Tyndall sought to revive Mayer's work and promote him as codiscoverer, along with James Joule, of energy conservation. The ensuing priority dispute between Mayer and Joule, which was largely fought out between their proxies—Tyndall on the one hand and William Thomson and Peter Guthrie Tait on the other—was carried on in increasingly acrimonious terms in private correspondence and then in the public arena.

Miller, George (ca. 1854–72), was Tyndall's younger assistant at the RI. He was the son of Mr. W. Miller of Park Road, Dalston, London. Miller accompanied Tyndall on his American lecture tour in 1872, along with Tyndall's other assistant, John Cottrell. While there, however, he contracted typhoid fever and died in Philadelphia on 30 November. His body was embalmed and brought back to England ("Births, Deaths, Marriages and Obituaries," *PMG*, 20 December 1872, 3; "The United States," *Glasgow Herald*, 19 December 1872, 5; *Nottingham Journal*, 27 December 1872, 2, col. 5; *Portland Daily Press*, 6 December 1872, 2).

Milnes, Richard Monckton (1809–85), 1st Baron Houghton, was a British traveler, poet, author, and politician. Milnes was a member of the famous Apostles Club, an intense group of undergraduates at Trinity College, Cambridge, who attempted to make sense of the religious changes that were taking place in mid-Victorian Britain in a spirit of freedom and openness. Members—which then included Alfred Tennyson, Arthur Hallam, Richard Chenevix Trench, Joseph Williams Blakesley, and many others—worried over natural science and natural theology. Milnes lived at Fryston Hall, in Castleford, West Yorkshire, where he entertained guests, including Tyndall.

Moigno, François-Napoléon-Marie (1804–84), was a French Catholic priest, physicist, writer, and editor. Educated at a Jesuit seminary, he studied science alongside theology and was given the mathematical chair at Jesuit House in Paris in 1836. After leaving the order in 1843 he became the science editor of *L'Époque*, and later became editor of the scientific journal *Cosmos* (1852–62). He founded the journal *Les Mondes* in 1863 and undertook the translation and publication of many of Tyndall's works in French, including *Radiation* (1867), *Le son* (1869), *La chaleur* (1864), and *Faraday inventeur* (1868) ("L'Abbé Moigno," *Nature* 30 [1884]: 291–92).

Newton, Isaac (1642–1727), was an English natural philosopher and mathematician of outstanding ability who rose to become Luscasian Professor of Mathematics at Cambridge (1669), master of the mint (1699), and president of the RS (1723). Newton's celebrated *Philosophiae Naturalis Principia Mathematica* (1687) examined mechanics, including concepts such as force, momentum, mass, and laws of motion. Considered the apogee of the scientific revolution, *Principia* paved the way for modern science. In his final years, Newton devoted his time to theology and the study of prophecies. His death was marked by a state funeral, with his remains interred in the nave of Westminster Abbey.

Odling, William (1829–1921), was an English chemist who had studied medicine at Guy's Hospital in 1846, before switching to chemistry. He was appointed director of the chemical laboratory at Guy's, then lecturer in chemistry, and finally professor of practical chemistry (1856). Elected a fellow of the Chemical Society in 1848, he served in several capacities, including president from 1873 to 1875. In 1859 he was elected FRS and became a fellow of the Royal College of Physicians. He was appointed joint professor of chemistry at Saint Bartholomew's Hospital, assuming the role of sole lecturer in 1864. In 1868 he succeeded Michael Faraday as Fullerian Professor of Chemistry at the RI. In 1872, during this volume, he was appointed Waynflete Professor of Chemistry at the University of Oxford.

Owen, Richard (1804–92), was a British paleontologist and comparative anatomist. Owen studied at the Royal College of Surgeons, intending to become a medical practitioner, but his career path changed when he was offered a post at the Hunterian Museum in 1827. Thus began his long and fruitful career as a natural scientist. In 1835, Owen married Caroline Amelia Clift (1801–73). The following year he was appointed Hunterian Professor of Comparative Anatomy and Physiology at the Royal College of Surgeons. At the RS, Owen was elected FRS (1834) and received the society's Royal Medal (1846) and Copley Medal (1851). In 1856 he was appointed superintendent of the natural history departments at the British Museum. Part of the older generation of scientists who Huxley and Darwin replaced, Owen was strongly opposed to the theory of evolution by natural selection. The Natural History Museum, in South Kensington, which Owen designed and opened in 1881, was considered his greatest achievement, although he made many other, less-celebrated scientific contributions. As letters in this volume show, Owen was critical of Joseph Hooker's management of the Royal Botanic Gardens, Kew, and his disapproval was aired publicly in the pages of *Nature* in May and November of 1872.

Peel, Emily (née Hay, 1836–1924), was the seventh daughter of George Hay (1787–1876), 8th Marquess of Tweeddale, and Susan Montagu (1797–1870). She married politician Robert Peel (1822–95), 3rd baronet, in 1856. The grand ceremony was held at the Royal Chapel at Whitehall, one of the bridesmaids being Emily's sister, the Lady Elizabeth Wellesley, Duchess of Wellington (1820–1904). They had four daughters, twins Gwendoline Cecilia and Agnes Helen, Victoria Alexandrina Julia, and Evelyn Emily, and one son,

Robert (1867–1925), 4th baronet. The Peels divided their time between their London home in Whitehall Gardens and Drayton Manor, in Tamworth, Staffordshire. In later years, Peel spent most of her time at Lammermoor, their villa on Lake Geneva, Switzerland. She was a regular attendee of Tyndall's lectures, and was regarded as a great hostess and philanthropist ("Births, Deaths, Marriages and Obituaries," *Tamworth Herald*, 12 April 1924, 5).

Playfair, Lyon (1818–98), 1st Baron Playfair, was a Scottish chemist and Liberal politician born in Bengal. He studied under eminent chemists such as Thomas Graham, Justus Liebig, and James P. Joule. Playfair was offered a chair in chemistry at King's College, Toronto, by Michael Faraday in 1842, a position he was persuaded to turn down by Prime Minister Robert Peel. Instead, he became professor of chemistry at the Royal Manchester Institution. In 1845 he became chemist to the Geological Survey of Great Britain and professor of chemistry at the Royal School of Mines, London, and in 1848 he was elected a fellow of the RS. Playfair assisted Prince Albert with the planning of the Great Exhibition of 1851, and from 1853 served as secretary of science in the Department of Science and Art, in which role he sought to increase the quality of science education in Britain. In 1858 he became professor of chemistry at Edinburgh University. He was elected Liberal member of Parliament for the Universities of Edinburgh and St Andrews (1868), and postmaster general in Gladstone's administration (1873–74). Playfair maintained a close relationship with the royal family and especially Albert, the Prince Consort, and for a time educated the queen's sons Albert and Alfred. He had three marriages; the first, in 1846, was to Margaret Oakes (d. 1855); the second, in 1857, to Jean Ann Millington (d. 1877); and the third, in 1878, to Edith Russell (d. 1832). Playfair served as vice president of the RS from 1864–70, and was created KCB in 1883.

Pole, William (1814–1900), was a British engineer and musician. With Tyndall and Augustus George Vernon Harcourt, Pole was one of the metropolitan gas referees, who were appointed by the Board of Trade under the City of London Gas Act (1868) to work with the city's suppliers of gas to maintain safety.

Pollock, Frederick (1845–1937), 3rd baronet, was the eldest son of William Frederick and Juliet Pollock. He attended Eton, where he was King's Scholar, and Trinity College, Cambridge, where he became a Cambridge apostle and was elected fellow of Trinity in 1868. Frederick became Corpus Professor of Jurisprudence at Oxford (1883–1903). He was editor of the *Law Reports* and, with Frederic William Maitland, author of *The History of English Law*, 2 vols.

(Cambridge: Cambridge University Press, 1895). He was known among family and friends as Fred.

Pollock, Juliet (née Creed, 1819–99), was the daughter of the Reverend Henry Creed, vicar of Corse, Gloucestershire, and wife of William Frederick Pollock, lawyer and author. They married in 1844 and had three sons, Frederick (1845–1937), 3rd baronet; Walter Herries (1850–1926); and Maurice Emilius (1857–1932). She and her husband were staunch supporters of the RI and frequently entertained Tyndall to dinner at their London home at Montagu Square. Pollock wrote several articles and books on literature and theater, including *Julian and his Playfellows* (London: Grant and Griffith, 1852), *New Friends* (London: John W. Parker and Son, 1858), and *Macready As I Knew Him* (London: Remington, 1885). In this volume, Tyndall refers to her article "The Imaginative Literature of America," published in the *Contemporary Review* in June 1873.

Pollock, William Frederick (1815–88), 2nd baronet, was a lawyer and author. The eldest son of Jonathan Frederick Pollock (1783–1870) and Frances Rivers (d. 1827), he married Juliet Creed in 1844. He was admitted to Trinity College, Cambridge, in 1832, where he obtained his BA (1836) and MA (1840). Admitted to the Inner Temple in 1833 and called to the bar in 1838, he then became a master of the court of exchequer in 1846 and was appointed queen's remembrancer in 1874. Pollock's two-volume *Personal Remembrances* (London: Macmillan, 1887) illustrates how the family circulated among London's elite and entertained at their Montagu Square home. Both William and Juliet Pollock were loyal supporters of the RI and Tyndall was frequently among their guests.

Pugh, Thomas Burnett (1829–84), was an American abolitionist, publisher, bookseller, and lecture manager. He was well known for his Star Course, established in 1869, in which he organized ten lectures annually at the Academy of Music, a concert hall and opera house in Philadelphia. Enormously popular, the star lectures not only encouraged intellectual discourse on a wide variety of subjects but also provided a form of entertainment, featuring distinguished and fashionable speakers from throughout America and abroad (V. Fischer and M. B. Frank, eds., *Mark Twain's Letters*, vol. 4, *1870–1871* [Berkeley: University of California Press, 1995]).

Rawlinson, Henry Creswicke (1810–95), was a British East India Company army officer, politician, and Assyriologist. He was president of the Royal Geographical Society (1871–73; 1874–76).

Reade, Thomas Mellard (1832–1909), was an English architect and geologist based in Liverpool. From 1853 to 1860, Reade was chief draftsman for the London and North Western railway company. He established a private architectural and civil engineering practice in 1860. A prolific writer, Reade published over two hundred scientific papers and an important and elaborately illustrated volume, *The Origin of Mountain Ranges* (1886). He became chief architect to the Liverpool School Board (1870–1902), joined the Liverpool Geological Society (1870), was elected fellow of the Geological Society (1872), was awarded its Murchison Medal (1896), and was made fellow of the Royal Institute of British Architects (1878). In 1886 he married Emma Eliza Taylor, widow of Alfred Taylor (DCP; "T. Mellard Reade," *Nature* 80 [3 June 1909]: 404).

Rhees, William Jones (1830–1907), was an American government administrator and archivist from Philadelphia, Pennsylvania. For most of his career, he worked at the Smithsonian Institution, first as private secretary to Joseph Henry, then as chief clerk and eventually becoming head archivist. In 1879 Rhees published *The Smithsonian Institution: Documents Relative to Its Origin and History*, a testament to his thoroughness as a writer and file keeper. A revised version was updated by Rhees and published in 1900. He married twice, first to Laura O. Clarke (d. 1864), and then in 1866 to Romenia Fontanette Ellis (1846–1923). His extensive collection of autographs of prominent individuals, known as the Rhees Collection, is housed at the Huntington Library in California (N. Reingold, "The Anatomy of a Collection: The Rhees Papers," *American Archivist* 27, no. 2 [1964]: 251–59).

Ripon, Lord: see Robinson, George Frederick Samuel.

Robinson, George Frederick Samuel (1827–1909), 1st Marquess of Ripon, was a politician, born at 10 Downing Street, London, during his father's, Frederick John Robinson (1782–1859), short term as prime minister (1827–28). His mother, Sarah Albinia Louisa (1793–1867), was the daughter of the 4th earl of Buckinghamshire. Considered a radical for most of his life, Lord Ripon became involved with the Christian Socialists in the late 1840s. He married his cousin Henrietta Anne Theodosia Vyner (1833/4–1907) in 1851, and the following year saw his election to Parliament as a member for Hull, and the birth of their only son, Frederick Oliver. As lord president of the council in the Gladstone administration, he contributed to the safe passage of the landmark elementary education legislation, a measure of justice for Ireland, the treaty of Washington, and passage of the secret ballot. During his life he was known by three titles: Viscount Goderich (1853–59); 3rd Earl De

Grey, following the death of his uncle in 1859; and finally Lord Ripon in 1871, when he was elevated to a marquessate.

Russell, (Mary) Agatha (1853–1933), was a daughter of John Russell and Frances Russell. She was coeditor of her mother's *Lady John Russell: A Memoir* (1910). Russell never married and lived at Pembroke Lodge until the death of her mother in 1898.

Russell, Frances (Fanny) Anna Maria (née Elliot, 1815–98), was the second wife of Lord John Russell. She was daughter of Gilbert Elliot Murray Kynynmound (1782–1859), 2nd Earl of Minto, and Mary Brydone (1786–1853). They lived at Pembroke Lodge in Richmond Park, a home provided for them by the queen, where they entertained the political, social, and scientific elite, including Tyndall, "one of the most genial and delightful of her guests" (Desmond MacCarthy and Agatha Russell, eds., *Lady John Russell: A Memoir with Selections from Her Diaries and Correspondence*, 2nd ed. [London: Methuen, 1910], 261). Russell had a keen interest in politics and strong religious beliefs evidenced in her publication *Family Worship*, which contained selections from the Bible and prayers for daily use.

Russell, Francis Albert Rollo (1849–1914), called Rollo by family and friends, was the third son of John and Frances Russell. He graduated with distinction in natural science from Oxford in 1872. He went on to make his scientific reputation not in physics but in meteorology, looking especially at the link between air pollution and poor health. (His 1892 book *Epidemics, Plagues, and Fevers: Their Causes and Prevention* was positively received.) He was an important influence on his nephew Bertrand Russell (1872–1970), who was the son of his eldest brother, John Russell (1842–76), and went on to become one of the most prominent philosophers of the twentieth century.

Russell, John (1792–1878), 1st Earl Russell, was an author and prime minister of the United Kingdom. While abroad in 1813 he was returned to Parliament as a Whig member of Parliament for the borough of Tavistock. In addition to his political duties he became a prolific author. Lord Russell was at the forefront of the campaign for parliamentary reform, hoping to enfranchise the large industrial towns at the expense of the small boroughs whose vote was given in exchange for government patronage. He served as prime minister from 1846 to 1852 and from 1865 to 1866. In 1866 he attended Tyndall's opening discourse, "On the Influence of Colour and Mechanical Condition on Radiant Heat," at the RI. After his first wife died in 1839, Russell was remarried in 1841 to Frances Russell. Their children were: John Russell (1842–76),

George Gilbert William Russell (1848–1933), Francis Albert Rollo Russell (1849–1914), and Agatha Russell (1853–1933).

Russell, Lord: see Russell, John.

Rutherford, Elizabeth (dates unknown), resided in an apartment at Dalhousie Castle, Lasswade, Scotland, which was the estate of the British politician Fox Maule-Ramsay (1801–74), 11th Earl of Dalhousie. In late 1872, however, she wrote to Tyndall that her "son-in-law, seeing that from different circumstances his family could only be resident there for 2 or 3 months during the year, and that the place was very expensive, resolved to give it up," thus requiring Rutherford to move out of Dalhousie Castle. It is unclear exactly who her son-in-law was. In another letter, she references her fifteen-year-old son, Thomas Rutherford, sharing that despite his interest in chemistry he would go into law. Their new residence was 13 Shandwick Place in Edinburgh. Rutherford appears to have been friends with the Pollocks and the Herries family.

Salisbury, Lord: see Gascoyne-Cecil, Robert Arthur Talbot.

Savage, Sarah (dates unknown), was housekeeper at the RI.

Shairp, John Campbell (1819–85), was a Scottish poet, critic, and academic. Shairp's formal education began at Edinburgh Academy (1829–36), followed by the University of Glasgow (1836–39), where he studied classics, philosophy, and poetry. In 1840 he was awarded the Snell Exhibition scholarship to undertake postgraduate work at Balliol College, Oxford. In 1853 Shairp married Eliza Douglas, the sister of his college friend Henry Alexander Douglas. At the University of St Andrews, he served as assistant to the professor (1857), chair of humanities (1861), and principal (1868). At Oxford, Shairp was professor of poetry. His publications included *Kilmahoe: A Highland Pastoral* (1864); *Studies in Poetry and Philosophy* (1868); a collection of lectures titled *Culture and Religion* (1870); and *The Life of Robert Burns* (1879) (W. A. Knight and C. M. St. John, *Principal Shairp and His Friends* [London: John Murray, 1888]) and, much-discussed in this volume, the co-edited *Life of James David Forbes*.

Smith, John (1798–1888), was a British botanist and horticulturist who specialized in ferns. Known as "Old Jock," Smith began working for the Royal Botanic Gardens, Kew, sometime after 1820, and was soon put in charge of the propagation department and hothouses. While there he increased Kew's fern collection fivefold and proposed a new classification of the world's fern genera. In 1841 he was given the official title of curator at Kew, although he had been working in that capacity for many years prior. Smith's publications included *Cultivated Ferns* (1857), *Ferns, British and Foreign* (1866), and *Historia filicum* (1875).

Smith, John Lawrence (1818–83), was an American chemist and mineralogist. Smith was educated at the University of Virginia, followed by the Medical College of South Carolina, where he received a medical degree in 1840. He continued his studies abroad, in Paris and in Giessen. He returned to the United States and began practicing medicine in 1844, but his interests in analytical chemistry led him away from medicine. Smith was appointed to investigate the mineral resources of Turkey from 1846 to 1850. In 1852 he became the chair of chemistry at the University of Virginia, which he followed in 1854 as the chair and professor of medical chemistry and toxicology at the University of Louisville. He retired from that position in 1866, and studied again in Europe, taking up his passion for researching and collecting meteorites (his extensive collection is now at Harvard). Smith was elected to the American Philosophical Society in 1857, and served as president of the AAAS (1872) and the American Chemical Society (1877).

Spencer, Herbert (1820–1903), was an English biologist, philosopher, journalist, and sociologist. He was baptized in the Methodist tradition, despite the misgivings of his father, who rejected institutional religion in favor of "rational" religion. Unsurprisingly, Spencer himself at an early age rejected all possibility of belief in an immanent or personal god, and later became a self-professed agnostic. Through family connections, at twenty-eight he secured a post as subeditor at the *Economist*, which put him at the heart of literary London. He published his first book, *Social Statics*, in 1850 and his second book, *Principles of Psychology*, in 1855. Spencer was a respected philosopher, but one of his most lasting contributions was his *Principles of Biology* (1864) in which he coined the phrase "survival of the fittest." He made evolution his guiding principle in his multivolume synthesis of all knowledge. In spite of his agnosticism, however, Spencer nevertheless affirmed the existence of the "Unknowable," a shadowy deity behind the evolutionary process that, he believed, would ultimately reconcile the growing conflict between science and religion. Along with Tyndall and seven others, Spencer was a member of the X Club, a monthly dining club the members of which promoted science in British society (Spencer was the only member who was not FRS).

Spottiswoode, Eliza (née Taylor Arbuthnot, 1837–94), was born in Madras, India. Eliza was the eldest daughter of William Urquhart Arbuthnot, member of the council for India. She married William Spottiswoode in 1861 and they lived at 41 Grosvenor Place in London, where they hosted gatherings that were attended by the scientific, political, and social elite and frequently included scientific demonstrations. Their country residence was in Combe Bank, Sundridge, near Sevenoaks. They had two sons, William Hugh (1864–1915), who became a partner and later director in his father's publishing firm, and Cyril Andrew (1867–1915), who also joined the family business.

Spottiswoode, William (1825–83), was an English mathematician and physicist. He was also a partner of Eyre & Spottiswoode, the queen's printer. In 1861 he married Eliza Taylor Arbuthnot. Parties at the Spottiswoode home were considered the height of the London season, attracting men of science, politics, and society. These events quite often featured scientific demonstrations in their home laboratory. In 1864 Spottiswoode, along with Tyndall, Thomas Huxley, Herbert Spencer, and five others, formed the X Club, a monthly dining club the members of which promoted science in British society. He was treasurer of the BAAS (1861–74), the RI (1865–73), and the RS (1871–8), and served as president of the BAAS (1878) and the RS (1878–83).

Stanhope, Lord: see Stanhope, Philip Henry.

Stanhope, Philip Henry (1805–75), 5th Earl Stanhope, was an English antiquarian and Tory politician.

Stanley, Edward Henry (1826–93), 15th Earl of Derby, was a politician and diarist. He was elected to Parliament as a Tory for King's Lynn in December 1848 and retained the seat until he took his seat in the Lords in 1869. His lack of faith alienated him from High Church Anglicans but he was still an important figure within the Tory party, being offered but declining leadership of the Lords in February 1870. That same year he married Mary Catherine Sackville (1824–1900). He returned to the Foreign Office, where he served as secretary for the second time (1874–78) and as colonial secretary (1882–85). Letters in this volume between Tyndall and Derby discuss the memorial in support of Joseph Hooker in his dispute with Acton Ayrton over the management of the Royal Botanic Gardens, Kew.

Steuart [Stewart], Elizabeth Dawson (née Duckett, 1802–93), was born into a very wealthy family with an estate of twelve thousand acres and a gothic mansion (Duckett's Grove, now a ruin) situated several miles northeast of County Carlow, in Ireland. In 1843 she married William Richard Steuart and they lived at Steuart's Lodge, Leighlinbridge. She took a keen interest in

Tyndall's career and was probably his most consistent and long-lived patron. Letters in this volume show that Steuart frequently helped Tyndall arrange financial support for his relatives and acquaintances in Ireland.

Stokes, George Gabriel (1819–1903), was an English mathematician and physicist. He graduated from the University of Cambridge as senior wrangler (1841) and was Lucasian Professor of Mathematics at the University of Cambridge (1849–1903). He served as secretary of the RS (1854–85), during which time he was responsible for refereeing papers for the *Phil. Trans.*, a task for which he frequently sought Tyndall's assistance. In the 1850s Stokes turned his attention to optics, and in 1852 published on fluorescence, a term he coined, showing that the phenomenon was the result of changes in the wavelength of light. For this work he was awarded the RS's Rumford Medal (1852), and later the RS's Copley Medal (1893) for his contributions to science.

Tait, Peter Guthrie (1831–1901), was a Scottish mathematician and physicist, who was senior wrangler in the mathematics Tripos at the University of Cambridge in 1852. He was professor of mathematics at Queen's College, Belfast (1854–60), and held the chair of natural philosophy at Edinburgh University from 1860 until shortly before his death. During the 1860s, Tait forged a friendship with William Thomson and began writing their *Treatise on Natural Philosophy* (1867). During the 1860s, Tait and Thomson had been in a dispute with Tyndall over his claims about Julius Mayer's priority over Joule in developing the mechanical equivalent of heat. This heated dispute spilled over into popular journals before they resolved their differences at the Dundee meeting of the BAAS in 1867, where, Tyndall wrote, they "seemed like brothers together" (letter 2646, *Tyndall Correspondence*, vol. 10). By 1873 Tyndall's feelings for Tait had soured again following a critique of Tyndall's *Forms of Water* regarding James David Forbes in Tait's coedited *Life of James David Forbes*. Bitter exchanges between the two men were printed in the correspondence columns of *Nature* and Tyndall ended the matter to his satisfaction with "Principal Forbes and His Biographers," published in the *Contemporary Review* in August 1873 (and as a standalone pamphlet by his publisher Longmans).

Thompson, Benjamin (1753–1814), Count Rumford, was an American physicist, inventor, and social reformer. During the American Revolutionary War, Rumford worked for the British military in England, providing information on colonial affairs. At the same time, he worked on scientific experiments with gunpowder and cannons. This research led to his first publication and fellowship in the RS, where he later established the Rumford Medal for notable research on heat or light. Rumford's challenges to established physical theory were part of the nineteenth-century development of thermodynamics. He was one of the founders of the RI, where he lived briefly before settling in France. As letters in this volume note, an account of his life, *Memoir of Sir Benjamin Thompson, Count Rumford, with Notices of His Daughter* (1871), was written by George E. Ellis of Boston.

Thomson, William (1824–1907), 1st Baron Kelvin, was a distinguished Irish-Scottish mathematician, physicist, and inventor. He was educated at the University of Glasgow and Peterhouse College, University of Cambridge, where he was second wrangler and Smith's Prizeman in 1841. He was professor of natural philosophy in Glasgow from 1846, paying frequent visits to London. In 1870, Thomson purchased a yacht called *Lalla Rookh*, which he turned into a floating laboratory, used for experiments on a new sounding apparatus and a new design of a dry-card magnetic compass. He first met Tyndall at the Edinburgh meeting of the BAAS in 1850, and from the start the two physicists had a strained relationship. For much of his career Tyndall regarded Thomson as a scientific rival; however, the two physicists resolved their differences by the Dundee meeting of the BAAS in 1867, where, Tyndall wrote, they "seemed like brothers together" (letter 2646, *Tyndall Correspondence*, vol. 10). As letters in this volume indicate, Tyndall was worried about renewed tensions between them with the publication of Tyndall's *Forms of Water*.

Tocqueville, Alexis de (1805–59), was a French diplomat, political philosopher, and historian who authored *Democracy in America* (1835–40), a book that Tyndall consulted for and referred to during his American lecture tour.

Trevor, Charles Cecil (1830–1921), was an English civil servant who was assistant secretary to the Board of Trade, Harbour Department (1867–95). Trevor was a student of Lincoln's Inn (1851) and was called to the bar in 1855. In 1863 he married Mary Weston (d. 1892). He was conservator of the River Thames (1895–1906) and delegate to the North Sea Fishery Conference (1881–82), the Submarine Cables Conference (1882–83), the North Sea Liquor Traffic Conference (1886), and the Channel Fisheries Negotiations. Trevor was also the author of *A Treatise on the Taxes on Succession*, which was published in several editions. In 1867 Trevor wrote to Tyndall to discuss the details of Tyndall taking over the Board of Trade duties from Faraday. Letters in this volume between Trevor and Tyndall discuss the use of gas in lighthouses (Joseph Foster, *Men-at-the-Bar: A Biographical Hand-List of the Members of the Various Inns*, 2nd ed. [London: Hazell, Watson and Viney,

1885]; R. M. MacLeod, "Science and Government in Victorian England: Lighthouse Illumination and the Board of Trade, 1866–1886," *Isis* 60, no. 1 [1969]: 5–38; "Sir Cecil Trevor," *Times*, 20 September 1921, 20).

Tunstall, Jane (1811–?), was a widowed housekeeper who worked for Tyndall and Hirst when they lived at 14 Waverley Place in London. Tunstall continued on as Tyndall's housekeeper when he moved into Faraday's rooms at the RI on 25 January 1868, following Faraday's death in August 1867.

Tyndale, Hector (1821–80), was an American businessman, Union general, and cousin of John Tyndall. He was the son of Tyndall's father's brother, Robinson Tyndall (1775–1842), who had immigrated to Philadelphia from Leighlinbridge around 1804 and founded a glass and ceramics import business. Hector inherited the business and first met his physicist cousin in 1854 on one of his frequent business trips to Europe. They immediately struck up an intellectual rapport and began an occasional correspondence. During the American Civil War (1861–65), Tyndale rose to the rank of brigadier general, and after his resignation in 1864 due to health, he was promoted to brevet major general in acknowledgment of his service (1865). Tyndale was instrumental in organizing his cousin's lecture tour in the United States and, with Joseph Henry and Edward L. Youmans, was named a trustee of the fund established with Tyndall's lecture profits for pure science research in America.

Tyndale, Julia (née Nowlen, 1823–97), was born in New York to Edward and Julia Nowlen and married Hector Tyndale in 1842. Upon her death, her and her husband's significant collection of ceramics was bequeathed to the Philadelphia Museum of Art, and is known as the General Hector Tyndale Memorial Collection.

Tyndall, Emily (Emma, 1817–1904), was John Tyndall's sister in Ireland, whom he called Emma. She does not appear to have married and later in life suffered from a mental illness that was manifested as an extreme form of religiosity. Tyndall regularly sent money for his sister and, often through Elizabeth Dawson Steuart, helped with managing her care.

Tyndall, John (ca. 1815–75), was a relative of Tyndall's in Gorey, County Wexford, Ireland. A doctor and member of the Royal Historical and Archaeological Society of Ireland, he died in 1875 of acute bronchitis ("Obituaries," *Lancet*, 1 May 1875, 633).

Valentin, William George (1829–79), was a German chemist. Born in Neuenbürg, Valentin arrived in England in 1855 to study at the Royal College of

Chemistry, where he quickly became senior assistant in the laboratory. He also taught at the Science Schools, South Kensington. Valentin held the post of gas examiner for the Great Western Railway, and finally as the chemical adviser to Trinity House. He died suddenly of apoplexy in 1879 ("William George Valentin," *Nature* 20 [8 May 1879]: 39).

Vieweg, Heinrich Rudolf (1826–90), was a German publisher. He was the son of Eduard Vieweg (1797–1869) and the grandson of Friedrich Vieweg (1761–1835), who had founded the family publishing business in Berlin (1786) and later Brunswick (1799). Vieweg worked and studied natural sciences in Heidelberg before he joined his father as a partner in the family firm in 1853 and took over the firm in 1866. On taking over the firm Vieweg continued his father's emphasis on scientific works and published German translations of a number of Tyndall's books. Letters in this volume discuss the German translation of a new edition of Tyndall's *Heat: A Mode of Motion* and *Six Lectures on Light*.

Wheatstone, Charles (1802–75), was an English experimental physicist, inventor, and businessman. He had been a maker of musical instruments before becoming professor of experimental philosophy at King's College, London (1834–75), where he was a reluctant lecturer. There he developed the electric telegraph with William Fothergill Cooke (1806–79) as well as a number of optical instruments, including the stereoscope, which led him into controversy with David Brewster (1781–68). He married Emma West in 1847 and together they had five children by 1855. The family settled in Hammersmith, where Wheatstone frequently entertained Tyndall, whom he first met in 1852.

Wiedemann, Clara Laura (née Mitscherlich, 1827–1914), the daughter of the chemist Eilhard Mitscherlich (1794–1863), was a German translator who married Gustav Wiedemann in 1851. With Anna Helmholtz, she translated Tyndall's *Heat* (1863) and *Sound* (1867) for their German editions in 1867 and 1869, respectively. Although the two women translated the work, only their husbands were credited (*HLS*; B. Meijer et. al., eds., *Nordisk familjebok*, 2nd ed. [Stockholm: Nordisk familjeboks, 1904–26]). When her husband became the editor of *Annalen der Physik und Chemie*, which subsequently took on the name *Wiedemann's Annalen*, Wiedemann assisted by translating the scientific papers. Gustav and Clara Wiedemann had two sons. Eilhard Ernst Gustav (1852–1928) pursued a career in physics, like his father, eventually becoming a professor at Erlangen. Their younger son, Alfred (1856–1936), became an Egyptologist.

Wiedemann, Gustav Heinrich (1826–99), was a German physicist and editor. Wiedemann was trained at the University of Berlin by Heinrich Gustav Magnus and came to know Hermann Helmholtz while there. He held professorships of physics at Basel (1854–63), Brunswick (1854–66), and Karlsruhe (1866–71) before taking the chair of physical chemistry at Leipzig in 1871. Together with Rudolph Franz, Wiedemann developed the so-called Wiedemann-Franz law in 1853, a fundamental law of physics that holds that the "ratio of the thermal conductivity of any pure metal to its electrical conductivity is approximately constant at a given temperature" (*Oxford Dictionary of Physics*). In 1851, Wiedemann married Clara Laura Mitscherlich (1827–1914). With Clara, Gustav translated some of Tyndall's books into German, although he often received the credit alone. When Johann Christian Poggendorff died in 1877, Wiedemann became the editor of *Annalen der Physik und Chemie*, which subsequently took on the name *Wiedemann's Annalen*.

Wigham, John Richardson (1829–1906), born in Edinburgh, went to Dublin in 1843 where he served an apprenticeship in a hardware and manufacturing business under his brother-in-law Joshua Edmundson. After Edmundson's death, the Edmundson's Electricity Corporation grew rapidly under Wigham's stewardship and focused on the provision of gas lighting to private and public buildings. Wigham was engineer to the Commercial Gas Company of Ireland and served as director of the Alliance and Dublin Consumers' Gas Company from 1866 and director and vice-chairman of the Dublin United Tramways Company from 1881, and was a member of the council of the Dublin Chamber of Commerce, as well as serving as its secretary (1881–93) and president (1894–96). He was best known as an inventor of powerful gas burners that were adopted by the Commissioners for Irish Lights. A system invented by him, with an on-site gas works, was installed in Baily Lighthouse in Howth, County Dublin.

Wills, Thomas (1850–79), was a chemist specializing in illumination. In 1868 he became William Odling's assistant, first at Saint Bartholomew's Hospital and later that year at the RI. He was elected a fellow of the Chemical Society during his time at the RI. In 1873 he accepted the position of demonstrator in chemistry at the Royal Naval College, Greenwich. Wills received two RS silver medals, the first for his paper "Recent Processes for the Manufacture of Gas for Illuminating Purposes" (1873) and the second for a series of lectures on "Explosions in Coal Mines." He was also appointed secretary to the RS Chemical Section. He died after a brief illness with typhoid fever (*Roy. Inst. Proc.* 6 [1870–72]: ii; "Science Gossip," *Athenaeum*, 17 May 1879, 636).

Winthrop, John (1588–1649), was the first governor and chronicler of Massachusetts Bay. Considered a preeminent figure of early New England, Winthrop was responsible not only for the survival of the colony but also the formation of its institutions and for maintaining order. Winthrop married four times and fathered sixteen children. His third marriage was to Margaret Tyndal (ca. 1591–1647), the daughter of Sir John Tyndal.

Winthrop, John (1714–79), was an American scientist and a fifth-generation descendant of Governor John Winthrop (1588–1649). Winthrop graduated from Harvard with a BA in 1732, received an MA in 1735, and was appointed Hollisian Professor of the Mathematics and of Natural and Experimental Philosophy in 1738. He was a gifted lecturer, and gave the first in-class electricity demonstrations in 1746. That same year he married Rebecca Townsend (d. 1753), and together they had five children. Three years after Rebecca's death, he married Hannah Fayerweather Tollman; there were no children from this marriage. Winthrop made important contributions in the fields of seismology, astronomy, and physics and was one of the most esteemed American scientists of his time. His fame extended overseas through his publications on a variety of topics such as the aberration of light, the nature of comets and earthquakes, and planetary transits. He was elected FRS (1766), and was awarded the first honorary LLD from Harvard (1774), among other honors.

Winthrop, Margaret Tyndal (ca. 1591–1647), was born in Essex County, England, to Sir John Tyndal and Anne Egerton. She was raised as a proper Puritan, but was also highly literate. In 1618 she married twice-widowed John Winthrop and they had five children. When her husband was elected governor of Massachusetts (1829), the family emigrated to America; their youngest daughter died on the voyage (1831). Correspondence between Margaret and John Winthrop can be found in the Winthrop Papers, held by the Massachusetts Historical Society.

Winthrop, Robert Charles (1809–94), was an American senator and representative from Massachusetts. After several terms as representative in Congress, as Speaker of the House of Representatives in the Thirtieth Congress (1847–49), and finally as senator, Winthrop's political career was over by age forty-three. He devoted the rest of his life to literary, historical, and philanthropic pursuits and became a noted orator. As letters in this volume show, Winthrop was a sixth-generation descendant of Governor John Winthrop.

Youmans, Edward Livingston (1821–87), was an American scientific popularizer, editor, and lecturer. Born in Coeymans, New York, he was raised in a home where books and ideas were highly valued. An attack of ophthalmia in 1835 damaged his eyesight, which prompted him to move to New York City for treatment in 1839. There he formed a lasting professional partnership with leading American publisher William Henry Appleton (1814–99), son of Daniel Appleton (1785–1849), and through this connection Youmans was crucial for arranging the US publication of works by Spencer, Tyndall, and others. In New York Youmans also studied chemistry under John William Draper and completed a medical degree through the University of Vermont. Youmans produced a stream of popular science books, including *A Class-book of Chemistry* (1851), *Alcohol and the Constitution of Man* (1853), and *The Handbook of Household Science* (1857). He possessed a genuine talent for rendering vast amounts of scientific material into clear, understandable prose. In *The Correlation and Conservation of Forces* (1864) and *Culture Demanded by Modern Life* (1867), Youmans insisted that science is not materialistic but leads to an ever higher plane of knowledge and experience. To this end, Youmans founded the International Scientific Series in 1871 and the *Popular Science Monthly* in 1872, which he "conducted" until his death, when his brother William Jay Youmans (1838–1901) took it over. Youmans also served as vice president of the Free Religious Association, which was composed of a diverse assortment of Unitarians, Universalists, spirituals, transcendentalists, scientific theists, and other disaffected religious minorities, and advanced the principles of free thought and moral philosophy without any reference to institutional Christianity. Youmans helped to organize Tyndall's lecture tour in the United States, and, with Joseph Henry and Hector Tyndale, was named a trustee of the fund established with Tyndall's lecture profits for pure science research in America.

Young, Thomas (1773–1829), was an English physician and natural philosopher whose work helped to establish the wave theory of light. Young held the RI professorship of natural philosophy from 1801 to 1803.

Zöllner, Johann Karl Friedrich (1834–82), was a German astrophysicist. He studied with Heinrich Magnus and Gustav Wiedemann, receiving his PhD in 1857 with a thesis on photometry. He became a professor at the University of Leipzig in 1866. His research included the development of several instruments for astrophysical research as well as more theoretical work, including a theory of comets. In the 1870s and 1880s he began using his published work to attack other scientific practitioners. This, along with his growing commitment to spiritualism, led to his increasing isolation from the scientific community; some considered that he suffered from mental illness.

INDEX

The index refers to the letters using their serial numbers. The letter 'n' denotes a footnote reference and appears if an item is not otherwise mentioned in the letter to which the footnote pertains. A number in **bold** signifies a letter to or from a correspondent.